PRINCIPLES OF

Electrical Engineering

McGraw-Hill Series in Electrical Engineering

Consulting Editor
Stephen W. Director, Carnegie-Mellon University

CIRCUITS AND SYSTEMS

COMMUNICATIONS AND SIGNAL PROCESSING

CONTROL THEORY

ELECTRONICS AND ELECTRONIC CIRCUITS

POWER AND ENERGY

ELECTROMAGNETICS

COMPUTER ENGINEERING

INTRODUCTORY

RADAR AND ANTENNAS

VLSI

INTRODUCTORY

Consulting Editor
Stephen W. Director, Carnegie-Mellon University

Belove, Schachter, and Schilling: Digital and Analog Systems, Circuits and Devices
Fitzgerald, Higginbotham, and Grabel: Basic Electrical Engineering
Hammond and Gehmlich: Electrical Engineering
Paul, Nasar, and Unnewehr: Introduction to Electrical Engineering
Peebles and Giuma: Principles of Electrical Engineering
Piel and Truxal: Technology: Handle with Care

PRINCIPLES OF Electrical Engineering

Peyton Z. Peebles, Jr., Ph.D.
Tayeb A. Giuma, Ph.D.

Department of Electrical Engineering
University of Florida

McGRAW-HILL, INC.

New York St. Louis San Francisco Auckland Bogotá
Caracas Hamburg Lisbon London Madrid Mexico
Milan Montreal New Delhi Paris San Juan São Paulo
Singapore Sydney Tokyo Toronto

Dedicated to our wives and children.

Principles of Electrical Engineering

1 2 3 4 5 6 7 8 9 0 DOC DOC 9 0 9 8 7 6 5 4 3 2 1

ISBN 0-07-049252-2

This book was set in Times Roman by Better Graphics, Inc.
The editors were Roger Howell and Eleanor Castellano;
the design was done by Caliber Design Planning, Inc.;
the production supervisor was Kathryn Porzio.
R. R. Donnelley & Sons Company was printer and binder.

Library of Congress Cataloging-in-Publication Data

Peebles, Peyton Z.
Principles of electrical engineering / Peyton Z. Peebles, Jr., Tayeb A. Giuma.
p. cm.—(McGraw-Hill series in electrical engineering. Introductory)
Includes bibliographical references and index.
ISBN 0-07-049252-2
1. Electric engineering. 2. Electronics. I. Giuma, Tayeb A. II. Title. III. Series.
TK146.P38 1991
621.3—dc20 90-45402

ABOUT THE AUTHORS

Peyton Z. Peebles, Jr.

Dr. Peebles is Professor and formerly Associate Chairman of the Department of Electrical Engineering at the University of Florida. His twenty-two years of teaching experience include time spent at the University of Tennessee and the University of Hawaii, as well as the University of Florida. He earned his Ph.D. degree from the University of Pennsylvania where he held a David Sarnoff Fellowship from RCA for 2 years.

Professor Peebles has published over 50 journal articles and conference papers as well as three textbooks prior to this one: *Probability, Random Variables, and Random Signal Principles, 2nd edition* (McGraw-Hill, 1987); *Digital Communication Systems* (Prentice-Hall, 1987); and *Communication System Principles* (Addison-Wesley, 1976).

Dr. Peebles is a member of Tau Beta Pi, Eta Kappa Nu, Sigma Xi, Sigma Pi Sigma, Phi Beta Chi, and is a Fellow of the IEEE.

Tayeb A. Giuma

Dr. Giuma is currently on the faculty of the Electrical Engineering Department at the University of Florida. He received the BSEE, MSEE, and the Ph.D. degrees from the University of Miami, Coral Gables, Florida, in 1979, 1980, and 1987, respectively. Prior to joining the University of Florida, he was on the faculty of the Electrical and Computer Engineering Department at the University of Miami. Dr. Giuma has taught a wide spectrum of electrical and computer engineering courses and won three awards for excellence in teaching. His research interest includes microprocessor-based systems, computer architecture, programming languages, and multivalued logic. He has published many technical papers in refereed journals and proceedings.

Dr. Giuma is a member of Tau Beta Pi, Eta Kappa Nu, Sigma Xi, IEEE, the Computer Society, and ASEE.

CONTENTS

PART 2

Electronic Devices and Circuits

CHAPTER 8 Operational Amplifiers, Feedback, and Control 259

CHAPTER 11 Digital System Components 421

CHAPTER 12 Computer Systems 457

PART 4

Electrical Communications Systems

APPENDIX A Software Package 707

APPENDIX B PSPICE 714

APPENDIX C USEFUL MATHEMATICAL QUANTITIES 731

PREFACE

This book was written as a textbook on electrical engineering topics for non-electrical engineers. Because the needs of aerospace, chemical, civil, environmental, industrial, materials, mechanical, ocean, and other types of engineering are so diverse, no optimum list of topics for one would necessarily serve another. When course length and emphasis variations from school to school are included, it became clear that the book had to be written to give the instructor of each specific course some freedom of choice in the topics covered.

Thus, a wide selection of topics in electrical engineering is included in the book so that each instructor can choose those that best fit each individual course. These topics fall broadly into five areas that form the five parts into which the book is subdivided. Part 1, "Basic Concepts and Circuits," contains the most elementary material (five chapters), most of which must be studied before any other topics are attempted. Part 2, "Electronic Devices and Circuits," discusses diodes and transistors (Chapters 6 and 7), operational amplifiers and controls (Chapter 8), and some specialized circuits (Chapter 9). Part 3, "Computers and Digital Devices," includes four chapters on various aspects of computers. This comprehensive integrated introduction to digital computers features an up-to-date treatment of fundamentals of digital logic design (Chapters 10 and 11), microprocessors and assembly language (Chapter 12), and computer networks (Chapter 13). Part 4, "Electrical Communications Systems," discusses wave propagation, antennas, and noise in systems (Chapter 14), analog radio systems, including AM and FM radio as well as TV (Chapter 15), and digital communications systems (Chapter 16). Part 5 considers power systems (Chapter 17) and electric machinery (Chapter 18).

Generally, Chapters 1–4 of Part 1 must be included in any first course. Parts 2–5 have almost no dependence on each other and can be selected in any order. Chapters in any of Parts 3–5 should generally be taken in sequence for best effect. Within Part 2 Chapter 8 can stand alone but has a weak dependence on Chapter 5 of Part 1; Chapter 7 requires Chapter 6 be taken first; and Chapter 9 has a weak dependence on Chapter 6.

Although the book has sufficient material to support a sequence of two one-semester courses, most applications are expected to be in single courses of one-quarter or one-semester length. The material can be selected to give various course lengths and emphasis of content. Several choices are listed in the table below. All choices assume three hours of class exposure per week.

Type of Course	**Length**	**Chapters Included**
Basics + some computer	1 quarter	1–4, 10, 11
Basics + some communications	1 quarter	1–4, 14, 15
Basics + devices + controls	1 quarter	1–6, 8
Basics + controls + specialized circuits	1 quarter	1–6 (diodes only), 8, 9
Basics + controls + machines	1 quarter	1–4, 8, 18
Basics + devices + controls + specialized circuits + some power	1 semester	1–9, 17
Basics + computers	1 semester	1–5, 10–13
Basics + communications	1 semester	1–5, 14–16
Basics + devices + controls + power	1 semester	1–6, 8, 17, 18
Basics + computer + some power	1 semester	1–5, 10–12, 17

An effort has been made to keep the level of presentation reasonably consistent throughout the book. Each topic has enough detail for the reader to have profitable learning, but not so much, it is hoped, that insight is lost. We believe that a course for nonelectrical engineers should give a learning experience and not just a shallow exposure to a short list of meaningless topics to be discussed one day and forgotten the next.

The background needed to use the book is only that typical of junior or senior engineering students. Basic courses are assumed in calculus and physics, where quantities of electric and magnetic fields, work, energy, and sound (waves) are covered.

To aid the student, worked examples are scattered throughout the book, and exercise problems are provided at the ends of all chapters. The more advanced or time-consuming problems are keyed by a star (★). A complete solutions manual is available to instructors from the publisher. Where feasible, the chapters also include both open-ended design exercises and exercises that require simple use of the computer. To aid in working computer exercises, a special diskette is contained in the instructor's solution manual. Software (programs) on the diskette supports the computer exercises. Readers may obtain a copy of the diskette from their instructor who is authorized by the publisher to freely make copies as required for student use. Each program has been written to be user-friendly so that the student's time is spent actually using the program instead of learning the "overhead" needed to make the program function.

The authors gratefully acknowledge the assistance of several people who have helped in various phases of the book's preparation. Mr. Robert Moore, one of our students, helped prepare much of the software for the computer diskette. Drs. T. Bullock, D. P. Carroll, L. W. Couch, II, R. Fox, S. Miller, and R. L. Sullivan, all of the University of Florida, read various parts of the manuscript and offered improvements. Helpful improvements were also obtained from the following reviewers: Alvin Day, Iowa State University; Anthony England, University of Michigan; Sergio Franco, San Francisco State University; Raj Misra, New Jersey Institute of Technology; Gerald Parks, Michigan State University; Ralph Santoro, U.S. Naval Academy; and Rolf Schanmann, University of Minnesota. Dr. D. Weiner of Syracuse University independently worked all of the book's problems; his efforts have materially reduced the chances of errors occurring in the solutions manual. Mr. G. Dean taught from an early version of the manuscript and provided helpful comments. Ms. Nancy Mishoe typed the bulk of the manuscript and did a superb job. To all these people, we record our sincere thanks. Finally, our wives, Barbara and Lourdes, were most helpful in proofreading assistance, and we are very grateful for their work.

PEYTON Z. PEEBLES, JR.
TAYEB A. GIUMA

PART 1

Basic Concepts and Circuits

CHAPTER 

Elementary Principles

1.0 INTRODUCTION

This book was written for the nonelectrical engineer. Its purpose is to provide aerospace, chemical, civil, industrial, mechanical, and other engineering readers with the basic principles from electrical engineering that will help them in their individual fields. Since the needs of one area of engineering can be quite different from another, it is difficult to satisfy all needs. However, by carefully subdividing electrical engineering into its most important subdisciplines and covering each of these in reasonable depth, it is possible to satisfy most needs.

Nearly all topics in electrical engineering can be classified into one of five categories that form the five parts into which this book has been subdivided. Part 1 (''Basic Concepts and Circuits'') and Part 2 (''Electronic Devices and Circuits'') contain the most fundamental principles, much of which must be studied before proceeding to other material.

Part 3 (''Computers and Digital Devices'') is dedicated entirely to the digital computer. The importance of the computer has grown to the point that it is now indispensable in nearly every type of engineering. The coverage given in this book is a bit more than the typical nonelectrical engineer will need on first reading. However, it is the authors' belief that these readers will require more than cursory knowledge as time goes on, so some effort has been made to be more complete than usual in a book of this type.

Part 4 (''Electrical Communications Systems'') discusses ways in which information is transferred from one place to another by radio means (commercial radio, television, etc.). It also discusses antennas, wave propagation, and noise in electrical systems. Although aerospace and some mechanical and industrial engineers may be interested in several topics of Part 4, and most nonelectrical engineers will find the noise topics beneficial, others may wish to omit this part. It is included so that the needs of all types of nonelectrical engineers are served.

Part 5 (''Power Systems and Machinery'') should prove valuable to all engineers. It gives an overview of how power is generated, distributed, and used in the United States. It also discusses the most common forms of motors and generators. Many engineers should have special interests in motors, either because they have applications in low-power instrumentation servomechanisms or because they are used in high-power applications such as chemical process control, production lines, or industrial machinery.

Broadly defined, electrical engineering is the study and application of *electrical systems*, *networks*, or *circuits* that are formed from an interconnection of electric devices and components. The name *component* usually refers to a resistor, capacitor, inductor, or transformer (passive elements); the name *device* often refers to active devices, such as transistors, vacuum tubes (today used mainly in high-power applications, such as radio and television transmitters, and at microwave frequencies), motors, and generators. However, the names may, and often are, used interchangeably. In a similar manner, a system, network, and circuit often mean similar things. Some intuitive distinction between the three can be gained if a circuit is viewed as a small-scale interconnection of a few devices and components but a network is viewed as a large-scale interconnection, perhaps including some of what we define as circuits. Finally, a system can be viewed as the largest-scale interconnection. Certainly, the power grid in the United States is an electrical system that is made up of many local power networks, each comprising many circuits. However, a local power engineer also works for a power ''system.'' Thus, we see that only an intuitive meaning can be attached to some terms in electrical engineering.

In this chapter we begin Part 1, ''Basic Concepts and Circuits,'' by introducing the most elementary principles that are needed to continue through the book. The first of these relates to electric charge and current.

1.1 ELECTRIC CHARGE AND CURRENT

The reader may recall from elementary physics that the electron is the smallest indivisible quantity of electricity and it is present in all matter. The electrical properties of the electron arise mainly from its *charge* rather than from its physical mass. R. A. Millikan (1868–1953), an American physicist, experimentally determined the magnitude of the electron's charge, which is 1.602×10^{-19} coulomb† (unit abbreviation is C); it was arbitrarily defined as negative. Of

† Named for Charles A. de Coulomb (1736–1806), a French army officer and scientist.

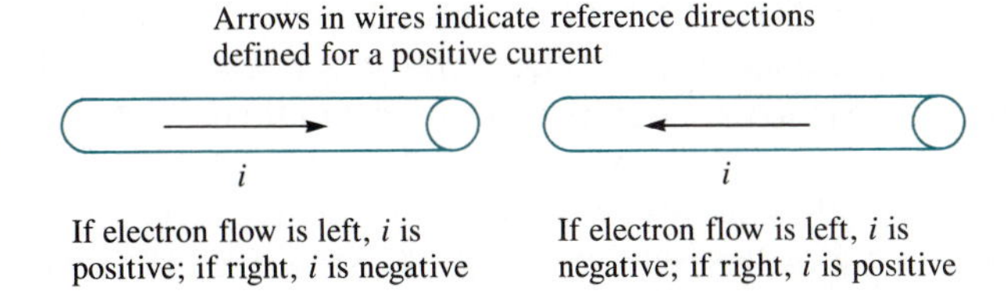

FIG. 1.1-1. Charge flow possibilities in defining currents in a wire.

course, the proton, which is present in the nuclei of all matter, has a charge equal in magnitude to that of the electron but is positive.

We now define an *electric current* as the rate at which charge moves past a point. If i denotes the current and q denotes charge, then

$$i = \frac{dq}{dt} \text{ amperes (A)} \tag{1.1-1}$$

The unit of current is the ampere† (abbreviated A) when q is in coulombs. These units are part of the Système International d'Unités, or SI units, which are adopted for use in this book. Additional details on units are given in Sec. 1.8.

For the most part, currents of interest in electrical systems flow in the components and devices that are interconnected in the system, as well as in the wires that provide the interconnections. Since charge can be positive or negative and can flow in either of two directions in a component or wire, a convention for current must be established. Current is said to be positive in a particular direction, called the *reference direction*, when there is a net flow of positive charges in that direction. This definition establishes *conventional current*, which is well-rooted in historical use.

In real metallic wires current is known to be due to flow of electrons (negative charges) in a direction opposite to the reference direction for conventional current. However, we note that a positive current resulting from a net positive charge flowing in the reference direction is the same as having a net negative charge of the same magnitude but flowing in the opposite direction. Thus, if the reference direction for a wire's current i is to the right, as in Fig. 1.1-1(a), the numerical value of i is positive if electron flow is to the left and negative for flow to the right. If the reference direction is reversed, as in Fig. 1.1-1(b), the sign of i is also reversed for the same electron flow directions as before.

EXAMPLE 1.1-1

In a wire running left to right the reference direction for a current I‡ is to the right. Assume that actual current is due to a flow of 10^{14} electrons per second to the right and find I.

† Named for André Ampère (1775–1836), a French mathematician and physicist.

‡ We shall use capital letters to imply quantities that are constant with time and lowercase letters to imply possible variation with time. A constant current I is called a *direct current* (dc).

The magnitude of the current is the product of the number of electrons flowing per second and the magnitude of the electron's charge: $(10^{14})(1.602 \times 10^{-19}) = 1.602 \times 10^{-5}$ A. Since the charge is negative and flows in the reference direction, the current is negative. Thus, $I = -1.602 \times 10^{-5}$ A.

If the electron flow had been to the left instead, it would have been equivalent to positive charges to the right and I would have been positive.

From the above discussions and example it should be clear that the reference direction can be chosen arbitrarily; the actual (true) current will be either positive or negative, depending on which choice is made.

The mechanical analogy of water flowing through a pipe is sometimes helpful to visualize. The water and water flow are analogous to charge and electric current, respectively.

1.2 ELECTRIC POTENTIAL AND VOLTAGE

Let us continue the analogy between water flow and current that was mentioned in the preceding section as a means of introducing the concepts of electric potential and voltage. Consider a waterwheel used to power a grain mill. Water from a pipe at the top of the wheel (a point of high potential energy in the field of gravity) enters buckets on the wheel's periphery and falls with the turning of the wheel. The water falls through a vertical distance equal to the wheel's diameter. At the bottom (a point of lower potential energy) the buckets empty the water into an exit pipe. Clearly, the water is converting its loss in potential energy into mechanical (kinetic) energy that turns the wheel.

Next, consider a typical two-terminal electric element, as depicted in Fig. 1.2-1. Presume initially that the element (waterwheel) is absorbing energy provided by the current i (water rate) that enters the element at point a (top of waterwheel). According to our convention on current, positive current (water rate) enters point a and exits the element at point b (bottom of waterwheel). Let a

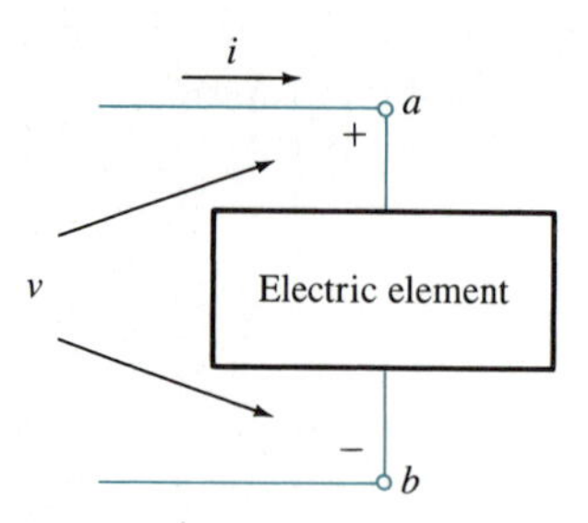

FIG. 1.2-1. An electric element used in defining current and voltage conventions.

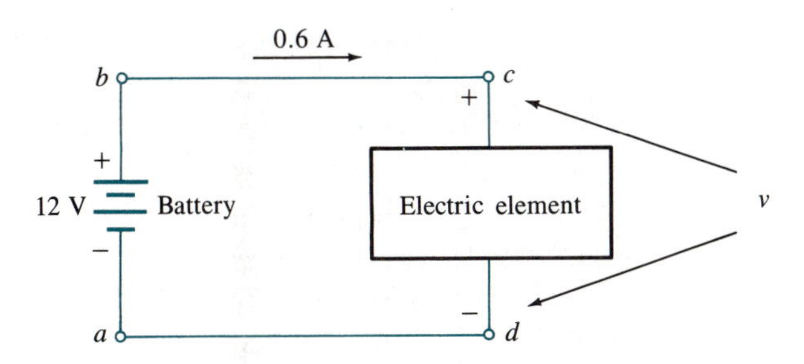

FIG. 1.2-2. An electric circuit containing a 12-V battery.

small charge of magnitude dq pass through the element and release a small amount of energy dw to the element. We associate this energy with a drop in electric potential energy from point a (point of higher potential energy) to point b (lower in potential energy). Finally, we define the energy released per unit charge as the electric *potential difference*, or *voltage*, denoted by v, that exits between points a and b. Thus,

$$v = \frac{dw}{dq} \text{ volts (V)} \tag{1.2-1}$$

The unit of v is the volt† (abbreviated V) when dw is in joules‡ (abbreviated J); both are SI units. The sign convention used in Fig. 1.2-1 is that + is assigned to the point of higher potential.

The polarity marks in our sign convention establish a voltage reference only. With the convention, a positive current entering the positive terminal of the voltage across an element corresponds to that element absorbing energy. A positive current entering the negative voltage terminal means the element is a source of energy. The voltage across an energy source is sometimes called an *electromotive force* (abbreviated emf); this name is not inappropriate since voltage is analogous to pressure in hydraulic systems and is the force that causes current flow.

A good example of a source of electric energy is the ordinary 12-V automobile battery, for which the electrical symbol is shown in Fig. 1.2-2. Since the battery is an energy source, positive current moving from point a to point b corresponds to energy gain, since it enters the negative and leaves the positive terminals, according to our voltage convention. This current traveling from point b to point c constitutes a current with reference direction to the right, as shown by the arrow that indicates the direction of positive current. It can be shown that no potential (voltage) change can occur along an *ideal* wire, so the potentials at b and c are the same, as are those at a and d. Thus, the voltage across the element has to be the same as that of the battery. Finally, since the positive current enters the positive

---

† Named for Alessandro Volta (1745–1827), an Italian physicist.

‡ Named for James P. Joule (1818–1889), a British physicist.

terminal of the element, it must absorb energy, that which is being generated in the battery.

The elementary circuit of Fig. 1.2-2, for which the current that flows is 0.6 A, serves to illustrate the topics of this section by means of an example.

EXAMPLE 1.2-1

For the circuit of Fig. 1.2-2 we determine the energy absorbed by the element during any 2-second (abbreviation is s for time in seconds) time interval.

By solving (1.1-1) for dq and substituting into (1.2-1), we have

$$dw = vi\,dt$$

Since $v = 12$ V and $i = 0.6$ A are constants, we can set $dt = 2$ s to obtain

$$dw = (12)(0.6)(2) = 14.4 \text{ J}$$

which is the energy absorbed in 2 s. Obviously, this energy is being supplied by the battery.

1.3 ELECTRIC ENERGY AND POWER

After having established the fundamental quantities of current and voltage, we may turn our attention to developments of energy and power in electrical systems.

Energy

From (1.2-1) the relation between incremental energy dw, voltage v, and incremental charge can be written as

$$dw = v\,dq = v\frac{dq}{dt}\,dt = vi\,dt \qquad (1.3\text{-}1)$$

after (1.1-1) is used. On integration of (1.3-1) over a time interval from a time t_1 to a later time t_2, we have

$$w = \int_{t_1}^{t_2} vi\,dt \qquad (1.3\text{-}2)$$

as the energy absorbed or provided during a time interval $t_2 - t_1$ by an electric element through which current i flows and across which a voltage v exists.

Power

Electric power is defined as the rate at which electric energy is transferred. If p denotes this power, then (1.3-1) gives

$$p = \frac{dw}{dt} = vi \text{ watts (W)} \qquad (1.3\text{-}3)$$

The unit of power is the watt† (abbreviated W), which is the same as joules per second.

EXAMPLE 1.3-1

We revisit Example 1.2-1 and find the power being absorbed by the electric element.

Here, $v = 12$ V and $i = 0.6$ A. From (1.3-3) $p = 12(0.6) = 7.2$ W.

It is worth noting that the power in (1.3-3) is instantaneous power when v and i are changing with time. In many practical cases it is desirable to determine an average amount of power that occurs between a time t_1 and some later time t_2. If this average power is denoted by P_{av}, then it is defined by

$$P_{av} = \frac{1}{t_2 - t_1} \int_{t_1}^{t_2} vi \, dt \qquad (1.3\text{-}4)$$

In Example 1.3-1 instantaneous and average power (for any times t_1 and t_2) are equal because v and i are constant in time.

1.4 ELEMENTARY ELECTRIC COMPONENTS

Electric circuits are formed by interconnecting various devices and components. In this section we define some of these and give their most important characteristics. These components may then be used in the next section to construct various real circuits.

Resistors

A resistor is a passive component that develops a voltage across its two terminals that is proportional to the current through it:

$$v = iR \qquad (1.4\text{-}1)$$

where R is the proportionality constant called the *resistance* of the resistor. R has the unit ohm,‡ for which the unit symbol Ω is used. A resistor for which (1.4-1) applies is called *linear*. Most resistors used in practice are good approximations to linear resistors for very large ranges of current. Figure 1.4-1(a) illustrates the symbol used to represent a resistor.

---

† Named for James Watt (1736–1819), a Scottish inventor.

‡ Named for George Simon Ohm (1787–1854), a Bavarian mathematician and physicist.

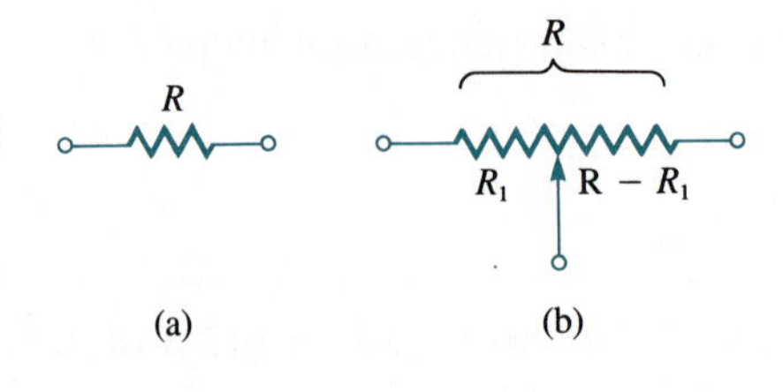

FIG. 1.4-1. Circuit symbols used to represent (a) a fixed resistor of resistance R and (b) a resistor R with a variable tap.

Equation (1.4-1) is known as *Ohm's law*, which first appeared in publication in 1827. Another version of the law is sometimes convenient:

$$i = \frac{v}{R} = Gv \tag{1.4-2}$$

where

$$G = \frac{1}{R} \tag{1.4-3}$$

is called *conductance*. Conductance has the unit siemens† (abbreviation S).

The value of a resistor is determined mainly by the physical dimensions and the resistivity of the material of which it is composed. For a bar of resistive material of length l and cross-sectional area A, R is given by

$$R = \frac{\rho l}{A} \tag{1.4-4}$$

where ρ is the *resistivity* of the material with the unit $\Omega \cdot$m. Table 1.4-1 illustrates

TABLE 1.4-1. Resistivity of Several Materials at 293.15 K (20°C), Except Germanium and Silicon at 300 K

Material	ρ ($\Omega \cdot$m)
Silver	1.63×10^{-8}
Copper, annealed	1.72×10^{-8}
Gold	2.44×10^{-8}
Aluminum, commercial	2.83×10^{-8}
Tungsten	5.52×10^{-8}
Germanium	4.6×10^{-1}
Nichrome (65 Ni, 12 Cr, 23 Fe)	100×10^{-8}
Silicon	2300

† Prior to around 1970 the unit of conductance was the mho (ohm spelled backward) and the abbreviation was ℧ (upside-down Ω). Although there is still considerable use of the old notation, we shall use the siemens, named in honor of Karl Wilhelm Siemens (1823–1883), a British inventor born in Germany, and his older brother Ernst Werner von Siemens (1816–1892), a German inventor.

values of ρ for several materials. The reciprocal of resistivity is called the *conductivity* of the material.

EXAMPLE 1.4-1

An audio enthusiast has a high-fidelity speaker that appears to be a 4-Ω resistance to the circuit supplying power to its terminals. The enthusiast has a pair of 18-gauge (American Wire Gauge, AWG) wires that can be used to connect the speaker to the driving audio unit from a remote distance of 75 m. The wire is annealed copper and has a diameter of 1.024×10^{-3} m. We determine if the resistance of the long connecting wires is large enough to suspect some loss in performance of the speaker.

From (1.4-4) we calculate the resistance R of one wire of 75-m length:

$$R = \frac{(1.72 \times 10^{-8})(75)}{\pi (1.024 \times 10^{-3})^2/4} \approx 1.57 \ \Omega$$

Since two wires are involved (the driving current must pass over one wire to the speaker, through the speaker, and then return over the second wire), the total resistance of the wires is 3.14 Ω. This value is quite significant with respect to the 4 Ω of the speaker, so we suspect that a significant power loss due to the wires will occur, and a loss in performance is likely.

Practical resistors are manufactured in standard values, various resistance tolerances, various power ratings, and in many forms of construction. The standard resistance values are listed in Table 1.4-2 for resistors from 1 to 9.1 Ω; other values are available (from about 1 Ω to about 22×10^6 Ω) in factors of 10. Thus, resistors such as 8.2 Ω, 82 Ω, 820 Ω, etc., are standard values. All values shown are available in a ±5% tolerance (accuracy of resistance), but only certain values in the table are available in other tolerances. For example, only the values in rows 1 and 3 are available in ±10% tolerance. For a ±20% resistor, only the values in row 1 are available.

In many precision resistors of ±1% accuracy or better the resistance value is marked directly on the unit. For the common carbon composition type illustrated in Fig. 1.4-2, standard color-coded bands are used to define resistance. Figure

TABLE 1.4-2. Standard Values of Resistors Having ±5% Resistance Tolerance

1.0	1.5	2.2	3.3	4.7	6.8
1.1	1.6	2.4	3.6	5.1	7.5
1.2	1.8	2.7	3.9	5.6	8.2
1.3	2.0	3.0	4.3	6.2	9.1

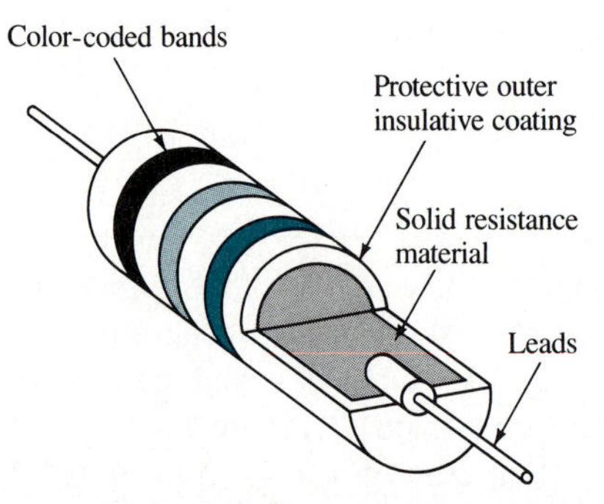

FIG. 1.4-2. A common carbon composition resistor.

1.4-3 gives the color code. According to the code, if bands 1 through 4 have brown, grey, red, silver colors, respectively, the resistor is $18 \times 10^2 = 1800\ \Omega$ and has a $\pm 10\%$ tolerance. Sometimes, a fifth band is present to define reliability, with black being the least reliable and orange being 1000 times more reliable than black.

If we substitute (1.4-1) into (1.3-3), the power dissipated in a resistor is

$$p = i^2 R = \frac{v^2}{R} \quad \text{W} \tag{1.4-5}$$

The energy, which is lost in the form of heat, becomes

$$w = \int_{t_1}^{t_2} p\, dt = R \int_{t_1}^{t_2} i^2\, dt = \frac{1}{R} \int_{t_1}^{t_2} v^2\, dt \tag{1.4-6}$$

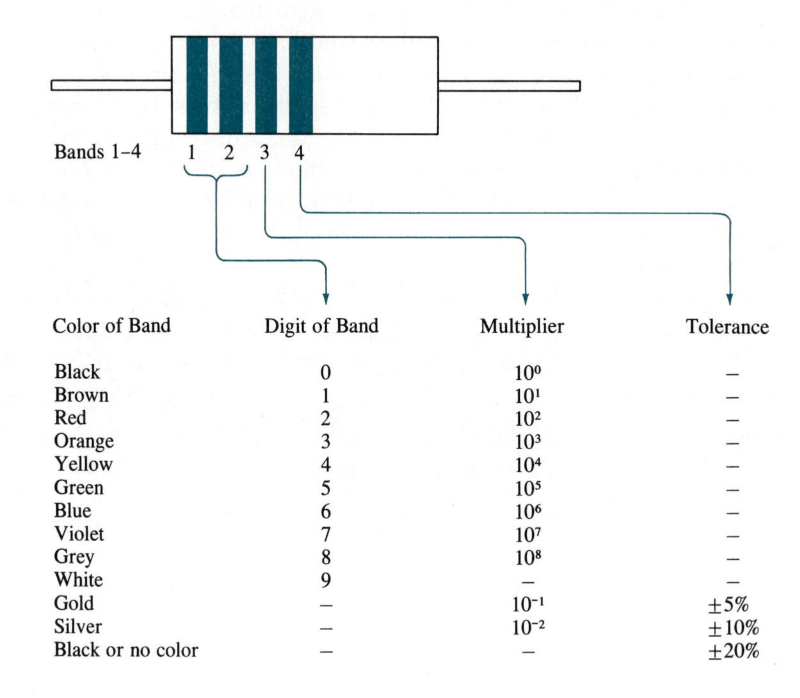

Color of Band	Digit of Band	Multiplier	Tolerance
Black	0	10^0	–
Brown	1	10^1	–
Red	2	10^2	–
Orange	3	10^3	–
Yellow	4	10^4	–
Green	5	10^5	–
Blue	6	10^6	–
Violet	7	10^7	–
Grey	8	10^8	–
White	9	–	–
Gold	–	10^{-1}	$\pm 5\%$
Silver	–	10^{-2}	$\pm 10\%$
Black or no color	–	–	$\pm 20\%$

FIG. 1.4-3. The standard color codes for defining resistance.

TABLE 1.4-3. Common Prefixes

Prefix	Symbol	Meaning
atto	a	10^{-18}
femto	f	10^{-15}
pico	p	10^{-12}
nano	n	10^{-9}
micro	μ	10^{-6}
milli	m	10^{-3}
centi	c	10^{-2}
deci	d	10^{-1}
deka	da	10^{1}
hekto	h	10^{2}
kilo	k	10^{3}
mega	M	10^{6}
giga	G	10^{9}
tera	T	10^{12}
exa	E	10^{15}
peta	P	10^{18}

Example: 4,300,000 Ω = 4.3 MΩ = 4300 kΩ = 43,000 hΩ = 4.3×10^{8} cΩ

from (1.3-2). Practical resistors are limited in their ability to dissipate heat, so they are assigned power ratings, usually $\frac{1}{8}$, $\frac{1}{4}$, $\frac{1}{2}$, 1, and 2 W, although larger ratings are also available. Since heat-dissipation capability is related to surface area, resistors with larger power ratings are physically larger than those with smaller ratings. A $\frac{1}{8}$-W common-composition resistor has a diameter and length of about 1.6 and 3.7 mm, respectively; the same dimensions for a 2-W unit are 7.9 and 17.5 mm. Here, mm stands for millimeter where *milli* is a prefix meaning 10^{-3}. Hence, 1 mm = 10^{-3} m. Other standard prefixes are given in Table 1.4-3.

EXAMPLE 1.4-2

An engineer desires to "bleed" a current of 62.5 mA ± 3.5 mA from a 15-V dc (constant-voltage) source by placing a resistor across its terminals. We determine the resistor needed, its resistance tolerance, and power rating required.

From Ohm's law the nominal resistance needed is 15 V/0.0625 A = 240 Ω. The maximum and minimum currents allowed correspond to resistance extremes of 15/0.066 = 227.27 Ω and 15/0.059 = 254.24 Ω. These resistance extremes are −5.30% and +5.93%, respectively, relative to 240 Ω. Thus, a 240-Ω, ±5% tolerance resistor is adequate. The largest power consumed by the resistor occurs if it is at its extreme low-tolerance value. This power is $v^2/R = (15)^2/228.0 =$ 0.987 W. At least a 1-W resistor is required to handle this power. However, for a margin of safety, the engineer might select a 2-W unit in practice.

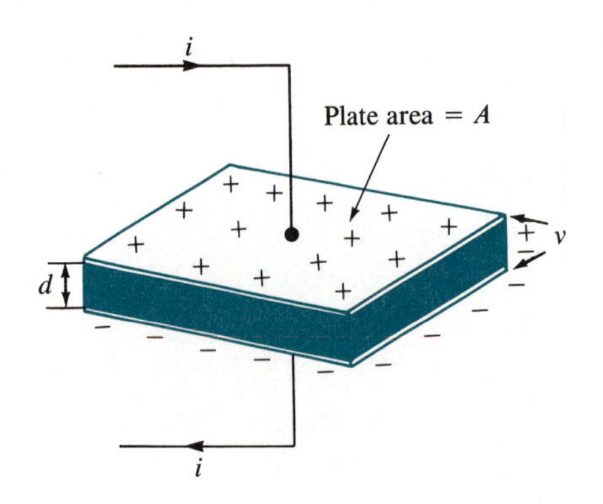

FIG. 1.4-4. Parallel plates forming a capacitor.

Various methods are used to manufacture resistors. The *composition resistor* mentioned above is a low-cost device that uses molding techniques. It has enjoyed widespread applications in past years but is increasingly being displaced by the *diffusion resistor* that is made by using integrated-circuit technology. *Metal-film resistors* employ an insulating substrate on which a resistive film is deposited. These devices can be made to precise tolerances by trimming the film with a laser. *Wire-wound resistors* are based on the resistance of a length of wire wound on a form. Wire-wound components are best used in lower-resistance, higher-power, lower-frequency applications. Their performance as a resistor decreases rapidly for frequencies above a few tens of kilohertz.†

In addition to fixed types, resistors can be adjustable. Figure 1.4-1(b) depicts a resistor designed to have a movable tap. If the tap is connected to either of the other terminals, the resistor becomes variable, ideally from zero to its maximum value R. Three-terminal (adjustable) resistors are usually called *potentiometers*, or *pots* for short. They may be single-turn or multiple-turn rotary, or multiple-turn linear motion. They may be obtained in large, high-power configurations or small, low-power configurations.

Capacitors

Another basic electric component is the *capacitor*. Its operation and definition can best be described by referring to Fig. 1.4-4, which illustrates two parallel metal plates, each with area A and an attached conducting wire, that are separated a distance d by a dielectric (insulator). Now, suppose by some method a *constant* current i flowing into the top plate is established. For such a constant current a steady flow of positive charges is directed to the top plate. However, because the dielectric cannot pass these charges, they "pile up," or accumulate, at the top plate. Because current must be continuous through an electric component, which

---

† Frequency has the unit hertz (same as cycles per second), which is abbreviated Hz. The unit is named in honor of Heinrich R. Hertz (1857–1894), a German physicist.

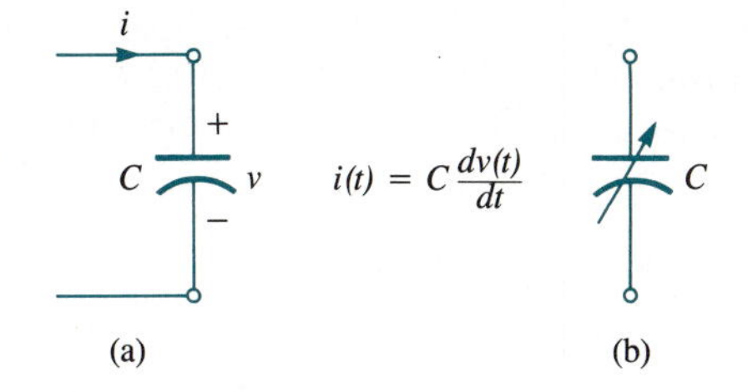

FIG. 1.4-5. (a) The electrical symbol and sign conventions for a fixed capacitor and (b) the symbol for a variable capacitor.

will become clear as we continue in later work, a similar condition must exist at the bottom plate. That is, positive charges must leave the bottom plate to establish the proper exiting current. As these charges exit at a constant rate, they leave behind a net negative charge on the lower plate. The magnitudes of the accumulating charges on the two plates are equal. The charge unbalance between plates causes a difference in electric potential energy, which means a voltage between the plates is established that increases in proportion to the charge increase. In other words, the voltage *rate* is proportional to the charge *rate* (current), so mathematically, we have

$$i = \frac{dq}{dt} = C\frac{dv}{dt} \tag{1.4-7}$$

Here, C is the constant of proportionality called *capacitance*. The unit of C is the farad† (abbreviated F). From (1.4-7) we also find that

$$q = Cv \tag{1.4-8}$$

that is, the charge on a capacitor is proportional to its capacitance and the voltage across it.

The electrical symbol for a capacitance is displayed in Fig. 1.4-5(a), where the proper sign conventions for current and voltage are also given. The symbol for a variable, or adjustable, capacitor is shown in Fig. 1.4-5(b).

By explicitly writing the dependence of i and v on time and integrating (1.4-7), we obtain an alternative form:

$$v(t) = \frac{1}{C}\int_{-\infty}^{t} i(\xi)\,d\xi \tag{1.4-9}$$

The lower limit in (1.4-9) is $-\infty$ because the entire past history of current flow affects the value of the voltage at any time t. By choosing some "initial" time t_0 of interest, we can restate (1.4-9) as

$$v(t) = \frac{1}{C}\int_{-\infty}^{t_0} i(\xi)\,d\xi + \frac{1}{C}\int_{t_0}^{t} i(\xi)\,d\xi = v(t_0) + \frac{1}{C}\int_{t_0}^{t} i(\xi)\,d\xi \tag{1.4-10}$$

† Named for Michael Faraday (1791–1867), a British physicist.

This expression shows that voltage at time t is the voltage at some prior time t_0 plus a contribution due to the current from all time after t_0.

For the parallel-plate capacitor of Fig. 1.4-4, capacitance is approximately given by†

$$C = \frac{\epsilon A}{d} \tag{1.4-11}$$

where ϵ is the *permittivity* of the dielectric medium separating the plates. For vacuum space ϵ is given the symbol ϵ_0 and its value is

$$\epsilon_0 = \frac{10^{-9}}{36\pi} \frac{\text{farad}}{\text{meter}} \left(\frac{\text{F}}{\text{m}}\right) \tag{1.4-12}$$

For other media $\epsilon > \epsilon_0$ is common and the ratio

$$\epsilon_r = \frac{\epsilon}{\epsilon_0} \tag{1.4-13}$$

is a useful quantity called *relative permittivity* (sometimes also called the *relative dielectric constant*). Table 1.4-4 lists the values of ϵ_r that apply at lower frequencies and room temperature (ϵ_r changes with temperature and frequency) for some selected dielectric materials.

EXAMPLE 1.4-3

We use (1.4-11) to compute the capacitance of two round metal plates of diameter 5 mm when separated 0.2 mm by a dielectric for which $\epsilon_r = 6.0$.

$$C = \frac{(6.0 \times 10^{-9})(\pi)(0.0025)^2}{36\pi(0.0002)} \approx 5.21(10^{-12}) = 5.21 \text{ pF}$$

The example above helps to illustrate that the smaller practical values of capacitors range in the low picofarads. The practical upper end ranges in the hundreds of thousands of microfarads.

Ideally, a capacitor does not dissipate any energy in the form of heat because there is no conductive current through an ideal dielectric. However, there is energy exchange between the capacitor and the external circuit because a current flows and a voltage exists. This energy is maintained in the component as *stored* energy. To obtain a relationship for this energy, we develop (1.3-3) by using (1.4-7). The instantaneous power flowing to the capacitor is

$$p(t) = v(t)i(t) = Cv(t)\frac{dv(t)}{dt} \tag{1.4-14}$$

† The approximation is good and becomes best when d is small in relation to the dimensions that establish area A.

TABLE 1.4-4. Relative Permittivity of Selected Materials at Frequencies Near 10^6 Hz and Approximately Room Temperature

Material Class	Material	ϵ_r
Ceramics	Aluminum oxide	8.80
	Steatite 410	5.77
	Tantalum oxide (Ta_2O_5)	27.60
Glass	Soda borosilicate	4.84
	Fused quartz	3.78
Plastics	Polyvinyl-chloride	2.88
	Polystryene	2.56
	Polyethylene	2.26
	Teflon	2.10
Rubbers	Neoprene	6.26
	Silicone	3.20
Other	Mica, various types	5.40–9.00
	Paper	2.99
	Water, distilled	78.20
	Air	1.00

Integration over all past time up to the present time gives the energy stored in the capacitor:

$$w(t) = \int_{-\infty}^{t} p(\xi)\, d\xi = C \int_{-\infty}^{t} v(\xi) \frac{dv(\xi)}{d\xi}\, d\xi \tag{1.4-15}$$

This integral evaluates directly by parts to give

$$w(t) = \tfrac{1}{2} C v^2(t) \tag{1.4-16}$$

if $v(t = -\infty) = 0$, which can be assumed. Alternatively,

$$w(t) = \frac{q^2(t)}{2C} \tag{1.4-17}$$

from substitution of (1.4-8) into (1.4-16). These last two expressions show that the instantaneous energy being stored in the capacitor is proportional to the square of either its instantaneous voltage or its charge on one plate.

Practical capacitors come in a wide range of values, shapes, sizes, voltage ratings, and constructions. Both fixed and adjustable devices are available [see Fig. 1.4-5(b) for the electrical symbol of an adjustable capacitor.] Capacitance values follow those listed in Table 1.4-2 for the most part, but some differences occur. The most popular values are those listed in the first row of the table. Most

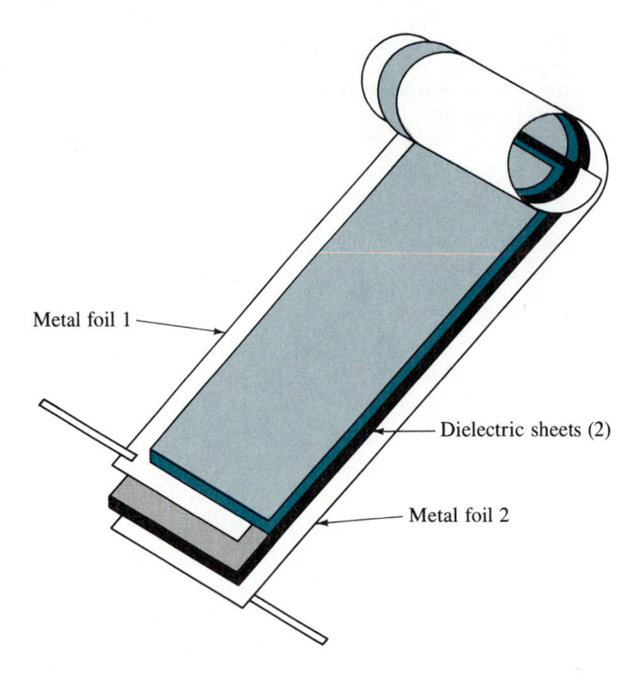

FIG. 1.4-6. The rolling of parts to form a paper or plastic-film capacitor.

capacitors have the capacitance value marked on them, but a few use a code of colored dots, with the colors defined as they were previously for resistors. Voltage ratings are based on the electrical breakdown of the dielectric used in manufacture and range from only a few volts to thousands of volts for capacitors used in high-power applications.

The *ceramic capacitor* finds heavy application in the low to middle range of capacitances (1 pF to 0.1 μF). Typically, one or more layers of thin ceramic dielectric of high permittivity are interlaced with layers of metal. Alternate layers of the conductors are connected to form one "side" of the capacitor. The others form the other "side." The entire unit is then encapsulated in a suitable insulator. Versions with wire leads or with soldering pads for circuit board mounting are both available.

The low to middle range of capacitance is also available from the *mica capacitor*. Its construction is somewhat similar to the ceramic capacitor, except that layers of metallic foil are separated by the dielectric mica. The *silvered-mica capacitor* is a variation, where the conducting metal is a thin layer of silver deposited on the mica.

In the middle range of capacitance values (0.001 to 1 μF), *paper* and the *plastic-film capacitors* are often used. Both are based on the same construction principle. Metal-foil conductors are rolled together with interlaced dielectric sheets (wax- or resin-impregnated kraft paper in the former case and a suitable plastic film, such as polystyrene, in the latter case), as shown in Fig. 1.4-6. The result is a capacitor of tubular form.

Larger capacitors (0.1 to 10^6 μF) are of the *electrolytic* type. In the *polarized*

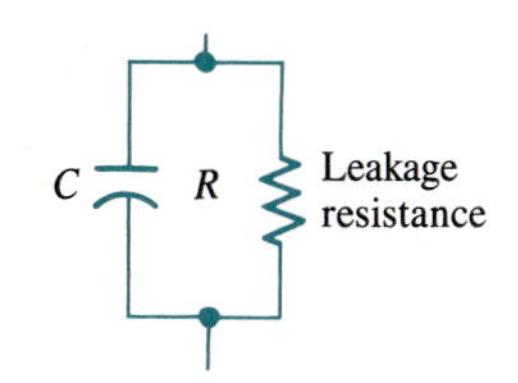

FIG. 1.4-7. A model of a more practical capacitor having losses accounted for by a leakage resistance.

variety the two metal plates are separated by an electrolyte that oxidizes the inside of one of the plates. The oxide acts as the dielectric, and the electrolyte in contact with the other plate acts as one of the capacitor terminals. When the oxide is being formed, a voltage of specific polarity is applied across the capacitor, and any voltage applied in any application must be such as to not reverse the specified polarity. *Nonpolarized electrolytic capacitors* have been developed that do not have to maintain constant polarity of voltage. When time-varying voltages are encountered that may have sign reversals, the nonpolarized electrolytic capacitor should be used.

Real capacitors are not perfectly lossless. Resistive losses can occur in both the metal plates, because of nonzero resistance, and in the dielectrics. A model to account for losses often uses a *leakage resistance* in parallel with an ideal capacitor, as shown in Fig. 1.4-7. This resistance is not constant with frequency, in general. At lower frequencies it is usually quite large (hundreds of megohms) and represents no problem in most applications, so it can be neglected. At higher frequencies its effects may need to be included.

Inductors

When current flows in a conductor, a magnetic field is set up around, and is proportional to, the current. These fields are visualized as lines of magnetic flux, as shown in Fig. 1.4-8(a) for a current i flowing in a metal wire. The density of flux lines indicates the strength of the magnetic field in a region near the wire. Field strength is largest near the wire and decreases inversely with distance from the wire. The validity of such fields was established in 1820 by Hans Christian Oersted (1775–1851), a Danish scientist, when he observed that a magnetic compass needle was deflected and aligned with the field (much as it does in the earth's magnetic field).

The direction of flux lines ("north") is established by the *right-hand rule*. If the wire is grasped in the right hand with the thumb outstretched and pointing in the direction of current flow, the fingers wrap around the wire pointing in the direction of the flux. The reader may want to check this rule against Fig. 1.4-8(a).

If the conductor is wound into a loop, an elementary *inductor* is formed, as shown in Fig. 1.4-8(b). Magnetic flux lines become concentrated inside the loop, and a certain amount of flux will permeate the area bounded by the loop. If the

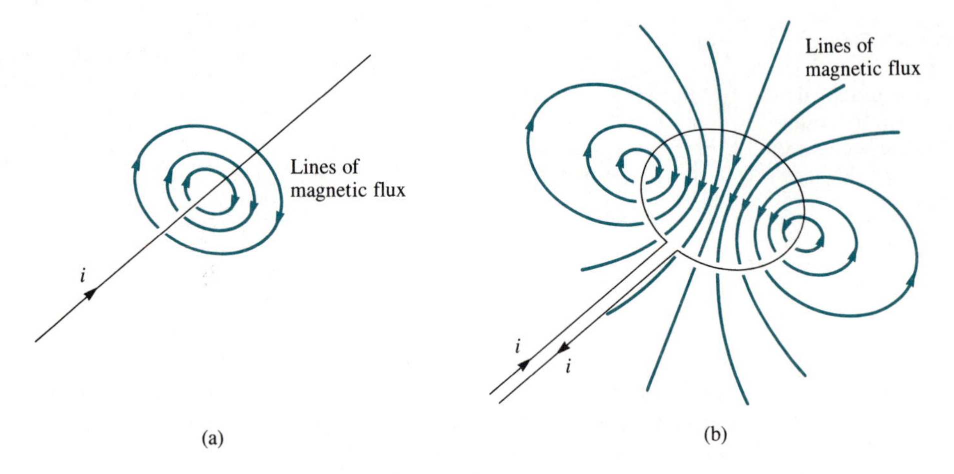

FIG. 1.4-8. Magnetic lines of flux around conductors: (a) around a straight wire and (b) around a wire loop.

number of turns (loops) is increased to N, the flux through the area will increase to a level as much as N times its single-turn value. The exact amount of increase depends on some geometrical factors. Ideally, all the total flux would pass through the area of all turns. Flux coupling is called linking. The total linkage, called *flux linkage* and given the symbol λ, can be as large as N times the total flux. The unit of flux linkage is the *weber*† (abbreviated Wb) after Wilhelm E. Weber (1804–1891), a German physicist.

The importance of flux linkages is rooted in some early experiments performed around 1831 by Michael Faraday that showed that the voltage drop v across the terminals of a *coil* (conductor of multiple turns), caused by a current i that produces a flux linkage λ, is

$$v = \frac{d\lambda}{dt} \tag{1.4-18}$$

Equation (1.4-18) is called *Faraday's law* of induction. It is also found that v is proportional to the rate at which current is changing:

$$v = L\frac{di}{dt} \tag{1.4-19}$$

where L is a constant of proportionality called *inductance*.‡ The unit of inductance is the *henry* (abbreviated H), named in honor of Joseph Henry (1797–1878), an American inventor. Joint consideration of the last two expressions gives

$$\lambda = Li \tag{1.4-20}$$

which shows that inductance is the proportionality constant relating current to

---

† Occasionally, the unit weber-turns (Wb • t) is used to account for the product of flux (webers) and the number of turns linked to get flux linkages.

‡ More precisely, L is the *self-inductance* of the coil. This distinction becomes important later when coupled magnetic circuits (transformers) are introduced.

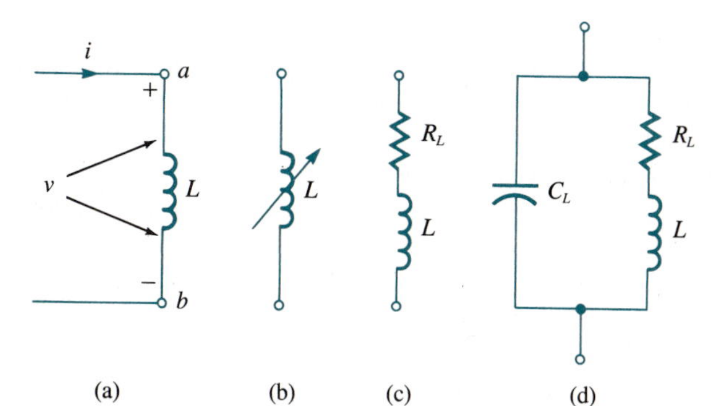

FIG. 1.4-9.
(a) The circuit symbol for an ideal fixed inductor of inductance L, (b) a variable inductor, (c) a more exact low-frequency model accounting for coil resistance, and (d) a still more exact model accounting for distributed capacitance of turns.

flux linkages. On integration, (1.4-19) becomes

$$i(t) = \frac{1}{L} \int_{-\infty}^{t} v(\xi)\, d\xi \tag{1.4-21}$$

If we are most interested in the current after some time t_0, then

$$i(t) = \frac{1}{L} \int_{-\infty}^{t_0} v(\xi)\, d\xi + \frac{1}{L} \int_{t_0}^{t} v(\xi)\, d\xi = i(t_0) + \frac{1}{L} \int_{t_0}^{t} v(\xi)\, d\xi \tag{1.4-22}$$

This last expression shows that current at time t equals current at time t_0 plus a contribution due to voltage at all times after t_0.

The circuit symbol for an arbitrary linear ideal fixed inductor is given in Fig. 1.4-9(a). An adjustable, or variable, inductor is symbolized in Fig. 1.4-9(b). The sign of the voltage across the inductor in Fig. 1.4-9(a) is also shown. If a current i is flowing into one terminal (terminal a in the figure) and is *increasing*, the voltage of that terminal is positive with respect to the other and is given by (1.4-19).

The idealized inductor assumes no resistance in the conductor. Energy flowing into the device from the external circuit that provides the current is stored in the magnetic fields of the inductor. By use of (1.3-3) we calculate the instantaneous power flowing into the inductor to be

$$p(t) = v(t)i(t) = Li(t)\frac{di(t)}{dt} \tag{1.4-23}$$

after substitution of (1.4-19). On integration of (1.4-23) to obtain energy, we have

$$w(t) = \int_{-\infty}^{t} p(\xi)\, d\xi = L \int_{-\infty}^{t} i(\xi)\frac{di(\xi)}{d\xi}\, d\xi = \frac{Li^2(t)}{2} \tag{1.4-24}$$

where $i(-\infty) = 0$ has been assumed (a reasonable assumption). On substitution of (1.4-20) we have an alternative energy relationship.

$$w(t) = \frac{\lambda^2(t)}{2L} \tag{1.4-25}$$

Thus, the instantaneous energy stored in the magnetic field of an inductor is proportional to the square of either the instantaneous current or the flux linkages.

In some applications the resistance of the wire from which the inductance is wound cannot be ignored. A better model in these cases is shown in Fig. 1.4-9(c). Another practical effect derives from capacitance occurring between conductors in the multiple turns. This capacitance is a distributed quantity and cannot be modeled exactly. However, the model given in Fig. 1.4-9(d) is quite good in many cases over a limited range of frequencies.

Practical inductors range from about 0.1 μH to hundreds of millihenries. Some, usually for special applications such as *filter chokes* in power supplies, can even have values as large as several henries. In general, the larger the inductance, the lower its frequency is in use. The smallest inductances tend to be most used at radio frequencies. Many inductors are available in standard values, somewhat like resistors, except that the number of values per decade is smaller. Similarly, units in several standard inductance tolerances may be found.

Inductors are wound in many ways. For lower inductance values, a single layer of turns forming a cylindrical shape called a *solenoid* is often used. The layer is typically wound on a low-loss dielectric form. For larger inductances multiple-layer solenoids are common. The volume inside the inductor's turns is called the *core*. Air-core coils give good performance (low loss, constancy of inductance) versus frequency but have low inductance. If a ferromagnetic material, such as iron, steel, powdered iron, or ferrite, is inserted into the core, inductance can be increased. If the position of the material is adjustable along the axis of a coil, inductance can be made adjustable. Some small fixed inductors are encapsulated in an insulating material and appear somewhat like resistors.

As an illustration of a real inductor, we modify slightly a formula given by Terman† for a single-layer solenoid of length l, with core area A (radius r) and n turns of round wire, as shown in Fig. 1.4-10(a):

$$L = \frac{\mu n^2 A}{l + \frac{9}{10} r} \tag{1.4-26}$$

Here, L is in henries when A is in square meters and l and r are in meters. The constant μ is called the *permeability* of the core material. For a vacuum μ is given the symbol μ_0 and its value is $4\pi \times 10^{-7}$ H/m (same as Wb/A·m). For air $\mu \approx \mu_0$ can be assumed. For other materials a *relative permeability,* denoted by μ_r, is defined by

$$\mu_r = \frac{\mu}{\mu_0} \tag{1.4-27}$$

Generally, $\mu_r > 1$ and can be as high as several thousand (typically) or even several hundred thousand for some materials.

---

† Equation (37), p. 55, of F. E. Terman, *Radio Engineers' Handbook*, McGraw-Hill, New York, 1943, is modified for our notation and its extension to include non-air cores.

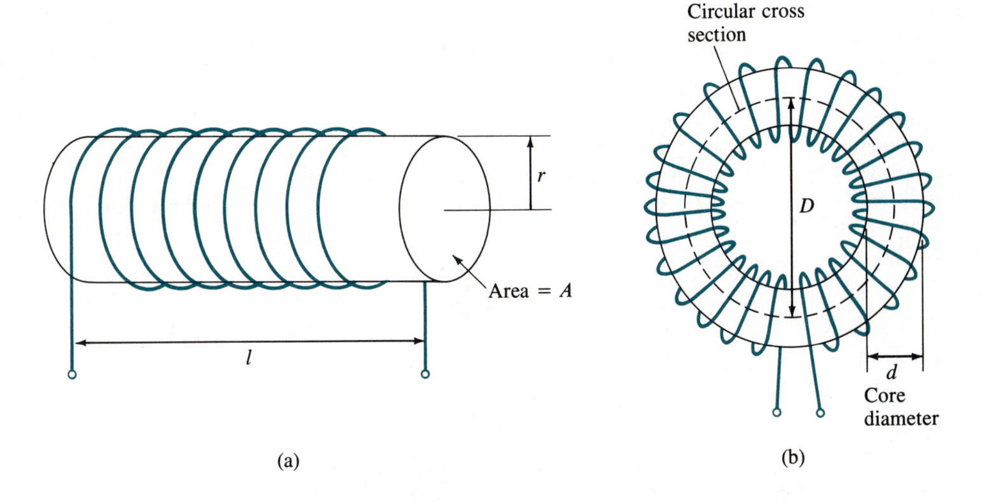

FIG. 1.4-10. Construction of (a) a single-layer solenoid and (b) a toroidal inductor.

EXAMPLE 1.4-4

To illustrate inductance, we determine how many turns wound on a 6-mm-diameter air core are needed to form a solenoidal inductor of 5-cm length and 2.0-μH inductance. Here, $\mu = 4\pi \times 10^{-7}$, $A = \pi(0.003)^2$ m^2, $l = 0.05$ m, and $r = 0.003$ m. From (1.4-26)

$$n = \left\{\frac{(2 \times 10^{-6})[0.05 + (\frac{9}{10})(0.003)]}{(4\pi \times 10^{-7})(\pi)(0.003)^2}\right\}^{1/2} \approx 54.47 \text{ turns}$$

For the length specified the wire diameter cannot exceed $0.05/54.47 = 0.918$ mm.

As a final example of a real inductor, we consider the *toroid* of Fig. 1-4-10(b) having a diameter of revolution D (in meters) and coil diameter d (in meters). Inductance is†

$$L = \frac{\mu n^2}{2}\left(D - \sqrt{D^2 - d^2}\right) \tag{1.4-28}$$

in henries.

Transformers—General

A transformer is created when two inductors are arranged so that some of the flux of either of the two is linked with the other. Figure 1.4-11(a) depicts this effect for two inductors of individual inductances L_1 and L_2. Both L_1 and L_2 are now most

† Equation (42), p. 57, of F. E. Terman, *Radio Engineers' Handbook*, McGraw-Hill, New York, 1943, has been modified slightly.

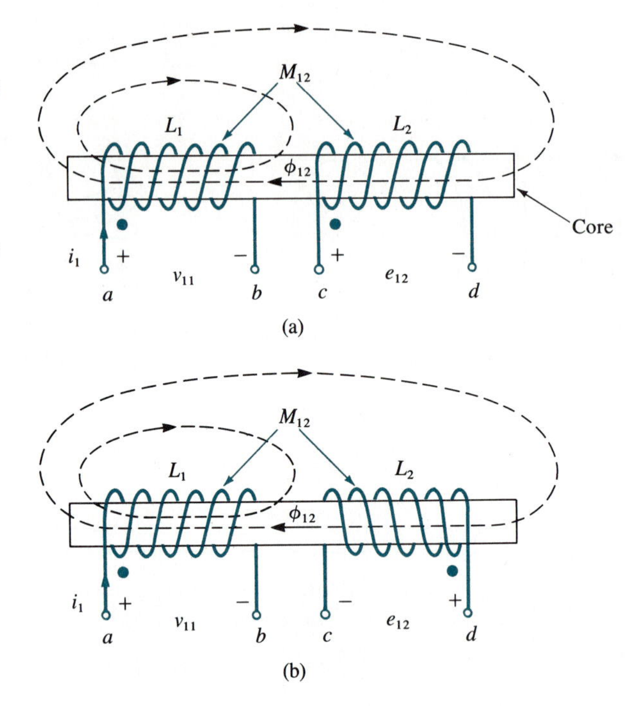

FIG. 1.4-11. Transformers with windings (a) in the same direction and (b) in opposite directions. The dots identify voltage polarities (see the text).

appropriately called *self-inductances*, although we shall continue to refer to them as simply inductances. The core, as shown, can simply be a hollow dielectric form for an air-core transformer, or it can be a magnetic material such as ferrite. If an increasing current i_1 is flowing into terminal a, two things happen. First, a voltage drop v_{11} occurs across inductance L_1 in the usual manner, given by

$$v_{11} = L_1 \frac{di_1(t)}{dt} \tag{1.4-29}$$

Second, some of the flux, denoted ϕ_{12}, couples, or links, the windings of inductance L_2. For the winding directions shown, an induced voltage, denoted by e_{12}, occurs in the second winding that is given by

$$e_{12} = M_{12} \frac{di_1(t)}{dt} \tag{1.4-30}$$

Its polarity is indicated in the figure. Here, M_{12} is called the *mutual inductance* between windings 1 and 2. In a similar manner, if an increasing current flows into terminal c, call it i_2, a voltage, denoted by e_{21}, is induced into winding 1 that is positive at terminal a and given by

$$e_{21} = M_{21} \frac{di_2(t)}{dt} \tag{1.4-31}$$

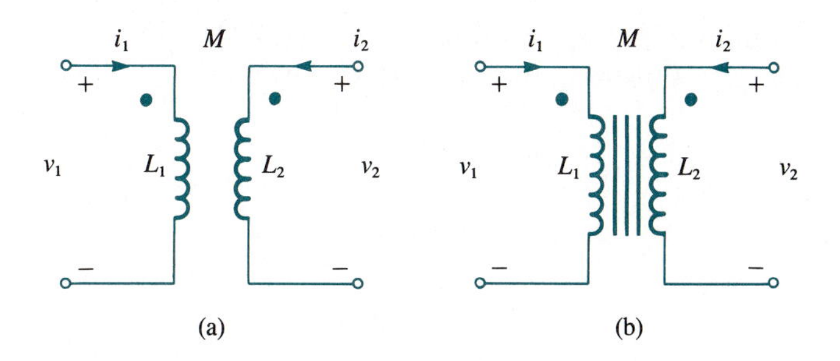

FIG. 1.4-12. (a) The circuit symbol and polarity notation for an idealized transformer and (b) another symbol occasionally seen.

Here, M_{21} is the mutual inductance between windings 2 and 1. Ordinarily, $M_{21} = M_{12}$; so subscripts are unimportant, and we can define

$$M = M_{12} = M_{21} \tag{1.4-32}$$

The normal voltage drop, denoted by v_{22}, across terminals c and d is

$$v_{22} = L_2 \frac{di_2(t)}{dt} \tag{1.4-33}$$

If we place dots on terminals a and c, it is clear that *an increasing current entering either dotted terminal induces a voltage that is positive at the other dotted terminal.* The dots provide a convention for keeping track of voltage polarities that occur in transformers.

To further illustrate the dot convention, let us suppose that one winding (the second) is reversed, as shown in Fig. 1.4-11(b). The induced voltage e_{12} is now positive at terminal d, but there is no ambiguity as long as the dot is placed at terminal d, as shown. The dot convention removes the necessity of specifying winding directions. It becomes a simple matter to model an idealized transformer by using the circuit symbol of Fig. 1.4-12(a). Sometimes the symbol of Fig. 1.4-12(b) is used to imply a magnetic core such as iron, steel, or ferrite.

One side of a transformer is called the *primary;* it is normally the side to which a source is connected. The other side is called the *secondary*. The total voltage on one side of a transformer is the sum of the drop due to the self-inductance of that side plus the voltage induced from action of the current in the other side. Thus, total voltages v_1 and v_2 in Fig. 1.4-12(a) are

$$v_2(t) = v_{22}(t) + e_{12}(t) = L_2 \frac{di_2(t)}{dt} + M \frac{di_1(t)}{dt} \tag{1.4-34}$$

$$v_1(t) = v_{11}(t) + e_{21}(t) = L_1 \frac{di_1(t)}{dt} + M \frac{di_2(t)}{dt} \tag{1.4-35}$$

A nonidealized transformer is one in which winding resistance and distributed capacitances are not negligible, including capacitance between the two windings. In this book we shall not need to consider these complications further.

The instantaneous power flowing into the transformer is

$$p(t) = v_1(t)i_1(t) + v_2(t)i_2(t) \tag{1.4-36}$$

After substituting for $v_2(t)$ and $v_1(t)$ from (1.4-34) and (1.4-35), we integrate to obtain the energy flowing into the transformer:

$$\begin{aligned} w(t) &= \int_{-\infty}^{t} p(\xi)\, d\xi \\ &= L_1 \int_{-\infty}^{t} i_1(\xi) \frac{di_1(\xi)}{d\xi}\, d\xi + M \int_{-\infty}^{t} i_1(\xi) \frac{di_2(\xi)}{d\xi}\, d\xi \\ &\quad + L_2 \int_{-\infty}^{t} i_2(\xi) \frac{di_2(\xi)}{d\xi}\, d\xi + M \int_{-\infty}^{t} i_2(\xi) \frac{di_1(\xi)}{d\xi}\, d\xi \end{aligned} \tag{1.4-37}$$

The first and third integrals readily evaluate to $L_1 i_1^2(t)/2$ and $L_2 i_2^2(t)/2$, respectively. The second and fourth integrals evaluate together by parts to give $Mi_1(t)i_2(t)$. Hence,

$$w(t) = \frac{L_1 i_1^2(t)}{2} + \frac{L_2 i_2^2(t)}{2} + Mi_1(t)i_2(t) \tag{1.4-38}$$

is the instantaneous energy. Equation (1.4-38) can also be written as

$$w(t) = \frac{1}{2}\left\{\left[\sqrt{L_1}\, i_1(t) + \frac{M}{\sqrt{L_1}}\, i_2(t)\right]^2 + \left(L_2 - \frac{M^2}{L_1}\right) i_2^2(t)\right\} \tag{1.4-39}$$

The energy flowing into the transformer cannot be negative; otherwise, it would *generate* energy. To guarantee $w(t) \geq 0$ in (1.4-39) for all values of currents and time requires

$$M \leq \sqrt{L_1 L_2} \tag{1.4-40}$$

A quantity, denoted by k and defined by the ratio

$$k = \frac{M}{\sqrt{L_1 L_2}} \tag{1.4-41}$$

is called the *coupling coefficient*, which is bounded according to

$$0 \leq k \leq 1 \tag{1.4-42}$$

Transformer—Ideal

An *ideal transformer* results from the special case where the coupling coefficient is at its maximum possible value $k = 1$. This situation is approached in practice by placing the primary and secondary windings close together and winding them on a high-permeability core that closes on itself to form a continuous magnetic path. A good example of such a transformer would result if a second winding were to be added to the toroidal inductor of Fig. 1.4-10(b). The goal of all these techniques is to cause all the flux of any one winding to fully link the other winding. For the ideal transformer, therefore,

$$M = \sqrt{L_1 L_2} \tag{1.4-43}$$

If (1.4-34) is solved for $di_2(t)/dt$ and then substituted into (1.4-35), we get

$$v_1(t) = \left(L_1 - \frac{M^2}{L_2}\right)\frac{di_1(t)}{dt} + \frac{M}{L_2}v_2(t) \tag{1.4-44}$$

Since (1.4-43) applies to an ideal transformer, (1.4-44) reduces to

$$v_1(t) = \sqrt{\frac{L_1}{L_2}}\,v_2(t) = \frac{N_1}{N_2}v_2(t) \tag{1.4-45}$$

The last form in (1.4-45) recognizes that L_1 and L_2 are closely approximated as being proportional to the squares of the numbers of turns, denoted N_1 and N_2, respectively, in their windings. Another form of (1.4-45) is

$$n = \frac{N_1}{N_2} = \frac{v_1(t)}{v_2(t)} \tag{1.4-46}$$

where n is defined as the *turns ratio* of the windings.

Next, observe that an ideal transformer is lossless (or nearly so, in practice), so that any power into one winding must leave the other. In other words, the net power flowing *into* the transformer must be zero. From (1.4-36) this constraint gives

$$i_1(t) = -\frac{v_2(t)}{v_1(t)}i_2(t) = -\frac{N_2}{N_1}i_2(t) \tag{1.4-47}$$

after use of (1.4-46).

The principal results for ideal transformers are (1.4-46) and (1.4-47). If $N_1 < N_2$, then $v_1(t) < v_2(t)$, and the transformer is called a *step-up transformer*. If $N_1 > N_2$, the opposite is true, and it is called a *step-down transformer*. The ability to step voltage up or down is very valuable in the design of power supplies, in commercial power distribution systems, and in many industrial processes. The common 110-V power line feeding a typical home is the result of the power company's nearby transformer stepping down voltage from a higher-voltage distribution line.

Along with voltage step-up/down, there is an attendant step-down/up in current from (1.4-47). This fact must be true to maintain equal power in and out of the transformer.

1.5 ELECTRIC SOURCES

An electric source is defined as any device capable of generating power. In this section sources are briefly discussed, and their circuit symbols and mathematical models are defined.

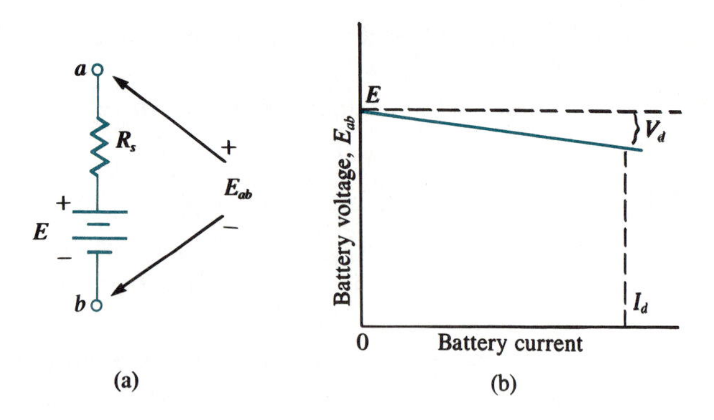

FIG. 1.5-1. (a) The symbol for an ideal battery with internal resistance to approximate a real battery and (b) its terminal voltage under current loads.

Batteries

A *battery* is a device ideally capable of providing a constant voltage independent of the current flowing through it. The symbol for a battery has already been given in Fig. 1.2-2. Ideal batteries do not exist. If one did, it would sustain an infinite current for an indefinite period if its terminals were connected together (*short-circuited*). Real batteries have both maximum current ratings and *internal resistance*, which helps limit the current that can be drawn. The effect of the internal resistance is to cause the battery's voltage to decrease slightly under load. A circuit symbol for a battery with open-circuit voltage E and internal resistance R_s is shown in Fig. 1.5-1(a). Its terminal voltage, E_{ab}, decreases linearly under current load, as shown in Fig. 1.5-1(b). The voltage droop, V_d, occurring at current I_d, is the drop that occurs across R_s:

$$V_d = I_d R_s \tag{1.5-1}$$

EXAMPLE 1.5-1

Assume a transistor radio's battery is rated at 9 V (open-circuited) but droops to 8.5 V when providing a current of 200 mA. We find its internal resistance from (1.5-1) to be

$$R_s = \frac{9.0 - 8.5}{0.2} = 2.5\ \Omega$$

Practical batteries come in a variety of sizes, voltages, and constructions. Voltage is typically generated by chemical action of an electrolyte placed between two electrodes of dissimilar metals. The electrolyte is usually a liquid or a paste, but in some newer devices it is solid. The chemical process leads to what is called a voltage *cell*, with a voltage from about 1 V to over 3.5 V, depending on material used. Larger batteries are made by adding several cells in series. A 12-V automobile battery, for example, is often made up from six 2-V cells, where the negative electrode is lead, the positive electrode is lead peroxide, and the electrolyte is water-diluted sulfuric acid.

Ordinary flashlight batteries, called dry cells, have a voltage of 1.5 V and come in several physical sizes (AA, C, and D are examples) and capacities. Standard-duty, heavy-duty, and long-life versions differ mainly in the type of electrolyte used. The long-life type is usually called an *alkaline cell*, because of the strongly alkaline solution of potassium hydroxide used as the electrolyte.

Other long-life batteries used in cameras, cardiac pacemakers, and some microcomputer memory circuits are called *lithium cells*, because lithium is used in its construction. Some lithium cells can last over 10 years on the shelf and for several years in circuit use, depending on the current required.

When energy is extracted from a battery, it is *discharging*. When a point is reached where the battery can no longer provide rated currents and voltage, the unit is said to be discharged. Most flashlight batteries when discharged must be replaced. Others, like the automobile battery, are designed to be *recharged*. The recharging process calls for a specified current to flow for some period of time through the battery opposite to its normal direction in *providing* energy. Thus, charging current flows into the battery's positive terminal. After the charging period of time the battery can again be used.

Ideal Independent Sources

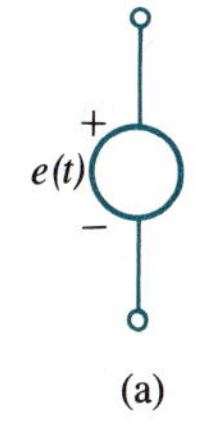

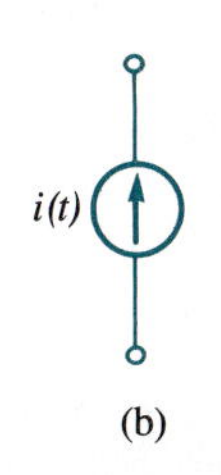

FIG. 1.5-2. Circuit symbols for (a) a time-varying source of voltage $e(t)$ and (b) a time-varying source of current $i(t)$.

A battery is an independent dc (direct current) voltage source. Many examples of other voltage sources exist. Generators, to be studied more in Chapter 18, are electromechanical devices that produce either dc or time-varying (sinusoidal) voltages in response to mechanical rotation of a shaft. The generator in a modern automobile generates an alternating (sinusoidal) voltage and is called an *ac* (alternating current) source; these devices are also called *alternators*. Special circuits convert the ac voltage to dc voltage to keep the battery charged. The circuit symbol for an ideal time-varying source of voltage $e(t)$ is illustrated in Fig. 1.5-2(a). Note that the voltage of an ideal source does not change regardless of the current passing through it. Since it does not "impede" current flow, it has zero internal *impedance*, a quantity to be defined later.

Other voltage sources exist. *Thermocouples* produce small dc voltages that are related to the temperature difference between two points. A *photovoltaic* (solar) *cell* produces a dc voltage related to the intensity of incident light. *Piezoelectric devices*, which are often some form of crystal (such as quartz) or ceramic (such as barium titanate), produce voltages related to pressure or stress applied to the crystal. A *microphone* produces a voltage proportional to sound pressure. Finally, a dc *power supply* is a collection of electronic circuits that converts an ac voltage to a dc voltage. Power supplies are found in nearly every electronic system, from commercial radios, tape recorders, stereos, remote-control toys, automobile electronics, computers, and electronic games, to military missiles, radar, navigation devices, control systems, and aircraft altimeters.

Power supplies can even be designed as *current sources*. An ideal indepen-

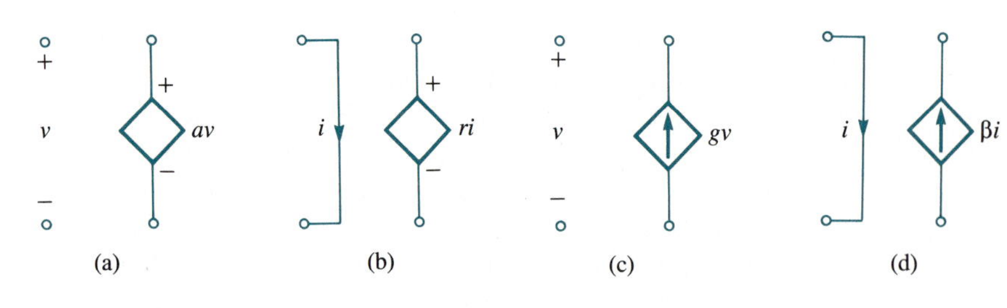

FIG. 1.5-3. Ideal voltage sources dependent on (a) another voltage and (b) a current. Ideal current sources dependent on (c) a voltage and (d) another current.

dent current source provides a given current between its two terminals regardless of the voltage that may exist across its terminals. The circuit symbol for such a source is shown in Fig. 1.5-2(b) for a possibly time-varying current $i(t)$. The direction of the arrow is the direction of flow when $i(t)$ is positive. When voltages are caused to occur across the terminals of a current source due to other sources in a network, the voltages have no effect on the value of the current produced by the current source. The current source, therefore, appears as an infinite impedance (between its terminals) to the other sources.

Ideal Dependent Sources

Quite often in electronic circuits, a source of either voltage or current is dependent on some other quantity, itself often a voltage or a current. There are four possible *ideal dependent sources*. The circuit symbols defining these cases are shown in Fig. 1.5-3. Here, a diamond-shaped symbol is used to imply a dependent source. In every case either a voltage or a current source is dependent on, and proportional to, another current or voltage. The constants of proportionality are a, r, g, and β. Both a and β are unitless and may be considered as *voltage-gain* and *current-gain* constants, respectively. Constant r has the unit of ohms and is called *transresistance*, after *transfer resistance*. Constant g is called *transconductance* for a similar reason, since its unit is the siemens.

We shall encounter dependent sources often in the later studies of electronic devices such as transistors.

1.6 ELECTRIC CIRCUITS

In this section some basic definitions involved with electric networks are given and some elementary networks are introduced. The more powerful and general methods of analyzing networks are given in Chapters 3 (dc networks) and 4 (ac networks).

The following definitions are needed:

Circuit. An interconnection of electric components, sources, and devices.

Network. A connection of two or more circuit components, sources, and devices (the terms *network* and *circuit* are synonymous).

Node. A point at which two or more components or devices are connected together.

Branch. A part of a circuit containing only one component, source, or device between two nodes.

Loop. A closed path through a circuit in which no electric element or node is encountered more than once.

Mesh. A loop that contains no other loops.

Some example circuits that illustrate the definitions above are given in Fig. 1.6-1. These circuits all have independent sources and no dependent sources. The simple circuit of Fig. 1.6-1(a) has 4 branches; 4 nodes, as shown; 1 loop; and 1 mesh. The slightly more complicated circuit of Fig. 1.6-1(b) has 7 branches; 6 nodes, as marked; 3 loops, marked $L1$, $L2$, and $L3$; and 2 meshes. The network in Fig. 1.6-1(c) may appear at first glance to have 4 nodes. However, on redrawing, as shown in Fig. 1.6-1(d), it is clear that it has 4 branches, 2 nodes, 6 loops (the

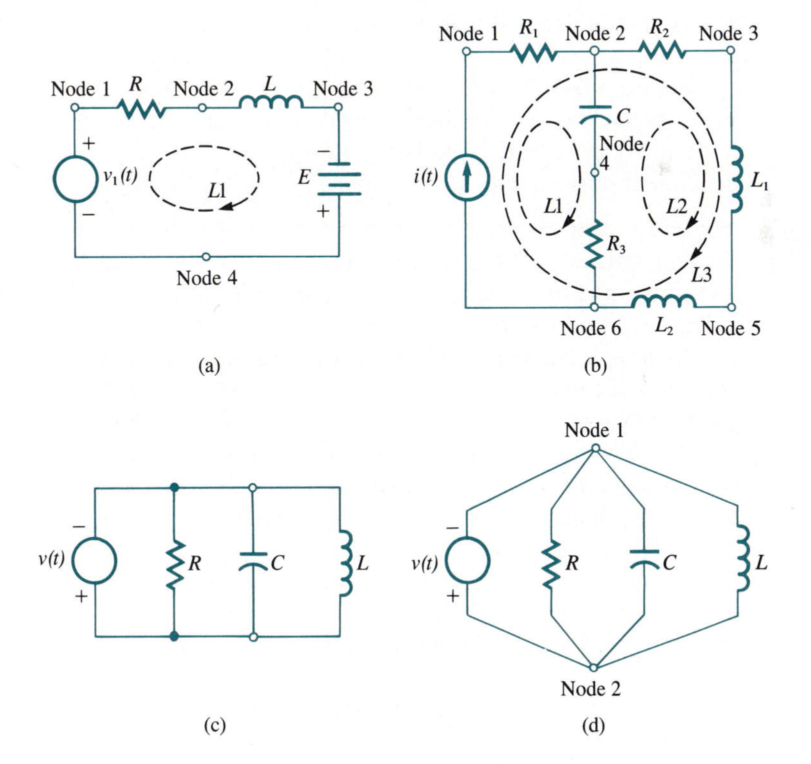

FIG. 1.6-1. Several circuits with different numbers of branches, nodes, loops, and meshes.

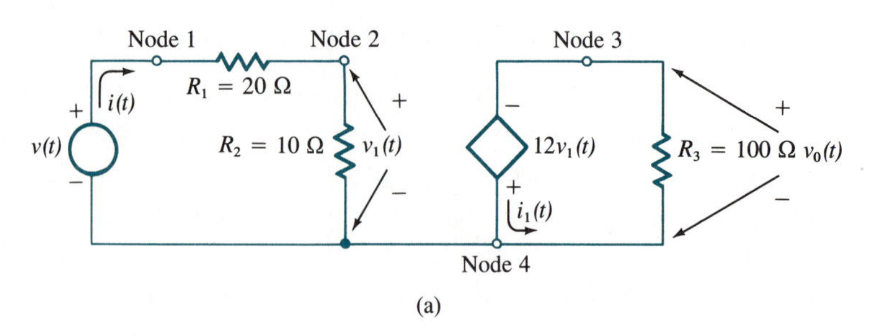

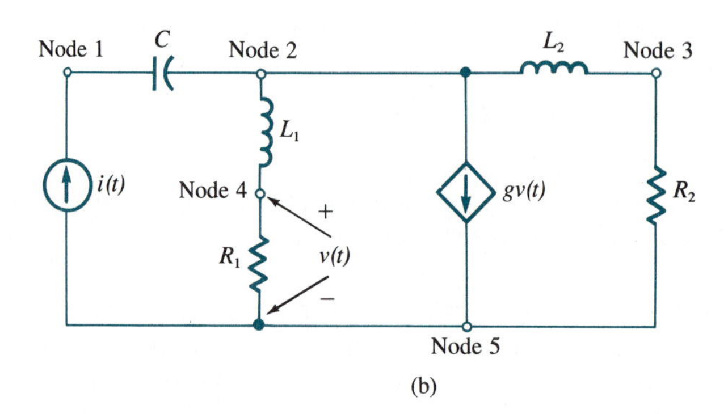

FIG. 1.6-2. Two circuits with dependent sources present: (a) a dependent voltage source and (b) a dependent current source.

reader should verify this fact as an exercise), and 3 meshes. In these figures nodes are represented by open dots and other connections by solid dots.

Figure 1.6-2 illustrates two circuits with dependent sources. The circuit of Fig. 1.6-2(a) has one voltage-dependent voltage source, and it has 5 branches, 4 nodes, 2 loops, and 2 meshes. In Fig. 1.6-2(b) the circuit has one voltage-dependent current source, 7 branches, 5 nodes, 6 loops, and 3 meshes.

EXAMPLE 1.6-1

To demonstrate that the concepts presented to this point will allow some elementary circuits to be solved, we consider the circuit of Fig. 1.6-2(a) and solve for the voltage $v_0(t)$ across R_3.

Assume a current $i(t)$ flows in the direction shown when $v(t)$ is positive. This current will cause a voltage drop in R_1 of $i(t)R_1 = 20i(t)$ such that node 1 is more positive than node 2. Similarly, a drop in R_2 equal to $10i(t)$ occurs such that node 2 is positive with respect to node 4. This drop also equals $v_1(t)$, so $v_1(t) = 10i(t)$. The total voltage from node 1 to node 4 is $(10 + 20)i(t)$. Since the total drop must also equal the voltage $v(t)$, we have $v(t) = 30i(t)$, or

$$i(t) = \frac{v(t)}{30}$$

By Ohm's law we now readily obtain

$$v_1(t) = i(t)R_2 = \frac{v(t)}{30}(10) = \tfrac{1}{3}v(t)$$

The voltage of the dependent source in the rightmost loop becomes

$$12v_1(t) = 4v(t)$$

This source voltage must appear across R_3, with node 3 negative with respect to node 4, which means that $v_0(t)$ is the *negative* of the source voltage. Hence,

$$v_0(t) = -12v_1(t) = -4v(t)$$

The entire circuit behaves as nothing more than an amplifier that produces an output voltage 4 times larger than the input in magnitude and has its polarity reversed. The overall "gain" of the circuit is -4.

1.7 SAFETY

The primary concern of anyone working with electric networks at any level must be the safety and the welfare of people. Circuits, instruments, and other physical facilities, although important, must receive consideration secondary to personal safety. The greatest threat to individuals is electrical shock.

Shock

An electrical shock can occur whenever the body forms a path through which current can pass. Generally, this means coming in contact with a voltage source and providing a path for current from the source. Depending on the voltage level and current, shock can be no more than a tingle in the body, or it can cause pain, fainting, heart fibrillation, respiratory paralysis, and death. An example of a very dangerous situation is accidentally touching a power line in a home workshop while standing barefoot on a wet concrete floor; the resulting shock could easily prove fatal.

In another common example, someone working on an electronic circuit could accidentally touch the "hot" wire of a power supply while resting an arm on the grounded circuit housing. Depending on the voltage, the shock may or may not be dangerous. If the circuit is a high-power transmitter with a 1500-V supply, it could be lethal. If it is a low-power analog circuit with a 24-V supply, the shock is much less likely to be lethal. In general, the higher the voltage, the greater the danger is.

Even when a piece of electric equipment is switched off from the power line, it can still be a shock hazard. Capacitors in equipment can often hold their voltages for a significant time after the power is turned off and can cause shock. The problem is most severe in circuits with large capacitors at high voltages, as in transmitters and some older tube-type circuits. Such circuits should always be considered dangerous until *proven* safe.

The susceptibility to shock varies tremendously from individual to individual and with conditions. A measurement of the resistance from one index finger through the body to the other index finger can be over 1 MΩ for some and less for others, both with relatively dry fingers. For sweaty fingers these values can easily decrease to 60 kΩ or less. Because only about 1 mA or less of current is needed to produce perceptible shock, it is clear that shock can occur for voltages significantly below 60 V. In fact, voltages as low as about 12 V can be perceived by some people.

The effect of shock also varies greatly between people. No absolute statements can be made about a person's susceptibility to shock or about the current level above which the effect is certain death. Even relatively healthy people can be killed by relatively low currents if the currents are sustained long enough and are large enough so that the person cannot break the contact. This fact emphasizes why it is important for people to not work alone when dealing with voltages above about 20 V. If a sustained shock occurs, others can often turn off the power source in time to save a life. In the event that shock causes unconsciousness or heart stoppage, another person properly trained in cardiopulmonary resuscitation (CPR) can mean the difference between life and death.

In summary, electrical shock can be dangerous even at low voltages (20 V) and low currents (5–10 mA). When voltages above about 20 V can be encountered, special efforts to prevent shock may be required. These precautions may include wearing rubber-soled shoes and insulating clothing and gloves.

Other Injuries

Proper attention to safety means more than prevention of shock. Overall work habits must be good, which includes being familiar with the circuits and instruments to be used, being aware of all points where dangerous voltages occur, allowing adequate work space for all activities, knowing how and where to obtain emergency treatments, keeping all instruments and facilities in proper repair, and using the correct devices and instruments for the tasks involved. In short, common sense, good training, and carefulness are the best defenses against injury when working with electric devices.

1.8 SUMMARY OF NOTATION AND UNITS

It is helpful to summarize the various circuit symbols, with attendant notation, that have been discussed in this chapter. Figure 1.8-1(a) shows the various ideal sources that have been defined, and ideal components are shown in Fig. 1.8-1(b).

In a similar way, Table 1.8-1 summarizes the various quantities and their SI units that have been used in this chapter. A few additional quantities are also given for later reference. The SI system of units is founded on seven base units

FIG. 1.8-1.
The notation and circuit symbols for (a) ideal sources and (b) ideal components.

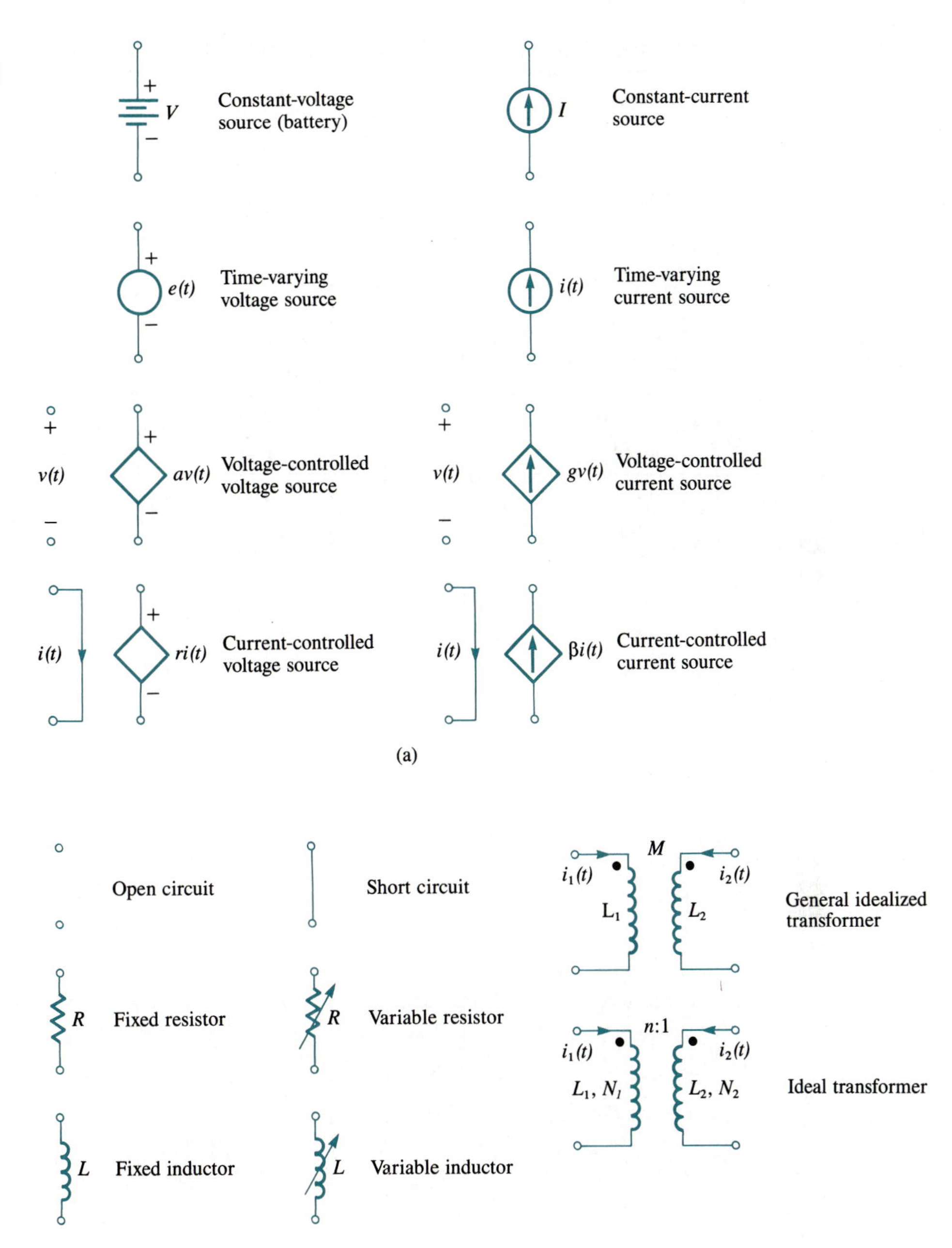

TABLE 1.8-1 Table of Various Quantities and Their SI Units with Standard Abbreviations

	Quantity		SI Unit		
Type of Unit	Name	Symbols Often Used	Name	Abbreviation	Derived Equivalent Unit
Base	Electric current	i, I	ampere	A	—
	Length	l	meter	m	—
	Mass	m, M	kilogram	kg	—
	Temperature	T	kelvin	K	—
	Time	t	second	s	—
Supplementary	Angle, plane	θ, ϕ	radian	rad	—
	Angle, solid	several	steradian	sr	—
Derived with special names	Admittance	Y, y	siemens	S	A/V
	Capacitance	C	farad	F	C/V
	Charge	q, Q	coulomb	C	A·s
	Conductance	g, G	siemens	S	A/V
	Energy	w, W	joule	J	W·s
	Force	F	newton	N	$kg \cdot m/s^2$
	Frequency	f	hertz	Hz	s^{-1}
	Impedance	Z	ohm	Ω	V/A
	Inductance	L	henry	H	Wb/A
	Magnetic flux	ϕ	weber	Wb	V·s
	Magnetic flux density	B	tesla	T	Wb/m^2
	Magnetic flux linkage	λ	weber	Wb	V·s
	Power	p, P	watt	W	J/s
	Reactance	x, X	ohm	Ω	V/A
	Resistance	r, R	ohm	Ω	V/A
	Susceptance	b, B	siemens	S	A/V
	Voltage	v, V, e, E	volt	V	W/A
Derived with no special names	Electric field intensity	$\mathcal{E}$	volt per meter	V/m	V/m
	Electric flux density	D	coulomb per meter squared	C/m^2	C/m^2
	Inertia	J	kilogram-meter squared	$kg \cdot m^2$	$kg \cdot m^2$
	Magnetic field intensity	H	ampere per meter	A/m	A/m
	Permeability	μ	henry per meter	H/m	H/m
	Permittivity	ϵ	farad per meter	F/m	F/m
	Resistivity	ρ	ohm-meter	Ω·m	Ω·m
	Torque	T	newton-meter	N·m	N·m

and two supplementary units; derived units apply to all other quantities. Two base quantities were omitted from Table 1.8-1 because they do not relate to the work of this book.

PROBLEMS

1-1. Charge flows into an electric component as follows:

$$q(t) = 26.4 + 0.87t$$

What current flows in the component?

1-2. Work Problem 1-1 except assume that charge varies with time according to Fig. P1-2.

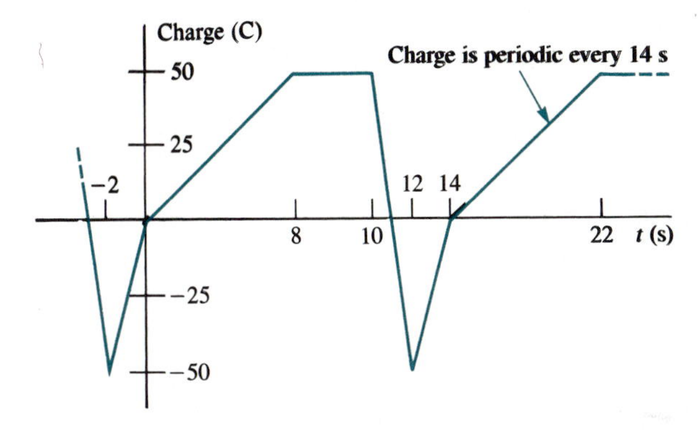

FIG. P1-2.

1-3. The current flowing through an electric component is $i(t) = 17.5 \cos 50t$. What charge is flowing?

1-4. Find the number of electrons per second that are passing some point in a wire if current in the wire is defined as in Problem 1-3.

1-5. *Coulomb's law* states that the force between two charges, Q and q, separated in air by a distance r is

$$F = \frac{Qq(9 \times 10^9)}{r^2} \text{ newtons (N)}$$

If Q is a fixed negative charge of 1 C and q is the charge of one electron, how much work must be done to bring q from infinity to 1 m from Q?

1-6. For charges as defined in Problem 1-5, what is the potential of a point 0.5 m from charge Q relative to another point 1.0 m from Q?

1-7. *Electric field intensity*, denoted by $\mathscr{E}$, is defined as the force exerted at a point having a unit positive charge. Write Coulomb's law of Problem 1-5 in terms of $\mathscr{E}$, and define $\mathscr{E}$.

1-8. Use Coulomb's law of Problem 1-5 and compute the force between a charge of 0.6 C and another of -0.2 C separated by 10 m. Is the force attractive or repulsive?

1-9. A charge of 0.2 C passes through an electric source of 15 V from its negative to its positive terminals. What change in energy did the charge receive? Did the charge gain or lose energy? What sign should be given to the energy change?

1-10. A current $i(t) = -4 \sin 30\pi t$ flows into terminal a of an electric component. The voltage at terminal a relative to terminal b is $v(t) = 20 \cos 30\pi t$. (a) What is the total energy flowing into the component from a time t_1 to another time t_2? (b) If $t_2 = t_1 + \frac{1}{15}$, what energy was absorbed?

1-11. What instantaneous power flows into the component of Problem 1-10? Interpret the meaning of the polarity of the power.

1-12. For constant current and voltage in a device, prove that average power during any time period is equal to the instantaneous power at any one time.

1-13. An electric meter used to measure average power to a home supplied with a voltage 156 cos $120\pi t$ has a current of 18 cos $120\pi t$ flowing through it. What does the meter read? Assume the meter averages power over some multiple of $\frac{1}{60}$ s.

1-14. The home of Problem 1-13 can be modeled as a resistor that is absorbing the power supplied by the power line. What is the resistance?

1-15. A dc current of 5.3 mA flows through a conductance of 62.5×10^{-6} S. What is the voltage across the conductance? What is the equivalent resistance?

1-16. Compute the current I in Fig. P1-16.

FIG. P1-16.

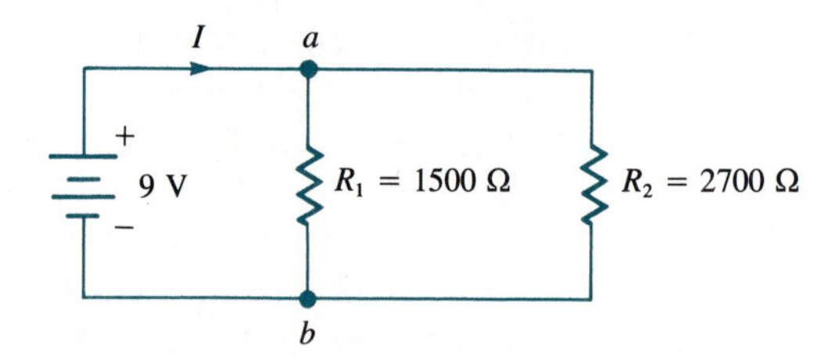

1-17. A spool contains 500 m of very fine copper wire (AWG 40 gauge) with a diameter of 0.0799 mm. What is the resistance of the whole spool of wire?

1-18. A special shunting bar of copper is placed across the terminals of a meter to keep the voltage across the meter from being larger than 0.1 V when the shunt is drawing 100 A of current. If the bar is 1 mm thick, has rectangular cross section, and is 80 mm long, what is the smallest width the shunt bar can have?

FIG. P1-20.

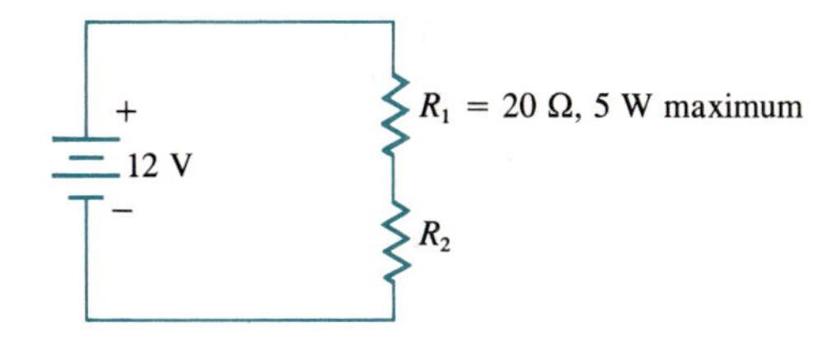

1-19. A resistor has color bands yellow (nearest the end), orange, red. What is the resistor's resistance? What color is the fourth color band likely to be?

1-20. A 12-V battery is to supply precisely 5 W of power to resistor R_1 in Fig. P1-20 to heat a small chemical solution at 5 J/s. The resistance of R_2 is to be precisely chosen to set current I so that R_1 dissipates the correct power. (a) What is I? (b) What is the resistance of R_2? (c) What power is dissipated in R_2?

1-21. A sinusoidal current $I_0 \cos 2\pi t$ flows in a resistance R. Sketch the current and the instantaneous power supplied to the resistance on the same time scale. Interpret the results in a physical sense.

1-22. A large capacitor in a power supply has a capacitance of 100 μF and a maximum allowable voltage of 100 V. At its maximum voltage, how much charge is on one of its plates?

1-23. A sinusoidal voltage $v(t) = 50 \cos 120\pi t$ exists across a 0.1-μF capacitor. Find the current in the capacitor. If the circuit remains the same, except that the frequency of the voltage is increased by a factor of 10, that is, $v(t) = 50 \cos 1200\pi t$ now, to what value has the current changed? Think about the consequences of this result. Note also that at both frequencies the current leads the voltage in phase by $\pi/2$ rad.

1-24. Figure P1-24 illustrates an air-dielectric variable capacitor often used in the frequency-tuning portion of a radio. Semicircular metal plates are all attached to a metal shaft that can rotate. Together, they form one side of the capacitor. These plates are interlaced between rectangular plates that are all electrically connected to form the other side of the capacitor. If the radius of the movable plates is 2.5 cm and their separation from the adjacent rectangular plate is 0.8 mm, how many plates are required to give a maximum capacitance of 260 pF? Neglect the cutout areas in the rectangular plates that allow the shaft to pass.

1-25. The voltage across a capacitor is $v(t) = V \cos \omega t$, where V is the peak voltage and $\omega = 2\pi f$ is called the *angular frequency* of the cosine function. (a) Find and sketch the power flowing into the unit. (b) Find and sketch the energy in the capacitor versus time. Use the same time scale as in (a) and interpret the sketches.

FIG. P1-24.

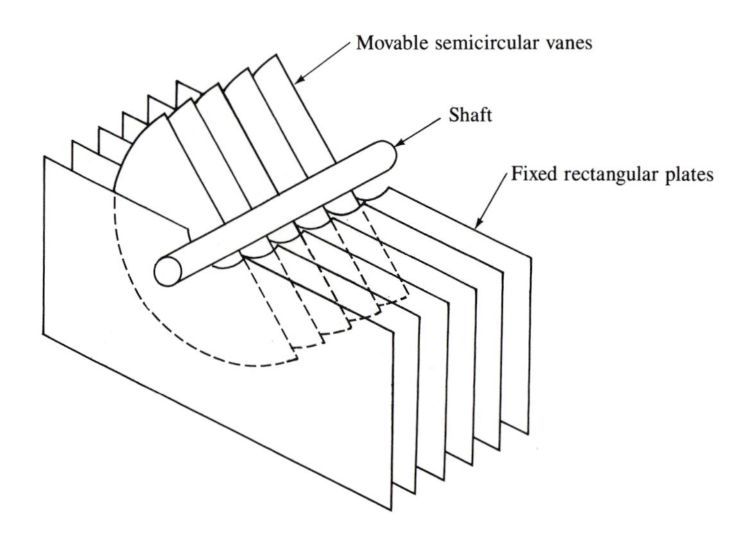

1-26. Analyze the circuit of Fig. P1-26 to determine the charge on the capacitor plate connected to point a.

FIG. P1-26.

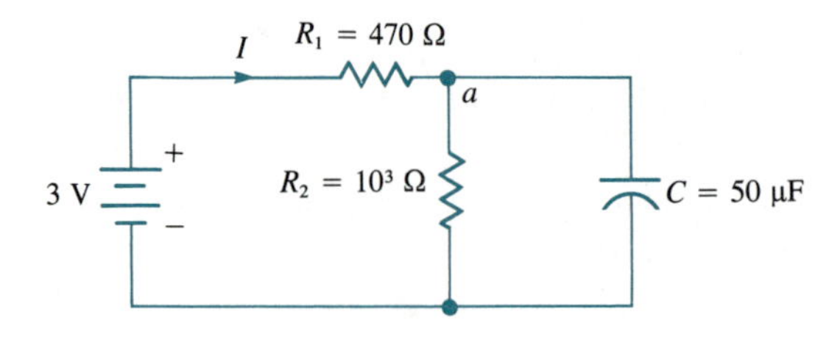

1-27. The current through an inductor is $i(t) = I \cos 2\pi t$, where I is the peak current. (a) Find the voltage across the inductor. (b) What phase relationship exists between the voltage and current? (c) If the angular frequency of the current is arbitrary, that is, if $i(t) = I \cos 2\pi f t$, what is the phase relationship?

1-28. A single-layer inductor is to be wound on an air core of 0.5-cm radius and 2-cm length, using 100 turns of wire. What is its inductance?

1-29. A single-layer solenoidal inductor of 15 mH is to be wound on a cylindrical piece of ferrite for which $\mu_r = 125$. The ferrite's diameter is 1.2 cm, and 200 turns are used. What is the length of the coil?

1-30. If l/r is large for the single-layer solenoidal inductor defined by (1.4-26), and if D/d is large in (1.4-28) for the toroidal inductor, show that these two equations become equal if a core length for the toroid is defined as $l = \pi D$.

1-31. A ferrite toroid for which $\mu_r = 2000$ has an inner diameter of 8 mm, an outer diameter of 24 mm, and a circular cross section. It is wound with 35 turns of wire to form an inductor. What is its inductance?

FIG. P1-36.

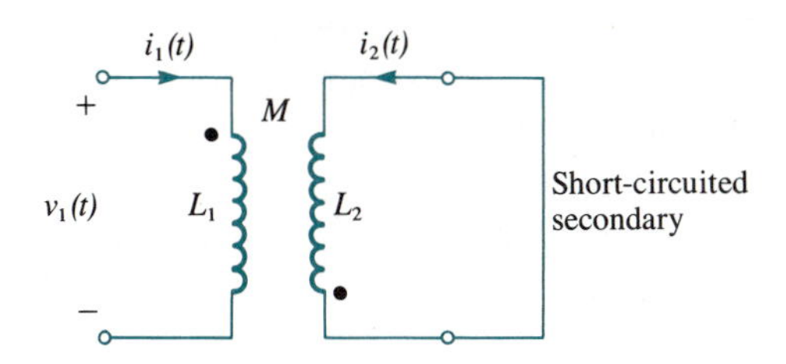

1-32. A toroidal inductance for which (1.4-28) applies has $\mu_r = 500$, $n = 75$ turns, $D = 25$ mm, and $L = 3.7$ mH. Find the cross-sectional diameter d.

1-33. A transformer is wound as in Fig. 1.4-11(a), and $L_1 = 1.0$ mH, $L_2 = 0.8$ mH, and $M_{12} = M_{21} = M = 0.5$ mH. If points b and c are connected together to form one inductor between terminals a and d, what is its inductance?

1-34. Work Problem 1-33 except assume that the windings are as shown in Fig. 1.4-11(b).

1-35. For the transformer of Fig. 1.4-11(a) with $M_{12} = M$, show that

$$e_{12}(t) = \frac{M}{L_1} v_{11}(t)$$

★1-36. In the circuit of Fig. P1-36 the transformer's secondary is short-circuited. Find general equations for $i_1(t)$ and $i_2(t)$ in terms of $v_1(t)$.

★1-37. Repeat Problem 1-36, but this time, place the dot on the secondary side of the transformer on the other end of inductance L_2.

1-38. Find the coupling coefficient for the transformer in Problem 1-33.

★1-39. Use (1.4-46) and (1.4-47) to show that the ideal transformer of Fig. P1-39(a), which has a resistive load on its secondary, can be replaced by a resistance n^2R_L, as shown in Fig. P1-39(b).

1-40. In a step-down ideal transformer $N_1/N_2 = 4$. If a load resistor of 16 Ω is tied across the secondary winding and $v_1(t) = 90 \cos 2000\pi t$, find (a) $i_1(t)$, (b) $i_2(t)$, (c) $v_2(t)$, (d) the power in the load resistor, and (e) the power supplied to the primary.

FIG. P1-39.

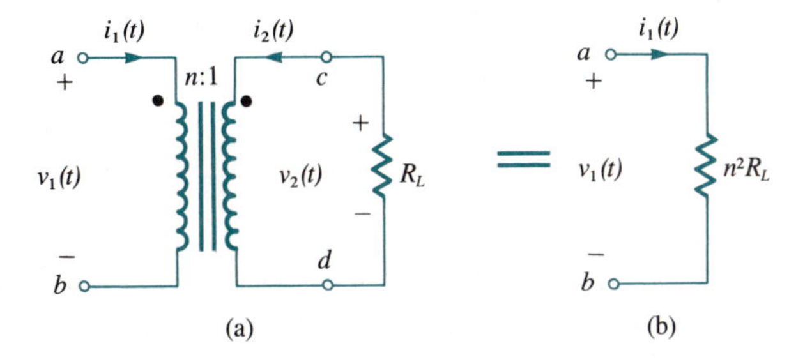

1-41. Show how one ideal voltage-dependent voltage source plus one ideal current-dependent current source can together be used to represent an ideal transformer of turns ratio n.

1-42. An independent voltage source v_1 drives a current-dependent current source connected as in Fig. P1-42. (a) Solve for v_0 in terms of v_1. (b) Solve for v_2 in terms of v_0.

FIG. P1-42.

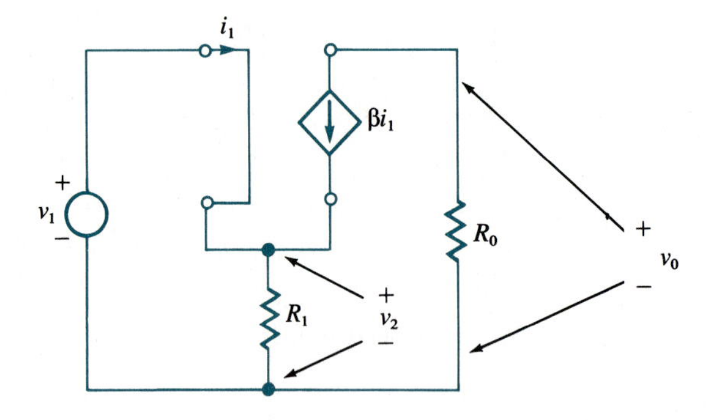

1-43. Find v_4 in terms of v_1 for the circuit of Fig. P1-43. If $R_1 = 10\ \text{k}\Omega$, $R_2 = 47\ \text{k}\Omega$, $R_3 = 150\ \Omega$, $R_4 = 1\ \text{k}\Omega$, and $g = 5 \times 10^{-3}$ S, what is v_4 in terms of v_1? If $R_1 \to 0$, how is v_4 affected? If $R_3 \to 0$, how is v_4 affected?

FIG. P1-43.

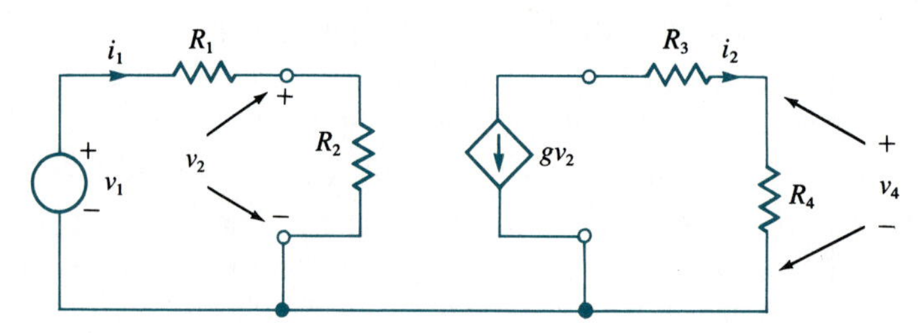

1-44. Find the number of branches, nodes, loops, and meshes in the circuit of Fig. P1-44.

FIG. P1-44.

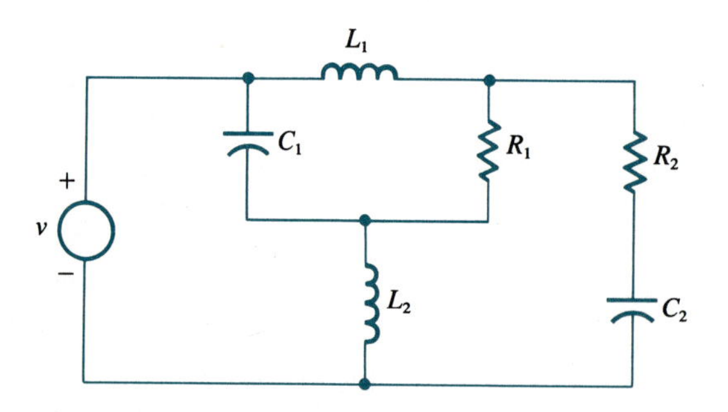

1-45. Work Problem 1-44 except for the circuit of Fig. P1-45.

FIG. P1-45.

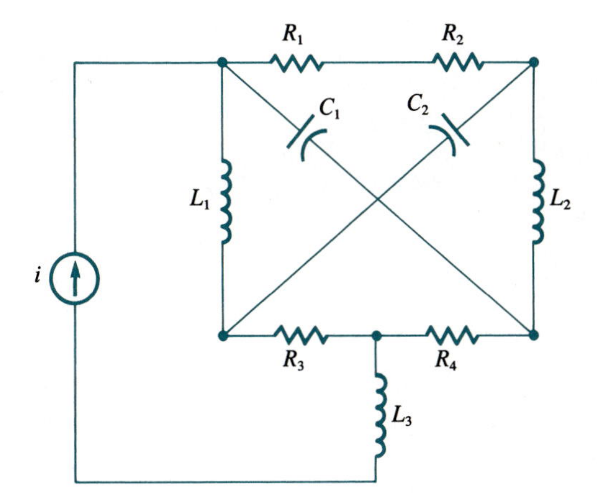

CHAPTER 2

Electric Signals

2.0 INTRODUCTION

In the broad study of topics in electrical engineering various waveforms, or signals, are encountered. These signals are usually currents or voltages, but they may also be other quantities such as charge, power, or frequency. The signals may be constant with time, as with direct current (dc), or may vary in a complicated manner with time. For example, instrumentation can be designed to produce a voltage proportional to the instantaneous displacement of the punch in an industrial machine that periodically punches out parts from a ribbon of metal stock. The voltage in this case would be sinusoidal. If one part was punched for each repetition of the machine, the frequency of the sinusoid would equal the rate at which parts were being made. In another application, the output (voltage) of a vibration analyzer attached to a jet engine under test would also be a periodic function, but with a much more complicated waveform.

Applications where waveforms suddenly change form are also common. The output (voltage) of a depth monitor for gasoline in an empty storage tank that begins to fill at a constant rate will change from zero to a linearly increasing (ramp) value, if the tank is an upright cylinder. The voltage at a panel meter will range from zero to some value (a step function) if the output signal from a temperature-monitoring thermocouple is switched to the meter. The voltage from a strain gauge designed to measure the force of a bullet striking a metal target will abruptly change from zero to a very large value and then quickly return to zero (an approximate impulse).

In this chapter we shall define the electric signals that are most useful and most commonly encountered in electrical systems. These signal definitions, when combined with the introductory principles of the preceding chapter, will form a good foundation on which we may proceed to study electric circuits in the following chapters.

2.1 ELEMENTARY SIGNALS

Two of the most basic and useful of the elementary electric signals are the dc waveform and the step function.

DC Waveform and Step Function

A dc waveform is one that is constant for all time, that is, for $-\infty < t < \infty$. The voltage of an ideal battery is taken as a good example of a dc signal, even though we must recognize that no real battery can provide a constant voltage forever, nor will we be able to verify that the voltage is constant forever.

It may seem somewhat inconsistent to the reader that the dc signal used to represent the voltage of a battery may not represent the real device for all time and cannot even be verified as a good model for all time. However, the reader should remember that mathematical definitions of signals serve only as *models* for the real world. In all engineering problems we are interested in what is happening during a given time span. As long as our models are accurate during the times of interest, their precise behavior at distant times (near ∞ and $-\infty$, for example) is of little concern. These same observations extend to all signals used in electrical engineering.

If we use the notation $v(t)$ to represent an arbitrary electrical signal,† then the dc signal can be written as

$$v(t) = a_0 \qquad (2.1\text{-}1)$$

where a_0 is a real constant.

As given above, (2.1-1) is a signal that presumably is nonzero for all values of t. If $v(t)$ is nonzero, but constant, only for some specific time interval, say $t > 0$, the result is the very useful signal called a *step function*, defined by

$$v(t) = \begin{cases} 0 & t < 0 \\ a_0 & t > 0 \end{cases} \qquad (2.1\text{-}2)$$

where the constant a_0 has the unit of volts if $v(t)$ is a voltage.

The step function is very valuable in electrical engineering; it is called a unit

† The chosen notation implies a voltage. However, the intent is to be general so that $v(t)$ can be used to represent current, power, etc.

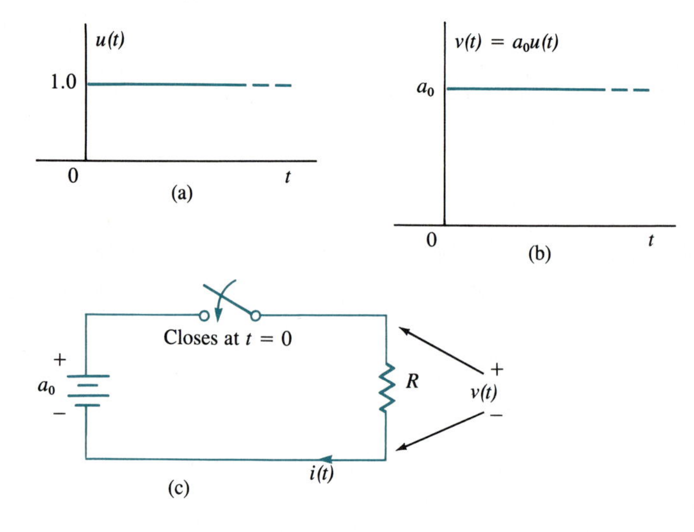

FIG. 2.1-1. (a) A unit step function, (b) a step function of voltage, where $a_0 > 0$, and (c) a circuit where $v(t)$ and $i(t)$ are step functions.

step function when $a_0 = 1$ and is given the special notation $u(t)$. Thus,

$$u(t) = \begin{cases} 0 & t < 0 \\ 1 & t \geq 0 \end{cases} \tag{2.1-3}$$

so the general step function becomes

$$v(t) = a_0 u(t) \tag{2.1-4}$$

Figure 2.1-1(a) illustrates the unit step function. The general step function is shown in Fig. 2.1-1(b), and a practical circuit in which step functions of both voltage and current occur is shown in Fig. 2.1-1(c).

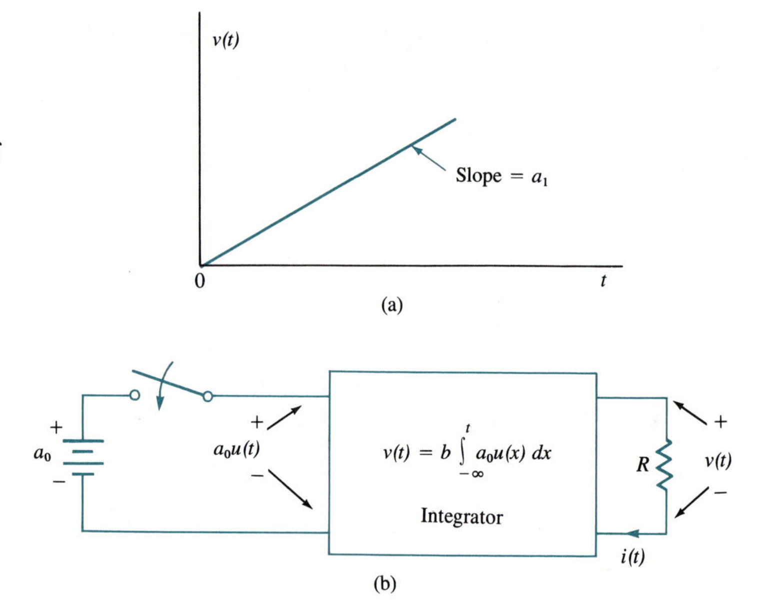

FIG. 2.1-2. (a) A ramp function and (b) a circuit where $v(t)$ and $i(t)$ are ramp functions of time.

Ramp

A *ramp* is a linearly changing signal. In most engineering applications the ramp is restricted to positive time only. Thus, we define a ramp signal by

$$v(t) = \begin{cases} 0 & t < 0 \\ a_1 t & t \geq 0 \end{cases} \tag{2.1-5}$$

If $v(t)$ represents a voltage, the constant a_1 has the unit volts per second. Figure 2.1-2(a) depicts a ramp of slope a_1.

EXAMPLE 2.1-1

The simple circuit of Fig. 2.1-1(c) is extended, as shown in Fig. 2.1-2(b), to illustrate the generation of ramps. A device called an *integrator* has been added; it is an idealized network with an output (voltage here) that is the time integral of its input (voltage here). The integrator has a gain constant b (unit is seconds^{-1}). If the integrator's input is a step function because of the action of the switch, the output becomes a ramp:

$$v(t) = b\int_{-\infty}^{t} a_0 u(\xi)\, d\xi = \begin{cases} 0 & t < 0 \\ b\displaystyle\int_0^t a_0\, d\xi & t \geq 0 \end{cases}$$

$$= \begin{cases} 0 & t < 0 \\ ba_0 t & t \geq 0 \end{cases} = \begin{cases} 0 & t < 0 \\ a_1 t & t \geq 0 \end{cases}$$

Constants are related by $a_1 = ba_0$.

The integrator of Example 2.1-1 is obviously an idealized device since the output voltage would become infinite as time becomes infinite. Practical integrators, to be discussed later in this book, are capable of generating the ramp output for only a finite time interval. Thereafter, the output either saturates (becomes constant) or decreases.

Steps and ramps are especially useful signals in the evaluation of feedback and control networks (Chapter 8).

2.2 SINUSOIDAL SIGNALS

One of the most useful and fundamental signals in electrical engineering is the sinusoidal signal, defined by

$$v(t) = A \cos(\omega t + \theta) \tag{2.2-1}$$

Here, A is called the *peak amplitude*, ω is the *angular frequency*, and θ is the *phase angle* of $v(t)$. The unit of A is the volt if $v(t)$ represents voltage, that of ω is

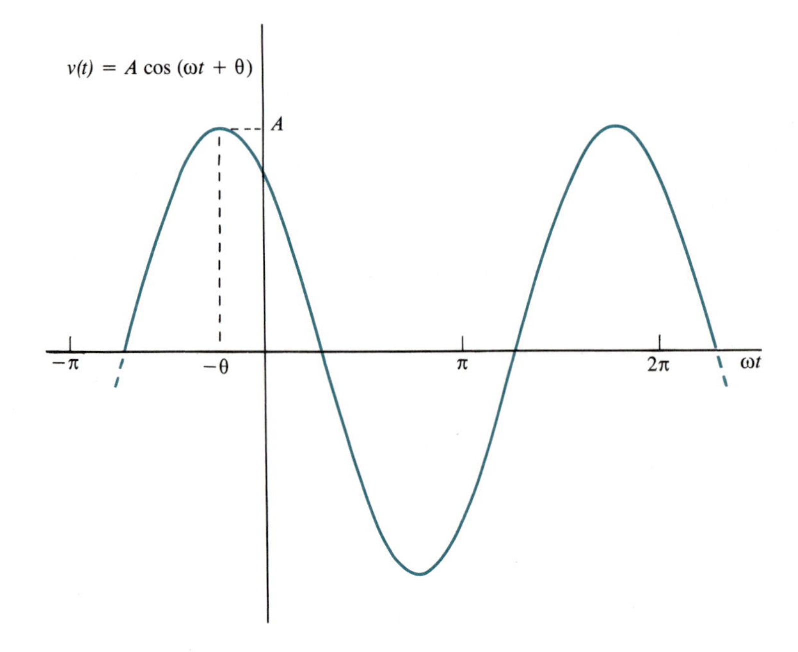

FIG. 2.2-1. A sketch of the cosine function of (2.2-1).

the radian per second (rad/s), and the unit of θ is the radian. Angular frequency is related to frequency, denoted by f, having the unit hertz (Hz), by

$$\omega = 2\pi f \tag{2.2-2}$$

Clearly, (2.2-1) represents all sinusoidal signals by choice of θ. If $\theta = 0$, $v(t)$ becomes a simple cosine function. If $\theta = -\pi/2$, then $v(t) = A \sin \omega t$. Figure 2.2-1 illustrates the waveform of (2.2-1).

Special observations should be made about (2.2-1). It is a signal that cycles steadily at a *single* frequency $\omega/2\pi$. Because of its repetitive nature, we say (2.2-1) is *periodic*. Its *period* of repetition, denoted by T, is given by

$$T = \frac{2\pi}{\omega_T} = \frac{1}{f_T} \tag{2.2-3}$$

if we use the notations $\omega_T = 2\pi f_T$ to refer to specific values of ω and f. The periodic waveform

$$v(t) = A \cos (\omega_T t + \theta) \tag{2.2-4}$$

can be viewed as a waveform $v(t)$ that has all its signal contributions at the *one* frequency $f_T = \omega_T/2\pi$. By adding a second signal at twice the frequency ($2f_T$), we create a waveform with signal contributions from *two* frequencies, f_T and $2f_T$:

$$v(t) = A \cos (\omega_T t + \theta) + B \cos (2\omega_T t + \theta) \tag{2.2-5}$$

This waveform is still periodic with period T, but its shape is different than it was before because of the component of peak amplitude B at frequency $2f_T$. Obviously, by adding other components to $v(t)$ at frequencies that are higher multiples of f_T, we can create a periodic signal with contributions from many *discrete* frequencies. The general case of this type of signal is developed in Sec. 2.3.

It is also possible for signals to have waveform content due to all possible frequencies, not just discrete frequencies as in periodic signals. Such signals are not periodic and are discussed in detail in Sec. 2.5.

EXAMPLE 2.2-1

We find the instantaneous and average powers delivered to a resistor R by a voltage defined in (2.2-4). By Ohm's law $i = v/R$, so instantaneous power, from (1.3-3), is

$$p = vi = \frac{v^2}{R} = \frac{A^2}{R}\cos^2(\omega_T t + \theta)$$

$$= \frac{A^2}{2R}[1 + \cos(2\omega_T t + 2\theta)]$$

We use (1.3-4) and find the average power over one period. With $t_2 = t_1 + T$ we develop

$$P_{av} = \frac{1}{T}\int_{t_1}^{t_1+T} \frac{A^2}{2R}[1 + \cos(2\omega_T t + 2\theta)]\,dt$$

$$= \frac{A^2}{2R} + \frac{A^2}{2RT}\int_{t_1}^{t_1+T} \cos(2\omega_T t + 2\theta)\,dt = \frac{A^2}{2R}$$

The second right-side integral readily evaluates to zero because it is the area under two cycles of the integrand's cosine function. If the averaging period had been any integral multiple of T, the average power would remain $A^2/2R$. Thus, we may say that average power, taken over all time, is $A^2/2R$.

The *root-mean-squared* (rms) value of a sinusoidal voltage (or current) is often used in defining the voltage (or current). The average squared voltage (over any period T) from (2.2-4) is

$$\frac{1}{T}\int_{t_1}^{t_1+T} A^2\cos^2(\omega_T t + \theta)\,dt = \frac{1}{T}\int_{t_1}^{t_1+T}\left(\frac{A^2}{2}\right)[1 + \cos(2\omega_T t + 2\theta)]\,dt = \frac{A^2}{2} \tag{2.2-6}$$

for any arbitrary time t_1. Thus, the rms value of the sinusoidal waveform of (2.2-4) is

$$V_{\text{rms}} = \frac{A}{\sqrt{2}} \tag{2.2-7}$$

2.3 ARBITRARY PERIODIC SIGNALS

The periodic sinusoidal signal discussed in the preceding section forms the basis for describing any arbitrary periodic signal.

Fourier Series—Real Form

A periodic waveform $v(t)$ with period T and fundamental frequency $f_T = 1/T$ can be expressed in terms of an infinite series of sinusoidal signals according to

$$v(t) = \frac{a_0}{2} + \sum_{n=1}^{\infty} a_n \cos(n\omega_T t) + \sum_{n=1}^{\infty} b_n \sin(n\omega_T t) \tag{2.3-1}$$

The series is called a *Fourier series*, after Jean Baptiste Joseph Fourier (1768–1830), a brilliant French mathematician. The constants a_n and b_n are given by

$$a_n = \frac{2}{T}\int_{-T/2}^{T/2} v(t)\cos(n\omega_T t)\,dt \qquad n = 0, 1, 2, \ldots \tag{2.3-2}$$

$$b_n = \frac{2}{T}\int_{-T/2}^{T/2} v(t)\sin(n\omega_T t)\,dt \qquad n = 1, 2, 3, \ldots \tag{2.3-3}$$

Note that the only angular frequencies present in a periodic waveform are $n\omega_T$, $n = 0, 1, 2, \ldots$. However, some of these components may be absent in specific signals if a_n and b_n are both zero for any values of n. An example is helpful in illustrating these points and the Fourier series.

EXAMPLE 2.3-1

Consider the square-wave signal of Fig. 2.3-1. Here, the actual signal over the central period is

$$v(t) = \begin{cases} 0 & \dfrac{-T}{2} < t < \dfrac{-T}{4} \\ A & \dfrac{-T}{4} < t < \dfrac{T}{4} \\ 0 & \dfrac{T}{4} < t < \dfrac{T}{2} \end{cases}$$

From (2.3-2) and (2.3-3) we find the series' coefficients to be

$$a_0 = \frac{2}{T}\int_{-T/4}^{T/4} A\,dt = A$$

$$a_n = \frac{2}{T}\int_{-T/4}^{T/4} A\cos\left(\frac{n2\pi t}{T}\right)dt = \frac{2A}{n\pi}\sin\frac{n\pi}{2} \qquad n = 1, 2, \ldots$$

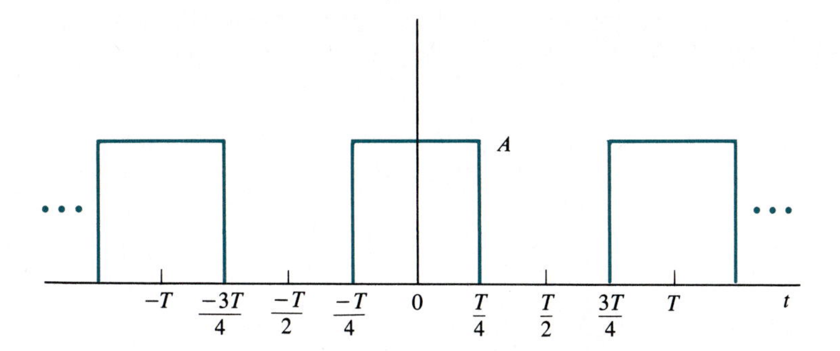

FIG. 2.3-1. A square-wave signal.

$$b_n = \frac{2}{T}\int_{-T/4}^{T/4} A \sin\left(\frac{n2\pi t}{T}\right) dt = 0 \qquad n = 1, 2, \ldots$$

The square-wave signal becomes

$$v(t) = \frac{A}{2} + A \sum_{n=1}^{\infty}\left[\frac{\sin (n\pi/2)}{n\pi/2}\right] \cos n\omega_T t$$

Because $\sin n\pi/2 = 0$ for even values of n, we see that only terms at odd multiples of ω_T are present in the given square wave.

The Fourier series of (2.3-1) is called the *real form* because all its terms are real functions.

Fourier Series—Complex Form

Another form of Fourier series can be obtained that makes use of the identities

$$\cos x = \tfrac{1}{2}(e^{jx} + e^{-jx}) \tag{2.3-4}$$

$$\sin x = \frac{1}{2j}(e^{jx} - e^{-jx}) \tag{2.3-5}$$

where $j = \sqrt{-1}$ is the unit-imaginary. After substituting (2.3-4) and (2.3-5) into (2.3-1), we obtain

$$v(t) = \sum_{n=-\infty}^{\infty} C_n e^{jn\omega_T t} \tag{2.3-6}$$

where the coefficients C_n are given by

$$C_n = \frac{1}{T}\int_{-T/2}^{T/2} v(t)e^{-jn\omega_T t}\, dt \qquad n = 0, \pm 1, \pm 2, \ldots \tag{2.3-7}$$

These coefficients are related to the previous coefficients a_n and b_n by

$$C_0 = \frac{a_0}{2} \tag{2.3-8}$$

$$C_n = \frac{1}{2}(a_n - jb_n) \qquad n = 1, 2, \ldots \tag{2.3-9}$$

$$C_{-n} = \frac{1}{2}(a_n + jb_n) = C_n^* \qquad n = 1, 2, \ldots \tag{2.3-10}$$

where the asterisk represents complex conjugation.

The series of (2.3-6) is known as the *complex form* of the Fourier series because its terms are complex functions. As an example of the use of (2.3-6) and (2.3-7), we shall again return to the waveform of Example 2.3-1.

EXAMPLE 2.3-2

For the square-wave signal defined in Fig. 2.3-1, we compute the coefficients C_n from (2.3-7) as follows:

$$C_n = \frac{1}{T}\int_{-T/4}^{T/4} Ae^{-jn\omega_T t}\,dt = \frac{A}{T}\left.\frac{e^{-jn\omega_T t}}{-jn\omega_T}\right|_{-T/4}^{T/4} = \frac{A}{2}\frac{\sin n\pi/2}{n\pi/2}$$

The Fourier series of (2.3-6) becomes

$$v(t) = \frac{A}{2}\sum_{n=-\infty}^{\infty}\frac{\sin n\pi/2}{n\pi/2}e^{jn\omega_T t}$$

In this example we observe that $C_{-n} = C_n$ because coefficients b_n are zero [see (2.3-9) and (2.3-10)].

As an exercise, the reader may wish to use (2.3-4) in the Fourier series of Example 2.3-1 to verify that it can be put in the form of the Fourier series of this example.

In many problems the complex form of the Fourier series is more useful than the real form. The usefulness of (2.3-6) derives to some extent from its compact form and to a larger extent from the presence of the exponential waveforms exp $jn\omega_T t$. We shall subsequently find that use of signals of the form exp $j\omega t$, which we shall call *complex signals*, are fundamental to the analysis of electric networks and greatly simplify analysis.

The complex signals exp $jn\omega_T t$ of (2.3-6) cannot be generated in the real world. They are mathematical conveniences that are used *to describe* real signals [as in (2.3-4) and (2.3-5)] and that make the solutions of the mathematical equations of networks easier. For positive values of n we think of the signal exp $jn\omega_T t$ as having the *positive* angular frequency $n\omega_T$. For negative values of n we consider the signal's frequency to be *negative*. Thus, from (2.3-6) a periodic signal will have series contributions from both positive and negative frequencies; the amplitudes of the complex signals are themselves complex and are the coefficients C_n.

Properties of Fourier Series

In this subsection we state some properties of Fourier series without proofs.†

Existence. If a bounded function $v(t)$ of period T has at most a finite number of maxima, minima, and discontinuities in any one period, then $v(t)$ will have a Fourier series. The series will converge to $v(t)$ at all points where $v(t)$ is continuous and to the average of the right- and left-hand limits of $v(t)$ at each point where $v(t)$ is discontinuous. These conditions on existence are known as the *Dirichlet conditions*, after Peter Gustov Lejeune Dirichlet (1805–1859), a German mathematician.

Delay. If a periodic function $v(t)$ is delayed by any multiple of its period T, the function is unchanged. That is,

$$v(t - nT) = v(t) \qquad n = \pm 1, \pm 2, \pm 3, \ldots \tag{2.3-11}$$

Symmetry. A periodic signal $v(t)$ with even symmetry such that $v(-t) = v(t)$ will have a Fourier series with no sine terms; that is, all coefficients b_n are zero. If $v(t)$ has odd symmetry such that $v(-t) = -v(t)$, its Fourier series will have no cosine terms, and $a_n = 0$ for all n.

Decomposition. An arbitrary periodic signal $v(t)$ can be expressed in terms of a part denoted by $v_e(t)$, with even symmetry, and a part $v_o(t)$, having odd symmetry, according to

$$v(t) = v_e(t) + v_o(t) \tag{2.3-12}$$

The parts may be individually structured from the signal by‡

$$v_e(t) = \tfrac{1}{2}[v(t) + v(-t)] \tag{2.3-13}$$

$$v_o(t) = \tfrac{1}{2}[v(t) - v(-t)] \tag{2.3-14}$$

Integration. The integral of a periodic function that has a valid Fourier series can be found by termwise integration of the function's Fourier series.

Differentiation. If a periodic function $v(t)$ is everywhere continuous and its derivative has a valid Fourier series, then wherever it exists, the derivative of $v(t)$ can be found by termwise differentiation of the Fourier series of $v(t)$.

---

† Justification of these properties can be found in *Advanced Engineering Mathematics* by C. R. Wylie, Jr., McGraw-Hill, New York, 1960, or in most advanced mathematics books.

‡ These results also apply to arbitrary signals in general, both periodic and nonperiodic.

2.4 IMPULSE FUNCTION

In this section we introduce what is popularly known as the *unit impulse function*, sometimes called the *delta function*, denoted by $\delta(t)$. Several ways exist for defining $\delta(t)$. In one, which is probably the most sound, it is defined through its integral property. If $\phi(t)$ is a function continuous at the point $t = t_0$ but otherwise arbitrary, then $\delta(t)$ is defined by

$$\int_{-\infty}^{\infty} \phi(t)\delta(t - t_0)\, dt = \phi(t_0) \tag{2.4-1}$$

The impulse function behaves as though it has an area of one, zero duration, and infinite amplitude at its point of occurrence, which is t_0 in (2.4-1). All these properties can be deduced from the definition (2.4-1).

Symbolically, the impulse function is shown as a vertical arrow (implying infinite amplitude) at its point of occurrence, as illustrated in Fig. 2.4-1(a). If the impulse function carries a multiplying factor, such as the 2 in $2\delta(t - t_0)$, it is sketched with an amplitude proportional to the factor, as shown in Fig. 2.4-1(b).

The unit impulse function and unit step function are related:

$$\delta(t) = \frac{du(t)}{dt} \tag{2.4-2}$$

$$u(t) = \int_{-\infty}^{t} \delta(\xi)\, d\xi \tag{2.4-3}$$

The unit impulse function is not realizable. It can be approximated in practice by a large-amplitude, short-duration pulse with an area of unity. However, the utility of the impulse function does not rest on its generation and use as a real signal. Its utility is in its mathematical help in describing certain phenomena in electrical systems. One of these phenomena will be discussed in the following section.

Another definition for $\delta(t)$ tends to carry a physical picture with it that may give the reader some additional insight. We can also define $\delta(t)$ as the limit of a sequence of functions that each has an area of unity and, in the limit, leads to

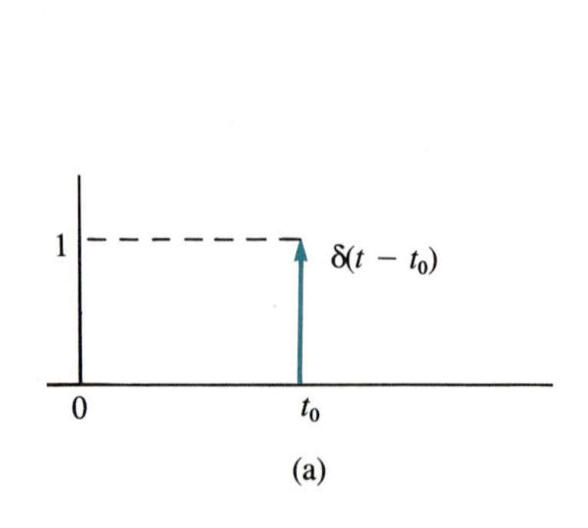

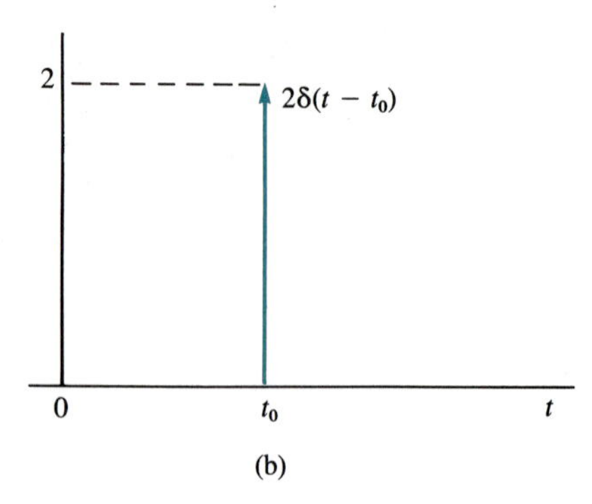

FIG. 2.4-1. (a) An impulse function occurring at time t_0 and (b) the same impulse function with a multiplying scale factor of 2.

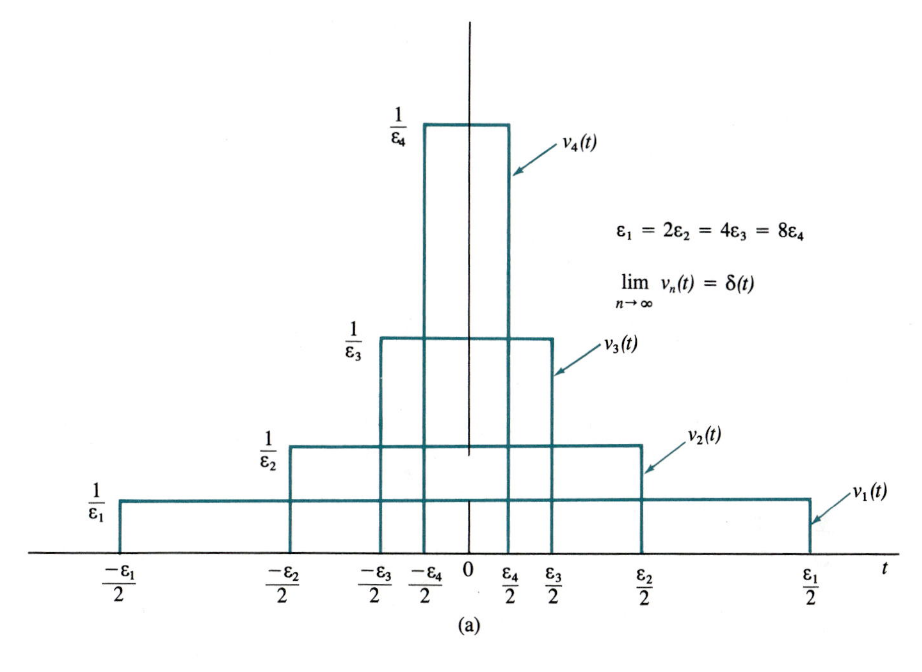

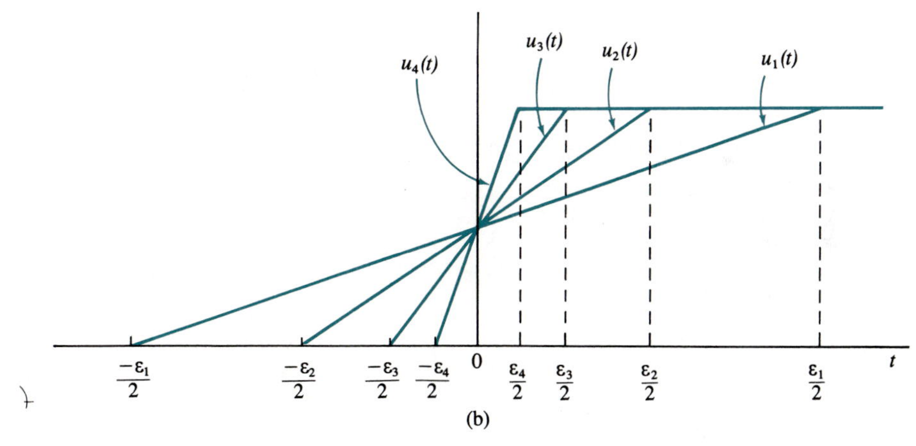

FIG. 2.4-2.
(a) A sequence of rectangular functions that are the derivatives of the step functions of (b).

infinite amplitude and zero duration. For example, the sequence of rectangular functions shown in Fig. 2.4-2(a) has such properties. In the limit we have

$$\lim_{n\to\infty} v_n(t) = \delta(t) \tag{2.4-4}$$

The rectangular functions are the derivatives of the step functions $u_n(t)$ of Fig. 2.4-2(b):

$$v_n(t) = \frac{du_n(t)}{dt} \tag{2.4-5}$$

Clearly, in the limit, the sequence of $u_n(t)$ becomes the unit step function, so that these functions also emphasize the validity of (2.4-2):

$$\delta(t) = \lim_{n\to\infty} v_n(t) = \lim_{n\to\infty} \frac{du_n(t)}{dt} = \frac{du(t)}{dt} \tag{2.4-6}$$

EXAMPLE 2.4-1

We examine the rectangular functions of Fig. 2.4-2 (a) to show that they satisfy (2.4-1). The functions are of the form

$$v(t) = \begin{cases} \frac{1}{\epsilon} & |t| < \frac{\epsilon}{2} \\ 0 & |t| > \frac{\epsilon}{2} \end{cases}$$

where

$$\delta(t) = \lim_{\epsilon\to 0} v(t)$$

From (2.4-1) we have

$$\int_{-\infty}^{\infty} \phi(t)v(t - t_0)\, dt = \int_{t_0-\epsilon/2}^{t_0+\epsilon/2} \phi(t)\frac{1}{\epsilon}\, dt$$

But if $\epsilon \to 0$, then $\phi(t) \to \phi(t_0)$ for all t near t_0, and

$$\lim_{\epsilon\to 0} \int_{-\infty}^{\infty} \phi(t)v(t - t_0)\, dt = \lim_{\epsilon\to 0} \frac{\phi(t_0)}{\epsilon} \int_{t_0-\epsilon/2}^{t_0+\epsilon/2} dt = \phi(t_0)$$

Thus

$$\lim_{\epsilon\to 0} \int_{-\infty}^{\infty} \phi(t)v(t - t_0)\, dt = \int_{-\infty}^{\infty} \phi(t) \lim_{\epsilon\to 0} v(t - t_0)\, dt$$

$$= \int_{-\infty}^{\infty} \phi(t)\delta(t - t_0)\, dt = \phi(t_0)$$

2.5 SPECTRA OF SIGNALS

In the preceding sections various electric signals have been discussed, both periodic and nonperiodic. Although some observations were made about the frequencies present in periodic waveforms, our discussions were limited to time-domain descriptions. The discussions about frequencies were with respect to only *time* signals in Fourier series that could be identified as having a specific frequency. In this section we introduce another method of signal analysis and definition that is based on the frequency, or spectral, content of a time waveform. Basically, we introduce a new domain, the frequency domain, for describing waveforms. When a signal is defined in one domain, time or frequency, it is also defined in the other, frequency or time. The mathematical devices that allow the movement from one domain to the other are transforms.

Fourier Transforms

Let $v(t)$ be an arbitrary signal defined in the time domain. A function $V(\omega)$ given by

$$V(\omega) = \int_{-\infty}^{\infty} v(t)e^{-j\omega t}\, dt \tag{2.5-1}$$

describes the frequency, or spectral, content of $v(t)$. The function $V(\omega)$ is called the *Fourier transform* of $v(t)$; it is the frequency-domain description of $v(t)$. In general, $V(\omega)$ is a complex function of ω for $-\infty < \omega < \infty$ even if $v(t)$ is a real function. If $V(\omega)$ is known, the *inverse Fourier transform* allows the recovery of $v(t)$:

$$v(t) = \frac{1}{2\pi}\int_{-\infty}^{\infty} V(\omega)e^{j\omega t}\, d\omega \tag{2.5-2}$$

Together, (2.5-1) and (2.5-2) form a Fourier transform *pair*. A transform pair is often represented by use of a double-ended arrow:†

$$v(t) \leftrightarrow V(\omega) \tag{2.5-3}$$

Some example pairs are listed in Chapter 5 (Table 5.1-1).

The existence of $V(\omega)$ is guaranteed for signals with bounded variation (where a bounded signal has at most a finite number of maxima, minima, and discontinuities in any finite time interval) and for which

$$\int_{-\infty}^{\infty} |v(t)|\, dt < \infty \tag{2.5-4}$$

However, these conditions are only sufficient and are not necessary to guarantee the existence of the Fourier transform. All signals of engineering interest will have transforms.

If $v(t)$ is a voltage, then its Fourier transform $V(\omega)$ is a voltage *density function* (unit is volts per hertz), as can be seen from (2.5-2). It describes the relative way in which the complex voltages present in the signal are distributed versus angular frequency ω.

To gain some insight into the physical meaning of a Fourier transform, or *spectrum*, as it is sometimes called, we consider some examples.

---

† The notation $\mathscr{F}\{\ \}$ will also be used to represent taking the Fourier transform of the quantity in the braces. Similarly, $\mathscr{F}^{-1}\{\ \}$ represents an inverse transform.

EXAMPLE 2.5-1

We find the Fourier transform and inverse Fourier transform of impulse functions. First, for a time impulse we have

$$\int_{-\infty}^{\infty} v(t)e^{-j\omega t}\,dt = \int_{-\infty}^{\infty} \delta(t)e^{-j\omega t}\,dt = e^{j0} = 1$$

from use of (2.4-1), so

$$\delta(t) \leftrightarrow 1$$

This result means that

$$\delta(t) = \frac{1}{2\pi}\int_{-\infty}^{\infty} e^{j\omega t}\,d\omega$$

from (2.5-2).

Next, for a frequency-domain impulse we have

$$\frac{1}{2\pi}\int_{-\infty}^{\infty} V(\omega)e^{j\omega t}\,d\omega = \frac{1}{2\pi}\int_{-\infty}^{\infty} \delta(\omega)e^{j\omega t}\,d\omega = \frac{1}{2\pi}e^{j0} = \frac{1}{2\pi}$$

so that

$$\frac{1}{2\pi} \leftrightarrow \delta(\omega)$$

From (2.5-1) we must have

$$\delta(\omega) = \frac{1}{2\pi}\int_{-\infty}^{\infty} e^{-j\omega t}\,dt$$

The third and last expressions are useful forms for representing impulse functions.

The example above illustrates two important points. First, an impulse $\delta(t)$ is a signal having equal-amplitude spectral content at *all* frequencies from $-\infty$ to ∞. Second, a constant $1/2\pi$ has a transform (spectrum) that is an impulse $\delta(\omega)$. This fact means that a dc voltage ($1/2\pi$ V here) has a spectrum that is infinite at $\omega = 0$ and has no values at any other angular frequency. In other words, the spectrum has a value only at a dc voltage (zero frequency) where it is infinite. A physical interpretation of this fact is readily obtained by remembering that the spectrum is a voltage *density* function. If a finite amount of voltage is assigned to a single frequency (a point on the ω axis having no width), then the density must be infinite, as indicated by the impulse's infinite amplitude.

EXAMPLE 2.5-2

A simple cosine signal of amplitude A and angular frequency ω_T is

$$v(t) = A\cos\omega_T t = \frac{A}{2}\left(e^{j\omega_T t} + e^{-j\omega_T t}\right)$$

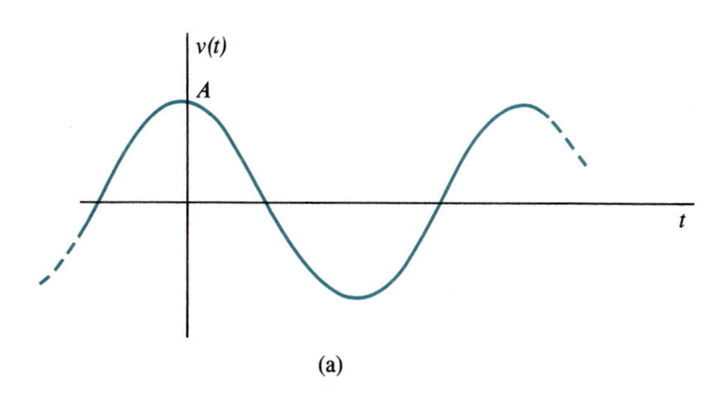

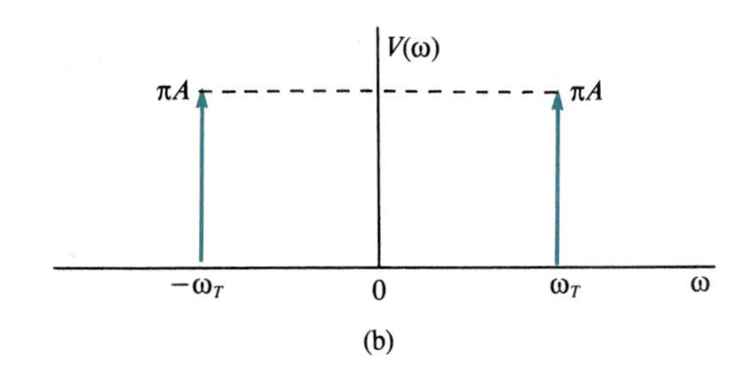

FIG. 2.5-1.
(a) A cosine signal and (b) its Fourier transform (or spectrum).

Earlier, we interpreted this waveform as having frequency content only at angular frequencies ω_T and $-\omega_T$. We now develop the spectrum of $v(t)$ to find the exact spectral content. From (2.5-1)

$$
\begin{aligned}
V(\omega) &= \int_{-\infty}^{\infty} \frac{A}{2}(e^{j\omega_T t} + e^{-j\omega_T t})e^{-j\omega t}\, dt \\
&= \frac{A}{2}\int_{-\infty}^{\infty} e^{-j(\omega-\omega_T)t}\, dt + \frac{A}{2}\int_{-\infty}^{\infty} e^{-j(\omega+\omega_T)t}\, dt \\
&= A\pi[\delta(\omega - \omega_T) + \delta(\omega + \omega_T)]
\end{aligned}
$$

This last form derives from use of the last expression in Example 2.5-1. The signal and its Fourier transform are illustrated in Fig. 2.5-1. Clearly, the voltage density is infinite at the expected angular frequencies ω_T and $-\omega_T$, as it should be for this signal at real angular frequency ω_T.

As a third and final example of the Fourier transform of an arbitrary signal, we consider a single rectangular pulse.

EXAMPLE 2.5-3

We find the Fourier transform (spectrum) of the voltage pulse of Fig. 2.5-2(a) that has amplitude A and duration τ. Here,

$$
v(t) = \begin{cases} A & \dfrac{-\tau}{2} < t < \dfrac{\tau}{2} \\ 0 & \text{elsewhere} \end{cases}
$$

so from (2.5-1),

$$
V(\omega) = \int_{-\infty}^{\infty} v(t)e^{-j\omega t}\, dt = \int_{-\tau/2}^{\tau/2} Ae^{-j\omega t}\, dt
$$

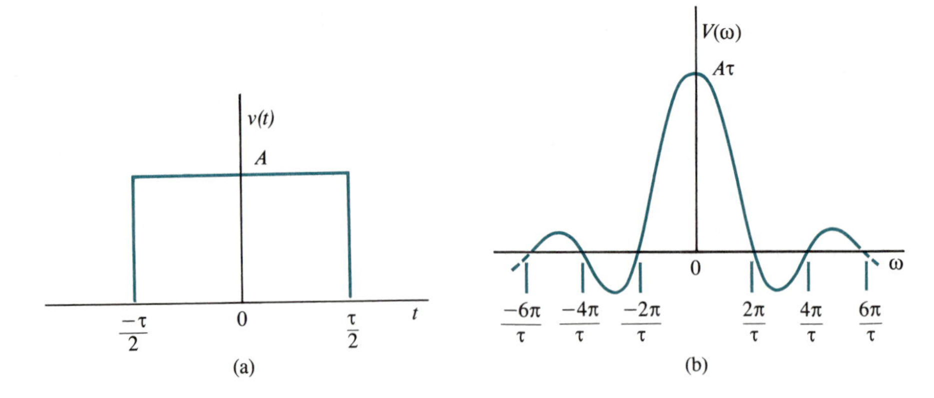

FIG. 2.5-2. (a) A voltage pulse and (b) its Fourier transform.

$$= \frac{Ae^{-j\omega t}}{-j\omega}\Bigg|_{-\tau/2}^{\tau/2} = (A\tau)\,\frac{\sin \omega\tau/2}{\omega\tau/2}$$

This spectrum is plotted in Fig. 2.5-2(b). Voltage components at *all* angular frequencies are present in the voltage pulse, although the amplitudes of the components at periodic frequencies n/τ, $n = \pm 1, \pm 2, \pm 3, \ldots$, are zero. Components at higher angular frequencies ($\omega \rightarrow \infty$) also tend to zero. Frequencies near dc are the most significant, since they have the largest spectral amplitudes. This fact might have been guessed at the start, because the pulse is just a cutout piece of a dc voltage waveform, which, of course, has *all* its spectrum at dc (zero frequency). The process of cutting up the dc voltage has created new frequencies. Although this analogy is crude, it does help give a physical interpretation to the spectrum.

Properties of Fourier Transforms

Fourier transforms have many properties that aid in the solution of many problems. We state a few of the most important ones without proof. The proofs of some of these and other properties are suggested for the reader in the problems at the end of this chapter.

Linearity. If signals $v_n(t)$ have respective Fourier transforms $V_n(\omega)$ for $n = 1, 2, \ldots, N$, then

$$v(t) = \sum_{n=1}^{N} a_n v_n(t) \leftrightarrow \sum_{n=1}^{N} a_n V_n(\omega) = V(\omega) \tag{2.5-5}$$

where N is a positive integer and the a_n are constants (possibly complex). In words (2.5-5) states that a signal $v(t)$, comprised of a linear sum of signals $v_n(t)$, has a Fourier transform $V(\omega)$ that is a linear sum of the tranforms $V_n(\omega)$ of the signals.

Time and Frequency Shifting. If t_0 and ω_0 are real constants and a signal $v(t)$ has a transform $V(\omega)$, then

$$v(t - t_0) \leftrightarrow V(\omega)e^{-j\omega t_0} \tag{2.5-6}$$

$$v(t)e^{j\omega_0 t} \leftrightarrow V(\omega - \omega_0) \tag{2.5-7}$$

Equation (2.5-6) gives the spectrum of a time-shifted version of $v(t)$, and (2.5-7) gives the time waveform of a frequency-shifted version of $V(\omega)$.

Differentiation. The spectrum of a time-differentiated signal is given by the pair

$$\frac{d^n v(t)}{dt^n} \leftrightarrow (j\omega)^n V(\omega) \tag{2.5-8}$$

for the nth derivative. If differentiation is with respect to ω we have the pair

$$(-jt)^n v(t) \leftrightarrow \frac{d^n V(\omega)}{d\omega^n} \tag{2.5-9}$$

Integration. For integration in either the time or frequency domains

$$\int_{-\infty}^{t} v(\xi)\, d\xi \leftrightarrow \pi V(0)\delta(\omega) + \frac{V(\omega)}{j\omega} \tag{2.5-10}$$

$$\pi v(0)\delta(t) - \frac{v(t)}{jt} \leftrightarrow \int_{-\infty}^{\infty} V(\xi)\, d\xi \tag{2.5-11}$$

apply.

Convolution. The products of spectra or signals have transforms according to

$$v(t) = \int_{-\infty}^{\infty} v_1(\xi)v_2(t - \xi)\, d\xi \leftrightarrow V_1(\omega)V_2(\omega) = V(\omega) \tag{2.5-12}$$

$$v(t) = v_1(t)v_2(t) \leftrightarrow \frac{1}{2\pi}\int_{-\infty}^{\infty} V_1(\xi)V_2(\omega - \xi)\, d\xi = V(\omega) \tag{2.5-13}$$

where the integrals are called *convolution integrals* because of their form. Engineers encounter the convolutional integral quite often in circuit work.

2.6 BANDWIDTH

A measure of the significant spectral content of an electrical waveform is *bandwidth*. There are many ways of defining bandwidth. However, relatively little difference usually exists in the various definitions, so we shall give here only the definition most often encountered.

Figure 2.6-1 sketches the absolute magnitude, $|V(\omega)|$, of the spectrum $V(\omega)$ of

FIG. 2.6-1.
The absolute magnitudes of signal spectra for (a) a lowpass waveform and (b) a bandpass signal.

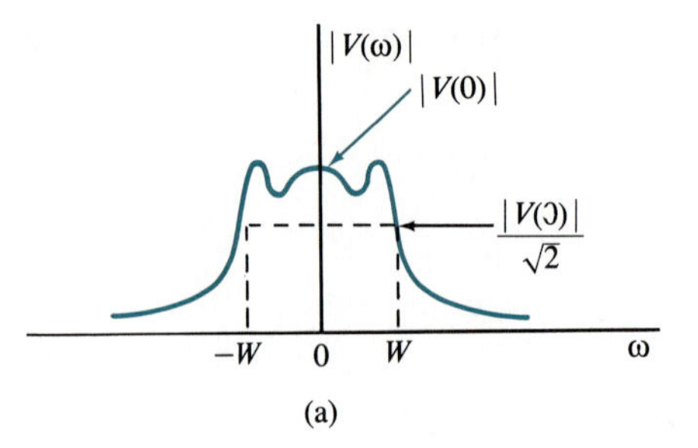

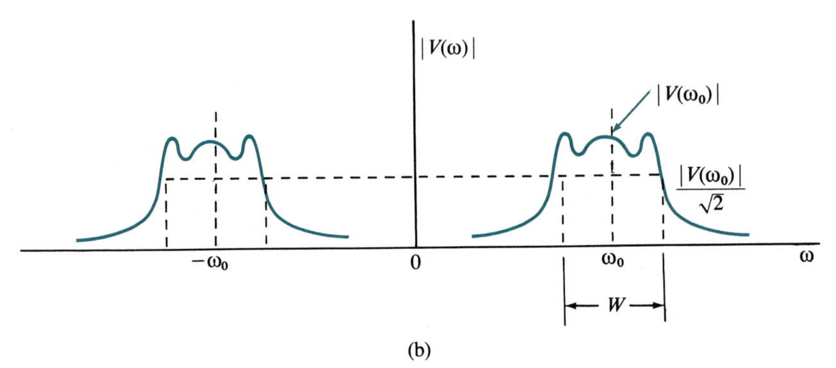

FIG. 2.6-2.
(a) A bandpass signal and (b) its spectrum.

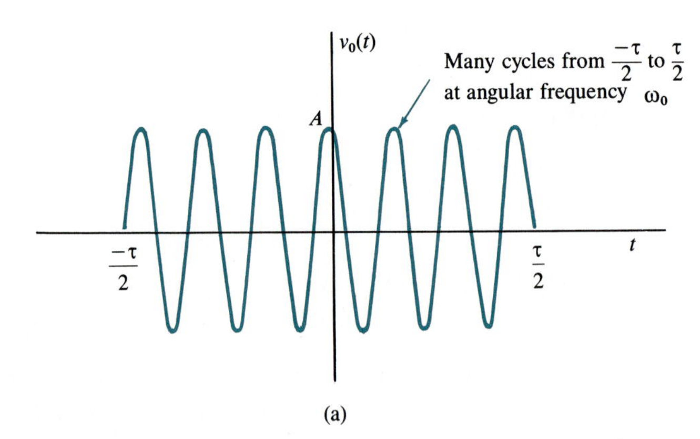

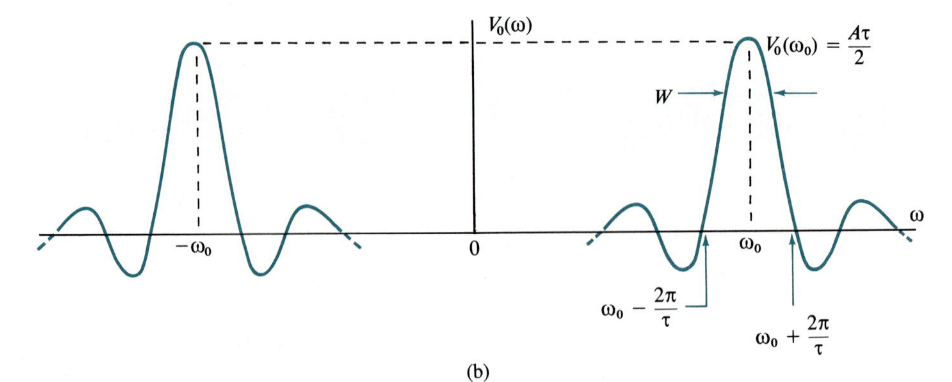

some signal $v(t)$. The sketch in Fig. 2.6-1(a) is for a possible *lowpass signal*, one in which its spectral components are most significant at frequencies near dc. Here, the bandwidth of $v(t)$ is that band of *positive* frequencies over which $|V(\omega)|$ remains above $1/\sqrt{2}$ times its value at a convenient frequency near dc (usually the frequency *is* dc). In the figure, bandwidth W (in radians per second) is the frequency band where $|V(0)|/|V(\omega)| \leq \sqrt{2}$. In terms of *decibels*,† $\sqrt{2}$ as a voltage ratio is 3 decibels (abbreviated dB), so W is usually called the 3-dB bandwidth.

Figure 2.6-1(b) illustrates $|V(\omega)|$ for a *bandpass signal*, one in which its spectral components are most significant near some "higher" angular frequency ω_0. Again, 3-dB bandwidth is the band of positive frequencies over which $|V(\omega)|$ remains above $1/\sqrt{2}$ times its value at a convenient frequency near center-band. Usually, the convenient frequency ω_0 is *at* center-band. We take an example of a bandpass signal.

EXAMPLE 2.6-1

We shall find first the spectrum of the signal in Fig. 2.6-2(a) and then determine its 3-dB bandwidth. Here,

$$v_0(t) = \begin{cases} A \cos \omega_0 t & \dfrac{-\tau}{2} < t < \dfrac{\tau}{2} \\ 0 & \text{elsewhere} \end{cases}$$

The spectrum $V_0(\omega)$ of $v_0(t)$ can be found by using Fourier transform property (2.5-7) if it is recognized that we can write $v_0(t)$ in terms of $v(t)$ of Example 2.5-3. Here,

$$v_0(t) = v(t) \cos \omega_0 t = \frac{v(t)}{2}(e^{j\omega_0 t} + e^{-j\omega_0 t})$$

From (2.5-7) and the prior example [where $V(\omega)$ is given] we have

$$V_0(\omega) = \tfrac{1}{2}[V(\omega - \omega_0) + V(\omega + \omega_0)]$$

$$= \frac{A\tau}{2}\left\{\frac{\sin[(\omega - \omega_0)\tau/2]}{(\omega - \omega_0)\tau/2} + \frac{\sin[(\omega + \omega_0)\tau/2]}{(\omega + \omega_0)\tau/2}\right\}$$

This function is sketched in Fig. 2.6-2(b). The 3-dB bandwidth is readily found as the value of W where $\omega = \omega_0 + (W/2)$ causes the first right-side term to decrease to $A\tau/(2\sqrt{2})$. Computation gives $W = (0.8845)(2\pi)/\tau$ rad/s.

The preceding example not only gave a good example of calculating 3-dB bandwidth but also illustrated the powerful utility of the properties of Fourier transforms.

†If two positive quantities V_1 and V_2 are voltages (or voltage densities here), the ratio V_1/V_2 expressed in decibels is given by dB $= 20 \log (V_1/V_2)$.

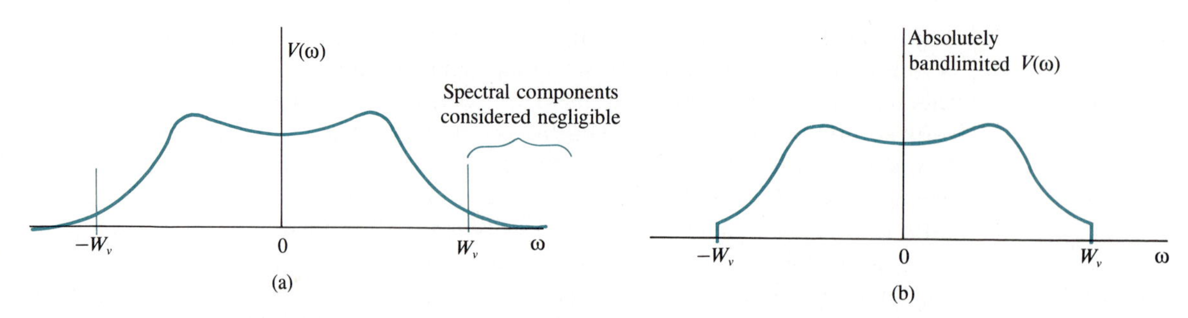

FIG. 2.7-1.
(a) The spectrum of a real signal and (b) the absolutely bandlimited approximation to $V(\omega)$, which assumes that negligible spectral components are zero.

2.7 SAMPLED SIGNALS AND THE SAMPLING THEOREM

In most analog circuits signals are processed in their entirety. However, in many modern electrical systems, especially those that convert waveforms for processing by digital circuits such as digital computers, only samples of signals are used for processing.

Sampling Theorem

A very powerful theorem that relates to the ability to sample a waveform and to reconstruct the entire waveform from its samples alone is called the *sampling theorem*. Although theorems can be stated for bandpass and random signals as well as for lowpass waveforms, we discuss only the latter since it is the most important case. The sampling theorem for lowpass signals can be stated as follows:

> Sampling Theorem: A lowpass signal $v(t)$, bandlimited such that its Fourier transform $V(\omega)$ is zero for all frequencies outside the interval $-W_v \leq \omega \leq W_v$, is uniquely determined by its values (samples) taken at equally spaced points in time separated by $T_s \leq \pi/W_v$ seconds.

The quantity W_v is the *absolute bandwidth* (in radians per second) of $v(t)$. Real waveforms are never absolutely bandlimited as required in the theorem. However, for all real waveforms there is always some angular frequency above which the signal's spectral components can be considered negligible; this angular frequency can be taken as W_v. Figure 2.7-1(a) illustrates a spectrum for a possible real signal. The absolutely bandlimited approximation of the spectrum is shown in Fig. 2.7-1(b).

The quantity T_s is the period between samples of $v(t)$. The sampling theorem states that there is a *maximum* allowed period, called the *Nyquist period*,† given

† After Harry Nyquist (1889–1976), a brilliant American scientist and engineer.

by T_s (maximum) $= \pi/W_v$. Alternatively, the *minimum* allowed sampling *rate* is $f_s = 1/T_s = W_v/\pi$ (samples per second). The minumum rate is called the *Nyquist rate*. Samples may be taken at a higher rate than the Nyquist rate but cannot be taken slower, if a bandlimited waveform is to be completely reconstructed from its samples alone.

EXAMPLE 2.7-1

An audio signal $v(t)$ is bandlimited to 15 kHz. It is to be sampled at a rate 80% higher than the minimum allowed rate, and the samples are to be stored and later used to reconstruct $v(t)$. We find the sampling rate. Since $W_v/2\pi = 15 \times 10^3$ Hz, the Nyquist (minimum) rate is $2(15 \times 10^3) = 30$ kHz. The actual rate is then $f_s = 1.80(30 \times 10^3) = 54 \times 10^3$ Hz.

Practical Sampling

One practical way of generating samples of a waveform $v(t)$ is to form the product of $v(t)$ with a "sampling" pulse train $s_p(t)$, as shown in Fig. 2.7-2(a). This form of sampling is usually called *natural sampling*. A possible signal and its spectrum are shown in Fig. 2.7-2(b). The pulse train is a periodic function, as shown in Fig. 2.7-2(c), that is represented by its Fourier series

$$s_p(t) = \frac{K\tau}{T_s} \sum_{k=-\infty}^{\infty} \frac{\sin(k\pi\tau/T_s)}{k\pi\tau/T_s} e^{jk\omega_s t} \tag{2.7-1}$$

where

$$\omega_s = \frac{2\pi}{T_s} \tag{2.7-2}$$

and τ is the duration of pulses occurring with period T_s. The sampled signal is the product

$$v_s(t) = v(t)s_p(t) = \frac{K\tau}{T_s} \sum_{k=-\infty}^{\infty} \frac{\sin(k\pi\tau/T_s)}{k\pi\tau/T_s} v(t)e^{jk\omega_s t} \tag{2.7-3}$$

which, from (2.5-7), has the Fourier transform

$$V_s(\omega) = \frac{K\tau}{T_s} \sum_{k=-\infty}^{\infty} \frac{\sin(k\pi\tau/T_s)}{k\pi\tau/T_s} V(\omega - k\omega_s) \tag{2.7-4}$$

A sketch of $V_s(\omega)$ is shown in Fig. 2.7-2(d).

An examination of (2.7-4) shows that the spectrum of the sampled signal is comprised of a central term for $k = 0$ that is proportional to the original signal's spectrum $V(\omega)$ plus replicas of $V(\omega)$ located at angular frequencies $k\omega_s$, $k = \pm 1, \pm 2, \ldots$. It is clear from Fig. 2.7-2(d) that these replicas do not overlap as long as $\omega_s > 2W_v$, which is just the Nyquist requirement on sampling rate in the sampling

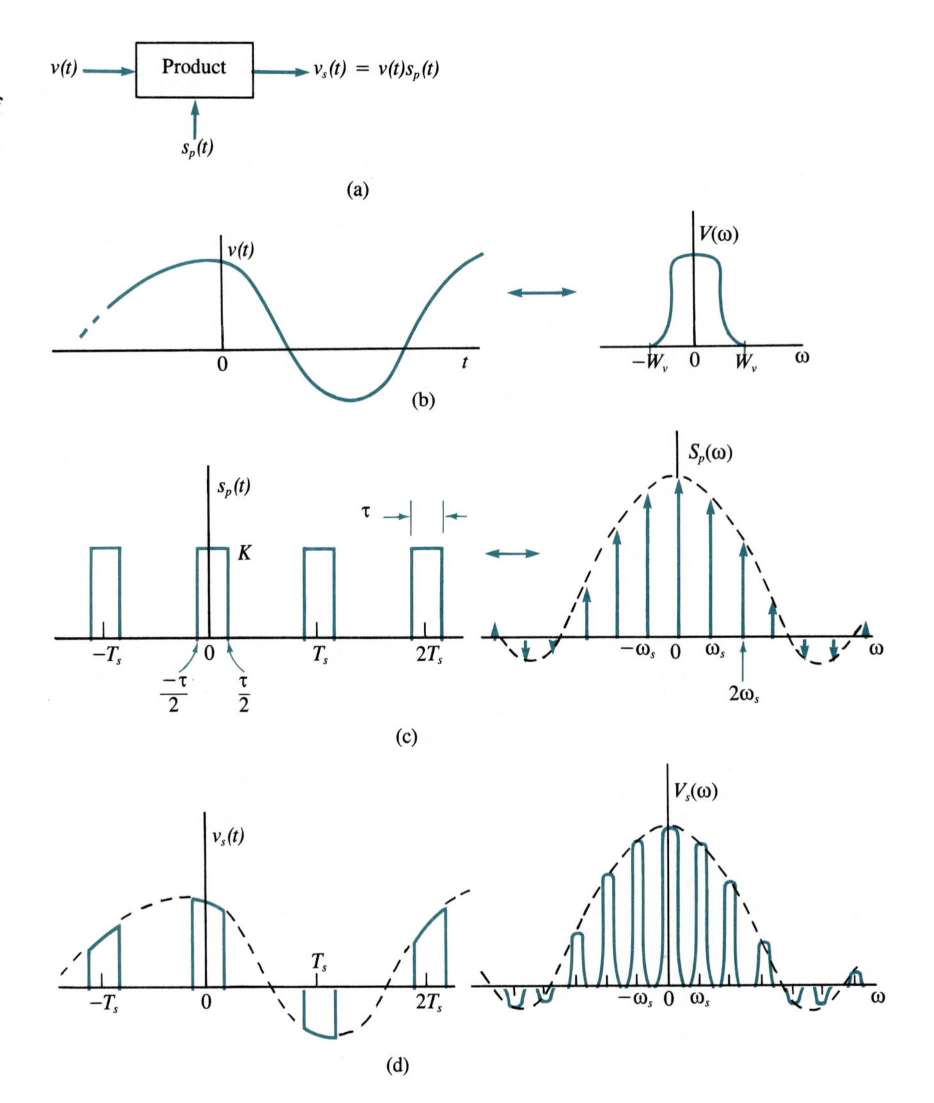

FIG. 2.7-2. (a) The product of the signals of (b) and (c). (d) The product signal and its Fourier transform.

theorem. The condition $\omega_s > 2W_v$ allows complete signal recovery, because a simple lowpass electrical filter can be used to pass only the central term in (2.7-4) while rejecting the replicas. The filter's output, or response, denoted by $v_0(t)$, has the spectrum

$$V_0(\omega) = \frac{K\tau}{T_s} V(\omega) \tag{2.7-5}$$

so

$$v_0(t) = \frac{K\tau}{T_s} v(t) \tag{2.7-6}$$

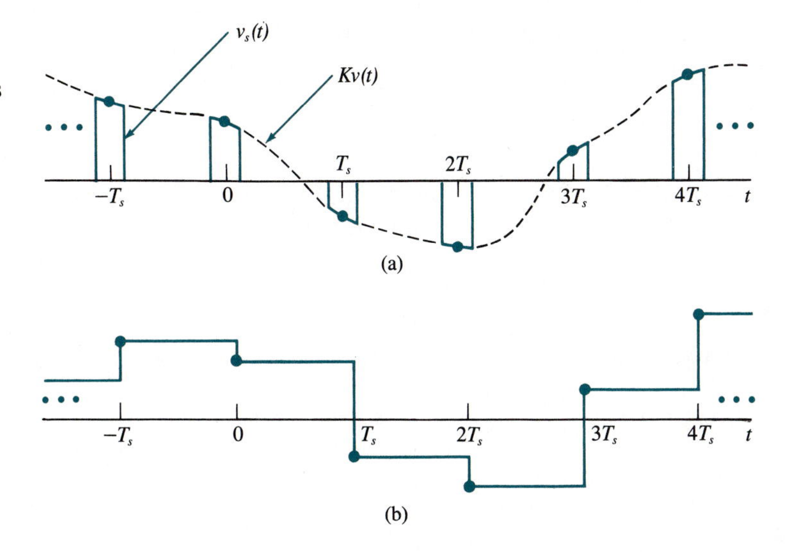

FIG. 2.7-3. (a) A sampled signal and (b) its sample-hold version.

The lowpass filter is a simple way to recover a signal from its samples, but it is not very efficient, because practical samples usually have short duration (small values of τ), which leads to low output in (2.7-6) due to the factor τ/T_s. The *sample-hold* circuit is an alternative method for recovering $v(t)$ with greater efficiency. Figure 2.7-3 illustrates its action. The sampled signal $v_s(t)$ is shown in Fig. 2.7-3(a). The sample-hold circuit takes "instantaneous" samples of $v_s(t)$ and "holds" these samples as shown in Fig. 2.7-3(b), where the samples occur at times kT_s, $k = 0, \pm 1, \pm 2, \ldots$, as illustrated by heavy dots in Fig. 2.7-3(a) and (b). In essence, the sample-hold circuit creates a new sampled signal, where now $\tau = T_s$. When the new sampled (and held) signal is passed through a lowpass filter, the response is

$$v_0(t) = Kv(t) \tag{2.7-7}$$

which is (2.7-6) with $\tau = T_s$.

There are many other aspects to sampling waveforms. The reader interested in additional detail and other references should consult the literature.†

Ideal Sampling

As a final topic in sampled signals we shall briefly discuss *ideal sampling*, which is sampling by use of a train of impulses. This condition can be realized from the preceding work by letting the areas of the pulses in the pulse train of Fig. 2.7-2(c) be unity, that is, let $K\tau = 1$ and allow $\tau \rightarrow 0$. All pulses become unit impulses.

† See, for example, Peyton Z. Peebles, Jr., *Digital Communications Systems*, Prentice-Hall, Englewood Cliffs, N.J., 1987, for more detailed discussions and other references.

Under these assumptions the spectrum of the sampled signal, given by (2.7-4), becomes

$$V_s(\omega) = \frac{1}{T_s} \sum_{k=-\infty}^{\infty} V(\omega - k\omega_s) \tag{2.7-8}$$

Two main points are now made. First, since $V(\omega)$ is assumed bandlimited to W_v and $\omega_s \geq 2W_v$ is also assumed, then

$$V_s(\omega) = \frac{1}{T_s} V(\omega) \qquad -W_v \leq \omega \leq W_v \tag{2.7-9}$$

Second, $V_s(\omega)$ is a periodic function in ω with period ω_s, so it has a Fourier series

$$V_s(\omega) = \sum_{n=-\infty}^{\infty} C_n e^{jn2\pi\omega/\omega_s} \tag{2.7-10}$$

where

$$\begin{aligned} C_n &= \frac{1}{\omega_s} \int_{-\omega_s/2}^{\omega_s/2} V_s(\omega) e^{-jn2\pi\omega/\omega_s}\, d\omega \\ &= T_s \frac{1}{2\pi} \int_{-\omega_s/2}^{\omega_s/2} \frac{1}{T_s} V(\omega) e^{-jn2\pi\omega/\omega_s}\, d\omega = v(-nT_s) \end{aligned} \tag{2.7-11}$$

Thus, from (2.7-10),

$$V_s(\omega) = \sum_{n=-\infty}^{\infty} v(-nT_s) e^{jnT_s\omega} \tag{2.7-12}$$

Because (2.7-12) is a valid representation for $V_s(\omega)$ and because (2.7-9) is true, we take the inverse Fourier transform of $V(\omega)$ to obtain

$$\begin{aligned} v(t) &= \frac{1}{2\pi} \int_{-\infty}^{\infty} V(\omega) e^{j\omega t}\, d\omega = \frac{1}{2\pi} \int_{-W_v}^{W_v} T_s V_s(\omega) e^{j\omega t}\, d\omega \\ &= \frac{T_s}{2\pi} \int_{-W_v}^{W_v} \sum_{n=-\infty}^{\infty} v(-nT_s) e^{j\omega(t+nT_s)}\, d\omega \\ &= \frac{T_s}{2\pi} \sum_{n=-\infty}^{\infty} v(-nT_s) \int_{-W_v}^{W_v} e^{j\omega(t+nT_s)}\, d\omega \\ &= \frac{W_v T_s}{\pi} \sum_{k=-\infty}^{\infty} v(kT_s) \frac{\sin [W_v(t - kT_s)]}{W_v(t - kT_s)} \end{aligned} \tag{2.7-13}$$

This expression shows that $v(t)$ is proportional to an infinite sum of terms. Each term has an amplitude factor $v(kT_s)$ that is the sample value of $v(t)$ at the sample time kT_s, and each has a form set by the function

$$Sa(\xi) = \frac{\sin \xi}{\xi} \tag{2.7-14}$$

called a *sampling function*. In terms of sampling funtions (2.7-13) is

$$v(t) = \frac{W_v T_s}{\pi} \sum_{k=-\infty}^{\infty} v(kT_s) Sa[W_v(T - kT_s)] \tag{2.7-15}$$

For the special case of Nyquist rate sampling $W_v T_s = \pi$. Sampling functions are said to be *orthogonal* because they possess the following property:

$$\int_{-\infty}^{\infty} Sa\left[\frac{\omega_s}{2}(t - kT_s)\right] Sa\left[\frac{\omega_s}{2}(t - mT_s)\right] dt = \begin{cases} 0 & m \neq k \\ T_s & m = k \end{cases} \qquad (2.7\text{-}16)$$

EXAMPLE 2.7-2

We make use of (2.7-16) to compute the total energy dissipated in a resistor R due to $v(t)$ applied across its terminals. Energy over all time is

$$w = \int_{-\infty}^{\infty} v(t)i(t)\, dt = \int_{-\infty}^{\infty} \frac{v^2(t)}{R}\, dt$$

from (1.3-2). On substitution of (2.7-15) and assuming Nyquist rate sampling, we have

$$\begin{aligned} w &= \frac{1}{R}\int_{-\infty}^{\infty} \sum_{k=-\infty}^{\infty} v(kT_s) Sa\left[\frac{\omega_s}{2}(t - kT_s)\right] \sum_{m=-\infty}^{\infty} v(mT_s) Sa\left[\frac{\omega_s}{2}(t - mT_s)\right] dt \\ &= \frac{1}{R}\sum_{k=-\infty}^{\infty} \sum_{m=-\infty}^{\infty} v(kT_s)v(mT_s) \int_{-\infty}^{\infty} Sa\left[\frac{\omega_s}{2}(t - kT_s)\right] Sa\left[\frac{\omega_s}{2}(t - mT_s)\right] dt \\ &= \frac{1}{R}\sum_{k=-\infty}^{\infty} v^2(kT_s)T_s \end{aligned}$$

This expression shows that total energy is the sum of energies due to each sample effectively generating energy in its sampling interval. The actual value of w will depend on $v(t)$, which will determine whether the sum defining w converges.

PROBLEMS

2-1. A *polynomial signal* is given by $v(t) = a_0 + a_1 t + a_2 t^2$. If it is required that $v(-1) = 0.5$, $v(0) = 2$, and $v(2) = 1.0$, find a_0, a_1, and a_2.

2-2. A polynomial waveform is $v(t) = 6 - 2t + t^2$. At what rate is the signal changing at $t = 4$ s?

2-3. Is the waveform of Problem 2-2 accelerating in voltage? If so, what is its acceleration?

2-4. In the circuit of Fig. 2.1-2(b), find $i(t)$ if $a_0 = 3$ V and $b = 2\ \text{s}^{-1}$.

2-5. Sketch the waveform of (2.2-5) if $\theta = 0$, $A = 2$, and $B = 1$. Does $v(t)$ remain periodic with period T?

2-6. Use (1.3-4) and compute the power averaged over one period in the waveform of (2.2-5) across a resistor R. Is this power the simple sum of the average powers of the two components taken separately?

2-7. Determine the Fourier series coefficients a_n and b_n for the waveform of Fig. P2-7.

FIG. P2-7.

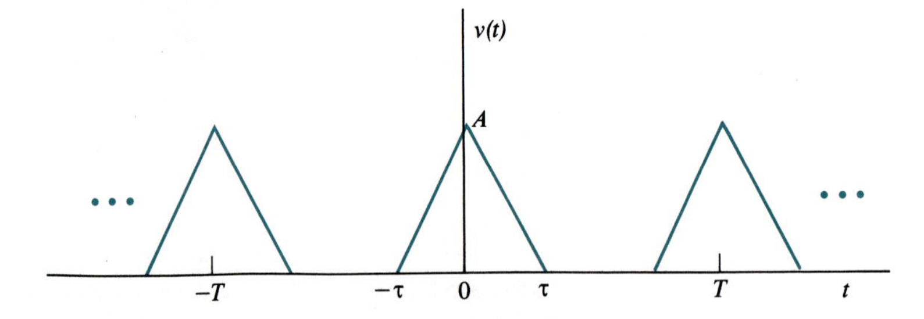

2-8. Work Problem 2-7 except for the waveform of Fig. P2-8.

FIG. P2-8.

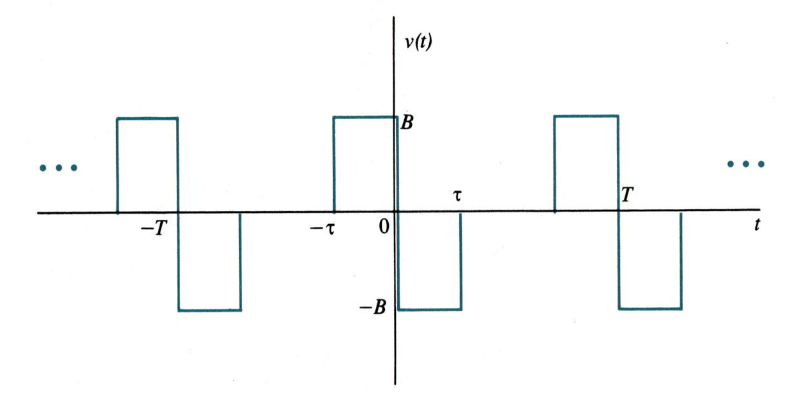

2-9. Work Problem 2-7 except for the waveform of Fig. P2-9.

FIG. P2-9.

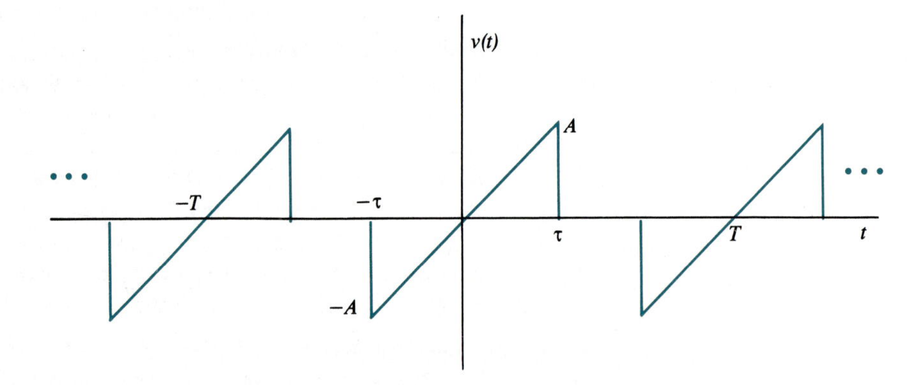

2-10. The waveform of Fig. P2-7 is delayed to a later time by an amount t_0. Find the complex Fourier series coefficients for the delayed signal.

2-11. Find the complex Fourier series of the periodic signal of Fig. P2-11. If $T = 2\tau$, find the resulting series.

FIG. P2-11.

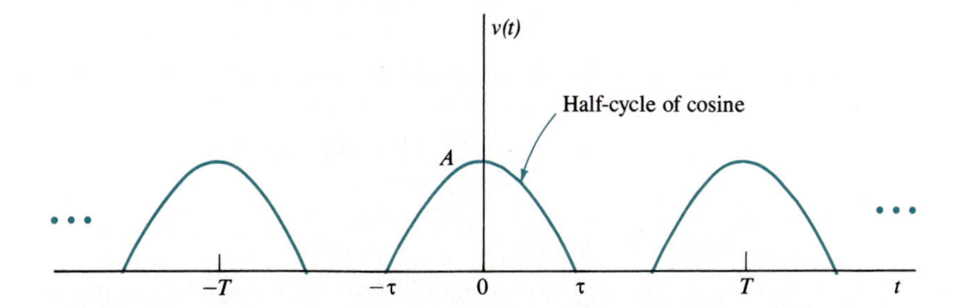

2-12. Find and sketch the even- and odd-symmetry components of the waveform in Fig. P2-12.

FIG. P2-12.

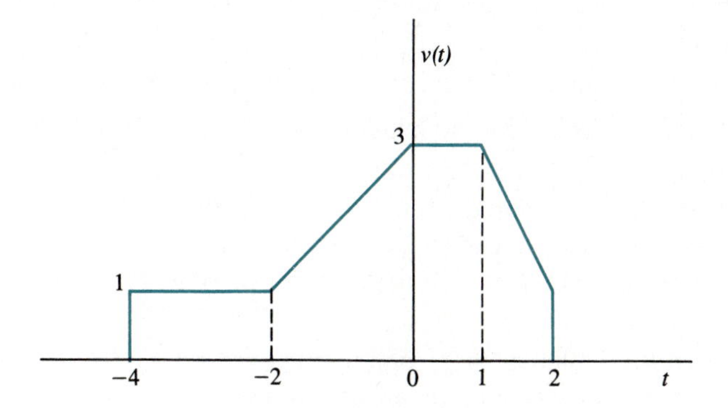

2-13. Evaluate each of the following integrals.

(a) $\displaystyle\int_{-\infty}^{\infty} \cos(6t)\delta(t+3)\,dt$

(b) $\displaystyle\int_{-\infty}^{\infty} \frac{\delta(t-4)}{t^2+6}\,dt$

(c) $\displaystyle\int_{-\infty}^{\infty} \delta(t+1)e^{-4t^2}\cos(9t)\,dt$

(d) $\displaystyle\int_{-\infty}^{\infty} \frac{e^{-7t}}{t^4+1}\,\delta(t-1)\,dt$

2-14. Find the Fourier transform of a single pulse defined by

$$v(t) = \begin{cases} At & 0 \le t < \tau \\ 0 & \text{elsewhere} \end{cases}$$

2-15. Find the Fourier transform of a single pulse of cosine-squared shape

$$v(t) = \begin{cases} A \cos^2 \left(\dfrac{\pi t}{2\tau}\right) & -\tau \leq t \leq \tau \\ 0 & \text{elsewhere} \end{cases}$$

★2-16. Find and sketch the Fourier transform of a pulse of carrier defined by

$$v(t) = \begin{cases} A \cos^2 \left(\dfrac{\pi t}{2\tau}\right) \cos (\omega_0 t) & -\tau \leq t \leq \tau \\ 0 & \text{elsewhere} \end{cases}$$

where $\omega_0 \gg 2\pi/\tau$.

2-17. Use the time integration property of Fourier transforms with the fact that an impulse $\delta(t)$ has a spectrum of unity and find the Fourier transform of a unit step function.

2-18. Find the 3-dB bandwidth of the spectrum $V(\omega) = 12/(\omega^6 + \alpha^6)$.

2-19. Find the 3-dB bandwidth of the spectrum defined by $|V(\omega)|^2 = A/(\omega^2 + \alpha^2)^3$.

2-20. Work problem 2-19 except for the function $|V(\omega)|^2 = A/(\omega^2 + \alpha^2)^4$.

2-21. A lowpass signal has absolute bandwidth 20 MHz. What is its Nyquist rate in hertz?

2-22. The television signal from a video camera is absolutely bandlimited to 6 MHz. Samples of the signal are to be transmitted to a distant point where complete reconstruction is to occur. At what minimum rate must samples be sent?

★2-23. Prove that even if the sampling pulses of Fig. 2.7-2 are not rectangular but have arbitrary shape, the process of natural sampling still allows a lowpass filter to recover $v(t)$ according to (2.7-6) if $p(t)$ represents the shape of the pulse in the train of pulses centered at $t = 0$, K is equal to $p(0)$, and τ is defined by

$$\tau = \frac{1}{p(0)} \int_{-T_s/2}^{T_s/2} p(t)\, dt$$

DESIGN EXERCISES

D2-1. One of the principal advantages of the sampling theorem is that it allows the sample trains of several signals to be time-interlaced and transmitted together using only one transmission path (line). Assume that $(N - 1)$ signals, where N is odd, are to be sampled by using natural sampling with rectangular pulses, as in Fig. 2.7-2. One signal is to be used to synchronize the receiver of the interlaced pulse train. Design a method of sampling and interlacing samples at the transmitter when $(N - 1)/2$ signals have absolute bandwidth of 5 kHz, the rest have absolute bandwidth of 40 kHz, and sampling is to be at twice the Nyquist rate for all signals. Also, design a method of separating the signals for reconstruction in the receiver. Finally, develop a means to synchronize the receiver so that it knows which message is being reconstructed.

D2-2. Design a way to construct a sample-hold circuit by using only ideal switches and capacitors. Do not be concerned with signal timing.

CHAPTER 3

Direct Current Circuits

3.0 INTRODUCTION

In this chapter we introduce electric circuits in a slightly more formal manner than was done in Chapter 1. However, since most topics in electrical engineering have circuits as their foundation, we shall cover the subject carefully in three chapters. Here, in the first of the three, we shall develop some important concepts using only the most elementary form of circuit, that containing only direct current (dc) sources and resistors. More generality is then introduced gradually in the following two chapters.

3.1 KIRCHHOFF'S VOLTAGE LAW

In Chapter 1 we found that either an increase or a decrease in voltage (electric potential difference) occurred as the path was traversed between the terminals of an element in a circuit. Let us now define such voltage increases or decreases as voltage *rises* and *drops*, respectively. One of the fundamental laws in circuit analysis, called *Kirchhoff's*† *voltage law* (abbreviated KVL), can now be stated:

† Gustav Robert Kirchhoff (1824–1887) was a German physicist.

> KVL: In a complete traversal of any loop in any circuit in a specified direction, the algebraic sum of the voltage rises is equal to the algebraic sum of the voltage drops.

KVL is general and applies to all our later work as well as our present discussion of dc circuits with resistors.

We illustrate the use of KVL by means of a simple example.

EXAMPLE 3.1-1

Consider the one-loop circuit of Fig. 3.1-1(a). To establish current and voltage references, assume current I flows clockwise (since the current is due to the one source, the chosen direction will also be the correct direction), and assume resistor voltages as shown. Note that V_3 is obviously defined to have the opposite polarity from the true voltage drop across R_3 (as the analysis will show). We use KVL to find the current I and all voltages. From KVL

$$\text{Voltage rises} = V + V_3 = \text{voltage drops} = V_1 + V_2$$

On use of Ohm's law

$$V - IR_3 = IR_1 + IR_2$$

so

$$I = \frac{V}{R_1 + R_2 + R_3} = \frac{24}{12} = 2 \text{ A}$$

and

$$V_1 = IR_1 = 2(3) = 6 \text{ V}$$

$$V_2 = IR_2 = 2(7) = 14 \text{ V}$$

$$V_3 = -IR_2 = -2(2) = -4 \text{ V}$$

Note from the third equation above that current I is the same as if the three resistors from terminal a to terminal b were replaced by a single resistor R equal to their sum (12 Ω), as shown in Fig. 3.1-1(b). This result can be generalized for any number of resistors in series.

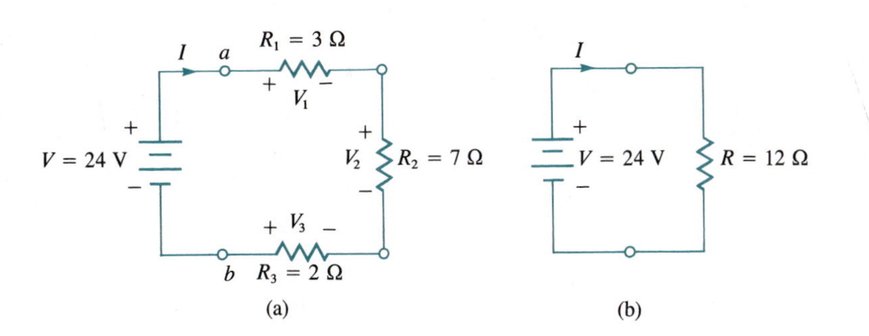

FIG. 3.1-1. (a) A simple circuit with one loop and (b) an equivalent circuit.

Resistors in Series

By repeating the procedures of Example 3.1-1, we can show that N resistors R_n, $n = 1, 2, \ldots, N$, all in series, are equivalent to a single resistor of resistance

$$R = R_1 + R_2 + \cdots + R_N \tag{3.1-1}$$

Thus, resistors in series add. Equation (3.1-1) also implies that conductances defined by $G = 1/R$ and $G_n = 1/R_n, = 1, 2, \ldots, N$, are related by

$$G = \frac{1}{R} = \frac{1}{R_1 + R_2 + \cdots + R_N} = \frac{1}{1/G_1 + 1/G_2 + \cdots + 1/G_N} \tag{3.1-2}$$

A Two-Mesh Circuit

We next take an example of a two-mesh circuit to illustrate the application of KVL to a bit more complicated circuit.

EXAMPLE 3.1-2

We use KVL to solve the circuit of Fig. 3.1-2 for its currents and voltages having the reference directions and polarities indicated. From KVL for the two loops shown,

$$\text{(rises)} \quad V + V_4 = V_1 + V_g \quad \text{(drops)}$$

$$\text{(rises)} \quad V_g = V_2 + V_3 + V_4 \quad \text{(drops)}$$

The first expression, after applying Ohm's law and known quantities, reduces as follows:

$$6 + (3)(6) = I_1(4) + V_g$$

$$24 = I_1(4) + V_g$$

The second expression similarly reduces:

$$V_g = I_2(12 + 3) + 3(6) = I_2(15) + 18$$

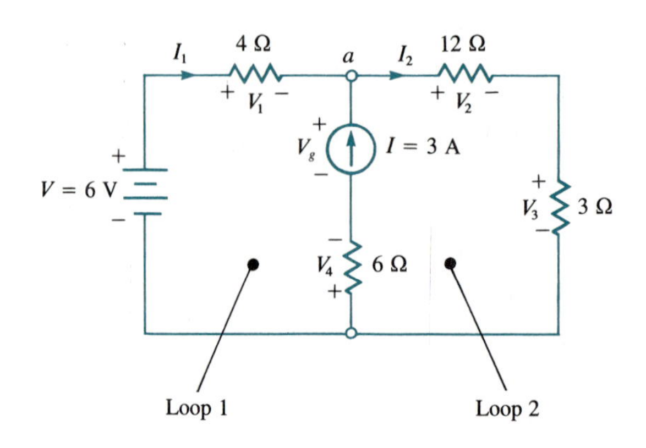

FIG. 3.1-2. A two-mesh circuit with two independent sources.

On substituting V_g from this last equation into the fourth, we have

$$4I_1 + 15I_2 = 6$$

For a final solution I_1 must be related to I_2. This relationship derives from the observation that the current leaving node a must equal the current entering the node:

$$I_2 = I + I_1 = 3 + I_1$$

On combining the last two expressions, we get

$$4I_1 + 15(3 + I_1) = 6$$

$$I_1 = \frac{6 - 45}{19} = \frac{-39}{19} \text{ A}$$

so

$$I_2 = 3 + \left(\tfrac{-39}{19}\right) = \tfrac{18}{19} \text{ A}$$

The remaining voltages follow easily from Ohm's law:

$$V_1 = I_1(4) = \frac{-39(4)}{19} = \frac{-156}{19} \text{ V}$$

$$V_2 = I_2(12) = \frac{18(12)}{19} = \frac{216}{19} \text{ V}$$

$$V_3 = I_2(3) = \frac{18(3)}{19} = \frac{54}{19} \text{ V}$$

$$V_4 = I(6) = 18 \text{ V}$$

$$V_g = V_2 + V_3 + V_4 = \frac{612}{19} \text{ V}$$

To completely solve the two-mesh circuit of Example 3.1-2 using KVL, we had to relate currents that enter or leave a circuit node. The relationship was just an application of the second fundamental law of circuit analysis, also due to Kirchhoff.

3.2 KIRCHHOFF'S CURRENT LAW

Kirchhoff's current law (abbreviated KCL) is as follows:

> KCL: At any node of any circuit the algebraic sum of all currents entering the node is equal to the algebraic sum of all currents leaving the node.

KCL is general and applies to all our later work as well as our present discussion of dc circuits with resistors.

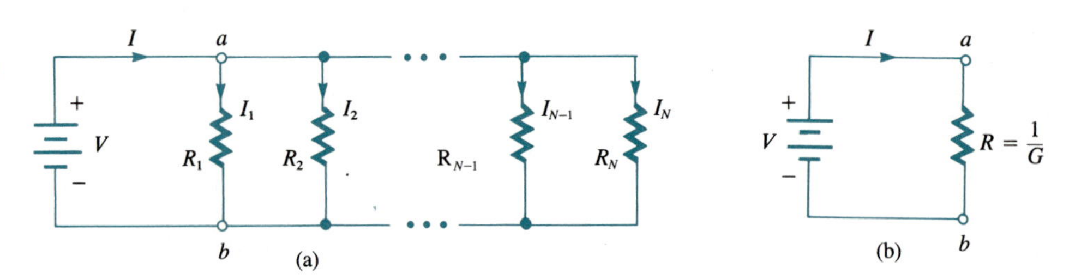

FIG. 3.2-1. (a) A circuit of N resistors in parallel and (b) the equivalent circuit.

Resistors in Parallel

We shall demonstrate the application of KCL by examining the circuit of Fig. 3.2-1(a) that consists of N resistors R_n, $n = 1, 2, \ldots, N$, all connected between nodes a and b. These resistors are said to be in parallel. From KCL at node a,

$$\text{(currents in)} \qquad I = I_1 + I_2 + \cdots + I_N \qquad \text{(currents out)} \tag{3.2-1}$$

Since $I_n = V/R_n$, $n = 1, 2, \ldots, N$, from Ohm's law, we have

$$I = \frac{V}{R_1} + \frac{V}{R_2} + \cdots + \frac{V}{R_N} = V\left(\frac{1}{R_1} + \frac{1}{R_2} + \cdots + \frac{1}{R_N}\right) \tag{3.2-2}$$

Again, from Ohm's law, we may consider V/I as an equivalent resistance to which current I is being supplied, as shown in Fig. 3.2-1(b). This equivalent resistance is

$$R = \frac{V}{I} = \frac{1}{1/R_1 + 1/R_2 + \cdots + 1/R_N} \tag{3.2-3}$$

from (3.2-2). For the special case of two resistors, where $N = 2$, (3.2-3) gives

$$R = \frac{1}{1/R_1 + 1/R_2} = \frac{R_1 R_2}{R_1 + R_2} \tag{3.2-4}$$

In terms of conductances defined by $G = 1/R$, $G_n = 1/R_n$, $n = 1, 2, \ldots, N$, (3.2-3) becomes

$$G = G_1 + G_2 + \cdots + G_N \tag{3.2-5}$$

In words, the conductance of N resistors in parallel is equal to the sum of the conductances of the individual resistances.

A Three-Node Circuit

We consider an example to illustrate the application of KCL to slightly more complicated networks.

EXAMPLE 3.2-1

We shall analyze the network of Fig. 3.2-2 to find all voltages and currents using KCL. For the two nodes a and b

$$\text{(currents in)} \qquad I = I_1 + I_2 + I_5 \qquad \text{(currents out)}$$

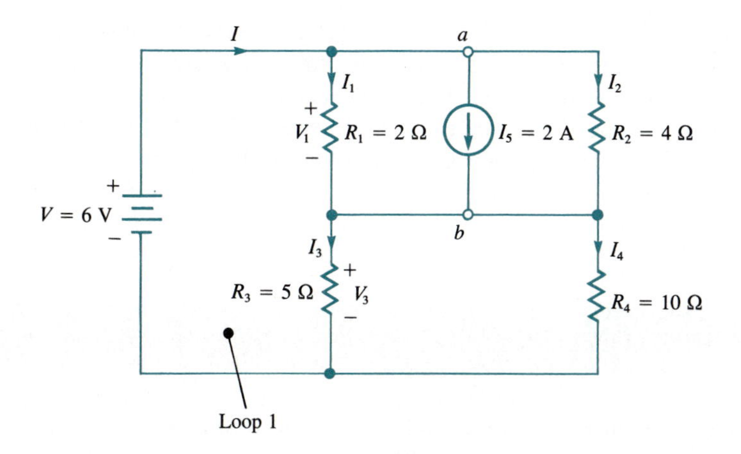

FIG. 3.2-2. The network analyzed in Example 3.2-1.

$$= \frac{V_1}{R_1} + \frac{V_1}{R_2} + 2 = V_1\left(\frac{1}{2} + \frac{1}{4}\right) + 2 = \frac{3V_1}{4} + 2$$

$$\text{(currents in)} \qquad I = I_3 + I_4 \qquad \text{(currents out)}$$

$$= V_3\left(\frac{1}{R_3} + \frac{1}{R_4}\right) = V_3\left(\frac{1}{5} + \frac{1}{10}\right) = \frac{3V_3}{10}$$

On equating these two expressions, we obtain

$$\frac{3V_1}{4} + 2 = \frac{3V_3}{10}$$

To completely solve the circuit, we must relate V_1 to V_3 through a KVL equation. For loop 1 defined in the figure

$$V = 6 = V_1 + V_3$$

By solving the last two equations simultaneously, we find

$$V_1 = \tfrac{-4}{21}\text{ V} \qquad V_3 = \tfrac{130}{21}\text{ V}$$

Thus, from Ohm's law

$$I_1 = \frac{V_1}{R_1} = \frac{-2}{21}\text{ V}$$

$$I_2 = \frac{V_1}{R_2} = \frac{-1}{21}\text{ V}$$

$$I_3 = \frac{V_3}{R_3} = \frac{26}{21}\text{ V}$$

$$I_4 = \frac{V_3}{R_4} = \frac{13}{21}\text{ V}$$

so

$$I = I_1 + I_2 + I_5 = \tfrac{39}{21}\text{ A}$$

In the preceding example an equation derived from KVL was needed along with two from KCL to solve the given network. In Section 3.6 we shall introduce an analysis method using mainly KCL equations. Similarly, in Example 3.1-2 an equation using KCL was required to support the two KVL equations for circuit solution. An analysis method using mainly KVL equations will be developed in Section 3.5.

3.3 THE SUPERPOSITION PRINCIPLE

When we solve for a particular response (voltage or current) in a linear circuit containing several sources, the solution can always be obtained by applying the superposition principle for linear networks.

Circuits with Independent Sources

For a network that has several independent sources, the *superposition principle* is as follows:

> Superposition Principle: In any linear network that contains several independent sources, the current through, or the voltage across, any element is the algebraic sum of the currents, or voltages, produced in the element by the sources acting separately.

This principle is general and is not restricted to the dc circuits with resistors discussed in this chapter.

Superposition implies that we solve for the *components* of a desired response due to each independent source alone and then algebraically add the components to find the desired response. In the use of this procedure all sources except the one of interest are set to zero. The zero requirement means all other voltage sources are assumed to generate no voltage, and they are replaced by their internal resistance, which is zero for an ideal source. Current sources are assumed to generate no current and are replaced by their internal resistance, which is an open circuit (infinite resistance) for an ideal source. We illustrate these points by two examples.

EXAMPLE 3.3-1

To demonstrate the use of superposition we analyze the circuit of Fig. 3.3-1(a) to find the currents I_1 and I_2.

First, we replace the parallel combination of 3-Ω and 6-Ω resistors with their equivalent 2-Ω resistor, using (3.2-4). Next, we find the currents caused by the

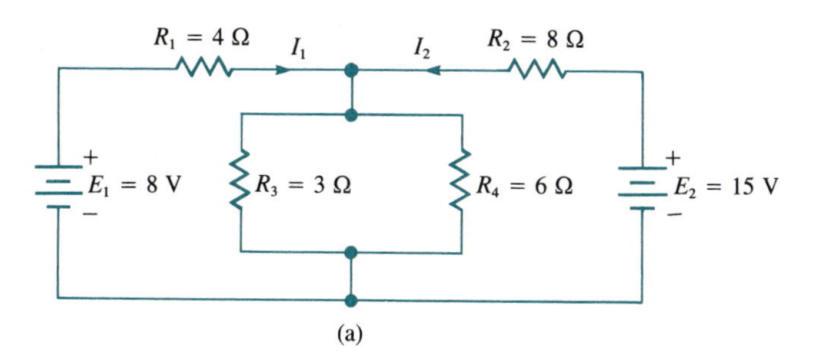

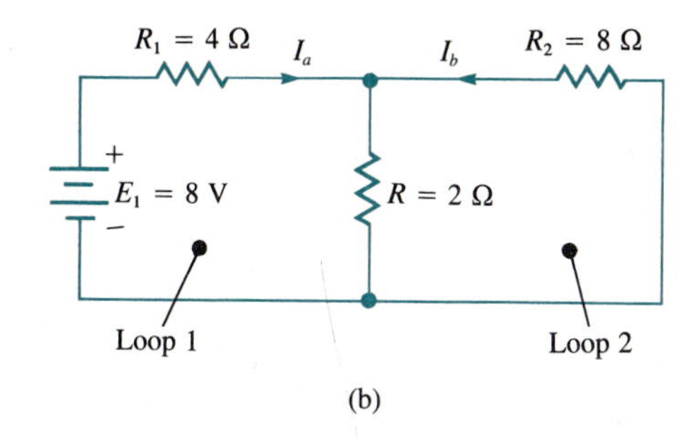

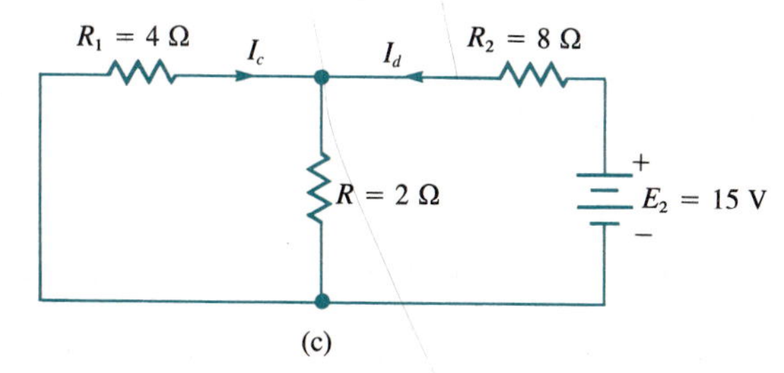

FIG. 3.3-1. (a) A two-source network and (b) and (c) its equivalent networks used to calculate the currents I_1 and I_2 by superposition of component currents due to 8- and 15-V sources, respectively.

8-V source alone, with the help of Fig. 3.3-1(b), which applies here. After writing KVL equations for loops 1 and 2, we have

$$E_1 = 8 = I_a R_1 + (I_a + I_b)R = 6I_a + 2I_b$$

$$0 = (I_a + I_b)R + I_b R_2 = 2I_a + 10I_b$$

These two equations solve simultaneously to give

$$I_a = \tfrac{10}{7} \text{ A}$$

$$I_b = \tfrac{-2}{7} \text{ A}$$

The current components of I_1 and I_2 due to the second source are found from solving the circuit of Fig. 3.3-1(c). Again, we write KVL equations:

$$15 = 2I_c + 10I_d$$

$$0 = 6I_c + 2I_d$$

which readily solve to give

$$I_c = \tfrac{-15}{28} \text{ A}$$

$$I_d = \tfrac{45}{28} \text{ A}$$

Finally, we combine current components from the two sources by superposition:

$$I_1 = I_a + I_c = \tfrac{25}{28} \text{ A}$$

$$I_2 = I_b + I_d = \tfrac{37}{28} \text{ A}$$

In another example we take the case where one of two sources is a constant-current source.

EXAMPLE 3.3-2

We apply superposition to the circuit of Fig. 3.3-2(a) to find I_1 and I_2. For the battery source alone the circuit of Fig. 3.3-2(b) applies, and we readily find

$$I_a = \frac{E_1}{R_1 + R_2} = \tfrac{1}{2} \text{ A}$$

For the current source alone we solve the circuit of Fig. 3.3-2(c). From a KCL equation at node a and a KVL equation for loop 1, we get

$$I_b = I_c + I_0 = I_c + 6$$

$$0 = I_b R_1 + I_c R_2 = 12 I_b + 6 I_c$$

The solutions for I_b and I_c are

$$I_b = 2 \text{ A}$$

$$I_c = -4 \text{ A}$$

Thus,

$$I_1 = I_a + I_b = \tfrac{5}{2} \text{ A}$$

$$I_2 = I_a + I_c = \tfrac{-7}{2} \text{ A}$$

Circuits with Dependent Sources

If a circuit has both independent and dependent sources in it, the principle of superposition still applies if the dependent sources are properly considered. The procedure is to determine the response components, as described above, by setting all independent sources to zero except the one of interest. Any dependent source will have zero value only if its controlling voltage or current is zero.

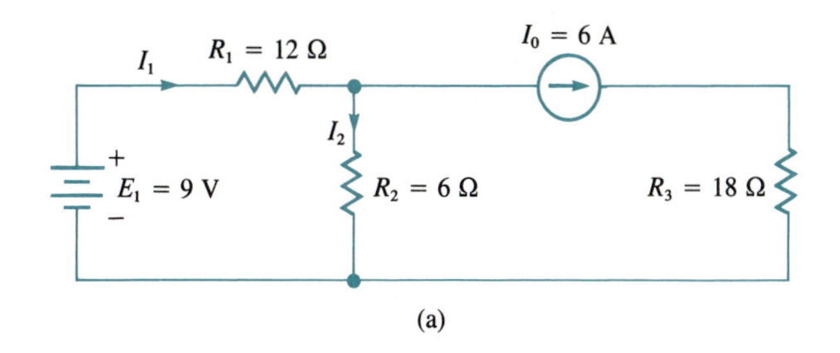

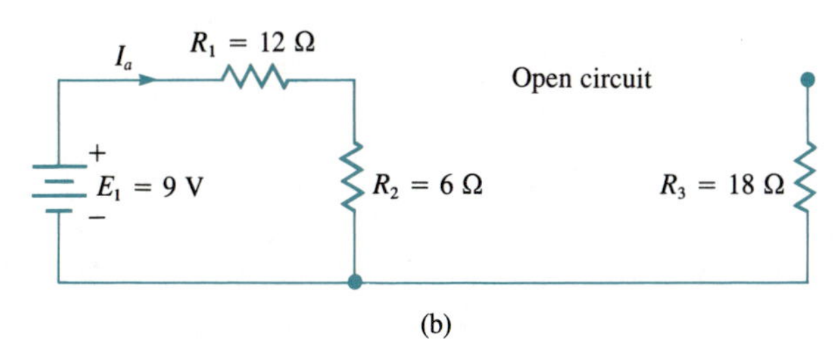

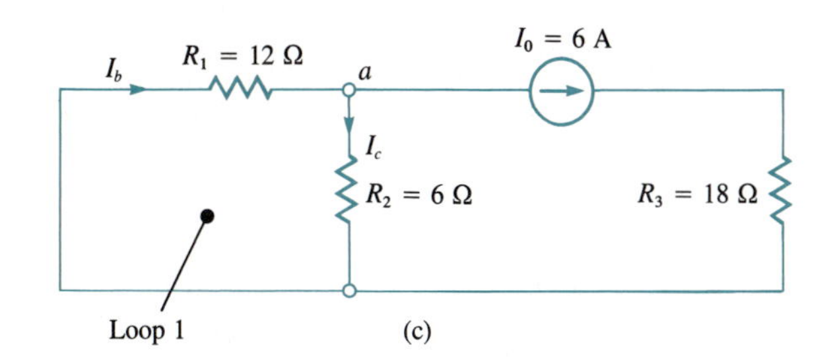

FIG. 3.3-2. (a) A two-source circuit and (b) and (c) its equivalent circuits used in applying superposition to find currents I_1 and I_2 (see Example 3.3-2).

EXAMPLE 3.3-3

Consider solving for I_0 in the circuit of Fig. 3.3-3(a), which has one dependent and two independent sources. First, we set the 9-V independent source to zero, as shown in Fig. 3.3-3(b). The 10-Ω and 12-Ω resistors in parallel are equivalent to a single resistor of 10(12)/(10 + 12) = 60/11 Ω. Current I_a easily is found to be

$$I_a = \frac{3}{5 + \frac{60}{11}} = \frac{33}{115}\text{ A}$$

Thus,

$$V_{3a} = \left(\tfrac{33}{115}\right)\left(\tfrac{60}{11}\right) = \tfrac{36}{23}\text{ V}$$

so

$$I_{0a} = \frac{6V_{3a}}{50} = \frac{6(36)}{23(50)} = \frac{108}{575}\text{ A}$$

Second, we set the 3-V independent source to zero, as shown in Fig. 3.3-3(c).

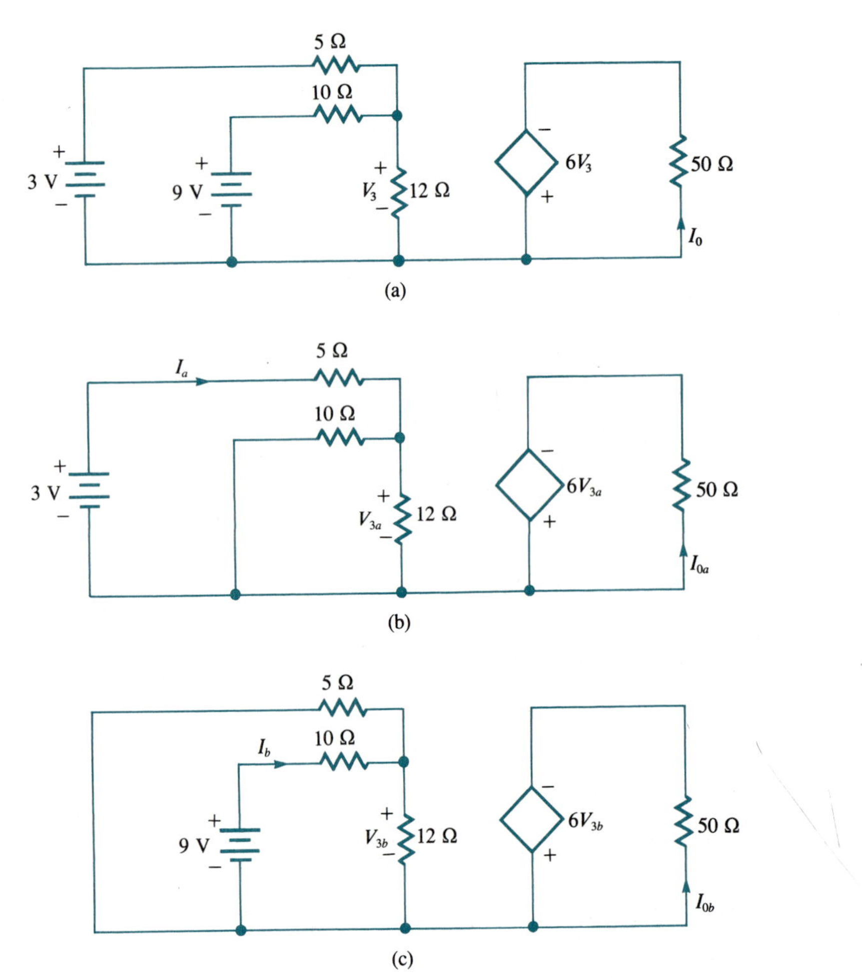

FIG. 3.3-3. (a) A circuit with a dependent source and (b) and (c) its equivalents, used in Example 3.3-3.

The form of this circuit is identical to that in Fig. 3.3-3(b), so the solution procedure is the same. We find

$$I_b = \frac{9}{10 + \frac{60}{17}} = \frac{153}{230} \text{ A}$$

$$V_{3b} = \left(\tfrac{153}{230}\right)\left(\tfrac{60}{17}\right) = \tfrac{54}{23} \text{ V}$$

$$I_{0b} = \frac{6(54)}{50(23)} = \frac{162}{575} \text{ A}$$

Finally, we add the component currents:

$$I_0 = I_{0a} + I_{0b} = \tfrac{108}{575} + \tfrac{162}{575} = \tfrac{54}{115} \text{ A}$$

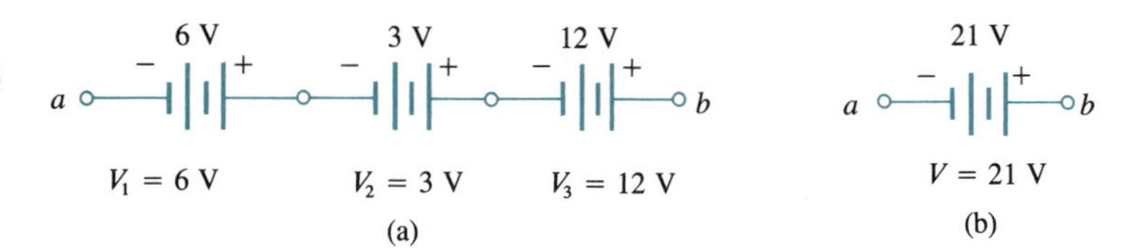

FIG. 3.4-1. (a) Three sources in series and (b) their equivalent source.

Observe that the dependent source in the preceding example never became zero, even when either one of the independent sources was zero. This situation occurred because the dependent source's control voltage V_3 was dependent on *both* independent sources.

3.4 EQUIVALENT SOURCES

Often in network analysis it is helpful to replace an actual network with a simpler one that performs in an identical way. This problem was addressed in part, in Secs. 3.1 and 3.2, where equivalent series and parallel resistors were developed. In this section we discuss several ways of developing equivalent (and usually simpler) representations of sources.

Series and Parallel Sources

When several sources, say N, of voltages V_n, $n = 1, 2, \ldots, N$, are in series, the series combination can be replaced by a single source V, according to

$$V = V_1 + V_2 + \cdots + V_N \tag{3.4-1}$$

If all sources have the same polarity as assumed for V, voltages simply add, as shown in Fig. 3.4-1. If some voltages have the opposite polarity as assumed for V, they are taken as negative in (3.4-1), as illustrated in Fig. 3.4-2.

When several sources, say M, of currents I_m, $m = 1, 2, \ldots, M$, are in parallel, the parallel combination can be replaced by a single source I, according to

$$I = I_1 + I_2 + \cdots + I_M \tag{3.4-2}$$

where the reference directions are the same for all currents.

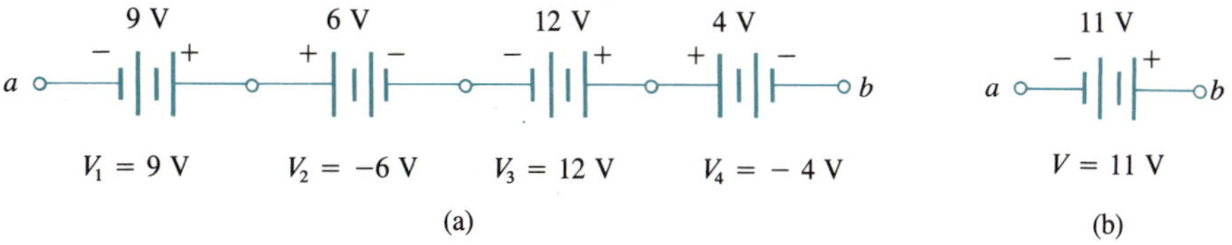

FIG. 3.4-2. (a) Four sources in series and (b) their equivalent source.

Thévenin's Theorem

Although we shall omit the proof, the superposition principle can be used to prove the following extremely valuable theorem, called *Thévenin's theorem*.†

> Thévenin's Theorem: If any linear network is partitioned into two parts that are connected only through a pair of terminals a and b, then either of the parts can be replaced by an equivalent network comprised of a source of voltage V_{oc} in series with a resistance R_{Th}, where V_{oc} is the voltage across terminals a and b caused by the network to be replaced when the other network part is removed (terminals a, b open-circuited), and $R_{Th} = V_{oc}/I_{sc}$, where I_{sc} is the current caused by the network to be replaced when terminals a, b are short-circuited.

Thévenin's theorem is general and is not restricted to the dc networks with resistors considered in this chapter. It is also valid when dependent as well as independent sources are contained in the network parts. Where dependent sources are present, the source and its control variable must be in the same part of the network. If only independent sources are present in the replaced network part, R_{Th}, called the *Thévenin equivalent resistance*, can be found as the resistance looking back into the replaced network from terminals a, b when all sources in the network are set to zero.

EXAMPLE 3.4-1

We apply Thévenin's theorem to replace network 1 in Fig. 3.4-3(a) by an equivalent network. Since the network has only one independent source, we find V_{oc} and R_{Th} directly. For V_{oc}, we easily solve the circuit of Fig. 3.4-3(b) for $I_1 = \frac{15}{10}$ A, so

$$V_{oc} = I_1(6) = 9 \text{ V}$$

Resistance R_{Th} becomes the resistance seen looking back into terminals a, b when the 15-V source is set to zero, as shown in Fig. 3.4-3(c). We derive

$$R_{Th} = \frac{6(4)}{6 + 4} = 2.4 \ \Omega$$

The final network with the Thévenin's equivalent substituted is shown in Fig. 3.4-3(d). Notice that the 5-Ω resistor does not enter into the results, because it is directly across the 15-V source and its current drain in no way affects the 20-Ω load.

† Named for M. L. Thévenin (1857–1926), a French engineer.

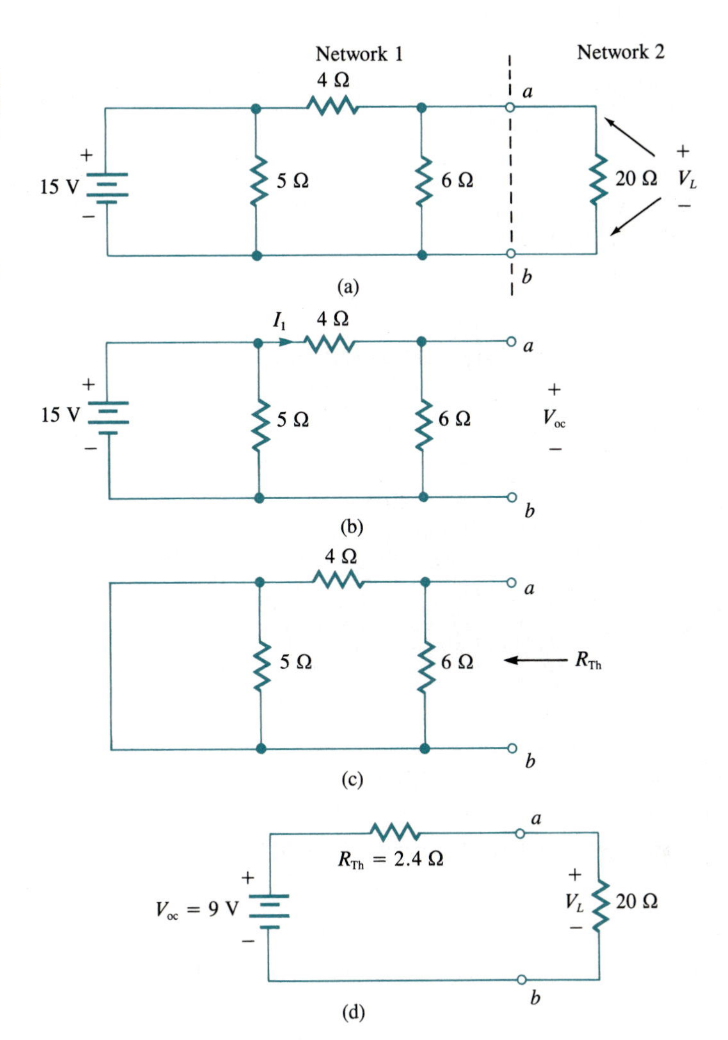

FIG. 3.4-3. (a) A partitioned network, (b) and (c) its variations used in Thévenin's theorem in Example 3.4-1, and (d) the final equivalent network.

EXAMPLE 3.4-2

As an example of Thévenin's theorem applied to a network with a dependent source, we consider the circuit of Fig. 3.4-4(a) and find the Thévenin equivalent of network 1. The open-circuit voltage is derived from a KVL equation around the loop containing the 10- and 14-Ω resistors in Fig. 3.4-4(b):

$$\text{(voltage rises)} \qquad 30 = \frac{V_{oc}}{2}(10) + V_{oc} \qquad \text{(voltage drops)}$$

so

$$V_{oc} = 5 \text{ V}$$

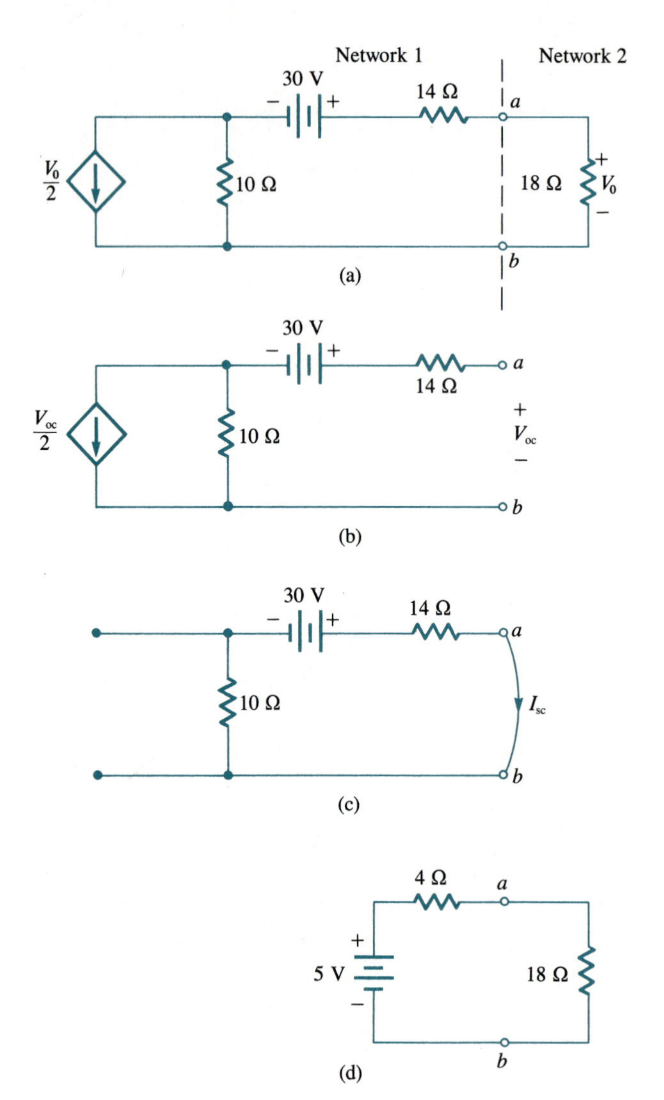

FIG. 3.4-4. (a) A network with a dependent source, and (b) and (c) networks used to compute the Thévenin equivalent network of (d) in Example 3.4-2.

To compute the short-circuit current I_{sc}, we apply the network of Fig. 3.4-4(c):

$$I_{sc} = \frac{30}{10 + 14} = \frac{5}{4}\text{ A}$$

Finally,

$$R_{Th} = \frac{V_{oc}}{I_{sc}} = 4\ \Omega$$

The Thévenin generator and network 2 are shown in Fig. 3.4-4(d).

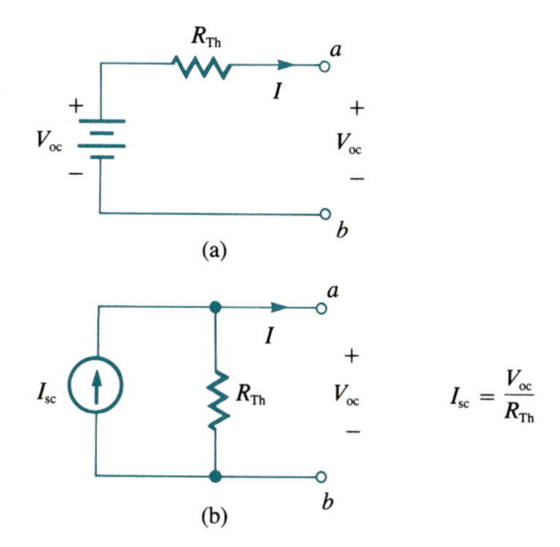

FIG. 3.4-5. (a) A Thévenin equivalent source and (b) the equivalent Norton source.

$$I_{sc} = \frac{V_{oc}}{R_{Th}}$$

Norton's Theorem

Whereas Thévenin's theorem allows us to replace a linear network by a voltage source in series with a resistance, *Norton's theorem*† allows us to replace the same network with a current source in parallel with a resistance. The current source's current is I_{sc}, and the resistance is $R_{Th} = V_{oc}/I_{sc}$, where V_{oc}, I_{sc}, and R_{Th} are defined exactly as for the Thévenin generator. Figure 3.4-5 depicts Thévenin and Norton generators.

The power and utility of both Thévenin's and Norton's theorems rest in their ability to reduce a large part of a linear network to a simpler equivalent network. Any other network being driven by the equivalent network experiences voltage, current, and power flow that are identical to those caused by the actual network. In any given network, whether one chooses to use a voltage or current form of an equivalent network would depend, at least in part, on the form of the network being driven by the equivalent network.

Source Conversions

Although the Thévenin and Norton equivalent sources were developed to replace a portion of a network, the final results (Fig. 3.4-5) may be used in an alternative manner.‡ We may view the Norton source as a *conversion* from a Thévenin source and vice versa. With this viewpoint in mind, if we have any source consisting of a voltage E in series with a resistance R, we may convert to a current

† Named for Edward L. Norton (1898–   ), an American scientist who worked at the Bell Telephone Laboratories.

‡ Because the Thévenin or Norton equivalent *network* appears in the form of an equivalent voltage or current *source* to the network being driven, it is sometimes convenient to refer to the equivalent networks as equivalent sources.

source of current $I = E/R$ in parallel with a resistance R. In a similar manner, a source of current I in parallel with a resistance R can be converted to a source of voltage $E = IR$ in series with a resistor R. These changes are called *source conversions*.

3.5 MESH ANALYSIS

In this section we introduce an analysis method that will permit the solution of relatively complicated circuits. The method, called *mesh analysis*, is most applicable to networks containing mainly voltage sources and is based on solving a set of simultaneous KVL equations. The method will apply to any *planar network*, which is one that can be diagramed on a plane without having any branch cross another.

Although mesh analysis can allow both current and voltage sources to be contained in the network, we shall initially discuss the more readily understood case where no current sources are present.

Circuits with No Current Sources

Consider the simple solved circuit of Fig. 3.5-1(a) that has two meshes. The circuit is redrawn in Fig. 3.5-1(b), except that it is assumed unsolved and we show currents circulating in each mesh that we call *mesh currents*. Mesh current I_1 is

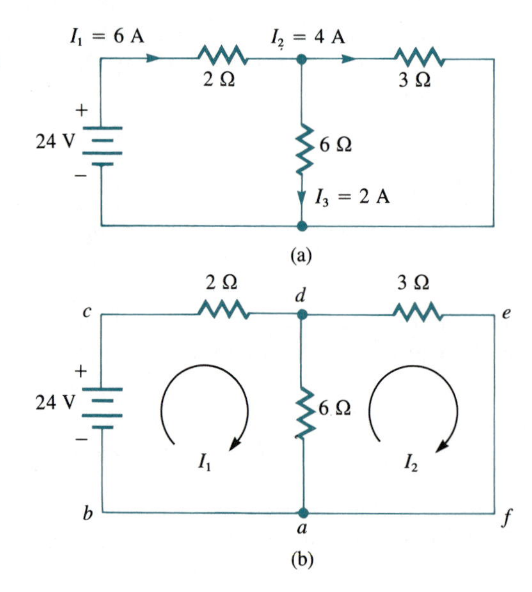

FIG. 3.5-1. (a) A circuit with two meshes and (b) the circuit redrawn to illustrate mesh currents.

clearly equal to the true current of 6 A over the path *abcd*; it is not equal to the real current of 2 A over the path *da*. Similarly, mesh current I_2 equals the real current of 4 A over *defa* but does not equal the real current over *ad*. However, the difference in mesh currents $I_1 - I_2$ does equal the real current of Fig. 3.5-1(a) over *da*. These similarities between real and mesh currents suggest that the circuit can be analyzed by using the mesh currents. We now show that this is a fact.

Suppose we write one KVL for each mesh as follows:

$$24 = I_1(2) + (I_1 - I_2)(6) = 8I_1 - 6I_2 \tag{3.5-1}$$

$$0 = (I_2 - I_1)(6) + I_2(3) = -6I_1 + 9I_2 \tag{3.5-2}$$

The simultaneous solution of (3.5-1) and (3.5-2) is facilitated if matrices are used. In matrix form we write

$$\begin{bmatrix} 24 \\ 0 \end{bmatrix} = \begin{bmatrix} 8 & -6 \\ -6 & 9 \end{bmatrix} \begin{bmatrix} I_1 \\ I_2 \end{bmatrix} \tag{3.5-3}$$

The solution for I_1 and I_2 becomes†

$$\begin{bmatrix} I_1 \\ \\ I_2 \end{bmatrix} = \begin{bmatrix} 8 & -6 \\ \\ -6 & 9 \end{bmatrix}^{-1} \begin{bmatrix} 24 \\ \\ 0 \end{bmatrix} = \begin{bmatrix} 9 & 6 \\ \\ 6 & 8 \end{bmatrix} \begin{bmatrix} 24 \\ \\ 0 \end{bmatrix} \left(\frac{1}{36}\right) \tag{3.5-4}$$

or $I_1 = 6$ A and $I_2 = 4$ A. These currents are the correct values, and for the chosen circuit, the mesh analysis method is shown to be valid.

We may state a procedure for mesh analysis in a more generalized network of many meshes:

1. For a planar network of resistors and voltage sources (dependent and/or independent), draw the network's diagram to clearly identify all meshes.
2. Assign a mesh current to each mesh. The usual convention is to show all currents clockwise, although this choice is not a requirement.
3. Write one KVL equation for each mesh in the direction of the mesh's current.
4. For each dependent voltage source, write a constraint equation that defines the source's dependence on the defined mesh currents. Substitute the constraint equations so that the dependent sources are explicit functions of the mesh currents.
5. Group the KVL equations and solve for the mesh currents by any convenient method. For N meshes there are N simultaneous equations in N unknown mesh currents. For $N > 2$ matrix methods are helpful.

We next demonstrate the mesh analysis procedure by using two examples, one with only independent voltage sources and one with both independent and dependent voltage sources.

† We use the notation [ ]$^{-1}$ to imply the inverse of a matrix [ ].

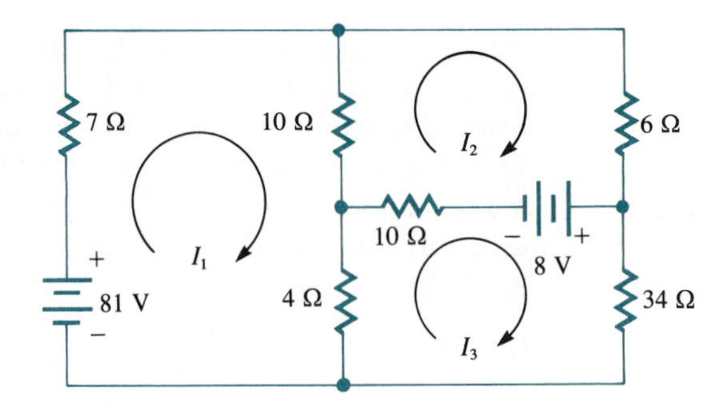

FIG. 3.5-2. The three-mesh network analyzed in Example 3.5-1.

EXAMPLE 3.5-1

We use mesh analysis to set up the equations for solution of the circuit of Fig. 3.5-2. The three required KVL equations are

$$81 = I_1(7) + (I_1 - I_2)(10) + (I_1 - I_3)(4)$$
$$-8 = I_2(6) + (I_2 - I_3)(10) + (I_2 - I_1)(10)$$
$$8 = I_3(34) + (I_3 - I_1)(4) + (I_3 - I_2)(10)$$

or on reduction

$$81 = 21I_1 - 10I_2 - 4I_3$$
$$-8 = -10I_1 + 26I_2 - 10I_3$$
$$8 = -4I_1 - 10I_2 + 48I_3$$

On solving these equations directly or through matrices, we find

$$I_1 = 5 \text{ A} \qquad I_2 = 2 \text{ A} \qquad I_3 = 1 \text{ A}$$

All other voltages follow use of these currents in Ohm's law. For example, the voltage drop across the 4-Ω resistor in the direction of I_1 is $(I_1 - I_3)(4) = 4(4) = 16$ V.

EXAMPLE 3.5-2

Mesh analysis will be used to solve the circuit of Fig. 3.5-3(a). It may first appear that mesh analysis is inappropriate because of the current source. However, we first convert the 15-A source in parallel with the 4-Ω resistor to a voltage source, as shown in Fig. 3.5-3(b). The network now has only voltage sources, one of which is a dependent source. The three required KVL equations are

$$60 = I_1(4) + (I_1 - I_3)(18) + (I_1 - I_2)(12)$$
$$0 = (I_2 - I_1)(12) + (I_2 - I_3)(12) + 15I$$
$$0 = (I_3 - I_1)(18) + I_3(28) + (I_3 - I_2)(12)$$

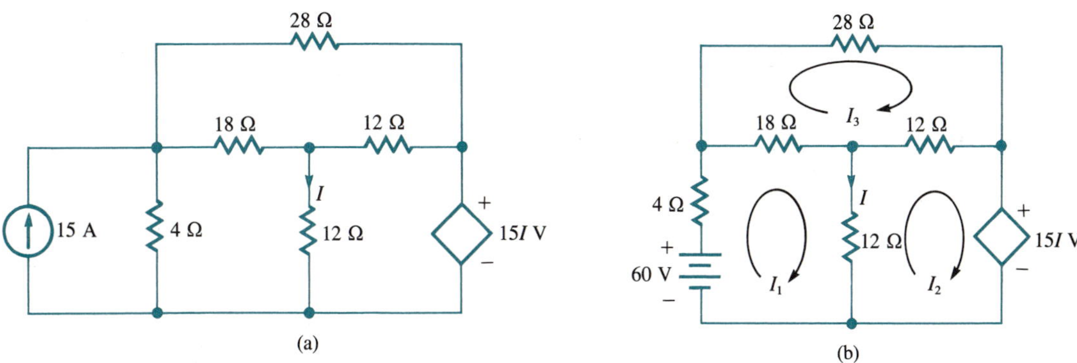

FIG. 3.5-3.
(a) A network analyzed by mesh analysis in Example 3.5-2 and (b) the same network after conversion of the current source to a voltage source.

We must now use the constraint equation

$$I = I_1 - I_2$$

to place the KVL equation of mesh 2 in the necessary form (a function of I_1, I_2, and I_3). After grouping terms, we have

$$60 = 34I_1 - 12I_2 - 18I_3$$

$$0 = 3I_1 + 9I_2 - 12I_3$$

$$0 = -18I_1 - 12I_2 + 58I_3$$

After solving these simultaneous equations, we find

$$I_1 = 2.25 \text{ A} \qquad I_2 = 0.25 \text{ A} \qquad I_3 = 0.75 \text{ A}$$

All voltages may be found from Ohm's law and these currents.

Circuits with Current Sources

If a circuit that is to be analyzed by mesh analysis contains some current sources (independent and/or dependent) in addition to voltage sources (independent and/or dependent), the analysis is usually simpler; but some special considerations need to be made. These considerations are now stated.

First, if there are dependent current sources, their constraint equations need to be used to make the source currents explicit functions of the mesh currents.

Next, if the current source is in a branch that borders only one mesh (an outer edge of the circuit), as illustrated in Fig. 3.5-4(a), the source's current is equal to, or is the negative of, the mesh current. In other words, the mesh current is *known*, and the number of equations required to solve the circuit is reduced by one for each such current source. For the example circuit the equations that give a solution are

$$\text{(mesh 1)} \qquad 19 = I_1(11) + (I_1 - I_2)(2) \tag{3.5-5}$$

$$\text{(mesh 2)} \qquad -19 = (I_2 - I_1)(2) + (I_2 - I_3)(5) + I_2(2) \tag{3.5-6}$$

$$\text{(mesh 3)} \qquad -25 = (I_3 - I_2)(5) + (I_3 - I_4)(6) \tag{3.5-7}$$

$$\text{(mesh 4)} \qquad I_4 = 3 \tag{3.5-8}$$

On substitution of the last result, the other three become

$$19 = 13I_1 - 2I_2 \tag{3.5-9}$$

$$-19 = -2I_1 + 9I_2 - 5I_3 \tag{3.5-10}$$

$$-7 = -5I_2 + 11I_3 \tag{3.5-11}$$

When solved, these equations yield $I_1 = 1$ A, $I_2 = -3$ A, and $I_3 = -2$ A.

Lastly, if a current source is in a branch that borders two meshes (interior branch), it forms a common link between the two meshes. If we assign a voltage across the source, the voltage can be eliminated by writing the KVL equations for the two meshes. One equation then results, which, when combined with the equation defining the dependence of the current source on its two mesh currents and the other circuit KVL equations, will give a complete solution. Figure 3.5-4(b) shows a circuit with an interior current source.

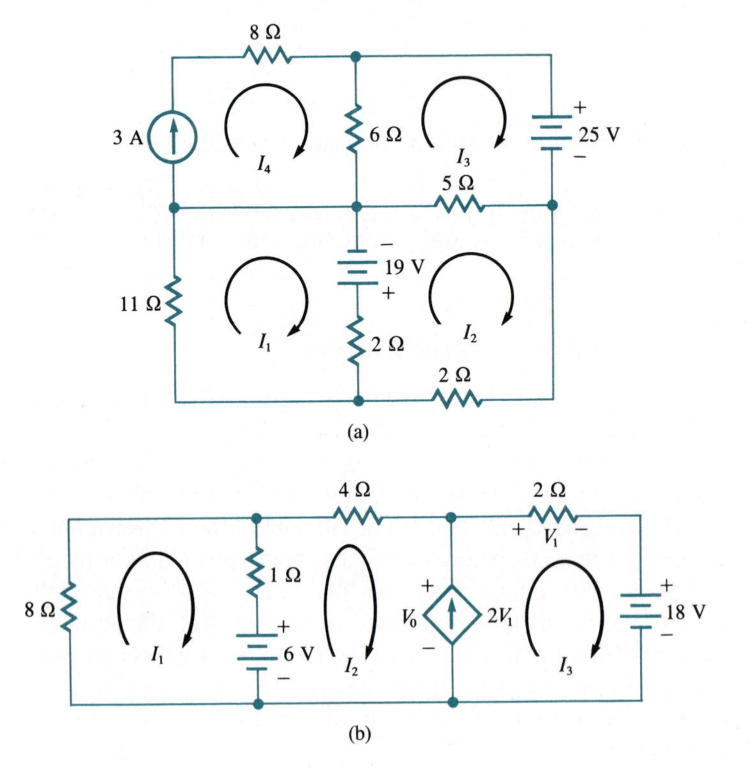

FIG. 3.5-4. (a) A circuit where a current source borders only one mesh and (b) a circuit where a current source borders two meshes.

EXAMPLE 3.5-3

We use mesh analysis to solve for the currents I_1, I_2, and I_3 in Fig. 3.5-4(b). Here, we have an interior dependent current source. We assign voltage V_0 across the source and write KVL equations for meshes 2 and 3.

$$\text{(mesh 2)} \qquad 6 = (I_2 - I_1) + I_2(4) + V_0$$

$$\text{(mesh 3)} \qquad -18 = -V_0 + I_3(2)$$

On elimination of V_0 only one equation results:

$$-12 = -I_1 + 5I_2 + 2I_3 \tag{A}$$

Another equation comes from that defining the current source:

$$2V_1 = 4I_3 = I_3 - I_2$$

or

$$0 = I_2 + 3I_3 \tag{B}$$

The remaining KVL equation is

$$\text{(mesh 1)} \qquad -6 = I_1(8) + (I_1 - I_2) \tag{C}$$

The required three equations in three unknown currents are (A), (B), and (C). On solving, we obtain

$$I_1 = -1\text{ A} \qquad I_2 = -3\text{ A} \qquad I_3 = 1\text{ A}$$

3.6 NODAL ANALYSIS

Nodal analysis is another rather powerful method of solving relatively complicated circuits. It can be viewed as a sort of corollary method to mesh analysis and it applies to any linear network. Whereas in mesh analysis a system of KVL equations were solved for unknown mesh currents, nodal analysis requires that we solve a system of KCL equations for unknown node voltages. In a circuit with N nodes, one node is designated the reference node, and the voltages at the other $N - 1$ nodes are defined relative to the reference node. After writing $N - 1$ KCL equations, one for each of the $N - 1$ nodes, we solve them to obtain the $N - 1$ node voltages. From these voltages and Ohm's law, other quantities of interest, such as currents in elements, may be determined.

Because nodal analysis is based on Kirchhoff's current law, networks that have only current sources are conceptually the most easily understood. We shall discuss this case first. However, the analysis procedure does not exclude the presence of voltage sources, and this case is discussed second.

Circuits with No Voltage Sources

For a network that has no voltage sources, the following steps form the nodal-analysis procedure.

1. For a network of resistors and independent and dependent current sources, draw the network's diagram to clearly identify all nodes.
2. Choose one node as the reference node. The choice can be arbitrary and made for convenience or to fit a physical condition. For example, the reference node might correspond to a circuit point that is grounded in practice. It is customary to associate zero potential with the reference node.
3. Assign a voltage to each remaining node. This voltage is, of course, relative to the reference node.
4. Write one KCL equation for each nonreference node.
5. For each dependent current source, write a constraint equation that defines the source's dependence on the assigned voltage variables. Substitute the constraint equations so that the dependent sources are explicit functions of the voltage variables.
6. Group the KCL equations and solve for the node voltages by any convenient method. For $N - 1$ nonreference nodes there are $N - 1$ simultaneous equations in $N - 1$ unknown node voltages. For $N - 1 > 2$ matrix methods may be helpful.

EXAMPLE 3.6-1

We demonstrate the nodal-analysis method by solving the circuit of Fig. 3.6-1(a) for the node voltages V_1, V_2, and V_3. The three KCL equations are

$$\text{(node 1)} \qquad 2 = \tfrac{1}{15} V_1 + (V_1 - V_2)(\tfrac{1}{3})$$

$$\text{(node 2)} \qquad 4I_a = V_2(\tfrac{1}{4}) + (V_2 - V_1)(\tfrac{1}{3}) + (V_2 - V_3)(\tfrac{1}{9})$$

$$\text{(node 3)} \qquad -4I_a = V_3(\tfrac{1}{3}) + (V_3 - V_2)(\tfrac{1}{9})$$

Next, we substitute the constraint equation defining the dependent source, which is

$$I_a = (V_1 - V_2)(\tfrac{1}{3})$$

and obtain

$$30 = 6V_1 - 5V_2 \qquad \text{(A)}$$

$$0 = -60V_1 + 73V_2 - 4V_3 \qquad \text{(B)}$$

$$0 = 12V_1 - 13V_2 + 4V_3 \qquad \text{(C)}$$

When these are solved, we find

$$V_1 = 15 \text{ V} \qquad V_2 = 12 \text{ V} \qquad V_3 = -6 \text{ V}$$

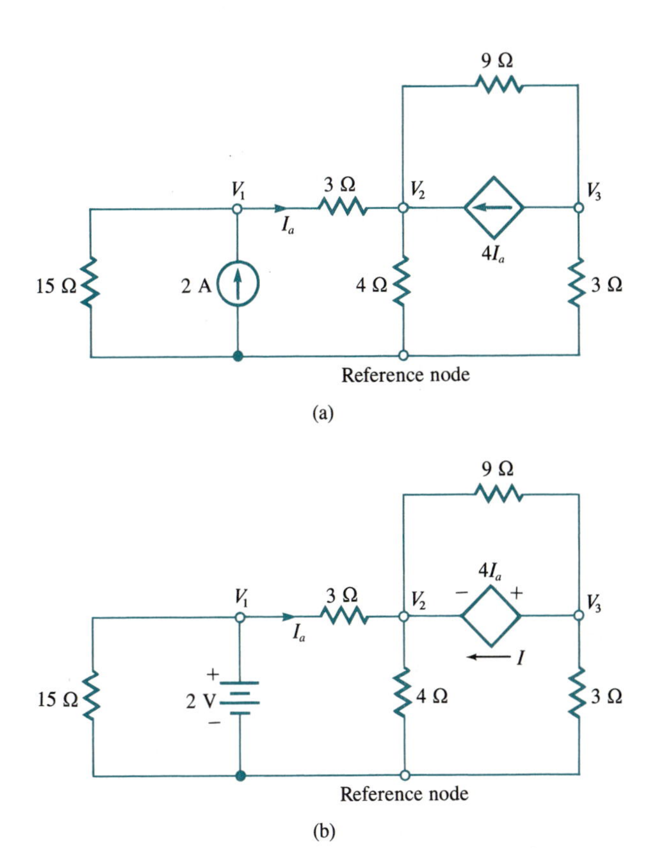

FIG. 3.6-1. (a) The network solved in Example 3.6-1 and (b) the revised network solved in Example 3.6-2.

Circuits with Voltage Sources

When independent and/or dependent voltage sources are contained in a network and nodal analysis is to be used, some modifications need to be made in the preceding procedure.

First, if there are any dependent voltage sources, their constraint equations are used to make the source's voltage an explicit function of the node voltages.

Next, for any voltage source that has the reference node as one of its connecting points, the source's voltage determines the node voltage at the non-reference node to which it is connected. This condition means that a KCL equation is not required at the nonreference node. The required number of KCL equations is reduced by one for each such voltage source.

Finally, if a voltage source is connected between two nonreference nodes, it establishes a direct relationship between the nonreference-node voltages. By assigning a current through the source and solving the two KCL equations derived at the two nonreference nodes to eliminate the current, we obtain a single equation. This equation, the other KCL equations, and the equation defining the dependence of the voltage source on its two node voltages will allow all node voltages to be found. We take an example to solidify these points.

EXAMPLE 3.6-2

Suppose we use the circuit of Fig. 3.6-1(a), except that we change the current sources to voltage sources, as shown in Fig. 3.6-1(b). Thus, we have a circuit with *no* current sources, one independent voltage source, and one current-controlled dependent voltage source. Since the independent 2-V source connects to the reference node, it establishes V_1 immediately,

$$V_1 = 2\text{ V}$$

and no KCL equation is needed at node 1. Because the dependent source is connected between nonreference nodes, we assign a current I as shown, write the KCL equations at the two nodes, which are

(node 2) $\quad I = (V_2 - V_3)(\tfrac{1}{9}) + (V_2 - V_1)(\tfrac{1}{3}) + V_2(\tfrac{1}{4}) = \tfrac{1}{36}(25V_2 - 4V_3 - 24)$

(node 3) $\quad -I = (V_3 - V_2)(\tfrac{1}{9}) + V_3(\tfrac{1}{3}) = \tfrac{1}{9}(-V_2 + 4V_3)$

and add the two to eliminate I. The result is

$$24 = 21V_2 + 12V_3 \qquad \text{(A)}$$

We only require the equation relating the controlled source to its node voltages to complete the analysis. It is

$$V_3 - V_2 = 4I_a = 4(V_1 - V_2)(\tfrac{1}{3}) = 4(2 - V_2)(\tfrac{1}{3})$$

or

$$8 = V_2 + 3V_3 \qquad \text{(B)}$$

Simultaneous solution of (A) and (B) gives

$$V_2 = -\tfrac{8}{17}\text{ V} \qquad V_3 = \tfrac{48}{17}\text{ V}$$

The reader may wish to use these node voltages to compute the currents in the circuit and verify that all quantities are valid.

PROBLEMS

3-1. Write a KVL equation for the circuit of Fig. P3-1 and solve for the voltages across each resistor.

FIG. P3-1.

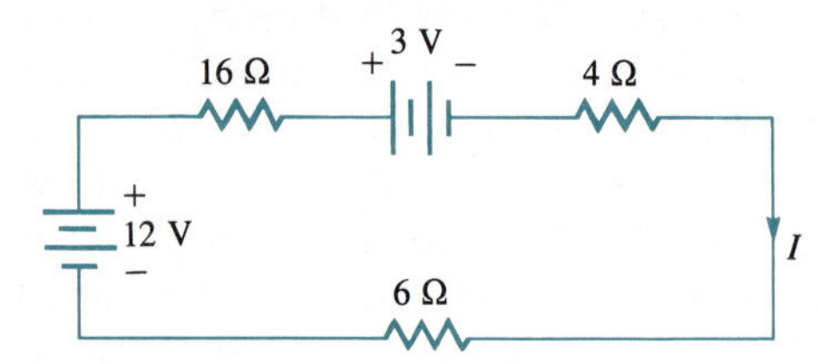

3-2. Use a KVL to solve the circuit of Fig. P3-2 for the voltage across the current source. Which of the terminals, *a* or *b*, is the more positive relative to the other?

FIG. P3-2.

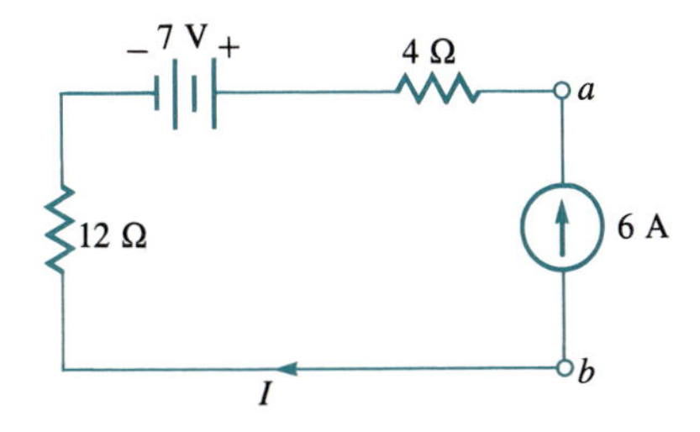

3-3. Five resistors of resistances 1.8, 3.6, 10, 22, and 33 Ω are in series between two terminals *a* and *b*. What is the equivalent resistance between the terminals?

3-4. A series of six resistors is defined in Fig. P3-4. (a) What value of *R* will cause a battery current of 3 mA to flow? (b) Find the voltage that occurs at each of the points, labeled 1 through 6, relative to the reference node. Do any of the voltages depend on the value of *R*?

FIG. P3-4.

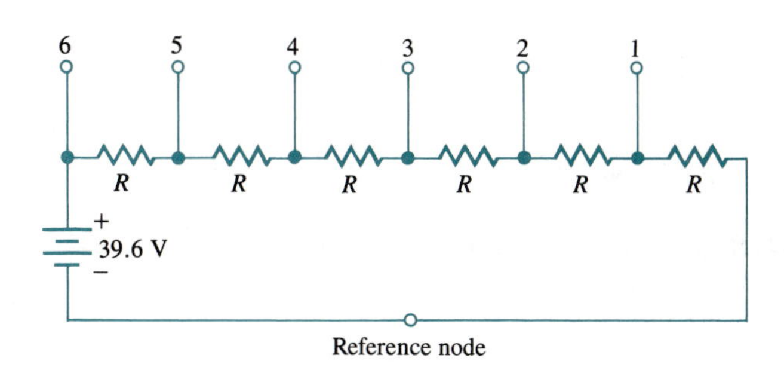

3-5. Write KVL equations for the circuit of Fig. P3-5 and solve for all currents and the voltages across the resistors. Are both batteries providing power to the resistors?

FIG. P3-5.

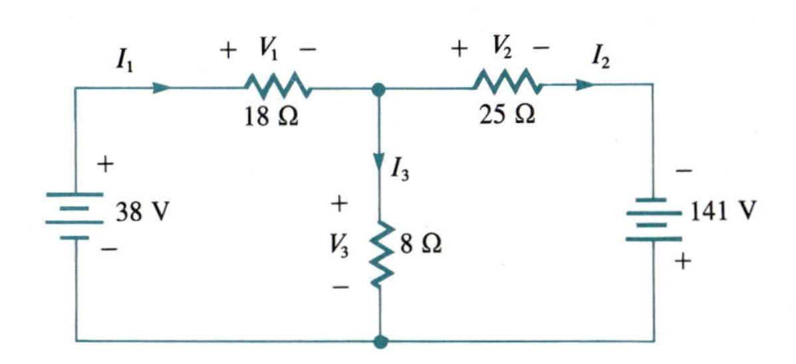

3-6. Use KVL equations in the circuit of Fig. P3-6 to find the value of R that will cause $I_1 = 0.5$ A when V_1 is larger than V_2 by 24 V.

FIG. P3-6.

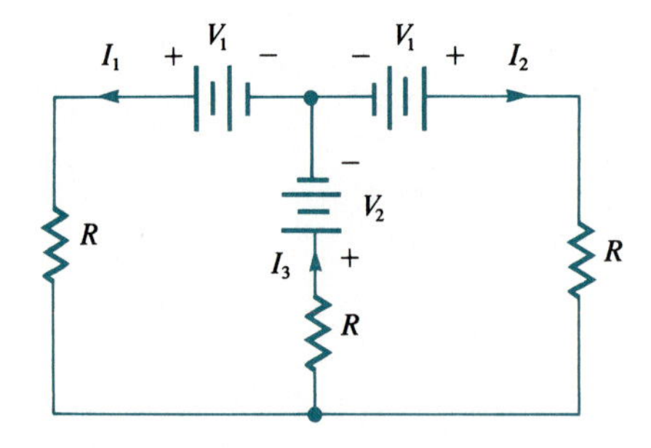

3-7. Find the equivalent resistance of four 8-Ω resistors all in parallel.

3-8. Assume N identical resistors of resistance R are connected in parallel. Determine an equation giving the equivalent resistance of the parallel combination for any N and R.

3-9. Use KCL equations to solve for the voltages and currents in all resistors of Fig. P3-9.

FIG. P3-9.

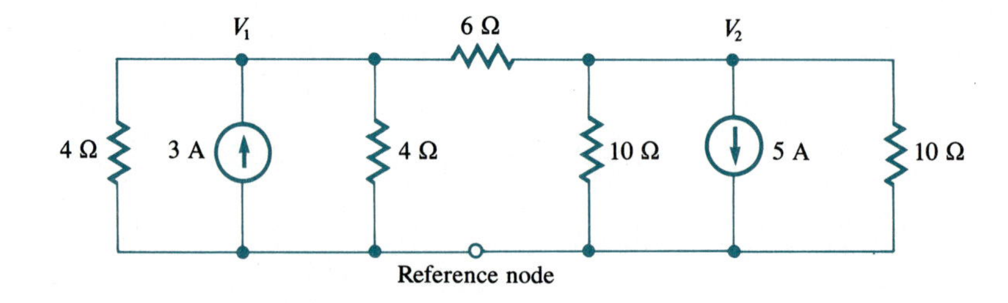

3-10. Apply superposition to the network of Fig. P3-10 to calculate the currents I_1 and I_2. What are the voltages at points a and b relative to the reference node?

FIG. P3-10.

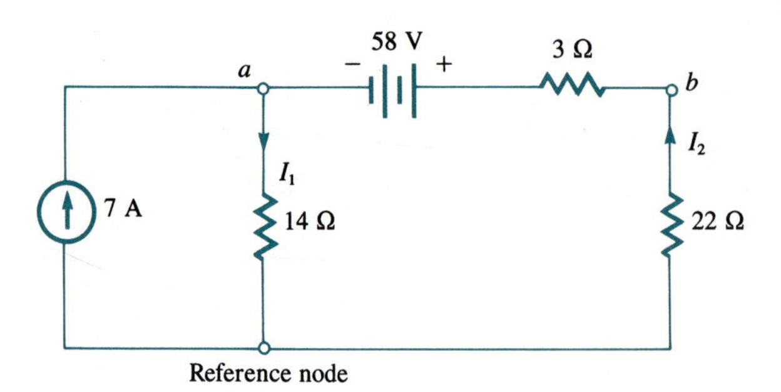

★ **3-11.** Use the superposition principle to find the five currents shown and the voltages at points a, b, c, and d, all relative to the reference node, in Fig. P3-11. Give the three component currents that constitute I_3.

FIG. P3-11.

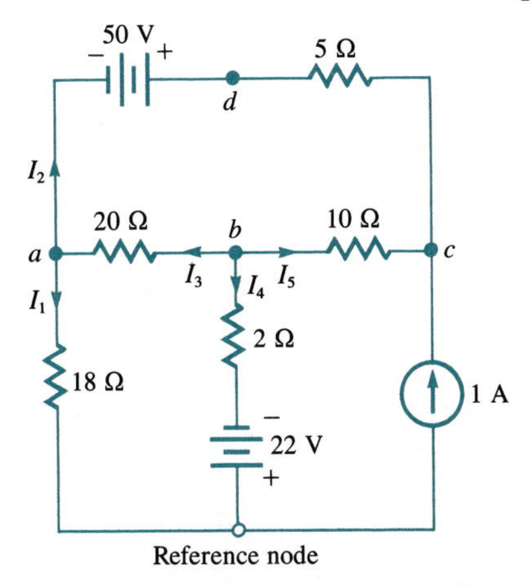

3-12. Use superposition to find currents I_1 and I_2 in the circuit of Fig. P3-12.

FIG. P3-12.

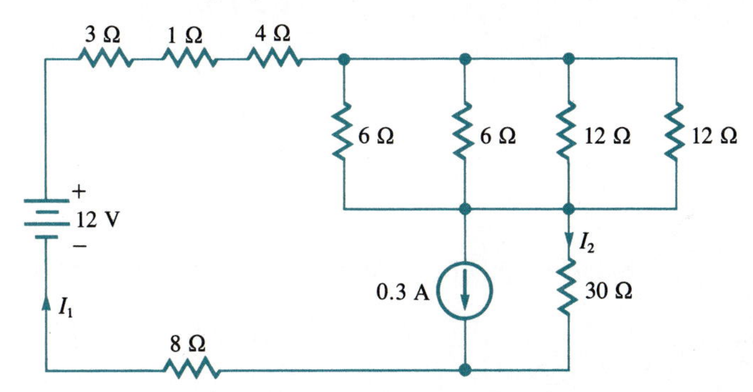

3-13. Work Problem 3-12 except double the resistance of each of the four parallel resistors and halve all other resistance values.

3-14. For the circuit of Fig. P3-14, find the node voltages V_1, V_2, and V_3 by using superposition.

FIG. P3-14.

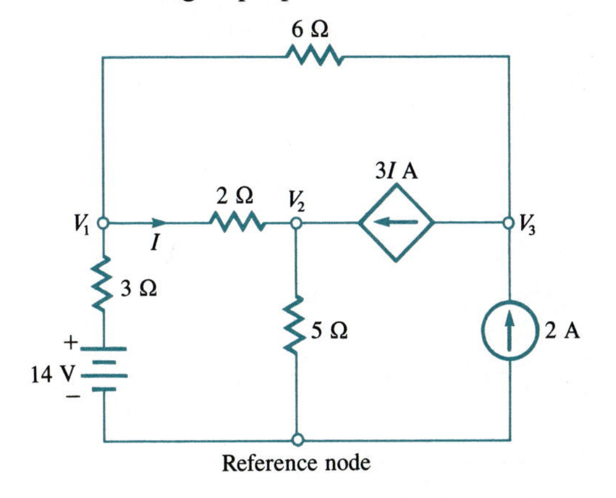

3-15. Use source conversions to show that the four-node network of Fig. P3-15 can be converted to a two-node network with two current generators. Find V_1.

FIG. P3-15.

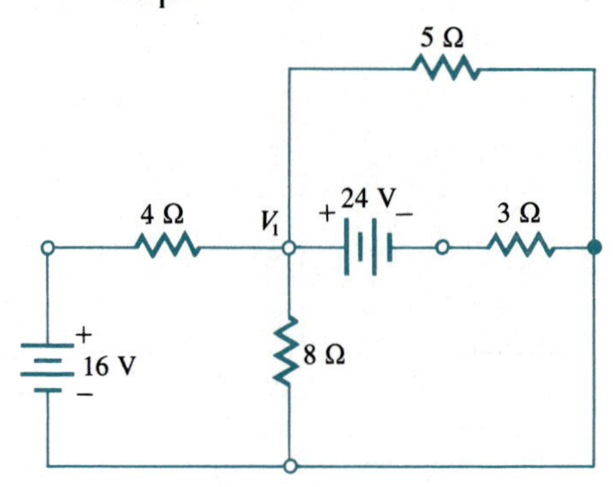

3-16. Apply superposition to determine V_0 in terms of V_a and V_b in Fig. P3-16.

FIG. P3-16.

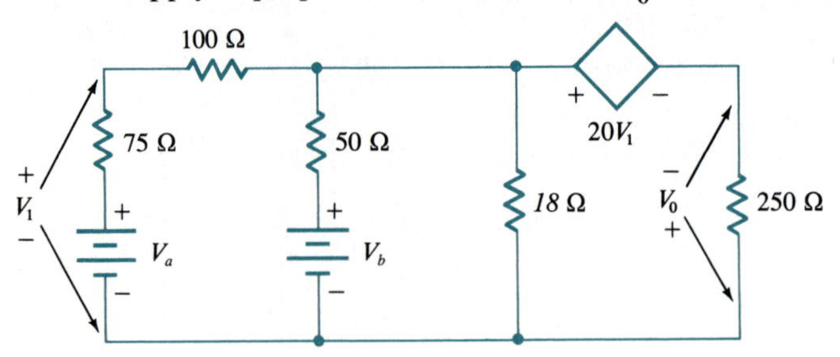

3-17. Find a single equivalent voltage source for the network of Fig. P3-17.

FIG. P3-17.

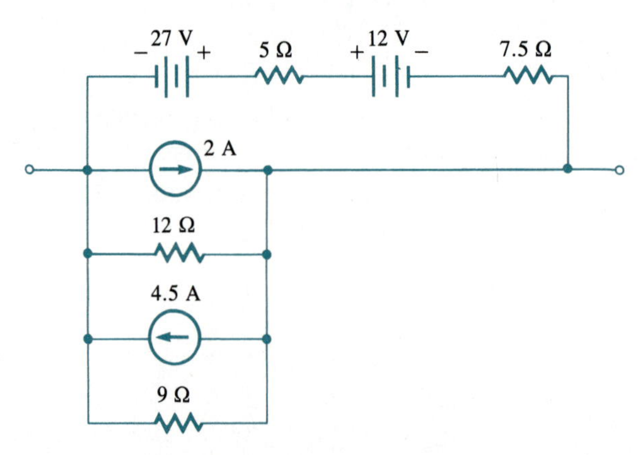

3-18. Work Problem 3-17 except find a current source.

3-19. Find a Thévenin equivalent source to connect between node V_1 and the reference node in Fig. P3-15 that will replace the loop containing the 16-V battery and the 4- and 8-Ω resistors.

3-20. Work Problem 3-19 except find a Norton equivalent source.

3-21. Draw a line through point *a* and the reference node in Fig. P3-11. Find a Thévenin equivalent source for the network on the right side of the partition line.

3-22. Work Problem 3-21 except find a Norton equivalent source.

3-23. Partition the circuit of Fig. P3-10 with a line through node *b* and the reference node. Find a Thévenin generator for the left part of the circuit.

3-24. Work Problem 3-23 except find a Norton generator.

3-25. Draw two vertical partition lines in the circuit of Fig. P3-9. One line is just left of the 6-Ω resistor and one is just to the right. Find a Thévenin source for the circuit left of the left line and another for the circuit right of the right line.

3-26. Work Problem 3-25 except find Norton sources.

3-27. Use mesh analysis to determine the resistance seen by the current source in Fig. P3-27. Check your result by using parallel and series equivalent resistances to reduce the network.

FIG. P3-27.

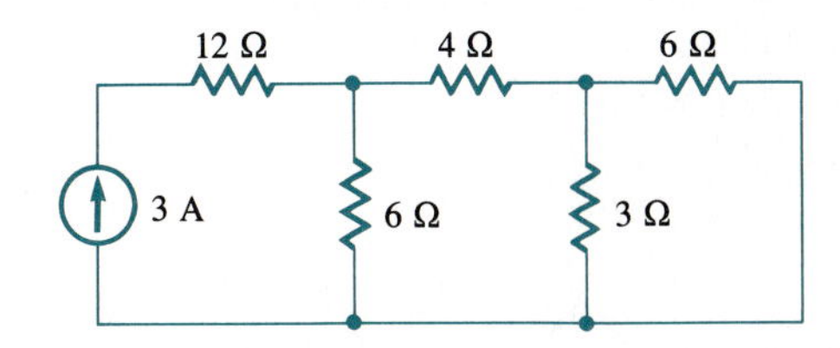

3-28. Solve the circuit of Fig. P3-28 for the mesh currents I_1, I_2, and I_3.

FIG. P3-28.

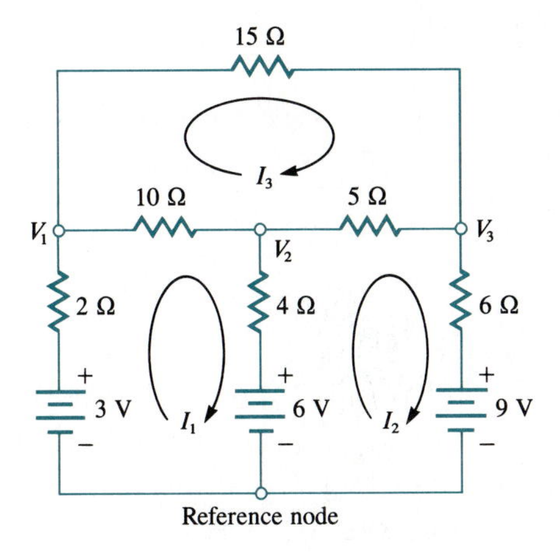

3-29. Use mesh analysis to solve the circuit of Fig. P3-29 for V_0. Find a Thévenin equivalent source for the entire network that drives the 20-Ω resistor. How much power is being delivered to the 20-Ω resistor?

FIG. P3-29.

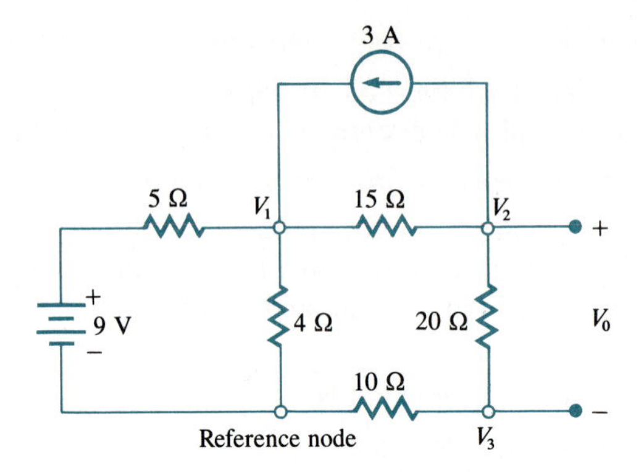

3-30. Use mesh analysis to solve for the node voltages V_1, V_2, and V_3 in the circuit of Fig. P3-30. (*Hint:* Reduce the independent voltage sources first to obtain a three-mesh circuit for analysis.)

FIG. P3-30.

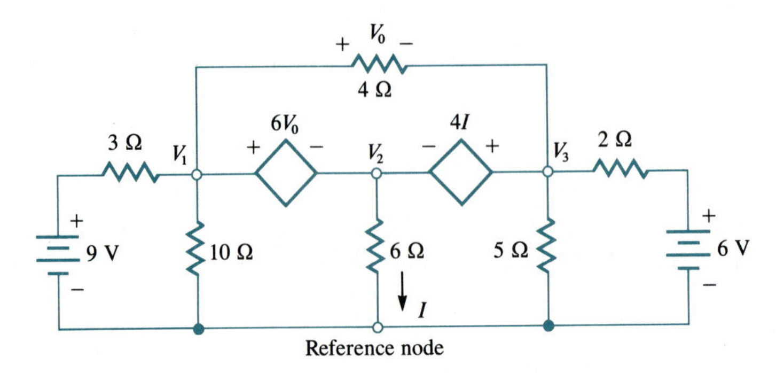

3-31. Solve the circuit of Fig. P3-31, using mesh analysis, to find V_0. This circuit is a model of a transistor amplifier.

FIG. P3-31.

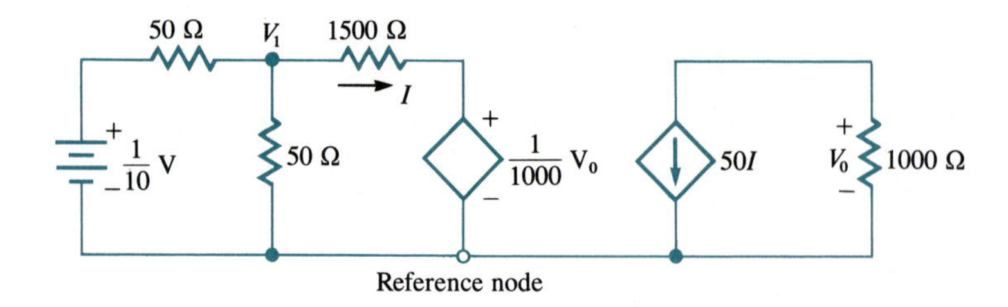

3-32. Use nodal analysis to work Problem 3-31 and find the node voltages V_0 and V_1.

3-33. Work Problem 3-30 except use nodal analysis.

3-34. Solve the circuit of Fig. P3-29 for V_1, V_2, V_3, and V_0 by using nodal analysis.

3-35. Work Problem 3-28 except use nodal analysis to find V_1, V_2, and V_3.

3-36. Use nodal analysis to write a set of equations that is sufficient to solve Fig. P3-36. It is not necessary to solve the equations.

FIG. P3-36.

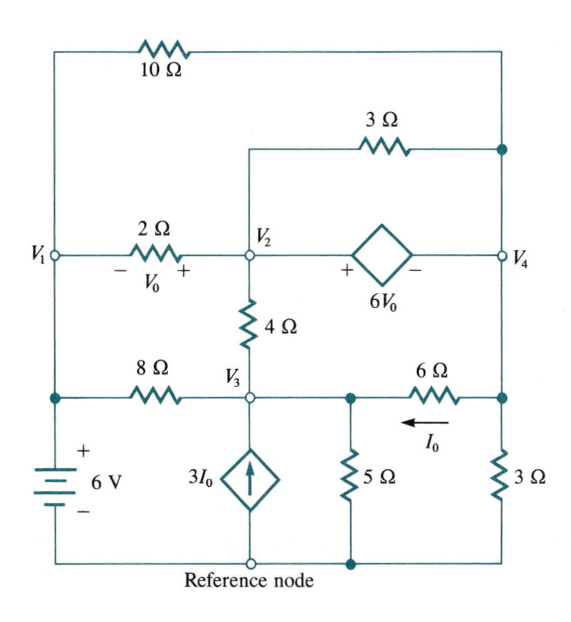

CHAPTER 4

Alternating Current Circuits

4.0 INTRODUCTION

Many electric networks involve more complexity than those described in the preceding chapter where only dc sources and resistors were present. In this chapter we turn our attention to the more general linear network. Mainly, emphasis will be placed on networks containing only resistors, capacitors, inductors, and transformers. This choice is only for ease of learning, and our developments will apply to networks that include devices other than passive components. Transistors and klystrons (a microwave device) are just two examples of components that may behave as a linear circuit element. They are usually called *active devices*. Such devices may be included in a linear circuit by modeling them as a dependent source.

Although we place no restrictions on the form of the linear network, one very important restriction will be placed on the sources considered in this chapter. It results that great insight into the behavior of circuits having arbitrary sources can be derived from knowledge of their behavior using sources containing only one frequency. Our restriction is, therefore, to consider here only single-frequency voltage and current sources. An example of a single-frequency source is the voltage

$$v(t) = V \cos(\omega t + \theta) \tag{4.0-1}$$

where V is the peak voltage (a constant), θ is an arbitrary phase angle, and ω is the (single) angular frequency of $v(t)$. Because sources such as $v(t)$ constitute an alternating (cyclic) voltage, they give rise to alternating currents in a circuit, so we refer to them as *alternating current* (abbreviated ac) *circuits*.

Although no effort will be made to offer proofs, the important concepts of Kirchhoff's laws, superposition, equivalent networks (Thévenin and Norton), and mesh and nodal analysis developed in Chapter 3 are all valid when applied properly to the circuits of this chapter.

As a final point before we proceed with details, it is important to note that our work will deal with ac *steady-state analysis*, because excitations such as (4.0-1) are presumed to exist in their fixed form for all time. These excitations produce circuit voltages and currents that do not change in peak amplitude, frequency, or phase angle as time varies, so they are considered to be in a steady state. More general excitations that may contain variations in amplitude, phase, or frequency may be called transient waveforms. Analysis with these types of signals is considered in Chapter 5.

4.1 COMPLEX SIGNALS AND BASIC ANALYSIS

Our general problem is to analyze linear networks when all voltage and current sources are sinusoidal, with the forms

$$v(t) = V \cos(\omega t + \theta) \tag{4.1-1a}$$

$$i(t) = I \cos(\omega t + \phi) \tag{4.1-1b}$$

Here V and I are the peak (constant) voltage and current amplitudes, respectively. Constants θ and ϕ are arbitrary phase angles, and ω is any constant angular frequency. The period T of (4.1-1) is given by

$$T = \frac{2\pi}{\omega} \tag{4.1-2}$$

Now, let us make a simple observation. In practice, we always have real networks excited by real voltage and current sources that generate real voltages and currents everywhere in the network. With this observation, the most obvious approach to solving a network would be to (1) specify all of the real sources, (2) establish the fundamental equations describing the network, and (3) solve the equations for the desired real voltages or currents. This procedure sounds simple but problems immediately arise. Since circuit elements are fundamentally described by differential equations [see (1.4-1) for a resistor, (1.4-7) or (1.4-9) for a capacitor, (1.4-19) or (1.4-21) for an inductor, and (1.4-34) and (1.4-35) for a transformer], and since several current or voltage variables may be unknown, the equations to be solved form a set of simultaneous integro-differential equations. Any reader who has had to deal with solving such equations knows the problems that occur.

Complex Signals

Fortunately, various techniques have been developed to solve the equations of linear networks,† especially those with steady-state excitations. For the special case where a *complex signal* is used to represent any source, these techniques reduce the integro-differential equations to a set of algebraic equations that can be solved by methods developed earlier for dc circuits. We formally define a complex signal, denoted by $v_c(t)$,‡ as having the form

$$v_c(t) = Ve^{j\omega t + j\theta} \tag{4.1-3}$$

where V and θ are called the amplitude and phase, respectively, of the complex signal and ω is its angular frequency.

By using Euler's§ identity

$$e^{jx} = \cos(x) + j\sin(x) \tag{4.1-4}$$

the complex signal can be put in the form

$$v_c(t) = V\cos(\omega t + \theta) + jV\sin(\omega t + \theta) \tag{4.1-5}$$

From this form it is easy to see how the complex signal can represent the real source of (4.1-1):

$$v(t) = \text{Re}\,[v_c(t)] = \tfrac{1}{2}[v_c(t) + v_c^*(t)] \tag{4.1-6}$$

Here Re [ ] represents taking the real part of the quantity in the brackets, and the asterisk represents forming the complex conjugate of the quantity on which it occurs. Alternatively,

$$v_c(t) = v(t) + jv\left(t - \frac{T}{4}\right) \tag{4.1-7}$$

A sort of physical significance can be attached to the complex signal

$$e^{j\omega t} = \cos(\omega t) + j\sin(\omega t) \tag{4.1-8}$$

by thinking of it as simply a complex number having a real part $\cos(\omega t)$ and an imaginary part $\sin(\omega t)$. In the complex plane having these real and imaginary components, $\exp(j\omega t)$ would correspond to a locus of (vector) points on a circle. As time progresses, angle ωt increases, and the complex point $\cos(\omega t) + j\sin(\omega t)$ rotates counterclockwise on a circle of unit radius. One rotation occurs for each

---

† Our results apply only to networks in which all values of resistance, capacitance, inductance, mutual inductance, and constraints on dependent sources are all constants with time.

‡ The subscript is used to maintain context in this section. Later when context is clear, complex signals may be denoted without a subscript. The chosen notation implies a voltage, but the form of (4.1-3) is not so restricted, and (4.1-3) can equally well represent a current with a minor notation change. When the term $j\omega t$ is suppressed (set to zero), the complex signal is called a *phasor*.

§ Leonard Euler (1707–1783) was a Swiss mathematician.

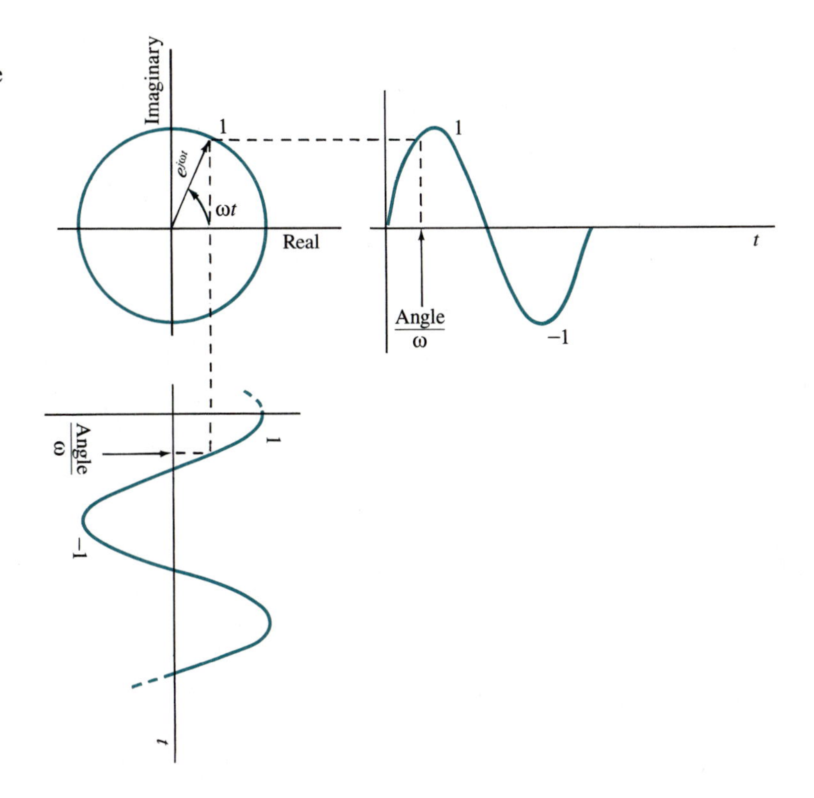

FIG. 4.1-1. A unit-amplitude complex signal showing its real and imaginary components.

period of the angle ωt. The *rate* of rotation of the vector is ω, the angular frequency of the complex signal. Figure 4.1-1 illustrates these points, where the real and imaginary components, which are $\cos(\omega t)$ and $\sin(\omega t)$, respectively, are shown as the projections with time of the complex signal.

A useful property of the complex signal is that a factor of j represents an advance in phase shift of $\pi/2$. The proof is straightforward:

$$\begin{aligned} e^{j(\omega t + \theta + \pi/2)} &= e^{j(\omega t + \theta) + j(\pi/2)} \\ &= e^{j(\omega t + \theta)}\left[\cos\left(\frac{\pi}{2}\right) + j\sin\left(\frac{\pi}{2}\right)\right] \\ &= je^{j(\omega t + \theta)} \end{aligned} \tag{4.1-9}$$

Of course, a factor of $-j$ represents a phase shift of $-\pi/2$.

It is often convenient to use a complex signal to represent the sum of two real ac signals with different amplitudes but having the same angle. If the amplitudes are denoted by A and B, we use trigonometric identities to write

$$A\cos(\omega t + \theta) + B\sin(\omega t + \theta) = D\cos(\omega t + \theta + \psi) \tag{4.1-10}$$

where

$$D = \sqrt{A^2 + B^2} \tag{4.1-11}$$

$$\psi = \tan^{-1} \frac{-B}{A} \tag{4.1-12}$$

or, alternatively,

$$A = D \cos \psi \tag{4.1-13}$$

$$B = -D \sin \psi \tag{4.1-14}$$

The complex signal representing $D \cos(\omega t + \theta + \psi)$ becomes $D \exp(j\omega t + j\theta + j\psi)$.

One of the most important properties of a complex signal is that it reproduces itself, within a factor that is constant in time, when integrated or differentiated. From (4.1-3) we have

$$\frac{dv_c(t)}{dt} = j\omega V e^{j\omega t + j\theta} = j\omega v_c(t) \tag{4.1-15}$$

$$\int_{-\infty}^{t} v_c(\xi)\, d\xi = \frac{V}{j\omega} e^{j\omega t + j\theta} = \frac{1}{j\omega} v_c(t) \tag{4.1-16}$$

where the integral of $v_c(t)$ at minus infinity is assumed to be zero. It is this unique property that allows the differential equations of a complicated circuit to be reduced to a set of linear equations that can be solved by algebraic methods.

Basic Analysis

The above discussions and the form of (4.1-6) help to suggest a general procedure for the analysis of ac linear networks:

1. Replace each real source voltage or current with a complex signal that has a real part equal to the real source's signal.
2. Perform the time differentiations or integrations indicated for all network elements. An algebraic system of equations with unknown currents and voltages will result.
3. Solve the system of equations in any convenient manner (such as by applying Kirchhoff's laws, for example). The solutions will be complex, in general.
4. Take the real solution of any desired variable as the real part of the complex solution. Alternatively, half the complex conjugate of the complex solution can be added to half the complex solution to achieve the real solution, as indicated in (4.1-6).

At this point in our discussions, step 2 requires additional development and discussion before a clear understanding can be achieved. These developments mainly come from solutions of some simple circuits as we proceed. Not only will these efforts serve to lead to a clearer understanding of the analysis procedure, they will also lead to some important new concepts.

4.2 EQUATIONS OF CIRCUIT ELEMENTS

An important part of the analysis procedure of the preceding section is step 2. In this section we show how to perform the time differentiations or integrations indicated for network elements.

Resistors

The current $i(t)$ through a resistance R and the voltage $v(t)$ across its terminals are defined by Ohm's law given previously as (1.4-1):

$$v(t) = i(t)R \tag{4.2-1}$$

In this relationship no derivative or integration is involved.

Capacitors

The voltage and current relationship for a capacitance C was defined either by (1.4-7),

$$i(t) = C\frac{dv(t)}{dt} \tag{4.2-2}$$

or by (1.4-9),

$$v(t) = \frac{1}{C}\int_{-\infty}^{t} i(\xi)\,d\xi \tag{4.2-3}$$

For a real capacitor voltage as defined by (4.1-1a), the current becomes

$$i(t) = C\frac{d}{dt}[V\cos(\omega t + \theta)] = -\omega CV\sin(\omega t + \theta) \tag{4.2-4}$$

For a complex signal such as (4.1-3), which represents the real voltage by having a real part equal to the real voltage, the current is

$$i_c(t) = C\frac{d}{dt}(Ve^{j\omega t + j\theta}) = j\omega CVe^{j\omega t + j\theta} = j\omega Cv_c(t) \tag{4.2-5}$$

These two expressions both show that the current in a capacitor leads (has a larger phase angle than) the voltage across it by a phase angle of $\pi/2$. This fact is obvious in (4.2-4) because $-\sin x = \cos(x + \pi/2)$. It is evident from (4.2-5) because the factor j represents a shift in phase by $\pi/2$.

Because of the form of (4.2-5), the ratio $v_c(t)/i_c(t) = 1/(j\omega C)$ can be taken as a kind of Ohm's law for capacitors. This result is made possible only through the special exponential form of the complex signal that allows the *ratio* of the voltage and current time functions to be independent of time. Thus,

$$v_c(t) = i_c(t)\left(\frac{1}{j\omega C}\right) \tag{4.2-6}$$

can be taken as Ohm's law for a capacitor, but *only* when currents and voltages are complex signals. Dimensionally, the unit of $1/(j\omega C)$ must be the ohm, although it is *not* a resistance. Because $j = \sqrt{-1}$ is unitless, it is customary to call $-1/(\omega C)$ the reactance, or *capacitive reactance* of C, denoted by X_C:

$$X_C = \frac{-1}{\omega C} \tag{4.2-7}$$

More will be said about reactance as we continue.

Inductors

The voltage-current relationship for an inductance L is established either by (1.4-19),

$$v(t) = L\frac{di(t)}{dt} \tag{4.2-8}$$

or by (1.4-21),

$$i(t) = \frac{1}{L}\int_{-\infty}^{t} v(\xi)\, d\xi \tag{4.2-9}$$

For a real current defined by (4.1-1b), we use (4.2-8) to write

$$v(t) = L\frac{d}{dt}[I\cos(\omega t + \phi)] = -\omega LI\sin(\omega t + \phi) \tag{4.2-10}$$

For a complex current

$$i_c(t) = Ie^{j\omega t + j\phi} \tag{4.2-11}$$

representing $i(t)$, the voltage is

$$v_c(t) = L\frac{d}{dt}[i_c(t)] = j\omega LIe^{j\omega t + j\phi} = j\omega Li_c(t) \tag{4.2-12}$$

Both (4.2-10) and (4.2-12) show that the current in an inductor lags (has a smaller phase angle than) the voltage across its terminals by $\pi/2$.

As with the capacitor when complex signals are used, the inductor also has a kind of Ohm's law defined by (4.2-12). The quantity ωL, denoted by X_L to imply an *inductive reactance*, has the ohm as its unit:

$$X_L = \omega L \tag{4.2-13}$$

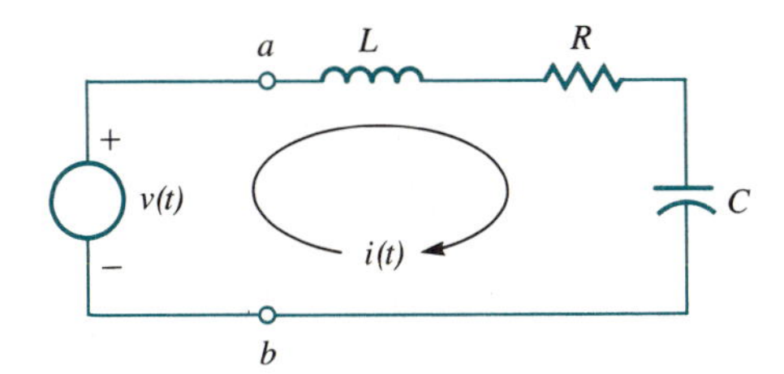

FIG. 4.2-1. A simple *R-L-C* circuit.

To demonstrate the use of complex signals in analyzing networks, we shall solve the circuit of Fig. 4.2-1 for the real current $i(t)$ that flows in response to the ac excitation of (4.0-1). First, we perform the analysis by using all real quantities.

EXAMPLE 4.2-1

To solve for $i(t)$ in Fig. 4.2-1 when

$$v(t) = V\cos(\omega t + \theta)$$

we write a KVL equation for the mesh:

$$v(t) = V\cos(\omega t + \theta) = L\frac{di(t)}{dt} + Ri(t) + \frac{1}{C}\int_{-\infty}^{t} i(\xi)\,d\xi \qquad \text{(A)}$$

Since the source is a sinusoid, we suspect that $i(t)$ must be a sinusoid, so we postulate the solution

$$i(t) = I\cos(\omega t + \theta + \psi)$$

where I and ψ are unknown. From (A)

$$\begin{aligned} V\cos(\omega t + \theta) &= -\omega LI\sin(\omega t + \theta + \psi) + RI\cos(\omega t + \theta + \psi) \\ &\quad + \frac{I}{\omega C}\sin(\omega t + \theta + \psi) \\ &= RI\cos(\omega t + \theta + \psi) + \left(\frac{1}{\omega C} - \omega L\right)I\sin(\omega t + \theta + \psi) \end{aligned}$$

On applying (4.1-10) this equation becomes

$$V\cos(\omega t + \theta) = I\left[R^2 + \left(\frac{1}{\omega C} - \omega L\right)^2\right]^{1/2}\cos(\omega t + \theta + \psi + \eta) \qquad \text{(B)}$$

where

$$\eta = \tan^{-1}\left[\frac{-1}{R}\left(\frac{1}{\omega C} - \omega L\right)\right]$$

For (B) to be true, we must have

$$I = \frac{V}{[R^2 + (\omega L - 1/\omega C)^2]^{1/2}} \qquad \text{and} \qquad \psi = -\eta = \tan^{-1}\left[\frac{1}{R}\left(\frac{1}{\omega C} - \omega L\right)\right]$$

so

$$\begin{aligned} i(t) &= I \cos(\omega t + \theta + \psi) \\ &= I \cos(\psi) \cos(\omega t + \theta) - I \sin(\psi) \sin(\omega t + \theta) \\ &= \frac{V}{R^2 + (\omega L - 1/\omega C)^2} \left[R \cos(\omega t + \theta) + \left(\omega L - \frac{1}{\omega C} \right) \sin(\omega t + \theta) \right] \end{aligned}$$

Next, we perform the analysis by using complex signals.

EXAMPLE 4.2-2

Again we solve the circuit of Fig. 4.2-1 except we replace $v(t)$ by its complex equivalent (4.1-3) and solve for the complex current $i_c(t)$. The final desired result $i(t)$ is the real part of $i_c(t)$. From (4.2-6) and (4.2-12) and a KVL equation for the loop:

$$\begin{aligned} v_c(t) &= L \frac{di_c(t)}{dt} + Ri_c(t) + \frac{1}{C} \int_{-\infty}^{t} i_c(\xi)\, d\xi \\ &= j\omega L i_c(t) + Ri_c(t) + \frac{1}{j\omega C} i_c(t) \end{aligned}$$

Thus,

$$i_c(t) = \frac{v_c(t)}{R + j(\omega L - 1/\omega C)} = \frac{Ve^{j\omega t + j\theta}[R - j(\omega L - 1/\omega C)]}{R^2 + (\omega L - 1/\omega C)^2} \tag{A}$$

so

$$\begin{aligned} i(t) &= \text{Re}\,[i_c(t)] \\ &= \frac{V}{R^2 + (\omega L - 1/\omega C)^2} \left[R \cos(\omega t + \theta) + \left(\omega L - \frac{1}{\omega C} \right) \sin(\omega t + \theta) \right] \end{aligned}$$

Although both analyses in the preceding two examples give the correct current, the analysis using the complex signals proved simpler.

Transformers

The terminal voltages of a transformer are related to its windings' currents through (1.4-34) and (1.4-35):

$$v_1(t) = L_1 \frac{di_1(t)}{dt} + M \frac{di_2(t)}{dt} \tag{4.2-14}$$

$$v_2(t) = M\frac{di_1(t)}{dt} + L_2\frac{di_2(t)}{dt} \tag{4.2-15}$$

Since each right-side term is just the form already discussed for an inductance, it's clear that the transformer only represents complication in the form of two currents and two voltages to be considered instead of one each, as in an inductor.

For the most important case to be demonstrated, where the voltages $v_1(t)$ and $v_2(t)$ and currents $i_1(t)$ and $i_2(t)$ are to be replaced by their complex-signal equivalents $v_{1c}(t)$ and $v_{2c}(t)$ and $i_{1c}(t)$ and $i_{2c}(t)$, respectively, (4.2-14) and (4.2-15) become

$$v_{1c}(t) = j\omega L_1 i_{1c}(t) + j\omega M i_{2c}(t) \tag{4.2-16}$$

$$v_{2c}(t) = j\omega M i_{1c}(t) + j\omega L_2 i_{2c}(t) \tag{4.2-17}$$

An example will help to show that complex signals lead to straightforward solution of networks with transformers.

EXAMPLE 4.2-3

We find the secondary voltage $v_2(t)$ in terms of the primary voltage $v_1(t)$ for the transformer of Fig. 4.2-2. The solution is the real part of the solution using complex signals. Since

$$v_{2c}(t) = -i_{2c}(t)R$$

(4.2-16) and (4.2-17) become

$$v_{1c}(t) = j\omega L_1 i_{1c}(t) - \frac{j\omega M}{R}v_{2c}(t)$$

$$v_{2c}(t) = j\omega M i_{1c}(t) - \frac{j\omega L_2}{R}v_{2c}(t)$$

After solving to eliminate $i_{1c}(t)$, we have

$$v_{2c}(t) = \frac{RMv_{1c}(t)}{RL_1 + j\omega(L_1L_2 - M^2)}$$

Next, we multiply numerator and denominator by the denominator's complex

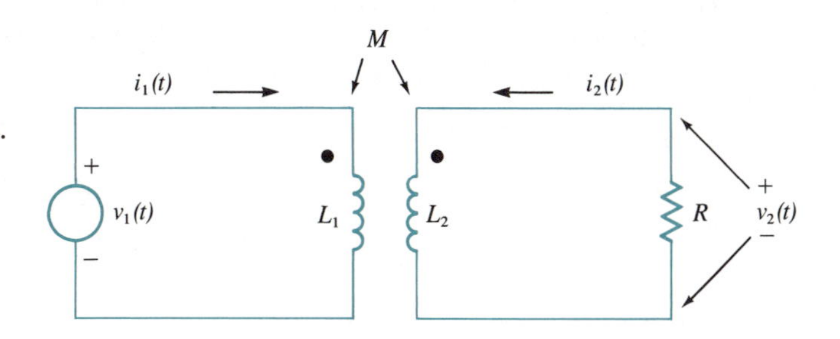

FIG. 4.2-2. A transformer circuit analyzed in Example 4.2-3.

conjugate and take the real part of both sides to obtain the real solution desired:

$$v_2(t) = \mathrm{Re}[v_{2c}(t)]$$
$$= \frac{MRV_1}{(RL_1)^2 + \omega^2(L_1L_2 - M^2)^2}\left[RL_1 \cos(\omega t + \theta) + \omega\left(L_1L_2 - M^2\right)\sin(\omega t + \theta)\right]$$

4.3 THE IMPEDANCE CONCEPT

We have found that a form of Ohm's law applies to resistors, capacitors, and inductors when complex signals are being used. Specifically, the ratio of $v_c(t)$, the complex voltage across the element's terminals, to $i_c(t)$, the complex current flowing into the element's most positive terminal, is

$$\frac{v_c(t)}{i_c(t)} = \begin{cases} R, & \text{resistor} \\ \dfrac{1}{j\omega C}, & \text{capacitor} \\ j\omega L, & \text{inductor} \end{cases} \tag{4.3-1}$$

We called $-1/(\omega C)$ the reactance of the capacitor and ωL the inductive reactance. But what shall we call the quantities $1/(j\omega C)$ and $j\omega L$? Of course, there is an answer ready.

We formally define *impedance*, denoted generally by Z, as the ratio of the complex voltage $v_c(t)$ across two terminals in a linear network to the complex current $i_c(t)$ flowing into its most positive terminal:

$$Z = \frac{v_c(t)}{i_c(t)} \tag{4.3-2}$$

Thus, $1/(j\omega C)$ is the impedance of a capacitor, which equals j times the capacitive reactance. An inductor's impedance is also j times its reactance.

Impedance can be real (resistor) or imaginary (inductor or capacitor) or have any complex value, in general. Because any complex number can be represented as the sum of a real part and an imaginary part, we may write Z in the form

$$Z = R + jX \tag{4.3-3}$$

Obviously, R is the resistive part of Z. The imaginary part X is called the reactance of Z. The form of (4.3-3) is called the *rectangular form* of expressing a complex number. The impedance also has the *polar form*

$$Z = |Z|e^{j\theta_Z} \tag{4.3-4}$$

where, respectively,

$$|Z| = \sqrt{R^2 + X^2} \tag{4.3-5}$$

$$\theta_Z = \tan^{-1}\frac{X}{R} \tag{4.3-6}$$

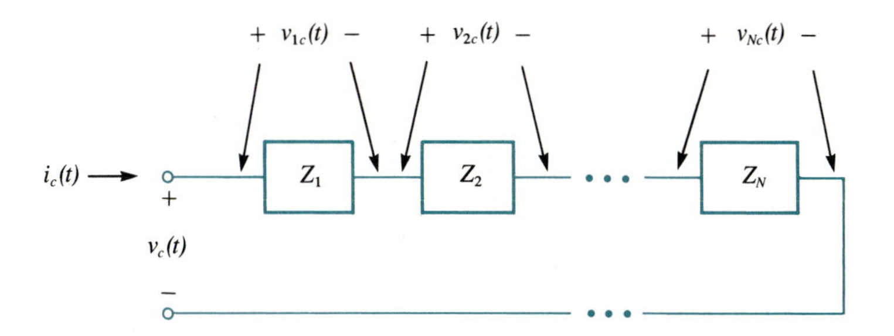

FIG. 4.3-1. N complex impedances in series.

are called the *magnitude* and *angle* of Z. Equations (4.3-5) and (4.3-6) allow us to convert from rectangular to polar form. To convert from polar to rectangular form, we use

$$R = |Z| \cos \theta_Z \tag{4.3-7}$$

$$X = |Z| \sin \theta_Z \tag{4.3-8}$$

Impedances in Series

To find the impedance of N (several) impedances in series, we discuss Fig. 4.3-1. From KVL and Ohm's law we have

$$\begin{aligned} v_c(t) &= v_{1c}(t) + v_{2c}(t) + \cdots + v_{Nc}(t) \\ &= i_c(t)Z_1 + i_c(t)Z_2 + \cdots + i_c(t)Z_N \end{aligned} \tag{4.3-9}$$

so

$$Z = \frac{v_c(t)}{i_c(t)} = Z_1 + Z_2 + \cdots + Z_N \tag{4.3-10}$$

This result states that the impedance of any number of individual impedances in series is the sum of the individual impedances.

Figure 4.2-1 shows a series connection of three elements. The impedance to the right of terminals a, b is found by forming the ratio of the complex voltage $v_c(t)$ to the complex current $i_c(t)$, which was determined in Example 4.2-2, equation (A). Hence, for the three elements of Fig. 4.2-1,

$$Z = \frac{v_c(t)}{i_c(t)} = R + j\omega L + \frac{1}{j\omega C} \tag{4.3-11}$$

is the sum of individual impedances.

Impedances in Parallel

The composite impedance of several impedances in parallel is found by using Fig. 4.3-2. From KCL and Ohm's law

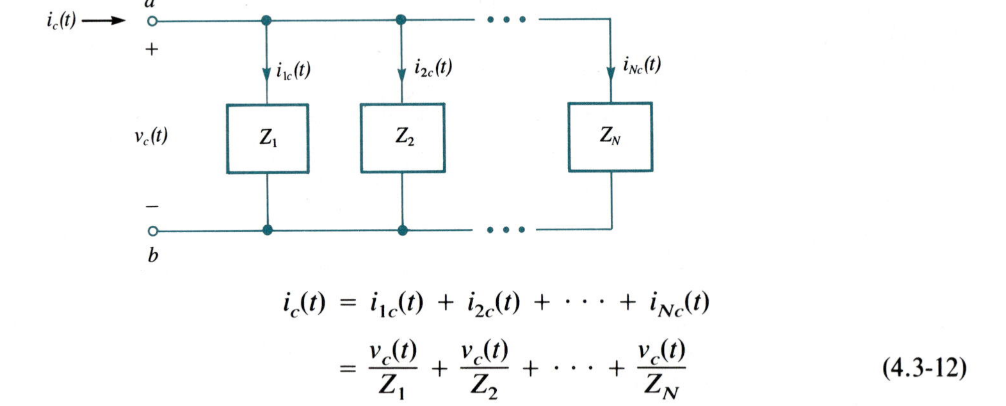

FIG. 4.3-2. *N* complex impedances in parallel.

$$i_c(t) = i_{1c}(t) + i_{2c}(t) + \cdots + i_{Nc}(t)$$
$$= \frac{v_c(t)}{Z_1} + \frac{v_c(t)}{Z_2} + \cdots + \frac{v_c(t)}{Z_N} \tag{4.3-12}$$

so

$$Z = \frac{v_c(t)}{i_c(t)} = \frac{1}{1/Z_1 + 1/Z_2 + \cdots + 1/Z_N} \tag{4.3-13}$$

If all impedances are just resistors, (4.3-13) reduces to (3.2-3), as it should.

EXAMPLE 4.3-1

As an example of series and parallel impedances, we solve for the impedance looking into terminals a, b of Fig. 4.3-3. If Z_2 is the impedance of the parallel combination of R and C, we have

$$Z_2 = \frac{1}{\frac{1}{4} + 1/(-j10)} = \frac{-j20}{2 - j5}$$

Hence,

$$Z = Z_1 + Z_2 = j7 + \frac{-j20}{2 - j5} = \frac{35 - j6}{2 - j5}$$
$$= \frac{(35 - j6)(2 + j5)}{(2 - j5)(2 + j5)} = \tfrac{100}{29} + j\tfrac{163}{29}$$

The network to the right of terminals a, b appears inductive, since the reactive component of Z is positive.

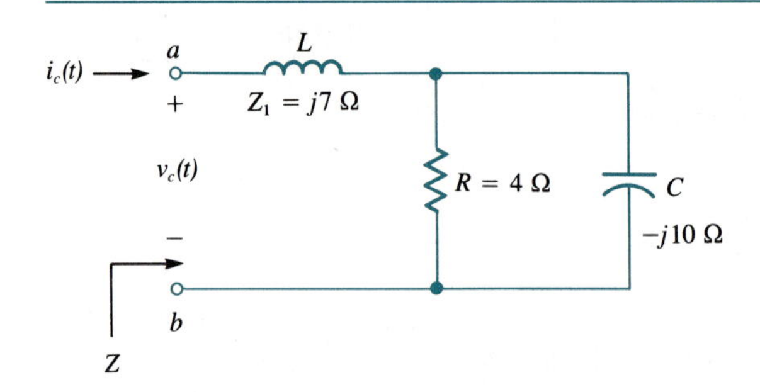

FIG. 4.3-3. A simple circuit analyzed in Example 4.3-1.

Admittance

The reciprocal of impedance is called *admittance*, denoted by Y. In general, Y is a complex quantity having a real part, denoted by G and called *conductance*, and an imaginary part, denoted by B and called *susceptance*:

$$Y = \frac{1}{Z} = G + jB \tag{4.3-14}$$

The siemens (S) is the unit of Y, G, and B.

Let $Y_n = 1/Z_n$, $n = 1, 2, \ldots, N$, be the admittances of the N impedances in series that resulted in (4.3-10). We take the reciprocal to find that

$$Y = \frac{1}{Z} = \frac{1}{Z_1 + Z_2 + \cdots + Z_N} = \frac{1}{1/Y_1 + 1/Y_2 + \cdots + 1/Y_N} \tag{4.3-15}$$

applies to admittances in series.

From (4.3-13) for impedances (admittances) in parallel, we get

$$Y = \frac{1}{Z} = \frac{1}{Z_1} + \frac{1}{Z_2} + \cdots + \frac{1}{Z_N} = Y_1 + Y_2 + \cdots + Y_N \tag{4.3-16}$$

EXAMPLE 4.3-2

Equation (4.3-16) will be used to find the admittance of the elements in parallel in Fig. 4.3-4. Here

$$Y_1 = 4 \text{ S}$$

$$Y_2 = j3 \text{ S}$$

$$Y_3 = \frac{1}{\frac{1}{2} + 1/(-j12)} = \frac{-j24}{2 - j12} = \frac{-j24(2 + j12)}{(2 - j12)(2 + j12)} = \tfrac{72}{37} - j\tfrac{12}{37}$$

By application of (4.3-16):

$$Y = 4 + j3 + \tfrac{72}{37} - j\tfrac{12}{37} = \tfrac{220}{37} + j\tfrac{99}{37} \text{ S}$$

This admittance has a capacitive component because the susceptance is positive.

For the special case where only two admittances are in parallel, (4.3-16) reduces to

$$Y = Y_1 + Y_2 = \frac{1}{Z_1} + \frac{1}{Z_2} = \frac{Z_1 + Z_2}{Z_1 Z_2} \tag{4.3-17}$$

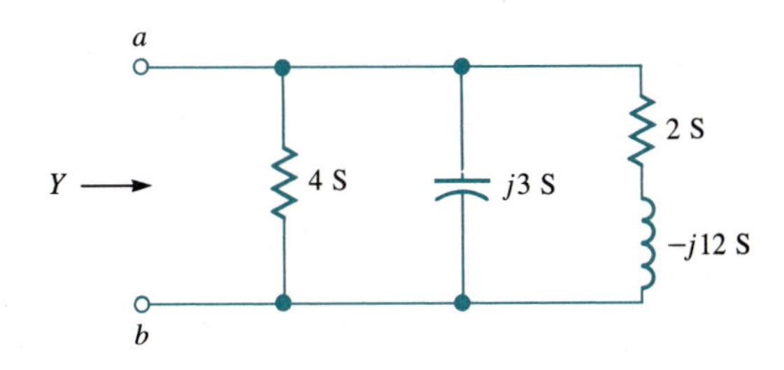

FIG. 4.3-4. A circuit applicable to Example 4.3-2.

or, equivalently,

$$Z = \frac{1}{Y} = \frac{Z_1 Z_2}{Z_1 + Z_2} \tag{4.3-18}$$

4.4 COMPLEX ANALYSIS OF AC NETWORKS

The name *complex analysis* will be attached to the process of analyzing ac circuits by using complex voltages and currents. The procedure for analysis was given in Sec. 4.1. Follow-up developments of Secs. 4.2 and 4.3 provided further detail on how to implement step 2 in the procedure. In this section additional and more complicated examples of complex analysis are given to demonstrate mesh and nodal analysis as well as equivalent (Thévenin and Norton) networks.

Equivalent Networks

For ac circuits, Thévenin's theorem stated in Sec. 3.4 remains valid, except that the open-circuited voltage and short-circuit currents are now complex signals and Thévenin's equivalent resistance now becomes Thévenin's equivalent impedance.

EXAMPLE 4.4-1

The circuit to the left of terminals a, b in Fig. 4.4-1(a) will be replaced by its Thévenin equivalent network. When the terminals are open-circuited, the current that flows in the capacitor is $v_c(t)/(5 + j20 - j6)$. The open-circuit terminal voltage is this current times the capacitor's impedance:

$$v_{oc}(t) = \frac{v_c(t)(-j6)}{(5 + j20 - j6)} = \frac{6e^{-j(\pi/2)}v_c(t)}{5 + j14} = \frac{6e^{-j(\pi/2) - j\tan^{-1}(14/5)}}{\sqrt{221}} v_c(t)$$

When terminals a and b are shorted, the short-circuit current that flows is

$$i_{sc}(t) = \frac{v_c(t)}{5 + j20}$$

Thévenin's equivalent impedance becomes

$$Z_{Th} = \frac{v_{oc}(t)}{i_{sc}(t)} = \frac{v_c(t)(-j6)}{5 + j14} \frac{(5 + j20)}{v_c(t)} = \tfrac{180}{221} - j\tfrac{1830}{221}$$

The equivalent network is shown in Fig. 4.4-1(b).

The Norton's equivalent current source follows directly from the procedures discussed for the Thévenin's source.

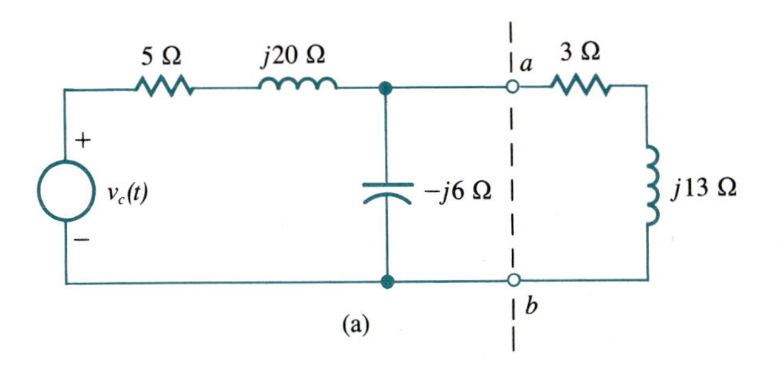

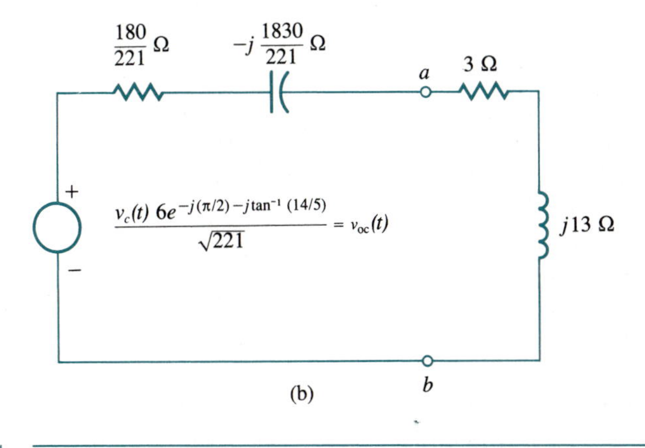

FIG. 4.4-1. (a) An ac circuit and (b) its equivalent using a Thévenin's generator to replace the circuit to the left of terminals a, b. See Example 4.4-1.

EXAMPLE 4.4-2

We replace the circuit of Fig. 4.4-1(a) to the left of terminals a, b with a Norton's current source by using the short-circuit current and Thévenin's impedance found in Example 4.4-1. The Norton current source becomes

$$i_{sc}(t) = \frac{v_c(t)}{5 + j20} = \frac{v_c(t)(5 - j20)}{425} = \frac{e^{-j\tan^{-1}(4)}}{\sqrt{425}}\,v_c(t)$$

The equivalent network is shown in Fig. 4.4-2.

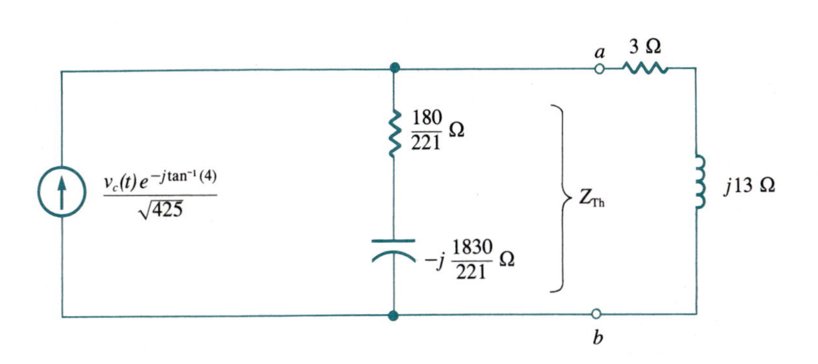

FIG. 4.4-2. A circuit applicable to Example 4.4-2.

Mesh Analysis

Mesh analysis with complex signals in ac circuits is performed according to the procedure listed in Sec. 3.5 if several minor changes are made in the steps listed. Step 1 no longer is restricted to a network of resistors and can now be any linear planar network of non-time-varying elements. Step 1 must also have the condition added that the voltage sources are all ac, and so that they apply to complex analysis, they are replaced by their complex-source representations. In step 2 the mesh currents to be assigned are now complex. In step 3 the KVL equations are to be written by using the complex impedances to represent each element in the network.

We use an example to illustrate the revised mesh-analysis procedure.

EXAMPLE 4.4-3

The circuit of Fig. 4.4-3 already has all elements and sources represented in complex form and will be analyzed by applying mesh analysis. Three KVL mesh equations are required:

$$v_{1c}(t) = 5i_{1c}(t) - j6[i_{1c}(t) - i_{2c}(t)]$$

$$0 = -j6[i_{2c}(t) - i_{1c}(t)] + j20i_{2c}(t) - j12[i_{2c}(t) - i_{3c}(t)]$$

$$-v_{2c}(t) = -j12[i_{3c}(t) - i_{2c}(t)] + 10i_{3c}(t)$$

These equations rearrange to

$$v_{1c}(t) = (5 - j6)i_{1c}(t) + j6i_{2c}(t)$$

$$0 = j6i_{1c}(t) + j2i_{2c}(t) + j12i_{3c}(t)$$

$$-v_{2c}(t) = j12i_{2c}(t) + (10 - j12)i_{3c}(\mathrm{t})$$

Although some complex algebra is involved, the solution of these equations is straightforward and is

$$i_{1c}(t) = \frac{(42 + j5)v_{1c}(t) + 18v_{2c}(t)}{5(66 - j67)}$$

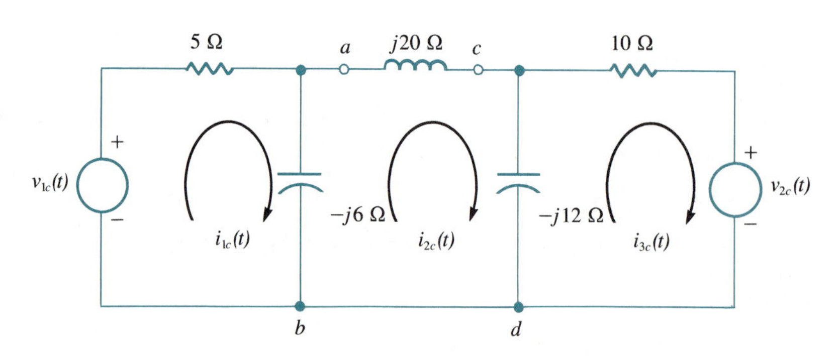

FIG. 4.4-3. An ac circuit analyzed in Example 4.4-3 by using mesh analysis.

$$i_{2c}(t) = \frac{3(6 + j5)[v_{2c}(t) - v_{1c}(t)]}{5(66 - j67)}$$

$$i_{3c}(t) = \frac{-18v_{1c}(t)}{5(66 - j67)} + \frac{(-150 + j119)v_{2c}(t)}{10(5 - j6)(66 - j67)}$$

The voltages across the circuit's elements may be found from the products of the complex impedances and the proper complex currents.

Nodal Analysis

The six-step nodal-analysis procedure given in Sec. 3.6 can be applied to ac circuits with complex representations if some minor changes are noted. Step 1 now applies to any linear network with elements that do not change values with time. The sources in step 1 are all ac, and so that complex analysis can be applied, they are all replaced by their complex-source representations. In step 3 the node voltages are assumed to be complex signals. In step 4 the KCL equations are written by using either the complex-impedance or the complex-admittance representations of each element in the network.

4.5 POWER IN AC CIRCUITS

The principal thrusts of this chapter are the introduction and use of complex analysis in ac circuits. The preceding concepts allow the easy solution of most circuits for their complex currents and voltages. The real currents and voltages become the real parts of the corresponding complex quantities. These real currents and voltages may be used to compute the instantaneous or average power dissipated in resistive components by using (1.3-3) or (1.3-4). However, how do we compute power when complex voltages and currents are analyzed in a network being represented by complex impedances? The purpose of this section is to answer this question. But first, it is desirable to review real power in real ac networks.

Real Power

Suppose the real ac voltage

$$v(t) = V \cos(\omega t + \theta_v) \tag{4.5-1}$$

appears across the terminals of some element of a linear network, while the current

$$i(t) = I \cos(\omega t + \theta_i) \tag{4.5-2}$$

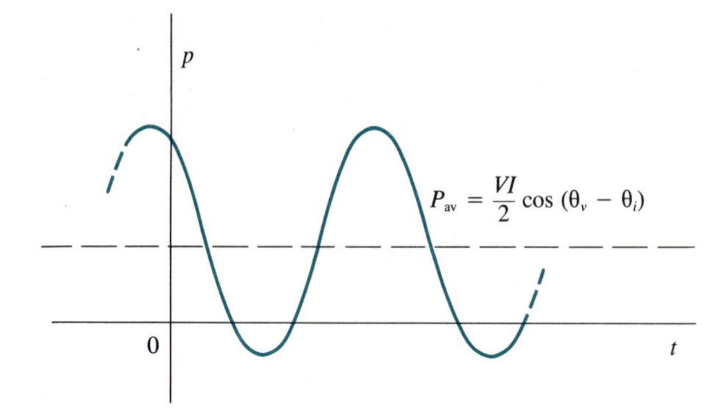

FIG. 4.5-1. Instantaneous power variations with time.

flows into the more positive of the two terminals. Clearly, V and I are the amplitudes of the voltage and current, respectively, while θ_v and θ_i are respective arbitrary voltage and current phase angles. The instantaneous power, from (1.3-3), that flows into the element is

$$p = v(t)i(t) = VI \cos(\omega t + \theta_v) \cos(\omega t + \theta_i)$$

$$= \frac{VI}{2} \cos(\theta_v - \theta_i) + \frac{VI}{2} \cos(2\omega t + \theta_v + \theta_i) \tag{4.5-3}$$

This function is sketched in Fig. 4.5-1. Since the second term's average value over any multiple of its period is zero, the first term represents the average power that is absorbed by the network's element.

$$P_{av} = \frac{VI}{2} \cos(\theta_v - \theta_i) \tag{4.5-4}$$

The quantity $\cos(\theta_v - \theta_i)$ depends on the difference in phases of the voltage and current. It is called the *power factor*. Power factor is largest when the voltage and current have the same phase angle, as in a resistor. When $\theta_v = \theta_i \pm \pi/2$, as with an inductor or capacitor, the power factor is zero and no average power is being absorbed. For this reason ideal inductors and capacitors are called *lossless elements*. An ideal transformer is also lossless.

The largest value that the real power can have is $VI/2$ from (4.5-4). This quantity is often called *apparent power*. It is most important in power distribution and generation systems. Here customers are usually charged only for the average real power used, but the system must cope with the largest possible real power. Strictly, the unit of apparent power $VI/2$ is the watt, but the *voltampere* (VA) unit is usually used in practice.

EXAMPLE 4.5-1

We shall compute the real and apparent powers supplied to terminals a, b in the circuit of Fig. 4.2-1 by the source of voltage $v(t) = V \cos(\omega t + \theta)$ when $R = 4\ \Omega$ and the inductive and capacitive reactances are $X_L = \omega L = 15\ \Omega$ and $X_C =$

$-1(\omega C) = -8\ \Omega$. The current, as found from Example 4.2-1, becomes

$$i(t) = \frac{V}{65}[4\cos(\omega t + \theta) + 7\sin(\omega t + \theta)] \tag{A}$$

after numerical substitutions are made. Instantaneous power is

$$p = v(t)\,i(t) = \frac{V^2}{65}[4\cos^2(\omega t + \theta) + 7\cos(\omega t + \theta)\sin(\omega t + \theta)]$$

$$= \frac{2V^2}{65} + \frac{2V^2}{65}\cos(2\omega t + 2\theta) + \frac{3.5V^2}{65}\sin(2\omega t + 2\theta)$$

The constant term is the average power:

$$P_{av} = \frac{2V^2}{65} \quad \text{W}$$

To obtain apparent power, we must first find the peak current for (A) by applying the conversion of (4.1-10):

$$i(t) = \frac{V}{\sqrt{65}}\cos[\omega t + \theta + \tan^{-1}(\tfrac{-7}{4})]$$

Apparent power is found to be $V(V/2\sqrt{65}) = V^2/2\sqrt{65}$ VA.

Observe that the power factor for this circuit is $\cos(\tan^{-1}\tfrac{7}{4}) = 4/\sqrt{65} \approx 0.496$.

Complex Power

When the complex signal (voltage or current) was defined, it was that exponential function having a real part equal to the real signal (voltage or current). In a similar manner, *complex power* is defined as that exponential function having a real part equal to the real *average* power. By expanding the real average power of (4.5-4) as

$$P_{av} = \frac{VI}{2}\cos(\theta_v - \theta_i) = \text{Re}\left[\frac{VI}{2}e^{j(\theta_v - \theta_i)}\right] \tag{4.5-5}$$

we conclude that the quantity within the brackets is the complex power, which we denote by P_c. We put complex power in a more useful form as follows:

$$P_c = \frac{VI}{2}e^{j(\theta_v - \theta_i)} = \frac{VI}{2}e^{j(\omega t + \theta_v - \omega t - \theta_i)}$$

$$= \tfrac{1}{2}Ve^{j\omega t + j\theta_v}Ie^{-j\omega t - j\theta_i} \tag{4.5-6}$$

or

$$P_c = \tfrac{1}{2}v_c(t)i_c^*(t) \tag{4.5-7}$$

This result shows that complex power is half the product of the complex voltage across a pair of circuit terminals and the complex conjugate of the complex current flowing into the most positive terminal. Complex power is customarily

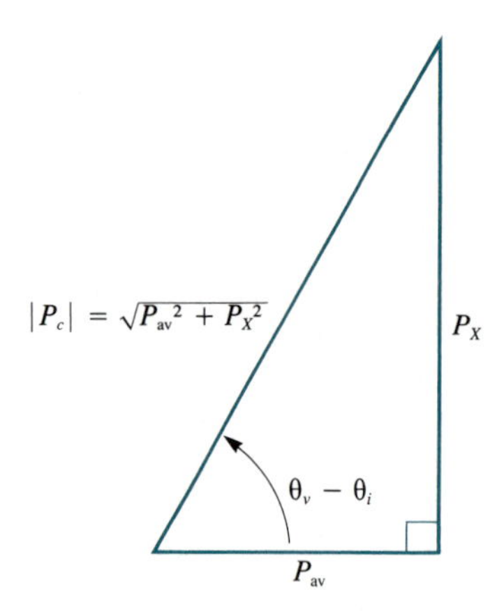

FIG. 4.5-2.
A power triangle.

said to have the unit of volt-amperes (VA) to keep it from being confused with real power (watts).

The imaginary part of complex power is called *reactive power*, denoted by P_X. Thus, we can write

$$P_c = P_{av} + jP_X \tag{4.5-8}$$

where†

$$P_X = \text{Im}\,[P_c] = \frac{VI}{2} \sin(\theta_v - \theta_i) \tag{4.5-9}$$

The unit of reactive power is the volt-ampere-reactive (abbreviated var); this choice (rather than its true unit, watt) helps reduce confusion between the three powers P_c, P_{av}, and P_X. Reactive power is due to the alternating storage and release of energy in the electric fields of capacitors or magnetic fields of inductors. The factor $\sin(\theta_v - \theta_i)$ in (4.5-9) is known as the *reactive factor*. It determines the type of load (the circuit between the terminals of interest) that is presented to the source. A load is said to have a lagging power factor if load current lags load voltage (inductive case), so that $\sin(\theta_v - \theta_i)$ is positive. A load with a leading power factor is one where current leads voltage such that $\sin(\theta_v - \theta_i)$ is negative (capacitive load).

A convenient way to visualize the relationship between $|P_c|$, P_{av}, and P_X is the *power triangle* depicted in Fig. 4.5-2. Since P_c is a complex number, with P_{av} its real part and P_X its imaginary part, a right triangle relates the parts. The angle $\theta_v - \theta_i$ of the power triangle is the same as the angle of the load's impedance. To show this fact, write

$$v_c(t) = i_c(t)Z \tag{4.5-10}$$

† Im [ ] represents taking the imaginary part of the quantity within the brackets. In some texts symbols S and Q are used for complex and reactive powers, respectively.

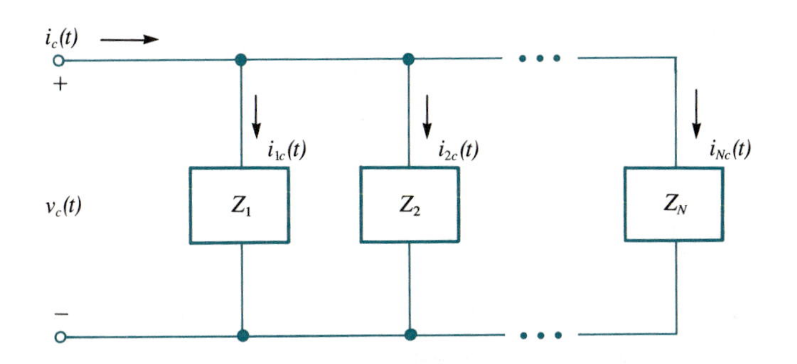

FIG. 4.5-3. Complex networks in parallel.

where Z is the load's impedance. From (4.5-7)

$$P_c = \tfrac{1}{2}v_c(t)i_c^*(t) = \frac{v_c(t)v_c^*(t)}{2\ Z^*} = \frac{|v_c(t)|^2 Z}{2|Z|^2} = \frac{V^2 Z}{2|Z|^2} \tag{4.5-11}$$

or, equivalently,

$$P_c = \tfrac{1}{2}v_c(t)i_c^*(t) = \frac{i_c(t)Zi_c^*(t)}{2} = \frac{|i_c(t)|^2 Z}{2} = \frac{I^2 Z}{2} \tag{4.5-12}$$

The last forms of (4.5-11) and (4.5-12) show that the angle of P_c is the angle of Z. Either form is obtainable from the other by using

$$|v_c(t)|^2 = V^2 = |i_c(t)Z|^2 = I^2|Z|^2 \tag{4.5-13}$$

from (4.5-10).

In terms of the rms value of a real sinusoidal voltage (see Sec. 2.2), the complex power of (4.5-11) equals the square of the rms voltage divided by the complex conjugate of the impedance across which the voltage occurs. Similarly, complex power is the product of the square of a real rms current times the impedance in which the current flows from (4.5-12).

In general, superposition of powers cannot be used in networks with multiple sources (unless all sources are at different frequencies). Superposition is used to find voltages and currents. Once these are found, the complex powers delivered to any number of interconnected loads, no matter how the interconnections are made, may be added to obtain the total complex power. In the circuit of Fig. 4.5-3, for example,

$$\begin{aligned} P_c &= \tfrac{1}{2}v_c(t)i_c^*(t) = \tfrac{1}{2}v_c(t)[i_{1c}^*(t) + i_{2c}^*(t) + \cdots + i_{Nc}^*(t)] \\ &= P_{1c} + P_{2c} + \cdots + P_{Nc} \end{aligned} \tag{4.5-14}$$

EXAMPLE 4.5-2

The real, reactive, and complex powers will be found for the circuit of Fig. 4.5-4. The total impedance across terminals a, b is

$$Z_{ab} = R + j\omega L$$

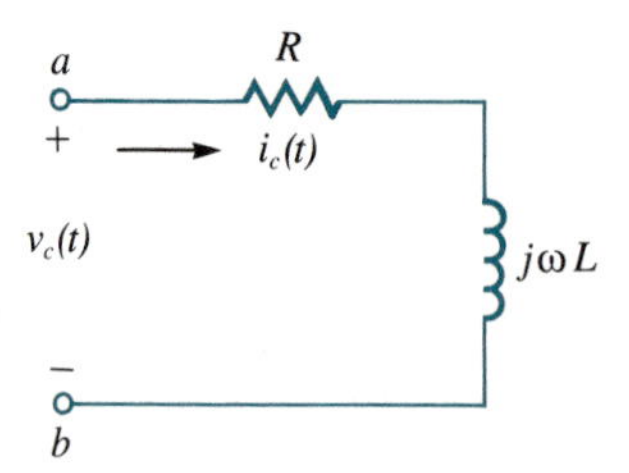

FIG. 4.5-4. An R-L network analyzed in Example 4.5-2.

From (4.5-11) the complex power is

$$P_c = \frac{V^2[R + j\omega L]}{2[R^2 + (\omega L)^2]}$$

so

$$P_{av} = \frac{V^2 R}{2[R^2 + (\omega L)^2]} \quad \text{and} \quad P_X = \frac{V^2 \omega L}{2[R^2 + (\omega L)^2]}$$

To get a better feeling for these results, suppose $V = 2$ V, $R = 20\ \Omega$, and $X_L = \omega L = 30\ \Omega$. The three powers calculate to $P_{av} = \frac{4}{130}$ W, $P_X = \frac{6}{130}$ var, and $P_c = (4 + j6)/130$ VA. The power factor is the cosine of the impedance's angle, which is $\tan^{-1} \frac{6}{4} \approx 0.983$ rad (or 56.31°), as can be determined from the power triangle.

Maximum Power Transfer Theorem

A problem of considerable interest to communications engineers, persons involved in the generation of power at utilities, and others, is defined in Fig. 4.5-5. A source, represented by its complex voltage $v_c(t)$, has an internal impedance Z_s (complex). The source drives a complex load Z_L connected across its terminals. The load's complex voltage is $v_{Lc}(t)$. The problem is: What load impedance will correspond to maximum real power in the load, if it is given that the source's impedance is fixed? To answer this question, we use

$$v_{Lc}(t) = i_c(t) Z_L \tag{4.5-15}$$

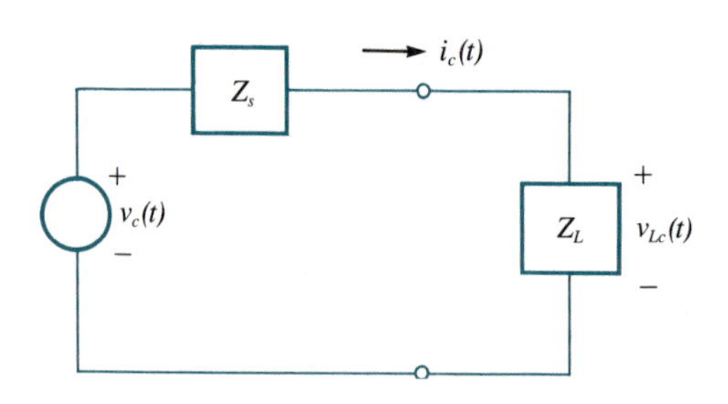

FIG. 4.5-5. The circuit for proving the maximum power transfer theorem.

$$i_c(t) = \frac{v_c(t)}{Z_s + Z_L} \tag{4.5-16}$$

in (4.5-7) applied to the load:

$$P_c(\text{load}) = \tfrac{1}{2}V_{Lc}(t)i_c^*(t) = \frac{|i_c(t)|^2 Z_L}{2} = \frac{|v_c(t)|^2 Z_L}{2|Z_s + Z_L|^2} \tag{4.5-17}$$

Next, let Z_s and Z_L have real and imaginary components defined by

$$Z_s = R_s + jX_s \tag{4.5-18}$$

$$Z_L = R_L + jX_L \tag{4.5-19}$$

Equation (4.5-17) becomes

$$P_c(\text{load}) = \frac{V^2(R_L + jX_L)}{2[(R_s + R_L)^2 + (X_s + X_L)^2]} \tag{4.5-20}$$

The real load power is the real part of (4.5-20):

$$P_{av}(\text{load}) = \frac{V^2 R_L}{2[(R_s + R_L)^2 + (X_s + X_L)^2]} \tag{4.5-21}$$

This real power is maximized first with respect to X_L by differentiation. The maximum occurs when $X_L = -X_s$. That a maximum occurs is checked by showing that the second derivative of (4.5-21) with respect to X_L is negative when $X_L = -X_s$. With $X_L = -X_s$ substituted into (4.5-21), the result is then found to be maximum when $R_L = R_s$. These facts may be summarized in the *maximum power transfer theorem*: Maximum real power is transferred to a load if its complex impedance Z_L is equal to the complex conjugate of the internal complex impedance Z_s of the voltage source driving the load

$$Z_L = Z_s^* \tag{4.5-22}$$

When (4.5-22) is satisfied, the load's real power is

$$P_{av}(\text{maximum}) = \frac{V^2}{8R_L} \tag{4.5-23}$$

as the reader may wish to prove as an exercise.

4.6 TRANSFER FUNCTIONS

Even though most of the preceding discussions have not stressed the point, all impedances, complex sources, and all currents and voltages in a linear network are functions of angular frequency ω. It is precisely this dependence that allows special networks such as filters, amplifiers, oscillators, and many other circuits to be designed for particular applications. An audio amplifier designed to pass frequencies only up to about 20 kHz would be useless for passing a video signal containing frequencies up to about 6 MHz, for example. In this section the frequency dependence of networks is discussed, and viewpoints are developed that aid in the interpretation of circuit behavior.

Transfer Functions

In a more or less arbitrary linear network, imagine that two pairs of terminals are of interest, as illustrated in Fig. 4.6-1. At one pair the network is being excited by a real source. We choose to illustrate a voltage source

$$v(t) = V \cos \omega t \tag{4.6-1}$$

but a current source could equally well have been selected. At the other terminal pair we are interested in a response. In the figure this response is taken as a voltage, but it could also have been the current through some network element connected between the terminals. The source terminals will be considered the "input," and the response is taken as the "output" of the network. Rigorous analysis shows that sinusoidal excitation causes only sinusoidal currents and voltages to occur anywhere in the network. This fact means that the output voltage, denoted by $v_o(t)$, can be altered only in amplitude and phase but not in form relative to the input $v(t)$. If α is a factor representing the network's effect on amplitude and β is an added phase shift due to the network, then $v_o(t)$ will have the form

$$v_o(t) = V\alpha(\omega) \cos [\omega t + \beta(\omega)] \tag{4.6-2}$$

where it is recognized that α and β may depend on angular frequency.

When the cosine is represented by its exponential (complex) signals,

$$v_o(t) = \frac{V}{2} \alpha(\omega)(e^{j\omega t + j\beta(\omega)} + e^{-j\omega t - j\beta(\omega)})$$

$$= \frac{V}{2} e^{j\omega t}\alpha(\omega)e^{j\beta(\omega)} + \frac{V}{2} e^{-j\omega t}\alpha(\omega)e^{-j\beta(\omega)} \tag{4.6-3}$$

Since superposition applies, we associate the first right-side term in (4.6-3) with the network's response to the complex signal $(V/2) \exp (j\omega t)$, which has a positive angular frequency ω. The quantity $\alpha(\omega) \exp [j\beta(\omega)]$ is entirely due to the network; it is defined as the *transfer function*, denoted by $H(\omega)$, of the network

$$H(\omega) = \alpha(\omega)e^{j\beta(\omega)} \tag{4.6-4}$$

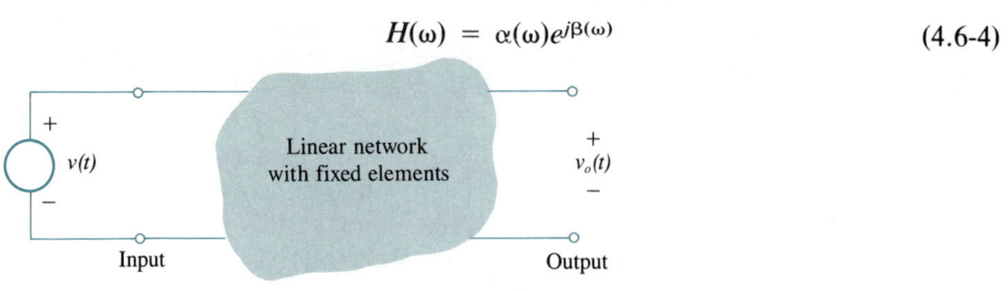

FIG. 4.6-1. A linear network having one input and one output.

TABLE 4.6-1. Transfer function types and units

Input	Output	Unit of $H(\omega)$
Voltage	Voltage	Unitless
Current	Voltage	Ohm (Ω)
Voltage	Current	Siemens (S)
Current	Current	Unitless

The second right-side term in (4.6-3) is taken as the response to the complex signal $(V/2)\exp(-j\omega t)$, which is at the negative angular frequency $-\omega$. The quantity $\alpha(\omega)\exp[-j\beta(\omega)]$ is interpreted as the effect of the network at this negative frequency. In other words,

$$H(-\omega) = \alpha(\omega)e^{-j\beta(\omega)} \tag{4.6-5}$$

or on combining with (4.6-4), we have

$$H(-\omega) = H^*(\omega) \tag{4.6-6}$$

for any real linear network expressed in terms of complex impedances. The transfer function is seen to be a function that is defined from, and to be used with, complex signals. It is a complex function, in general.

There are four types of transfer functions, as summarized in Table 4.6-1, which also defines the unit of $H(\omega)$. In a practical circuit $H(\omega)$ is found from the ratio of output-to-input complex signals found by complex analysis. An example will illustrate the procedure.

EXAMPLE 4.6-1

We find the voltage-to-voltage transfer function of the network of Fig. 4.6-2(a), where the output is taken across the resistor R. On using complex analysis and two mesh KVLs, we get

$$v_c(t) = i_{1c}(t)\left(j\omega L + \frac{1}{j\omega C}\right) - i_{2c}(t)\left(\frac{1}{j\omega C}\right)$$

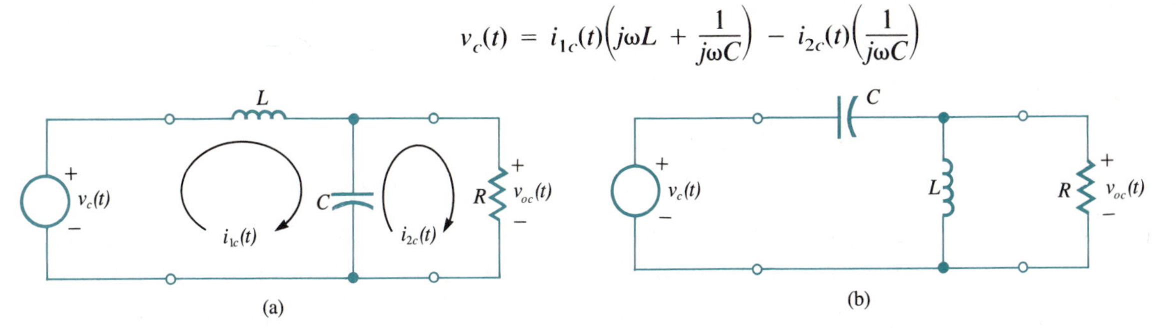

FIG. 4.6-2.
Circuits of (a) lowpass and (b) highpass types for which transfer functions are found in Examples 4.6-1 and 4.6-2, respectively.

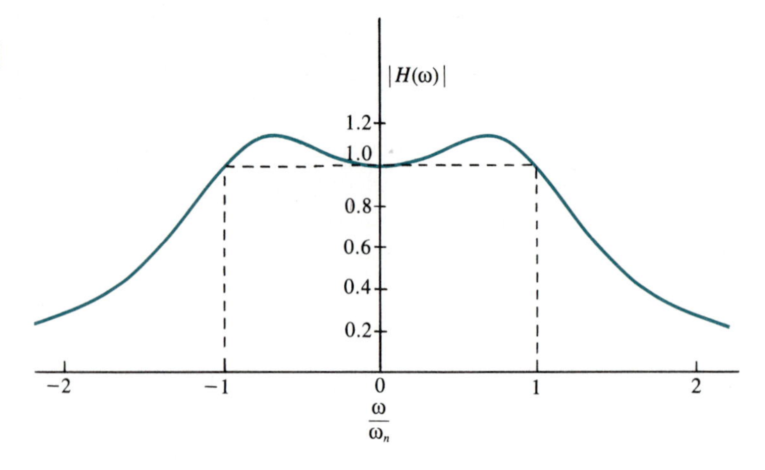

FIG. 4.6-3. The plot of $|H(\omega)|$ for the circuit of Fig. 4.6-2(a). (See Example 4.6-1.)

$$0 = -i_{1c}(t)\left(\frac{1}{j\omega C}\right) + i_{2c}(t)\left(R + \frac{1}{j\omega C}\right)$$

These equations are solved for $i_{2c}(t)$, and $v_{oc}(t)$ is computed to be

$$v_{oc}(t) = i_{2c}(t)\,R = \frac{v_c(t)}{(1 - \omega^2 LC) + j(\omega L/R)}$$

Thus,

$$H(\omega) = \frac{v_{oc}(t)}{v_c(t)} = \frac{1}{(1 - \omega^2 LC) + j(\omega L/R)}$$

If we define

$$\omega_n^2 = \frac{1}{LC}$$

and let

$$C = \frac{1}{\omega_n} \quad \text{and} \quad R = 1\ \Omega$$

then

$$H(\omega) = \frac{1}{1 - (\omega/\omega_n)^2 + j(\omega/\omega_n)} \tag{A}$$

To see the behavior of $H(\omega)$, we have plotted $|H(\omega)|$ in Fig. 4.6-3. The circuit is best described as a filter that passes all angular frequencies with about the same response out to about $1.25\omega_n$ and then attenuates the response rapidly for angular frequencies above $1.25\omega_n$.

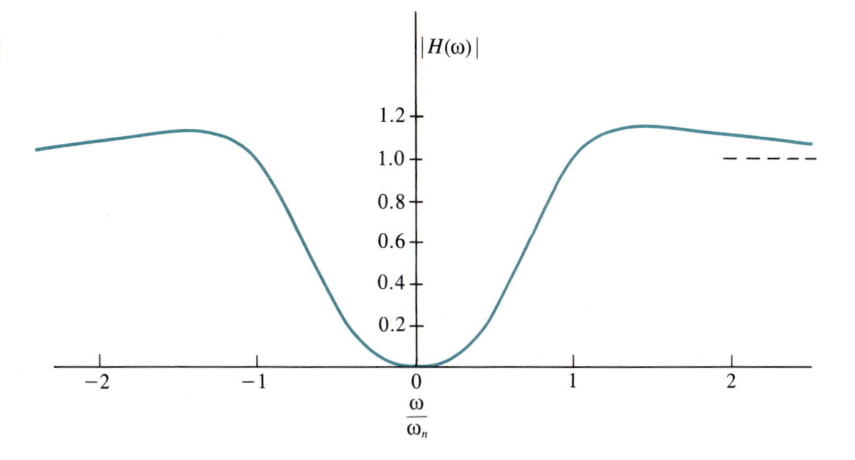

FIG. 4.6-4. The plot of $|H(\omega)|$ for the circuit of Fig. 4.6-2(b). (See Example 4.6-2.)

In the foregoing example the filter would be called a *lowpass filter*† in practice because it passes lower frequencies only. Filters may also be *bandpass*, where only a band of frequencies is passed while all others are attenuated (rejected). *Highpass* filters pass only frequencies higher than a particular value, as the following example illustrates.

EXAMPLE 4.6-2

The circuit of Fig. 4.6-2(b) is a highpass filter. By analysis that parallels the analysis used in Example 4.6-1, the transfer function is found to be

$$H(\omega) = \frac{v_{oc}(t)}{v_c(t)} = \frac{-(\omega/\omega_n)^2}{[1 - (\omega/\omega_n)^2] + j(\omega/\omega_n)}$$

where we define

$$\omega_n^2 = \frac{1}{LC}$$

and select

$$C = \frac{1}{\omega_n} \quad \text{and} \quad R = 1\ \Omega$$

as before. The highpass behavior of $|H(\omega)|$ is demonstrated in Fig. 4.6-4.

Many other types of transfer functions (filters) exist. Some are designed to suppress a specific band of frequencies and no others; they are called *stopband*

† Lowpass filters are also called *baseband filters*.

filters. Another, called a *notch filter*, is designed to eliminate a single frequency while not affecting all others as much as possible.

Circuit Bandwidth

A bandwidth is often associated with a network through its transfer function. Because there are many possible definitions of bandwidth, it can mean different things to different people. Generally, it refers to the band of frequencies over which a network's transfer function has its principal responses. It is nearly universal that the bandwidth, however defined, refers to the principal responses only for positive frequencies. Consequently, the bandwidth of a lowpass filter would contain frequencies from dc to some positive frequency (which is the bandwidth itself in this case). In a bandpass network the bandwidth would include frequencies between some positive value and a larger positive value. A highpass filter would theoretically have an infinite bandwidth by any definition.

One specific definition is the *3-dB bandwidth*, denoted by W (in radians per second). It refers to the band of frequencies over which $|H(\omega)|$ does not fall more than 3 dB below the value of $|H(\omega)|$ evaluated at a convenient reference value of ω near where $|H(\omega)|$ is largest. For a lowpass transfer function the reference value is often $\omega = 0$. For a bandpass transfer function the reference is usually at some value of ω equal to, or near, the center of the passband. These definitions are all analogous to the 3-dB-bandwidth definitions given in Sec. 2.6 for signals, only here we use $|H(\omega)|$ instead of the magnitude of the signal's spectrum.

Circuits in Cascade

When the output of one circuit provides the input of another, the two circuits are said to be *cascaded*. For the voltage-to-voltage or current-to-current transfer functions of Table 4.6-1, the overall network's transfer function can easily be related to those of the individual networks. Let $H_n(\omega)$, $n = 1, 2, \ldots, N$, be the transfer function of network n *when connected to its following network*. The transfer function of the overall cascade becomes

$$H(\omega) = H_1(\omega)H_2(\omega) \cdots H_N(\omega) = \prod_{n=1}^{N} H_n(\omega) \tag{4.6-7}$$

4.7 AC RESONANCE

An important electrical effect occurs when particular circuits are formed from a resistor, an inductance, and a capacitance. The effect, called *resonance*, is best understood by describing the two principal circuits in which it occurs.

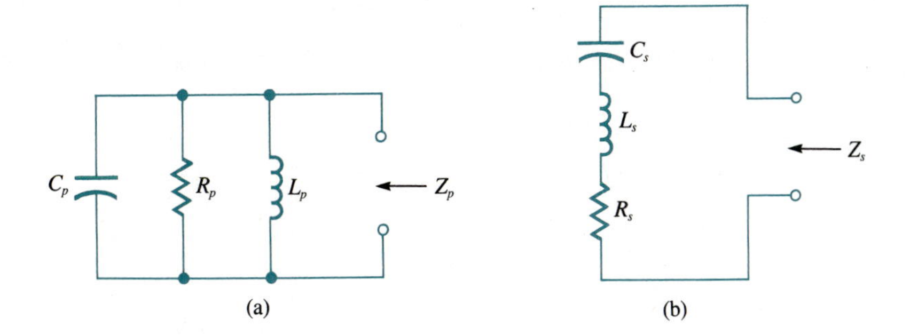

FIG. 4.7-1. Resonant circuits: (a) parallel and (b) series.

Consider first the parallel combination of elements shown in Fig. 4.7-1(a). We examine the impedance of the parallel circuit, which is

$$Z_p = \left(\frac{1}{R_p} + \frac{1}{j\omega L_p} + j\omega C_p\right)^{-1} = \left(\frac{R_p - \omega^2 R_p L_p C_p + j\omega L_p}{j\omega R_p L_p}\right)^{-1} \tag{4.7-1}$$

If we define parameters ω_0 and Q_0 by

$$\omega_0 = \frac{1}{\sqrt{L_p C_p}} \tag{4.7-2}$$

$$Q_0 = \frac{R_p}{\omega_0 L_p} \tag{4.7-3}$$

the impedance can be written as

$$\frac{Z_p}{R_p} = \frac{j\omega/Q_0\omega_0}{(1 - \omega^2/\omega_0^2) + j\omega/Q_0\omega_0} \tag{4.7-4}$$

Figure 4.7-2(a) sketches the behavior of the magnitude of (4.7-4) as a function of ω for Q_0 as a parameter. The peaked behavior of $|Z_p/R_p|$ at frequencies near ω_0, called the *resonant frequency*, is the important effect referred to as *resonance*. At the resonant frequency, where $\omega = \omega_0$, the reactances of the inductance and capacitance are equal in magnitude but opposite in sign so add to zero in (4.7-1). As a consequence, the parallel resonant circuit appears as a pure resistance R_p at resonance. The parameter Q_0, called the circuit's *quality factor*, determines the sharpness of the impedance's response near resonance (sharpness can be interpreted as a type of "bandwidth" for the circuit; a very sharp characteristic has a narrow "bandwidth"). For larger Q_0 the angular-frequency separation or "bandwidth" between points 3 dB below the peak of Z_p/R_p is ω_0/Q_0.

The phase response of the parallel resonant circuit is plotted in Fig. 4.7-2(b). As Q_0 becomes large, the phase makes a more rapid transition between its extremes of $\pi/2$ (as $\omega \to 0$) and $-\pi/2$ (as $\omega \to \infty$). This phase passes through zero at resonance (at $\omega = \omega_0$).

The parallel resonant circuit is sometimes called a *tank circuit* because, at resonance where $\omega = \omega_0$, relatively large amounts of energy may be stored in the circuit (tank). This energy is stored in, and transfers between, the capacitor and

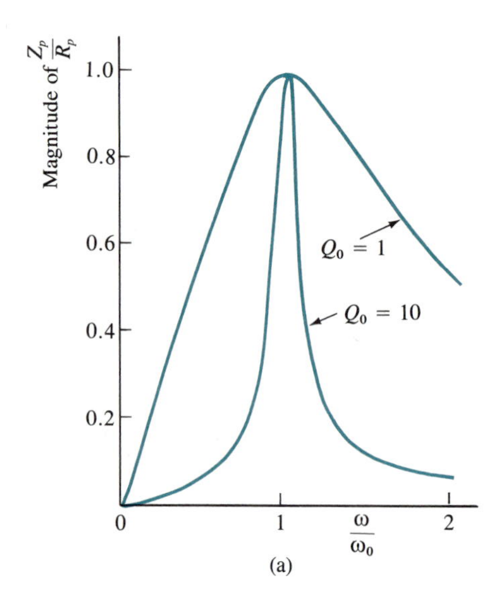

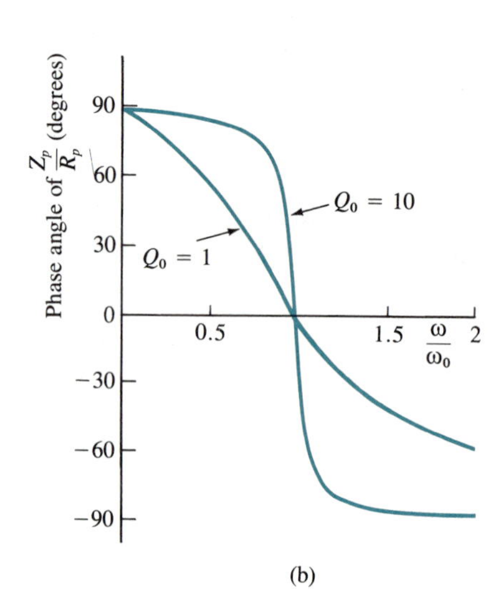

FIG. 4.7-2. (a) The magnitude and (b) the phase angle of Z_p/R_p for the circuit of Fig. 4.7-1(a).

inductor but remains in the circuit. The parallel resonant circuit finds considerable use as a simple bandpass filter and in amplifiers designed to pass a band of frequencies while rejecting others.

Next, we examine the admittance of the series connection of elements as shown in Fig. 4.7-1(b). It is

$$\frac{1}{Z_s} = Y_s = \left(R_s + j\omega L_s + \frac{1}{j\omega C_s}\right)^{-1} = \left(\frac{1 - \omega^2 L_s C_s + j\omega R_s C_s}{j\omega C_s}\right)^{-1} \tag{4.7-5}$$

If we now define ω_0 and Q_0 by

$$\omega_0 = \frac{1}{\sqrt{L_s C_s}} \tag{4.7-6}$$

$$Q_0 = \frac{\omega_0 L_s}{R_s} \tag{4.7-7}$$

(4.7-5) reduces to

$$\frac{R_s}{Z_s} = \frac{j\omega/Q_0\omega_0}{(1 - \omega^2/\omega_0^2) + j\omega/Q_0\omega_0} \tag{4.7-8}$$

Since the right sides of (4.7-8) and (4.7-4) have the same form, the behavior of R_s/Z_s for the series circuit is identical to that of Z_p/R_p for the parallel circuit, as shown in Fig. 4.7-2, provided Q_0 and ω_0 are the same for the two circuits. As in the parallel resonant circuit the series circuit resonates at an angular frequency ω_0 where the capacitive and inductive reactances are equal in magnitude. The resulting admittance at resonance is resistive and equals $1/R_s$.

The circuits of Fig. 4.7-1 are known as *tuned circuits* because the inductance and capacitance are chosen (tuned) to resonate at a specified angular frequency ω_0 according to (4.7-2) or (4.7-6).

PROBLEMS

4-1. A real voltage $v(t) = 6 \cos(\omega t + \pi/3) + 4 \sin(\omega t - \pi/3)$ is to be represented by a complex voltage $v_c(t)$. Find $v_c(t)$.

4-2. A complex signal is $12 \exp(j3000t + j\pi/6)$. (a) What real signal does it represent? (b) What is the signal's angular frequency? (c) If the complex signal is a voltage across a 4-Ω resistor, what real average power is dissipated in the resistor?

4-3. Two real voltages $v_1(t) = V_1 \cos \omega_1 t$ and $v_2(t) = V_2 \cos \omega_2 t$ are added. Find how ω_1 and ω_2 must be related if the power of the sum $v(t) = v_1(t) + v_2(t)$ averaged over a very long time must equal the sum of separate average powers. Assume all voltages exist across a resistance R.

4-4. What frequency will give a capacitive reactance of −1500 Ω if the capacitor's value is 0.06 μF?

4-5. A capacitance of 100 pF is in series with an inductance L. What value of L is required if the net reactance of the series combination must be zero at 20 MHz?

4-6. What inductance will give an impedance of $j600$ Ω at 3.5 MHz?

4-7. The current $i_c(t)$ in Fig. 4.2-1 is given by (A) of Example 4.2-2. At what value of ω (in terms of L, R, and C) will $|i_c(t)|$ be maximum?

4-8. Rework Example 4.2-3 for the resistor R in Fig. 4.2-2 replaced by a capacitance C. What value does $v_{2c}(t)/v_{1c}(t)$ have if $\omega = 1/\sqrt{L_2 C}$?

4-9. Equation (4.3-11) defines the impedance of the elements between terminals a, b in Fig. 4.2-1. Note that a value of ω exists such that the capacitive reactance and inductive reactance cancel each other. Find this value of ω, called the *resonant angular frequency* of the network. What determines the current at the resonant frequency?

4-10. Find the impedance Z_{ab} defined in Fig. P4-10.

FIG. P4-10.

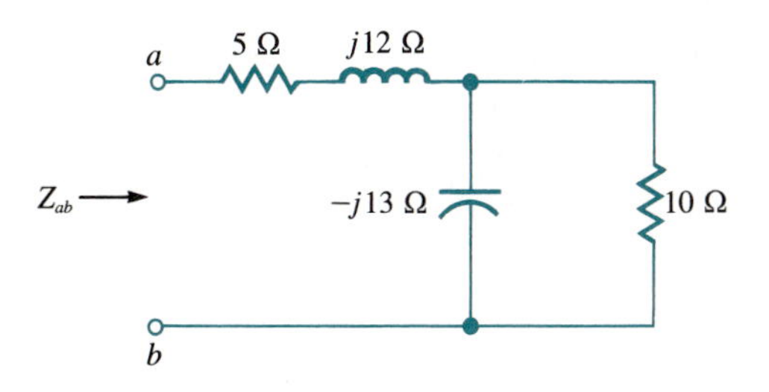

4-11. Find the impedance Z_{ab} defined in Fig. P4-11.

FIG. P4-11.

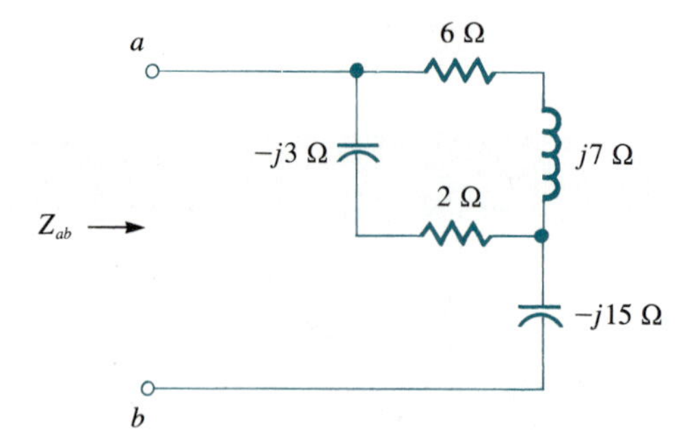

4-12. The impedance of a resistance $R = 12\ \Omega$ in series with an inductance L is $\sqrt{180}\exp(j\tan^{-1}0.5)$ when $\omega = 5000$ rad/s. What is the reactance of L? What is L?

4-13. A practical inductance $L = 10\ \mu\text{H}$ has a series resistance of 7 Ω. At what frequency will the impedance of the practical inductance have an angle of $89\pi/180$ rad? What are the magnitude of the impedance and reactance of the inductance at this frequency?

4-14. An impedance Z is placed in parallel with a 15-Ω resistor to produce an impedance of $12 - j6\ \Omega$. What is Z?

4-15. In Fig. 4.3-3, replace the inductance with a capacitor, the capacitor with an inductor, and keep the same impedance magnitudes as before. Find the impedance looking into terminals a, b. How has the impedance changed from that found in Example 4.3-1?

4-16. Find the admittance Y_{ab} shown in Fig. P4-16. Give Y_{ab} in both polar and rectangular forms.

FIG. P4-16.

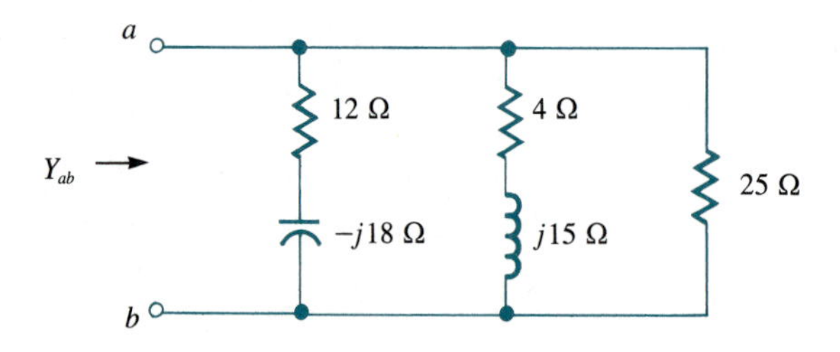

4-17. Two admittances $G_1 + jB_1$ and $G_2 + jB_2$ are in series. Show that the admittance of the series combination has a conductance G of

$$G = \frac{G_1(G_2^2 + B_2^2) + G_2(G_1^2 + B_1^2)}{(G_1 + G_2)^2 + (B_1 + B_2)^2}$$

and susceptance B of

$$B = \frac{B_1(G_2^2 + B_2^2) + B_2(G_1^2 + B_1^2)}{(G_1 + G_2)^2 + (B_1 + B_2)^2}$$

4-18. Find a Thévenin's equivalent source for the circuit of Fig. P4-18 to the left of terminals a, b.

FIG. P4-18.

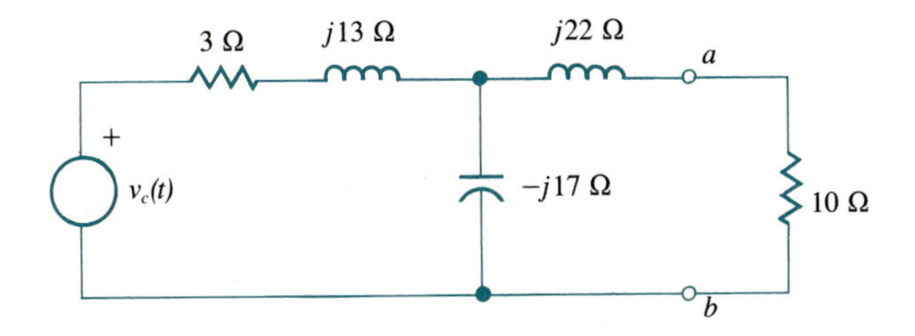

4-19. Work Problem 4-18 except for the circuit of Fig. P4-19.

FIG. P4-19.

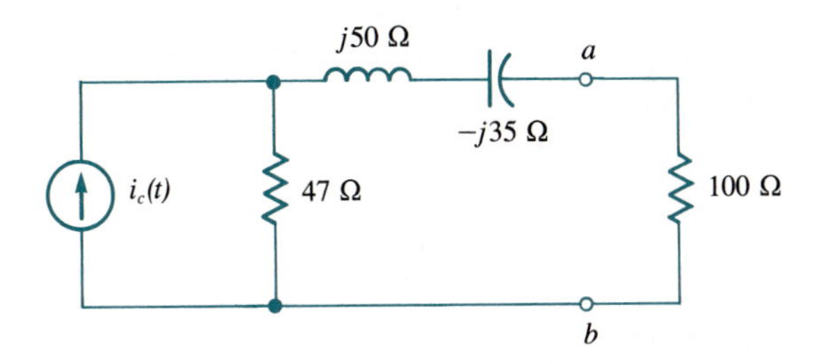

4-20. Work Problem 4-18 except find a Norton's source instead.

4-21. Find a Norton's equivalent source for the circuit of Fig. P4-19 to the left of terminals a, b.

4-22. Find a Thévenin's equivalent source for the circuit of Fig. P4-22 to the left of terminals a, b.

FIG. P4-22.

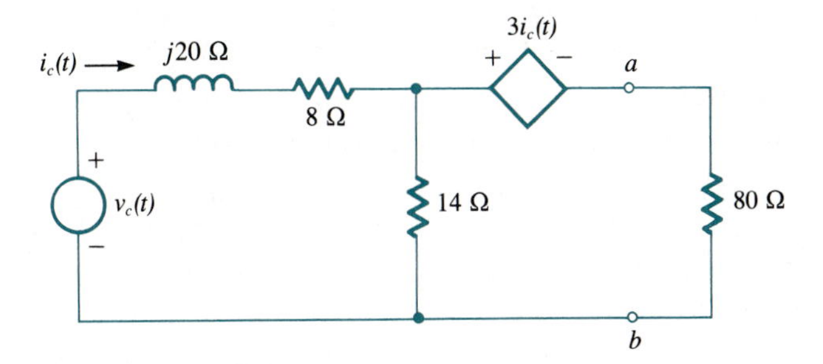

4-23. Work Problem 4-22 except find a Norton's source.

4-24. Use mesh analysis to solve for $v_{0c}(t)$ in the circuit of Fig. P4-24. Assume $v_{1c}(t) = 12 \exp j\omega t$ and $v_{2c}(t) = 5 \exp (j\omega t + j\pi/4)$.

FIG. P4-24.

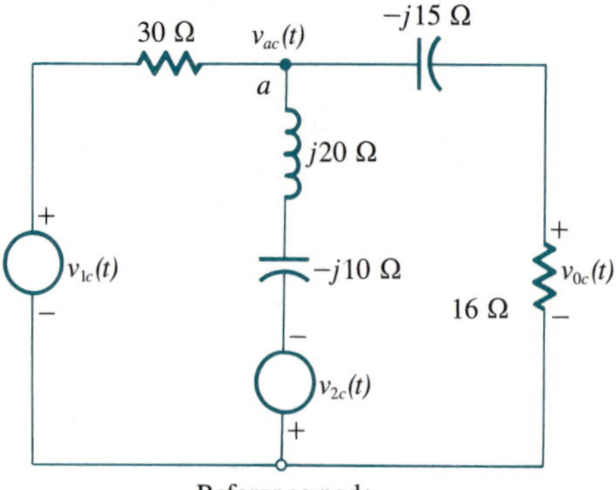

4-25. Use mesh analysis to solve the circuit of Fig. P4-25 for $v_{0c}(t)$ when $i_{1c}(t) = 3 \exp j\omega t$ and $i_{2c}(t) = 9 \exp j\omega t$. (*Hint*: Convert the current sources to equivalent voltage sources.)

FIG. P4-25.

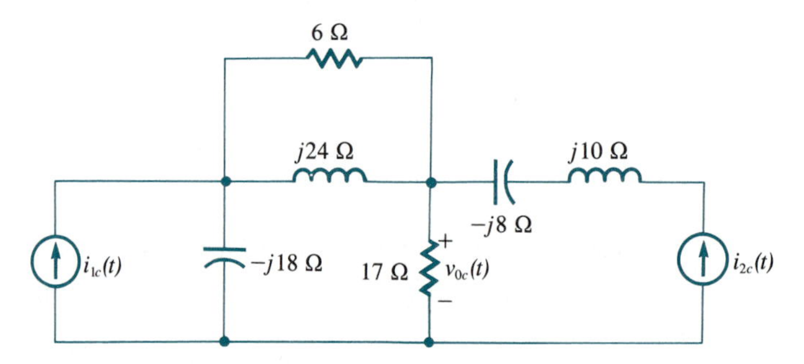

4-26. Use nodel analysis to find the voltage $v_{ac}(t)$ at node a in Fig. P4-24. Assume $v_{1c}(t) = 24 \exp j\omega t$ and $v_{2c}(t) = 8 \exp j\omega t$.

4-27. Redraw Fig. 4.4-3 after replacing the circuits to the left of terminals a, b and to the right of terminals c, d by Thévenin's voltage sources.

4-28. A complex voltage $v_c(t) = 10 \exp j\omega t$ is applied to terminals a, b in Fig. P4-10. What real power is delivered by the source?

4-29. A complex voltage $v_c(t) = 7 \exp j\omega t$ occurs across a complex impedance $Z = 15 - j8.5\ \Omega$. (a) What is the current through Z? (b) Find the apparent, complex, real, and reactive powers. (c) What is the power factor of this load? What is the reactive factor?

4-30. In the circuit of Fig. 4.2-2, find the complex power flowing into the resistor and the complex power flowing out of the source. Are they equal? If not, explain.

4-31. For the circuit of Fig. 4.3-3 that was analyzed in Example 4.3-1, what complex power flows into terminals a, b? What is the real power expended in the 4-Ω resistor?

4-32. If a source is connected to terminals a, b in Fig. 4.3-4 to provide a complex current $i_c(t) = 16 \exp j\omega t$ into terminal a, what are the real powers in the two resistors? What is the complex power into terminals a, b?

4-33. If the load in Fig. 4.5-5 is restricted to be only resistive, that is, $Z_L = R_L$, what value of R_L will correspond to maximum power in R_L when Z_s is fixed?

4-34. In the circuit of Fig. 4.5-5, assume that the complex load impedance Z_L is constant. What value of complex source impedance will maximize the real power delivered to the load?

4-35. In the circuit of Fig. 4.4-1(a) a reactance can be added in series with the 3-Ω load resistor to maximize its real power. What value should the reactance have?

4-36. Find the transfer function $H(\omega) = v_{oc}(t)/v_c(t)$ for the network of Fig. P4-36.

FIG. P4-36.

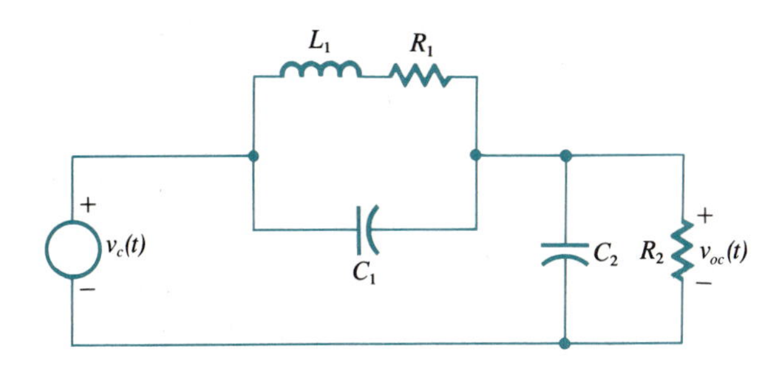

4-37. Find the exact value of the 3-dB bandwidth of the circuit of Fig. 4.6-2(a) for which the transfer function is (A) in Example 4.6-1.

4-38. Assume $\omega_0/2\pi = 10.7$ MHz, $C_p = 56$ pF, and $R_p = 15$ kΩ in the circuit of Fig. 4.7-1(a) and find Q_0 and L_p.

DESIGN EXERCISES

D4-1. Design a filter using only two resistors and one capacitor to have a voltage-in to voltage-out transfer function that has a magnitude that is nearly constant out to $\omega = 10^3$ rad/s, where it then decreases with ω until $\omega = 10^5$ rad/s, where it becomes nearly constant for $\omega > 10^5$ rad/s.

D4-2. Design a filter as in Design Exercise D4-1, but use two resistors and two capacitors. Also, the transfer function's magnitude must steadily decrease for $\omega > 10^7$ rad/s such as to approach zero as $\omega \rightarrow \infty$.

D4-3. Design a circuit to connect an ideal voltage source to a load resistance. The voltage-to-voltage transfer function of the circuit, when connected to source and load, is to have a transfer function that equals 0.8 for $\omega = 0$ and $\omega \rightarrow \infty$. The *circuit* can contain only one resistor, one inductor, and one capacitor, all of which may be considered ideal. The transfer function is to become *zero* at a specific angular frequency ω_0.

D4-4. Design a bandpass filter containing only one resistor, one inductor, and one capacitor, all of which may be taken as ideal, to connect an ideal voltage source to a 1500-Ω resistive load. The transfer function, when connected to both the source and the load, must be zero at $\omega = 0$ and $\omega \rightarrow \infty$, and be $0.5 + j0$ at $\omega_0 = 2\pi \times 10^7$ rad/s. The 3-dB bandwidth of the transfer function must be $\frac{5}{9}$ MHz.

CHAPTER 5

Transients in Circuits

5.0 INTRODUCTION

To complete the development of analysis methods for various networks, our attention is now turned to the general linear network that can contain arbitrary sources. By *arbitrary* we mean a source that provides a current or voltage of arbitrary waveshape. We are no longer restricted to the steady-state ac or dc signals of the preceding chapters. A waveform with no steady behavior is called a *transient signal.*

Transient signals are common in all areas of engineering. A mechanical engineer may be interested in the heat propagation along some structure in response to a sudden, limited-duration application of heat at one location. One of the concepts introduced in this chapter actually had its roots in heat-transfer problems. A chemical or industrial engineer could be concerned with the effects of a sudden pressure surge in a pipeline. An aeronautical engineer may be interested in the effects on wing flexures in an aircraft subject to landing shock to the wheels. The list could be easily extended. Although the mathematical theory required in these problems is similar, we shall be most concerned with transient excitations in electric networks.

Two principal methods are usually considered for general linear network analysis. One is based on the theory of Laplace transforms.† It is the most

† Named in honor of Pierre Simon Laplace (1749–1827), a brilliant French mathematician.

powerful approach but requires either that the reader have some background in the theory of complex variables or that this background be developed in the book. For the intended purpose and scope of this book both requirements are unrealistic. For our needs the second analysis method is best and the Laplace transform approach is left as an advanced topic in circuit theory for those readers interested in more background.

The analysis procedure that we shall use is based on the theory of Fourier transforms and the concept of a transfer function defined in Sec. 4.6. It is powerful enough for the analysis of most practical networks but does have some limitations. These limitations are partially overcome through the use of impulse functions. Others are overcome by assuming, as we shall in this book, that initial conditions are zero. This assumption means that prior to the application of any stimuli from any sources, the various capacitors in the network have no charges and all inductors, including transformers, have no currents flowing.

5.1 FUNDAMENTAL ANALYSIS USING FOURIER TRANSFORMS

When complex analysis, transfer functions, and Fourier transforms are combined to form an analysis procedure, a flexibility is created to work in either the time or frequency domain. We consider the frequency domain first.

Frequency-Domain Analysis

Let $v(t)$ represent a possibly complex waveform† constrained to have a Fourier transform denoted by $V(\omega)$; otherwise, $v(t)$ is arbitrary. From the inverse Fourier transform:

$$v(t) = \frac{1}{2\pi}\int_{-\infty}^{\infty} V(\omega)e^{j\omega t}\,d\omega \tag{5.1-1}$$

On recognizing that integration is just a summation of an infinite number of infinitesimal quantities, we may write (5.1-1) as

$$v(t) = \sum_{k=-\infty}^{\infty}\left[V(\omega_k)\frac{d\omega}{2\pi}\right]e^{j\omega_k t} = \sum_{k=-\infty}^{\infty} V_k e^{j\omega_k t} = \sum_{k=-\infty}^{\infty} v_k(t) \tag{5.1-2}$$

where

$$V_k = V(\omega_k)\frac{d\omega}{2\pi} \tag{5.1-3}$$

$$v_k(t) = V_k e^{j\omega_k t} \tag{5.1-4}$$

† Here complex means only that $v(t)$ has both real and imaginary parts; it does not restrict $v(t)$ to be a complex signal as defined in Chapters 2 and 4.

The first right-side term in (5.1-2) is the result of dividing the ω axis into contiguous infinitesimal intervals of width $d\omega$ and writing the integral as a sum. The center angular frequency of the kth interval is ω_k. Over any small interval $V(\omega)$ is nearly constant at its center-interval value $V(\omega_k)$. The second right-side term in (5.1-2) recognizes that $V(\omega_k)\ d\omega/2\pi$ is just a constant complex amplitude of an exponential signal $\exp j\omega_k t$, as defined in (5.1-3) to be V_k. The last right-side form demonstrates that $v(t)$ is an infinite sum of (infinitesimally small) complex signals, as defined in (5.1-4).

Next, suppose we are interested in a response,† denoted by $v_o(t)$, that occurs someplace in a linear network when $v(t)$ of (5.1-2) is applied as a source to two terminals someplace else in the network. The response to $v(t)$ is the sum of the responses to the signals $v_k(t)$ because of superposition. The response to a complex signal $v_k(t)$, which we denote by $v_{ok}(t)$, is given by the definition of the transfer function, $H(\omega)$,

$$v_{ok}(t) = H(\omega_k)v_k(t) \tag{5.1-5}$$

The output is the sum of all such responses. It is

$$\begin{aligned} v_o(t) &= \sum_{k=-\infty}^{\infty} v_{ok}(t) = \sum_{k=-\infty}^{\infty} H(\omega_k)v_k(t) = \sum_{k=-\infty}^{\infty} H(\omega_k)V_k e^{j\omega_k t} \\ &= \sum_{k=-\infty}^{\infty} H(\omega_k)V(\omega_k)\frac{d\omega}{2\pi}\ e^{j\omega_k t} \end{aligned} \tag{5.1-6}$$

Finally, we replace the sum in (5.1-6) by its integral to obtain

$$v_o(t) = \frac{1}{2\pi}\int_{-\infty}^{\infty} H(\omega)V(\omega)e^{j\omega t}d\omega \tag{5.1-7}$$

Equation (5.1-7) is the inverse Fourier transform that produces $v_o(t)$. Therefore $H(\omega)V(\omega)$ must be the Fourier transform of $v_o(t)$, which we denote by $V_o(\omega)$. Our principal results become

$$V_o(\omega) = H(\omega)V(\omega) \tag{5.1-8}$$

$$v_o(t) = \frac{1}{2\pi}\int_{\infty}^{\infty} V_o(\omega)e^{j\omega t}d\omega \tag{5.1-9}$$

Equation (5.1-8) forms the basis of frequency-domain analysis. It states that the spectrum of the response of a linear network at a pair of (output) terminals is the product of the spectrum of an arbitrary waveform that excites another pair of (input) terminals and the transfer function defined between the pairs of terminals. No time functions are involved. The output time function is found from inverse transformation, using (5.1-9). A simple example will demonstrate the use of frequency-domain analysis.

---

† We shall imply a voltage response by the choice of notation, but the response could equally well be a current. Similarly, $v(t)$ could be replaced by notation to imply that the excitation is a current.

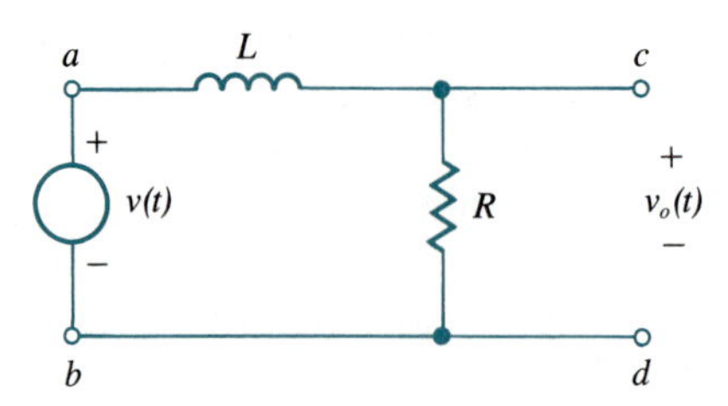

FIG. 5.1-1. A lowpass network for Example 5.1-1.

EXAMPLE 5.1-1

Frequency-domain analysis will be employed to solve for $v_o(t)$ in the circuit of Fig. 5.1-1 when

$$v(t) = u(t)e^{-\alpha t} \tag{A}$$

and $\alpha > 0$ is a constant. Although we omit the details, it can be shown that the spectrum $V(\omega)$ of $v(t)$ and the network's transfer function are

$$V(\omega) = \frac{1}{\alpha + j\omega} \tag{B}$$

$$H(\omega) = \frac{R/L}{R/L + j\omega}$$

respectively. From (5.1-8) the output signal's spectrum is

$$V_o(\omega) = V(\omega)H(\omega) = \frac{R/L}{(\alpha + j\omega)\,[R/L + j\omega]}$$

$$= \frac{R/(R - \alpha L)}{\alpha + j\omega} - \frac{R/(R - \alpha L)}{(R/L) + j\omega} \tag{C}$$

Because both terms in (C) are of the form of (B), which corresponds to the time function of (A), $v_o(t)$ follows easily

$$v_o(t) = \left(\frac{R}{R - \alpha L}\right)u(t)(e^{-\alpha t} - e^{-(R/L)t}) \tag{D}$$

A sketch of $v_o(t)$ is given in Fig. 5.1-2 for R/L equal to 3α, α, and $\alpha/3$. When $R/L = \alpha$ equation (D) yields an indeterminate form. On application of L'Hôpital's† rule we have

$$v_o(t) = \alpha t u(t)e^{-\alpha t} \qquad \frac{R}{L} = \alpha$$

We observe that as $(R/L) \rightarrow \infty$, which means the transfer function's bandwidth approaches infinity, the output $v_o(t)$ becomes proportional to the input $v(t)$. This

† Named for Guillaume François Antoine de L'Hôpital (1661–1704), a French mathematician.

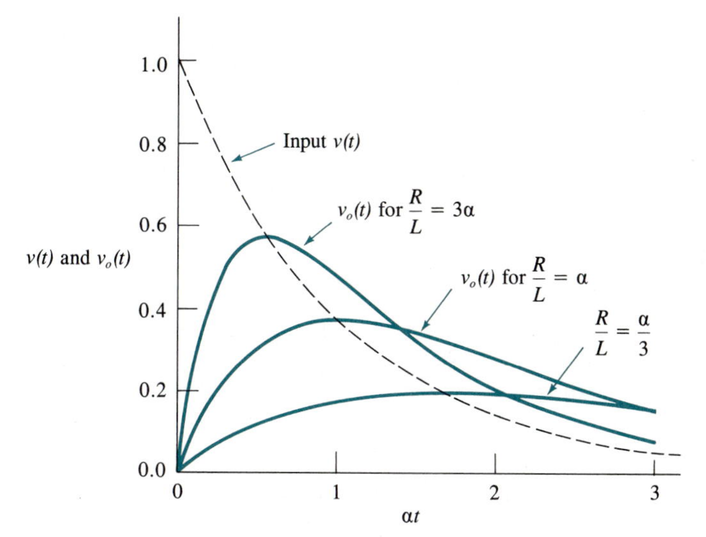

FIG. 5.1-2. Input and output signals of the circuit of Fig. 5.1-1.

behavior is characteristic of all transfer functions for which bandwidth is very large relative to the input signal's bandwidth. (Why?)

When we apply frequency-domain analysis, $V(\omega)$ and $H(\omega)$ must be determined. Complex analysis yields $H(\omega)$, but it is necessary to perform a Fourier transformation to obtain $V(\omega)$ when $v(t)$ is specified. Similarly, it is necessary to perform an inverse transformation if $v_o(t)$ is desired once $V_o(\omega) = V(\omega)H(\omega)$ is determined. Since Fourier transforms may sometimes be difficult to solve, it is helpful to establish a list, or table, of some known transforms to serve as a ready reference.

Fourier Transform Pairs

Table 5.1-1 gives a convenient list of some Fourier transform pairs. The second column lists the time signal $v(t)$. The third column gives the corresponding Fourier transform $V(\omega)$ of $v(t)$. The fourth column states conditions under which the transforms are valid. The utility of the table of pairs can be greatly increased by judicial use of the properties of Fourier transforms, some of which were given in Sec. 2.5 of Chapter 2. For example, if we wish to inverse transform the function $2/[j(\omega - \omega_0]$, where ω_0 is a constant, we do not find it under the column marked $V(\omega)$ in the table. However, by using (2.5-7), the frequency-shifting property, on pair 4 in the table, we have

$$v(t) = \text{Sgn}(t)e^{j\omega_0 t} \leftrightarrow \frac{2}{j(\omega - \omega_0)} = V(\omega) \tag{5.1-10}$$

The answer sought is $\text{Sgn}(t) \exp(j\omega_0 t)$, where $\text{Sgn}(t)$ is the *signum function* defined by

$$\text{Sgn}(t) = \begin{cases} 1 & t > 0 \\ -1 & t < 0 \end{cases} \tag{5.1-11}$$

TABLE 5.1-1. Fourier Transform Pairs

Pair	$v(t)$	$V(\omega)$	Remarks
1	$\delta(t)$	1	
2	1	$2\pi\delta(\omega)$	
3	$\|t\|$	$-\dfrac{2}{\omega^2}$	
4	$\mathrm{Sgn}(t)$	$\dfrac{2}{j\omega}$	
5	$u(t)$	$\pi\delta(\omega) + \dfrac{1}{j\omega}$	
6	$tu(t)$	$j\pi \dfrac{d\delta(\omega)}{d\omega} - \dfrac{1}{\omega^2}$	
7	$e^{-\alpha t}u(t)$	$\dfrac{1}{\alpha + j\omega}$	$\alpha > 0$
8	$te^{-\alpha t}u(t)$	$\dfrac{1}{(\alpha + j\omega)^2}$	$\alpha > 0$
9	$t^n e^{-\alpha t}u(t)$	$\dfrac{n!}{(\alpha + j\omega)^{n+1}}$	$\alpha > 0$
10	$e^{j\omega_0 t}$	$2\pi\delta(\omega - \omega_0)$	
11	$\cos \omega_0 t$	$\pi[\delta(\omega - \omega_0) + \delta(\omega + \omega_0)]$	
12	$\sin \omega_0 t$	$-j\pi[\delta(\omega - \omega_0) - \delta(\omega + \omega_0)]$	
13	$e^{j\omega_0 t}u(t)$	$\pi\delta(\omega - \omega_0) + \dfrac{1}{j(\omega - \omega_0)}$	
14	$\cos (\omega_0 t)u(t)$	$\dfrac{\pi}{2}[\delta(\omega - \omega_0) + \delta(\omega + \omega_0)] + \dfrac{j\omega}{\omega_0^2 - \omega^2}$	
15	$\sin (\omega_0 t)u(t)$	$-j\dfrac{\pi}{2}[\delta(\omega - \omega_0) - \delta(\omega + \omega_0)] + \dfrac{\omega_0}{\omega_0^2 - \omega^2}$	
16	$e^{-(\alpha + j\omega_0)t}u(t)$	$\dfrac{1}{\alpha + j(\omega - \omega_0)}$	$\alpha > 0$
17	$e^{-\alpha t}\cos (\omega_0 t)u(t)$	$\dfrac{\alpha + j\omega}{\omega_0^2 + (\alpha + j\omega)^2}$	$\alpha > 0$
18	$e^{-\alpha t}\sin (\omega_0 t)u(t)$	$\dfrac{\omega_0}{\omega_0^2 + (\alpha + j\omega)^2}$	$\alpha > 0$
19	$e^{-\alpha\|t\|}$	$\dfrac{2\alpha}{\alpha^2 + \omega^2}$	$\alpha > 0$
20	$e^{-\alpha t^2}$	$\sqrt{\dfrac{\pi}{\alpha}}\, e^{-\omega^2/4\alpha}$	$\alpha > 0$

Note: For an extensive list of pairs, see G.A. Campbell and R.M. Foster, *Fourier Integrals for Practical Applications*, Van Nostrand, Princeton, N.J., 1948.

Table 5.1-1 can be effectively doubled in length by the duality property of Fourier transforms. It states that

$$V(t) \leftrightarrow 2\pi v(-\omega) \tag{5.1-12}$$

In other words, if the given time function is equivalent to replacing ω by t in a function found in the $V(\omega)$ column of the table, then the desired transform is given by 2π times the corresponding function from the $v(t)$ column with t replaced by $-\omega$. For example, the pair

$$\frac{1}{\alpha + jt} \leftrightarrow 2\pi e^{\alpha\omega}u(-\omega) \tag{5.1-13}$$

derives from use of (5.1-12) with pair 7 of Table 5.1-1. Other Fourier transform properties are also helpful in extending the table (see Problems 5-8 through 5-12).

5.2 PARTIAL-FRACTION EXPANSIONS FOR INVERSE TRANSFORMS

For all but the simplest networks the problem of inverse-Fourier-transforming a given spectral function $V(\omega)$ to find the corresponding time function $v(t)$ is difficult by direct methods. It is helpful to put $V(\omega)$ in a simplified form that is more readily transformed. Such a form results from a *partial-fraction expansion* that expresses $V(\omega)$ as a sum of simple terms. The inverse transform of the complicated function $V(\omega)$ is thereby reduced to a sum of simple inverse transforms, each of which may typically be obtained from a table such as Table 5.1-1.

The partial-fraction expansion depends on the function $V(\omega)$ being expressible as a ratio of polynomials in the form

$$V(s) = \frac{N(s)}{D(s)} = \frac{b_m s^m + b_{m-1}s^{m-1} + \cdots + b_1 s + b_0}{a_n s^n + a_{n-1}s^{n-1} + \cdots + a_1 s + a_0} \tag{5.2-1}$$

where we have replaced $j\omega$ by the variable s for convenience,

$$s = j\omega \tag{5.2-2}$$

and the coefficients a_k and b_k are real numbers. Many practical problems will satisfy the form of (5.2-1). The numerator of (5.2-1) will have m roots because it is a polynomial of degree m. These roots are called *zeros* of $V(s)$ because when the variable s equals the value of a root, $V(s) = 0$. Similarly, the denominator's polynomial is of degree n and has n roots, but these roots are called *poles* of $V(s)$. When s equals the value of a pole, $V(s)$ becomes infinite, so the name is appropriate.

The partial-fraction representation of $V(s)$ depends on the polynomial degrees n and m and on the types of poles present. When $n > m$, $V(s)$ is said to be a *proper fraction*. We next examine the various cases that can occur.

Simple Poles Only

When $n > m$ and all poles are distinct, (5.2-1) can be written in the form

$$V(s) = \frac{K_1}{s - p_1} + \frac{K_2}{s - p_2} + \cdots + \frac{K_n}{s - p_n} \tag{5.2-3}$$

where p_k, $k = 1, 2, \ldots, n$, are the poles of $V(s)$ and the constants K_k, $k = 1,$

2, . . . , n, are called the *residues* of the poles. They may be found from

$$K_k = (s - p_k)V(s)\Big|_{s=p_k} \qquad k = 1, 2, \ldots, n \tag{5.2-4}$$

After inverse Fourier transformation of (5.2-3) we have

$$v(t) = u(t)[K_1e^{p_1t} + K_2e^{p_2t} + \cdots + K_ne^{p_nt}] \tag{5.2-5}$$

EXAMPLE 5.2-1

As an example of a spectrum with simple poles only, consider solving for $v_o(t)$ in Fig. 5.2-1 when

$$v(t) = 4u(t)e^{-3t} \leftrightarrow \frac{4}{3 + j\omega} = V(\omega)$$

It is easy to find the transfer function to be

$$H(\omega) = \frac{v_o(t)}{v(t)} = \frac{3}{(2 + j\omega)\,(\frac{1}{2} + j\omega)}$$

The output spectrum, which we wish to inverse-transform, is

$$V_o(\omega) = V(\omega)H(\omega) = \frac{12}{(3 + j\omega)(2 + j\omega)(\frac{1}{2} + j\omega)}$$

Since there are three simple poles, we let $s = j\omega$ and apply (5.2-3):

$$V_o(s) = \frac{12}{(s + 3)(s + 2)(s + \frac{1}{2})} = \frac{K_1}{s + 3} + \frac{K_2}{s + 2} + \frac{K_3}{s + \frac{1}{2}} \tag{A}$$

where

$$p_1 = -3 \qquad p_2 = -2 \qquad p_3 = -\tfrac{1}{2}$$

From (5.2-4):

$$K_1 = (s + 3)V(s)\Big|_{s=-3} = \frac{12}{(s + 2)(s + \frac{1}{2})}\Big|_{s=-3} = \tfrac{24}{5}$$

$$K_2 = (s + 2)V(s)\Big|_{s=-2} = \frac{12}{(s + 3)(s + \frac{1}{2})}\Big|_{s=-2} = -8$$

$$K_3 = (s + \tfrac{1}{2})V(s)\Big|_{s=-1/2} = \frac{12}{(s + 3)(s + 2)}\Big|_{s=-1/2} = \tfrac{16}{5}$$

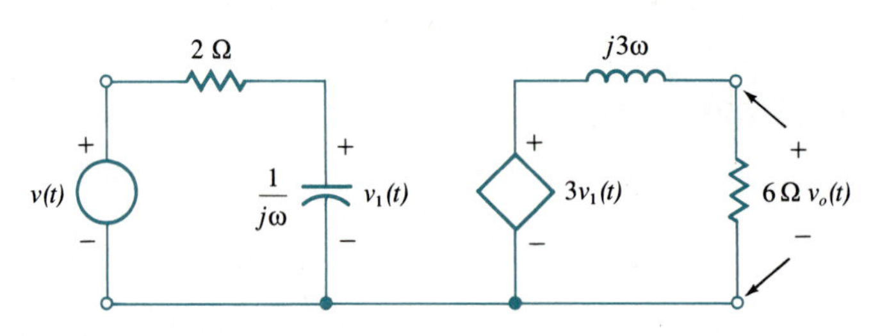

FIG. 5.2-1. A network analyzed in Example 5.2-1.

When we replace s with $j\omega$, (A) becomes

$$V_o(\omega) = \frac{\frac{24}{5}}{3 + j\omega} + \frac{-8}{2 + j\omega} + \frac{\frac{16}{5}}{\frac{1}{2} + j\omega}$$

Finally, we use either (5.2-5) or pair 7 of Table 5.1-1 to obtain the inverse transform of $V_o(\omega)$:

$$v_o(t) = u(t)\left(\tfrac{24}{5}e^{-3t} - 8e^{-2t} + \tfrac{16}{5}e^{-t/2}\right)$$

Repeated Poles

If the denominator of (5.2-1) contains a factor of the form $(s - p_q)^r$, it is said to have a repeated root p_q of multiplicity r when $r > 1$. Assume the other $n - r$ roots are simple and $n > m$. Then $V(s)$ can be written in the form

$$V(s) = \frac{K_1}{s - p_1} + \frac{K_2}{s - p_2} + \cdots + \frac{K_{n-r}}{s - p_{n-r}} + \frac{A_1}{s - p_q} + \frac{A_2}{(s - p_q)^2} + \cdots + \frac{A_r}{(s - p_q)^r} \tag{5.2-6}$$

where we have arbitrarily chosen to number the $n - r$ simple roots $1, 2, \ldots, (n - r)$. The repeated root is, therefore, labeled

$$q = n - r + 1 \tag{5.2-7}$$

The residues K_k, $k = 1, 2, \ldots, (n - r)$, in (5.2-6) are found from (5.2-4). The coefficients A_i are found from

$$A_i = \frac{1}{(r - i)!}\frac{d^{r-i}}{ds^{r-i}}[(s - p_q)^r V(s)]\Big|_{s=p_q} \qquad i = 1, 2, \ldots, r \tag{5.2-8}$$

The inverse transform of (5.2-6) is

$$v(t) = \sum_{k=1}^{n-r} u(t)K_k e^{p_k t} + \sum_{i=1}^{r} u(t)A_i \frac{t^{i-1}}{(i - 1)!} e^{p_q t} \tag{5.2-9}$$

An example will illustrate the use of these results:

EXAMPLE 5.2-2

Let the spectrum of the output of some system be

$$V_o(\omega) = \frac{2(2 + j\omega)^2}{(3 + j\omega)(5 + j\omega)^2}$$

This function has one simple pole and one repeated pole of multiplicity 2. On letting $s = j\omega$, we see that $n = 3$, $r = 2$, $q = 2$, $p_1 = -3$, and $p_2 = -5$:

$$V_o(s) = \frac{2(2 + s)^2}{(3 + s)(5 + s)^2} = \frac{K_1}{s + 3} + \frac{A_1}{s + 5} + \frac{A_2}{(s + 5)^2} \tag{A}$$

From (5.2-4)

$$K_1 = (s - p_1)V_o(s)\Big|_{s=p_1} = (s + 3)V_o(s)\Big|_{s=-3}$$
$$= \frac{2(2 + s)^2}{(5 + s)^2}\Bigg|_{s=-3} = \tfrac{1}{2}$$

From (5.2-8)

$$A_1 = \frac{1}{(2 - 1)!}\frac{d^{2-1}}{ds^{2-1}}[(s + 5)^2V_o(s)]\Bigg|_{s=-5} = \frac{d}{ds}\left[\frac{2(2 + s)^2}{3 + s}\right]\Bigg|_{s=-5} = \tfrac{3}{2}$$

$$A_2 = \frac{1}{(2 - 2)!}\frac{d^{2-2}}{ds^{2-2}}[(s + 5)^2V_o(s)]\Bigg|_{s=-5} = \frac{2(2 + s)^2}{3 + s}\Bigg|_{s=-5} = -9$$

On substitution into (A) with $s = j\omega$, we get

$$V_o(\omega) = \frac{\frac{1}{2}}{3 + j\omega} + \frac{\frac{3}{2}}{5 + j\omega} + \frac{-9}{(5 + j\omega)^2} \tag{B}$$

The corresponding time function $v_o(t)$ is the inverse transform of (B). From pairs 7 and 8 of Table 5.1-1 or from (5.2-9):

$$v_o(t) = \tfrac{1}{2}u(t)e^{-3t} + \tfrac{3}{2}u(t)e^{-5t} - 9u(t)te^{-5t}$$
$$= \tfrac{1}{2}u(t)e^{-3t} + (\tfrac{3}{2} - 9t)u(t)e^{-5t}$$

Complex-Pole Pairs

When a simple complex pole occurs in (5.2-1), there will be another pole that is the complex conjugate of the first. If we label these p_1 and p_2, where

$$p_1 = -\alpha - j\omega_0 \tag{5.2-10}$$

$$p_2 = -\alpha + j\omega_0 = p_1^* \tag{5.2-11}$$

the function $V(s)$ can be written as

$$V(s) = \frac{K_1}{s - p_1} + \frac{K_2}{s - p_2} + V_1(s) \tag{5.2-12}$$

The terms K_1 and K_2 are the residues of the poles p_1 and p_2, respectively, as before, and $V_1(s)$ is what is left of $V(s)$ after the first two right-side terms are removed. On using (5.2-4) to evaluate K_1 and K_2, we find that $K_2 = K_1^*$, so

$$V(s) = \frac{K_1}{s - p_1} + \frac{K_1^*}{s - p_1^*} + V_1(s) \tag{5.2-13}$$

where

$$K_1 = (s - p_1)V(s)\big|_{s=p_1} \tag{5.2-14}$$

By writing K_1 as

$$K_1 = a_1 + jb_1 \tag{5.2-15}$$

we can show by direct expansion that

$$\frac{K_1}{s - p_1} + \frac{K_1^*}{s - p_1^*} = \frac{2a_1(s + \alpha)}{\omega_0^2 + (s + \alpha)^2} + \frac{2b_1\omega_0}{\omega_0^2 + (s + \alpha)^2} \tag{5.2-16}$$

After replacing s by $j\omega$, we can use pairs 17 and 18 in Table 5.1-1 to evaluate the inverse transform of (5.2-12):

$$v(t) = 2u(t)e^{-\alpha t}[a_1 \cos(\omega_0 t) + b_1 \sin(\omega_0 t)] + v_1(t) \tag{5.2-17}$$

where $v_1(t)$ is the inverse Fourier transform of $V_1(\omega)$.

EXAMPLE 5.2-3

The response signal $v_o(t)$ from the network of Fig. 5.2-2 will be found when the input is a unit step function

$$v(t) = u(t) \leftrightarrow \pi\delta(\omega) + \frac{1}{j\omega} = V(\omega)$$

The network's transfer function is easily found to be

$$H(\omega) = \frac{17 + j\omega 2}{17 - \omega^2 + j\omega 2}$$

so the spectrum of the response is

$$V_o(\omega) = V(\omega)H(\omega) = \left[\pi\delta(\omega) + \frac{1}{j\omega}\right]\left[\frac{17 + j\omega 2}{17 - \omega^2 + j\omega 2}\right]$$

$$= \pi\delta(\omega) + \frac{17 + j\omega 2}{j\omega[17 - \omega^2 + j\omega 2]} = \pi\delta(\omega) + V_2(\omega)$$

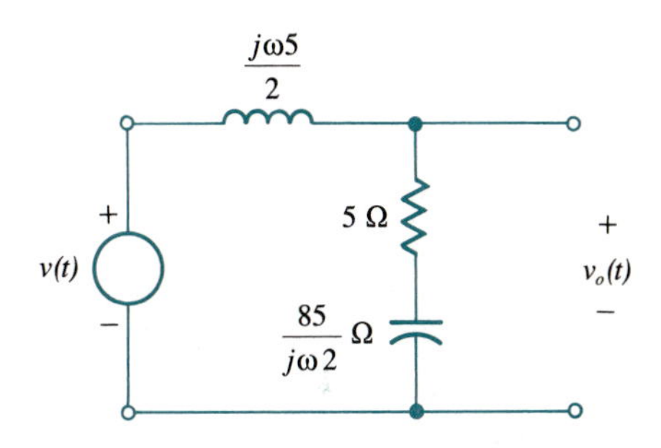

FIG. 5.2-2. A network analyzed in Example 5.2-3.

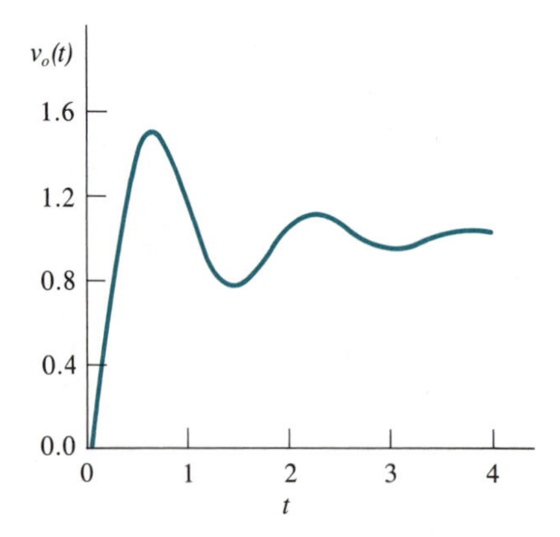

FIG. 5.2-3. The response of the network of Fig. 5.2-2 when the input is a unit step function.

We must inverse-Fourier-transform $V_o(\omega)$. The impulse is no problem, so we consider $V_2(\omega)$. On substituting $s = j\omega$, we have

$$V_2(s) = \frac{2(s + \frac{17}{2})}{s(s^2 + 2s + 17)}$$

which has roots (poles) at $s = 0$ and $s = -1 \pm j4$. Thus, from (5.2-10),

$$\alpha = 1 \qquad \omega_0 = 4$$

and $V_2(s)$ can be expanded as

$$V_2(s) = \frac{K_0}{s} + \frac{K_1}{s - p_1} + \frac{K_1^*}{s - p_1^*}$$

The residues are

$$K_0 = sV_2(s)\Big|_{s=0} = \frac{2(s + \frac{17}{2})}{s^2 + 2s + 17}\Big|_{s=0} = 1$$

$$K_1 = [s - (-1 - j4)]V_2(s)\Big|_{s=-1-j4} = \frac{2(s + \frac{17}{2})}{s[s - (-1 + j4)]}\Big|_{s=-1-j4} = -\tfrac{1}{2} + j\tfrac{1}{8}$$

so from (5.2-15),

$$a_1 = -\tfrac{1}{2} \qquad b_1 = \tfrac{1}{8}$$

After all substitutions are made and s is replaced by $j\omega$, we have

$$v_o(\omega) = \pi\delta(\omega) + \frac{1}{j\omega} + \frac{-(1 + j\omega)}{4^2 + (1 + j\omega)^2} + \frac{1}{4^2 + (1 + j\omega)^2}$$

and

$$\begin{aligned} v_o(t) &= \tfrac{1}{2} + \tfrac{1}{2}\text{Sgn}(t) + u(t)e^{-t}[-\cos(4t) + \tfrac{1}{4}\sin(4t)] \\ &= u(t)\{1 - e^{-t}[\cos(4t) - \tfrac{1}{4}\sin(4t)]\} \end{aligned}$$

This response is sketched in Fig. 5.2-3.

Nonproper Forms

If $n \leq m$ in (5.2-1), $V(s)$ has a nonproper form. By dividing the numerator by the denominator to obtain a quotient polynomial plus a remainder, we can express $V(s)$ as

$$V(s) = d_{m-n}s^{m-n} + d_{m-n-1}s^{m-n-1} + \cdots + d_1 s + d_0 + \frac{R(s)}{D(s)} \tag{5.2-18}$$

where the degree of $R(s)$ is less than n. The partial-fraction expansion of $R(s)/D(s)$ follows the preceding three methods for a proper fraction. The other right-side terms in (5.2-18) inverse-transform according to the pair

$$\frac{d^n\delta(t)}{dt^n} \leftrightarrow (j\omega)^n \tag{5.2-19}$$

for $n = 0, 1, \ldots$. Derivatives of the unit impulse function are defined through their integral property:

$$\int_{-\infty}^{\infty} \frac{d^n\delta(t)}{dt^n}\,\phi(t)\,dt = (-1)^n \left.\frac{d^n\phi(t)}{dt^n}\right|_{t=0} \tag{5.2-20}$$

EXAMPLE 5.2-4

The function

$$V(s) = \frac{3s^2 + 32s + 42}{s^2 + 9s + 14}$$

has nonproper form and poles at $s = -2$ and -7. Division gives

$$V(s) = 3 + \frac{5s}{s^2 + 9s + 14} = 3 + \frac{K_1}{s+2} + \frac{K_2}{s+7}$$

The residues are

$$K_1 = (s+2)\left.\frac{5s}{(s+2)(s+7)}\right|_{s=-2} = -2$$

$$K_2 = (s+7)\left.\frac{5s}{(s+2)(s+7)}\right|_{s=-7} = 7$$

so

$$V(s) = 3 + \frac{-2}{s+2} + \frac{7}{s+7}$$

Thus,

$$v(t) = 3\delta(t) - 2u(t)e^{-2t} + 7u(t)e^{-7t}$$

Finding Roots of Polynomials

The various partial-fraction expansions described in this section require that the roots (poles) of the denominator of (5.2-1) be known. This knowledge may not always be available; and when $n > 2$, it may be difficult to easily determine. Computers and special algorithms for finding roots of polynomials become valuable in these cases. Such a program has been provided on a diskette that is part of the solutions manual that is available to instructors using this book. Readers may obtain copies of the diskette from their instructor who is authorized by the publisher to freely make copies as required for student use. The program's name is FRESP and several computer exercises are given at the end of the chapter so that the reader can gain some experience from its use.

5.3 FUNDAMENTAL ANALYSIS USING CONVOLUTION

Impulse Response and Convolution

The spectrum $V_o(\omega)$ of the output signal of a linear network, with a transfer function $H(\omega)$, is related to the input signal's spectrum $V(\omega)$ by

$$V_o(\omega) = V(\omega)H(\omega) \tag{5.3-1}$$

as shown in Sec. 5.1. If we formally inverse-Fourier-transform $V_o(\omega)$, we obtain $v_o(t)$:

$$\begin{aligned} v_o(t) &= \frac{1}{2\pi}\int_{-\infty}^{\infty} V_o(\omega)e^{j\omega t}\,d\omega = \frac{1}{2\pi}\int_{-\infty}^{\infty} V(\omega)H(\omega)e^{j\omega t}\,d\omega \\ &= \frac{1}{2\pi}\int_{-\infty}^{\infty}\left[\int_{-\infty}^{\infty} v(\xi)e^{-j\omega\xi}\,d\xi\right]H(\omega)e^{j\omega t}\,d\omega \\ &= \int_{-\infty}^{\infty} v(\xi)\,\frac{1}{2\pi}\int_{-\infty}^{\infty} H(\omega)e^{j\omega(t-\xi)}\,d\omega\,d\xi \end{aligned} \tag{5.3-2}$$

The inner integral is an inverse transform of $H(\omega)$ and is a time function, which we define by using the notation $h(t)$ as follows:

$$h(t) = \frac{1}{2\pi}\int_{-\infty}^{\infty} H(\omega)e^{j\omega t}\,d\omega \tag{5.3-3}$$

We call $h(t)$ the *impulse response* of the network because $H(\omega)$ is the output signal's spectrum for an input unit impulse (remember that the spectrum of a unit impulse is unity). When (5.3-3) is substituted into (5.3-2), the main result of this section is achieved:

$$v_o(t) = \int_{-\infty}^{\infty} v(\xi)h(t - \xi)\,d\xi \tag{5.3-4}$$

Equation (5.3-4) is called a *convolution integral*. It is the basis of *time-domain analysis* of a linear network since all quantities are given in time. An alternative form of (5.3-4) is

$$v_o(t) = \int_{-\infty}^{\infty} h(\xi)v(t - \xi)\, d\xi \tag{5.3-5}$$

Sometimes, the analysis of a network problem is more successful if we use time-domain analysis via convolution than frequency-domain analysis. An example serves to demonstrate this point.

EXAMPLE 5.3-1

Consider the simple baseband pulse of Fig. 2.5-2(a), which is defined by the transform pair

$$v(t) = A\left[u\left(t + \frac{\tau}{2}\right) - u\left(t - \frac{\tau}{2}\right)\right] \leftrightarrow A\tau \frac{\sin(\omega\tau/2)}{\omega\tau/2} = V(\omega)$$

as proved in Example 2.5-3. We apply $v(t)$ to a lowpass filter, defined by the pair

$$h(t) = \alpha u(t)e^{-\alpha t} \leftrightarrow \frac{\alpha}{\alpha + j\omega} = H(\omega)$$

and seek to compute the response $v_o(t)$. To use frequency-domain analysis, we would have to inverse-transform the function

$$V_o(\omega) = V(\omega)H(\omega) = \frac{2A\alpha \sin(\omega\tau/2)}{\omega(\alpha + j\omega)}$$

which does not lend itself to easy solution. With time-domain analysis we use (5.3-5):

$$v_o(t) = \alpha \int_{-\infty}^{\infty} u(\xi)e^{-\alpha\xi}Au\left(t + \frac{\tau}{2} - \xi\right) d\xi - \alpha \int_{-\infty}^{\infty} u(\xi)e^{-\alpha\xi}Au\left(t - \frac{\tau}{2} - \xi\right) d\xi$$

$$= \alpha A \int_{0}^{\infty} u\left(t + \frac{\tau}{2} - \xi\right)e^{-\alpha\xi}\, d\xi - \alpha A \int_{0}^{\infty} u\left(t - \frac{\tau}{2} - \xi\right)e^{-\alpha\xi}\, d\xi$$

or

$$v_o(t) = A \int_{0}^{t+\tau/2} \alpha e^{-\alpha\xi}\, d\xi - A \int_{0}^{t-\tau/2} \alpha e^{-\alpha\xi}\, d\xi$$

The first integral is nonzero only for $t > -\tau/2$. The second is nonzero only for $t > \tau/2$. The final result is

$$v_o(t) = Au\left(t + \frac{\tau}{2}\right)(1 - e^{-\alpha(t+\tau/2)}) - Au\left(t - \frac{\tau}{2}\right)(1 - e^{-\alpha(t-\tau/2)})$$

which is sketched in Fig. 5.3-1 along with the input $v(t)$. The sketch assumes $\alpha = 2/\tau$. It can be shown that α is the 3-dB bandwidth (in radians per second) of the

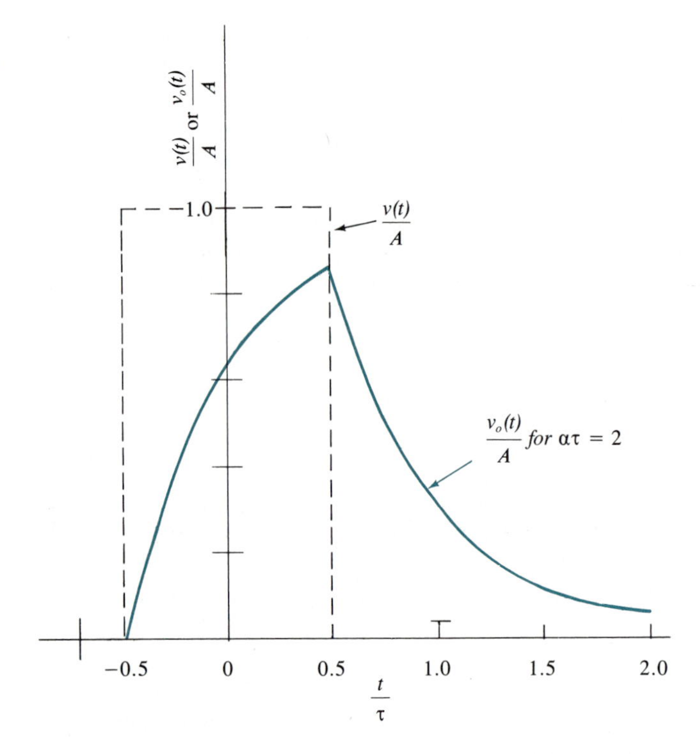

FIG. 5.3-1. Input and output signals for the circuit of Example 5.3-1.

network. If α were increased, the response would more closely approximate the input pulse.

A star between two time functions is often used as a short-form notation to imply the convolution of the two functions. Thus, for signals $v_1(t)$ and $v_2(t)$,

$$v_1(t) \bigstar v_2(t) = \int_{-\infty}^{\infty} v_1(\xi) v_2(t - \xi)\, d\xi \tag{5.3-6}$$

For the network's response of (5.3-4) we write

$$v_o(t) = v(t) \bigstar h(t) \tag{5.3-7}$$

Choosing Time- versus Frequency-Domain Analysis

Generally, there is no clear-cut way of deciding to use time-domain analysis (convolution) or frequency-domain analysis [product of $V(\omega)$ and $H(\omega)$] to solve for the response $v_o(t)$ of a network in the easiest way. Convolution may prove easiest if the input signal $v(t)$ and the network's impulse response $h(t)$ are given. On the other hand, if the signal's spectrum $V(\omega)$ and the network's transfer function $H(\omega)$ are specified, it is probably easier to first find the output spectrum

$V_o(\omega)$ by frequency-domain analysis and then inverse-transform to obtain $v_o(t)$. In many cases these procedures prove best because only one integration (transform), at most, is required.

Where *both* signal and network are not similarly specified, the solution method is the least clear. For example, if $v(t)$ and $H(\omega)$ are given, the frequency-domain method requires first transforming $v(t)$ to get $V(\omega)$ and then inverse-transforming $V_o(\omega)$. Two integrations are required. Alternatively, one can inverse-transform $H(\omega)$ to get $h(t)$ and use convolution to realize $v_o(t)$. Again two integrations are needed. The choice of which method to use will obviously depend on the difficulty encountered in solving the integrals.

PROBLEMS

5-1. Use frequency-domain analysis to solve for $v_o(t)$ in Fig. 5.1-1 when $v(t) = u(t)$.

5-2. Rework Example 5.1-1, but change the inductance L in Fig. 5.1-1 to a capacitance C.

5-3. Find $v_o(t)$ for the circuit of Fig. P5-3 when $v(t) = \delta(t)$. Use the frequency-domain-analysis method.

FIG. P5-3.

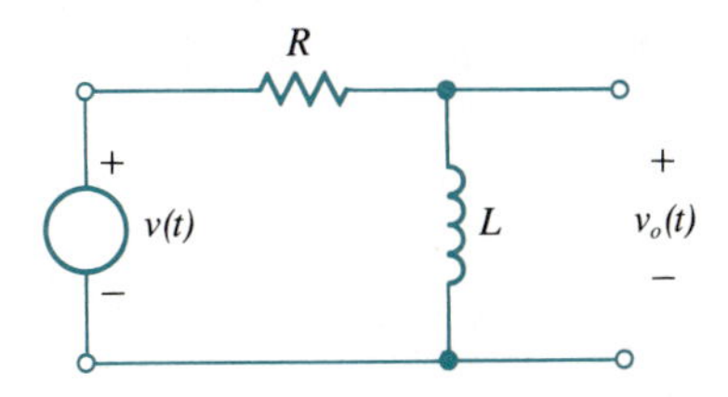

5-4. (a) Find the transfer function of the network in Fig. P5-4. (b) If $v(t) = \mathrm{Sgn}(t)$, find the spectrum of $v_o(t)$ and $v(t)$. (c) Sketch $v_o(t)$ and $v(t)$ when $R_1 = 10\ \Omega$, $R_2 = 12\ \Omega$, and $C = \frac{11}{30}$ F.

FIG. P5-4.

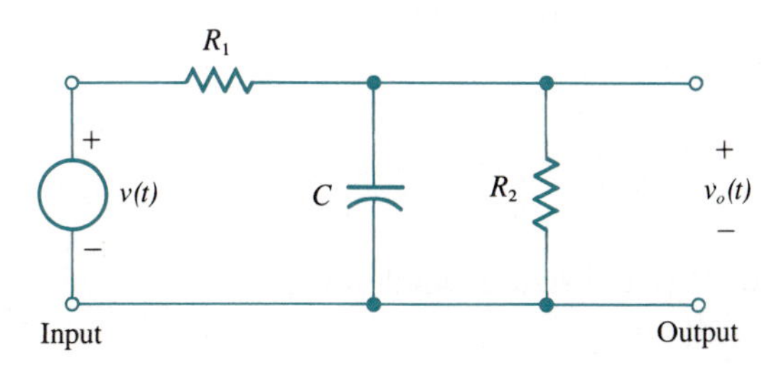

5-5. (a) For the circuit of Fig. P5-5, find the transfer function from input to output. (b) Write a general equation for the spectrum $V_o(\omega)$ of $v_o(t)$ when $v(t) = u(t)$.

FIG. P5-5.

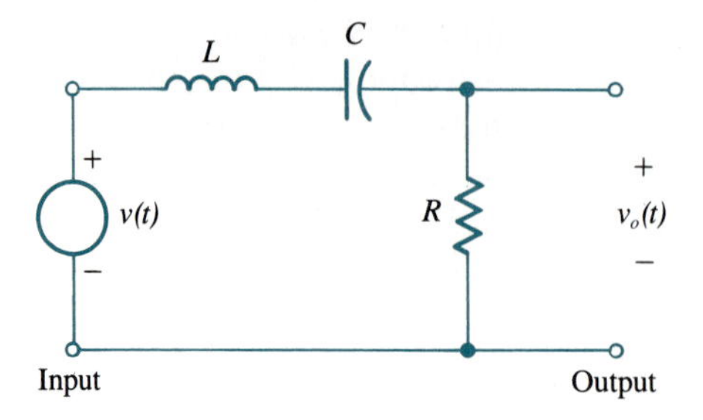

5-6. Use the time differentiation property of Fourier transforms [see (2.5-8)] to prove pair 12 in Table 5.1-1 by differentiating $\cos(\omega_0 t)$.

5-7. Write Sgn(t) as the sum of two waveforms, as defined in pair 7 of Table 5.1-1, and use a limiting operation to prove $\mathcal{F}\{\text{Sgn}(t)\} = 2/(j\omega)$.

5-8. Prove the *scaling property* of Fourier transforms, which is

$$v(\alpha t) \leftrightarrow \frac{1}{|\alpha|} V\left(\frac{\omega}{\alpha}\right)$$

where α is a real constant.

5-9. Prove the *conjugation property* of Fourier transforms, which is

$$v^*(t) \leftrightarrow V^*(-\omega) \qquad \text{and} \qquad v^*(-t) \leftrightarrow V^*(\omega)$$

5-10. Prove the *correlation property* of Fourier transforms, which is

$$v(t) = \int_{-\infty}^{\infty} v_1^*(\xi)v_2(\xi + t)\, d\xi \leftrightarrow V_1^*(\omega)V_2(\omega) = V(\omega)$$

$$v(t) = v_1^*(t)v_2(t) \leftrightarrow \frac{1}{2\pi}\int_{-\infty}^{\infty} V_1^*(\xi)V_2(\xi + \omega)\, d\xi = V(\omega)$$

where $v_1(t)$ and $v_2(t)$ are possibly complex signals with Fourier transforms $V_1(\omega)$ and $V_2(\omega)$, respectively.

5-11. Use the results from Problem 5-10 to prove *Parseval's theorem*, which is

$$\int_{-\infty}^{\infty} v_1^*(t)v_2(t)\, dt = \frac{1}{2\pi}\int_{-\infty}^{\infty} V_1^*(\omega)V_2(\omega)\, d\omega$$

or if $v_1(t) = v_2(t) = v(t)$,

$$\int_{-\infty}^{\infty} |v(t)|^2\, dt = \frac{1}{2\pi}\int_{-\infty}^{\infty} |V(\omega)|^2\, d\omega$$

Discuss the physical significance of this last expression. What interpretation can be given to $|V(\omega)|^2$?

5-12. Prove the *area properties* of Fourier transforms, which are

$$V(0) = \int_{-\infty}^{\infty} v(t)\,dt$$

$$v(0) = \frac{1}{2\pi}\int_{-\infty}^{\infty} V(\omega)\,d\omega$$

5-13. Find the Fourier transform of the signal

$$v(t) = u(t)t^3e^{-\alpha t^2}$$

where $\alpha > 0$ is a real constant.

5-14. Find the inverse Fourier transform of the spectrum

$$V(\omega) = u(\omega)\sin\omega\tau$$

5-15. Use the time differentiation property of Fourier transforms with pair 19 of Table 5.1-1 to prove that

$$\lim_{\alpha\to 0}\left\langle \mathcal{F}\left\{\frac{-1}{\alpha}\frac{d}{dt}\left[e^{-\alpha|t|}\right]\right\}\right\rangle = \frac{2}{j\omega}$$

5-16. (a) Find the partial-fraction expansion of the function

$$H(s) = \frac{2s}{s^2 + 10s + 16}$$

(b) Assume $H(s)$ defines the transfer function of a network when $s = j\omega$. Find the network's time response to a unit-impulse input.

5-17. Work Problem 5-16 except assume

$$H(s) = \frac{s^2 + 6s + 8}{(s^2 + 7s + 6)(s^2 + 14s + 45)}$$

5-18. For the network of Fig. 5.2-1, assume the input is $v(t) = u(t)$, a unit step function. (a) Find $V_o(\omega)$ and its partial-fraction expansion. (b) Find $v_o(t)$.

5-19. (a) Find the partial-fraction expansion of the function

$$H(s) = \frac{s^2 + 2s + 3}{(s + 2)^3}$$

(b) What is $h(t)$?

5-20. In the circuit of Fig. P5-4, let $R_1 = 3\ \Omega$, $R_2 = 6\ \Omega$, $C = \frac{1}{6}$ F, and

$$v(t) = 3u(t)t^2e^{-3t}$$

(a) Find $H(\omega)$ for the network. (b) What is the signal's Fourier transform? (c) Determine the spectrum $V_o(\omega)$ of $v_o(t)$. (d) Find $v_o(t)$.

5-21. An input signal

$$v(t) = 2u(t)e^{-2t}\sin(6t)$$

is applied to the network of Fig. 5.1-1 when $R = 1\ \Omega$ and $L = 0.5$ H. (a) Find $H(\omega)$. (b) Find $V_o(\omega)$ and its partial-fraction expansion. (c) What is $v_o(t)$? Sketch $v_o(t)$.

5-22. Assume the function

$$V(\omega) = \frac{4(2 + j\omega)}{(10 - \omega^2) + j\omega 2}$$

is the transform of a signal $v(t)$. (a) Find the partial-fraction expansion of $V(\omega)$. (b) Find the signal $v(t)$.

5-23. (a) Determine the transfer function of the network of Fig. P5-23. Is it of proper form? (b) If $v(t) = \delta(t)$, what is the partial-fraction expansion of $V_o(\omega)$? (c) What is $v_o(t)$?

FIG. P5-23.

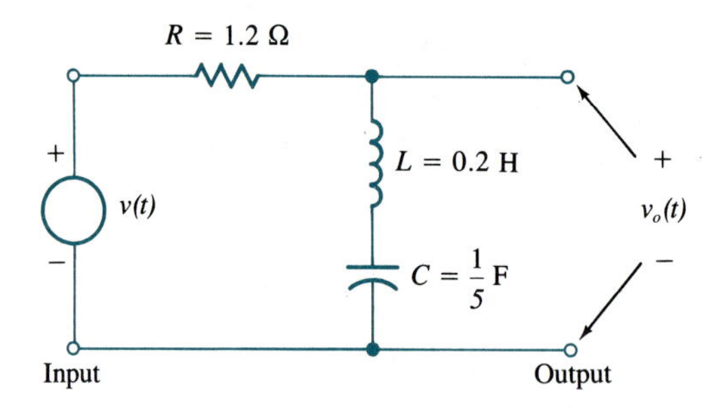

★ 5-24. If the signal

$$v(t) = u(t)e^{-2t} \sin 5t$$

is the input to the circuit of Fig. P5-23, determine (a) $V(\omega)$, (b) $V_o(\omega)$, (c) the partial-fraction expansion of $V_o(\omega)$, and (d) $v_o(t)$.

5-25. The derivative of a unit-impulse function is called a *unit doublet*. Evaluate the following integrals.

(a) $\displaystyle\int_{-\infty}^{\infty} \frac{d\delta(t)}{dt} \cos^2 (6t)\, dt$

(b) $\displaystyle\int_{-\infty}^{\infty} \frac{d\delta(t-3)}{dt} \frac{4}{24 + 5t^2}\, dt$

(c) $\displaystyle\int_{-\infty}^{\infty} \left[2\frac{d\delta(t)}{dt} + 4\delta(t)\right] \tfrac{1}{3} e^{-6t^2}\, dt$

5-26. For the circuit and signal $v(t)$ defined in Problem 5-20, solve for (a) the circuit's impulse response $h(t)$ and (b) the output $v_o(t)$, using convolution.

5-27. For the network and signal of Problem 5-4, find $v_o(t)$ by convolution when $R_1 = 10\ \Omega$, $R_2 = 12\ \Omega$, and $C = \frac{11}{30}$ F.

5-28. Determine $v(t) \bigstar v(t)$, where $v(t)$ is defined in Example 5.3-1. What is $\mathcal{F}\{v(t) \bigstar v(t)\}$?

★ 5-29. In a cascade of N networks the output of the first is the input of the second, the output of the second is the input of the third, and so forth. Show that the impulse response of the overall cascade is

$$h(t) = h_1(t) \star h_2(t) \star \cdots \star h_N(t)$$

where $h_n(t)$ is the impulse response of network n.

DESIGN EXERCISES

D5-1. Select constants a_1 and a_2 in the function

$$H(\omega) = \frac{1}{1 + a_1(j\omega) + a_2(j\omega)^2}$$

so that as many derivatives as possible of $|H(\omega)|^2$ at $\omega = 0$ can be made zero. In addition, the 3-dB bandwidth of the lowpass (filter) function is to be 10 rad/s. After defining a_1 and a_2, design any form of network with only passive elements to realize the transfer function.

D5-2. Design a network, by determining the necessary transfer function $H(\omega)$, to cause the output signal to have the largest possible peak power in a resistor R that can possibly occur at a specific time t_0 when the input signal is

$$v(t) = u(-t)e^{\alpha t} \cos(\omega_0 t)$$

Here $\alpha > 0$ and ω_0 are real constants. (*Hint:* Use Schwarz's inequality.)

COMPUTER EXERCISES

The following Exercises require use of the computer program named FRESP. The program is part of the software on a diskette included with the solutions manual that is available to instructors from the publisher. Readers may obtain the software from their instructor who is authorized by the publisher to freely make copies of the diskette as needed for student use.

C5-1. Use the computer program FRESP to determine the denominator's roots in the function

$$V(\omega) = \frac{1}{-ja\omega^3 - b\omega^2 + j2\omega + 1}$$

Determine the locations of all poles for several values of the constants a and b, beginning with $a = 6/2^3$ and $b = 4/2^3$ and increasing both by factors of 4 until $a = 6(2^3)$ and $b = 4(2^3)$.

C5-2. Determine the poles and zeros of the function

$$V(\omega) = \frac{-j\omega^3 - 7.5\omega^2 + j16.5\omega + 10}{\omega^4 - j7.2\omega^3 - 73.96\omega^2 + j168\omega + 784}$$

It may be necessary to use the computer program FRESP.

C5-3. Work Exercise C5-2 except for the function

$$V(\omega) = \frac{-j\omega^3 - 3\omega^2 + j18.25\omega + 32.5}{j\omega^5 + 5\omega^4 - j16\omega^3 - 28\omega^2 + j32\omega + 16}$$

C5-4. Use the computer program FRESP to generate data sufficient to plot $|V(\omega)|$ in decibels versus ω on a log ω scale for the function of Computer Exercise C5-2.

P A R T 2

Electronic Devices and Circuits

CHAPTER 6

Diode and Transistor Devices

6.0 INTRODUCTION

Electronics can loosely be defined as that area of science and engineering that is concerned with the processing of charged particles in various electrical materials, devices, and circuits. The usual charged particle of interest is the electron, but electronics is not restricted to only electrons. In this part of the book four chapters are devoted to the study of electronics. This, the first, chapter discusses the most important topics of materials and devices. The following three chapters then develop various circuit applications that use the devices.

Most people have some notion of what electronics is. Older readers may associate electronics with vacuum tubes, including diodes, triodes, pentodes, and various microwave tubes such as klystrons, magnetrons, traveling-wave tubes, and others. Many of these older devices made possible our early radio, television, and computer systems. Others are still around because they satisfy needs that cannot yet be met by more modern devices. For example, most high-power microwave radio, radar, and communications systems continue to use microwave vacuum-based tubes because of their high-power capabilities.

Most younger readers will associate electronics with the more recent products such as stereo amplifiers, compact-disk audio systems, solid-state radios and televisions, satellites, hand-held calculators, electronic typewriters, laser printers, desktop and large-scale digital computers, push-button, cellular, and other portable telephones, digital timepieces, video cassette recorders, electronic bank tellers, radar speed detectors, special automobile monitoring devices, and many more. Most of these products were made possible by the invention of the transistor in the late 1940s. In fact, one might say that the modern age of electronics began in about 1950 when the transistor moved out of the laboratory and practical devices became more readily available.

It is obvious that electronics is such a broad subject that a book of this scope cannot hope to cover even a small fraction of all applications. However, of the varied modern applications almost all are based on just a few key solid-state components, namely diodes and several types of transistors. Familiarity with these devices forms the basis for constructing most of the larger-scale devices and systems mentioned in the foregoing paragraph. We study diodes and transistors in this chapter and then apply them to several broad categories of system problems in the following three chapters.

6.1 ELEMENTARY PROPERTIES OF MATERIALS

All materials are made up of atoms. An atom consists of a positively charged nucleus around which a number of electrons orbit in elliptical paths. The electrons arrange themselves in *shells*, or rings. The number of shells depends on the number of electrons. A silicon (abbreviated Si) atom, for example, has a total of 14 electrons in orbit; 2 are in the innermost shell, 8 are in the second shell, and 4 are in the outer shell. Electrons in the outer shell of an atom are called *valence electrons*.

Atoms in a pure solid element arrange themselves in a precise pattern called a lattice, which on a larger scale becomes a *crystal* of the element's bulk material. The valence electrons determine the shape of the lattice within the crystal. A given atom can have several nearest-neighboring atoms with which valence electrons are shared in the form of *covalent bonds*. These bonds hold the lattice together.

Conductors and Insulators

The ability of valence electrons to move about in a material determines the classification of the material as an insulator, semiconductor, or conductor. In an insulator the valence electrons are all so tightly bound in the lattice that the application of stimuli such as heat or electric fields does not impart enough energy to cause them to break their covalent bonds and cause any significant conduction of current. Diamond, which is carbon in its crystalline form, is a good example of an element that is an insulator. Other materials that are good insulators and probably well known to the reader are many plastics and ceramics.

At the other extreme from the insulator is the conductor. In a conductor the valence electrons are so loosely associated with nuclei that they are able to move freely about within the material in response to an electric force such as an electric field. These electrons are called *conduction electrons*. Metals such as silver, copper, aluminum, gold, and platinum are conductors.

Semiconductors

Materials with conductivities too large to be classified as insulators (above about 10^{-6} S/m) or too small to be called conductors (conductivity below about 10^5 S/m) are classified as *semiconductors*. Semiconductors are the materials from which transistors and other solid-state devices are made today and are vital to the electronics industry. In the earliest days of the transistor germanium was the semiconductor used in fabrication. By the early 1960s, however, germanium was beginning to be displaced by silicon, which had superior temperature characteristics and was more economical (silicon comprises about 25% of the earth's crust in the form of silica and silicates and is second only to oxygen in abundance). At the present time *compound semiconductors*, especially gallium arsenide (GaAs), are receiving considerable research attention and have been used mainly in microwave and photonic applications. As the technology of these materials advances, we may begin to see more displacement of the silicon-based devices by compound semiconductors. For the present, however, silicon is the most important semiconductor for transistor construction and we shall emphasize mainly this material in the following work.

Silicon atoms have four valence electrons that are shared with four nearest-neighbor atoms located on the corners of a tetrahedron, as shown in Fig. 6.1-1(a).

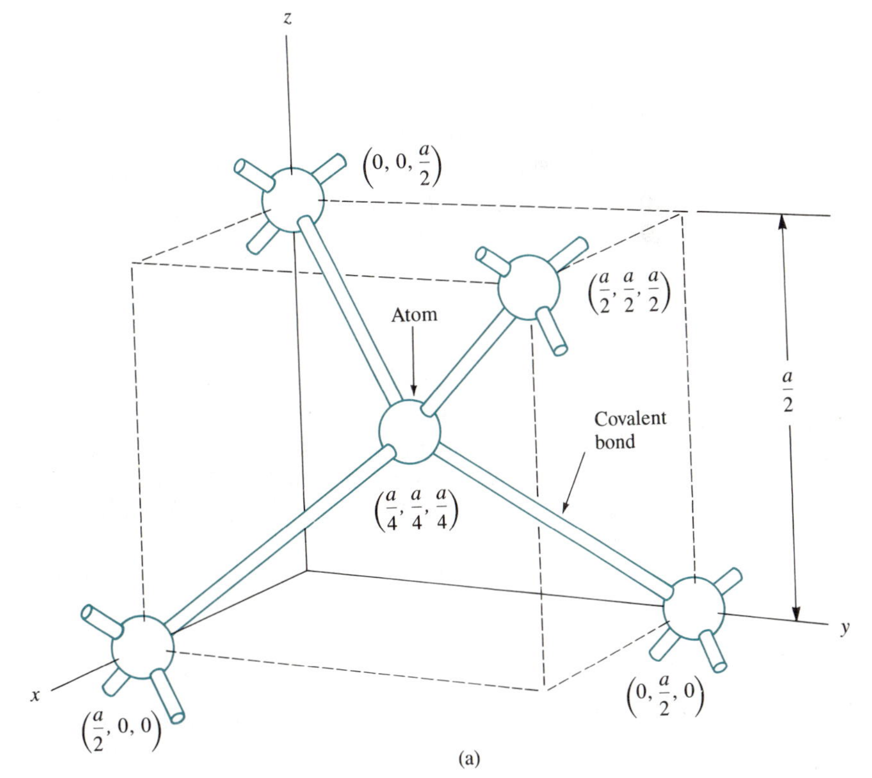

FIG. 6.1-1. Lattice representations for silicon: (a) three-dimensional and (b) (p. 170) planar.

FIG. 6.1-1.
(Continued)

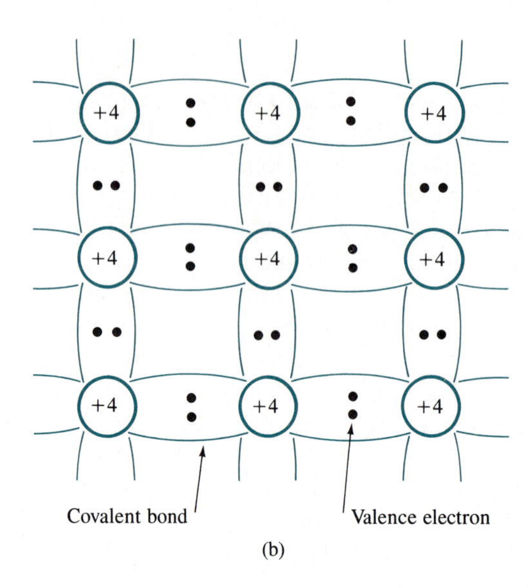

(b)

The constant a is called the *lattice constant* and equals 5.43×10^{-10} m for silicon at 300 K. Covalent bonds between atoms are illustrated by connecting links. Another lattice representation is shown in Fig. 6.1-1(b). Here valence electrons are depicted as heavy dots, and covalent bonds between pairs of atoms are shown by dashed lines. Two electrons, one from each atom of a pair, contribute to a bond. The remainder of an atom is illustrated by a circle with charge +4, which refers to 4 times the magnitude of the charge of an electron; this charge maintains the overall balance of charge in the atoms.

In silicon a *unit cell*, which is representative of the entire lattice, is a cube of length a per side. This cell is an interconnection of four complete tetrahedrons arranged such that there is one atom on every corner and one at the center of every face of the cubic cell.

EXAMPLE 6.1-1

We find the density of atoms in pure silicon at 300 K. There are four atoms internal to a unit cell (four complete tetrahedrons). There are eight at the corners, of which each is shared by eight adjacent unit cells. There are also six in the face centers, each of which is shared between two unit cells. The number of atoms per unit cell is, therefore, $4 + (8/8) + (6/2) = 8$. Since $a = 5.43 \times 10^{-10}$ m for silicon, it has

$$\frac{8}{a^3} = \frac{8 \text{ atoms}}{(5.43 \times 10^{-10} \text{ m})^3} \approx 50 \times 10^{27} \text{ atoms per cubic meter}$$

Almost all valence electrons in silicon occupy covalent bonds and cannot easily move about in the material. In fact, at a temperature of absolute zero (0 K)

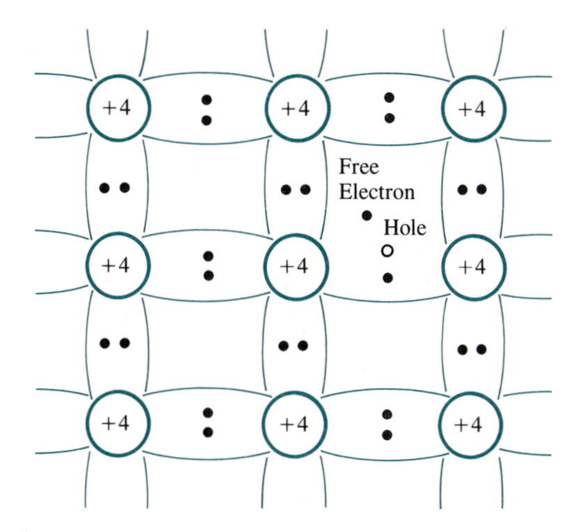

FIG. 6.1-2. The representation of a broken covalent bond in a silicon crystal.

they are unable to move at all, and silicon is an insulator. At any temperature above absolute zero some valence electrons can attain enough thermal energy to break a covalent bond and become a free charge for conduction. The concentration (density) of these free carriers increases exponentially with temperature. These comments are illustrated by Fig. 6-1.2. Under influence of an electric field the electron can move through the material to form a current. The position it vacates in the valence bond is called a *hole*. If some other valence electron is able to break its bond, but has insufficient energy to stay a free charge, it can move over to fill the hole. The effect is to move the *hole* to the electron's previous position. Because the atom at which the hole resides has a net positive charge (with a magnitude equal to the electron's charge), the movement of a hole represents a conventional current (opposite to electron flow). Total current in the material is the sum of the free-electron current and the hole current. For a given electric field the current is small in relation to conductors because, after all, silicon is still a semiconductor (conductivity is about 4.35×10^{-4} S/m for pure silicon). In an equilibrium condition, electron-hole pairs are removed by *recombination* at the same rate at which they are generated.

Doped Semiconductors

Through a process called *doping* the conductivity of a semiconductor can be greatly increased and precisely controlled. Doping consists in adding a small amount of either a *trivalent element* (such as boron, gallium, or indium) or a *pentavalent element* (such as antimony, phosphorus, or arsenic) to the semiconductor. The added element is referred to as an *impurity* that changes the *intrinsic* (pure) semiconductor to an *extrinsic* (impure) material.

Trivalent impurities alter the semiconductor's lattice as described by Fig. 6.1-3(a). An atom of the impurity has only three valence electrons and can form only three complete bonds in the lattice. The fourth bond is incomplete, or

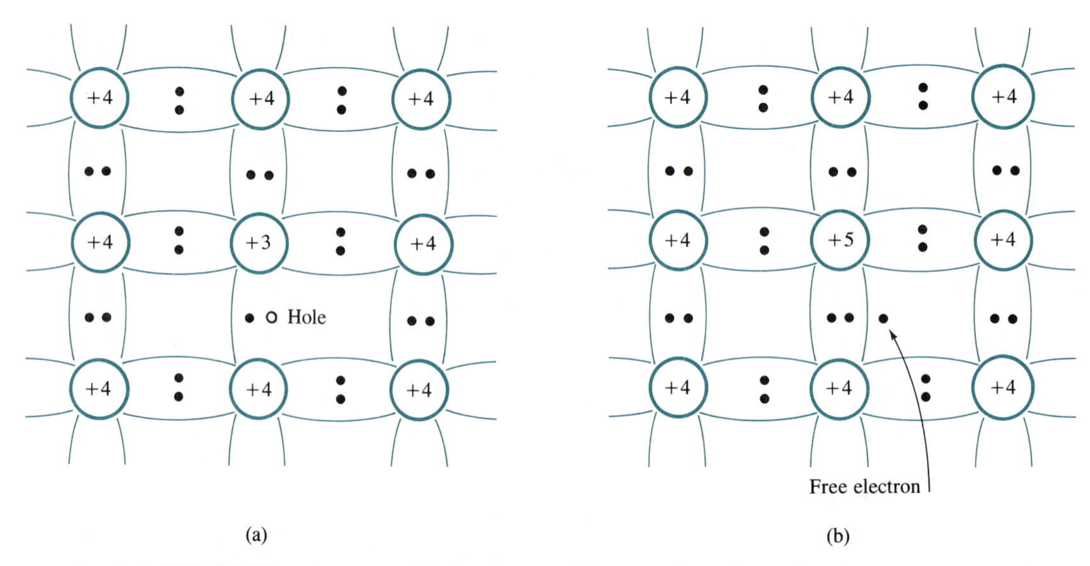

FIG. 6.1-3.
Silicon lattices with (a) trivalent impurity to create a p-type semiconductor and (b) pentavalent impurity to form n-type material.

broken, due to the hole. Holes form the *majority carriers* and electrons are the *minority carriers* in trivalent doping, and the extrinsic material is called *p-type* (p for positive majority carrier). The doping substance is called an *acceptor*.

The lattice structure near a pentavalent atom in a semiconductor is shown in Fig. 6.1-3(b). Here all four bonds are complete and one electron is left free. The free electrons become the majority carriers and the extrinsic semiconductor is called *n-type*. The doping substance is known as a *donor*. Holes are the minority carriers of current in n-type semiconductors.

EXAMPLE 6.1-2

The conductivity σ and resistivity ρ of n-type silicon are related to the donor concentration n (electrons per cubic meter) and *electron mobility* μ_n ($m^2/V\cdot s$) by

$$\sigma = \frac{1}{\rho} = qn\mu_n \qquad \text{S/m}$$

where $q = 1.602 \times 10^{-19}$ C is the magnitude of the charge on an electron. Mobility varies with temperature and donor concentration. At 300 K and a donor concentration of 10^{23} atoms per cubic meter, $\mu_n \approx 0.073$ $m^2/V\cdot s$. We compute conductivity to be

$$\sigma = (1.602 \times 10^{-19})(10^{23})(0.073) = 1169 \text{ S/m}$$

$$\rho = 8.55 \times 10^{-4}\ \Omega\cdot\text{m}$$

This calculation assumes that all donors contribute free-charge carriers. This semiconductor, after doping, is near the low end of the range of conductivity for conductors.

Both *p*- and *n*-type semiconductors are vitally important in solid-state-device technology because transistors, diodes, and other devices depend on the characteristics of a *pn junction* formed when the two materials are joined together as a single crystal. There are several ways in which such junctions may be created. However, we shall not be too concerned about manufacturing techniques. Rather, we shall concentrate on understanding junction operation. A single *pn* junction with appropriate contacts for connecting the junction to external circuits is called a *semiconductor- (pn-) junction diode*.

6.2 SEMICONDUCTOR-JUNCTION DIODES

Figure 6.2-1(a) depicts two pieces of semiconductor crystal, one *n*-type and one *p*-type. Each crystal has a metallic contact, called an *ohmic contact*, that allows a wire to be connected to an external circuit. Both crystals are electrically neutral. In Fig. 6.2-1(b) the two are joined at the junction to form a single crystal.

The *pn* Junction without Bias

To describe what exists near the junction, we imagine what occurs immediately after the junction is formed.† Because the concentration of holes is high in the *p* region, they diffuse across the junction and combine with free electrons in the *n* material near the junction. This combination neutralizes some electrons and "uncovers" bound positive charges in the *n* region. In a similar manner, some electrons in the *n* region diffuse across into the *p* material and combine with holes near the junction to "uncover" bound negative charges. The total *diffusion current* is the sum of the hole and electron currents and is in the direction toward the *n* side from the *p* side. As these majority carriers diffuse across the junction and uncover more and more bound charges, there develops a region near the junction that becomes depleted of majority carriers. It is called the *depletion region*. Because of the charge distribution through the depletion region, there are developed an electric field and a potential, denoted by V_0, across the junction.

---

† Obviously, junctions are not formed abruptly at an instant in time in a manufacturing process. Our descriptive approach is simply a plausible way to explain how the junction conditions evolve.

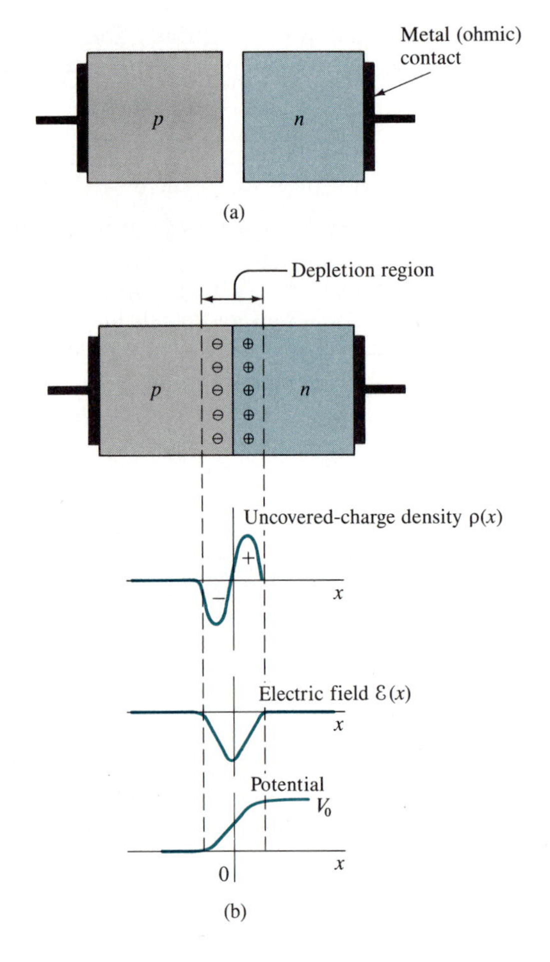

FIG. 6.2-1. (a) Materials of *p*- and *n*-type needed to form the *pn* junction of (b).

This voltage, called the *contact potential* or *barrier voltage*, acts to reduce the diffusion of majority carriers.

The barrier voltage has the opposite effect on minority carriers. Minority holes generated in the *n* region near the depletion region are acted on by the electric field and swept into the *p* region. Similarly, minority electrons from the *p* side are swept into the *n* region. These two minority-carrier currents add to give the *drift current*, which is in the direction toward the *p* region from the *n* material. Since the drift and diffusion currents are in opposite directions, they must be equal in magnitude at equilibrium, because the crystal has no external path in which a net current can flow.

For a *pn* junction at thermal equilibrium there is, in fact, no external current flow, even if an external path is available to support current as long as the path has no sources. If this were not a fact, a current would heat the external circuit (wires), and the required energy would have to be provided by the crystal, which must therefore release energy and cool down. Simultaneous heating of the exter-

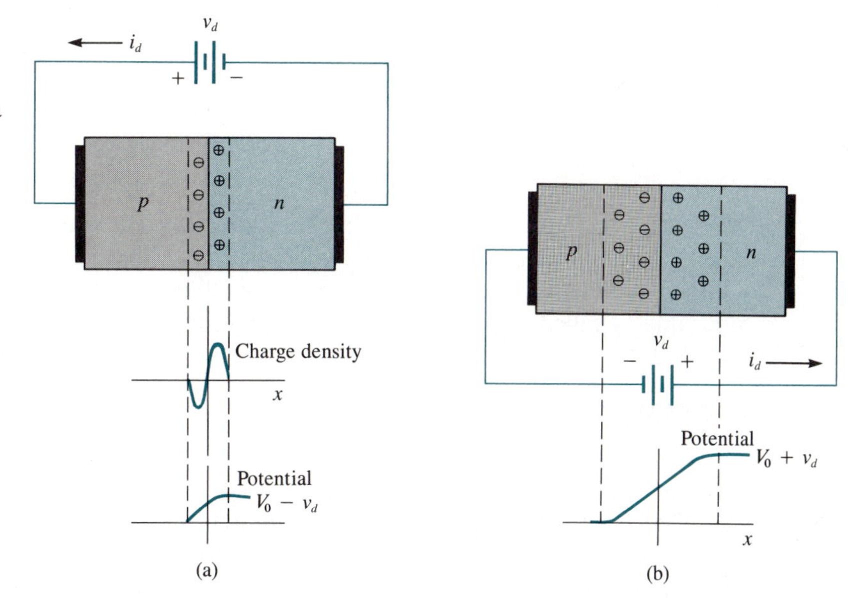

FIG. 6.2-2. (a) A forward-biased *pn* junction and (b) a reversed-biased *pn* junction.

nal circuit and cooling of the crystal are not possible under thermal equilibrium, so the current must be zero. From a simple KVL equation around the loop we conclude that zero current occurs because the junction's contact potential is exactly compensated by contact potentials at the ohmic contacts. As a consequence, no open-circuit voltage can occur across the crystal between the ohmic contacts, and the junction's barrier voltage cannot be measured by a voltage meter, no matter how perfect or ideal the meter is.

The *pn* Junction with Bias

When an external voltage, often called a *bias*,† v_d is applied across the crystal such as to make the *p* region positive with respect to the *n* region, the junction is *forward-biased*. As shown in Fig. 6.2-2(a), the barrier voltage is reduced by the applied voltage v_d. The effect is to cause a diffusion current of majority carriers (holes from *p* to *n* regions and electrons from *n* to *p* regions) to flow. This current increases exponentially with v_d and can be quite significant for v_d only slightly above about 0.6 V for silicon.

The opposite occurs when the polarity of the external voltage is reversed, as illustrated in Fig. 6.2-2(b). The barrier voltage is raised, majority-carrier current quickly (exponentially) becomes negligible, and the principal current is due to minority carriers (electrons going from *p* to the *n* regions, holes from *n* to *p*

† Our lower-case notation implies that v_d and i_d are total quantities that may include time-varying as well as dc components. For purposes of this discussion v_d and i_d can be viewed as dc quantities.

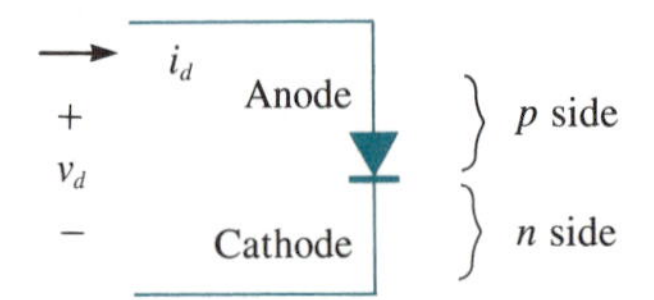

FIG. 6.2-3. The symbol for a *pn*-junction diode.

materials). This drift current quickly saturates (becomes constant) at a very low level, because minority carriers are generated only through thermal agitation and at the usual operating temperatures are in small numbers. The small current that does flow is called the *reverse saturation current*, denoted by I_0; the *pn* junction is said to be *reverse-biased* to cause this current.

Junction Diode

A *pn* junction with ohmic contacts forms a *junction diode*, and the preceding discussions define its behavior. Figure 6.2-3 defines the circuit symbol commonly accepted for a diode. The *p* side of the junction is called the *anode*; the other side is the *cathode*. If a voltage v_d is applied across the diode with the polarity shown, the current i_d that flows in the direction shown is accurately described by

$$i_d = I_0(e^{qv_d/\eta kT} - 1) \tag{6.2-1}$$

where $q = 1.602 \times 10^{-19}$ C is the magnitude of the charge of one electron, $k = 1.381 \times 10^{-23}$ J/K is *Boltzmann's*† *constant*, and T is absolute temperature (in kelvin). The current I_0 is called the *reverse saturation current*. It is usually not more than a few nanoamperes for low-power silicon diodes. The quantity η is called the *ideality factor*. It is a number from 1 to 2 that depends on the semiconductor material, manufacturing methods, and even to some extent on the current that flows. For germanium and silicon the values of η are often taken as 1 and 2, respectively. For diodes in integrated circuits η is near 1. Figure 6.2-4 illustrates the behavior of (6.2-1) for some assumed parameters. These curves indicate that current is very small until v_d reaches a *threshold* or *cut-in voltage*, which we denote as V_γ. For $v_d > V_\gamma$ current abruptly and dramatically increases. For germanium where $\eta = 1$, the cut-in voltage is about 0.2 to 0.3 V, and current increases rapidly for $v_d > 0.2$ V. For silicon where $\eta = 2$, current also rises rapidly as voltage exceeds the cut-in value of about 0.6 to 0.7 V.

Cut-in voltage is a decreasing function of temperature. For constant current a rough rule of thumb for silicon is that V_γ decreases about 2 mV/K.‡ Temperature also affects I_0. It is found experimentally that I_0 approximately doubles for every 10-K rise in temperature.

An *ideal diode* would have zero reverse-biased current and would conduct abruptly for $v_d > 0$ with a current limited only by the external circuit's impedance.

---

† Named for Ludwig Boltzmann (1844–1906), an Austrian physicist who was a founder of modern statistical physics.

‡ In terms of °C this constant is 2.0 mV/°C, since K = °C + 273.15.

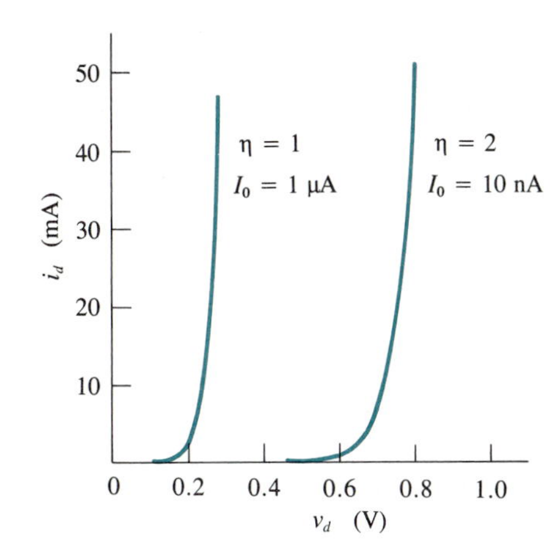

FIG. 6.2-4. Diode characteristic curves for T = 300 K.

As found in Fig. 6.2-4, these characteristics are only approximated in real diodes. All real diodes have other limitations, too. For example, the forward current is limited to a maximum specified value. To exceed the value may cause diode burnout. Reverse biases are also limited as indicated in Fig. 6.2-5. If v_d is more negative than a value $-V_Z$, called the *breakdown voltage*, the diode fails to operate properly and a large negative current occurs. If this current is allowed to be large enough, burnout occurs. Most diodes are not intended to be operated in the breakdown region. An exception is the zener diode.

Zener Diode

Through control of semiconductor processes a junction diode can be purposely designed to have a nearly constant voltage in the breakdown region for a large range of reverse current. Called *zener diodes*, these devices are intentionally operated in the breakdown region.

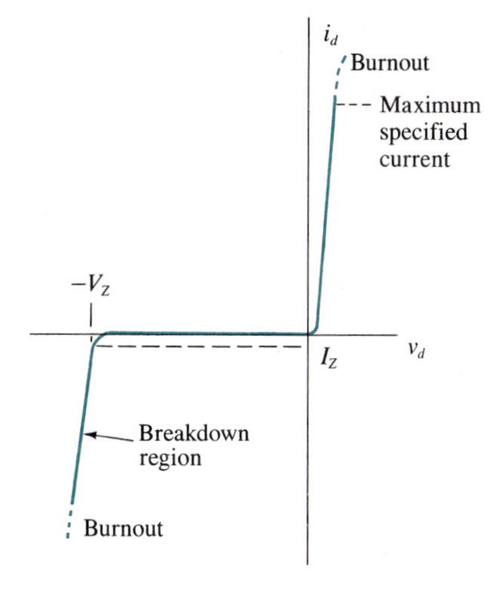

FIG. 6.2-5. Extreme voltage-current characteristics of a junction diode.

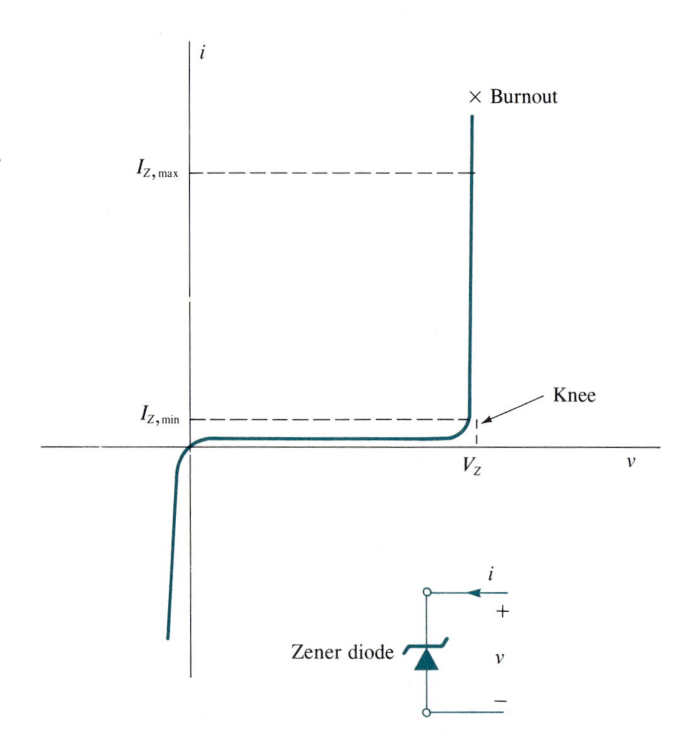

FIG. 6.2-6. The circuit symbol and the current-voltage characteristic for a zener diode.

Since the principal operating region for a zener diode is the negative (both voltage and current) of that for a regular diode, we shall define reference directions (both voltage and current) for the zener diode to be opposite to those of the normal diode. Figure 6.2-6 illustrates the circuit symbol for a zener diode along with its current-voltage characteristic. The voltage at which the onset of breakdown occurs is called the *zener voltage*, denoted by V_Z, which defines the *knee* of the curve in the figure.

One of the main uses of a zener diode is in a regulator circuit to maintain a constant voltage for a load. One simple circuit is given in Fig. 6.2-7. We may analyze this circuit in several ways. We give one to emphasize the use of a zener diode. We shall assume that the diode is *ideal* in that currents between $I_{Z,\min}$ and $I_{Z,\max}$ correspond to a truly *constant* voltage V_Z.

Generally, the regulator of Fig. 6.2-7 attempts to maintain the load voltage equal to V_Z for a source voltage $V_{s,\min} \le V_s \le V_{s,\max}$, even under variations in load resistance R_L. The result can be achieved provided that the zener's current

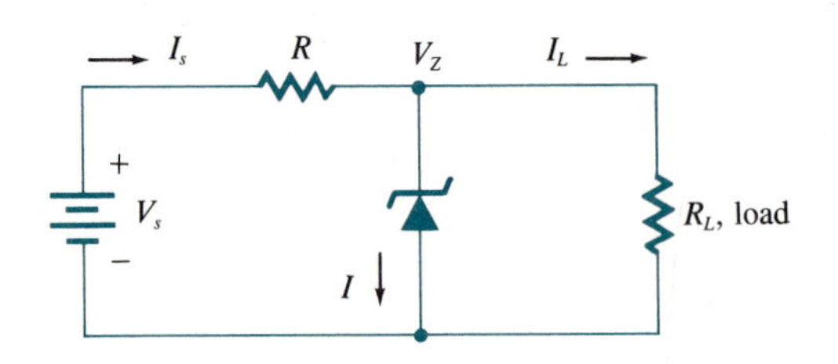

FIG. 6.2-7. A voltage regulator using a zener diode.

always falls between its allowable extremes $I_{Z,\min}$ and $I_{Z,\max}$. The largest diode current will occur when the load is open-circuited and the voltage is maximum:

$$I = I_s = \frac{V_{s,\max} - V_Z}{R} \leq I_{Z,\max} \tag{6.2-2}$$

so we require

$$R \geq \frac{V_{s,\max} - V_Z}{I_{Z,\max}} \tag{6.2-3}$$

The smallest diode current occurs for the largest load current and the smallest source voltage:

$$I = I_s - I_L = \frac{V_{s,\min} - V_Z}{R} - \frac{V_Z}{R_{L,\min}} \geq I_{Z,\min} \tag{6.2-4}$$

so

$$R_{L,\min} \geq \frac{V_Z R}{V_{s,\min} - V_Z - RI_{Z,\min}} \tag{6.2-5}$$

is required.

EXAMPLE 6.2-1

We select R and find the smallest load resistance allowed in Fig. 6.2-7 when V_Z = 12 V and the source is 25 V ± 20%. We shall assume a maximum desired diode current of 20 mA and a minimum of 1 mA. Here $V_{s,\max}$ = 30 V and $V_{s,\min}$ = 20 V. From (6.2-3)

$$R \geq \frac{30 - 12}{20 \times 10^{-3}} = 900\ \Omega$$

From (6.2-5)

$$R_{L,\min} \geq \frac{12(900)}{20 - 12 - (900 \times 10^{-3})} = 1521\ \Omega$$

In practice, the nearest resistor to 900 Ω is 910 Ω, which, if used instead of 900 Ω, would correspond to a load $R_{L,\min} \geq 1540\ \Omega$.

Practical zener diodes can be obtained with zener voltages from a few volts to several hundred volts at power ratings up to tens of watts. The breakdown region of a practical diode is also not flat (not parallel to the current axis in Fig. 6.2-6). There is a finite positive slope. At a given operating (nominal) point the reciprocal of the slope is called the *dynamic resistance*, denoted by R_Z,

$$R_Z = \frac{dv}{di} \tag{6.2-6}$$

The resistance R_Z will vary as a function of the operating point. For large currents R_Z may be only several ohms. At small currents R_Z may be several hundred ohms. As current approaches the diode's knee, R_Z may become quite large.

Equation (6.2-6) indicates that a small change in current (Δi) from an initial operating point corresponds to a change in diode voltage (Δv) of

$$\Delta v = R_Z \, \Delta i \tag{6.2-7}$$

This change in v relates directly to the practical voltage regulation abilities of a given diode.

A useful equivalent circuit for a zener diode is a source of voltage V_Z in series with a resistance R_Z. The equivalent model is valid for operation in the breakdown region at currents near the operating value for which R_Z is defined through (6.2-6).

6.3 JUNCTION DIODE MODELS

So that we can use junction diodes in practical circuits, it is helpful to define some additional characteristics and develop some circuit models.

Diode Resistances

Let a diode have a current flow I_Q when its voltage is V_Q, as shown in Fig. 6.3-1(a). We call point Q the *operating point*. For purposes of dc circuit analysis the diode *static resistance*

$$R_s = \frac{V_Q}{I_Q} \tag{6.3-1}$$

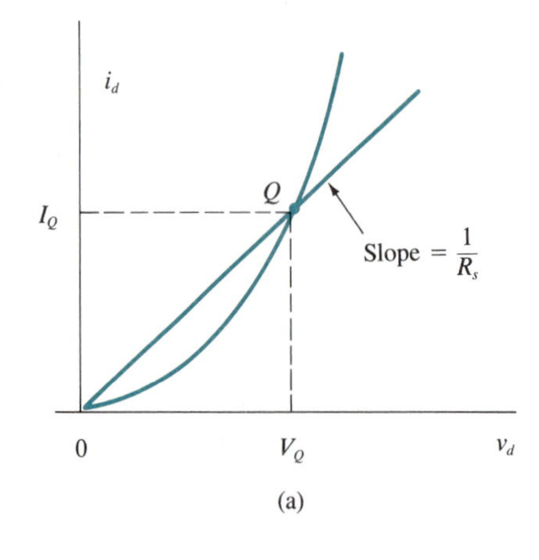

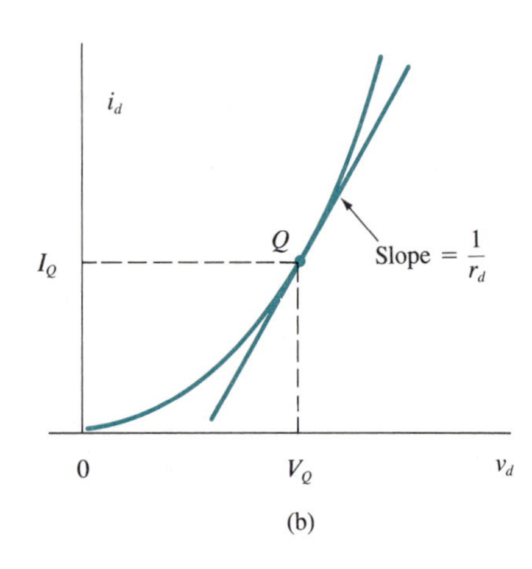

FIG. 6.3-1. Diode characteristics defining (a) static resistance R_s and (b) dynamic resistance r_d.

is useful. However, for most other applications the static resistance has little value.

A more useful diode quantity is its *dynamic resistance*, defined as the reciprocal of the slope of the diode's current-voltage characteristic at the operating point. Dynamic resistance, denoted by r_d, is illustrated in Fig. 6.3-1(b). After differentiation of (6.2-1) to obtain the slope, we have

$$r_d = \frac{1}{\partial i_D/\partial v_d}\bigg|_Q = \frac{\eta kT}{q(I_Q + I_0)} \tag{6.3-2}$$

Since r_d varies as the reciprocal of diode current, it is small for voltages well above cut-in and large for operation below cut-in.

EXAMPLE 6.3-1

Assume a silicon diode for which Fig. 6.2-4 applies. We shall find r_d at cut-in ($v_d \approx 0.6$ V) and at $v_d = 0.8$ V. Since $\eta = 2$, $T = 300$ K, and $I_0 = 10^{-8}$ A are assumed, then $q/\eta kT = (1.602 \times 10^{-19})/[2(1.381 \times 10^{-23})(300)] \approx 19.33$. Equation (6.2-1) gives

$$I + I_0 = I_0 \exp\left(\frac{qv_d}{\eta kT}\right) = (10^{-8}) \exp 19.33 v_d$$

For $v_d = 0.6$ V and $v_d = 0.8$ V:

$$r_d = \frac{1}{(19.33 \times 10^{-8}) \exp[19.33(0.6)]} \approx 47.5\ \Omega$$

$$r_d = \frac{1}{(19.33 \times 10^{-8}) \exp[19.33(0.8)]} \approx 1.0\ \Omega$$

It is interesting to observe that if v_d is only 25% below cut-in ($v_d = 0.45$ V), r_d rises to 863 Ω, or over 18 times larger than at cut-in.

Diode Capacitances

Two types of capacitances are associated with a *pn* diode: the *junction capacitance* (also known as depletion capacitance, space-charge capacitance, and other names) and the *diffusion capacitance*.

With no external bias a *pn* junction has a barrier voltage V_0 that corresponds to a depletion region of uncovered charges that has some width [Fig. 6.2-1(b)]. With a forward external bias v_d the barrier potential decreases to $V_0 - v_d$, and the depletion region decreases in width [Fig. 6.2-2(a)]. The smaller width means that fewer uncovered charges are in the depletion region. The change in charge caused by a change in voltage can be taken as a capacitive effect. This interpretation is also supported by what happens with an externally applied reverse bias. Here the barrier potential increases to $V_0 + v_d$ (v_d positive), and the depletion region widens to uncover more charges [Fig. 6.2-2(b)]. The capacitance that accounts for these effects is the junction capacitance C_J.

For a reverse-bias diode C_J is the dominant capacitance, and its value can range from about 1 pF to several hundred picofarads, depending on bias level and details of diode design and manufacture. In most applications it is desirable that C_J be small, but not always. The *varactor diode*, for example, makes use of the variation of C_J with the level of the reverse bias to develop frequency tuning of oscillators. Oscillator frequency can be made proportional to the inverse square root of a capacitance (C_J). Diodes with large C_J can then be used at lower frequencies.

The second type of diode capacitance is called *diffusion capacitance*. It is usually negligible for a reverse-biased diode and is most significant for the forward-bias condition. Under a forward bias holes from the p region diffuse across the depletion layer and are injected into the n region, where they combine with electrons as they move away from the depletion-layer boundary. Although the concentration of holes decreases exponentially with distance, their presence corresponds to a positive charge near the n-region edge of the depletion region. A similar situation occurs with electrons injected into the p region, where they correspond to negative charges. Since these charges are voltage-sensitive, they are accounted for by the diffusion capacitance, denoted by C_D. It is found that C_D is proportional to diode current:

$$C_D = K_D I \tag{6.3-3}$$

where K_D is a constant. Under forward bias and heavy current C_D can become very large compared with C_J (typically in the picofarad range).

Diode Models

Various diode models may be defined that depend on the degree to which the real diode must be approximated and on its application. For applications where currents and voltages change slowly enough that diode capacitances are of no concern, the models of Fig. 6.3-2 are useful. The model and current-voltage (i_d-v_d) characteristic for an ideal diode are shown in Fig. 6.3-2(a). An ideal diode is simply a switch that is closed with zero resistance when forward-biased and is open with infinite resistance when reverse-biased. The ideal-diode model can be improved, as shown in Fig. 6.3-2(b), by adding a voltage to prevent conduction until the bias exceeds the cut-in voltage V_γ.

A third simple model is called the *piecewise-linear model*. It is illustrated in Fig. 6.3-2(c) and is an extension of the model of (b) to allow for more realistic i_d-v_d slopes. For $v_d < V_\gamma$ the diode's *reverse resistance*, denoted by r_r, is given by (6.3-2) when evaluated at a convenient value of v_d. It could be $r_r = r_d$ when $v_d = 0$, for example. When $v_d > V_\gamma$, the diode's *forward resistance*, denoted by r_f, is again given by (6.3-2), but the evaluation is at a convenient point where $v_d > V_\gamma$.

For applications where diode capacitance is important, the small-signal equiv-

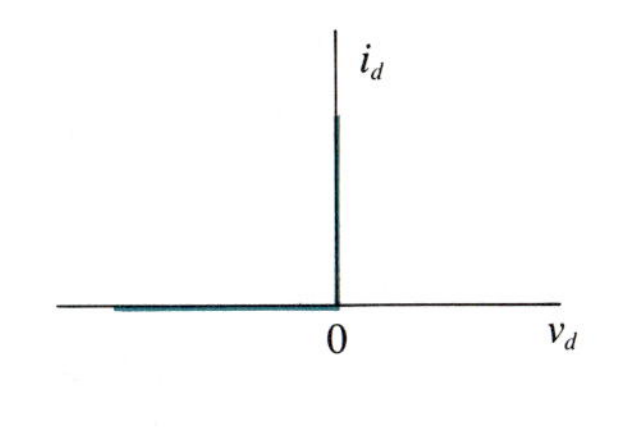

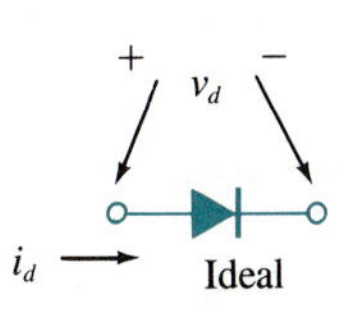

(a)

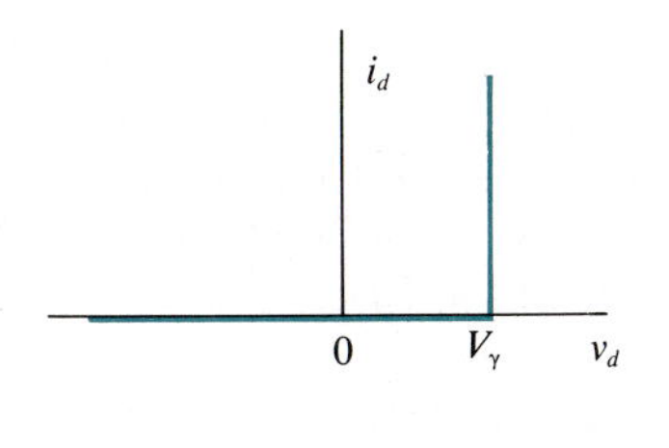

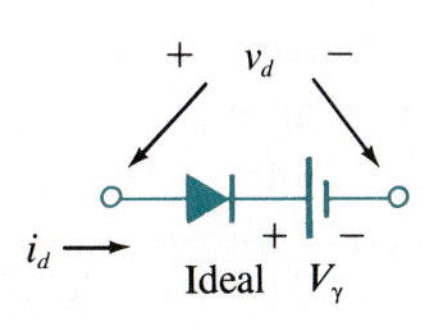

(b)

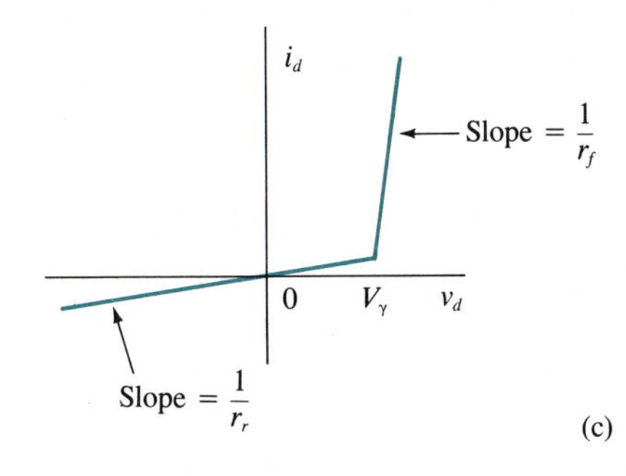

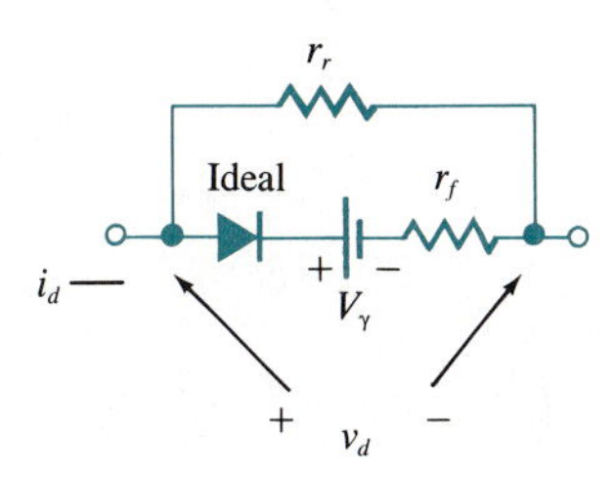

(c)

FIG. 6.3-2. Simple diode models: (a) ideal, (b) ideal with cut-in voltage taken into account, and (c) linear piecewise.

alent circuit under back-biased operation is r_r in parallel with C_J. For forward bias it is the parallel combination of r_f, C_J, and C_D.

6.4 ELEMENTARY DIODE CIRCUITS

Semiconductor diodes are used in a wide variety of applications. They are found in nearly every electrical system, such as radio (mixers, automatic gain-control circuits, message detectors), radar (phase detectors, gain-control circuits, power detectors, parametric amplifiers), computers (clamps, clippers, logic gates), communications systems (limiters, gates, clippers, mixers), and television (clamps, limiters, phase detectors, etc.). In this section several simple diode circuits are discussed to serve only as examples. Other diode circuits are encountered at many other places in this book.

Load Lines

A simple circuit is shown in Fig. 6.4-1, where it is desired to find the diode's voltage v_d and current i_d. From a voltage equation around the loop

$$v = v_d + i_d R_L \tag{6.4-1}$$

When plotted in coordinates (v_d, i_d), (6.4-1) is a linear function with a slope $-1/R_L$, as shown by the *load line* in Fig. 6.4-2. Clearly, any point (v_d, i_d) on the line satisfies (6.4-1). However, since only one current and one voltage can occur, more information is needed than that provided by (6.4-1). The diode's characteristic of (6.2-1) completes the solution. The actual values of i_d and v_d must satisfy both curves in Fig. 6.4-2, which happens at the intersection of the two at point Q, where $v_d = v_Q$ and $i_d = i_Q$.

The idea of a load line to establish an operating point has broader applications than in the diode circuit demonstrated above. It can be drawn independent of the device's (diode's) characteristic. When the device's characteristic is added, even for more complicated devices such as transistors, the operating point will be the point of intersection.

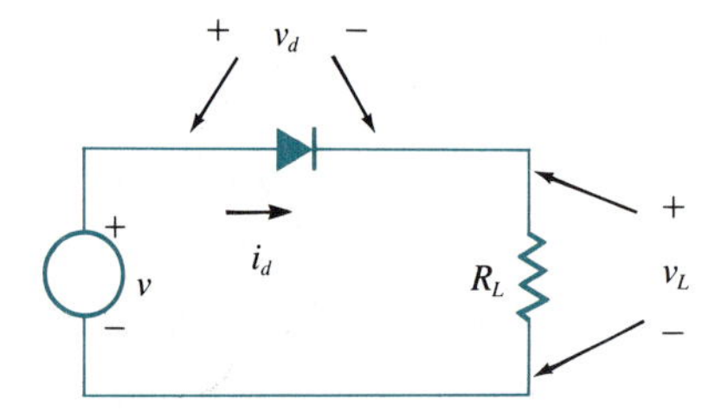

FIG. 6.4-1. A simple diode circuit.

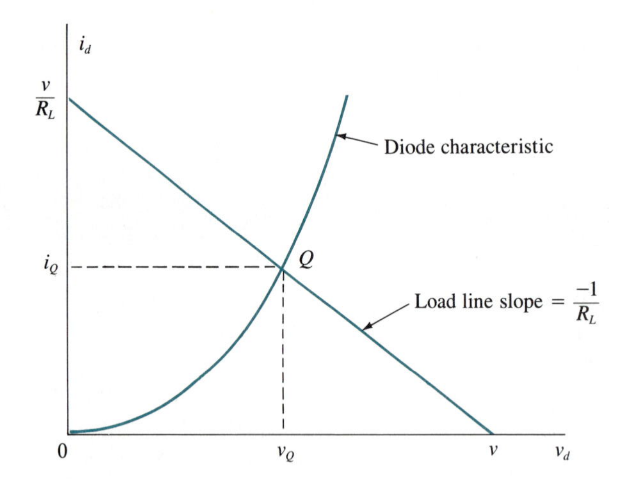

FIG. 6.4-2. Static characteristics that define the operating point of the circuit in Fig. 6.4-1.

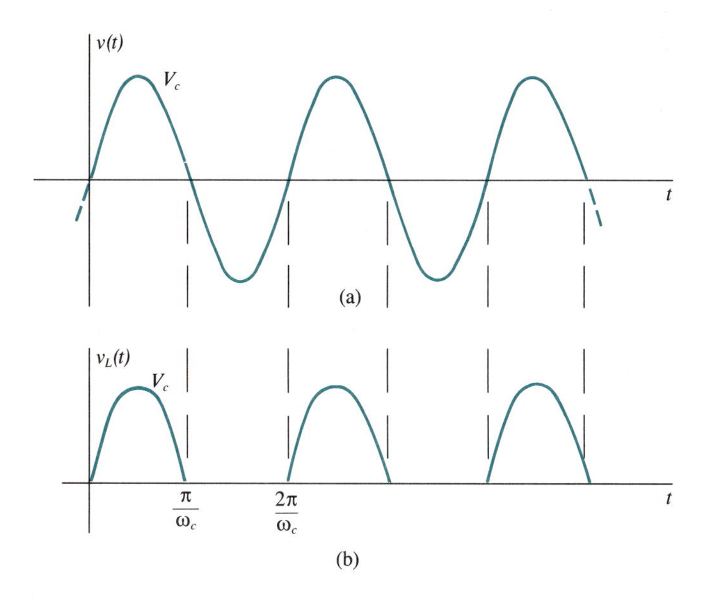

FIG. 6.4-3. Half-wave rectifier waveforms for (a) input and (b) output.

Rectifiers

If the applied voltage v in Fig. 6.4-1 is large compared with the diode's cut-in voltage V_γ, then $v_L \approx v$ whenever $v > 0$, and the ideal-diode model is good enough for most applications. With this assumption the circuit is called a *half-wave rectifier* when the input signal has the form

$$v(t) = V_c \cos \omega_c t \tag{6.4-2}$$

The name derives from the form of the output, or load, voltage $v_L(t)$, as sketched in Fig. 6.4-3. Only the positive half-cycles of the input voltage occur in the output because of the switching action of the diode.

From another viewpoint the half-wave rectifier can be considered as a *clipper* that reproduces the input signal everywhere except when it is negative. Negative voltages are clipped off.

A *full-wave rectifier* uses two or more diodes to pass the positive portions and rectify (invert the negative parts of) an input signal. One of many possible circuits is shown in Fig. 6.4-4(a). It uses a highly coupled transformer to feed two half-wave rectifiers, where each operates on one polarity of the input signal. Their combined outputs occur across the common load resistor R_L. The waveforms are shown in Fig. 6.4-4(b).

Rectifiers are mainly used in power supplies where an ac signal is to be converted to a dc waveform. The dc voltage is realized by passing the rectifier's output through a lowpass filter to remove the *ripple* (ac components). In the simplest case a capacitance C can be placed across the load resistor R_L in Fig. 6.4-4(a). The load's waveform becomes that shown in Fig. 6.4-5. Operation is explained by starting at $t = t_1$. Diode D_1 conducts from t_1 to t_2 and the output follows the input. At $t = t_2$ the input-voltage peak is reached. For $t > t_2$ the input

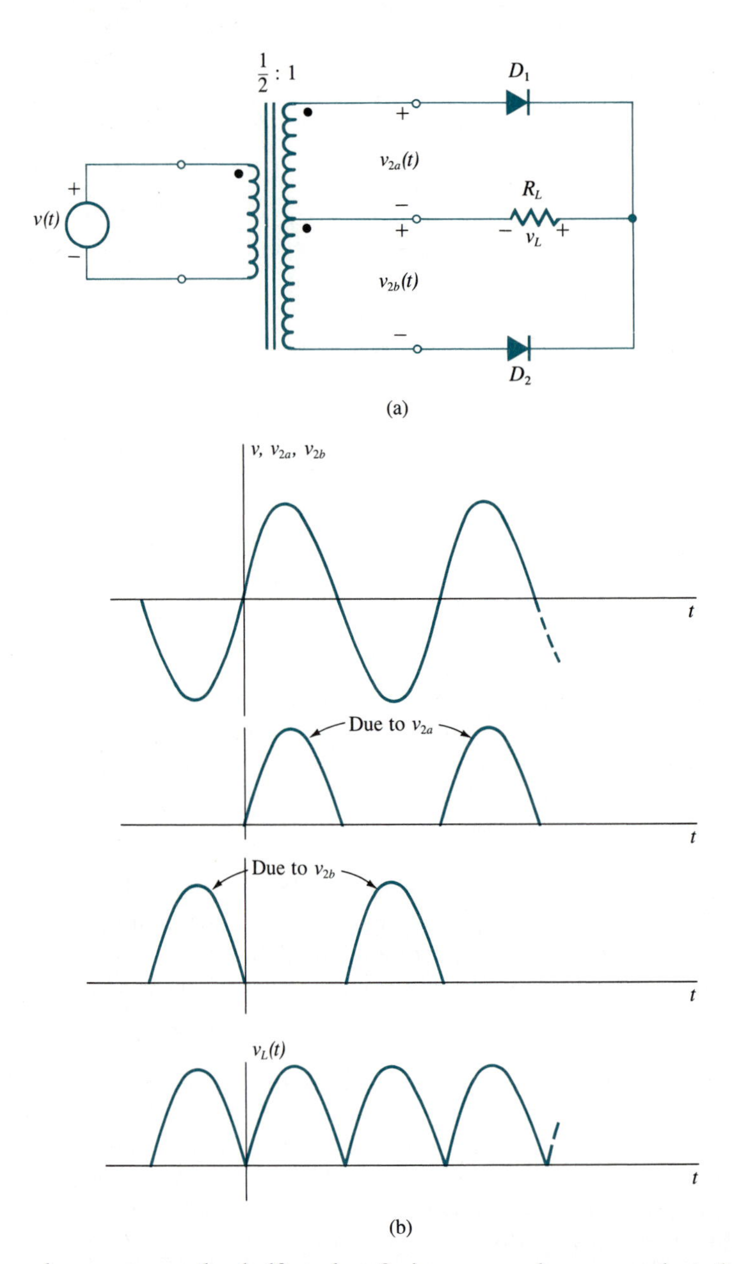

FIG. 6.4-4. (a) A full-wave rectifier and (b) its applicable waveforms.

decreases as the half-cycle of sine wave decreases, but the voltage across C cannot discharge as fast as the sine wave decreases. The result is that $v_L(t)$ back-biases D_1 for $t > t_2$ and the diode cuts off. Diode D_1 remains off until two half-cycles later (third half-cycle). During the second half-cycle D_2 turns on but only after the input voltage exceeds $v_L(t)$ at time $t = t_3$. Diode D_2 conducts from t_3 to t_4 and then cuts off, as diode D_1 did in the first half-cycle interval. The process repeats in all other half-cycle intervals. During times when the diodes are cut off,

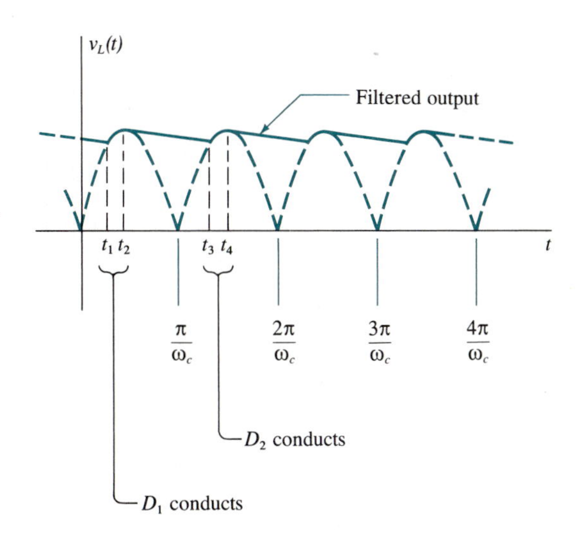

FIG. 6.4-5. The output waveform of the circuit of Fig. 6.4-4(a) when a capacitance C is added across R_L.

the capacitor discharges only through R_L; and $v_L(t)$ decreases exponentially as $\exp(-t/R_LC)$, where R_LC is called the *time constant* of voltage decay. If

$$R_LC \gg \frac{\pi}{\omega_c} \tag{6.4-3}$$

the "droop" or "ripple" in $v_L(t)$ will be very small, and $v_L(t)$ becomes approximately constant.

Signal Detector

A rectifier can also be used as a *signal detector*. We consider the response of the circuit of Fig. 6.4-6(a) to the signal

$$v(t) = [A_c + A_m \cos(\omega_m t)] \cos \omega_c t \tag{6.4-4}$$

Equation (6.4-4) has the form of the *amplitude-modulation* (AM) *signal* broadcast over commercial radio. The bracketed factor is called the amplitude of the signal. The constant ω_c is the station's broadcast angular frequency (called the *carrier frequency*). The unmodulated carrier's peak amplitude is A_c, and $A_m \cos \omega_m t$ represents a sinusoidal audio message of peak amplitude A_m and angular frequency ω_m. The input signal sketched in Fig. 6.4-6(b) produces the response of (c). As long as the decay time constant R_LC is small enough for $v_L(t)$ to follow the most rapid changes in the input signal's amplitude, the output is a reasonable recovery of a dc term A_c plus the message $A_m \cos \omega_m t$. This requirement becomes

$$R_LC < \frac{1}{\omega_{m,\max}} \tag{6.4-5}$$

where $\omega_{m,\max}$ is the maximum value of ω_m.

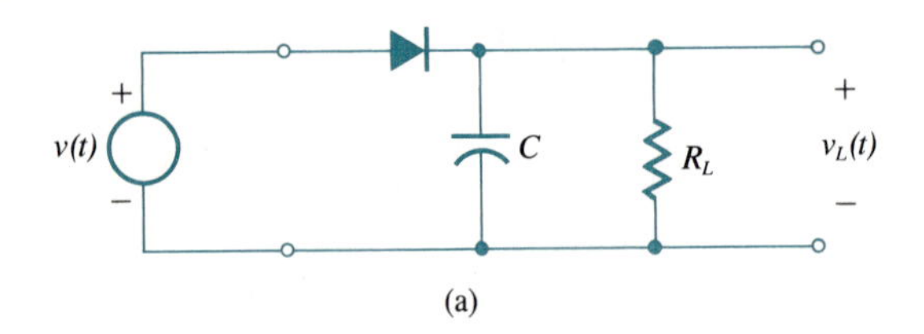

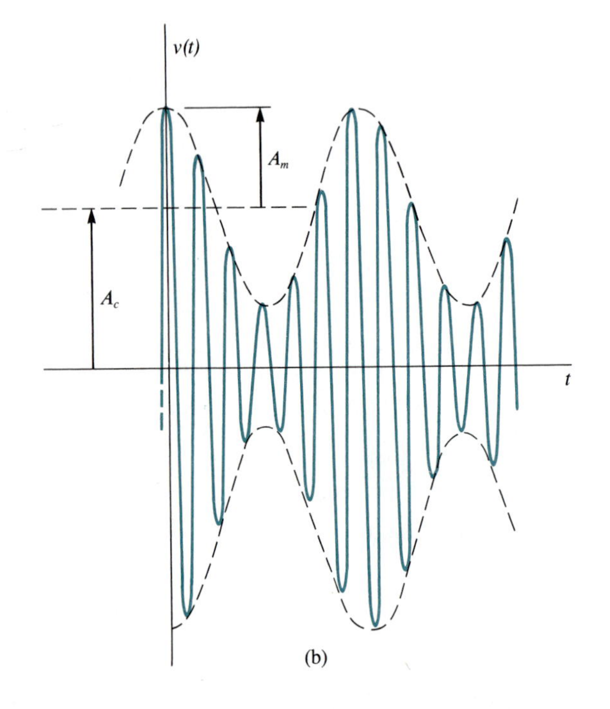

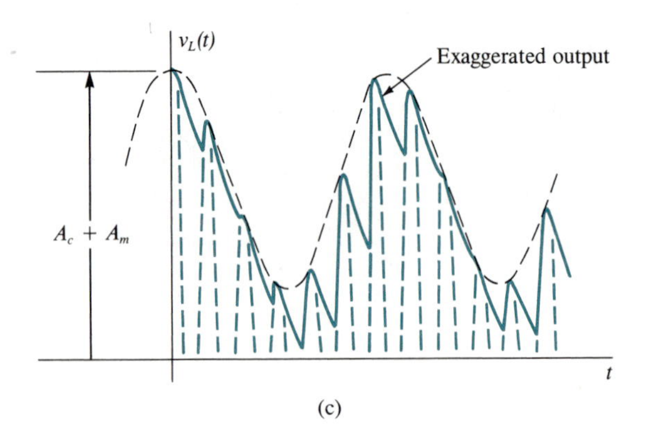

FIG. 6.4-6.
(a) A signal detector for which the input signal of (b) produces the response of (c).

EXAMPLE 6.4-1

In broadcast radio the highest frequency in the message broadcast is about 5 kHz. Thus, $\omega_{m,\max} = 2\pi(5000)$ rad/s. The required receiver (detector) time constant is

$$R_L C < \frac{10^{-4}}{\pi} = 31.83\ \mu\text{s}$$

A value of $R_L = 10\ \text{k}\Omega$ is reasonable, so $C < 3183$ pF is needed.

Of course, $R_L C$ cannot be made too small since it must still satisfy (6.4-3). The combined requirement

$$\frac{\pi}{\omega_c} \ll R_L C < \frac{1}{\omega_{m,\max}} \tag{6.4-6}$$

is usually no problem since $\omega_c \gg \omega_{m,\max}$ in the broadcast system.

The signal detector of Fig. 6.4-6(a) is also known as an *envelope detector* or *amplitude detector*.

6.5 BIPOLAR-JUNCTION TRANSISTOR (BJT)

Where two *pn* junctions are combined to form a single crystal, the resulting three-terminal device is the *bipolar-junction transistor* (BJT). Two realizations (types) are possible, as shown in Fig. 6.5-1(a). They depend on the *npn* or *pnp* order of materials that form the device. The middle region is called the *base* (*B*), one end region is called the *collector* (*C*), and the other end is the *emitter* (*E*). The difference between collector and emitter occurs because the emitter material is more heavily doped than the collector. The base region is typically lightly doped and is designed to be very thin so that the distance between the two junctions is small.

Practical fabrication of transistors can take many forms. Two are illustrated in

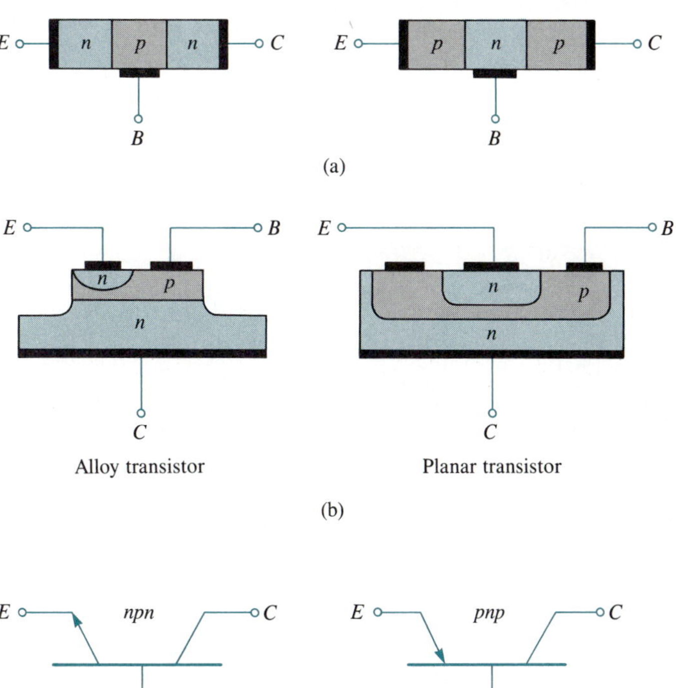

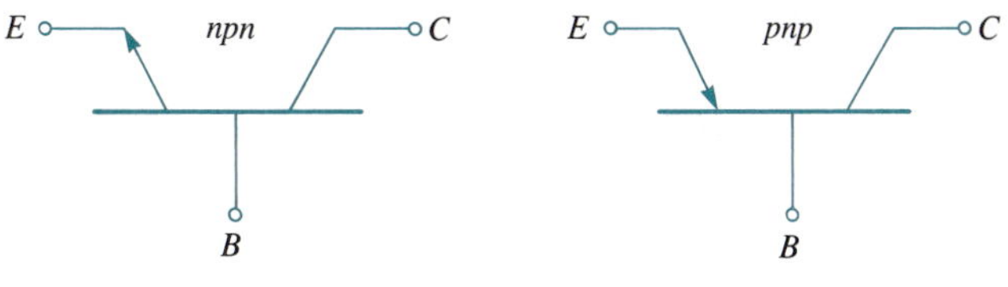

FIG. 6.5-1. (a) Realizations of *npn*- and *pnp*-type transistors, (b) practical transistor constructions, and (c) circuit symbols.

Fig. 6.5-1(b). Circuit symbols for both *npn* and *pnp* transistors are given in Fig. 6.5-1(c). The emitter is identified by the lead having the arrowhead. The arrow points in the direction of conventional emitter current flow when the base-emitter junction is forward-biased.

Device Operation

Figure 6.5-2 is helpful in understanding the operation of a bipolar transistor. We choose to describe an *npn* device. Operation of the *pnp* transistor is similar except for minor changes (holes and electrons for electrons and holes, respectively, and reversal of biases and current directions). A transistor can operate in three modes, cutoff, saturation, and active. We describe the active mode first. As developments proceed, the cutoff and saturation modes will be defined as a natural consequence of the work.

In the active mode the base-emitter junction (BEJ) is forward-biased by a voltage v_{BE}, as shown in Fig. 6.5-2. The collector-base junction (CBJ) is reverse-biased by the voltage v_{CB}. The positive directions of current are chosen to be into the device for base and collector and out of the device for the emitter.

At the forward-biased BEJ current flow is due to majority carriers. Electrons diffuse into the base from the emitter, and holes flow from the base to the emitter. Since the emitter is more heavily doped than the base, the electron flow is by far the more dominant part of the emitter current. These electrons become *minority carriers* in the base region, and because the base is very thin, they are quickly accelerated into the collector by action of the reverse bias on the CBJ. However, while the electrons are traversing the base region, some are removed by recom-

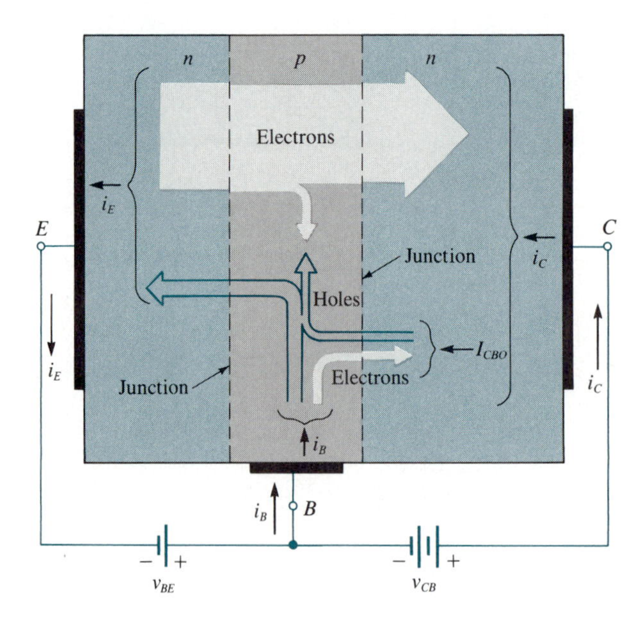

FIG. 6.5-2. The carrier flow in an *npn* transistor.

bination with (majority-carrier) holes. The number lost through recombination is typically less than 5% of the total.

There is also a small current flow across the CBJ due to the usual minority-carrier drift current at a reverse-biased *pn* junction. This current is denoted by I_{CBO} and is called the *reverse saturation current*. Its value is typically less than a few microamperes. It is the collector's current when the emitter is open-circuited, which explains the subscript O.

Transistor Characteristics

It can be shown that the currents in a transistor are approximately given by

$$i_E = I_{SE} e^{v_{BE}/V_T} \tag{6.5-1}$$

$$i_C = \alpha i_E + I_{CBO} \tag{6.5-2}$$

$$i_B = (1 - \alpha)\, i_E - I_{CBO} \tag{6.5-3}$$

Here I_{SE} is the reverse saturation current of the BEJ, α is the fraction of i_E that contributes to collector current and is called the *common-base current gain*, I_{CBO} is the reverse saturation current of the CBJ, and

$$V_T = \frac{kT}{q} \tag{6.5-4}$$

is called the *thermal voltage*. Voltage V_T is the voltage equivalent of temperature and has a value of 25.861×10^{-3} V when $T = 300$ K.

Another important transistor parameter is the *common-emitter current gain*, denoted by β and given by

$$\beta = \frac{\alpha}{1 - \alpha} \tag{6.5-5}$$

The parameter α is approximately constant at a value near, but less than, unity. It typically ranges from about 0.9 to 0.998, so β ranges from about 9 to 500, typically. Changes in β are very sensitive to changes in α.

The foregoing models and equations apply to the ideal transistor. In practice, devices depart somewhat from ideal in several ways, which are best observed from the transistor's *static characteristics*. Figure 6.5-3 illustrates some static characteristics for a hypothetical *npn* transistor. The *emitter characteristic*, as shown in Fig. 6.5-3(a), defines emitter current versus emitter forward bias v_{BE} for several values of v_{CB}. There is some dependence of the i_E-v_{BE} curve on v_{CB}. The dependence is less pronounced for v_{CB} larger than a few volts. The i_E-v_{BE} curve is also sensitive to temperature, and v_{BE} for silicon decreases about 2 mV for each 1 K rise in temperature where i_E is held constant.

The *collector characteristic* of Fig. 6.5-3(b) demonstrates the behavior of collector current versus collector-base voltage for various constant values of i_E. As expected from (6.5-2), collector current is nearly independent of collector voltage in the active region and depends only on i_E. Ideally, $i_C = i_E$ if α were

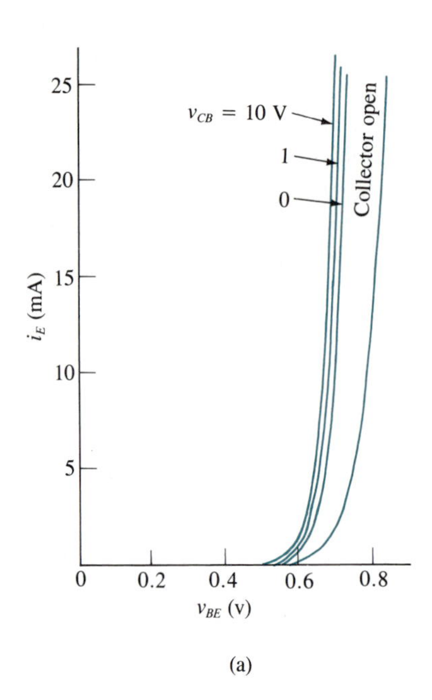

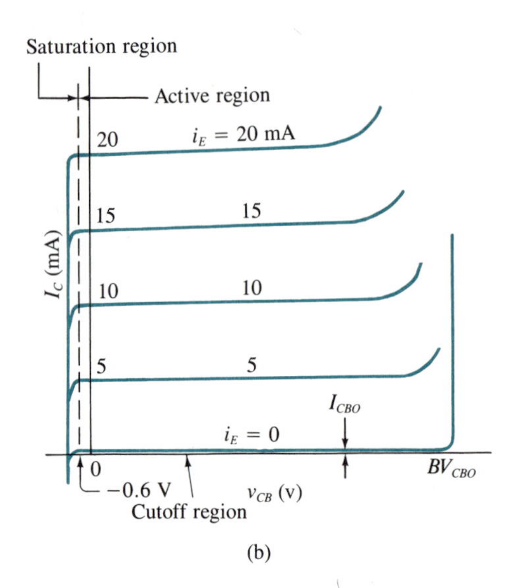

FIG. 6.5-3. Common-base static curves for a hypothetical *npn* silicon transistor, showing (a) emitter and (b) collector characteristics.

unity. In practical devices $i_C \approx i_E$, and the curves have a small slope, indicating a weak dependence on v_{CB}. For large values of v_{CB} the dependence becomes strong and a breakdown voltage BV_{CBO} is usually given on data sheets, above which operation is not possible. In breakdown i_C rises rapidly. It is not destructive to operate in breakdown provided the current and temperature limits of the device are not exceeded.

Figure 6.5-3(b) also illustrates other modes of transistor operation. In the active mode $v_{CB} > 0$ to keep the CBJ back-biased. However, if $v_{CB} < 0$, the CBJ becomes forward-biased (cut-in at about $v_{CB} = -0.6$ V); the current caused by the emitter is rapidly overcome and i_C decreases. It can even go negative as the CBJ acts more like a regular diode. Operation as described, where $v_{CB} < -0.6$ V, corresponds to the *saturation mode* of operation. The other mode, the *cutoff mode*, occurs when $i_E = 0$ (with $v_{CB} > -0.6$ V), corresponding to $i_C = I_{CBO} \approx 0$ and the BEJ cut off.

In some electric circuits the transistor is operated as a two-port (input, output) device, with the base forming the common terminal. In this so-called *common-base* configuration, the characteristics of Fig. 6.5-3(a) and (b) are called *input* and *output* characteristics, respectively.

The most common use of transistors is in a *common-emitter* configuration, where the input is to the base and the output is from the collector. If the input is taken as the base-emitter voltage, the output characteristic is as sketched in Fig. 6.5-4. The active region corresponds to v_{CE} above about 0.6 to 0.8 V, where the curves are approximately linear until they near the breakdown point. If the linear curves are extrapolated back to the v_{CE} axis, they meet at a point $-V_A$ called the

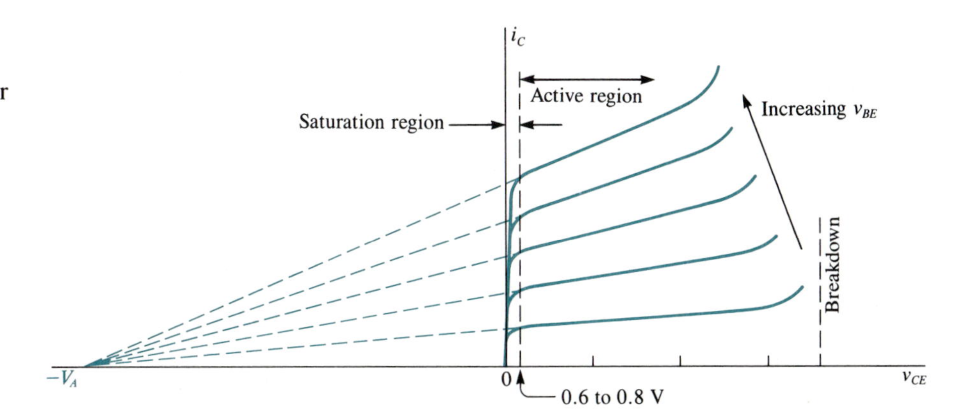

FIG. 6.5-4. The collector characteristic for a hypothetical *npn*-silicon transistor with base-emitter voltage as a parameter.

Early voltage. Voltage V_A is typically 50–100 V for a BJT. The Early effect causes the nonzero slope and is due to the fact that increasing v_{BE} makes the width of the depletion region of the CBJ larger, thereby reducing the effective width of the base. It results that I_{SE} in (6.5-1) is inversely proportional to the base width, so i_C increases according to (6.5-2). The increase in i_C can be accounted for by adding a factor to I_{SE} as follows:†

$$i_C = \alpha I_{SE} e^{v_{BE}/V_T}\left(1 + \frac{v_{CE}}{V_A}\right) + I_{CBO} \tag{6.5-6}$$

Another set of static characteristics applies to the common-emitter configuration, as illustrated in Fig. 6.5-5 for a hypothetical *npn* silicon transistor. The input characteristic of Fig. 6.5-5(a) is the device's i_B-v_{BE} behavior. A dependence on v_{CE}, the collector-to-emitter voltage, is seen (it is traceable to the Early effect). The output characteristics of Fig. 6.5-5(b) are somewhat analogous to Fig. 6.5-4, except that the parameter applicable to each curve is base current instead of voltage. Breakdown in the common-emitter configuration is more complex than in a common-base situation and will not be described in detail. However, there is a breakdown voltage, denoted by BV_{CEO} and sometimes called a *sustaining voltage*, above which the transistor should not be operated. Typically, BV_{CEO} is about half BV_{CBO}.

Small-Signal Common-Emitter Equivalent Circuits

One of the principal uses of a transistor is as an amplifier for small time-varying signals. In this and the following subsection we develop the two most important transistor equivalent circuits for such signals that will be used in the amplifier designs of Chapter 7. Our work is a small-signal linearization of the device's equations to apply specifically (and only) to small varying signals.

---

† A. S. Sedra and K. C. Smith, *Microelectronic Circuits*, 2d ed., Holt, Rinehart and Winston, New York, 1987; see p. 413.

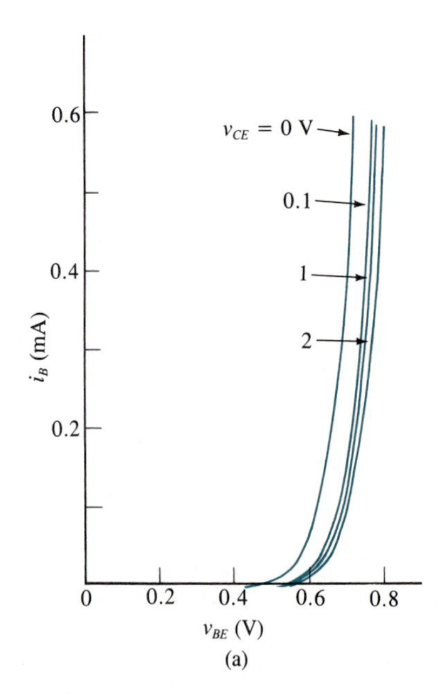

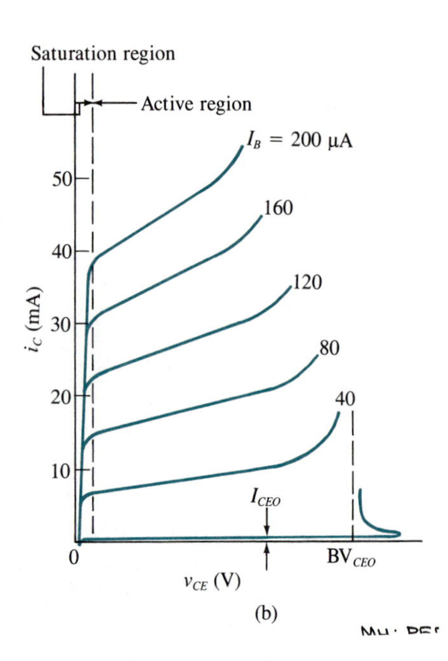

FIG. 6.5-5. Common-emitter static curves for a hypothetical *npn* silicon transistor, showing (a) input and (b) output characteristics.

We assume that the transistor is biased to a particular set of dc currents and voltages that we call the operating point Q. This assumption means that the "input" base-to-emitter voltage has a static (dc) value, denoted by V_{BEQ}, about which a small change, denoted by Δv_{BE}, can occur. An external generator applied across the base-emitter terminals can easily generate Δv_{BE}. In a similar manner, we shall be interested in the change in collector current, denoted by Δi_C, that occurs about an operating point value I_{CQ}. The change Δi_C is caused by the change Δv_{BE}. From (6.5-16) i_C is a function of both v_{BE} and v_{CE}, so Δi_C is found to be

$$\Delta i_C = \left.\frac{\partial i_C}{\partial v_{BE}}\right|_Q \Delta v_{BE} + \left.\frac{\partial i_C}{\partial v_{CE}}\right|_Q \Delta v_{CE} \tag{6.5-7}$$

where the notation $[\quad]|_Q$ implies that the bracketed quantity is evaluated at the operating point Q. We define the first derivative as the transistor's *transconductance* g_m, and the second derivative as the reciprocal of the *output resistance* r_o. From (6.5-6) these quantities are found to be

$$g_m = \left.\frac{\partial i_C}{\partial v_{BE}}\right|_Q = \frac{I_{CQ}}{V_T} \tag{6.5-8}$$

$$\frac{1}{r_o} = \left.\frac{\partial i_C}{\partial v_{CE}}\right|_Q \approx \frac{I_{CQ}}{V_A} \tag{6.5-9}$$

Thus, (6.5-7) can be written as

$$\Delta i_C = g_m \, \Delta v_{BE} + \frac{\Delta v_{CE}}{r_o} \tag{6.5-10}$$

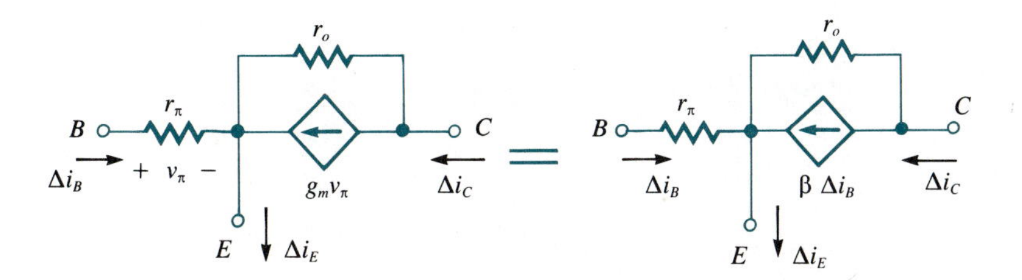

FIG. 6.5-6. Small-signal equivalent circuits for a transistor.

Next, consider what a small base-current change, denoted by Δi_B, occurs due to Δv_{BE}. We define a resistance r_π to relate the two according to

$$r_\pi = \left.\frac{\Delta v_{BE}}{\Delta i_B}\right|_Q = \left.\frac{\Delta i_C}{\Delta i_B}\frac{\Delta v_{BE}}{\Delta i_C}\right|_Q \approx \left.\frac{\partial i_C}{\partial i_B}\right|_Q \frac{1}{g_m} = \frac{\beta}{g_m} \tag{6.5-11}$$

where (6.5-2), (6.5-3), (6.5-5), and (6.5-8) have been used. Finally, we define v_π by

$$v_\pi = \Delta v_{BE} = r_\pi \, \Delta i_B \tag{6.5-12}$$

A combined consideration of (6.5-10) and (6.5-12) shows that the circuits of Fig. 6.5-6 are the small-signal equivalents of the transistor. These circuits apply to both *npn* and *pnp* transistors and are valid at lower frequencies, where effects of transistor capacitances can be ignored.

EXAMPLE 6.5-1

Assume a transistor has $\beta = 61.5$, an operating point defined by $I_{CQ} = 2.46$ mA, and an Early voltage of $V_A = 50$ V. We shall find the equivalent circuit parameters g_m, r_o, and r_π. We have

$$g_m = \frac{I_{CQ}}{V_T} = \frac{2.46 \times 10^{-3}}{25.681 \times 10^{-3}} = 95.79 \times 10^{-3} \text{ S}$$

from (6.5-8). From (6.5-9):

$$r_o \approx \frac{V_A}{I_{CQ}} = \frac{50}{2.46 \times 10^{-3}} \approx 20{,}325 \ \Omega$$

The value of r_π is found from (6.5-11):

$$r_\pi \approx \frac{\beta}{g_m} = \frac{61.5}{95.79 \times 10^{-3}} \approx 642 \ \Omega$$

At higher frequencies other components must be added, as shown in Fig. 6.5-7. Capacitor C_π is the sum of the junction and diffusion capacitances of the forward-biased BEJ evaluated at the operating point. It typically ranges in value up to a few tens of picofarads. Capacitor C_μ is the junction capacitance of the reverse-biased CBJ. It typically ranges up to a few picofarads in value. Resistor r_b

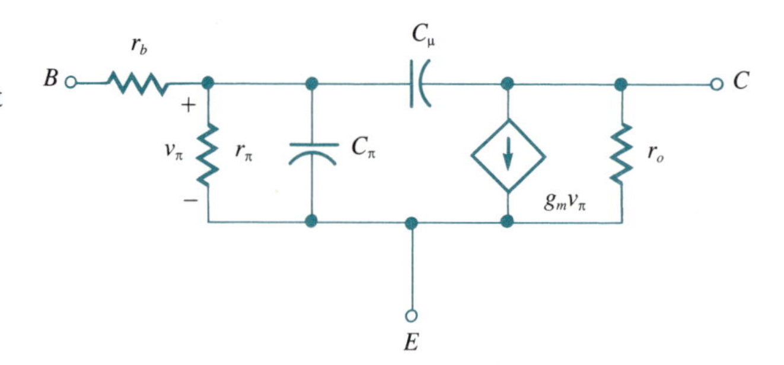

FIG. 6.5-7. High-frequency equivalent circuit to extend the circuit of Fig. 6.5-6.

represents the bulk resistance of the transistor material between the base lead and the base side of the base-emitter junction. At lower frequencies r_b (which has a value of a few tens of ohms) can be neglected by defining r_π to include r_b. At higher frequencies, where the effects of C_π and C_μ are important, it must be included separately.

6.6 JUNCTION FIELD-EFFECT TRANSISTOR (JFET)

We have seen that the bipolar-junction transistor (BJT) is principally a three-terminal, current-controlled current device. In this section we discuss the *junction field-effect transistor* (JFET), which is a three-terminal, *voltage-controlled* current device. As we shall see, the principles of JFET operation are different from those of the BJT, and these differences give it some special advantages over the BJT. Among the advantages are much higher input resistance (usually over $10^7\ \Omega$ and often over $10^{10}\ \Omega$), lower noise, easier fabrication, and in some JFETs, ability to handle higher currents and powers. Some disadvantages that often occur are slower speeds in switching circuits and, for a given gain in an amplifier, smaller bandwidth.

Device Operation

A typical method of constructing a JFET is shown in Fig. 6.6-1(a). Three terminals, called the *source* (S), *drain* (D), and *gate* (G), are attached through metal (ohmic) contacts to either n- or p-type semiconductor material. The lower gate is not a fourth terminal because it is electrically connected to the upper gate inside a typical device and is not normally accessible. We may consider the JFET's source, drain, and gate as roughly analogous to the emitter, collector, and base of a BJT, respectively. The semiconductor material connecting source and drain defines the JFET's *channel*, which is the region of length L, width Z, and depth a†

† Note that a is half the physical separation between the two pn junctions.

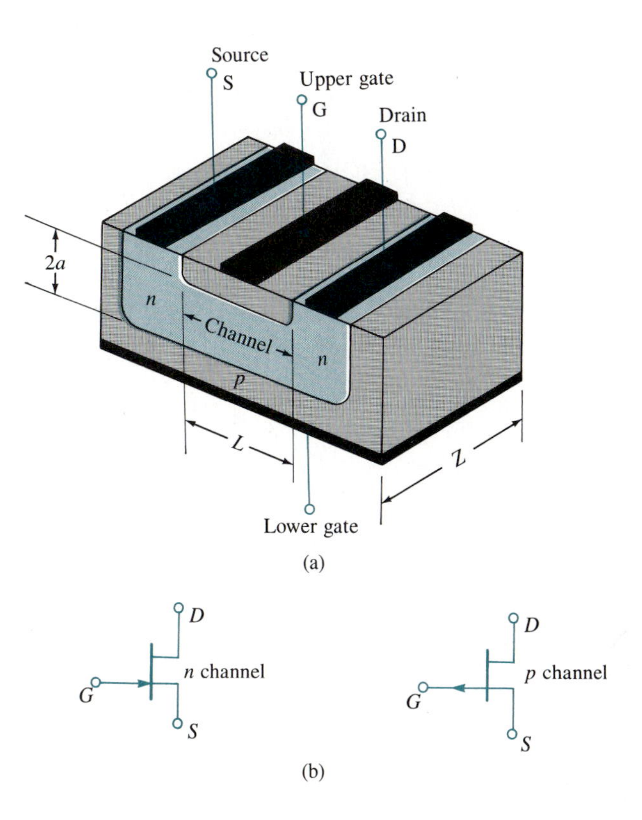

FIG. 6.6-1. (a) Construction and (b) circuit symbols for junction field-effect transistors.

that separates the two *pn* junctions. An *n*-channel device is illustrated, but *p*-channel devices are also available. In a typical JFET the channel-doping level is low compared with the gate's semiconductor material. This fact means that the depletion region is much wider on the channel side of the *pn* junctions than on the other (gate) side. Thus, when we refer to the width of a depletion region, we refer mainly to the part in the channel.

Various circuit symbols are found in the literature. We shall adopt those shown in Fig. 6.6-1(b). The gate terminal carries an arrow that points in the direction of conventional gate current flow if the gate-source *pn* junction were to be forward-biased.

For purposes of discussing JFET operation it is helpful to imagine that the transistor is constructed as sketched in Fig. 6.6-2 for an *n*-channel device. In typical operation the gate-source junction is reverse-biased by a voltage v_{GS}, and a voltage v_{DS} from drain to source causes drain current to flow. We adopt the reference polarities for v_{GS} and v_{DS} as shown in Fig. 6.6-2(a). Thus, for an *n*-channel JFET, v_{GS} will have a negative value while v_{DS} will be positive. Since the principal current (drain-source) flows in only one type of semiconductor (*n* in the figure), a JFET is often called a *unipolar device*, in contrast to the *bipolar*-junction transistor, where currents flow through two types of material.

To understand a JFET's operation, assume first that $v_{GS} = 0$. When v_{DS} is

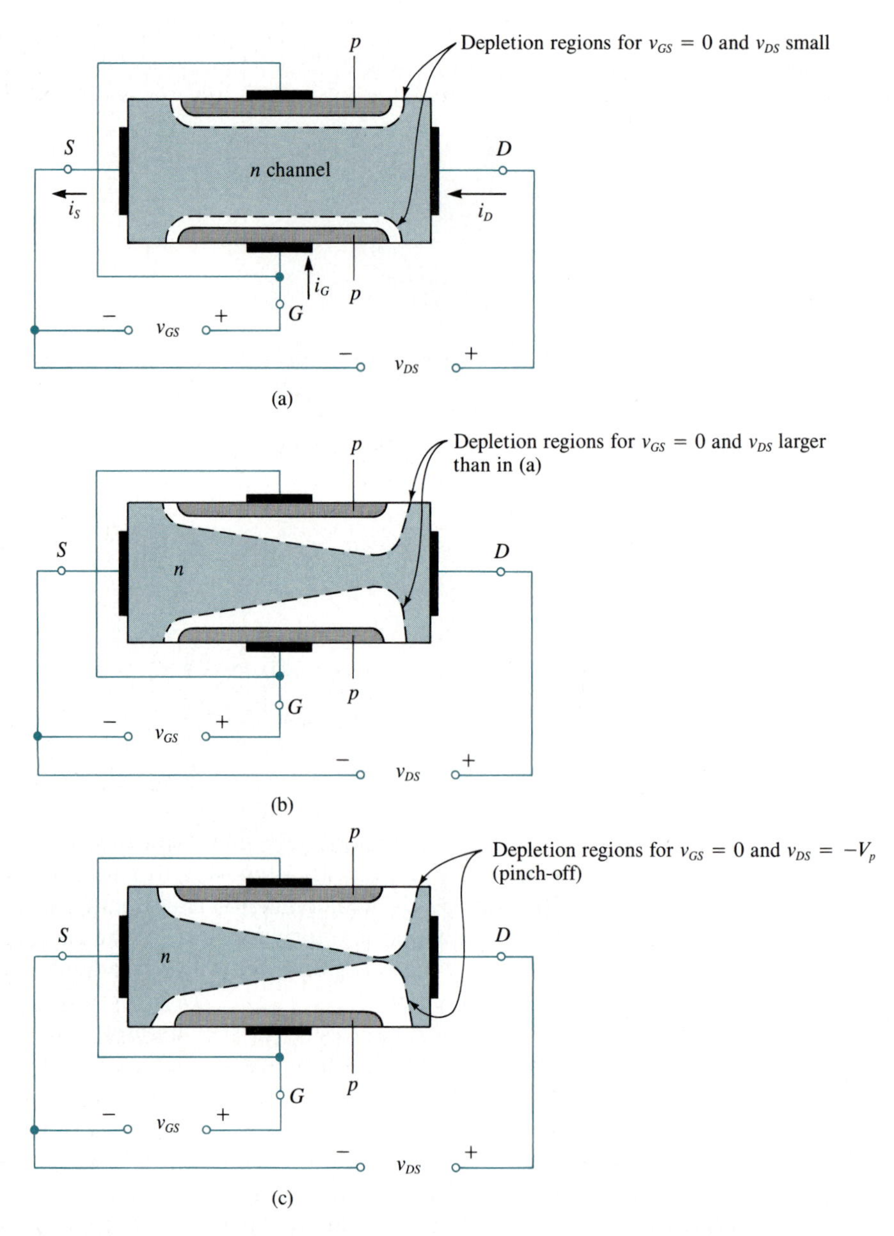

FIG. 6.6-2. Models useful in describing the operation of a JFET.

very small, the depletion regions of Fig. 6.6-2(a) are relatively thin and a small current i_D flows from D to S through the channel. To this current the channel appears simply as a resistance, and Ohm's law relates i_D to v_{DS}; so as v_{DS} increases from zero, i_D increases linearly. As the current moves through the channel, it causes an almost uniform rate of decrease in voltage, so that the voltage at a channel point near S is smaller than that at a point near D. Since $V_{GS} = 0$, the voltage in the channel acts as a reverse bias to the *pn* junction and

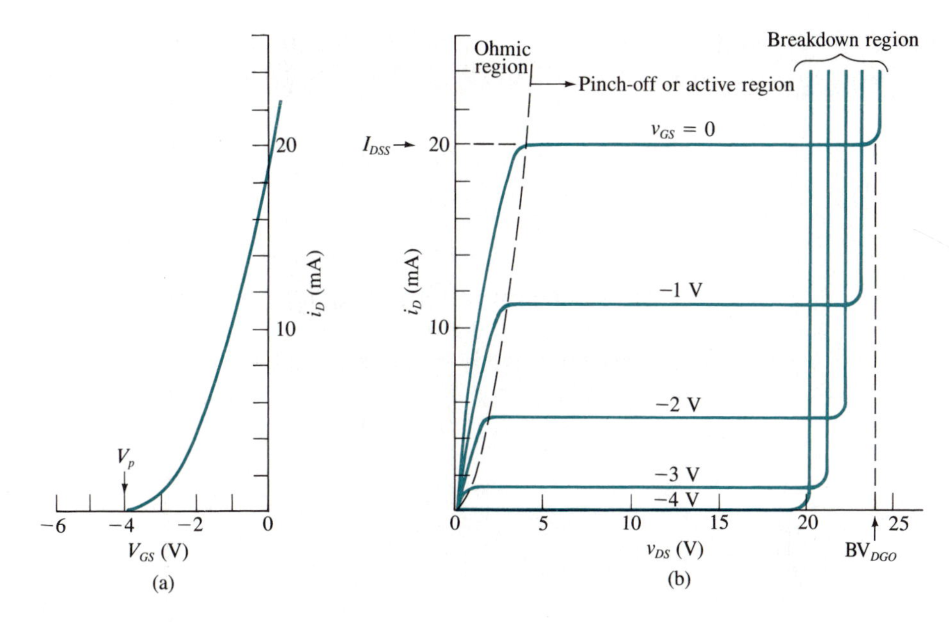

FIG. 6.6-3. JFET characteristics: (a) drain current versus v_{GS} for the pinch-off region and (b) i_D versus v_{DS} for v_{GS} as a parameter.

widens the depletion region. As the voltage decreases through the channel, the depletion region is widened more toward the *D* end than at the *S* end, as shown in Fig. 6.6-2(b). As v_{DS} increases, the widening becomes a significant part of the channel's width. Since no current flows in the depletion region, the widening decreases channel area and its effective resistance increases. The overall effect is to reduce the amount of current that can flow. Ohm's law now does not apply, and current i_D no longer increases linearly with v_{DS}. If v_{DS} increases further, a point is reached where the depletion regions widen to occupy almost all the channel, as shown in Fig. 6.6-2(c). At this point the current I_D becomes constant with further increases in v_{DS}, and the value of v_{DS} where this point occurs is called the *pinch-off voltage*, denoted by $-V_p$.† Thus, at the pinch-off point $v_{DS} = -V_p$. At and above pinch-off, where $v_{DS} \geq -V_p$, we have $i_D \approx$ constant $= I_{DSS}$, where we define I_{DSS} as the *drain-source saturation current*. The behavior just described for i_D versus v_{DS} is illustrated in Fig. 6.6-3(b) by the curve for $v_{GS} = 0$. Also shown is a breakdown voltage, denoted by BV_{DGO}, at which breakdown in the drain-gate junction occurs in the channel near the drain (highest reverse-bias region). Values of BV_{DGO} range from about 20 to 50 V in most JFETs.

When $v_{GS} \neq 0$, we mainly need to consider only $v_{GS} < 0$, because JFETs are normally operated with the gate-source junction reverse-biased. Because v_{GS} is a reverse bias, it acts to widen the depletion region along the whole channel, which, in turn, reduces i_D for a given value of v_{DS}. It also causes the value of v_{DS} at which the pinch-off point is reached to be lowered. In addition, v_{GS} acts to increase the

† For an *n*-channel JFET, V_p is defined as a negative quantity, so $-V_p$ is positive. Typical values of V_p range from about -2 to -6 V.

total voltage drop from drain to gate, so the value of v_{DS} at which breakdown occurs is reduced by $|v_{GS}|$. Mathematically, breakdown occurs when $v_{DG} = BV_{DGO}$, so

$$v_{DG} = v_{DS} - v_{GS} = BV_{DGO} \tag{6.6-1}$$

and therefore,

$$v_{DS}\ \text{(at breakdown)} = BV_{DGO} + v_{GS} = BV_{DGO} - |v_{GS}| \tag{6.6-2}$$

These effects are illustrated by the i_D-v_{DS} characteristics given in Fig. 6.6-3(b). In these curves $I_{DSS} = 20$ mA, $V_p = -4$ V, and $BV_{DGO} = 24$ V were assumed.

In the *ohmic region* (sometimes called the *triode region*), where the JFET behaves much like a voltage-variable resistance, the drain current is approximated by

$$i_D = I_{DSS}\left[2\left(1 - \frac{v_{GS}}{V_p}\right)\left(\frac{v_{DS}}{-V_p}\right) - \left(\frac{v_{DS}}{V_p}\right)^2\right] \tag{6.6-3}$$

For small v_{DS}

$$i_D \approx \frac{2I_{DSS}}{-V_p}\left(1 - \frac{v_{GS}}{V_p}\right)(v_{DS}) = \frac{v_{DS}}{r_{DS}} \tag{6.6-4}$$

where r_{DS} is the equivalent resistance of the JFET channel:

$$r_{DS} = \left.\frac{V_{DS}}{i_D}\right|_{v_{DS}(\text{small})} = \frac{-V_p^2}{2I_{DSS}(V_p - v_{GS})} \tag{6.6-5}$$

The ohmic region ends when v_{DS} is large enough to cause pinch-off, which occurs when the reverse bias between the drain and the gate at the drain end of the channel equals $-V_p$. This condition means $v_{DG} = -V_p$ and since

$$v_{DS} = v_{DG} + v_{GS} \tag{6.6-6}$$

the ohmic region ends when

$$v_{DS} = v_{GS} - V_p \tag{6.6-7}$$

From (6.6-3), the value of i_D becomes

$$i_D = I_{DSS}\left(\frac{v_{DS}}{V_p}\right)^2 \qquad \text{(ohmic-region boundary)} \tag{6.6-8}$$

This equation defines the dashed boundary shown in Fig. 6.6-3(b). The values of i_D at the boundary between the ohmic and active regions must equal the values of i_D in the active region, since i_D is ideally constant there. Thus, from (6.6-8) with (6.6-7) substituted, the drain current in the active region is

$$i_d = I_{DSS}\left(1 - \frac{v_{GS}}{V_p}\right)^2 \tag{6.6-9}$$

The current is plotted in Fig. 6.6-3(a).

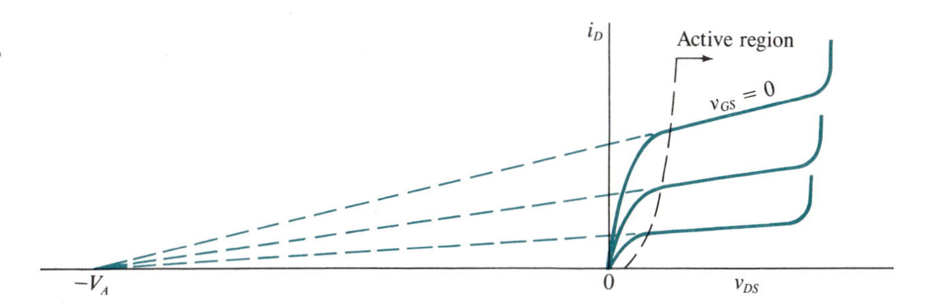

FIG. 6.6-4. A practical JFET static characteristic.

EXAMPLE 6.6-1

A JFET has $I_{DSS} = 32$ mA and $V_p = -5$ V. It is biased to produce $i_D = 27$ mA at $v_{DS} = 4$ V. We determine v_{GS} and the region in which the device is operating. From (6.6-8) with $v_{DS} = 4$ V, the boundary of the ohmic region is

$$i_D = (32 \times 10^{-3})\left(\frac{4}{-5}\right)^2 = 20.48 \text{ mA}$$

Since i_D is larger than 20.48 mA when $v_{DS} = 4$ V, the transistor is operating in the ohmic region. From (6.6-3)

$$v_{GS} = V_p + \frac{V_p^2}{2v_{DS}}\left[\frac{i_D}{I_{DSS}} + \left(\frac{v_{DS}}{V_p}\right)^2\right]$$

$$= -5 + \left[\frac{25}{2(4)}\right]\left[\frac{27}{32} + \left(\frac{4}{-5}\right)^2\right] = \frac{-93}{256} \approx -0.363 \text{ V}$$

JFET Static Characteristics

Figure 6.6-3 constitutes the static characteristics of an idealized JFET. In a more practical transistor the curves of i_D versus v_{DS} are not flat in the active region but tend to increase slightly with v_{DS}, as shown in Fig. 6.6-4. When extended, these curves tend to intersect at a point of $-V_A$ on the v_{DS} axis, where V_A typically has a positive value of about 100 V† for integrated-circuit JFETs. A first-order correction can be applied to (6.6-9) to account for the dependence of i_D on v_{DS} as follows:

$$i_D = I_{DSS}\left(1 - \frac{v_{GS}}{V_p}\right)^2\left(1 + \frac{v_{DS}}{V_A}\right) \tag{6.6-10}$$

† A. S. Sedra and K. C. Smith, *Microelectronic Circuits*, 2d ed., Holt, Rinehart and Winston, New York, 1987, p. 270.

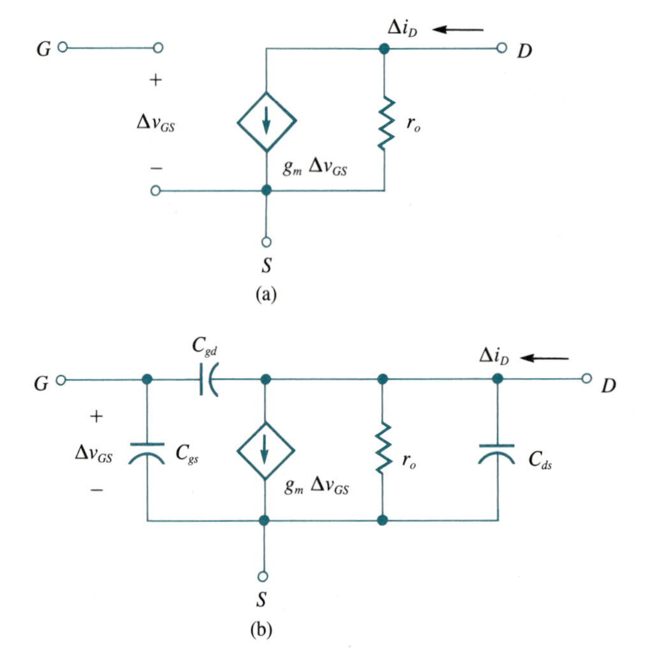

FIG. 6.6-5. JFET equivalent circuits for (a) low-frequency small signals and (b) high-frequency small signals.

Small-Signal Equivalent Circuits

As with the bipolar transistor earlier, let Q represent an operating point about which small-signal changes may occur. For a JFET, point Q is defined by dc values of v_{GS}, i_D, and v_{DS}, which we denote by V_{GSQ}, I_{DQ}, and V_{DSQ}, respectively. The small changes that occur in v_{GS}, i_D, and v_{DS} are denoted by Δv_{GS}, Δi_D, and Δv_{DS}, respectively. From (6.6-10) we have

$$\Delta i_D = \left.\frac{\partial i_D}{\partial v_{GS}}\right|_Q \Delta v_{GS} + \left.\frac{\partial i_D}{\partial v_{DS}}\right|_Q \Delta v_{DS} = g_m \,\Delta v_{GS} + \frac{1}{r_o}\,\Delta v_{DS} \qquad (6.6\text{-}11)$$

where

$$\begin{aligned} g_m &= \left.\frac{\partial i_D}{\partial_{GS}}\right|_Q = 2I_{DSS}\left(1 - \frac{v_{GS}}{V_p}\right)\left(1 + \frac{v_{DS}}{V_A}\right)\left.\left(\frac{-1}{V_p}\right)\right|_Q \\ &= \frac{-2I_{DSS}}{V_p}\left[\left(\frac{I_{DQ}}{I_{DSS}}\right)\left(1 + \frac{V_{DSQ}}{V_A}\right)\right]^{1/2} \approx \left(\frac{-2}{V_p}\right)\left[I_{DSS}I_{DQ}\right]^{1/2} \end{aligned} \qquad (6.6\text{-}12)$$

and

$$\frac{1}{r_o} = \left.\frac{\partial i_D}{\partial v_{DS}}\right|_Q = I_{DSS}\left(1 - \frac{v_{GS}}{V_p}\right)^2\left.\left(\frac{1}{V_A}\right)\right|_Q = \frac{I_{DQ}/V_A}{1 + v_{DSQ}/V_A} \approx \frac{I_{DQ}}{V_A} \qquad (6.6\text{-}13)$$

From (6.6-11) the small-signal equivalent circuit of Fig. 6.6-5(a) can be constructed. This circuit is valid at frequencies where capacitances can be neglected.

At higher frequencies the small-signal equivalent circuit must include capacitances from gate to source (C_{gs}), from gate to drain (C_{gd}), and from drain to source (C_{ds}), as shown in Fig. 6.6-5(b). These capacitors generally range from a few tenths of a picofarad to a few picofarads. Proper values should be obtained from manufacturers' data sheets, since they depend on the operating point chosen.

6.7 METAL-OXIDE-SEMICONDUCTOR FET

Field-effect transistors can be constructed in ways other than those used to make a JFET. We shall discuss two types of FET that use a special layer of oxide material (usually silicon dioxide, SiO_2, which is quartz) to insulate the metal gate (such as aluminum) from the semiconductor part of the FET. This metal-oxide-semiconductor construction leads to the name MOSFET. MOSFETs are also known as insulated-gate FETs, or IGFETs. One type of construction leads to the *depletion* MOSFET. The second type is called an *enhancement* MOSFET. The names derive from the way in which channels are formed and operate in the two MOSFET types.

Because of the gate's insulating layer, the MOSFET's input resistance (typically 10^{10} to 10^{15} Ω) is even higher than that of the JFET. This fact means, as in the JFET, that conductive gate current can be assumed to be negligible in most applications. This advantage is not achieved without an attendant disadvantage, however. The insulating oxide layer can easily be damaged by buildup of static charges. Since the oxide layer acts as the very thin dielectric of a capacitor, a small static charge can create a large enough voltage to puncture the layer and ruin the transistor. Thus, MOSFET devices are often shipped with leads conductively tied together to neutralize static charges. Users, too, must be careful in handling to prevent damage to MOSFETs from static electricity.

Operation and Characteristics—Depletion MOSFET

Typical construction of a depletion MOSFET is depicted in Fig. 6.7-1(a). On a lightly doped p-type-semiconductor *substrate* (also called the *body*) a layer of n-type material forms a channel of length L (and width W) between two heavily doped n^+-type regions.† The source (S) connects to one of these regions, and the drain (D) attaches to the other. Channel doping is heavier than that of the body but much less than in the source and drain regions. An insulating layer (usually SiO_2) of thickness d is placed above the channel. This layer separates the gate (G) from the channel. There is also a connection to the body (B). In many MOSFETs B and S are internally connected, and we shall consider only this form of device. The

† The plus sign is often used to imply a heavily doped semiconductor.

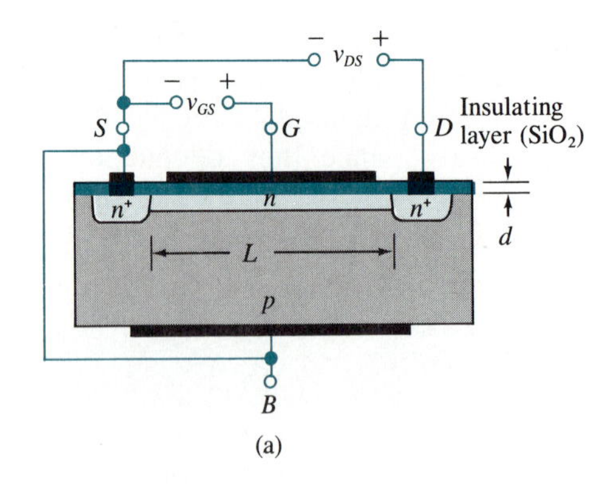

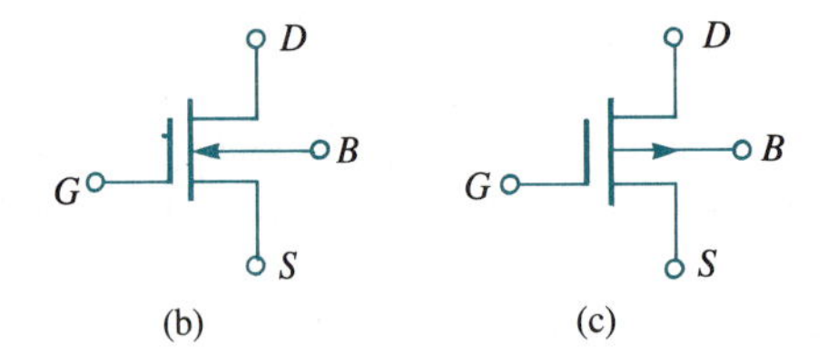

FIG. 6.7-1. (a) Construction of an n-channel depletion-type MOSFET and (b) its circuit symbol. (c) The circuit symbol for a p-channel depletion-type MOSFET.

electrical symbol for an n-channel MOSFET is given in Fig. 6.7-1(b). The name NMOS is often used to refer to the technology of n-channel MOS devices.

A p-channel MOSFET is also possible; it has an n-type substrate, p-type semiconductor for the channel, and p^+ material for the source and drain connections. Its electric circuit symbol is given in Fig. 6.7-1(c). The name PMOS is often used to refer to the technology of p-channel MOS devices. Except for voltage polarities and current directions, which are all reversed in the p-channel device compared with the n-channel MOSFET, behavior is similar to the n-channel transistor. Therefore, only the n-channel MOSFET is selected for further discussion.

MOSFET operation is briefly described by assuming that the body is connected to the source. For some positive bias applied to the drain relative to the source, assume a negative bias at the gate ($v_{GS} < 0$). The depletion region near the pn junction widens and affects drain current much as it does in a JFET. However, the electric field just under the insulating layer acts to drive electrons out of the top part of the channel, thereby creating another depletion region that widens as v_{GS} becomes more negative. Both depletion regions act to reduce the channel's current until pinch-off occurs, when v_{GS} is sufficiently negative (equal to the pinch-off voltage V_p). In fact, for $v_{DS} > 0$ and $v_{GS} < 0$, the equations describing drain current i_D have the same form as for the JFET. In the ohmic region

$$i_D = I_{DSS}\left[2\left(1 - \frac{v_{GS}}{V_p}\right)\left(\frac{v_{DS}}{-V_p}\right) - \left(\frac{v_{DS}}{V_p}\right)^2\right] \tag{6.7-1}$$

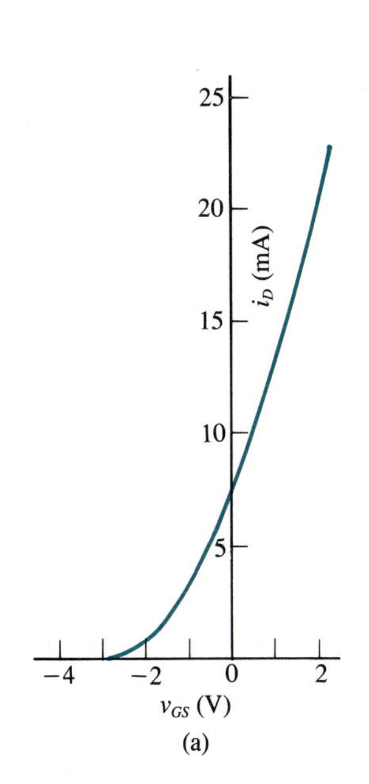

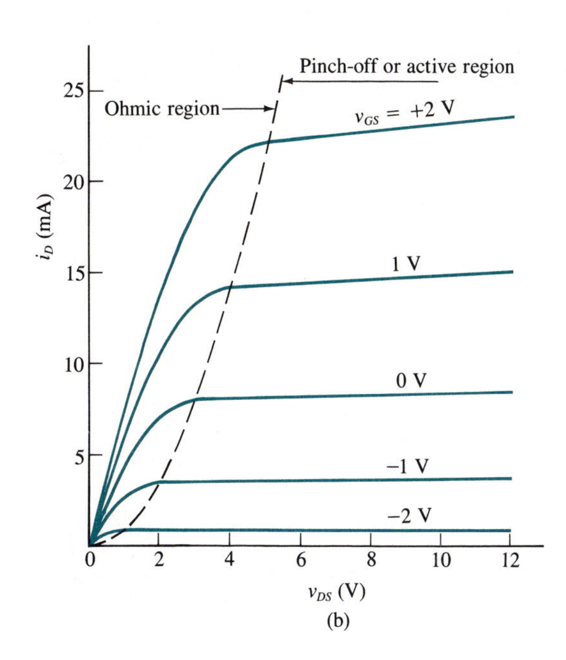

FIG. 6.7-2. Depletion-type MOSFET characteristics: (a) drain current at the start of the pinch-off region and (b) drain current versus v_{DS} for various values of v_{GS}.

when

$$v_{DS} < v_{GS} - V_p \tag{6.7-2}$$

In the pinch-off or active region, where

$$v_{DS} \geq v_{GS} - V_p \tag{6.7-3}$$

i_D is†

$$i_D = I_{DSS}\left(1 - \frac{v_{GS}}{V_p}\right)^2\left(1 + \frac{v_{DS}}{V_A}\right) \tag{6.7-4}$$

where V_A and I_{DSS} are positive constants.

Because of the insulating layer, no conductive gate current can occur, and v_{GS} can be positive as well as negative. For $v_{GS} > 0$ the electric field in the channel just below the insulating layer acts to attract and concentrate electrons in a thin layer at the top of the channel. These negative charges effectively increase the channel's conductivity and increase current flow. Drain current is still given by (6.7-1) and (6.7-4). Figure 6.7-2 illustrates characteristic curves for a hypothetical depletion-type MOSFET. When $v_{GS} \leq 0$, the MOSFET is said to operate in the *depletion mode*. It is in the *enhancement mode* for $v_{GS} > 0$, because the electric field caused by the gate's voltage enhances the channel's conductivity due to the electron concentration below the insulating layer.

---

† Again, the factor $[1 + (v_{DS}/V_A)]$ is added to approximately account for the nonzero slope of the i_D-v_{DS} curves of a practical device, as was done in (6.6-10).

EXAMPLE 6.7-1

An n-channel depletion-type MOSFET, for which $I_{DSS} = 6.8$ mA and $V_p = -3.6$ V, is known to operate in the ohmic region with drain current $i_D = 1.007$ mA when $v_{DS} = 0.8$ V. If the affect of v_{DS} on i_D is negligible, what is v_{GS}? We solve (6.7-1) for v_{GS}:

$$v_{GS} = -V_p\left\{\left[\frac{i_D}{I_{DSS}} + \left(\frac{v_{DS}}{V_p}\right)^2\right]\left(\frac{-V_P}{2v_{DS}}\right) - 1\right\}$$

$$= 3.6\left\{\left[\frac{1.007}{6.8} + \left(\frac{0.8}{3.6}\right)^2\right]\left(\frac{3.6}{1.6}\right) - 1\right\} = -2.00 \text{ V}$$

To check that ohmic-region operation is true, we find that $v_{DS} = 0.8$ V is less than $v_{GS} - V_p = -2.0 + 3.6 = 1.6$ V, as required from (6.7-2).

Operation and Characteristics—Enhancement MOSFET

Figure 6.7-3(a) sketches the typical construction of an enhancement-type MOSFET. Two heavily doped regions of n^+ material are placed on a p-type body (substrate). Source (S) and drain (D) connections are made to these n^+ regions. The gate (G) is attached above a layer of insulating oxide (SiO_2). Ideally, it has no conducting path to the rest of the device. Observe that this device has no channel between S and D. In fact, there are two back-to-back diodes between S and D formed by the two pn^+ junctions. For normal operation, where D is biased positive relative to S, negligible drain current flows because the drain's pn^+ junction is back-biased; this condition remains true until G is biased sufficiently positive with respect to S that current can flow from S to D.

The circuit symbol for the enhancement MOSFET is shown in Fig. 6.7-3(b). The line between D, B, and S is broken to imply the absence of a channel. For reasons subsequently made clear, the device is called an n channel. The circuit symbol for the complementary device (n substrate, p^+ drain and source regions)

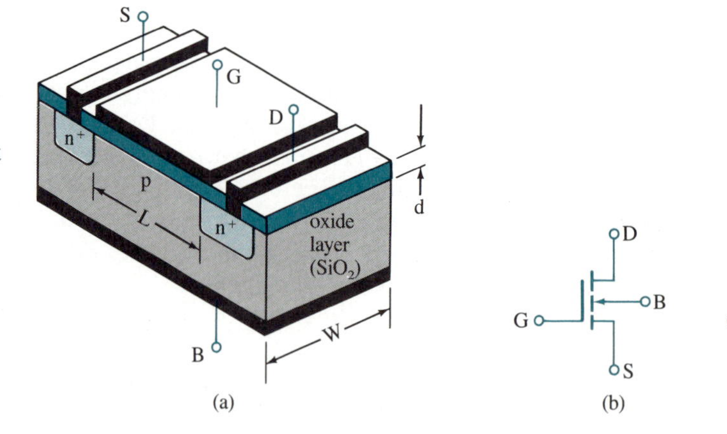

FIG. 6.7-3. (a) Construction of an n-channel enhancement-type MOSFET and (b) its circuit symbol. (c) The circuit symbol for a p-channel enhancement-type MOSFET.

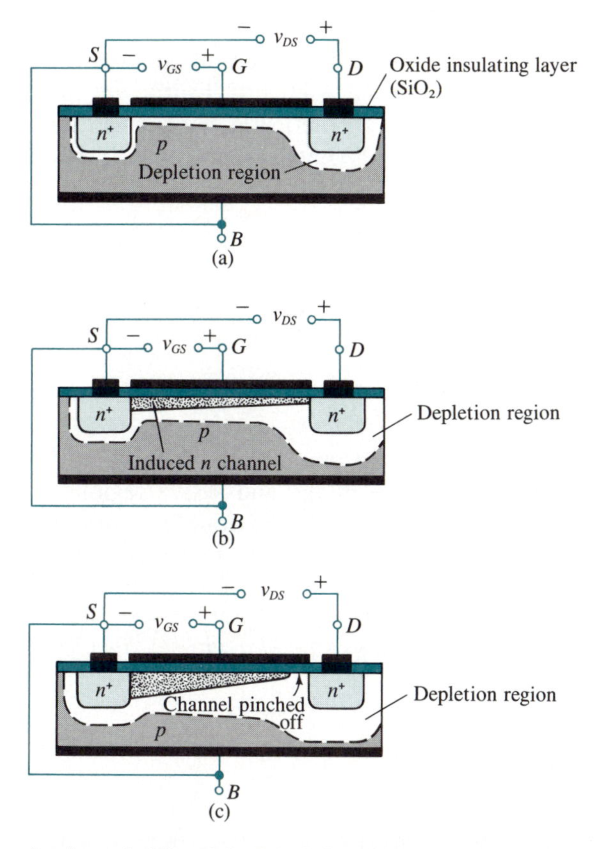

FIG. 6.7-4. Enhancement-type n-channel MOSFET structures when (a) $v_{GS} < V_t$, (b) $v_{GS} > V_t$ and $v_{DS} < v_{GS} - V_t$, and (c) $v_{GS} > V_t$ and $v_{DS} \geq v_{GS} - V_t$.

is given in Fig. 6.7-3(c); it is called a p-channel device. For many practical reasons such as smaller size, faster switching speeds, and compatibility with popular digital-logic circuits, the n-channel MOSFET is preferred over the p-channel device. For these reasons, we shall discuss only the n-channel device below.

A channel has to be *induced* into the enhancement MOSFET for drain current to flow. How the channel is induced and how the MOSFET behaves are best described with the help of Fig. 6.7-4. We shall assume B is connected to S. In Fig. 6.7-4(a) a small positive bias (v_{DS}) is applied to the drain relative to the source. No drain current flows because of the depletion region at the back-biased pn^+ junction at the drain. For v_{GS} negative the back-bias is even larger, and no drain current flows. If v_{GS} is a small positive voltage, it creates an electric field that moves the free holes away from the region below the insulating layer. The effect is to create a depletion region under the insulating layer that extends between the n^+ regions, as shown in Fig. 6.7-4(a). If v_{GS} is increased to a sufficiently large positive value called the *threshold voltage*, denoted by V_t, the electric field it produces becomes strong enough to pull free electrons to the layer under the insulator. This layer of electrons creates the induced channel. Current now flows in the channel because the channel's electrons flow under the action of the voltage v_{DS}. The induced channel is shown in Fig. 6.7-4(b). The described behavior corresponds to operation in the ohmic region as long as $v_{DS} < v_{GS} - V_t$.

If $v_{DS} \geq v_{GS} - V_t$, the channel becomes pinched off, as illustrated in Fig. 6.7-4(c). Since $v_{DS} = v_{GS} - v_{GD}$, the pinch-off condition is equivalent to $v_{GD} \leq V_t$, which says that the gate-drain voltage is insufficient to maintain the induced channel (at the drain end). As a practical consequence, increases in v_{DS} above $v_{GS} - V_t$ will not change the channel's shape, and i_D becomes approximately constant.

In the ohmic region where

$$v_{GS} > V_t \tag{6.7-5}$$

$$v_{DS} < v_{GS} - V_t \tag{6.7-6}$$

drain current is given by

$$i_D = K[2(v_{GS} - V_t)v_{DS} - v_{DS}^2] \tag{6.7-7}$$

where K is a constant having the unit amperes per square volt.

The boundary between the ohmic and active regions occurs, from (6.7-6), when $v_{DS} = v_{GS} - V_t$. For larger v_{DS} the drain's current is ideally constant at a value $K(v_{GS} - V_t)^2$ from (6.7-7). However, if a factor is added to account for the effect of v_{DS} on i_D, we have

$$i_D = K(v_{GS} - V_t)^2\left(1 + \frac{v_{DS}}{V_A}\right) \tag{6.7-8}$$

when

$$v_{GS} > V_t \tag{6.7-9}$$

$$v_{DS} \geq v_{GS} - V_t \tag{6.7-10}$$

which define the active region. Here V_A is a constant, usually in the range of 30 to 200 V.

Figure 6.7-5 illustrates characteristics of a hypothetical *n*-channel enhancement MOSFET for which $V_t = 4$ V, $V_A = 200$ V, and $K = 0.4$ mA/V^2.

Small-Signal Equivalent Circuits

The small-signal equivalent circuits for both the depletion and enhancement MOSFETs are identical in *form* to those of Fig. 6.6-5(a) and (b) for the JFET. In fact, because the drain-current expression, (6.6-10), for the JFET and that for the depletion MOSFET, (6.7-4), are identical in form, the equations defining g_m and r_o for the JFET apply to the depletion MOSFET. These equations were given in (6.6-12) and (6.6-13).

For the enhancement MOSFET (6.7-8) is used to find g_m and r_o:

$$g_m = \left.\frac{\partial i_D}{\partial v_{GS}}\right|_Q = 2K(v_{GS} - V_t)\left.\left(1 + \frac{v_{DS}}{V_A}\right)\right|_Q \approx 2\sqrt{KI_{DQ}} \tag{6.7-11}$$

$$r_o = \left[\left.\frac{\partial i_D}{\partial v_{DS}}\right|_Q\right]^{-1} = \left.\frac{V_A}{K(v_{GS} - V_t)^2}\right|_Q \approx \frac{V_A}{I_{DQ}} \tag{6.7-12}$$

Here all definitions are the same as used previously for the JFET.

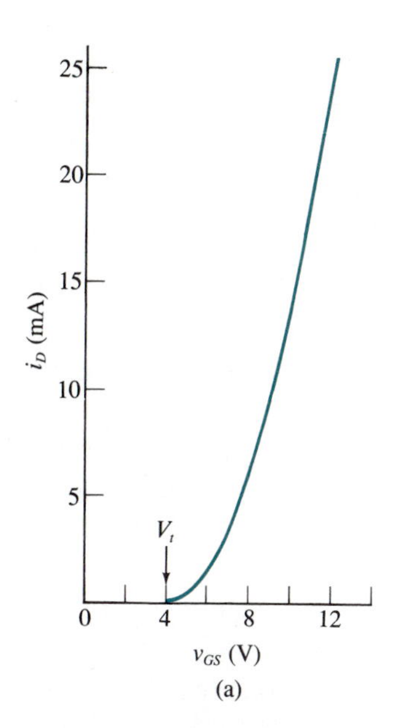

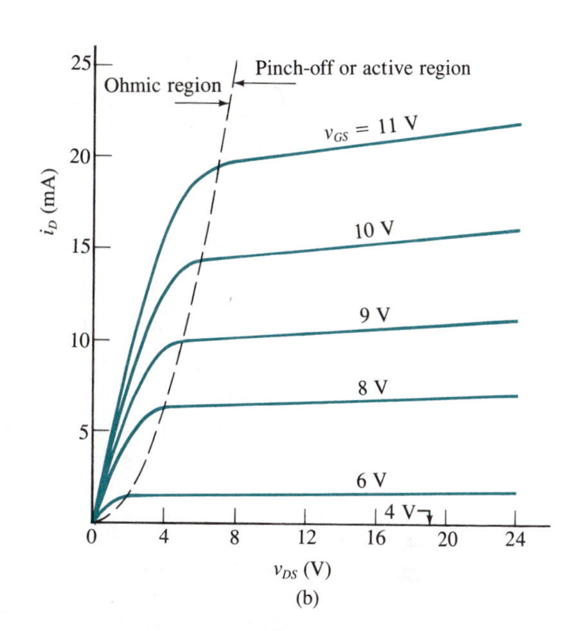

FIG. 6.7-5. Enhancement-type MOSFET characteristics: (a) drain current at the start of the pinch-off region and (b) drain current versus v_{DS} for various values of v_{GS}.

PROBLEMS

6-1. Measured data given by Sze† for hole and electron mobilities μ_n and μ_p, respectively, in silicon at 300 K can be fitted approximately by the following functions:

$$\mu_n \approx (1630 \times 10^{-4})\left(1 + \frac{N_n}{3.2 \times 10^{22}}\right)^{-1/2} \qquad \text{m}^2/\text{V}\cdot\text{s}$$

$$\mu_p \approx (460 \times 10^{-4})\left(1 + \frac{N_p}{10^{22}}\right)^{-1/3} \qquad \text{m}^2/\text{V}\cdot\text{s}$$

where N_n and N_p are electron and hole donor concentrations, respectively, in atoms per cubic meter. The two functions apply for N_n and $N_p \leq 5 \times 10^{24}$ atoms per cubic meter. If all impurity atoms contribute to conduction, find the conductivity and resistivity of n-type silicon at 300 K when $N_n = 2 \times 10^{24}$ atoms per cubic meter. Compare results with those of Example 6.1-2.

6-2. For p-type silicon conductivity σ and resistivity ρ are related to the density of free holes p (holes per cubic meter) and hole mobility μ_p (see Problem 6-1) by

$$\sigma = \frac{1}{\rho} = qp\mu_p$$

† S. M. Sze, *Semiconductor Devices, Physics and Technology*, Wiley, New York, 1985, p. 34.

If all donors contribute to conduction, so that $N_D \approx p$, find σ and ρ for $N_p = 10^{23}$ atoms per cubic meter. Compare the results with those of Example 6.1-2.

6-3. If donor concentration in p-type silicon at $T = 300$ K is 7×10^{22} atoms per cubic meter, what donor concentration is required in a piece of n-type silicon if the two materials are to have the same conductivity? Assume all donor atoms contribute to conduction.

6-4. A silicon-junction diode at 300 K produces a current of 9 mA when $v_d = 0.7$ V. What is the diode's reverse saturation current?

6-5. A silicon diode, for which $\eta = 2$ and $I_0 = 10^{-8}$ A, is forward-biased by a constant dc voltage of 0.725 V. Its current is to be measured and used as an indication of temperature. If I_0 is assumed constant, find the current that flows at temperatures of 275, 300, 325, 350, and 375 K. Plot current versus T. Is the diode a good sensor?

6-6. Two identical silicon diodes, for which $I_0 = 10^{-9}$ A and $\eta = 2$, are connected as indicated in Fig. P6-6. Find and plot the output voltage v_0 versus v for $-15 \text{ V} \le v \le 15$ V. Assume the diodes are at temperature $T = 300$ K.

FIG. P6-6.

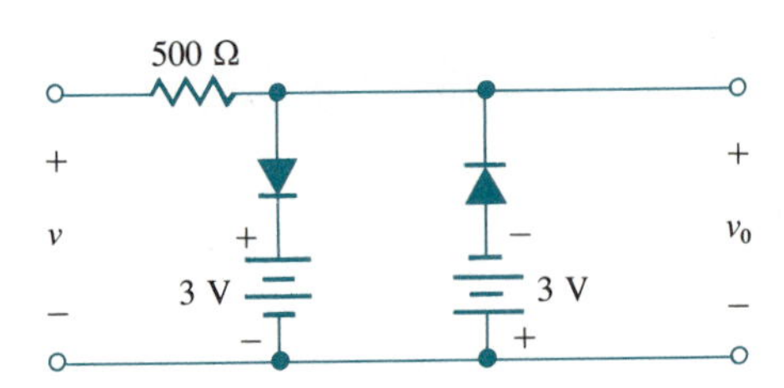

6-7. In a silicon diode cut-in voltage V_γ is to be defined as the value of v_d for which i_d/I_0 has a chosen value. Assume $\eta = 2$ and $T = 300$ K, and find V_γ for $i_d/I_0 = 10^2$, 10^4, and 10^6. What value of i_d/I_0 will give $V_\gamma = 0.6$ V?

6-8. Work Problem 6-7 except assume a diode for which $\eta = 1$, and find i_d/I_0 that gives $V_\gamma = 0.2$ V.

6-9. A constant voltage of 1.5 V is applied across a string of n identical diodes in series all at temperature 300 K. For each diode $\eta = 1$, $I_0 = 10^{-6}$ A, and the maximum allowable current is 100 mA. What is the smallest number (n_{min}) that can be used?

6-10. A string of seven identical silicon diodes ($\eta = 2$), all at 300 K, are in series and used as a voltage regulator to maintain 4.95 V across a 330-Ω resistor R_L, as shown in Fig. P6-10. (a) What is the reverse saturation current of the diodes? (b) If V_s increases by a small change $\Delta V_s = 2$ V, how much will v_L change?

6-11. Work Example 6.2-1 except assume $V_Z = 6$ V and the source is 18 V ± 15%.

FIG. P6-10.

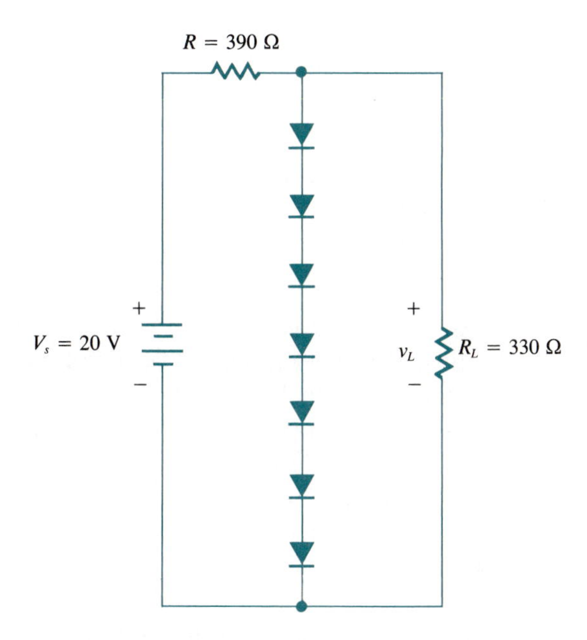

6-12. Assume a first-order practical representation of a zener diode's voltage for $i \geq I_{Z,\min}$ is

$$v = V_Z + \frac{(\Delta V)(i)}{I_{Z,\max}}$$

where ΔV is a constant. Find replacement equations for (6.2-3) and (6.2-5) when the practical diode is used.

6-13. For the circuit of Fig. 6.2-7, assume $R = 1500\ \Omega$ and the zener diode to be the practical device of Problem 6-12 with $\Delta V = 1$ V, $I_{Z,\max} = 15$ mA, and $V_Z = 8.5$ V. (a) What value of R_L will correspond to a voltage across R_L of $V_L = 9.0$ V when $V_s = 28$ V? (b) If V_s fluctuates between 24 and 32 V, what is the fluctuation in V_L? (c) What is the maximum power dissipated in the diode under worst-case assumptions (R_L can become open- or short-circuited)?

6-14. Two zener diodes are connected as shown in Fig. P6-14. For each diode $V_Z = 5$ V. Reverse saturation currents are 2 μA for D_1 and 4 μA for D_2. (a) Find v_1 and v_2 when $V_s = 4$ V. (b) If V_s is raised to 8 V, find v_1 and v_2.

★ 6-15. In the circuit of Fig. 6.2-7 $V_s = 94$ V, $V_Z = 12$ V, $R = 820\ \Omega$, and $R_L = 220\ \Omega$ nominally. The zener diode's reverse saturation current is so small it can be assumed to be zero, and its dynamic resistance can be assumed to be constant at 25 Ω over the breakdown region. (a) Find the load's voltage, current, and power. (b) What are the powers dissipated in R and the diode? (c) If the load is accidentally open-circuited, find the diode's

FIG. P6-14.

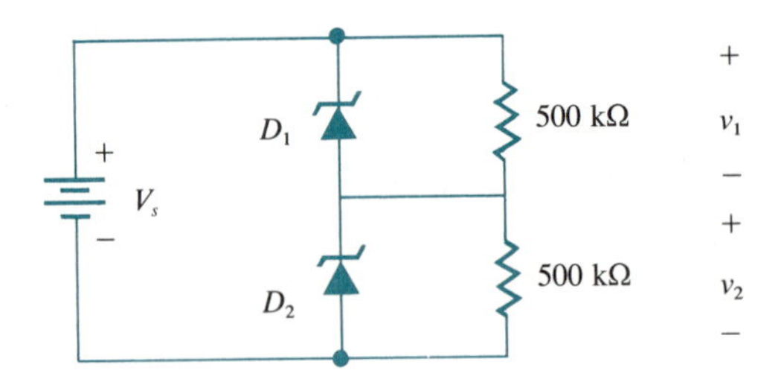

current and power and the power in R. (d) Repeat part (c) for a shorted load. (e) What minimum power ratings must R and the diode have if the circuit must anticipate any type of resistive load?

6-16. For the nominal circuit of Problem 6-15, what load-voltage variation (as a percentage of nominal) occurs for a $\pm 10\%$ variation in V_s?

6-17. A silicon diode has a reverse saturation current of 6.5×10^{-12} A and $\eta = 2$ when $T = 260$ K. (a) What current flows if a forward bias of 0.6 V is applied? (b) What is the diode's static resistance? (c) What is its dynamic resistance?

6-18. (a) Find the ratio of r_d/R_s for a junction diode. (b) If $qv_d/\eta kT \geq 3$ is true, so that $i_d \gg I_0$ can be assumed, what does the ratio become? (c) Above what value of $qv_d/\eta kT$ does r_d become small (less than one-tenth) relative to R_s?

6-19. In a diode for which $\eta = 2$, $T = 300$ K, and $I_0 = 3 \times 10^{-8}$ A, V_γ is defined as the diode's forward bias (v_d) when its current (i_d) is 0.5 mA. (a) Determine r_f for a piecewise-linear diode model if r_f is the average value of the diode's dynamic resistance over the range of v_d from V_γ out to the value of v_d where $i_d = 100$ mA. (b) Determine r_r for the model if it corresponds to the straight line through the origin and the point ($i_d = 0.5$ mA, V_γ).

★ 6-20. For the circuit of Fig. P6-20, assume V is small and the diode is defined by $I_0 = 10^{-7}$ A, $\eta = 2$, $T = 300$ K, and the total diode forward-biased capacitance ($C_J + C_D$) is 8000 pF. (a) Find the total diode current i_d for

FIG. P6-20.

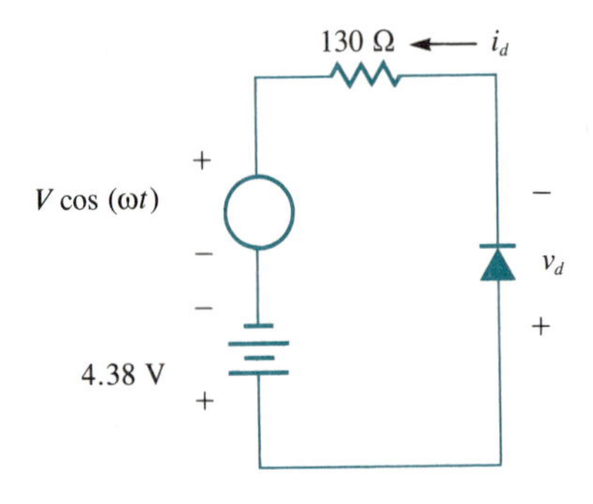

any ω. (b) Find the value of ω where the peak value of the currents through the diode's resistance and capacitance are equal. For ω above this value, which component of diode current is largest?

6-21. Neglect the ac voltage in Problem 6-20 and plot the diode characteristic and load line (as in Fig. 6.4-2). What are the dc operating current and voltage? On the same graph, plot the load line corresponding to a dc source voltage of 4.70 V and observe the change in the current and voltage of the operating point of the diode.

6-22. The output $v_L(t)$ given in Fig. 6.4-3(b) for a half-wave rectifier can be assumed to be half-cycles of a sine wave at a frequency of 60 Hz. If $v_L(t)$ is applied to a simple lowpass filter, as shown in Fig. P6-22, what product RC is needed if the lowest-frequency ac term in the Fourier series of $v_o(t)$ is to be reduced to a peak amplitude of one-hundredth of the dc component in $v_o(t)$?

FIG. P6-22.

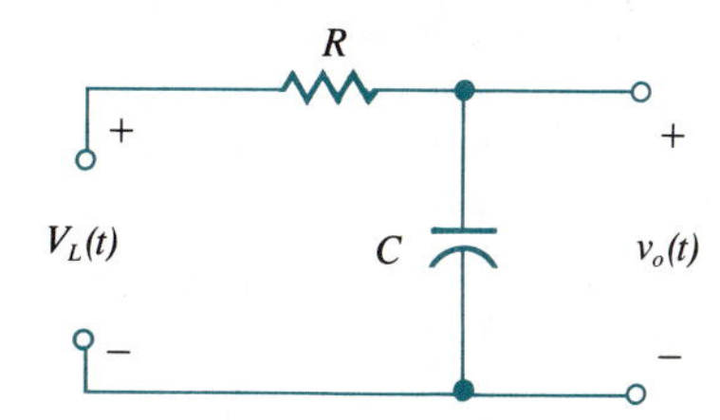

6-23. Work Problem 6-22 except assume $v_L(t)$ is for a full-wave rectifier [Fig. 6.4-4(b)].

6-24. In Fig. 6.4-6(a) the signal is

$$v(t) = \{10 + 3 \cos [\pi(10^4)(t)]\} \cos [1.08\pi(10^6)(t)]$$

and $R_L = 10\ \text{k}\Omega$. What should be the value of C?

6-25. In a transistor that has a base current $i_B = 24\ \mu\text{A}$, $\alpha = 0.985$. Find β, i_E, and i_C. Neglect I_{CBO}.

6-26. In a silicon transistor at 300 K the emitter's current is 4.5 mA when the BEJ is forward-biased by $v_{BE} = 0.67$ V. What is the reverse saturation current of the BEJ? If I_{CBO} is assumed negligible, what are i_C, β, and i_B if $\alpha = 0.99$?

6-27. In a particular transistor $\alpha = 0.99$ nominally. (a) What is the nominal β? (b) If α can easily change $\pm 1\%$, what percentage changes occur in β?

6-28. A transistor is known to have the parameters $\alpha = 0.975$, $I_{CBO} = 80$ nA, and $i_C = 7.2$ mA. Find (a) β, (b) i_B, and (c) i_E.

6-29. If $I_{CBO} = 3.8$ nA, $i_E = 0.8$ mA, and $i_C = 0.72$ mA in a transistor when $v_{BE} = 0.71$ V, find (a) α, (b) i_B, (c) I_{SE}, and (d) β.

6-30. A transistor can actually be operated in a fourth mode that is normally not useful. It corresponds to the BEJ reverse-biased and the CEJ forward-biased. What is the physical meaning of this mode?

6-31. (a) For the circuit of Fig. P6-31, assume $\beta = 85$ for the silicon transistor. Compute i_B, i_C, and i_E. Is the transistor in the active mode? (b) Let β be increased by 20% to 102, and explain what has happened. (c) If β is reduced by 10%, what happens? Neglect I_{CBO} in every case.

FIG. P6-31.

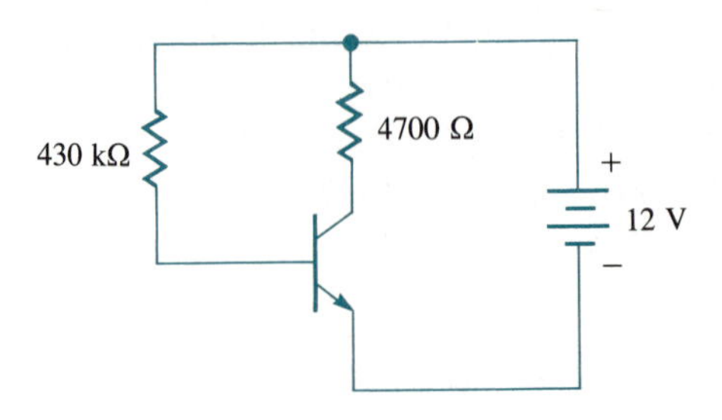

6-32. A transistor, for which $\beta = 75$, $V_A = 65$ V, and $g_m = 0.03$ S, is biased to have a dc collector current of 6 mA. The small-signal equivalent circuit is of the form of Fig. 6.5-6, except that a load resistor R_L is connected from the collector to the emitter and a source of voltage $\Delta v_{BE} = 0.05$ V is connected from B (positive) to E. (a) What change in voltage exists across R_L due to the small change Δv_{BE} if $R_L = 10{,}000\ \Omega$? (b) What base current change Δi_B occurs?

6-33. Pinch-off voltage in an n-channel JFET is given by†

$$V_p = -\frac{qa^2N_D}{2\epsilon_r\epsilon_0}$$

where $q = 1.602 \times 10^{-19}$ C is the magnitude of the electron's charge, a is the channel's depth, $\epsilon_0 = 10^{-9}/36\ \pi$ F/m is the permittivity of free space, ϵ_r is the relative permittivity of the JFET's material ($\epsilon_r = 11.9$ for silicon), and N_D is the donor concentration in the channel material. If $a = 0.5\ \mu$m for a silicon JFET, what value of N_D will give $V_p = -5$ V?

6-34. For an n-channel JFET with channel depth $a = 0.8\ \mu$m, find the donor concentrations that give (a) $V_p = -1.5$ V, (b) $V_p = -3$ V, and (c) $V_p = -6$ V. (*Hint*: Use the results given in Problem 6-33.)

6-35. In an n-channel JFET, $V_p = -3$ V and $I_{DSS} = 6$ mA. (a) What is the smallest value v_{DS} can have when $v_{GS} = -2$ V if operation is to be in the active region? What is i_D for the smallest v_{DS}? (b) Repeat part (a) for $v_{GS} = -1$ V.

6-36. What polarities of voltages would be needed to operate a p-channel JFET in its active region? In what directions do conventional currents flow at the source and drain terminals?

---

† M. S. Ghausi, *Electronic Devices and Circuits, Discrete and Integrated*, Holt, Rinehart and Winston, New York, 1985, p. 127.

6-37. A silicon n-channel JFET has $I_{DSS} = 12.5$ mA and $V_p = -4.75$ V. When $v_{GS} = -3.2$ V, $i_D = 0.5$ mA. (a) Is the device operating in the ohmic or active mode? (b) What is v_{DS}?

6-38. A JFET has $I_{DSS} = 26$ mA and $V_p = -2.9$ V. What is the channel's resistance r_{DS} for $v_{GS} = -0.5, -1.0$, and -2.0 V? Assume v_{DS} is small.

6-39. If $V_A = 350$ V, $I_{DSS} = 10$ mA, and $V_p = -3$ V for an n channel JFET, what value of v_{DS} will cause $i_D = 10.5$ mA when $v_{GS} = 0$?

6-40. A JFET, for which $V_A = 300$ V, $V_p = -2$ V, and $I_{DSS} = 8$ mA, is to be operated in the active region. (a) If the device is an n channel, what is the current i_D when $v_{DS} = 15$ V and $v_{GS} = -0.5$ V? (b) What does i_D become if v_{DS} decreases to 5 V?

6-41. An n-channel JFET has $I_{DSS} = 10$ mA, $V_p = -3$ V, and $V_A = 100$ V. (a) Plot the device's g_m versus v_{GS} for v_{DS} constant at 10 V. (b) Repeat part (a) for $V_A = 500$ V and $V_A = \infty$.

6-42. Work Problem 6-41 except compute and plot r_o.

6-43. In a depletion-mode MOSFET for which $V_p = -3.2$ V and $I_{DSS} = 11.2$ mA, drain current is 2.8 mA when v_{DS} is set at the largest value that will maintain operation in the ohmic region. What is v_{GS} if V_A is very large?

6-44. In an NMOS depletion-type FET, $V_p = -2.8$ V, $I_{DSS} = 4.3$ mA, $v_{DS} = 4.5$ V, $v_{GS} = 1.2$ V, and V_A can be assumed to be large. (a) Is the MOSFET in the active or pinch-off region of operation? (b) What is i_D? (c) Is the device operating in the depletion mode or enhancement mode?

★ 6-45. An n-type depletion MOSFET for which $V_p = -3.6$ V has $i_D = 3.611$ mA when $v_{DS} = 2$ V and operation is in the ohmic region. When v_{GS} is held constant but v_{DS} is raised to 8 V to operate well into the pinch-off region, i_D increases to 3.932 mA. Assume V_A is large, and find (a) v_{GS} and (b) I_{DSS}.

6-46. In an NMOS enhancement-type FET, V_A is very large, $v_{GS} = 6$ V, $V_t = 4.2$ V, and $i_D = 0.972$ mA. Operation is in the active region. What is K?

CHAPTER

Transistor Amplifiers for Small Signals

7.0 INTRODUCTION

A voltage amplifier is an electronic circuit that accepts an input voltage at a pair of "input" terminals and provides an output voltage at another pair of "output" terminals. The output voltage is often taken across a load resistor connected between the output terminals, as sketched in Fig. 7.0-1, but the load could also be the input terminals of another network. When the input and output signals are sinusoids (represented by their complex signals v_1 and v_L, respectively), the *voltage gain* of the amplifier is defined as the ratio v_L/v_1. For most voltage amplifiers that are used in instrumentation applications, voltage gain is a real number that can be greater than or less than unity but is approximately constant over a band of frequencies (its bandwidth). Outside the frequency band, gain is a complex number (the network's transfer function).

Other forms of amplifier are possible. In a *current amplifier* both the input and the output signals are currents. Obviously, voltage-in current-out and current-in voltage-out amplifiers are also possible. Since amplifier gain is just the network's transfer function, the four cases described above are just those listed in Table 4.6.-1.

Amplifiers find extensive use in instrumentation applications. In a common problem some current or voltage being generated by a sensor (such as a temperature-sensitive thermocouple) may be too small to use directly. If we amplify the sensor's output, a more useful signal level is established (to drive a panel meter, possibly).

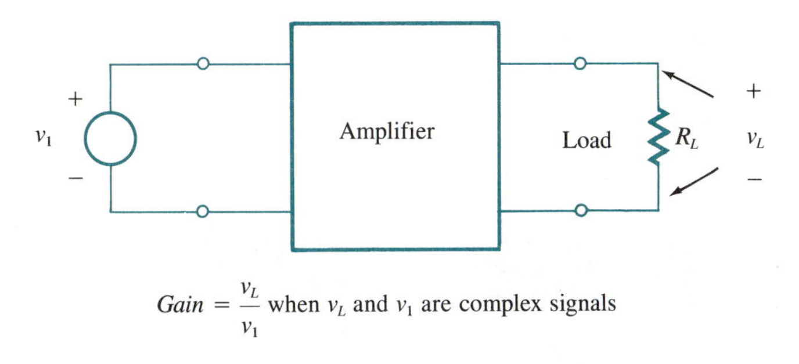

$$Gain = \frac{v_L}{v_1} \text{ when } v_L \text{ and } v_1 \text{ are complex signals}$$

FIG. 7.0-1. A voltage-to-voltage amplifier.

Sometimes, amplifiers are used for reasons other than gain alone. An amplifier can be designed to have high input impedance so that it does not affect the output of a sensor, while at the same time giving a low output impedance so it can drive large currents into its load (perhaps a lamp or heating element). In other applications an amplifier having a low input impedance might be desirable.

Fortunately, most instrumentation applications encountered by nonelectrical engineers can be solved by using already-designed, general-purpose amplifiers that are available off the shelf in standard integrated-circuit packages. These devices, called *operational amplifiers*, can be combined properly with a few resistors and capacitors to achieve a wide variety of amplifier functions. They can even be used to synthesize electrical filters. Operational amplifiers are discussed extensively in the following chapter. Nonelectrical engineering readers that are mostly interested in applications can omit this chapter and proceed directly to Chapter 8 with only minor loss.

For those readers interested in some details on *how* amplifiers are constructed, this chapter will serve as a bridge to show how amplifiers can be implemented from the components and devices described in the preceding chapters. In the course of this chapter we also discuss a few amplifiers that are not easily (or cannot be) realized from operational amplifiers.

Our work will focus mainly on small-signal amplifiers constructed from transistors. By *small-signal* we mean that all currents and voltages that occur anywhere in the amplifier are small enough that no distortions occur. In essence, this requirement means that the output signal's waveform is an amplified replica of the input signal.

Amplifiers may be structured in so many ways that it is necessary to consider only some of the more important and representative structures. We shall limit the discussions to amplifiers that use only a single bipolar-junction transistor (BJT), junction field-effect transistor (JFET), or metal-oxide-semiconductor FET (MOSFET) as its active device.

There are two principal concerns in amplifier design and analysis. One is the actual calculation of the response for a given input signal. This calculation, which uses the small-signal equivalent circuits developed in the preceding chapter, can be called *ac analysis*. However, before these calculations can be done, it is necessary to conduct a *dc design and analysis*, which is needed to establish an *operation*, *bias*, or *quiescent point*, called a *Q point*, about which the ac signals can occur without significant distortion.

7.1 BASIC TRANSISTOR AMPLIFIER EXAMPLE

The basic idea behind an amplifier is illustrated in Fig. 7.1-1, which assumes an *npn*, low-current, bipolar transistor, although the concepts hold for other devices. The circuit of Fig. 7.1-1(a) has the collector characteristic of (b). For the specified collector dc source of $V_{CC} = 15$ V and collector resistance $R_C = 100$ kΩ as shown, selection of the bias point Q is established by choosing a dc base current. In the figure $I_{BQ} = 0.5$ μA gives a Q point of $V_{CEQ} = 6.5$ V, $I_{CQ} = 85$ μA. This point must fall on the load line defined by the two points ($v_{CE} = 15$ V, $i_C = 0$) and ($v_{CE} = 0$, $i_C = V_{CC}/R_C = 150$ μA).

With the Q point established in Fig. 7.1-1, suppose now that a triangular-base-current *variation* Δi_B is added to I_{BQ} such as to swing the total base current i_B between extremes of 0.4 to 0.6 μA (or ±0.1 μA about I_{BQ}). As shown in Fig. 7.1-1(b), the total collector current i_C swings between extremes of 69 and 102 μA (+17 μA and −16 μA about $I_{CQ} = 85$ μA). The shape of the changes Δi_C in i_C are triangular. The triangularly shaped current change Δi_C gives rise to a voltage drop $\Delta i_C R_C$ in R_C. The resulting voltage change Δv_{CE} in v_{CE} is also triangular and varies between −1.6 V and +1.7 V (about $V_{CEQ} = 6.5$ V). If the stage is to be a current amplifier, its current gain can be defined as the ratio of the peak-to-peak *change* in collector current (102 − 69 = 33 μA) to the peak-to-peak *change* in base current (0.2 μA), or 33/0.2 = 165. This current gain is sometimes called the *ac beta* of the transistor; it is nearly equal to the *dc beta*, which is $\beta = I_{CQ}/I_{BQ} = 85/0.5 = 170$. Because the ac and dc betas are often nearly equal, we have made no real distinction between the two in this book.

The amplifier of Fig. 7.1-1(a) can be made to give voltage gain through a simple change. If Δi_B is provided instead by a voltage source Δv_B feeding the transistor's base through a large resistor R_B (so very little of I_{BQ} is diverted through R_B), then $\Delta i_B \approx \Delta v_B/R_B$. Since $\Delta v_{CE} = -\Delta i_C R_C \approx -\beta \Delta i_B R_C$, then $\Delta v_{CE} \approx -\beta R_C \, \Delta v_B/R_B$, and voltage gain is $\Delta v_{CE}/\Delta v_B \approx -\beta R_C/R_B$. Numerically, suppose $R_B = 100$ kΩ, corresponding to Δv_B varying between extremes of ±10 mV; the voltage gain is −165.

The amplifier described above has a small amount of distortion because the extremes of Δi_C (+17 μA and −16 μA) are not symmetric. For a smaller input Δi_B the distortion is reduced (small-signal case). For larger Δi_B distortion is minimized by selecting Q to be about midway in the range of possible collector currents. In Fig. 7.1-1(b) this range is from 0 to 150 μA as Δi_B ranges from −0.5 μA to +0.5 μA. In a given amplifier, if the current swings are known to be small, Q can

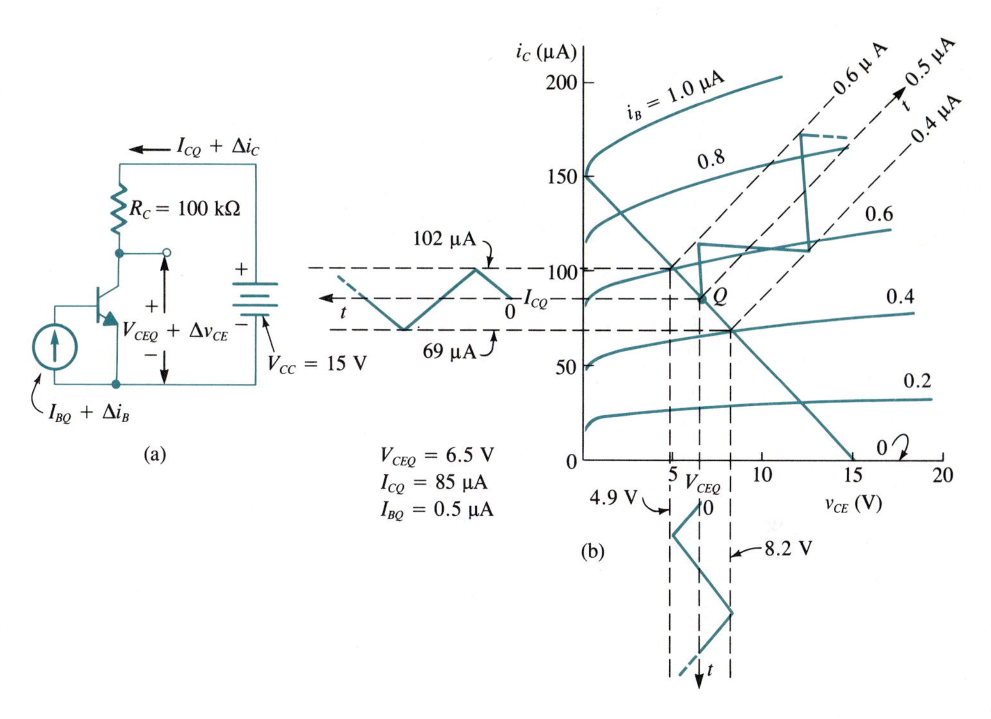

FIG. 7.1-1. (a) An *npn* bipolar transistor amplifier and (b) its collector characteristic with load line.

be placed at a smaller collector current to reduce the current drain on the power source V_{CC}.

In discussing Fig. 7.1-1, we have been somewhat vague about exactly how the bias current I_{BQ} is established. In the following sections we discuss ways of biasing various types of transistors. Because many biasing methods are possible, we limit the discussion to only the most useful, popular, and stable schemes.

7.2 BIASING THE BJT

One of the most popular methods of biasing the BJT is sketched in Fig. 7.2-1. Biasing consists mainly of choosing values of V_{CEQ}, I_{CQ}, and I_{BQ}, which define the operating (or Q) point, when some supply voltage V_{CC} is specified. However, some consideration is usually given to operating-point stability, that is, keeping collector current I_{CEQ} constant with variations in transistor parameters such as β and I_{CBO}.

Operating-point stability tends to be better as R_E increases and as the parallel combination of R_1 and R_2 decreases. The latter condition is increasingly satisfied as I_2 in Fig. 7.2-1 is made larger than I_{BQ}. On the other hand, increasing R_E for Q-

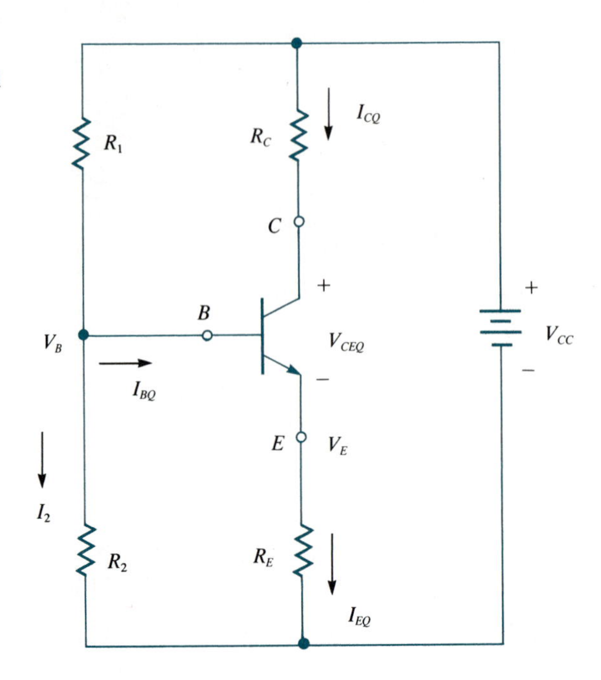

FIG. 7.2-1. A biasing method for a BJT.

point stability can be shown to reduce ac gain, so a compromise is required.

There is no general procedure for biasing a BJT that will work in all cases. It is often necessary to make an initial bias design and then make adjustments as needed to satisfy ac gain requirements. One reasonable approach to the initial bias design is to assign one-half of V_{CC} as the drop across the transistor and $3V_{CC}/8$ to the drop across R_C to allow an adequate ac voltage swing capability in the collector circuit. This rule-of-thumb approach allows the drop across R_E to be $V_{CC}/8$, so that R_E will have a value of about $R_C/3$. Thus, for V_{CC}, I_{CQ}, and $I_{BQ} = I_{CQ}/\beta$ specified, we select

$$R_C = \frac{3V_{CC}}{8I_{CQ}} \tag{7.2-1}$$

$$R_E = \frac{V_{CC}}{8(I_{CQ} + I_{BQ})} = \frac{V_{CC}\beta}{8(1 + \beta)I_{CQ}} \tag{7.2-2}$$

To choose R_1 and R_2, note that V_B in Fig. 7.2-1 is equal to $V_E + V_{BE}$, which is approximately $(V_{CC}/8) + 0.7$ V for silicon. If we select $I_2 = 5I_{BQ}$, then

$$R_2 \approx \frac{0.7 + (V_{CC}/8)}{5I_{BQ}} \tag{7.2-3}$$

$$R_1 \approx \frac{V_{CC} - V_B}{6I_{BQ}} = \frac{(7V_{CC}/8) - 0.7}{6I_{BQ}} \tag{7.2-4}$$

We illustrate the bias design with an example.

EXAMPLE 7.2-1

We find the rule-of-thumb dc design for a silicon *npn* BJT with $\beta = 65$ when the Q-point is defined by $I_{CQ} = 13.0$ mA and $I_{BQ} = 0.2$ mA. The dc supply voltage is to be $V_{CC} = 12.0$ V. The required resistors are found from (7.2-1) through (7.2-4):

$$R_C = \frac{3(12)}{8(13 \times 10^{-3})} = \frac{4.5}{13}\ \text{k}\Omega \approx 346\ \Omega$$

$$R_E = \frac{65(12)}{8(66)(13 \times 10^{-3})} \approx 113.6\ \Omega$$

$$R_1 \approx \frac{[7(12)/8] - 0.7}{6(0.2 \times 10^{-3})} \approx 8167\ \Omega$$

$$R_2 \approx \frac{[0.7 + (12/8)]}{5(0.2 \times 10^{-3})} = 2200\ \Omega$$

Other bias methods are possible for BJTs (see Problems 7-1 and 7-2).

7.3 BIASING THE FET

In this section we first consider biasing of the junction field-effect transistor (JFET) and then illustrate how to bias metal-oxide-semiconductor FETs (MOSFETs).

Biasing of JFET Devices

One of the most popular and practical biasing methods for a JFET is sketched in Fig. 7.3-1. The circuit is similar in form to that used in Fig. 7.2-1 for the BJT, but operation is different. Because the gate current of a JFET is usually negligible we have

$$V_G = \frac{V_{DD}R_2}{R_1 + R_2} \tag{7.3-1}$$

The bias circuit's dc behavior is best examined by using the JFET's transfer characteristic. It is the plot of drain current i_D versus gate-source voltage v_{GS},

$$i_D = I_{DSS}\left(1 - \frac{v_{GS}}{V_p}\right)^2 \tag{7.3-2}$$

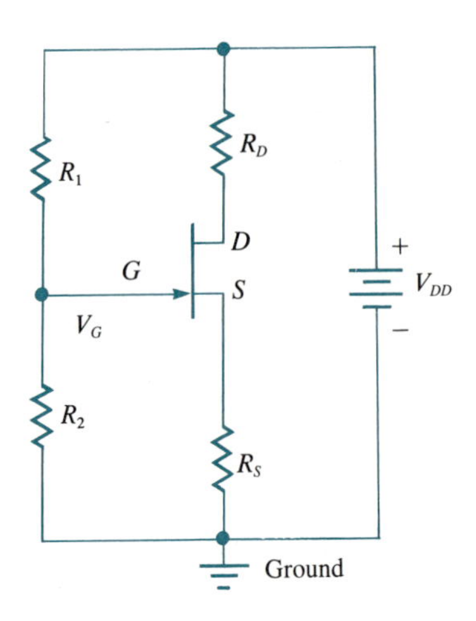

FIG. 7.3-1. A biasing method for a JFET.

as given by (6.6-9).† It is plotted in Fig. 7.3-2. To locate the operating point on the transfer characteristic, we must construct a load line. The equation for the load line is found from applying Kirchhoff's voltage law to the loop in Fig. 7.3-1 that contains R_2 and R_S:

$$i_D = \frac{V_G - v_{GS}}{R_S} \tag{7.3-3}$$

This function is plotted in Fig. 7.3-2. The operating point Q is the intersection of the load line and the transfer characteristic. A good dc design attempts to maintain a nearly constant Q point so that drain current changes are small. This condition is achieved by making V_G and R_S both large (for a given I_{DQ}) to give a small load-line slope through Q.

A systematic bias-point design procedure does not exist that serves all applications. However, we can outline a simple rule-of-thumb procedure that will serve many problems. For a given transistor (I_{DSS} and V_p specified) and a given source voltage V_{DD} the procedure must determine V_G, R_S, R_D, R_2, and R_1. First, we select V_G and R_S.

Both V_G and R_S are determined once point Q (Fig. 7.3-2) is specified in some logical and reasonable way. To establish point Q, we note that large I_{DQ} corresponds to less dc stability and, as can be shown, larger ac gain. A reasonable compromise between high stability and high gain is to choose

$$I_{DQ} = \frac{I_{DSS}}{3} \tag{7.3-4}$$

---

† Ideally, this characteristic is independent of the drain-source voltage v_{DS}, as given. However, i_D does depend on v_{DS} to a first approximation as given in (6.6-10). In the present discussions we assume that the effect of v_{DS} is small enough to be ignored.

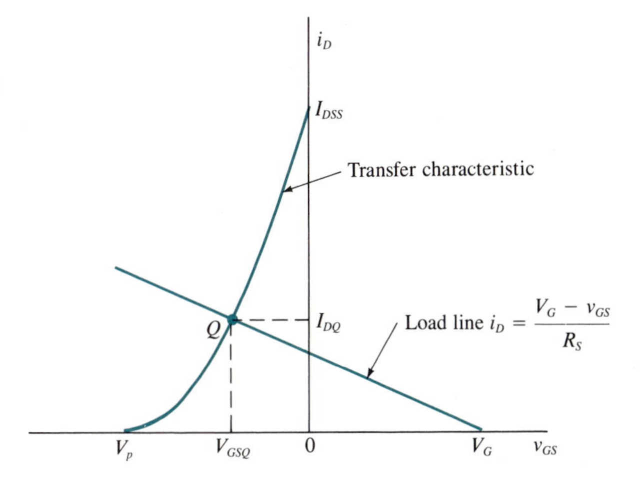

FIG. 7.3-2. Transfer characteristic and the load line for the biasing of a JFET.

On substituting (7.3-4) into (7.3-2), we determine V_{GSQ}, the gate-source voltage at the operating point,

$$V_{GSQ} = \left(\frac{\sqrt{3} - 1}{\sqrt{3}}\right) V_p \approx 0.423 V_p \tag{7.3-5}$$

Next, we observe from Fig. 7.3-2 that choosing

$$V_G = -1.5 V_p \tag{7.3-6}$$

will give a reasonably low slope to the load line. Other values can be selected but (7.3-6) is a reasonable compromise. The voltage drop across R_S is now $V_G - V_{GSQ} \approx -1.5V_p - 0.423V_p = -1.923V_p$, so

$$R_S = \frac{-1.923 V_p}{I_{DQ}} = \frac{-1.923 V_p}{I_{DSS}/3} = \frac{-5.768 V_p}{I_{DSS}} \tag{7.3-7}$$

To select R_2 and R_1, we observe that R_2 can be chosen arbitrarily. If we select

$$R_2 = 100 R_S \tag{7.3-8}$$

to keep the resistance large that is across the gate, we can solve (7.3-1) for R_1:

$$R_1 = \frac{R_2(V_{DD} - V_G)}{V_G} = -100 R_S\left(\frac{V_{DD}}{1.5 V_p} + 1\right) \tag{7.3-9}$$

To find R_D, we must establish how the voltage drops across R_D and the transistor (V_{DSQ}) are to be assigned. The total of these two drops is V_{DD} minus the drop across R_S. We choose to make the transistor's drop equal to that across R_D plus the pinch-off voltage (necessary to maintain active-region operation):

$$V_{DD} - I_{DQ} R_S = -V_p + 2 I_{DQ} R_D \tag{7.3-10}$$

or

$$R_D = \frac{V_{DD} - I_{DQ} R_S + V_p}{2 I_{DQ}} = \frac{3(V_{DD} + 2.923 V_p)}{2 I_{DSS}} \tag{7.3-11}$$

When we use the dc design procedure, the source's voltage V_{DD} must be larger than the minimum necessary to maintain adequate voltage swings for ac signals. A value of $-4.923V_p$ is a reasonable minimum that will allow peak collector voltage swings of $|V_p|$, although with some distortion.

We develop an example to illustrate the dc design procedure.

EXAMPLE 7.3-1

We select V_G, R_S, R_2, R_1, and R_D to bias a JFET by the given procedure. Assume $V_p = -4$ V and $I_{DSS} = 20$ mA for the JFET and a source of voltage $V_{DD} = 24$ V. From (7.3-7) through (7.3-9):

$$R_S = \frac{-5.768V_p}{I_{DSS}} = \frac{-5.768(-4)}{20 \times 10^{-3}} = 1153.6\ \Omega$$

$$R_2 = 100R_S = 115{,}360\ \Omega$$

$$R_1 = -100R_S\left(\frac{V_{DD}}{1.5V_p} + 1\right) = 115{,}360\left(\frac{-24}{1.5(-4)} - 1\right) = 346{,}080\ \Omega$$

From (7.3-11):

$$R_D = \frac{3(V_{DD} + 2.923V_p)}{2I_{DSS}} = \frac{3[24 + 2.923(-4)]}{2(20 \times 10^{-3})} = 923.1\ \Omega$$

We note that $V_{DD} = 24$ V is larger than $-4.923V_p \approx 19.69$ V, so the device is biased above the reasonable minimum, which should produce performance better than the minimum.

In most cases the above design procedure will need to be refined further to make use of available standard-component values and power-supply voltages. It is also necessary to ensure that the design does not exceed any transistor maximum ratings. In some instances it may be desirable to readjust the dc design after the ac design has been made.

Biasing of Depletion-Type MOSFET

The procedures developed for dc biasing of JFETs can be applied directly to biasing of depletion-type MOSFETs. It is only necessary to recognize that v_{GS} can exceed zero in a MOSFET. This fact means that it may be desirable to place the MOSFET's operating point at a higher current I_{DQ} for some applications than in a JFET circuit.

Biasing of Enhancement-Type MOSFET

Figure 7.3-3 illustrates a popular biasing method for the enhancement-type MOSFET. The method *looks* like that of Fig. 7.3-1 for the JFET. Operation is different, however, except for the voltage divider consisting of R_1 and R_2. Since

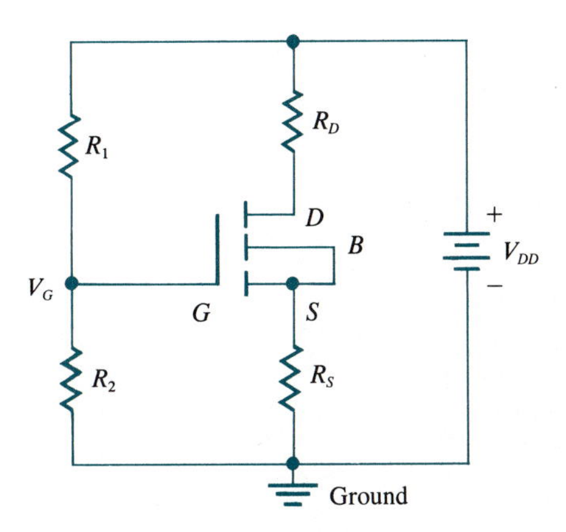

FIG. 7.3-3. A biasing method for an *n*-channel enhancement-type MOSFET.

the gate draws no current, V_G is given by

$$V_G = \frac{V_{DD}R_2}{R_1 + R_2} \tag{7.3-12}$$

In the JFET's bias circuit of Fig. 7.3-1 the voltage developed across R_S by the drain current has the correct polarity to aid in producing the correct gate-source voltage V_{GSQ}.† Thus, R_S helps provide bias as well as *Q*-point stability. In the MOSFET's circuit of Fig. 7.3-3 R_S is used *only* to provide *Q*-point stability. Its voltage drop is *not* of the correct polarity to establish a quiescent point.

Bias-circuit analysis and design are somewhat similar to that for the JFET in that a transfer characteristic and load line are used. For the (*n*-channel) enhancement-type MOSFET, however, the transfer characteristic applies only for v_{GS} positive because the threshold voltage V_t is positive, as shown in Fig. 7.3-4. The *Q*-point's stability improves as both R_S and V_G are made larger. If the small effect of v_{DS} on i_D is neglected, drain current in the active region is given by

$$i_D = K(v_{GS} - V_t)^2 \tag{7.3-13}$$

from (6.7-8). Here K can be determined from the transfer characteristic given in the manufacturer's data sheets, where V_t is also specified. For a desired *Q*-point current I_{DQ} the required gate-source voltage V_{GSQ} is

$$V_{GSQ} = V_t + \sqrt{\frac{I_{DQ}}{K}} \tag{7.3-14}$$

At this point the bias designer can select V_{DSQ} to give a desired operating point on the I_D-V_{DS} curves of the device. Next, for a specified V_{DD} the total voltage drop across R_D and R_S is $V_{DD} - V_{DSQ}$, so

$$R_S + R_D = \frac{V_{DD} - V_{DSQ}}{I_{DQ}} \tag{7.3-15}$$

---

† It is also possible to obtain V_{GSQ} entirely from the drop across R_S by setting $V_G = 0$, that is, by removing R_1. The resulting simple circuit is referred to as *self-bias*.

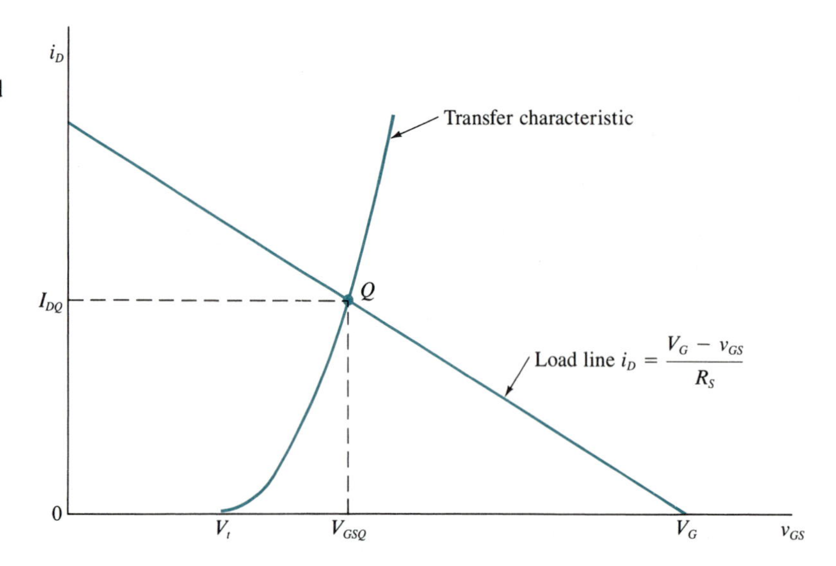

FIG. 7.3-4. The transfer characteristic and load line for the n-channel enhancement-type MOSFET.

The choice of individual values of R_D and R_S can be made by trading off ac gain (larger R_D) against dc stability (larger R_S). Once R_S is chosen, then V_G must be set by

$$V_G = V_{GSQ} + I_{DQ}R_S \tag{7.3-16}$$

Finally, R_1 and R_2 can be arbitrarily selected to produce V_G, but they should both be large to keep the gate's impedance large.

The dc design approach of the preceding paragraph is not very systematic but is probably more like what has to be done for good bias-point selection. We take an example that should illustrate some of the trade-offs involved.

EXAMPLE 7.3-2

A high-power n-channel enhancement MOSFET, the MFE 930, has $V_t = 3.0$ V, and K is found to equal 0.12 A/V^2 from the manufacturer's transfer characteristic. From the i_D-v_{DS} (output) characteristic it is desired to operate with $I_{DQ} = 400$ mA at $V_{DSQ} = 10.0$ V when $V_{DD} = 20$ V. We determine V_{GSQ}, V_G, R_D, R_S, R_1, and R_2. By solving (7.3-14), we obtain V_{GSQ}:

$$V_{GSQ} = V_t + \sqrt{\frac{I_{DQ}}{K}} = 3 + \sqrt{\frac{0.4}{0.12}} \approx 4.83 \text{ V}$$

Since the supply has 20 V and 10 V is across the transistor, we desire about 10 V for ac swing across R_D. However, to compromise, we allow some drop across R_S for dc stability, say 3 V. Thus,

$$R_D = \frac{7 \text{ V}}{0.4 \text{ A}} = 17.5\ \Omega \qquad \text{and} \qquad R_S = \frac{3 \text{ V}}{0.4 \text{ A}} = 7.5\ \Omega$$

From (7.3-16):

$$V_G = V_{GSQ} + I_{DQ}R_S = 4.83 + 3 = 7.83 \text{ V}$$

We arbitrarily select R_2 large to keep the gate's impedance large and then solve (7.3-12) for R_1:

$$R_2 = 10{,}000\ \Omega \qquad \text{(arbitrary choice)}$$

$$R_1 = \frac{R_2(V_{DD} - V_G)}{V_G} = 10^4\,\frac{20 - 7.83}{7.83} \approx 15.54\text{ k}\Omega$$

Final checks against manufacturer's data show that voltage, current, and power ratings of the device are not exceeded.

7.4 BJT AMPLIFIERS

In general, an amplifier can be made up of a cascade of several stages. A stage consists of an elementary amplifier that normally uses only one transistor. The cascade is formed by making the first stage's output be the input to the second stage, the second's output feeds the third's input, and so on. In this section we study three basic forms of amplifier stages that use a bipolar-junction transistor (BJT).

Common-Emitter Configuration

Figure 7.4-1(a) illustrates a BJT amplifier stage in which the emitter part of the circuit is common between the input and output circuits. It is therefore called a *common-emitter* (CE) *amplifier*. Resistors R_1, R_2, R_C, and R_E are mainly set by biasing. A source, modeled here by its Thévenin's generator voltage v_s and resistance R_s, provides an ac (changing) voltage to the transistor's base. This voltage is the input to be amplified. The amplified output ac voltage v_L appears across the load resistor R_L. Resistor R_L could represent the input resistance of the next stage if a cascade is involved; otherwise, it would represent the resistance of whatever device is to accept the output signal. Capacitors C_B, C_C, and C_E are designed to be short circuits at the lowest frequency of interest. Thus, C_E would be made large enough that $1/\omega C_E$ is small relative to R_E in parallel with the impedance looking into the emitter at the smallest ω of interest. Similarly, the reactances of C_C and C_B would be chosen small relative to the resistances in their parts of the circuit. Capacitors C_B and C_C appear as short circuits to the ac signals but block the dc voltages and currents from one part of the circuit from coupling into another part (such as previous or next stages in a cascade). Capacitor C_E is mainly used to increase gain. It is called a *bypass capacitor* because it bypasses the ac current around R_E so that no significant ac voltage is generated across R_E (which, as will be seen, decreases gain).

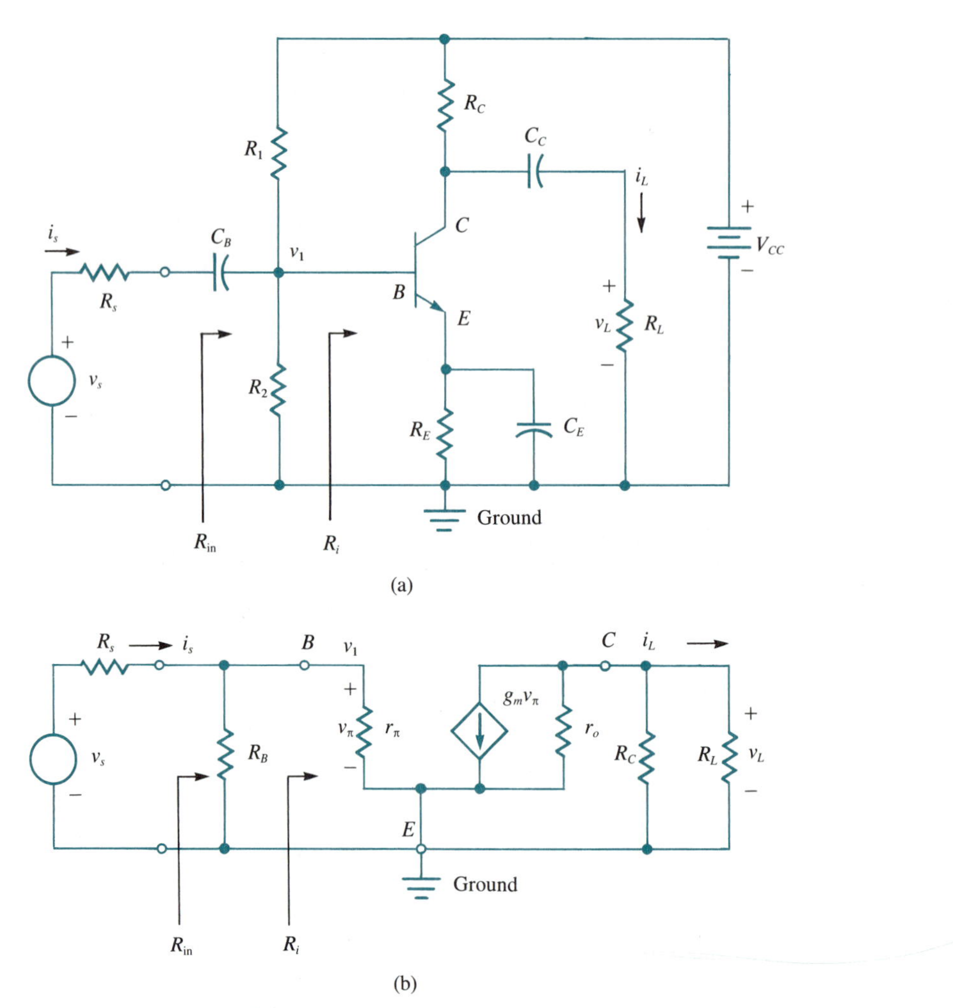

FIG. 7.4-1. (a) The common-emitter form of a BJT amplifier and (b) its small-signal ac equivalent circuit.

The small-signal ac equivalent circuit for Fig. 7.4-1(a) is given in (b). It uses the small-signal model of Fig. 6.5-6 for the transistor. In this circuit R_B represents R_1 and R_2 in parallel, a condition that is established because the dc source V_{CC} ideally has zero impedance to ac signals, and the model of Fig. 7.4-1(b) applies only to ac signals. Resistor R_i represents the resistance between the transistor's base and ground as seen looking into the base. Resistor R_{in} represents the total input resistance seen by the source. We shall omit details of analysis and only summarize the important results:†

$$R_{in} = R_1 \,||\, R_2 \,||\, R_i = R_B \,||\, R_i = \frac{R_B r_\pi}{R_B + r_\pi} \tag{7.4-1}$$

† The notation $R_1 \,||\, R_2$ implies the resistance of R_1 and R_2 in parallel. Similarly, $R_1 \,||\, R_2 \,||\, R_3$ is the resistance of three resistors R_1, R_2, and R_3, all in parallel.

$$R_B = R_1 \,||\, R_2 = \frac{R_1 R_2}{R_1 + R_2} \tag{7.4-2}$$

$$R_i = r_\pi \tag{7.4-3}$$

$$v_1 = v_\pi = \frac{v_s R_{\text{in}}}{R_s + R_{\text{in}}} \tag{7.4-4}$$

Voltage and current gains, denoted by A_{v1} and A_i, respectively, are defined by

$$A_{v1} = \frac{v_L}{v_1} = \frac{-g_m R_L \,(r_o \,||\, R_C)}{R_L + (r_o \,||\, R_C)} \tag{7.4-5}$$

$$A_i = \frac{i_L}{i_s} = \frac{-g_m (r_o \,||\, R_C)(r_\pi \,||\, R_B)}{R_L + (r_o \,||\, R_C)} \tag{7.4-6}$$

EXAMPLE 7.4-1

We find the ac voltage and current gains for the transistor biased in Example 7.2-1 when $R_L = 500\ \Omega$ is assumed and $V_A = 75$ V for the transistor. The amplifier is shown in Fig. 7.4-2. For this transistor

$$g_m \approx \frac{I_{CQ}}{V_T} = \frac{13 \times 10^{-3}}{25.861 \times 10^{-3}} \approx 0.503 \text{ S}$$

from (6.5-8). By using (6.5-9), we get

$$r_o \approx \frac{V_A}{I_{CQ}} = \frac{75}{13.0 \times 10^{-3}} \approx 5769\ \Omega$$

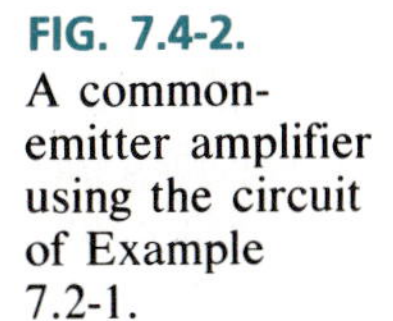

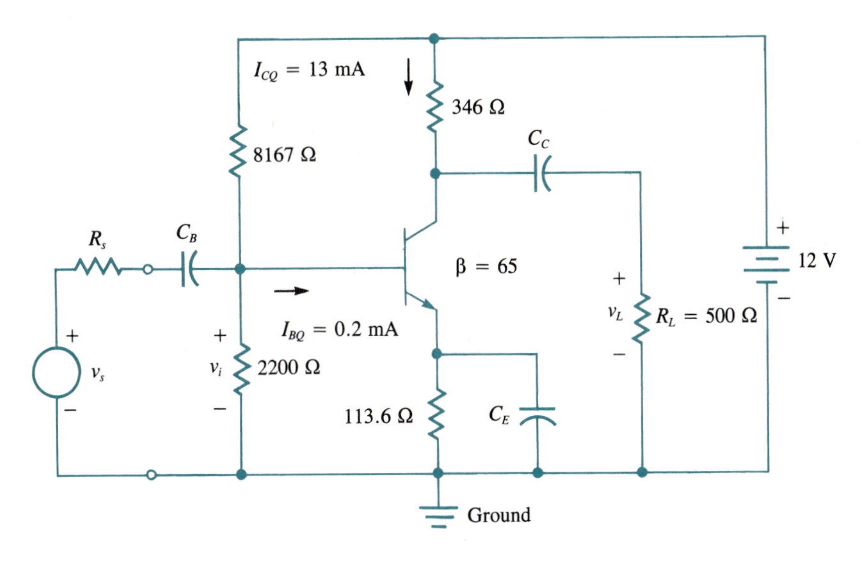

FIG. 7.4-2. A common-emitter amplifier using the circuit of Example 7.2-1.

Next,

$$r_o \,||\, R_C = \frac{5769(346)}{5769 + 346} \approx 326.4\ \Omega$$

so

$$A_{v1} = \frac{-0.503(500)(326.4)}{500 + 326.4} \approx -99.3$$

from (7.4-5). For current gain we first need r_π from (6.5-11):

$$r_\pi = \frac{\beta}{g_m} = \frac{65}{0.503} \approx 129.2\ \Omega$$

and R_{in}, as given by

$$R_{\text{in}} = (R_1 \,||\, R_2 \,||\, r_\pi) = \frac{1}{(1/R_1) + (1/R_2) + (1/r_\pi)}$$

$$= \frac{1}{(1/8167) + (1/2200) + (1/129.2)} \approx 120.2\ \Omega$$

Finally, from (7.4-6)

$$A_i = \frac{-0.503(326.4)(120.2)}{500 + 326.4} \approx -23.88$$

The CE amplifier is capable of giving both high voltage and high current gains. In both cases the gains are negative, indicating that the CE configuration produces signal inversion. Input resistance is relatively low in a CE amplifier; it typically ranges from a few hundred ohms to, at most, a few thousand ohms.

Input resistance of a CE amplifier can be increased, at the expense of reduced voltage and current gains, by leaving a portion of the emitter resistance unbypassed, as shown in Fig. 7.4-3(a) as R_{E1}. The small-signal ac equivalent circuit of Fig. 7.4-3(b) now applies, where it has been assumed for simplicity that r_o is large. Analysis now gives

$$R_i = r_\pi + R_{E1}(1 + g_m r_\pi) \qquad r_o \to \infty \tag{7.4-7}$$

$$A_{v1} = \frac{-g_m R_L R_C}{R_L + R_C}\left(\frac{r_\pi}{R_i}\right) \qquad r_o \to \infty \tag{7.4-8}$$

$$A_i = \frac{-g_m R_C R_{\text{in}}}{R_L + R_C}\left(\frac{r_\pi}{R_i}\right) \qquad r_o \to \infty \tag{7.4-9}$$

where

$$R_{\text{in}} = R_1 \,||\, R_2 \,||\, R_i \tag{7.4-10}$$

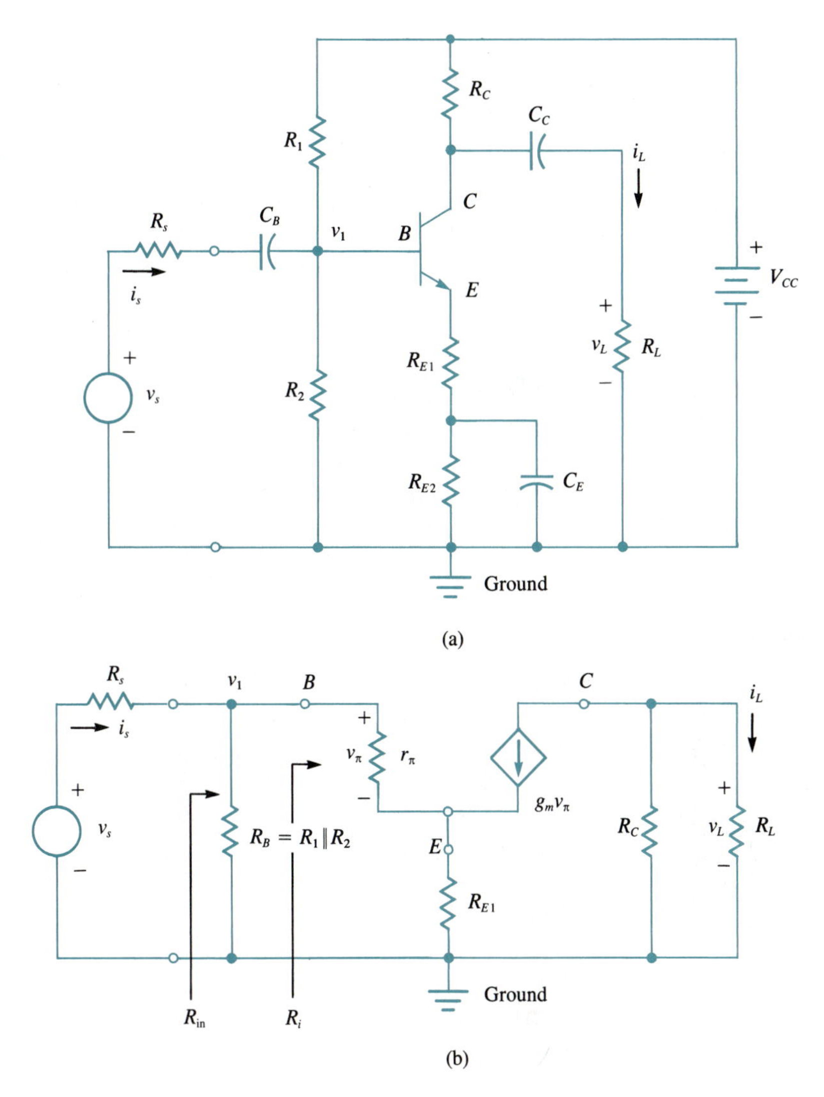

FIG. 7.4-3. (a) A CE amplifier having an unbypassed emitter resistance R_{E1} and (b) the ac equivalent circuit for small signals.

Since

$$\beta \approx g_m r_\pi \tag{7.4-11}$$

from (6.5-11), (7.4-7) indicates that the input resistance has increased by $R_{E1}(1 + \beta)$ over its value r_π, which occurs when the emitter resistance is fully bypassed. In other words, input resistance has increased by a factor $1 + [R_{E1}(1 + \beta)/r_\pi] = R_i/r_\pi$ compared with the input resistance when the full emitter's resistor is bypassed. On comparing (7.4-8) with (7.4-5), we see that voltage gain is reduced by the reciprocal of this factor. It might also appear that current gain is also reduced by this same factor when (7.4-9) is compared with (7.4-6). Indeed, such a result would be true except that R_{in} is different in the two cases. Closer examination reveals A_i with R_{E1} unbypassed is reduced by a factor equal to the reciprocal of $1 + [R_{E1}(1 + \beta)/(r_\pi + R_B)]$ relative to A_i with R_{E1} bypassed. Here R_B is given by (7.4-2).

Although the unbypassed emitter's resistance in a CE amplifier reduces current and voltage gains, which are disadvantages, there are also advantages that accrue. One is that the amplifier's bandwidth increases. Another is that the signal level that can be accommodated at the input is larger because a large fraction of the input v_1 appears across R_{E1} and not across the base-emitter junction. In fact, the base-emitter voltage is reduced by the same factor as is the voltage gain.

Common-Collector Configuration

A transistor-amplifier stage can be configured so that its collector forms a common terminal between the input and output circuits. The most common configuration, called a *common-collector* (CC) *amplifier* (also known as an *emitter-follower*), is shown in Fig. 7.4-4(a). Resistors R_1, R_2, and R_E are established by

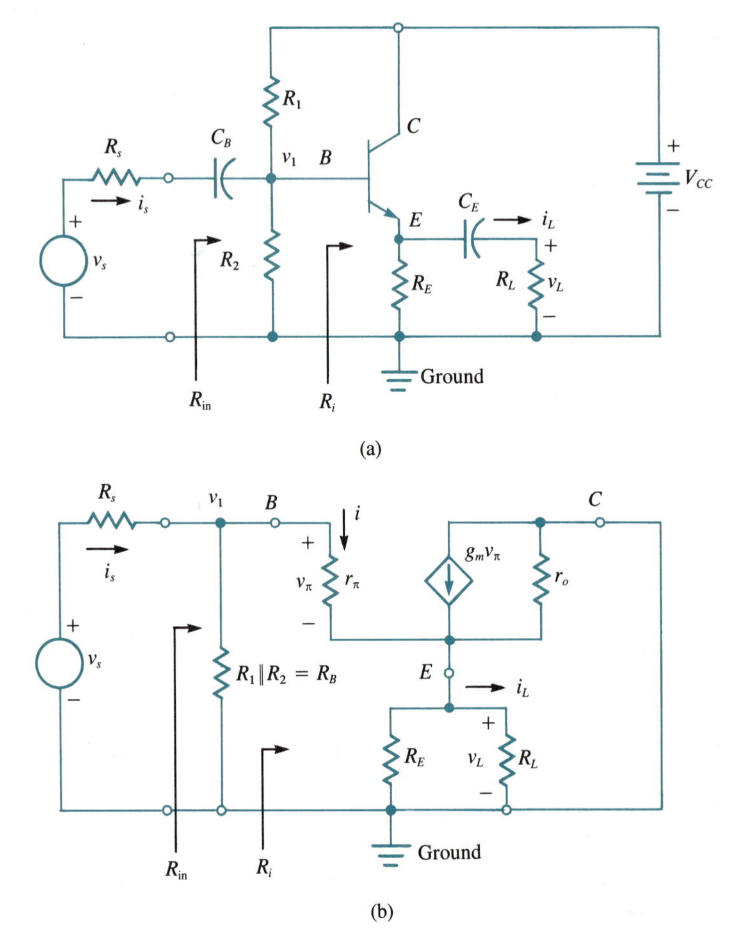

FIG. 7.4-4. (a) The common-collector (emitter-follower) form of a BJT amplifier and (b) its small-signal equivalent circuit.

biasing. The output voltage is taken across resistor R_L. In some cases R_E may serve as the output load. The input signal is shown applied to the base from a source having a Thévenin's voltage v_s and impedance R_s. Capacitors C_B and C_E are selected large enough to appear as short circuits at the lowest frequency of interest in the signal v_s. The amplifier's small-signal ac equivalent circuit is shown in Fig. 7.4-4(b).

Analysis of Fig. 7.4-4(b) gives the following important results:

$$R_i = r_\pi + R_W(1 + g_m r_\pi) \tag{7.4-12}$$

where

$$R_W = r_o \,||\, R_E \,||\, R_L \tag{7.4-13}$$

$$R_{\text{in}} = \frac{R_B R_i}{R_B + R_i} \tag{7.4-14}$$

$$R_B = \frac{R_1 R_2}{R_1 + R_2} \tag{7.4-15}$$

$$A_{v1} = \frac{v_L}{v_1} = \frac{R_W(1 + g_m r_\pi)}{r_\pi + R_W(1 + g_m r_\pi)} \tag{7.4-16}$$

$$A_i = \frac{i_L}{i_s} = \frac{v_L R_{\text{in}}}{R_L v_1} = \frac{R_{\text{in}}}{R_L} A_{v1} = \frac{R_{\text{in}} R_W(1 + g_m r_\pi)}{R_L[r_\pi + R_W(1 + g_m r_\pi)]} \tag{7.4-17}$$

It is often true that r_o is large enough to be ignored. When this condition is true, we have

$$R_i \approx r_\pi + (1 + g_m r_\pi)(R_E \,||\, R_L) \qquad r_o \to \infty \tag{7.4-18}$$

$$A_{v1} \approx \frac{(1 + g_m r_\pi)(R_E \,||\, R_L)}{r_\pi + (1 + g_m r_\pi)(R_E \,||\, R_L)} \approx 1 \qquad r_o \to \infty \tag{7.4-19}$$

$$A_i \approx \frac{(1 + g_m r_\pi)(R_E \,||\, R_L) R_B}{R_L[r_\pi + R_B + (1 + g_m r_\pi)(R_E \,||\, R_L)]} \qquad r_o \to \infty \tag{7.4-20}$$

In most CC designs $\beta = g_m r_\pi$ is large enough that $r_e = r_\pi/(1 + \beta) \ll R_W$, so $A_{v1} \approx 1$. Thus, we find that voltage gain of the CC amplifier is approximately unity, but never exceeds unity, from (7.4-19). On the other hand, current gain is large from (7.4-17), since $R_{\text{in}} \gg R_L$ is typical. Finally, input impedance can be relatively large, even for relatively small R_L, because of the factor $(1 + g_m r_\pi) = (1 + \beta)$ in (7.4-18).

Common-Base Configuration

If the base of a BJT is ac-grounded so as to become the common terminal for the input and output of an amplifier stage, it is called a *common-base* (CB) *amplifier*. Figure 7.4-5(a) depicts a CB amplifier where a source with Thévenin's resistance R_s and voltage v_s is applied to the emitter. The output is taken from the collector across resistor R_L. As usual, R_1, R_2, R_C, and R_E are selected through biasing; and capacitors C_B, C_C, and C_E are all chosen large enough to be short circuits at the

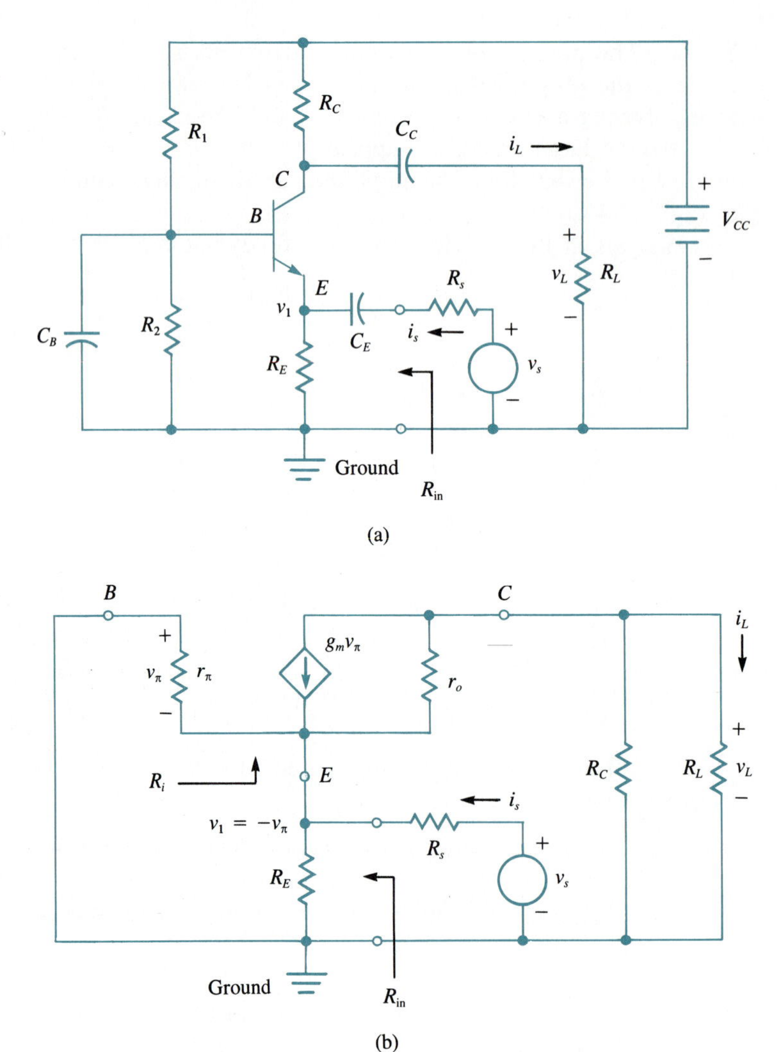

FIG. 7.4-5. (a) The common-base form of a BJT amplifier and (b) its small-signal ac equivalent circuit.

lowest frequency of interest in the signal v_s. The small-signal ac equivalent circuit is given in Fig. 7.4-5(b). Analysis of the equivalent circuit gives the following important quantities:

$$A_{v1} = \frac{v_L}{v_1} = \frac{v_L}{-v_\pi} = \frac{R_H(1 + g_m r_o)}{R_H + r_o} \tag{7.4-21}$$

$$R_H = R_C \,||\, R_L = \frac{R_C R_L}{R_C + R_L} \tag{7.4-22}$$

$$R_i = \frac{r_\pi(r_o + R_H)}{r_\pi + R_H + r_o(1 + g_m r_\pi)} \tag{7.4-23}$$

$$R_{\text{in}} = \frac{R_E R_i}{R_E + R_i} = \frac{(r_\pi \,||\, R_E)(r_o + R_H)}{r_o + R_H + (r_\pi \,||\, R_E)(1 + g_m r_o)} \tag{7.4-24}$$

$$A_i = \frac{R_{\text{in}}}{R_L} A_{v1} = \frac{R_H(r_\pi \,||\, R_E)(1 + g_m r_o)}{R_L[r_o + R_H + (r_\pi \,||\, R_E)\,(1 + g_m r_o)]} \tag{7.4-25}$$

If r_o is considered very large in relation to all other resistance values, we have the following special forms of the above results:

$$R_i \approx \frac{r_\pi}{1 + g_m r_\pi} \qquad r_o \to \infty \tag{7.4-26}$$

$$R_{\text{in}} \approx \frac{(r_\pi \,||\, R_E)}{1 + g_m(r_\pi \,||\, R_E)} \qquad r_o \to \infty \tag{7.4-27}$$

$$A_{v1} \approx g_m(R_C \,||\, R_L) \qquad r_o \to \infty \tag{7.4-28}$$

$$A_i \approx \frac{g_m(R_C \,||\, R_L)(r_\pi \,||\, R_E)}{R_L[1 + g_m(r_\pi \,||\, R_E)]} \qquad r_o \to \infty \tag{7.4-29}$$

Because the CB amplifier has a low input resistance, its usual application is as a current amplifier, accepting an input current into its low impedance and providing an output current into a high impedance. By analogy with the voltage-in voltage-out emitter-follower, the CB amplifier can be called a *current-follower*.

Comparison of CE, CC, and CB Configurations

To simplify a comparison of the three BJT-amplifier configurations, we shall assume r_o is large enough to be approximated as infinite.

To compare R_i, we bring together (7.4-3), (7.4-18), and (7.4-26):

$$R_i = \begin{cases} r_\pi & \text{CE} \\ r_\pi + (1 + g_m r_\pi)(R_E \,||\, R_L) & \text{CC} \\ \dfrac{r_\pi}{1 + g_m r_\pi} & \text{CB} \end{cases} \tag{7.4-30}$$

We see that R_i is largest for the CC configuration, smallest for the CB amplifier, and in between these extremes for the CE unit. A similar statement follows for the input resistance R_{in}.

Voltage-gain expressions to be compared are (7.4-5), (7.4-19), and (7.4-28):

$$A_{v1} \approx \begin{cases} -g_m(R_C \,||\, R_L) & \text{CE} \\ \dfrac{(1 + g_m r_\pi)(R_E \,||\, R_L)}{r_\pi + (1 + g_m r_\pi)(R_E \,||\, R_L)} \approx 1 & \text{CC} \\ g_m(R_C \,||\, R_L) & \text{CB} \end{cases} \tag{7.4-31}$$

These equations show that CE and CB configurations give the same magnitude of gain,† but there is a sign inversion with the CE amplifer. The magnitude of these gains can be larger than unity, whereas A_{v1} in the CC unit cannot exceed unity.

Current gains for the three configurations are

$$A_i \approx \begin{cases} \dfrac{-g_m(R_C \,||\, R_L)(r_\pi \,||\, R_B)}{R_L} & \text{CE} \\[2ex] \dfrac{(1 + g_m r_\pi)(R_E \,||\, R_L)R_B}{R_L[r_\pi + R_B + (1 + g_m r_\pi)(R_E \,||\, R_L)]} & \text{CC} \\[2ex] \dfrac{g_m(R_C \,||\, R_L)(r_\pi \,||\, R_E)}{R_L[1 + g_m(r_\pi \,||\, R_E)]} & \text{CB} \end{cases} \qquad (7.4\text{-}32)$$

from (7.4-6), (7.4-20), and (7.4-29). Since the resistance of any two resistors in parallel is less than the resistance of either of the two resistors taken separately, we may easily show that $|A_i|(\text{CE}) < g_m r_\pi = \beta$, $A_i(\text{CC}) < 1 + \beta$, and $A_i(\text{CB}) < 1$ from (7.4-32). Thus, both the CE and CC configurations are capable of giving large current-gain magnitudes, but the CB stage has a current gain less than unity.

7.5 FET AMPLIFIERS

Field-effect transistors (FETs) can be used to construct amplifier stages in three basic configurations analogous to the BJT amplifiers. These configurations are called common-source (CS), common-drain (CD), and common-gate (CG), and they are analogous to the CE, CC, and CB amplifiers, respectively. In this section we summarize the important equations for CS, CD, and CG amplifiers using junction FETs (JFETs). In the next section we indicate how the JFET results are modified when MOSFETs are used.

JFET in Common-Source Configuration

Figure 7.5-1(a) depicts an amplifier that uses an n-channel JFET in the CS configuration. As usual, capacitors C_G, C_D, and C_S are selected to be large enough that they are approximately short circuits at the lowest frequency of interest in the input signal v_s from the source of internal resistance R_s. The output signal is taken across R_L. All other resistors are established mainly by bias design. The amplifier's small-signal ac equivalent circuit is drawn in Fig. 7.5-1(b), where the JFET is replaced by its equivalent form as given in Fig. 6.6-5(a).

The most important parameters of interest are input resistance R_i of the JFET, input resistance R_{in} that is seen by the source, voltage gain $A_{v1} = v_L/v_1$,

---

† A more careful comparison shows that the CB amplifier's voltage gain can be slightly larger than for the CE amplifier.

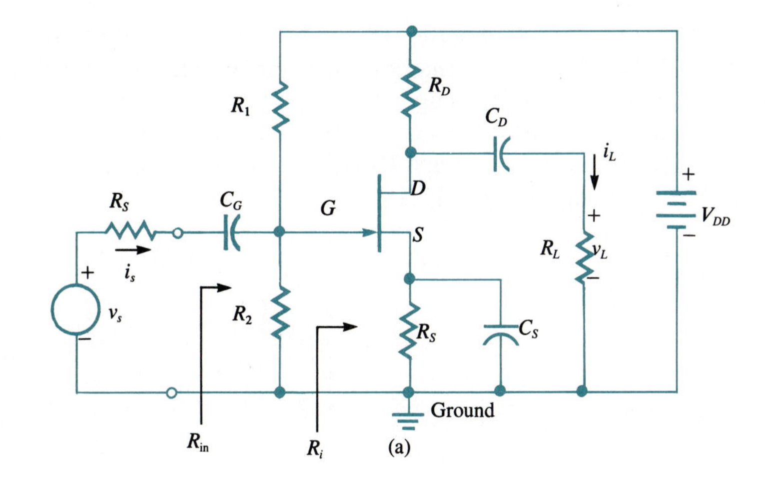

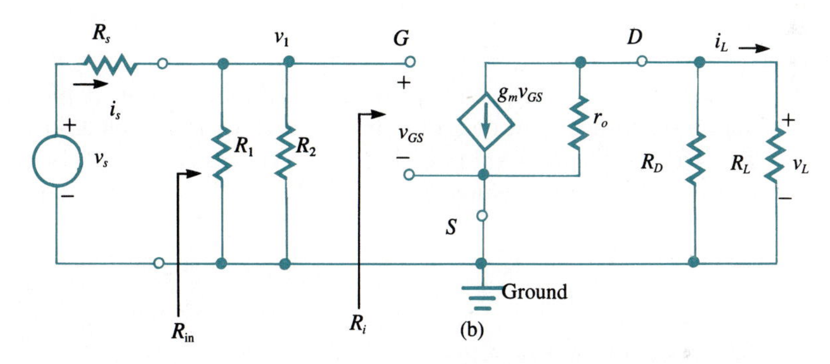

FIG. 7.5-1. (a) A common-source JFET amplifier and (b) its small-signal ac equivalent circuit.

and current gain $A_i = i_L/i_s$. Because the iput impedance of a JFET is very large, we have

$$R_i \approx \infty \tag{7.5-1}$$

$$R_{\text{in}} = R_1 \,||\, R_2 = \frac{R_1 R_2}{R_1 + R_2} \tag{7.5-2}$$

Further analysis gives

$$A_{v1} = \frac{-g_m r_o R_F}{r_o + R_F} \tag{7.5-3}$$

$$A_i = \frac{-g_m r_o R_F (R_1 \,||\, R_2)}{R_L (r_o + R_F)} \tag{7.5-4}$$

where we define

$$R_F = R_D \,||\, R_L = \frac{R_D R_L}{R_D + R_L} \tag{7.5-5}$$

These results indicate that the CS amplifier is capable of large voltage and current gains.

As in a CE amplifer using a BJT, a portion of the source resistance R_S in the JFET CS stage is sometimes left unbypassed. The result is lower amplifier gain, which is a disadvantage, but advantages of increased bandwidth, gain stability, and larger allowable signal amplitudes are realized. Let R_{S1} be the portion of R_S that is not bypassed and let R_{S2} be that part that is bypassed, where $R_S = R_{S1} + R_{S2}$. Analysis now gives

$$A_{v1} = \frac{-g_m r_o R_F}{r_o + R_F + R_{S1}(1 + g_m r_o)} \tag{7.5-6}$$

$$A_i = \frac{-g_m r_o R_F(R_1 \,||\, R_2)}{R_L[r_o + R_F + R_{S1}(1 + g_m r_o)]} \tag{7.5-7}$$

These results become equal to (7.5-3) and (7.5-4), as they must, when $R_{S1} = 0$.

EXAMPLE 7.5-1

We compute R_{in}, A_{v1}, and A_i for a CS JFET amplifier using the nominal device and bias arrangement of Example 7.3-1. Thus, $R_1 = 346{,}080\ \Omega$, $R_2 = 115{,}360\ \Omega$, $R_D = 923.1\ \Omega$, $R_S = 1153.6\ \Omega$, $I_{DQ} = 6.667$ mA, $V_p = -4$ V, and $I_{DSS} = 20.0$ mA. We also shall assume $R_{S1} = 153.6\ \Omega$, $V_A = 90$ V, and load resistor $R_L = 1000\ \Omega$. For R_{in}

$$R_{\text{in}} = R_1 \,||\, R_2 = \frac{346{,}080(115{,}360)}{346{,}080 + 115{,}360} \approx 86{,}520\ \Omega$$

From (6.6-13) and (6.6-12):

$$r_o \approx \frac{V_A}{I_{DQ}} = \frac{90}{6.667 \times 10^{-3}} \approx 13{,}449\ \Omega$$

$$g_m \approx \frac{-2I_{DSS}}{V_p}\sqrt{\frac{I_{DQ}}{I_{DSS}}} = \frac{-2(20 \times 10^{-3})}{-4}\sqrt{\frac{6.667}{20.0}} \approx 5.774 \times 10^{-3}\ \text{S}$$

From (7.5-3) through (7.5-5):

$$R_F = R_D \,||\, R_L = \frac{923.1(1000)}{923.1 + 1000} \approx 480\ \Omega$$

$$A_{v1} = \frac{-g_m r_o R_F}{r_o + R_F} = \frac{(-5.774 \times 10^{-3})(13{,}499)(480)}{13{,}499 + 480} \approx -2.68$$

$$A_i = \frac{R_{\text{in}}}{R_L} A_{v1} \approx \frac{86{,}520}{1000}(-2.68) \approx -232$$

This particular design gives relatively small voltage gain but very large current gain (because R_1 and R_2 were selected so large).

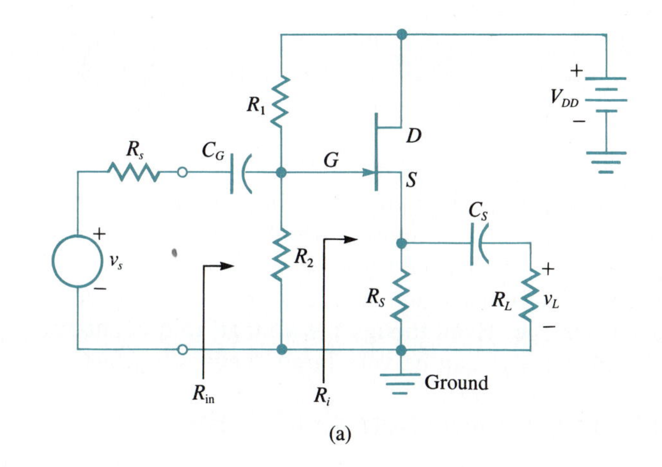

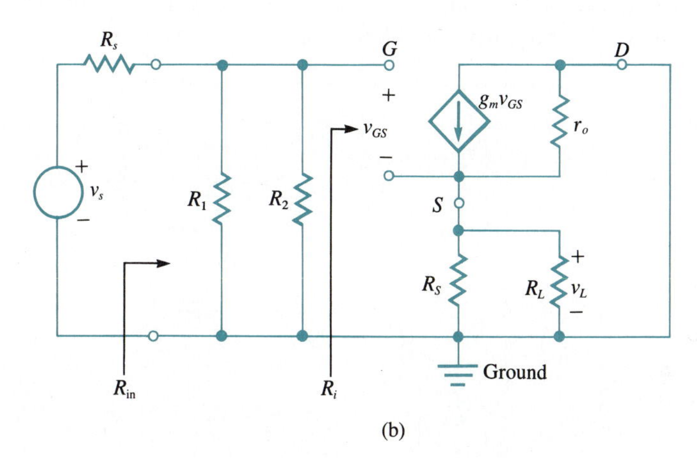

FIG. 7.5-2. (a) A common-drain JFET amplifier and (b) its small-signal ac equivalent circuit.

JFET in Common-Drain Configuration

An amplifier using a JFET in the common-drain (CD) configuration is shown in Fig. 7.5-2(a). Resistors R_1, R_2, and R_S are set in the bias design. Capacitors C_G and C_S are chosen large enough to be short circuits at frequencies in the band of interest (called the midband). A source of voltage v_s and internal resistance R_s is applied to the gate. The output is taken from the source terminal across resistor R_L. The drain is ac-grounded and forms a common terminal between the input and output circuits (hence the name CD). The small-signal ac equivalent circuit is sketched in Fig. 7.5-2(b). Analysis of this circuit produces

$$R_i \approx \infty \tag{7.5-8}$$

$$R_{\text{in}} = R_1 \,||\, R_2 = \frac{R_1 R_2}{R_1 + R_2} \tag{7.5-9}$$

$$A_{v1} = \frac{v_L}{v_1} = \frac{g_m r_o (R_S \,||\, R_L)}{r_o + (R_S \,||\, R_L)(1 + g_m r_o)} \tag{7.5-10}$$

$$A_i = \frac{i_L}{i_s} = \frac{g_m r_o (R_S \,||\, R_L)(R_1 \,||\, R_2)}{R_L[r_o + (R_S \,||\, R_L)(1 + g_m r_o)]} \tag{7.5-11}$$

In many cases $g_m(R_S \,||\, R_L) \gg 1$ and r_o is large, so $A_{v1} \approx 1$ and $v_L \approx v_1$. The load voltage at the source approximately "follows" the input. For this reason the CD amplifier is often called a *source-follower*. It is an excellent buffer device to couple a high-resistance source to a low-resistance load with nearly no loss in signal voltage. Even though the voltage gain is about unity in a source-follower, the current gain can be very large, leading to large *power gain*.

JFET in Common-Gate Configuration

In a common-gate amplifier the input signal is applied between the source and ground, while the output is taken from the drain, as shown in Fig. 7.5-3(a). The

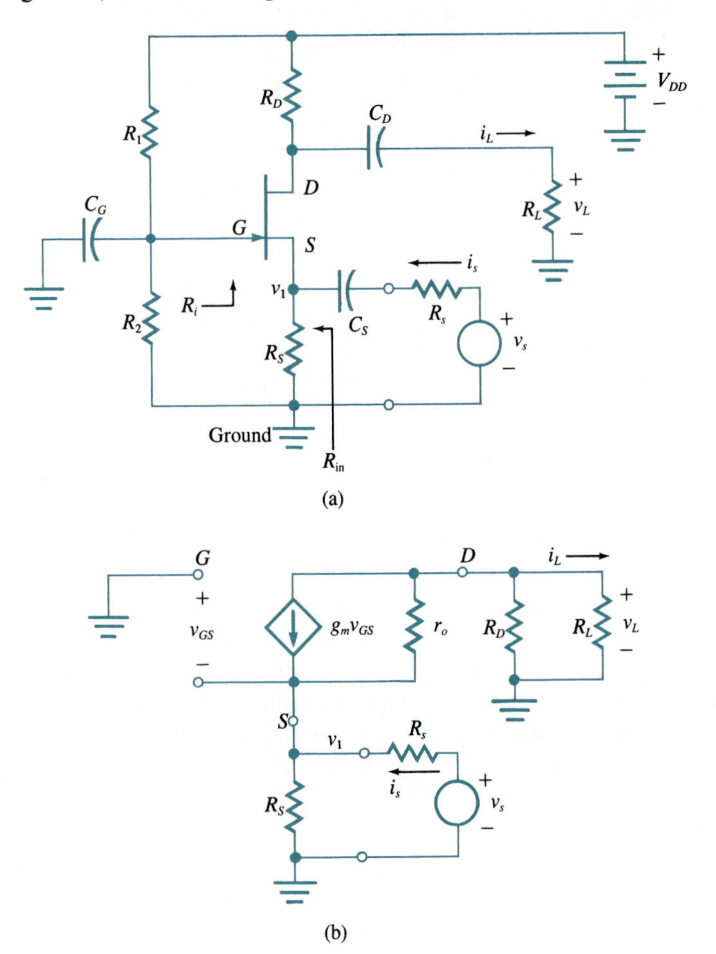

FIG. 7.5-3. (a) A common-gate JFET amplifier and (b) its small-signal ac equivalent circuit.

name *common-gate* derives from the gate's being ac-shorted to ground (through C_G), thereby forming the common terminal between the input and output signals. Resistors R_1, R_2, R_D, and R_S are mainly set by bias-point design. Capacitors C_D and C_S are chosen large enough to be short circuits at the lowest frequency of interest in the signal v_s. Straightforward analysis of the small-signal equivalent circuit of Fig. 7.5-3(b) yields

$$R_{\text{in}} = \frac{R_S(r_o + R_F)}{r_o + R_F + R_S(1 + g_m r_o)} \tag{7.5-12}$$

$$A_{v1} = \frac{v_L}{v_1} = \frac{R_F(1 + g_m r_o)}{r_o + R_F} \tag{7.5-13}$$

$$A_i = \frac{i_L}{i_s} = \frac{R_F(1 + g_m r_o)R_S}{R_L[r_o + R_F + R_S(1 + g_m r_o)]} \tag{7.5-14}$$

where R_F is given by (7.5-5).

Often $g_m R_S \gg 1$ and r_o is large, so $R_{\text{in}} \approx 1/g_m$, and $A_i \approx R_F/R_L = R_D/(R_D + R_L)$, which is less than unity.

Comparison of CS, CD, and CG Configurations

Only the CG configuration has an input resistance R_{in} of special interest, because it is the only case where R_{in} is not very large (often on the order of a few hundred ohms). For the CS and CD configurations $R_{\text{in}} = R_1 \,||\, R_2$, which can usually be selected to be large during the bias design.

Voltage gain can be compared by using (7.5-3), (7.5-10), and (7.5-13):

$$A_{v1} = \begin{cases} \dfrac{-g_m r_o R_F}{r_o + R_F} & \text{CS} \\[2ex] \dfrac{g_m r_o(R_S \,||\, R_L)}{r_o + (R_S \,||\, R_L)(1 + g_m r_o)} & \text{CD} \\[2ex] \dfrac{R_F(1 + g_m r_o)}{r_o + R_F} & \text{CG} \end{cases} \tag{7.5-15}$$

These results show that A_{v1} for the CG configuration is slightly larger than for the CS amplifier. In both cases A_{v1} can exceed unity. For the CD configuration $A_{v1} < 1$ but can be near unity for some parameter choices.

Since current gain equals $R_{\text{in}}A_{v1}/R_L$ for every configuration, it can be almost arbitrarily large for the CS and CD configurations, because R_{in} can be selected arbitrarily and is usually large. In the CG configuration A_i cannot be larger than unity, from (7.5-14). The reader might try to show, as an exercise, that this fact is true.

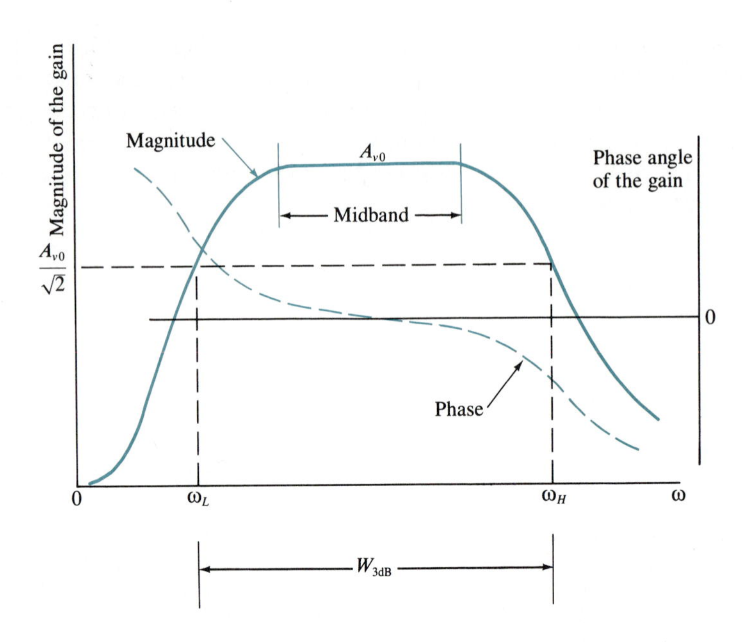

FIG. 7.7-1. The magnitude and phase angle of an amplifier. Also indicated is the 3-dB bandwidth W_{3dB} (rad/s).

7.6 MOSFET AMPLIFIERS

The small-signal ac equivalent circuit for MOSFETs is the same as in Fig. 6.6-5(a) for the JFET. In fact, the equations for g_m and r_o with a depletion MOSFET are the same as for the JFET, as given by (6.6-12) and (6.6-13). For the enhancement MOSFET g_m and r_o are given by (6.7-11) and (6.7-12), respectively.

Because the same small-signal equivalent-circuit form applies to both the JFET and the MOSFETs, all the equations developed in Sec. 7.5 for the JFET apply equally well to the MOSFETs, as long as g_m and r_o are computed properly.

7.7 FREQUENCY RESPONSE OF AMPLIFIERS

Amplifier gain is defined as the ratio of the output signal (voltage or current) to the input signal (voltage or current), when the input is a complex signal as defined in Sec. 4.1. It equals, therefore, the network's transfer function, which is complex in general and not necessarily constant at all frequencies. The variation in gain represents the *frequency response* of the amplifier. A typical response is shown in Fig. 7.7-1 for a voltage amplifier. The amplitude response (magnitude) is most often discussed, because it defines the frequency range of useful gain. However, the phase response is important in transient calculations.

Amplifiers normally are considered most useful at frequencies where gain is nearly constant.† The midband region shown in Fig. 7.7-1 corresponds to this condition. All the amplifiers discussed in the preceding sections of this chapter assumed operation in the midband region.

Amplifier Bandwidth

It is customary to define the bandwidth of an amplifier as the (useful) band between two frequency values, denoted by ω_L and ω_H in Fig. 7.7-1, that correspond to the gain falling to 3 dB below the midband gain. If $W_{3\text{dB}}$ denotes the 3-dB bandwidth, then

$$W_{3\text{dB}} = \omega_H - \omega_L \tag{7.7-1}$$

When ω_L is small, such as in an audio amplifier, where $\omega_L/2\pi$ can be less than 20 Hz, the amplifier is called either baseband or lowpass. In a lowpass unit where direct coupling between stages is used (no coupling capacitors), gain can remain constant all the way down to zero frequency (dc), and there is no low-frequency 3-dB frequency. These are often called *dc amplifiers*. If both ω_L and ω_H have large values, the amplifier is called bandpass. A narrowband bandpass amplifier is one in which $W_{3\text{dB}}$ is small relative the center frequency of the midband region.

Comments on General Analysis and Design

The general problem of *analyzing* any given amplifier to determine ω_L or ω_H is extremely difficult and well beyond the scope of this book. Even for single stages, exact results are cumbersome in most cases and require the development of the theory of poles and zeros in networks, a subject more advanced than is appropriate to this discussion. Similar statements apply to the general problem of *designing* amplifiers to satisfy values of ω_L and ω_H. It may appear that nothing more can profitably be achieved by continuing to study frequency response. However, there is much to be gained from studying several specific amplifier cases to gain a sense of what is involved in determining frequency response. We consider one configuration using an FET as an example.

Common-Source FET Amplifier

We consider the amplifier of Fig. 7.5-1(a). The small-signal ac equivalent circuits are sketched in Fig. 7.7-2. At lower frequencies near ω_L the circuit of Fig. 7.7-2(a) applies. Near ω_H the high-frequency circuit of Fig. 7.7-2(b) is used. In both cases the FET has been replaced by its equivalent circuit, as shown in Fig. 6.6-5.

---

† This condition is not always desired. Some amplifiers that are designed to approximate integrators or differentiators, as examples, have decreasing or increasing amplitude responses, respectively, with frequency.

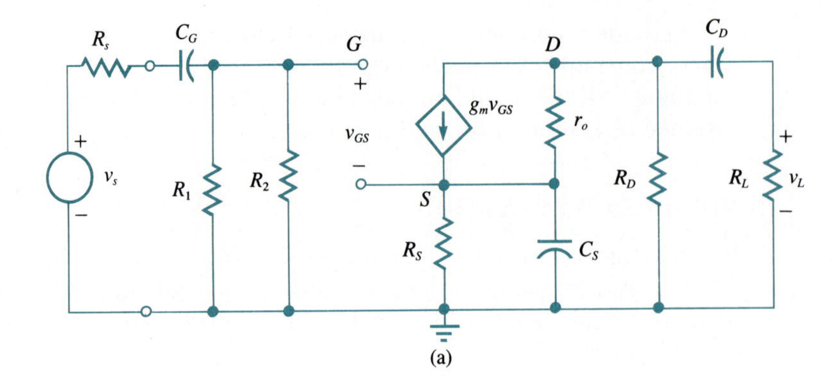

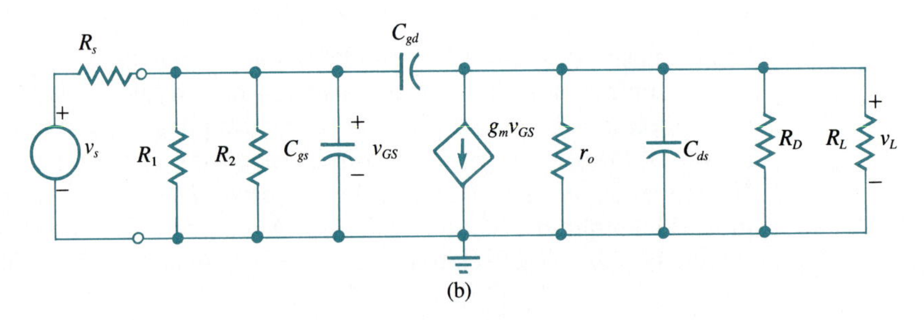

FIG. 7.7-2. Small-signal ac equivalent circuits for the FET amplifier of Fig. 7.5-1(a): (a) for low frequencies near ω_L and (b) for high frequencies near ω_H.

If r_o is assumed large for convenience, the gain of the low-frequency circuit of Fig. 7.7-2(a) is found to be

$$A_v = \frac{v_L}{v_s} = \frac{A_{v0}(j\omega)^2(\omega_Z + j\omega)}{(\omega_{L1} + j\omega)(\omega_{L2} + j\omega)(\omega_{L3} + j\omega)} \tag{7.7-2}$$

where

$$A_{v0} = \frac{-g_m R_G R_D R_L}{(R_s + R_G)(R_D + R_L)} \tag{7.7-3}$$

$$R_G = R_1 \,||\, R_2 = \frac{R_1 R_2}{R_1 + R_2} \tag{7.7-4}$$

$$\omega_Z = \frac{1}{R_S C_S} \tag{7.7-5}$$

$$\omega_{L1} = \frac{1}{R_{L1} C_S} \tag{7.7-6}$$

$$\omega_{L2} = \frac{1}{R_{L2} C_D} \tag{7.7-7}$$

$$\omega_{L3} = \frac{1}{R_{L3} C_G} \tag{7.7-8}$$

$$R_{L1} = \frac{R_S}{1 + g_m R_S} \tag{7.7-9}$$

$$R_{L2} = R_D + R_L \tag{7.7-10}$$

$$R_{L3} = R_s + R_G \tag{7.7-11}$$

The frequencies ω_Z, ω_{L1}, ω_{L2}, and ω_{L3} are called *break frequencies*. They represent values of ω where the behavior of $|A_v|$ changes. For ω above all break frequencies $A_v \approx A_{v0}$, the midband value of gain. The largest break frequency is equal to ω_L if the others are much smaller. This fact provides one of several possible approaches to design:

1. Determine which of R_{L1}, R_{L2}, or R_{L3} is the smallest resistance. Usually, the smallest is R_{L1}. From the corresponding equation (7.7-6)–(7.7-8), choose the capacitance so that the break frequency equals ω_L.
2. Choose the remaining two capacitors such that their break frequencies are at least 10 times smaller than ω_L.

Note that this procedure determines C_S so that break frequency ω_Z is determined. It will have some effect on performance, but from a comparison of (7.7-5) and (7.7-6), we see that $\omega_Z < \omega_{L1} = (1 + g_m R_s)\omega_Z$ and its effect is often small.

Amplifier gain at higher frequencies near ω_H is found from an analysis of Fig. 7.7-2(b). After one drops a small second-order term (in ω^2) in the denominator, the gain is found to be

$$A_v = \frac{v_L}{v_s} \approx \frac{A_{v0}\omega_H(\omega_Z - j\omega)}{\omega_Z(\omega_H + j\omega)} \tag{7.7-12}$$

where

$$A_{v0} = \frac{-g_m R_G R_{DL}}{R_s + R_G} \tag{7.7-13}$$

$$R_{DL} = r_o \,||\, R_D \,||\, R_L = \frac{r_o R_D R_L}{r_o R_D + r_o R_L + R_D R_L} \tag{7.7-14}$$

$$R_G = R_1 \,||\, R_2 = \frac{R_1 R_2}{R_1 + R_2} \tag{7.7-15}$$

$$\omega_Z = \frac{g_m}{C_{gd}} \tag{7.7-16}$$

$$\omega_H = \frac{1}{R_A[C_{gs} + C_{gd}(1 + g_m R_{DL} + R_{DL}/R_A)]} \tag{7.7-17}$$

$$R_A = R_s \,||\, R_G = \frac{R_s R_G}{R_s + R_G} \tag{7.7-18}$$

Clearly, the problem of *designing* for a specific value of ω_H means choosing mainly g_m, r_o, R_1, R_2, R_D, and R_L from (7.7-17). Since all these parameters except

R_L are established in the bias design, there is interaction between the ac and bias designs. Iteration back and forth between the two may be necessary in some cases. For a large desired value of ω_H, the choice of a transistor having small capacitances C_{gs} and C_{gd} is also important.

EXAMPLE 7.7-1

As an example of both low- and high-frequency designs, we again consider the amplifier of Examples 7.3-1 and 7.5-1. If we round off component values, the amplifier is specified by

$$R_1 = 346\ \text{k}\Omega \quad R_S = 1150\ \Omega \quad R_L = 1000\ \Omega \quad g_m = 5.77 \times 10^{-3}\ \text{S}$$

$$R_2 = 115\ \text{k}\Omega \quad R_D = 923\ \Omega \quad r_o = 13.5\ \text{k}\Omega$$

We shall also assume that $R_s = 2200\ \Omega$, $C_{gs} = 3.0$ pF, $C_{gd} = 1.0$ pF, and the desired value of $\omega_L/2\pi$ is 100 Hz.

For the low-frequency design we have

$$R_G = \frac{R_1R_2}{R_1 + R_2} = \frac{346(115 \times 10^3)}{346 + 115} = 86{,}312\ \Omega$$

$$R_{L1} = \frac{R_S}{1 + g_mR_S} = \frac{1150}{1 + 5.77(1.15)} = 150.6\ \Omega$$

$$R_{L2} = R_D + R_L = 923 + 1000 = 1923\ \Omega$$

$$R_{L3} = R_s + R_G = 2200 + 86{,}312 = 88{,}512\ \Omega$$

Thus, R_{L1} is smallest as expected, so (7.7-6) gives

$$C_S = \frac{1}{R_{L1}\omega_L} = \frac{1}{150.6(200\pi)} \approx 10.57\ \mu\text{F}$$

From (7.7-7) and (7.7-8)

$$C_D = \frac{1}{R_{L2}\omega_{L2}} = \frac{1}{R_{L2}\omega_L/10} = \frac{10}{(1.923 \times 10^3)(200\pi)} \approx 8.28\ \mu\text{F}$$

$$C_G = \frac{1}{R_{L3}\omega_{L3}} = \frac{1}{R_{L3}\omega_L/10} = \frac{10}{88{,}512(200\pi)} \approx 0.18\ \mu\text{F}$$

To determine ω_H, we first find

$$R_A = R_s \,||\, R_G = \frac{2200(88{,}512)}{2200 + 88{,}512} \approx 2147\ \Omega$$

$$R_D \,||\, R_L = \frac{923(1000)}{923 + 1000} \approx 480\ \Omega$$

$$R_{DL} = r_o \,||\, R_D \,||\, R_L = \frac{13{,}500(480)}{13{,}500 + 480} \approx 463.5\ \Omega$$

Finally, from (7.7-17):

$$\omega_H = \frac{10^{12}}{2147\,[3 + 1 + (5.77 \times 10^{-3})(463.5) + 463.5/2147]} \approx 67.6 \times 10^6 \text{ rad/s}$$

or

$$\frac{\omega_H}{2\pi} \approx 10.76 \text{ MHz}$$

In the midband we also find that gain is $A_v \approx -2.61$, from (7.7-13).

7.8 SOME OTHER TRANSISTOR AMPLIFIERS

The amplifiers discussed in this chapter up to this point have all been small-signal, linear stages that use capacitance coupling. Because they need only resistors and capacitors, they are sometimes called *RC* amplifiers. In this section we briefly examine some other transistor amplifiers that do not have the small-signal limitation and are not limited to the use of resistors and capacitors in their structure.

Amplifier Classifications

In all the amplifier discussions of the foregoing sections transistor currents that flowed in response to a small input signal were linearly related to the signal. Because of this relationship, such amplifiers are called *linear*. For larger input signals the linear dependence of current on signal level may not be maintained, but it may still be true that currents flow during all portions of the input signal. Such large-signal amplifiers are called *class A*.

In the classifying of amplifiers it is helpful to consider the input signal to be sinusoidal. Thus, a class A amplifier is one in which currents flow over the entire cycle of the sinusoid, as shown in Fig. 7.8-1(a). If the transistor is biased to cut off in the absence of the input signal, as shown in Fig. 7.8-1(b), the amplifier will have current flow over one-half of a cycle; this type of operation is called *class B*. For current flow over more than one half-cycle but less than a full cycle, the amplifier is said to be *class AB*. *Class C* corresponds to current flow over less than one half-cycle.

Class A amplifiers are usually employed in small-signal applications such as amplifying the small voltages emerging from microphones, from antennas at the ''front end'' of a radio receiver, from pressure and temperature transducers, etc. Amplifiers in classes B, AB, and C are usually found in high-power (large-signal) applications such as the driving of audio speakers, cathode-ray tubes in television sets, radio and television transmitters, and sonar and radar transmitters.

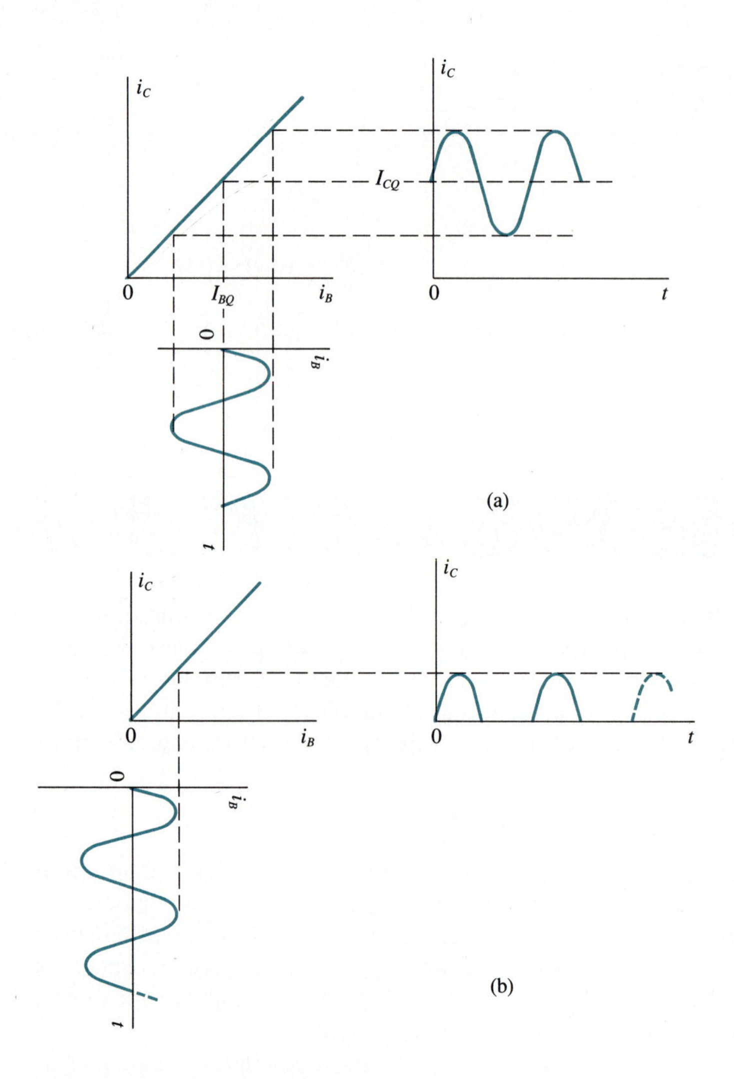

FIG. 7.8-1. Transistor currents for (a) class A and (b) class B amplifiers.

Some Class A Amplifiers

Figure 7.8-2(a) illustrates an amplifier that uses a bipolar transistor with a tuned resonant network in its collector circuit. Capacitors C_B, C_E, and C_C are selected large enough to approximate short circuits at all frequencies of interest, and resistors R_1, R_2, and R_E are set during bias selection. The impedance seen by the collector is due to C, L, and a resistance comprised of R_L in parallel with r_o (the transistor's output resistance), all in parallel. This impedance will have a maximum value equal to $r_o \,||\, R_L$ (resistive) at an angular frequency

$$\omega_0 = \frac{1}{\sqrt{LC}} \tag{7.8-1}$$

called the *resonant frequency* (see Sec. 4.7). For frequencies both above and

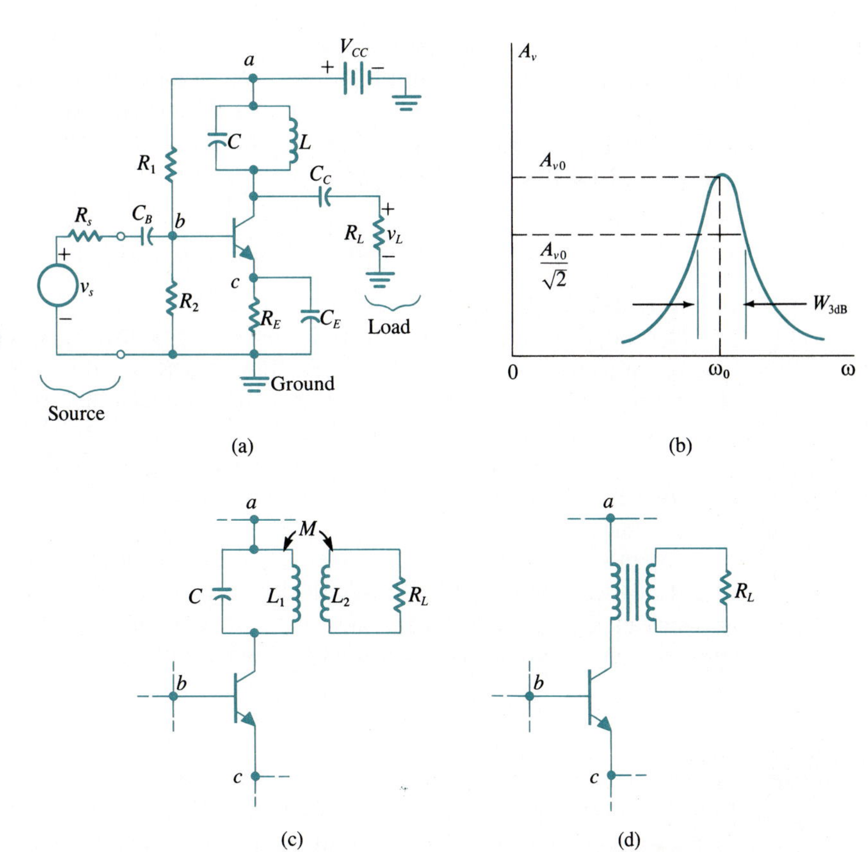

FIG. 7.8-2. Some linear (class A) amplifiers: (a) A single-tuned amplifier and (b) its voltage gain; (c) a single-tuned transformer-coupled amplifier; and (d) an untuned transformer-coupled amplifier.

below ω_0, the collector's impedance decreases, causing the voltage gain of the stage to decrease, as shown in Fig. 7.8-2(b). Gain is given approximately by

$$A_v = \frac{v_L}{v_s} \approx \frac{A_{v0}}{1 + j[2(\omega - \omega_0)/W_{3\text{dB}}]} \tag{7.8-2}$$

for ω near ω_0, where

$$A_{v0} = \frac{-g_m r_\pi R_B(r_o \,||\, R_L)}{(R_s + R_B)[r_b + r_\pi + (R_s \,||\, R_B)]} \tag{7.8-3}$$

$$R_B = R_1 \,||\, R_2 = \frac{R_1 R_2}{R_1 + R_2} \tag{7.8-4}$$

$$W_{3\text{dB}} = \frac{\omega_0}{Q} \tag{7.8-5}$$

$$Q = \frac{r_o \,||\, R_L}{\omega_0 L} \tag{7.8-6}$$

Of course, g_m, r_b, and r_π are parameters of the transistor. The quantity W_{3dB} is the 3-dB bandwidth of the amplifier, and Q is a measure of W_{3dB}.

Equation (7.8-2) assumes that $Q \gg 1$, which corresponds to a 3-dB bandwidth that is small relative to the amplifier's center (resonant) frequency. If this condition is not true, the more exact gain expression is

$$A_v = \frac{A_{v0}}{1 + j[\omega - \omega_0)(\omega + \omega_0)/\omega W_{3dB}]} \tag{7.8-7}$$

which holds for all $-\infty < \omega < \infty$.

The tuned amplifier of Fig. 7.8-2(a) uses a capacitor C_C to couple the amplified signal to the load. Other versions using magnetic (transformer) coupling are also possible. For example, if the network between terminals a, b, c is replaced by that shown in Fig. 7.8-2(c), a tuned transformer-coupled amplifier results. Here C is selected to resonate,† at angular frequency ω_0, with the reactance of the impedance looking into the primary (collector's side) of the transformer when the load is connected.

An untuned amplifer can also use a transformer for coupling. A typical circuit is that of Fig. 7.8-2(a) with the circuit between terminals a, b, c replaced by the network of Fig. 7.8-2(d). A popular application is in an audio circuit where the transformer has broad bandwidth and a high coupling coefficient (see Sec. 1.4).

EXAMPLE 7.8-1

We shall determine Q, L, and C in an amplifier like that of Fig. 7.8-2(a) when it is to serve at the intermediate frequency (IF) of a standard AM radio, which is $\omega_0/2\pi = 455$ kHz. Bandwidth is to be 10 kHz, and r_o is large enough to be neglected. The amplifier is to drive a load consisting of a 180-pF capacitor in parallel with 1000 Ω (input of the next stage).

From (7.8-5)

$$Q = \frac{\omega_0}{W_{3dB}} = \frac{2\pi(455 \times 10^3)}{2\pi \times 10^4} = 45.5$$

Next, we use (7.8-6) to find the required inductance:

$$L = \frac{r_o || R_L}{\omega_0 Q} \approx \frac{R_L}{\omega_0 Q} = \frac{1000}{2\pi(455 \times 10^3)(45.5)} \approx 7.69\ \mu\text{H}$$

From (7.8-1)

$$C = \frac{1}{\omega_0^2 L} \approx \frac{1}{4\pi^2(455)^2(10^6)(7.69 \times 10^{-6})} \approx 15{,}911\ \text{pF}$$

Since part of C is the 180 pF of the load, we require a circuit capacitance of $15{,}911 - 180 = 15{,}731$ pF to be added in the collector.

† Resonance occurs when the magnitude of the capacitor's reactance equals the magnitude of the inductor's reactance.

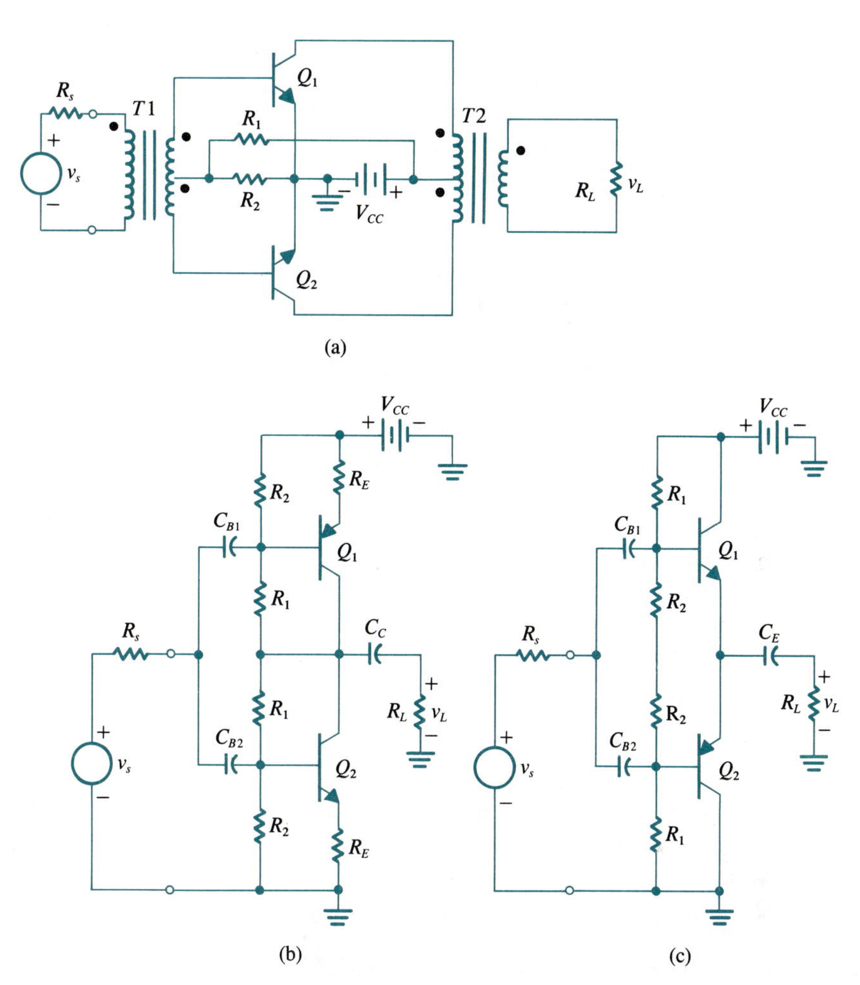

FIG. 7.8-3. Push-pull amplifiers: (a) Transformer-coupled; (b) common-emitter form using no transformers; and (c) emitter-follower form with no transformers.

Some Class B Amplifiers

Three class B amplifiers are shown in Fig. 7.8-3. All use two transistors and are called *push-pull amplifiers* because of their mode of operation. The circuit of Fig. 7.8-3(a) uses transformer coupling. Resistors R_1 and R_2 are present to bias transistors Q_1 and Q_2 to the cutoff point (in practice, the bias point is slightly above cutoff to reduce distortion due to the gradual turn-on characteristics of the transistors, but the finite no-signal currents that flow are small and operation is approximately class B). When v_s is positive, Q_1 conducts while Q_2 is cut off. When v_s is negative, Q_2 conducts and Q_1 is cut off. The sum of the currents due to Q_1 and Q_2 are such as to add in the load. Thus, the load sees a signal approximately proportional to v_s, and Q_1 and Q_2 conduct only when signal is present. No zero-signal quiescent current is necessary. As a consequence, efficiency, defined as the ratio of signal power delivered to the load to the total power required by the amplifier, can be much larger than in a class A amplifier. Ideally, the maximum efficiency of a class B amplifier is 78.5%, compared with 25% for a class A device.

The disadvantage of the push-pull amplifier is the need for driving signals to Q_1 and Q_2 that are 180° out of phase. Transformer $T1$ provides the drive signals in Fig. 7.8-3(a). It is also possible to derive these signals without a transformer by using a common-emitter amplifier, where the equal-amplitude voltages are taken from the collector and from an unbypassed-emitter resistor; such an amplifier is sometimes called a *phase splitter*.

The use of transistors with *complementary symmetry* (which means that two transistors, one *npn* and one *pnp*, have the same characteristics except for polarity of currents and voltages) allows a common-emitter form of class B push-pull amplifier to be constructed, as shown in Fig. 7.8-3(b). When v_s is positive, Q_1 conducts and v_L is proportional to the negative of v_s. When v_s is negative, Q_2 conducts (Q_1 is cut off) and v_L is again proportional to the negative of v_s. The advantage of the circuit is that it requires no transformers. The disadvantage is that it requires matching transistors. Of course, resistors R_1 and R_2 are used to set the bias point (at cutoff, ideally, and slightly above cutoff, in practice), and C_{B1}, C_{B2}, and C_C are coupling capacitors chosen to have negligible impedance at all frequencies of interest. The (small) resistors R_E are often added in practice to reduce distortion due to nonlinearities in the cut-on characteristics of the transistors. Although distortions are reduced as R_E is increased, gain is reduced; and some compromise is necessary in practice.

The push-pull amplifier of Fig. 7.8-3(b) is not as suited to driving low-impedance loads as that of (c), which is an amplifier using complementary-symmetric emitter-followers. As usual, C_{B1}, C_{B2}, and C_E are coupling capacitors designed to be approximately short circuits at all frequencies of interest, and R_1 and R_2 are used to establish the bias point.

PROBLEMS

7-1. A simple biasing method for a BJT, called *fixed bias*, is shown in Fig. P7-1. If I_{CBO} is small in relation to I_{BQ} and I_{CQ}, what value of R_B is required to place the operating point at I_{CQ} = 13.0 mA, V_{CEQ} = 6.0 V when V_{CC} = 12 V and the silicon BJT has a nominal β of 65? What is R_C?

FIG. P7-1.

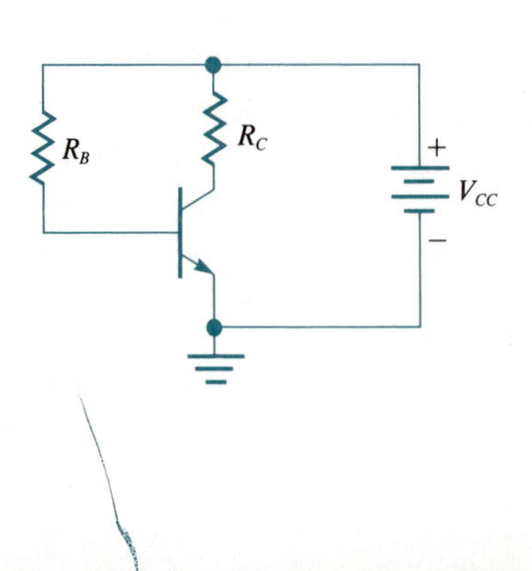

7-2. Work Problem 7-1 except for the *collector-to-base*-biasing method of Fig. P7-2.

FIG. P7-2.

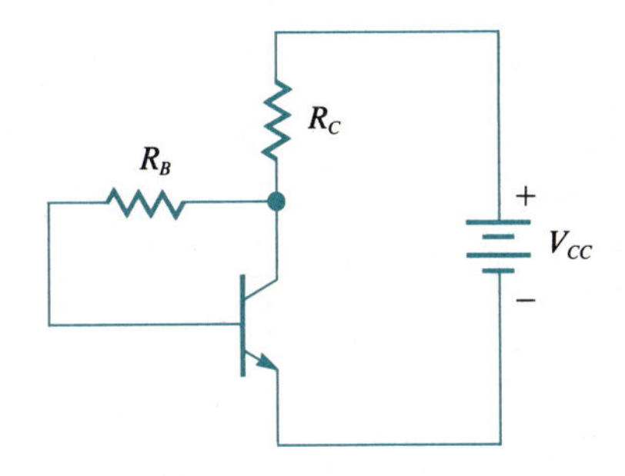

7-3. A silicon *npn* BJT is biased according to Fig. 7.2-1, with $R_E = 220\ \Omega$, $R_2 = 2700\ \Omega$, and $V_{CC} = 20$ V. If the Q point corresponds to $V_{BEQ} = 0.7$ V, $I_{BQ} = 100\ \mu$A, $V_{CEQ} = 12.0$ V, and $I_{CQ} = 9.9$ mA, find (a) R_C and (b) R_1.

7-4. Use the rule-of-thumb procedure of the text to find all resistors in the bias method of Fig. 7.2-1 for a silicon *npn* BJT for which $\beta = 150$ and the Q point is defined by $I_{CQ} = 4.8$ mA. Assume $V_{CC} = 9$ V.

7-5. Work Problem 7-4 except assume $\beta = 70$, $I_{CQ} = 84$ mA, and $V_{CC} = 15$ V.

7-6. The BJT biased as shown in Fig. P7-6 has $V_{BEQ} = 0.7$ V, $\beta = 108$, $V_{CEQ} = 11.0$ V, and $I_{CQ} = 6.0$ mA. Find I_1, I_2, and I_{EQ}.

FIG. P7-6.

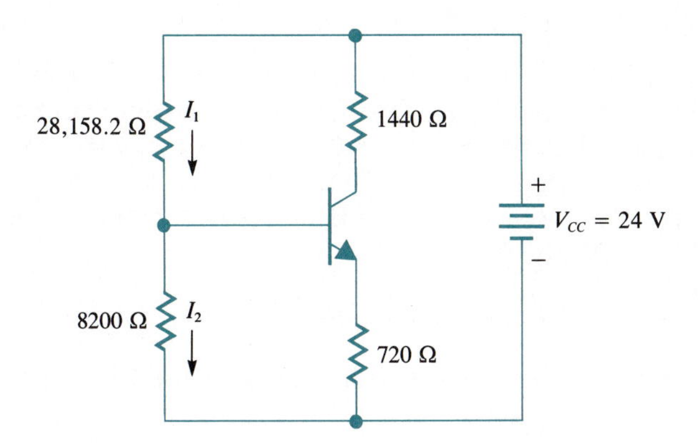

7-7. A JFET for which $V_p = -3.0$ V and $I_{DSS} = 24.0$ mA is to be biased by the network of Fig. P7-7 to operate at an active-region Q point defined by $I_{DQ} = 4.5$ mA and $V_{DSQ} = 7.0$ V. Find R_S and R_D so that this Q point is realized. This method is called *self-bias*.

FIG. P7-7.

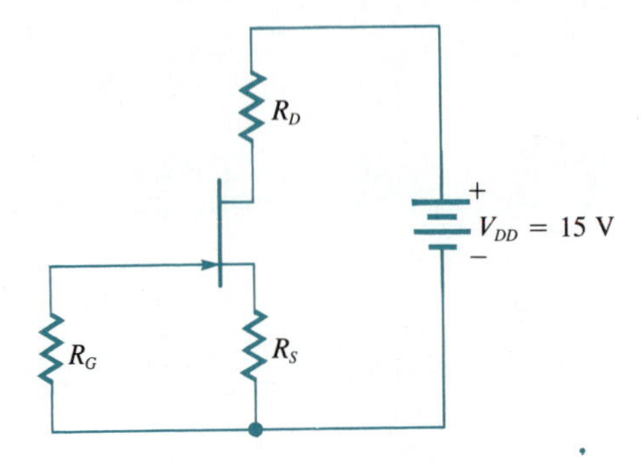

7-8. An n-channel JFET for which $I_{DSS} = 5$ mA and $V_p = -3.0$ V is biased by the circuit of Fig. 7.3-1, with $V_{DD} = 30$ V, $R_S = 3300\ \Omega$, and $R_2 = 100\ \text{k}\Omega$. The operating point is defined by $I_{DQ} = 2.0$ mA and $V_{DSQ} = 13.2$ V. (a) Find V_G. (b) What is R_D? (c) Find R_1.

7-9. In a certain low-noise JFET, $V_p = -3.5$ V. It has $I_{DSS} = 5.0$ mA. For the circuit of Fig. 7.3-1, use the design procedure to find V_G, R_S, R_D, R_2, and R_1 if $V_{DD} = 28$ V.

7-10. Work Problem 7-9 except assume $V_{DD} = 32$ V; and use a higher-power JFET for which $V_p = -4.0$ V and $I_{DSS} = 295$ mA.

7-11. A 2N3797 is an n-channel depletion-type MOSFET for which $V_p = -3.2$ V and $I_{DSS} = 2.9$ mA when $v_{DS} = 10.0$ V. Find R_S, R_D, V_G, and R_1 to bias the device to a Q point where $I_{DQ} = 2.9$ mA, $V_{DSQ} = 10.0$ V, and 40% of the voltage drop across R_D and R_S is across R_S. Assume $V_{DD} = 20.0$ V and $R_2 = 1\ \text{M}\Omega$, and use the bias circuit of Fig. 7.3-1.

7-12. A particular n-channel enhancement-type MOSFET has $V_t = 1.2$ V, $K = 5.33\ \text{mA/V}^2$, and a maximum continuous power dissipation of 0.3 W. The bias point is to be at $I_{DQ} = 9.0$ mA when $V_{DSQ} = 14$ V. (a) Find R_S, R_D, V_G, and R_1 for the circuit of Fig. 7.3-3 if $V_{DD} = 25$ V, the voltage across R_S is to be 3 V, and $R_2 = 1\ \text{M}\Omega$. (b) What continuous power is dissipated in the MOSFET?

7-13. Show that (7.4-6) is true by analysis of Fig. 7.4-1(b).

7-14. Determine voltage and current gains for the CE amplifier in Fig. P7-14.

FIG. P7-14.

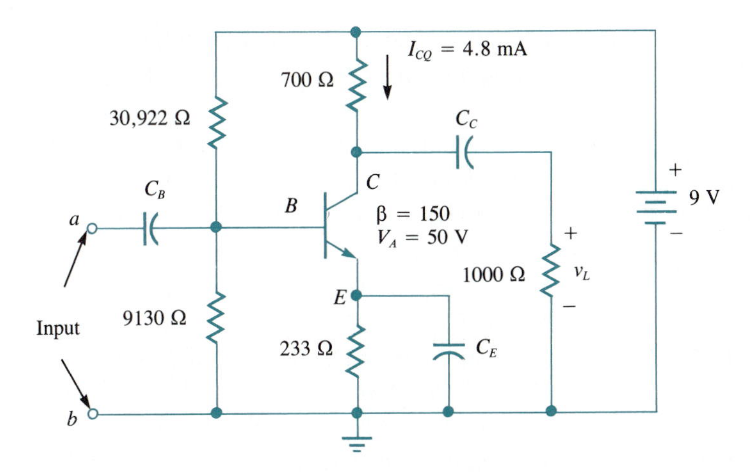

7-15. In the CE amplifier of Fig. 7.4-1(a), $R_1 = 1589\ \Omega$, $R_2 = 389\ \Omega$, $R_C = 67\ \Omega$, $R_E = 22\ \Omega$, and $R_L = 150\ \Omega$. The transistor has $\beta = 70$, $V_A = 50$ V, and $I_{CQ} = 84$ mA when $V_{CC} = 15$ V. Find the amplifier's voltage and current gains.

7-16. The CE amplifier of Fig. 7.4-1(a) is defined by $R_1 = 28{,}158\ \Omega$, $R_2 = 8200\ \Omega$, $R_C = 1440\ \Omega$, $R_E = 720\ \Omega$, $R_L = 2200\ \Omega$, $V_{CC} = 24$ V, and $I_{CQ} = 6.0$ mA when $\beta = 108$ and $V_A = 100$ V for the BJT. Find the voltage and current gains of the amplifier.

7-17. Work Problem 7-14 except assume that C_E is removed.

7-18. Work Problem 7-15 except assume that half of R_E is unbypassed. That is, assume $R_{E1} = 11\ \Omega$.

7-19. Work Problem 7-16, but assume that 220 Ω of R_E is not bypassed.

7-20. In a common-collector circuit, as shown in Fig. 7.4-4(a), $R_1 = 150.1$ kΩ, $R_2 = 150$ kΩ, $R_E = 2200\ \Omega$, $R_L = 3300\ \Omega$, and $V_{CC} = 9$ V. The BJT has $\beta = 80$, $I_{CQ} = 1.2$ mA, and $V_A = 70$ V. Find (a) R_i, (b) R_{in}, (c) A_{v1}, and (d) A_i.

7-21. Show that (7.4-12), (7.4-16), and (7.4-17) reduce to (7.4-18) through (7.4-20) if $r_o \to \infty$.

7-22. In the common-collector amplifier stage of Fig. 7.4-4(a), $R_1 = 34{,}175\ \Omega$, $R_2 = 23{,}359\ \Omega$, $R_E = 420\ \Omega$, $R_L = 270\ \Omega$, and $V_{CC} = 9$ V. For the BJT, $V_A = 60$ V, $\beta = 40$, and $I_{CQ} = 3.8$ mA. Find (a) R_i, (b) R_{in}, (c) A_{v1}, and (d) A_i.

7-23. Define power gain in an amplifier stage as the ratio of load power, given by $v_L i_L$, to the power delivered from the source, as given by $v_1 i_s$. Derive a formula for power gain, denoted by A_p, for a common-collector stage.

7-24. Work Problem 7-23 except for a common-emitter stage.

7-25. Work Problem 7-23 except for a common-base stage.

7-26. In the circuit of Fig. P7-14 terminals a and b are connected together and capacitor C_E is disconnected from ground and used to couple a signal from a source to the emitter. The circuit becomes a common-base amplifier. Find (a) R_i, (b) R_{in}, (c) A_{v1}, and (d) A_i.

★7-27. Complete the analysis of the text by proving that (7.4-23) is true.

7-28. In a common-base circuit $R_C = 1000\ \Omega$, $R_L = 5600\ \Omega$, and $R_{in} = 21.0\ \Omega$ when the transistor parameters are $\beta = 50$, $V_A = 75$ V, and $g_m = 0.03$ S. Find (a) R_i, (b) R_E, (c) A_{v1} and (d) A_i.

7-29. In a common-source amplifier $R_D = 1800\ \Omega$ and $R_L = 2700\ \Omega$. The JFET has $r_o = 15{,}000\ \Omega$ and gives a voltage gain $A_{v1} = -4.5$ in the circuit when all source resistance is bypassed. What is g_m?

7-30. A JFET for which $V_A = 80$ V, $V_p = -3.6$ V, and $I_{DSS} = 10$ mA has a quiescent drain current of 2.7 mA when used as a common-source amplifier where $R_D = R_S = 1500\ \Omega$ and $R_L = 2700\ \Omega$. If R_S is fully bypassed, what is the amplifier's voltage gain A_{v1}?

7-31. A common-source amplifier uses a JFET for which $V_A = 100$ V, $r_o = 2500\ \Omega$, and $V_p = -4.2$ V in a circuit where $R_L = 350\ \Omega$, $R_D = 150\ \Omega$, and $R_S = 100\ \Omega$ (fully bypassed). (a) If the voltage gain is -3.0, what are g_m and I_{DSS} for the JFET? (b) If the voltage gain is to be reduced to -1.5 by leaving part of R_S unbypassed, what is R_{S1}?

7-32. Show that alternative forms for (7.5-10) and (7.5-11) are

$$A_{v1} = \frac{g_m(r_o \,||\, R_S \,||\, R_L)}{1 + g_m(r_o \,||\, R_S \,||\, R_L)}$$

$$A_i = \frac{g_m(r_o \,||\, R_S \,||\, R_L)(R_1 \,||\, R_2)}{R_L[1 + g_m(r_o \,||\, R_S \,||\, R_L)]}$$

★7-33. For a JFET source-follower, find the output resistance seen by a load when looking back into the amplifier between the source and ground.

7-34. In a source-follower the voltage gain is to be 0.65 when $R_S = 270\ \Omega$ and a JFET with $r_o = 5000\ \Omega$ and $g_m = 25 \times 10^{-3}$ S is used. What load resistance R_L is required?

7-35. The circuit of Problem 7-30 is converted to a common-gate amplifier having the same component values. (a) Find the amplifier's voltage gain A_{v1}. (b) What is A_i? (c) What is R_{in}?

7-36. A common-gate JFET amplifier has $R_L = 15$ kΩ, $R_D = 7.5$ kΩ, $R_S = 6.2$ kΩ, $r_o = 100$ kΩ, and $g_m = 6 \times 10^{-3}$ S. (a) Find R_{in}. (b) What is A_{v1}? (c) Determine A_i.

7-37. Determine an equation for R_i in Fig. 7.5-3.

7-38. The voltage gain of an amplifier is

$$A_v = \frac{20\, j\omega}{(130\pi + j\omega)[1 + j\omega/(6\pi \times 10^{-4})]}$$

What are (a) ω_H, (b) ω_L, and (c) the midband gain?

7-39. A circuit for which $R_1 = 513$ kΩ, $R_2 = 141$ kΩ, $R_D = 1030$ Ω, and $R_S = 1410$ Ω uses an FET for which $g_m = 5.3 \times 10^{-3}$ S, $r_o = 22$ kΩ, and $C_{gs} = C_{gd} = 2.5$ pF. If the amplifier is in the CS configuration, $R_s = 500$ Ω, $R_L = 1500$ Ω, and the low-frequency 3-dB angular frequency is to be 40π rad/s, find (a) C_S, (b) C_D, (c) C_G, (d) A_{v0}, and (e) ω_H.

7-40. In a common-source FET amplifier $R_s = 1500\ \Omega \ll (R_1 \,||\, R_2)$, $R_L = R_D$, $\omega_Z/\omega_H = 77.4$, $\omega_L \ll \omega_H$, and $W_{3dB} = 25.84 \times 10^6$ rad/s. The transistor has $g_m = 6 \times 10^{-3}$ S, $r_o \gg (R_D \,||\, R_L)$, and $C_{gs} = C_{gd}$. Determine (a) R_L, (b) C_{gs}, and (c) A_{v0}.

7-41. An amplifier of the form of Fig. 7.8-2(a) is to have $\omega_0/2\pi = 10.7$ MHz and $W_{3dB}/2\pi = 200$ kHz when $R_L = 1000$ Ω and $r_o = 7$ kΩ for the transistor. What are the required values of Q, L, and C?

7-42. Work Problem 7-41 except assume, instead, that $R_L = 10$ kΩ.

DESIGN EXERCISES

★D7-1. A source has a peak-to-peak open-circuit voltage of 0.5 V and an internal resistance that is nominally 3000 Ω but can vary from 1000 to 10,000 Ω. A common-source JFET amplifier is to be designed to work with the source and drive a load of 150 Ω ± 5 Ω. The JFET to be used has $C_{gs} = 2$ pF, $C_{gd} = 3$ pF, $V_A = 100$ V, a maximum power-dissipation capability of 600 mW, and a maximum drain-source voltage rating of 20 V. The nominal JFET device has $I_{DSS} = 200$ mA and $V_p = -3$ V. A high-current device is defined by $I_{DSS} = 300$ mA when $V_p = -4$ V, and a low-current device is defined by $I_{DSS} = 100$ mA and $V_p = -2$ V. Power supplies for V_{DD} are available in voltages of 12, 14, 16, 18, 20, 22, and 24 V. Design an amplifier with the following characteristics:

1. The Q point varies ±15% or less from its nominal value over the range of devices.
2. The ac current swings always remain in the active region.
3. Variations in source resistance do not affect voltage gain v_L/v_s by more than ±10% from its nominal value. This gain shall be as large as possible, consistent with an upper 3-dB frequency of at least $\omega_H/2\pi = 1.0$ MHz under all variations in the source's resistance.
4. The lower 3-dB frequency shall be $\omega_L/2\pi = 100$ Hz.
5. All resistors are to have standard values in 5% tolerance. Capacitors are to have only the scale values given in the first row of Table 1.4-2.

For the final nominal design, determine A_v, average power dissipated in all resistors and in the JFET, the required voltage ratings of all capacitors, all break frequencies, and the power delivered to the nominal load if the source has a sinusoidal voltage.

D7-2. A common-emitter BJT amplifier for which $C_E = 0$ is to drive the base of a common-collector BJT amplifier through a dc connection (C_C replaced by a wire). The CC stage (emitter-follower) is to drive a 50-Ω load resistor through a *large*-coupling capacitor. The load voltage v_L is to be sinusoidal of peak value 1 V. The output (CC) BJT has $\beta = 60$, $V_A = 50$ V, and a maximum allowable collector current of 160 mA. For the CE stage $\beta = 120$, $V_A = 100$ V, and I_C (maximum) = 80 mA. A power supply of $V_{CC} = 9$ V is to be used. Design the two-stage amplifier, giving due consideration to current swings necessary for good linearity. Compute the overall midband gain v_L/v_1 for the amplifier. Use only standard resistance values, assuming 5% tolerance. Find the dc powers dissipated in all resistors and transistors.

D7-3. A common-emitter audio amplifier is to be designed to have the form of Fig. 7.8-2(d), where $R_L = 400$ Ω, and the transformer to be used is approximately ideal with turns ratio $n = 2$ (Sec. 1.4). A transistor with $\beta = 100$ and $V_A = 80$ V is to be used. Design the amplifier, using standard-value resistors and assuming a 5% tolerance. Gain v_L/v_1 is to be not less than 10.6, but as near 10.6 as possible, by allowing a part of R_E to be unbypassed. Assume that $V_{CC} = 12$ V for the power supply and that the input voltage v_1 is a sinusoid of peak voltage 188.7 mV. Design the amplifier, giving consideration to keeping the distortion low.

CHAPTER 8

Operational Amplifiers, Feedback, and Control

8.0 INTRODUCTION

By using modern integrated-circuit (IC) methods of manufacturing whole amplifier circuits on a single "chip," designers can create a sort of "building block" amplifier having flexible and almost universal design applications in low-frequency (baseband) circuits. These amplifiers are called *operational amplifiers*, or simply *op amps*. Op amps have been used for many years, but early versions were expensive. Current-day IC technology has made the cost of many general-purpose op amps less than a dollar.

When an op amp is combined properly with a few external components (resistors and capacitors), it is capable of performing many different operations—hence the name *operational*. Some early applications involved linear operations of integration, differentiation, addition, and subtraction that were needed in analog computers. Other linear applications are instrumentation amplifiers, voltage-to-current and current-to-voltage converters, voltage-followers, active filters, and many more. Most of these applications are discussed in this chapter. Op amps are also useful in *nonlinear* applications, such as in limiters, comparators, voltage regulators, signal rectifiers and detectors, logarithmic amplifiers, multipliers, and many digital circuits. Some of these applications are discussed in Chapter 9.

In general, an op amp is a high-gain, broad-bandwidth, direct-coupled (dc), two-input, single-output† amplifier with high input impedances and low output

† There are also two-output versions of op amps. These versions are usually discussed in the more advanced texts.

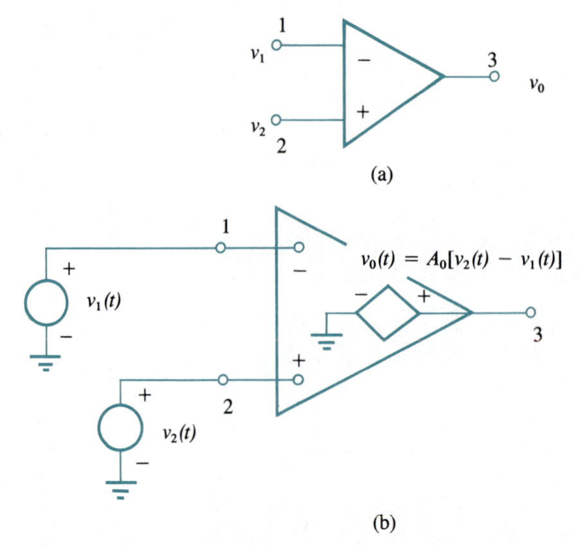

FIG. 8.0-1. (a) The circuit symbol for an op amp and (b) the equivalent circuit that is ideal except for having a finite voltage gain A_0.

impedance. Figure 8.0-1(a) illustrates the circuit symbol for an op amp. The input (v_1 and v_2) and output (v_0) voltages are all defined relative to a common reference point (usually ground), which is sometimes omitted in the circuit diagrams for simplicity. Operation is such that the component of v_0 caused by the input v_1 is inverted in sign. Because of this fact, terminal 1 is called the *inverting input terminal*. Terminal 2 is called the *noninverting input terminal* because no sign change occurs in the component of v_0 caused by the input v_2. Signs are commonly placed on the circuit symbol to define these terminals. When both input terminals see the same voltage gain to the output (say a gain of A_0), the equivalent circuit of Fig. 8.0-1(b) applies (the usual case). This circuit is discussed further below.

8.1 THE IDEAL OPERATIONAL AMPLIFIER

The ideal op amp has infinite voltage gain, infinite bandwidth, infinite input impedances at terminals 1 and 2 (to ground) and between terminals 1 and 2, zero output impedance, and no limitation on either input- or output-signal voltage or power levels. Of course, these ideal characteristics can only be approximated in practice. Fortunately, real op amps are available that behave almost as though they were ideal, at least in many lower-frequency (lowpass) applications. It is not uncommon for the less expensive, lower-power, general-purpose op amps to have voltage gains of 10^5 or more, a usable bandwidth of 1 MHz or more, input impedances of over 1 MΩ, and output resistances less than 100 Ω. More expensive op amps (up to several tens of dollars) can be found with higher-power capability (several volts across an output load of tens of ohms) and usable bandwidths up to several hundred megahertz.

Although the reader may find it reasonable to accept that bandwidth and input impedances may be made large enough, while output impedance may be made small enough, to approximate ideal in an op amp, it may be hard to accept any amplifier as having infinite gain, even in an ideal case. In reality, an op amp is almost never used alone. External elements are commonly connected from the output terminal back to the input terminals in a way that makes the overall *circuit's gain* independent of the op amp's gain, at least in the limit as its gain becomes infinite. Finiteness of gain matters little in these cases.

In the remainder of this section we analyze two types of amplifier to illustrate the preceding statements. Analyses assume both finite- and infinite-gain op amps that are otherwise ideal. For the case of finite voltage gain, denoted by A_0, the equivalent circuit of Fig. 8.0-1(b) is assumed to apply. The output voltage becomes

$$v_0(t) = A_0[v_2(t) - v_1(t)] \tag{8.1-1}$$

Inverting Amplifier

We analyze the circuit of Fig. 8.1-1, first assuming that the op amp's voltage gain is A_0 and then assuming that it is infinite.

When $A_0 < \infty$, we have

$$v_1 = -\frac{v_0}{A_0} \tag{8.1-2}$$

since $v_2 = 0$ because terminal 2 is grounded. Thus,

$$i_1 = \frac{v_{in} - v_1}{R_1} = \frac{v_{in} + (v_0/A_0)}{R_1} \tag{8.1-3}$$

$$i_2 = \frac{v_1 - v_0}{R_2} = \frac{-(v_0/A_0) - v_0}{R_2} \tag{8.1-4}$$

Since no current is drawn by terminal 1 of the ideal op amp, we have $i_2 = i_1$. Therefore, we can equate (8.1-3) and (8.1-4) to obtain the circuit's gain:

$$\frac{v_0}{v_{in}} = \frac{-R_2}{R_1} \frac{1}{1 + [(R_1 + R_2)/A_0R_1]} \tag{8.1-5}$$

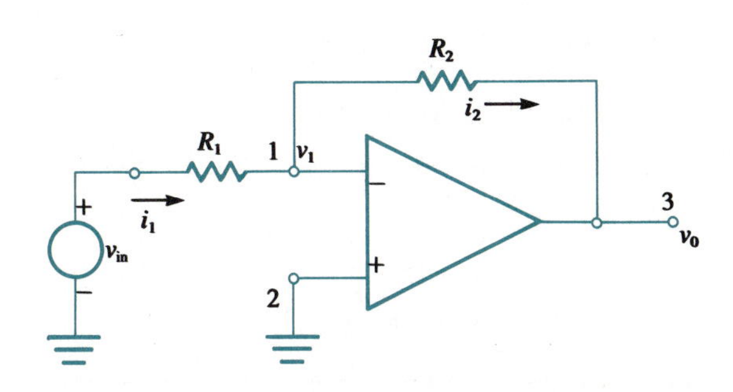

FIG. 8.1-1. The inverting amplifier.

Now if $A_0 \to \infty$, we have the ideal circuit gain:

$$\frac{v_O}{v_{in}} = -\frac{R_2}{R_1} \tag{8.1-6}$$

which means that the circuit behaves as an amplifier with voltage gain $-R_2/R_1$ that is independent of the op amp's gain A_0. We use an example to demonstrate that if A_0 is at least 200 times the ideal circuit gain's magnitude R_2/R_1, then the actual gain v_O/v_{in} will be within 1% of the ideal gain.

EXAMPLE 8.1-1

Consider what value of A_0 will make the circuit gain of (8.1-5) within 1% of the ideal gain of (8.1-6). We require

$$\frac{v_O/v_{in}}{-(R_2/R_1)} = \frac{1}{1 + [(1 + R_2/R_1)/A_0]} \geq 0.99$$

On solving for A_0, we obtain

$$A_0 \geq 99\left(\frac{R_2}{R_1}\right)\left(1 + \frac{R_1}{R_2}\right)$$

For R_2/R_1 more than 99 the required value of A_0 is only 100 times R_2/R_1. If R_2/R_1 is less than 99, the required value of A_0 slowly increases. Even for $R_2/R_1 = 1$, A_0 must be only 198 times R_2/R_1 to maintain the 1% criterion. Since $A_0 > 20{,}000$ for most practical op amps, we can safely assume infinite gain even for amplifiers with circuit gains up to 100.

Next, we demonstrate analysis assuming that the gain of the op amp is infinite. From Fig. 8.0-1(b) with a nonzero output signal and $A_0 \to \infty$, we see that the voltage between terminals 2 and 1, which is $v_2 - v_1$, must be zero. Thus, $v_1 = v_2$, in general. However, since $v_2 = 0$ for the circuit in question (Fig. 8.1-1), then $v_1 = 0$. Because terminal 1 is at ground potential (zero), it is said to be a *virtual ground*, although it is *not actually grounded*. Currents i_1 and i_2 are now

$$i_1 = \frac{v_{in} - v_1}{R_1} = \frac{v_{in} - 0}{R_1} = \frac{v_{in}}{R_1} \tag{8.1-7}$$

$$i_2 = \frac{v_1 - v_O}{R_2} = \frac{0 - v_O}{R_2} = -\frac{v_O}{R_2} \tag{8.1-8}$$

Again, the op amp draws no current, so $i_1 = i_2$. On equating the last two expressions, we have

$$\frac{v_O}{v_{in}} = -\frac{R_2}{R_1} \tag{8.1-9}$$

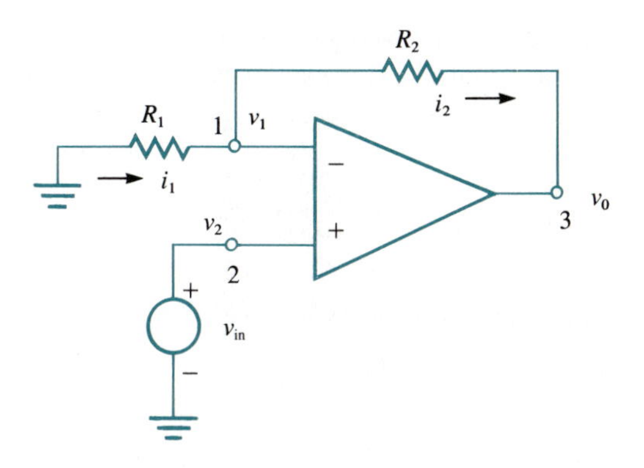

FIG. 8.1-2. The noninverting amplifier.

It is interesting to observe that when we set $R_2 = R_1$, the inverting amplifier becomes a *voltage-follower* with sign inversion and a gain magnitude of unity.

Noninverting Amplifier

The circuit of Fig. 8.1-2 is called a noninverting amplifier because it does not produce a sign inversion. Assume first an op amp with finite gain A_0. The voltage between terminals 2 and 1 is now $v_2 - v_1 = v_0/A_0$, so

$$v_1 = v_2 - \frac{v_0}{A_0} \tag{8.1-10}$$

Currents become

$$i_1 = \frac{-v_1}{R_1} = \frac{-v_2}{R_1} + \frac{v_0}{A_0 R_1} \tag{8.1-11}$$

$$i_2 = \frac{v_1 - v_0}{R_2} = \frac{v_2}{R_2} - \frac{v_0}{R_2}\left(\frac{1}{A_0} + 1\right) \tag{8.1-12}$$

Since the op amp draws no current, $i_1 = i_2$ and we equate (8.1-11) and (8.1-12) to obtain

$$\frac{v_0}{v_{in}} = \frac{v_0}{v_2} = \left(1 + \frac{R_2}{R_1}\right) \frac{1}{1 + [(R_1 + R_2)/A_0 R_1]} \tag{8.1-13}$$

If $A_0 \to \infty$, the ideal circuit gain becomes

$$\frac{v_0}{v_{in}} = 1 + \frac{R_2}{R_1} \tag{8.1-14}$$

The forms of these last two expressions are similar to those of (8.1-5) and (8.1-6). Analysis that follows the procedure of Example 8.1-1 shows that the required value of A_0 to maintain the circuit gain within 1% of the ideal gain is 99 times the ideal gain's magnitude.

For the ideal op amp with infinite gain the input voltage $v_2 - v_1$ is ideally zero, so $v_1 = v_2$. Analysis continues by finding expressions for i_1 and i_2 and noting that $i_1 = i_2$. When we solve the equated expressions, the (ideal) circuit gain is found; it is given by (8.1-14).

Note that if $R_2 = 0$ for any nonzero value of R_1 (8.1-14) gives

$$\frac{v_0}{v_{in}} = 1 \tag{8.1-15}$$

which is the gain of an ideal *voltage-follower*. The same result is achieved if $R_1 = \infty$ (open circuit) for any finite R_2.

8.2 SOME APPLICATIONS OF OP AMPS

In this section we give several examples of applications for which op amps perform admirably. In all cases the op amp is assumed to be ideal.

Inverting Summing Amplifier

The circuit of Fig. 8.1-1 can be extended by adding other input points, as shown in Fig. 8.2-1. Because of the virtual ground at terminal 1, the response of the circuit to any one input is not affected by the presence of the other $(N - 1)$ inputs. This means that the total output is just the sum of all responses, each having the form of (8.1-9):

$$v_0 = -\left(\frac{R_f}{R_1} v_{b1} + \frac{R_f}{R_2} v_{b2} + \cdots + \frac{R_f}{R_N} v_{bN}\right) = -R_f \sum_{n=1}^{N} \frac{v_{bn}}{R_n} \tag{8.2-1}$$

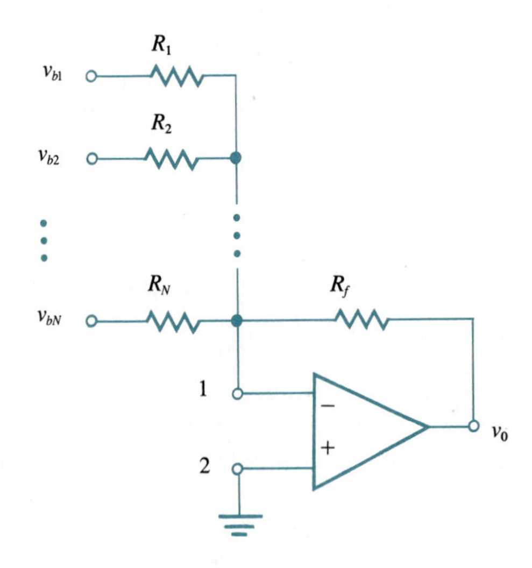

FIG. 8.2-1. An inverting summing amplifier.

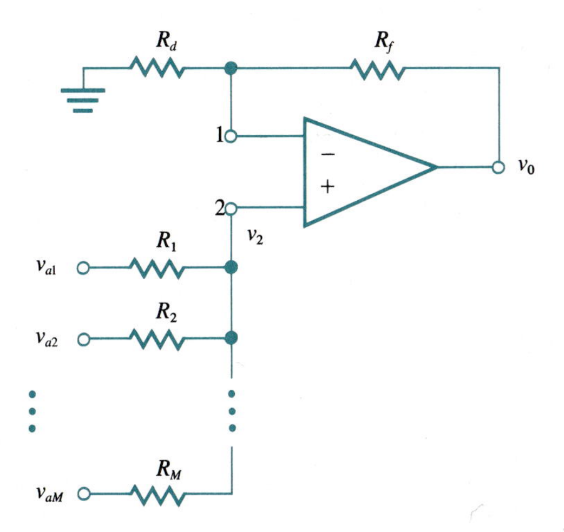

FIG. 8.2-2. A noninverting summing amplifier.

Clearly, if all resistors R_n are the same so that $R_n = R_1$, for all n, then

$$v_0 = -\frac{R_f}{R_1}\sum_{n=1}^{n} v_{bn} \qquad \text{if all } R_n = R_1 \tag{8.2-2}$$

which corresponds to an inverting summing amplifier with gain.

Noninverting Summing Amplifier

By generalizing the circuit of Fig. 8.1-2 as shown in Fig. 8.2-2, we can obtain an M-input noninverting summing amplifier. It can be shown (see Problem 8-11) that the output v_0 is given by

$$v_0 = \left(1 + \frac{R_f}{R_d}\right)(R_1 \,||\, R_2 \,||\, \cdots \,||\, R_M) \sum_{m=1}^{M} \frac{v_{am}}{R_m} \tag{8.2-3}$$

where $(R_1 \,||\, R_2 \,||\, \cdots \,||\, R_M)$ represents the resistance of all resistors R_1, $R_2, \ldots, R_M$ in parallel. For the special case where all these resistors are equal, say $R_m = R$ for all m, then

$$v_0 = \left(1 + \frac{R_f}{R_d}\right)\frac{1}{M}\sum_{m=1}^{M} v_{am} \tag{8.2-4}$$

Equation (8.2-4) can be interpreted two ways. First, it can be considered as a single-input noninverting amplifier of gain $[1 + (R_f/R_d)]$ that has an input equal to the average of the M inputs. Second, it can be considered as an M-input noninverting summing amplifier for which the gain seen by each input is $[1 + (R_f/R_d)]/M$.

Differential Amplifier

An op amp connected in the manner of Fig. 8.2-3 is called a weighted differencing amplifier. Use of the word *weighted* refers to the fact that the output has the form

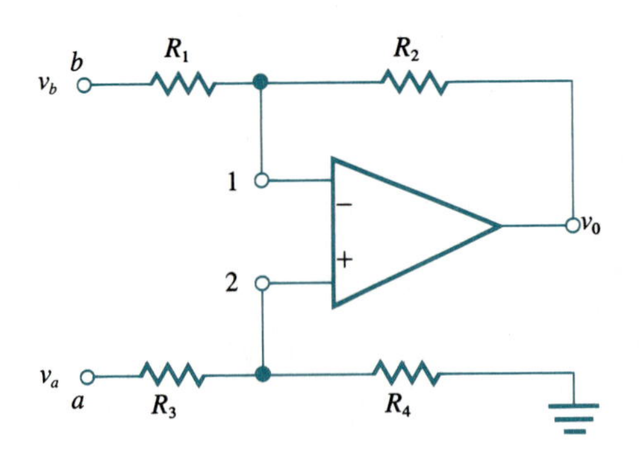

FIG. 8.2-3. A weighted differencing amplifier.

$v_0 = w_a v_a + w_b v_b$, where the weighting coefficients, w_a and w_b, are the voltage gains seen by the inputs. Analysis of the circuit is simple. We apply superposition since the network is linear.

First, let $v_a = 0$ by grounding the input terminal of R_3. Except for the presence of R_3 in parallel with R_4 between terminal 2 and ground, the circuit is just an inverting amplifier of gain $-R_2/R_1$ to the input v_b. Since no current flows at terminal 2, these resistors have no effect on the circuit, and we have

$$w_b = -\frac{R_2}{R_1} \tag{8.2-5}$$

Second, we let $v_b = 0$ by grounding the left terminal of R_1. With respect to the voltage on terminal 2, the circuit is just a noninverting amplifier with gain $[1 + (R_2/R_1)]$. Since the voltage divider consisting of R_3 and R_4 just corresponds to a "gain" of $R_4/(R_3 + R_4)$ from the input to terminal 2, the overall gain becomes

$$w_a = \left(1 + \frac{R_2}{R_1}\right)\frac{R_4}{R_3 + R_4} \tag{8.2-6}$$

Finally, the overall response is obtained by superposition:

$$\begin{aligned} v_0 = w_a v_a + w_b v_b &= \left(1 + \frac{R_2}{R_1}\right)\frac{R_4}{R_3 + R_4} v_a - \frac{R_2}{R_1} v_b \\ &= \frac{R_2}{R_1}\left[\left(\frac{1 + R_1/R_2}{1 + R_3/R_4}\right) v_a - v_b\right] \end{aligned} \tag{8.2-7}$$

In the special case where

$$\frac{R_1}{R_2} = \frac{R_3}{R_4} \tag{8.2-8}$$

the response becomes that of a *differential amplifier:*

$$v_0 = \frac{R_2}{R_1}(v_a - v_b) \tag{8.2-9}$$

For some practical reasons the differential amplifier is usually implemented with $R_3 = R_1$ and $R_4 = R_2$.

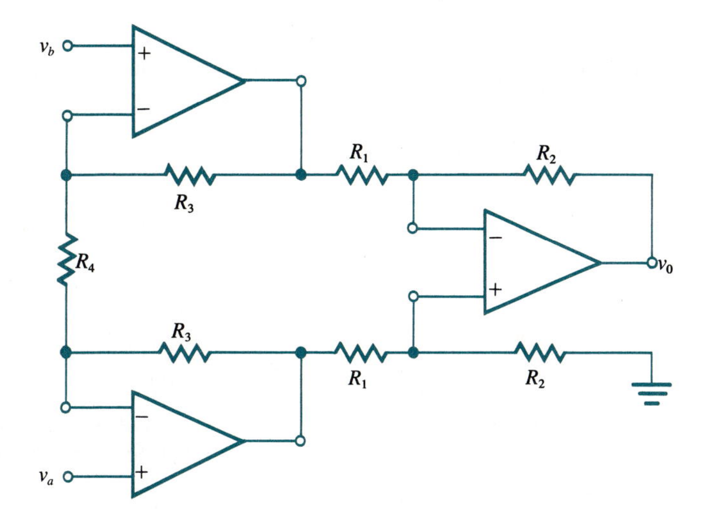

FIG. 8.2-4. An improved differential amplifier.

The input impedance of a differential amplifier can be small (see Problem 8-14), especially if its gain is large, which leads to a small value for R_1. Performance is badly affected if the sources v_a and v_b have large and unequal source impedances. For such a case buffers that make use of the high input impedance of the noninverting amplifier are often used. The modified differential amplifier is shown in Fig. 8.2-4. Its response is now

$$v_0 = \left(1 + \frac{2R_3}{R_4}\right)\frac{R_2}{R_1}(v_a - v_b) \tag{8.2-10}$$

Negative Impedance Converter

The op amp circuit of Fig. 8.2-5 produces a negative resistance R_{in} between the input terminal and ground. In the more general case where R is replaced by an impedance Z, the circuit gives a negative impedance, which leads to its being

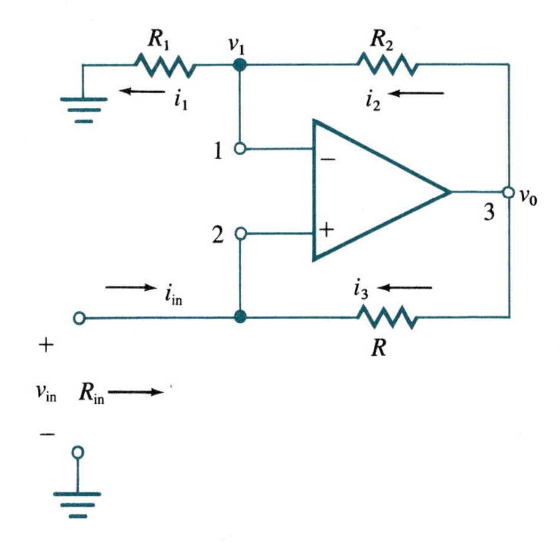

FIG. 8.2-5. A negative impedance converter.

called a *negative impedance converter*. Analysis is relatively straightforward. Because of the virtual short between terminals 1 and 2 of the op amp, we have $v_1 = v_{in}$, $i_1 = v_1/R_1 = v_{in}/R_1$. Since the op amp draws no current from terminal 1, $i_2 = i_1 = v_{in}/R_1$ and $v_O = i_2(R_1 + R_2) = v_{in}[1 + (R_2/R_1)]$. Finally,

$$i_3 = \frac{v_O - v_{in}}{R} = v_{in}\frac{R_2}{RR_1} \tag{8.2-11}$$

But $i_{in} = -i_3$, so

$$R_{in} = \frac{v_{in}}{i_{in}} = -R\frac{R_1}{R_2} \tag{8.2-12}$$

which is a negative resistance.

The negative impedance converter is useful in transforming a voltage source to a current source by means of a voltage-to-current converter (see Problem 8-17).

Integrators

If a resistor and capacitor are added to a negative impedance converter, as sketched in Fig. 8.2-6, the result is a *noninverting integrator*. Analysis parallels that given above for the converter, except that now

$$v_O = 2v_1 \tag{8.2-13}$$

and

$$i_3 = \frac{v_1}{R} \tag{8.2-14}$$

The total capacitor current is

$$i = i_{in} + i_3 = \frac{v_{in} - v_1}{R} + \frac{v_1}{R} = \frac{v_{in}}{R} \tag{8.2-15}$$

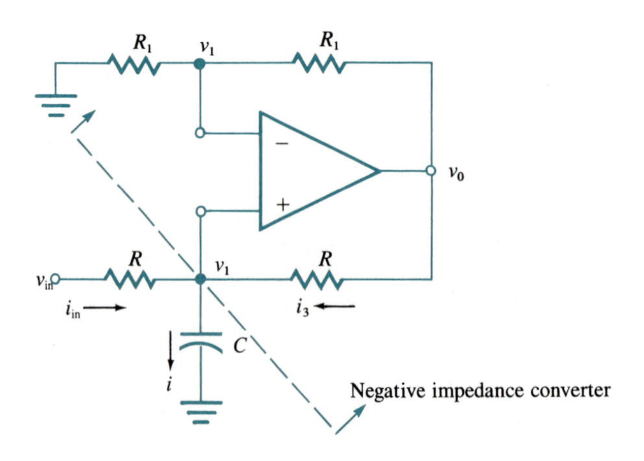

FIG. 8.2-6. A noninverting integrator.

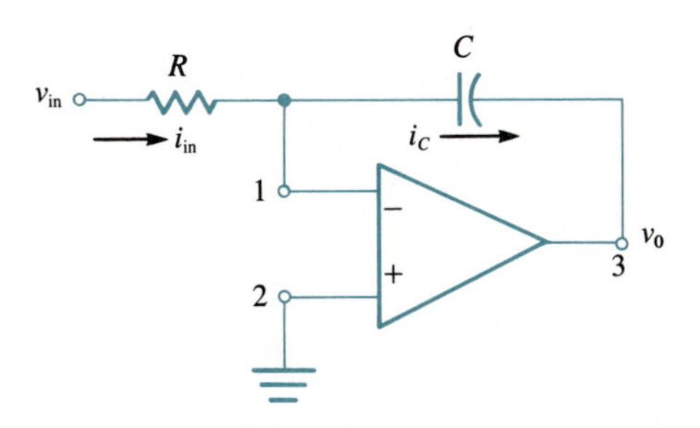

FIG. 8.2-7. An inverting integrator.

Capacitor voltage, from (1.4-9), becomes

$$v_1(t) = \frac{1}{C}\int_{-\infty}^{t} i(\xi)\, d\xi = \frac{1}{RC}\int_{-\infty}^{t} v_{in}(\xi)\, d\xi \tag{8.2-16}$$

Thus,

$$v_0(t) = \frac{2}{RC}\int_{-\infty}^{t} v_{in}(\xi)\, d\xi \tag{8.2-17}$$

which shows that the circuit functions as an integrator.

A somewhat simpler integrator is formed by replacing R_2 in the inverting amplifier of Fig. 8.1-1 by a capacitance C. The circuit, as shown in Fig. 8.2-7, is called an *inverting integrator*. Since terminal 1 is a virtual ground and draws no current in the ideal op amp, $i_C = i_{in} = v_{in}/R$. The voltage across C is just v_0, so

$$v_0(t) = \frac{-1}{C}\int_{-\infty}^{t} i_C(\xi)\, d\xi = -\frac{1}{RC}\int_{-\infty}^{t} v_{in}(\xi)\, d\xi \tag{8.2-18}$$

which shows the network to be an integrator with sign inversion.

Differentiator

A *differentiator*, as shown in Fig. 8.2-8, is obtained by replacing R_1 in the inverting amplifier by a capacitor C. Again, due to action of the ideal op amp, $i = i_C$ and

$$v_0 = -Ri = -Ri_C \tag{8.2-19}$$

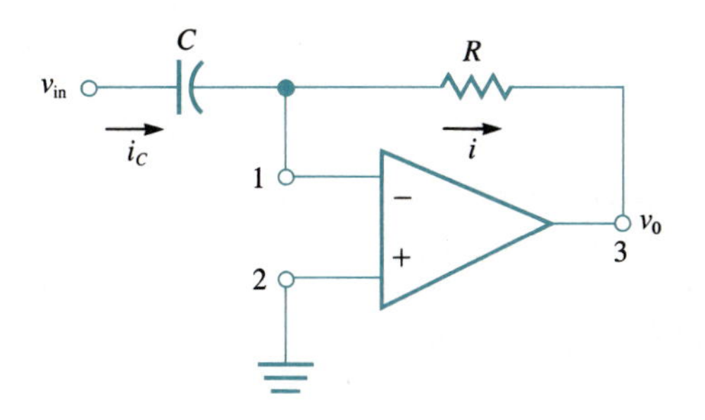

FIG. 8.2-8. A differentiator.

But since

$$i_C(t) = C\,\frac{dv_{\text{in}}(t)}{dt} \tag{8.2-20}$$

we have

$$v_0(t) = -RC\,\frac{dv_{\text{in}}(t)}{dt} \tag{8.2-21}$$

which corresponds to a differentiator with a "gain" of $-RC$.

The use of differentiators is usually avoided wherever possible in practice. The differentiator has a transfer function that increases with frequency, so it accentuates high-frequency noise. Besides being noisy, differentiators can have stability problems (they tend to oscillate).

Active Filters

In proper combination with other components an op amp provides the active element in structuring an *active* filter. We give two examples that use the generalized inverting amplifier of Fig. 8.2-9(a). The first example is the lowpass-filter circuit of Fig. 8.2-9(b). Here $Z_1(\omega) = R_1$ and $Z_2(\omega) = R_2 \,||\, (1/j\omega C) = R_2/(1 + j\omega R_2C)$. Thus the transfer function of this filter is

$$H(\omega) = \frac{v_0}{v_{\text{in}}} = -\frac{Z_2(\omega)}{Z_1(\omega)} = \frac{-(R_2/R_1)}{1 + j\omega R_2C} \tag{8.2-22}$$

which defines a lowpass filter with an angular break frequency (the same as the -3-dB frequency, in this case) of $1/R_2C$. A sketch of the behavior of $|H(\omega)|$ is also shown in Fig. 8.2-9(b). When $\omega \ll 1/R_2C$, the gain is nearly constant at $-R_2/R_1$. As ω exceeds $1/R_2C$, gain decreases. This decrease is approximately a linear function if the gain-frequency plot is on log-log scales, as shown. The break frequency defines the point of separation between the two types of behavior.

A second lowpass-type filter follows choosing the impedances as

$$Z_1(\omega) = R_1 \tag{8.2-23}$$

$$Z_2(\omega) = R_2 \,||\, \left(R_3 + \frac{1}{j\omega C}\right) = \frac{R_2(1 + j\omega R_3C)}{1 + j\omega(R_2 + R_3)C} \tag{8.2-24}$$

Thus,

$$H(\omega) = \frac{v_0}{v_{\text{in}}} = \frac{-(R_2/R_1)(1 + j\omega R_3C)}{1 + j\omega(R_2 + R_3)C} \tag{8.2-25}$$

This filter's circuit and a sketch of the absolute value of its transfer function are shown in Fig. 8.2-9(c). Two break frequencies are involved. At the lowest, which is $1/(R_2 + R_3)C$, the typical lowpass behavior is observed. However, at the second, given by $1/R_3C$, the decrease in circuit gain stops and gain becomes constant at $-R_2R_3/R_1(R_2 + R_3) = -(R_2 \,||\, R_3)/R_1$ for $\omega > 1/R_3C$.

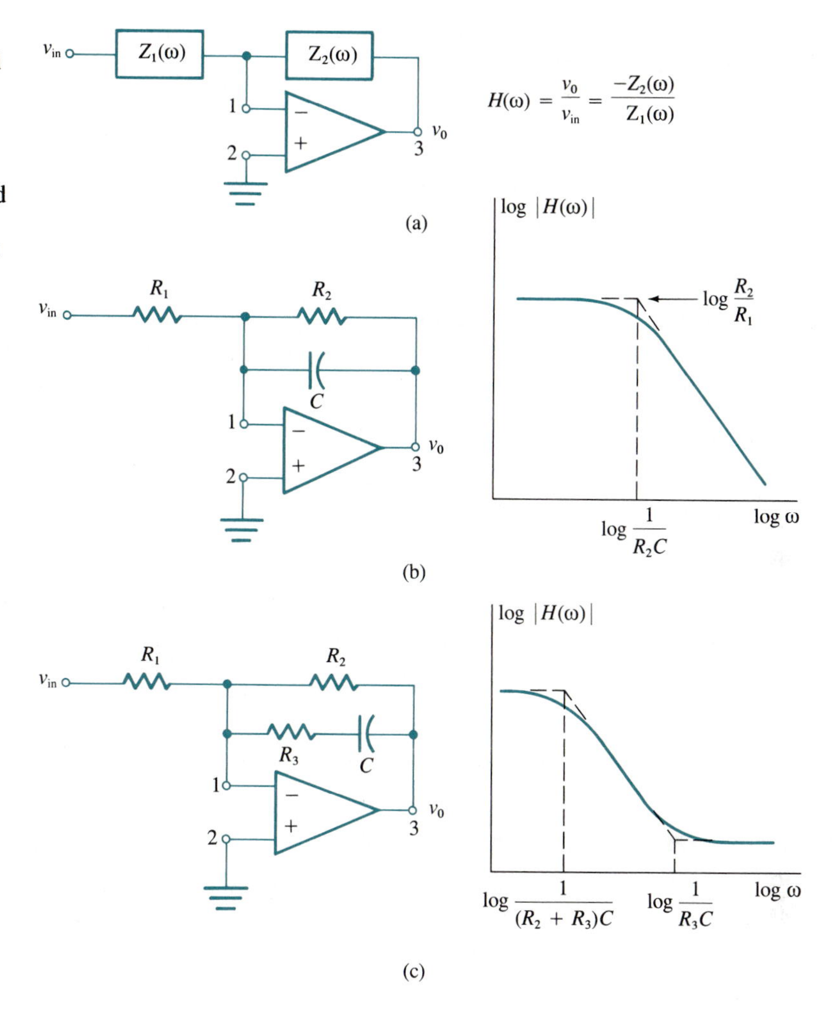

FIG. 8.2-9. (a) A generalized inverting amplifier and two filters derived from it: (b) a lowpass filter and (c) a lowpass filter with a limit on its attenuation.

Op amps have been used to implement all forms of filters: lowpass, highpass, bandpass, band-reject, and special cases. Many texts cover design methods in detail.

8.3 PRACTICAL OP AMPS

Although real op amps can give nearly ideal performance in some applications, there are many practical considerations to be made in a careful design. In this section we touch briefly on some of the more important considerations. However, to stay within the scope of this book, we shall make no effort to be all-inclusive.

Readers interested in greater detail should refer to the many specialized books that are available on op amps.

Maximum Ratings

Op amps are usually designed to use two power supplies of the same voltage magnitude but opposite polarities. Typical supply voltages range from a few volts up to a few tens of volts, with a typical range of about 5–25 V. A given op amp can usually operate within a range of supply voltages that does not exceed the absolute maximum ratings given in the manufacturer's data sheet. The popular μA741A op amp, for example, has absolute maximum supply voltages of ± 22 V, but it can operate from less than ± 5 to ± 22 V, with ± 15 V being a typical value.

Other absolute maximum ratings are given on the manufacturer's data sheet and they must be carefully followed. The maximum average power dissipated within the op amp often is stated for several forms of package and for a specified ambient temperature; care must be used to apply the correct value. Short-circuit durations for the output are usually stated in terms of where the short can occur (to ground or to the supply sources are common) and what ambient temperatures apply. The absolute maximum input terminal voltages and the maximum differential input voltage are often specified (in terms of the power supply voltages); these ratings are especially important in large-signal low-gain applications if output saturation is not desired. Finally, we note that maximum ratings are sometimes stated for output voltage, load current (or minimum impedance), storage-temperature range, operating-temperature range, pin-soldering temperature, etc.

Finite Gain and Bandwidth

Practical op amps have a finite maximum gain (at low frequencies), which we denote by A_0, and a lowpass frequency response. Thus, to a good approximation, a real op amp's voltage gain (transfer function) is given by

$$A(\omega) = \frac{A_0}{1 + j(\omega/\omega_b)} \tag{8.3-1}$$

where ω_b is the 3-dB bandwidth (in radians per second) when no external components are used. Gain $A(\omega)$ is called the *open-loop gain*, as opposed to the *closed-loop gain* that corresponds to any circuit using the op amp with external elements to form a feedback path between the output and one or more input terminals. Equation (8.3-1) is sketched in Fig. 8.3-1; the scales are drawn by assuming that $A_0 = 10^5$ and $\omega_b = 10$ rad/s. Note that ω_u is defined as the angular frequency at which the gain's magnitude is unity. It is related to ω_b by

$$\omega_u = \omega_b\sqrt{A_0^2 - 1} \tag{8.3-2}$$

or by

$$\omega_u \approx A_0\omega_b \tag{8.3-3}$$

if $A_0 \gg 1$.

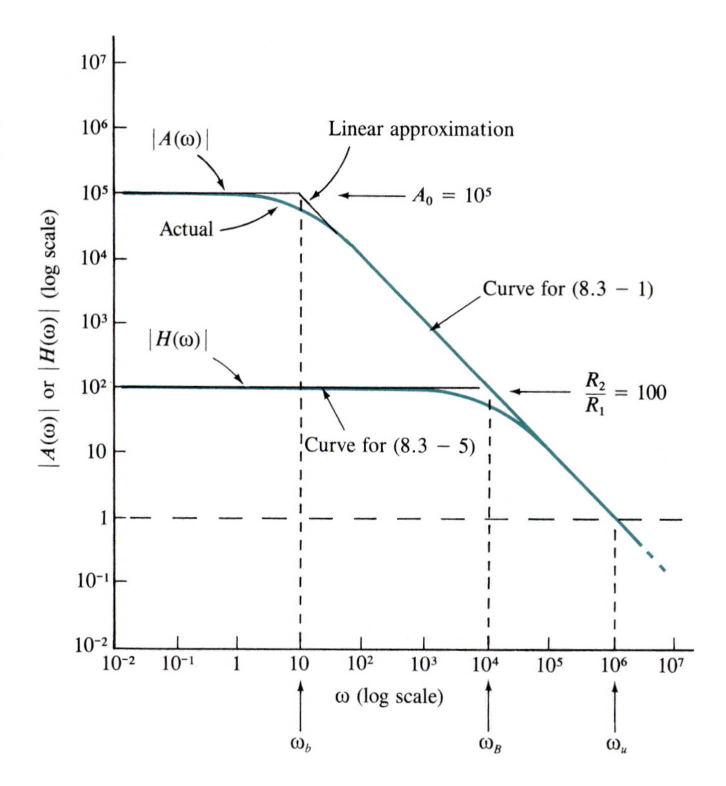

FIG. 8.3-1. Op amp transfer-function magnitudes when $A_0 = 10^5$ and $\omega_b = 10$.

Limited gain and bandwidth affect the performance of the op amp in practical circuits. We shall develop one example for illustration. Consider the inverting amplifier defined by (8.1-5), where now (8.3-1) applies. Its transfer function (gain), $H(\omega)$, now becomes

$$H(\omega) = \frac{v_0}{v_{in}} = -\frac{R_2}{R_1}\left\{\frac{1}{1 + [(R_1 + R_2)/A_0 R_1]}\right\}\left[\frac{1}{1 + j(\omega/\omega_b)/\{1 + [A_0 R_1/(R_1 + R_2)]\}}\right] \tag{8.3-4}$$

which is a lowpass function with different gain and bandwidth. If $A_0 > 200R_2/R_1$, as discussed in Example 8.1-1 following (8.1-5), (8.3-4) can be approximated by

$$H(\omega) \approx \frac{-R_2/R_1}{1 + j(\omega/\omega_b A_0)(1 + R_2/R_1)} \tag{8.3-5}$$

which is a lowpass function with a low-frequency gain $-R_2/R_1$, the desired circuit gain, and a 3-dB bandwidth, denoted by ω_B, of

$$\omega_B = \omega_b \frac{A_0}{1 + R_2/R_1} \approx \omega_b \frac{A_0}{R_2/R_1} \tag{8.3-6}$$

Equation (8.3-5) is plotted on Fig. 8.3-1 (assuming $A_0 = 10^5$, $\omega_b = 10$ rad/s, and $R_2/R_1 = 100$). It is noted that the bandwidth of the "closed-loop" circuit is greater than that of the open-loop op amp alone, but its performance curve is still limited by the op amp's open-loop gain *at the high-frequency end.* A gain-bandwidth trade-off is present, as confirmed by (8.3-6) when written in the form

$$A_0\omega_b = \frac{R_2}{R_1}\omega_B \tag{8.3-7}$$

The left side of (8.3-7) is the gain bandwidth product of the op amp; the right side is the same product for the final circuit. Thus, the gain-bandwidth product of an op amp is a measure of the maximum performance of amplifiers that use the op amp.

EXAMPLE 8.3-1

A μA741A op amp has $A_0 = 2 \times 10^5$ and $\omega_b = 16\pi$ rad/s when supply voltages are ±15 V. We find ω_u and the gain of an inverting amplifier designed for $R_2 =$ 47,000 Ω and $R_1 = 1000$ Ω. From (8.2-2)

$$\omega_u = 16\pi\sqrt{4 \times 10^{10} - 1} \approx 32\pi \times 10^5 \text{ rad/s}$$

or $\omega_u/2\pi = 1.6$ MHz. From (8.3-5)

$$H(\omega) \approx \frac{-47{,}000/1000}{1 + j(\omega/32\pi \times 10^5)(1 + 47{,}000/1000)} = \frac{-47}{1 + j(3\omega/2\pi)(10^{-5})}$$

The 3-dB bandwidth of the circuit is $2\pi \times 10^5/3$ rad/s.

Input Offset Voltage

In an ideal op amp if the two input voltages are zero, the output is zero. In a real op amp the output voltage is typically nonzero even if the inputs are zero (grounded). *Input offset voltage*, v_{io}, is defined as the voltage required at the input to force the output offset to zero. Typical values for v_{io} fall in the range 0.2 to 5 mV. The input offset voltage can be provided by the circuits of Fig. 8.3-2. By proper choices for R_4 and R_5 a small voltage in the range $\pm VR_4/(R_4 + R_5)$ is available at an input terminal to serve as v_{io}. The voltages $\pm V$ can be those of the op amp's power supplies. In the two circuits the resistance from terminal 3 to ground should be small relative to resistances R_3 in Fig. 8.3-2(a) and R_1 in (b). In both circuits R_3 should equal $R_1 \,||\, R_2$ (see below).

Input Offset Current

The ideal op amp draws no current at its input terminals. In a real op amp some current must be present. If the input stages internal to the op amp are bipolar transistors, this current is the required (small) base current. For FET inputs it is the smaller, but finite, gate current. For modeling purposes these currents can be considered as two dc sources of currents, I_{b1} and I_{b2}, connected between the input terminals 1 and 2, respectively, and ground. When resistors are connected

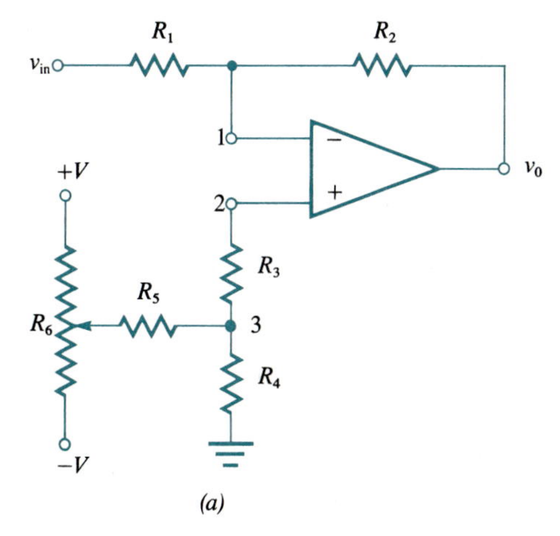

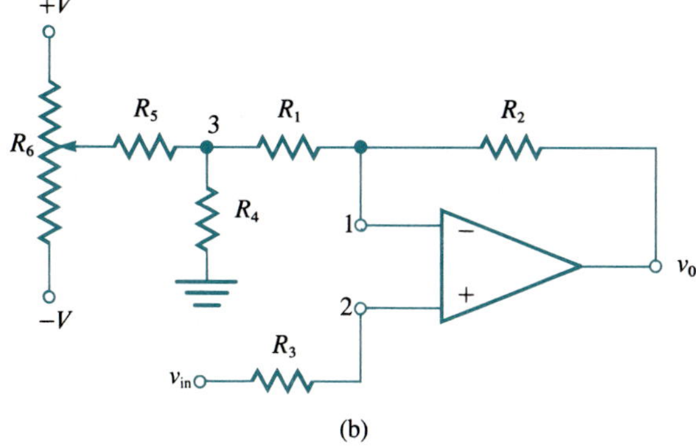

FIG. 8.3-2. Input-offset-voltage–nulling circuits for (a) inverting and (b) noninverting amplifiers.

externally to the input terminals, these bias currents lead to voltages that are amplified and affect the output as offset (unwanted) voltages.

If I_{b1} and I_{b2} are equal, and external resistances are chosen properly, the output offset can be made to be zero. If I_{b1} differs from I_{b2}, other approaches must be taken. Thus, analysis can be broken into two parts: one based on the portions of I_{b1} and I_{b2} that are equal, and the other based on their difference. The difference, denoted by I_{io}, is called the *input offset current*:†

$$I_{io} = I_{b2} - I_{b1} \tag{8.3-8}$$

Their average is called the *input bias current*:

$$I_b = \frac{I_{b1} + I_{b2}}{2} \tag{8.3-9}$$

Typical values for I_b range from 10 to 100 nA for bipolar inputs and 10 to 100 pA

† A manufacturer's data sheet normally gives $|I_{io}|$.

for FETs. Typical values of I_{io} usually range from 2 to 20 nA for bipolar inputs and 5 to 50 pA for FETs.

The definitions of (8.3-8) and (8.3-9) are equivalent to writing I_{b1} and I_{b2} as

$$I_{b1} = I_b - \frac{I_{io}}{2} \tag{8.3-10}$$

$$I_{b2} = I_b + \frac{I_{io}}{2} \tag{8.3-11}$$

which show that I_b is the common component of I_{b1} and I_{b2}, and $I_{io}/2$ is half their difference.

It can be shown that if the dc resistance of external components connected between one input terminal and ground is equal to the similarly computed resistance from the other terminal to ground, the output offset voltage due to I_b is zero. The remaining output offset voltage due to I_{io} can be nulled by use of the balancing circuits of Fig. 8.3-2.

In summary, to minimize output voltage offsets in practical op amp circuits, the designer should (1) provide a dc path from each input terminal to ground, (2) make each input terminal see the same external resistance to ground (as much as possible), and (3) use external balancing circuits (Fig. 8.3-2), if necessary, to null any remaining output offset voltage.

We develop an example.

EXAMPLE 8.3-2

We examine the effects of input bias currents on the inverting amplifier of Fig. 8.1-1, except that we add a resistance R_3 from terminal 2 to ground. The model is depicted in Fig. 8.3-3. We have

$$v_2 = -I_{b2}R_3$$

Since the op amp's gain is large, we can continue to assume a zero differential input voltage, so

$$v_1 = v_2$$

Next,

$$I_1 = -\frac{v_1}{R_1} = -\frac{v_2}{R_1} = \frac{I_{b2}R_3}{R_1}$$

$$I_2 = I_{b1} - I_1 = I_{b1} - \frac{I_{b2}R_3}{R_1}$$

and

$$v_0 = I_2R_2 + v_1 = R_2\left[I_{b1} - I_{b2}R_3\left(\frac{R_1 + R_2}{R_1R_2}\right)\right] \tag{A}$$

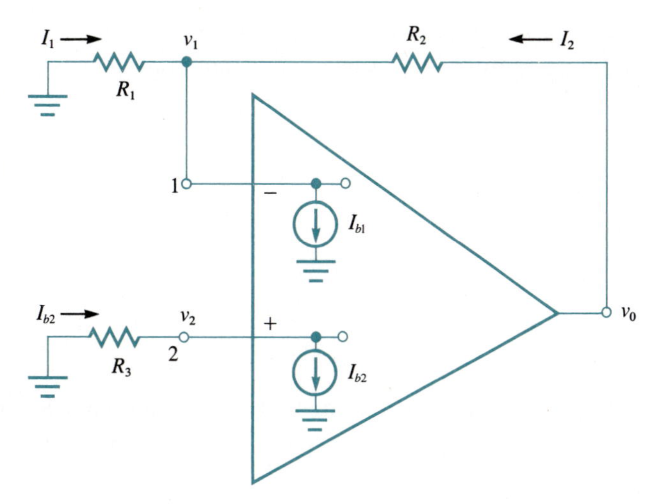

FIG. 8.3-3. An inverting amplifier showing the model for the input bias currents analyzed in Example 8.3-2.

Next, we use (8.3-10) and (8.3-11) in (A):

$$v_O = R_2\left\{I_b\left[1 - R_3\left(\frac{R_1 + R_2}{R_1R_2}\right)\right] - \frac{I_{io}}{2}\left[1 + R_3\left(\frac{R_1 + R_2}{R_1R_2}\right)\right]\right\} \tag{B}$$

Clearly, if

$$R_3 = \frac{R_1R_2}{R_1 + R_2} = R_1 \,||\, R_2$$

the component of v_O due to I_b is zero. The remaining component due to I_{io} is

$$v_O = -I_{io}R_2 \tag{C}$$

To compare this residual output to what occurs if R_3 were zero, we find from (B) and (C) that

$$\left|\frac{v_O(\text{for } R_3 = 0)}{v_O(\text{for } R_3 = R_1 \,||\, R_2)}\right| = \left|\frac{I_{b1}}{I_{io}}\right|$$

Since $|I_{b1}/I_{io}|$ can easily be 5 to 10 or more in general-purpose op amps, the reduction in offset voltage is significant.

Common-Mode Rejection

An op amp that has a finite voltage gain A_0, but is otherwise ideal, produces an output voltage v_O that is proportional to the difference in its two input voltages. This output voltage equals A_0v_d, where

$$v_d = v_2 - v_1 \tag{8.3-12}$$

Ideally, there is no output-voltage component related to the common voltage, denoted by v_{cm}, present in v_2 and v_1. This common part is given by

$$v_{\text{cm}} = \frac{v_2 + v_1}{2} \tag{8.3-13}$$

so we can write v_1 and v_2 as

$$v_1 = v_{\text{cm}} - \frac{v_d}{2} \tag{8.3-14}$$

$$v_2 = v_{\text{cm}} + \frac{v_d}{2} \tag{8.3-15}$$

In real op amps there *is* a component in the output voltage due to v_{cm} that equals $A_{\text{cm}}v_{\text{cm}}$. Here A_{cm} is a common-mode voltage gain. The total output voltage becomes

$$v_0 = A_0 v_d + A_{\text{cm}} v_{\text{cm}} \tag{8.3-16}$$

The ratio of the ideal-output part to the common-mode output for equal excitation (that is, for $v_d = v_{\text{cm}}$) is a measure of the quality of the real op amp. The ratio expressed in decibels is called the *common-mode rejection ratio*, CMRR:

$$\text{CMRR} = 20 \log \frac{A_0}{A_{\text{cm}}} \tag{8.3-17}$$

CMRR is a positive number, typically in the 50 to 100 dB range; CMRR $= \infty$ for an ideal op amp.

8.4 GENERAL FEEDBACK AND CONTROL CONCEPTS

The circuits discussed above use components in various configurations between the output of the op amp and one or both of its input terminals. The circuits are said to have *feedback*, and the components form a *feedback loop*. Generally, there are four basic topologies that are possible in a feedback loop, as illustrated in Fig. 8.4-1. These topologies depend on whether the output voltage or current is to be observed for feedback, and on whether the signal fed back to the input is a voltage in series with the input (voltage) source or is a current in shunt with the input (current) source. Examples of voltage-series feedback are (1) the noninverting op amp of Fig. 8.1-2 (with a load resistor added to the output) and (2) the voltage-follower circuits that use either a bipolar transistor or an FET.

It is not always easy to identify exactly which parts of a given network belong to the amplifier and feedback portions of the topologies of Fig. 8.4-1. There are, of course, general procedures for making these identifications, but the space and detail needed to develop them are beyond our scope. However, if we observe that

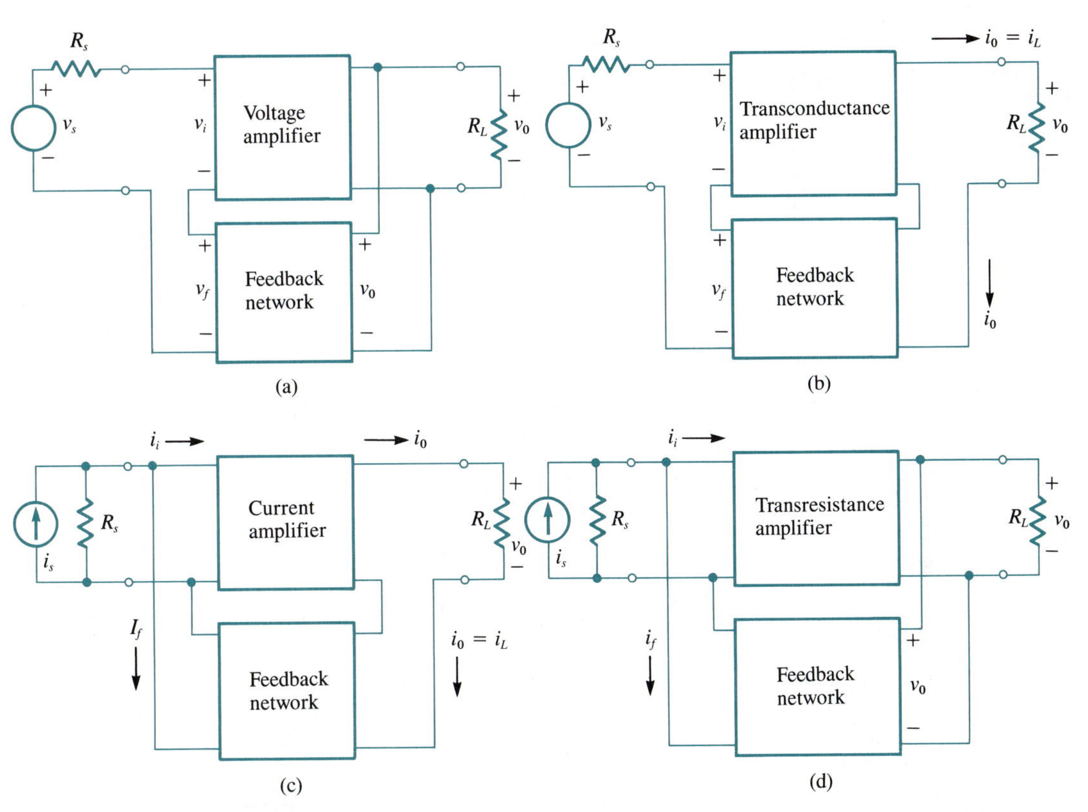

FIG. 8.4-1.
Feedback topologies: (a) voltage-series, (b) current-series, (c) current-shunt, and (d) voltage-shunt.

all four topologies can be drawn in the general form of Fig. 8.4-2(a), some very general and useful concepts can be developed about feedback and control systems through a study of the general form.

Closed-Loop Equations

Let $x(t)$ represent some input signal to the feedback system of Fig. 8.4-2(a). The purpose of the system is to force a feedback signal $x_f(t)$ to follow $x(t)$ as closely as possible so that the difference, called the *error signal*, $x_e(t)$, is as small as possible. The signals $x(t)$, $x_f(t)$, and $x_e(t)$ might represent voltages or currents or even parameters such as position or speed in a practical system. The error is passed through a network with transfer function $H_1(\omega)$ to produce a desired output, denoted by $y(t)$. The output can represent either a current, a voltage, or other parameters, depending on the practical system being modeled. The output is fed back through the feedback path's transfer function $H_2(\omega)$ to generate $x_f(t)$ and to close the loop. Analysis is most easily conducted in the frequency domain. If $X(\omega)$, $X_f(\omega)$, $X_e(\omega)$, and $Y(\omega)$ are the Fourier transforms of $x(t)$, $x_f(t)$, $x_e(t)$, and $y(t)$, respectively, then

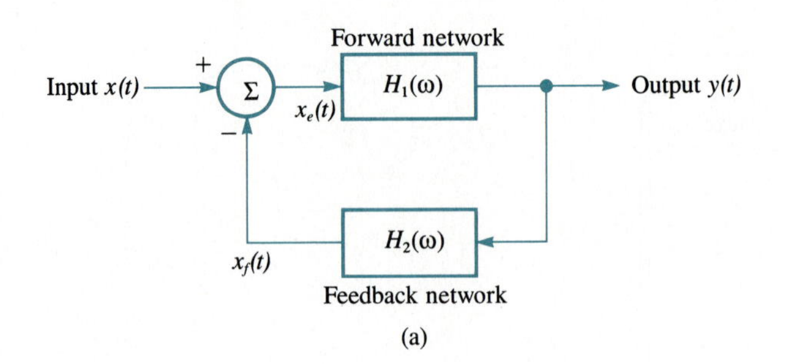

Input $x(t)$ → $H(\omega) = \dfrac{H_1(\omega)}{1 + H_1(\omega)H_2(\omega)}$ → Output $y(t)$

(b)

FIG. 8.4-2. (a) The general structure of a simple feedback system and (b) its equivalent form.

$$Y(\omega) = H_1(\omega)X_e(\omega) \tag{8.4-1}$$

$$X_e(\omega) = X(\omega) - X_f(\omega) \tag{8.4-2}$$

$$X_f(\omega) = H_2(\omega)Y(\omega) \tag{8.4-3}$$

On combining these three results, we solve for the ratio $Y(\omega)/X(\omega)$, which we denote by $H(\omega)$ and call the *closed-loop transfer function*:

$$H(\omega) = \frac{Y(\omega)}{X(\omega)} = \frac{H_1(\omega)}{1 + H_1(\omega)H_2(\omega)} \tag{8.4-4}$$

From this result, we see that the form of Fig. 8.4-2(b) is equivalent to the system of (a).

To illustrate the application of (8.4-4), consider the noninverting op amp sketched in Fig. 8.4-3(a). The finite-gain, finite-bandwidth op amp of (8.3-1) is assumed. Here

$$H_1(\omega) = A(\omega) = \frac{A_0}{1 + j(\omega/\omega_b)} \tag{8.4-5}$$

$$H_2(\omega) = \frac{R_1}{R_1 + R_2} \tag{8.4-6}$$

so (8.4-4) becomes

$$H(\omega) = \left(1 + \frac{R_2}{R_1}\right)\left[\frac{1}{1 + [(R_1 + R_2)/A_0R_1]}\right]\left[\frac{1}{1 + j(\omega/\omega_b)/\{1 + [A_0R_1/(R_1 + R_2)]\}}\right] \tag{8.4-7}$$

The first two right-side factors are just the low-frequency gain, as expected from (8.1-13). The third factor is the same as found in (8.3-4) for the inverting amplifier.

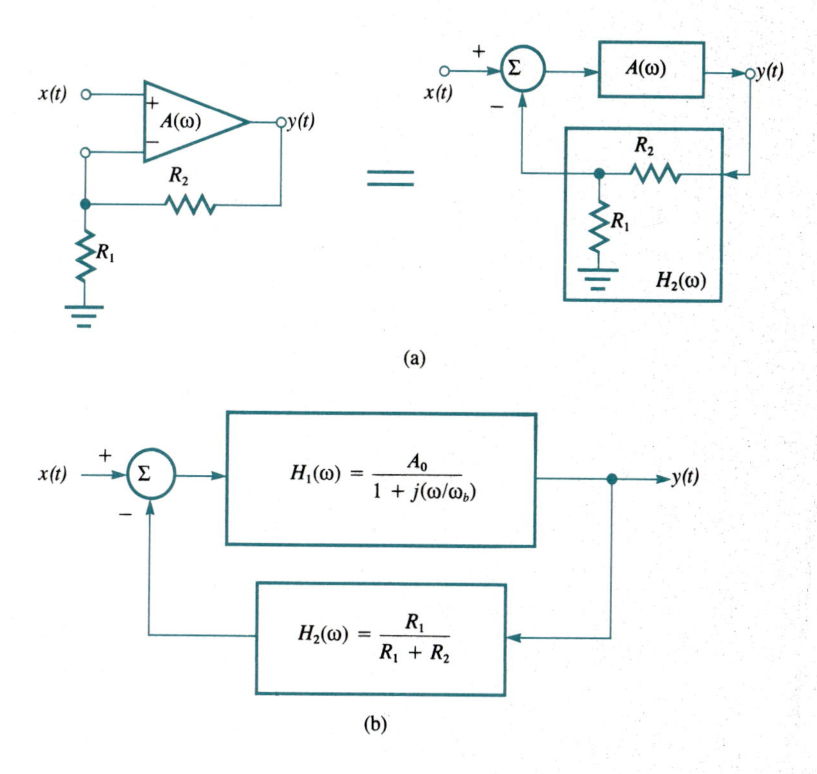

FIG. 8.4-3. (a) A noninverting op amp circuit and (b) its closed-loop representation.

Together these factors indicate that the op amp's gain-bandwidth product enters into the performance of the noninverting amplifier exactly as it did for the inverting amplifier.

Loop Gain

The product $H_1(\omega)H_2(\omega)$ is called the *loop gain*. Its behavior determines the behavior of the closed loop, at least to a great extent. For example, if loop gain is large, (8.4-4) becomes

$$H(\omega) \approx \frac{1}{H_2(\omega)} \qquad \text{loop gain large} \tag{8.4-8}$$

which indicates that the closed-loop gain is almost entirely set by the feedback network. This result is not surprising. We have encountered it many times already in disguised form. The case of an ideal op amp, for instance, where we can assume $A_0 \to \infty$ in (8.4-7), gives $H(\omega) = 1 + (R_2/R_1)$, the expected op amp gain, which depends only on the feedback components R_1 and R_2.

More generally, if $H_2(\omega)$ is due to passive elements and loop gain is large, (8.4-8) implies that closed-loop gain can be very stable. Stabilization of gain is one of the major advantages of using feedback.

FIG. 8.4-4.
Polar (Nyquist) plots for three loop-gain functions.

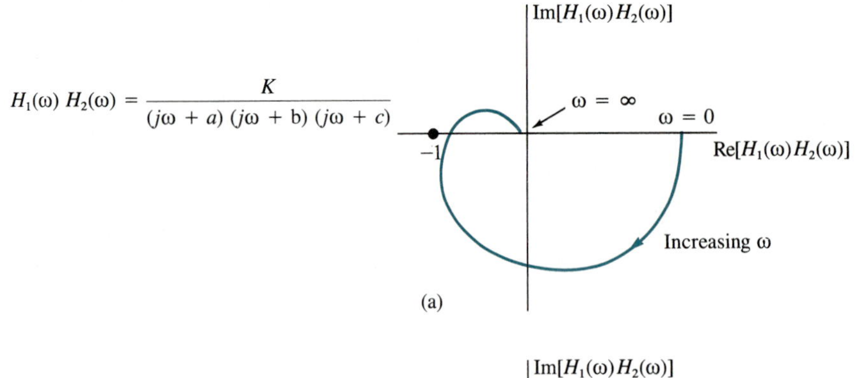

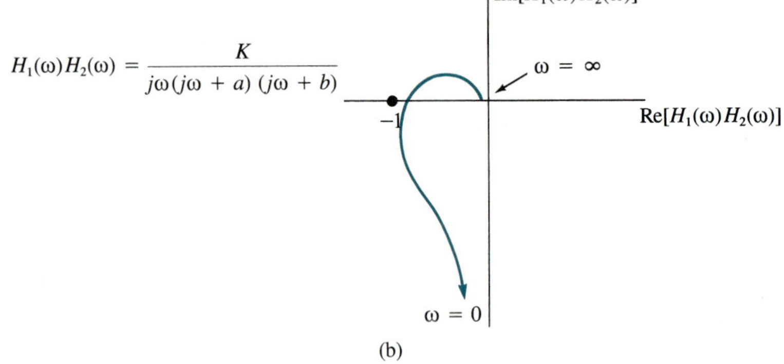

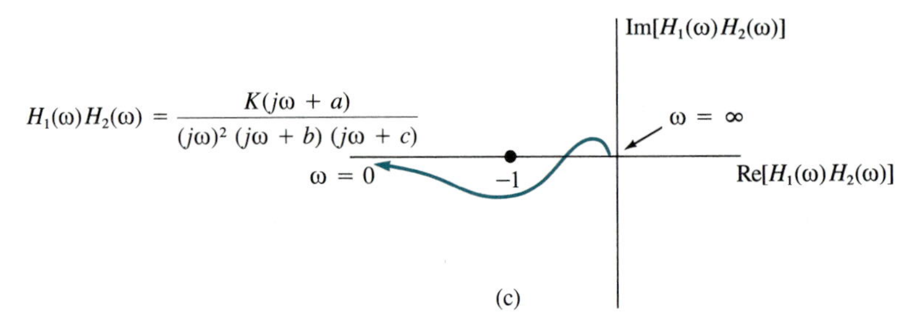

Stability and the Nyquist Plot

The closed loop of Fig. 8.4-2 may not be stable for some choices of the transfer functions $H_1(\omega)$ and $H_2(\omega)$, even if these functions are themselves perfectly stable. Stability is determined by the behavior of the loop gain $H_1(\omega)H_2(\omega)$. Because $H_1(\omega)H_2(\omega)$ is a complex quantity, it can be represented by a plot of its magnitude and phase (polar plot) at any given value of ω. This polar plot for all values of ω from $-\infty$ to ∞ is called the *Nyquist plot*. The part of the plot for $0 < \omega < \infty$ is especially convenient in determining closed-loop stability (the plot for negative ω is the mirror image about the real axis of the plot for positive ω). Some typical examples are shown in Fig. 8.4-4.

A simple interpretation of the polar plot of loop gain enables the determination of stability. For the usual case where $H_1(\omega)$ and $H_2(\omega)$ are stable functions (the only case we shall consider), the closed-loop system will be stable if the point -1 always lies in the area to the left of the polar plot when we move along the curve in the direction of increasing ω. The systems corresponding to Fig. 8.4-4 are all stable as illustrated but could become unstable if absolute gain (constant K) is made large enough to cause the point -1 to be encircled by (to the right of) the curve. When encirclement occurs, there will be a value of ω where the phase of the loop gain is $-\pi$ when $|H_1(\omega)H_2(\omega)| \geq 1$; these conditions lead to instability (oscillations). Intuition justifies that the polar plot of a system's loop gain should not come too close to the point -1. If it does, slight variations in power-supply voltages, aging effects, and changes in component values might cause instability. We next develop some measures of how close a loop-gain plot comes to this critical point.

Gain and Phase Margins

Let ω_π represent the angular frequency at which the loop gain reaches a phase of $-\pi$. *Gain margin* (GM) is defined by

$$\text{GM} = -20 \log |H_1(\omega_\pi)H_2(\omega_\pi)| \qquad (8.4\text{-}9)$$

which is the negative of the magnitude of the loop gain at $\omega = \omega_\pi$ when expressed in decibels (dB). In most cases a positive gain margin is required for a stable system. With large gain margin, a system is less likely to become unstable due to variations in the previously mentioned practical parameters of the system. The GM represents the increase in loop gain (in decibels) required to make the loop unstable. A negative gain margin usually corresponds to an unstable system.

EXAMPLE 8.4-1

Loop gain in a system is -0.35 when $\omega = 140\pi$. We find the system's gain margin. Here $\omega_\pi = 140\pi$ since loop gain is negative and real at this angular frequency. From (8.4-9):

$$\text{GM} = -20 \log |-0.35| = -20 \log 0.35 \approx 9.12 \text{ dB}$$

which is a reasonably good gain margin for a practical system.

If ω_u represents the value of ω where the loop gain has a magnitude of unity, then *phase margin* (PM) is defined as the phase of the loop gain at $\omega = \omega_u$ (which is a negative value) plus π:

$$\text{PM} = \pi + \tan^{-1} \frac{\text{Im}\,[H_1(\omega_u)H_2(\omega_u)]}{\text{Re}\,[H_1(\omega_u)H_2(\omega_u)]} \qquad (8.4\text{-}10)$$

A stable system will have a positive phase margin; PM is negative for an unstable system. Practical closed loops are usually designed to have a phase margin of $\pi/4$ or more. A smaller value causes undesirable "overshoot" in the transient (step) response of the loop.

EXAMPLE 8.4-2

Consider a closed loop for which

$$H_1(\omega) = \frac{80}{(1 + j5\omega)(1 + j\omega)[1 + j(\omega/20)]}$$

$$H_2(\omega) = \tfrac{1}{4}$$

Loop gain is

$$H_1(\omega)H_2(\omega) = \frac{20}{(1 + j5\omega)(1 + j\omega)[1 + j(\omega/20)]}$$

$$= \frac{20e^{-j\tan^{-1}(5\omega) - j\tan^{-1}(\omega) - j\tan^{-1}(\omega/20)}}{[(1 + 25\omega^2)(1 + \omega^2)(1 + \omega^2/400)]^{1/2}}$$

By trial and error we find that $|H_1(\omega)H_2(\omega)| = 1$ when $\omega = \omega_u = 1.869$ rad/s. At this angular frequency the phase of $H_1(\omega_u)H_2(\omega_u)$ is -0.839π (or $-151.1°$). Phase margin becomes $\pi - 0.839\pi = 0.161\pi$ (28.9°), which is a somewhat small value. Similarly, trial and error shows that the loop gain's phase is $-\pi$ at $\omega = \omega_\pi = 4.919$ rad/s. At this frequency $|H_1(\omega_\pi)H_2(\omega_\pi)| = 0.157$, so the gain margin is GM $= -20 \log 0.157 = 16.07$ dB, which is a fairly substantial amount.

Bode Plots

One convenient way to examine the behavior of the loop-gain function is to make separate plots of its magnitude (in decibels) and phase as a function of $\log \omega$. These are known as *Bode plots*. Piecewise-linear approximations to Bode plots are especially helpful in making rapid estimates of gain and phase margins, and these are called *idealized Bode plots*.

To visualize Bode plots, consider the function

$$H(\omega) = \frac{A}{1 + j(\omega/\omega_p)} \tag{8.4-11}$$

Its magnitude, expressed in decibels (dB), is

$$|H(\omega)|_{\text{dB}} = 20 \log |H(\omega)| = 20 \log (|A|) - 20 \log \left[\sqrt{1 + \left(\frac{\omega}{\omega_p}\right)^2}\right]$$

$$= 20 \log (|A|) - 10 \log \left[1 + \left(\frac{\omega}{\omega_p}\right)^2\right] \tag{8.4-12}$$

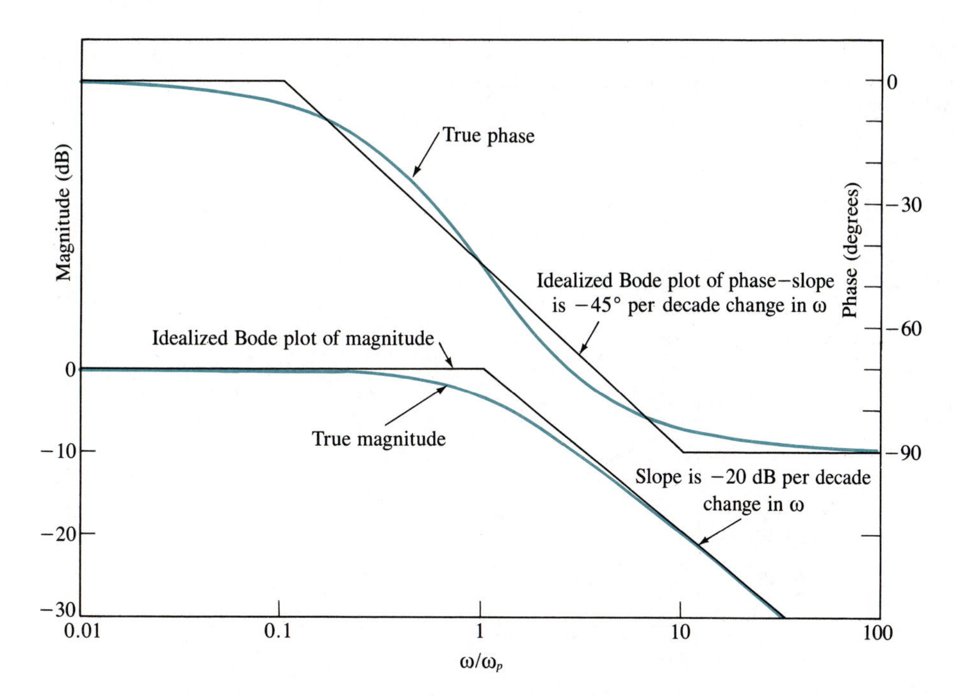

FIG. 8.4-5. Bode plots and idealized Bode plots of the magnitude and phase of the transfer function of (8.4-11).

For $\omega \ll \omega_p$, $|H(\omega)|_{\text{dB}} \approx 20 \log |A|$, which is a constant equal to the magnitude of the low-frequency gain A, expressed in decibels. For $\omega \gg \omega_p$, $|H(\omega)|_{\text{dB}}$ decreases linearly with log ω, with a slope of −20 dB per decade change in ω. The exact and piecewise approximations to $H(\omega)/A$ are shown in Fig. 8.4-5. To a good approximation $|H(\omega)|$ can be replaced by its idealized Bode plot at all frequencies. The largest error is 3 dB, which occurs at the break frequency ω_p. The error is negligible for $\omega > 10\omega_p$ and $\omega < 0.1\omega_p$.

The phase of $H(\omega)$ as given by (8.4-11) is

$$\text{Phase} = -\tan^{-1} \frac{\omega}{\omega_p} \tag{8.4-13}$$

which is plotted in Fig. 8.4-5. It is approximately constant at 0° for $\omega < 0.1\omega_p$ and at −90° for $\omega > 10\omega_p$. In between, phase is approximately linear with log ω. Thus, for approximate calculations, the idealized Bode plot of phase can be used. The maximum error in the approximation is about 5.71°.

The utility of Bode plots resides in the fact that general transfer functions can be expressed as proportional to products of factors of the form of $[1 + j(\omega/\omega_p)]$ in both the numerator and the denominator.† As an example, consider the slightly

† In some networks factors occur in the form $[1 - (\omega^2/\omega_n^2) + j2\zeta(\omega/\omega_n)]$, where ω_n and ζ are real constants. When $\zeta > 1$, the factor can be written as $(1 + j\omega a)(1 + j\omega b)$, where a and b are real numbers. When $\zeta = 1$, the factor can be written as $[1 + (j\omega/\omega_n)]^2$. For $\zeta < 1$, the exact Bode plots should be used in calculations.

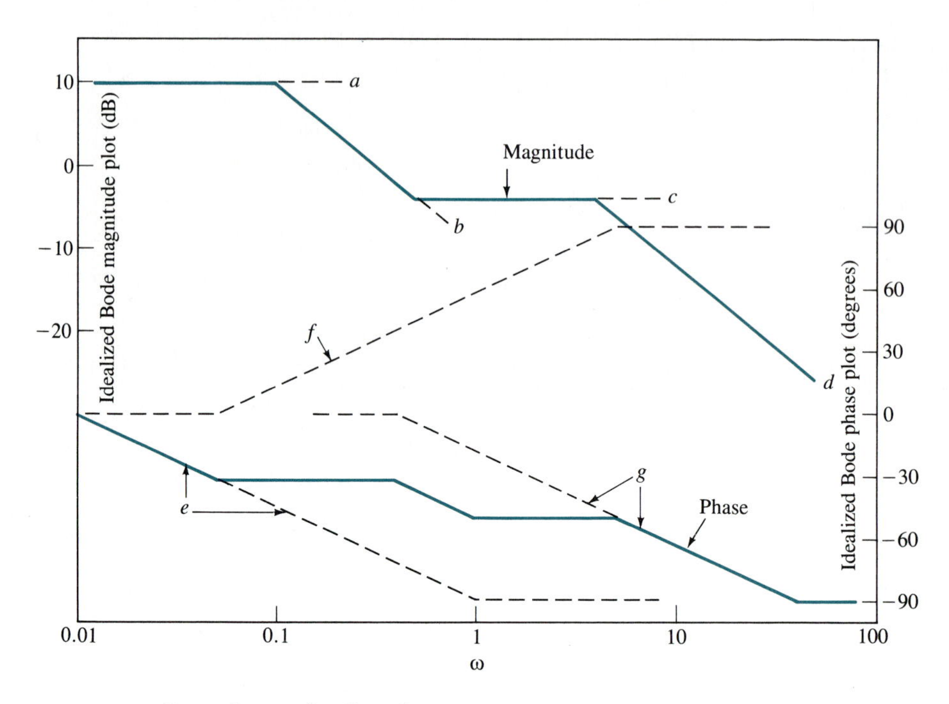

FIG. 8.4-6. Idealized Bode plots for the function in (8.4-14).

more complicated transfer function

$$H(\omega) = \frac{10(1 + j2\omega)}{(1 + j10\omega)(1 + j0.25\omega)} \tag{8.4-14}$$

Break frequencies here are $\frac{1}{10}$, $\frac{1}{2}$, and 1/0.25 rad/s. Since the logarithm of a product of factors is the sum of the logarithms of the factors, the Bode magnitude plot associated with the denominator of (8.4-14) is just the sum of the Bode magnitude plots of the two factors. With the correct interpretation, the Bode magnitude plot of the numerator factor in (8.4-14) can also be added. It is handled exactly as with a denominator factor, except that the curve breaks linearly *upward* above the break frequency, rather than downward. Fig. 8.4-6 shows the idealized Bode magnitude plot for (8.4-14). The numerator factor of 10 gives a constant gain of 10 dB (curve *a*). Above $\omega = 0.1$ the denominator factor $(1 + j10\omega)$ causes the gain to decrease from 10 dB by −20 dB per decade (curve *b*). However, above its break frequency of $\omega = 0.5$ rad/s the numerator factor $(1 + j2\omega)$ causes the gain to *increase* by 20 dB per decade, which compensates the decrease (curve *b*) to again give constant gain (curve *c*). Finally, above its break frequency of $\omega = 4$ rad/s the denominator factor $(1 + j0.25\omega)$ gives another decrease of −20 dB per decade to realize the final response (solid curve to point *d*).

A transfer function comprised of a product of factors, as in (8.4-14), has a phase equal to the sum of the phases of the individual factors. The phases of the numerator factors are positive, and those of the denominator are negative, as in (8.4-13). Thus, the denominator factors contribute the idealized Bode phase terms of curves *e* and *g* in Fig. 8.4-6, and the numerator contributes curve *f*. The sum of

FIG. 8.4-7. Idealized Bode plots for the loop gain of (8.4-15).

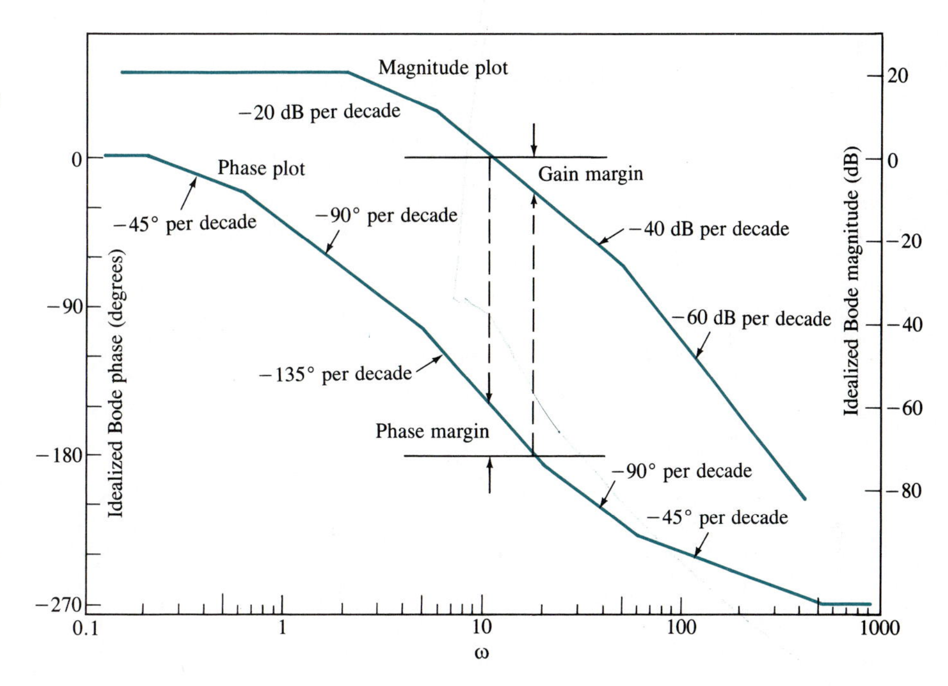

the three curves is the idealized Bode phase plot shown as the solid curve.

The transfer function of (8.4-14) has a phase that is never less than $-\pi/2$, so it is stable as a loop gain. To show how Bode plots can easily estimate gain and phase margin, consider the example loop-gain function

$$H_1(\omega)H_2(\omega) = \frac{10}{[1 + j(\omega/2)][1 + j(\omega/6)][1 + j(\omega/50)]} \tag{8.4-15}$$

Its idealized Bode plots are shown in Fig. 8.4-7. The magnitude plot is constant at 10 dB for ω less than the first break frequency at $\omega = 2$ rad/s. It then decreases by -20 dB per decade until the second break frequency at $\omega = 6$ rad/s, where it then decreases at -40 dB per decade. The effect of the last break frequency at $\omega = 50$ rad/s is to increase the slope to -60 dB per decade for $\omega > 50$ rad/s. The phase plot is the sum of three plots of the form given in Fig. 8.4-5. Each component is located at one of the three denominator break frequencies. For ω larger than 10 times the highest break frequency, the full phase shift of $3(-90°) = -270°$ is realized.

Gain and phase margins can be read directly from the idealized Bode plots of Fig. 8.4-7. Loop-gain magnitude is unity (0 dB) at $\omega_u = 10.8$ rad/s, where the phase margin is found to be 30°. The phase of the loop gain is $-\pi\,(-180°)$ at $\omega_\pi = 17.7$ rad/s, where the gain margin is 8.5 dB. Careful calculations give $\omega_u = 9.94$ rad/s, PM $= 31.25°$, $\omega_\pi = 20.3$ rad/s, and GM $= 11.8$ dB. These more exact results are, of course, found from (8.4-15) or can be obtained from the exact Bode plots, but these are not nearly as easy to construct as the idealized plots.

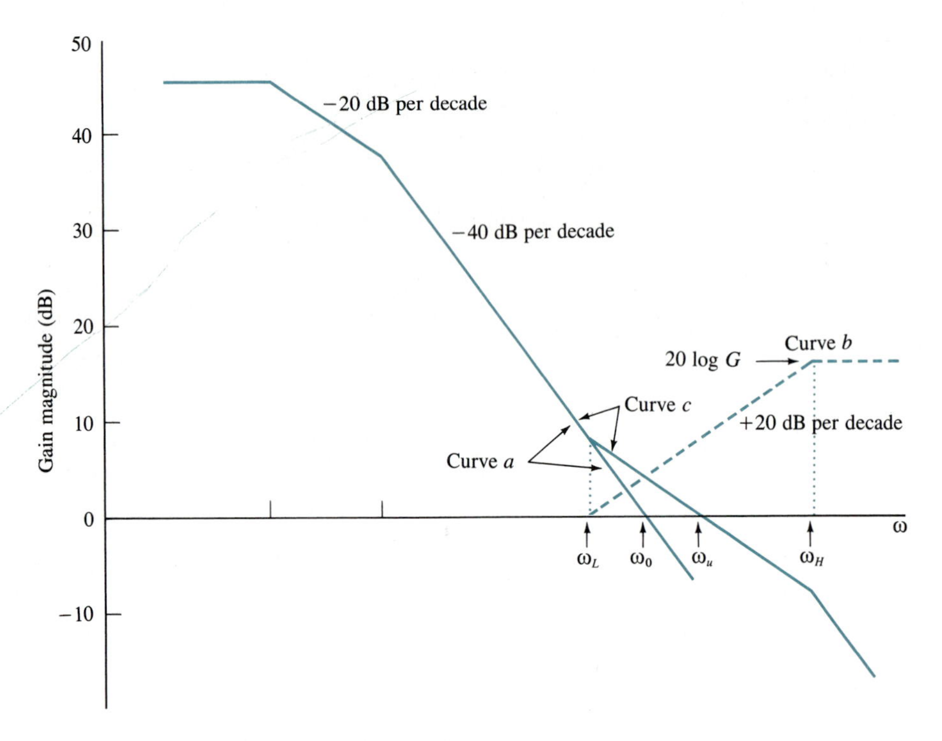

FIG. 8.4-8. The loop gains involved in loop compensation. Curve a: the uncompensated loop gain; curve b: the compensation network's gain; and curve c: the compensated loop's gain.

Compensation

Some loop-gain functions that lead to an unstable system when the loop is closed can be made stable by a procedure called *compensation*. The basic idea is sketched in Fig. 8.4-8. A reasonably common case corresponds to a loop-gain magnitude that crosses the 0-dB axis (unity gain) with a long run at −40 dB per decade (curve a). Phase is, therefore, almost $-\pi$ rad, and the closed loop would be, at best, marginally stable. In fact, there are almost always other, higher break frequencies present in practice that would easily add a few degrees to the phase to make the loop unstable. The idea of compensation is to adjust the loop gain at frequencies near the unity-gain (crossover) value so that an interval of −20 dB per decade gain decrease occurs. The phase will back up from $-\pi$ to a value approaching $-\pi/2$ (for a very long run at −20 dB per decade). A network having the gain magnitude of curve b will compensate curve a to obtain curve c.

If the compensation network has a total voltage-gain increase of G, which is $20 \log G$ in decibels, the frequency ω_L at which compensation starts should be set where the uncompensated loop gain's magnitude is $\sqrt{G}$ (or half of G, in decibels). Figure 8.4-9 gives a network that provides the compensation of curve b of Fig. 8.4-8. Its transfer function is

$$H(\omega) = \frac{v_0}{v_{\text{in}}} = \frac{1 + j\omega R_2C_2[R_1/(R_1 + R_2)][(C_1 + C_2)/C_2]}{1 + j\omega R_2C_2} \qquad (8.4\text{-}16)$$

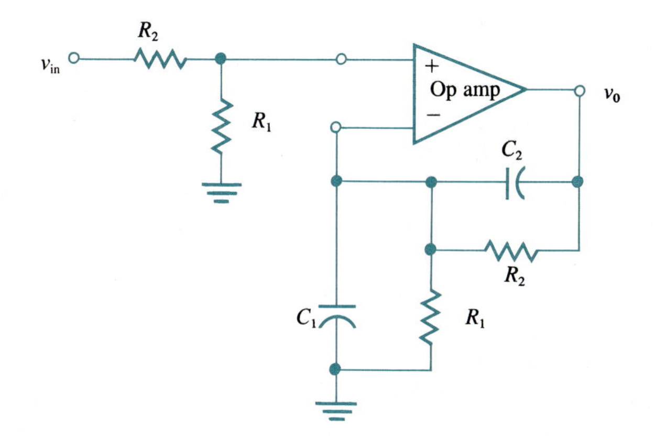

FIG. 8.4-9. A compensation network having the transfer function of (8.4-16).

in general. Since some of the parameters R_1, R_2, C_1, and C_2 can be selected arbitrarily in design, we choose convenient values of R_2 and C_2 so that

$$\omega_L = \frac{1}{GR_2C_2} \tag{8.4-17}$$

and then arbitrarily select R_1 and C_1 according to

$$R_1 = GR_2 \tag{8.4-18}$$

$$C_1 = GC_2 \tag{8.4-19}$$

In terms of these choices $H(\omega)$ can be written as

$$H(\omega) = \frac{1 + j(\omega/\omega_L)}{1 + j(\omega/G\omega_L)} \tag{8.4-20}$$

In the practical use of the compensation circuit of Fig. 8.4-9, G should be at least 6.3 (16 dB), which gives a phase margin of 46.4°. The gain margin is very large (theoretically infinite here) because the compensated loop gain's phase theoretically approaches $-\pi$ as $\omega \to \infty$ (is never less than $-\pi$). In practice, gain margin will be finite because of higher break frequencies that also will reduce the phase margin to less than 46.4°.

8.5 SOME CONTROL-SYSTEM COMPONENTS

As we have already observed, feedback loops have been used with op amps to provide a prescribed gain or transfer function. However, the general feedback concepts of Sec. 8.4 also apply to other forms of systems. They apply, for example, to systems that maintain the fluid level in a tank at a desired value, keep the speed of a motor at a desired value even in the presence of varying shaft loads, cause a mill machine to roll sheet steel to a prescribed thickness, direct the antenna of a radar to point to a designated point in the sky, or cause the acceleration of a rocket to follow a prescribed format with time. These are all examples of *control systems*, where the position, rate (speed), or acceleration of some parameter is controlled. Such systems are also often called *servomechanisms* (*servos*, for short). Many special components such as gears, motors, potentiometers, and tachometers are used in systems involving mechanical control. Other components, such as phase detectors, limiters, frequency discriminators, voltage-controlled oscillators, and mixers, are used in electronic control systems.

In this section the control-system components listed above are briefly defined and discussed so that some example control systems can be developed in the following section.

Gears

Let a gear (call it gear 1) have N_1 teeth around its circumference and come in contact (or mesh) with a second gear having N_2 teeth on its periphery. If the shaft to which gear 1 is attached rotates through an angle θ_1, the number of its teeth that move past the point of contact (see Fig. 8.5-1) is

$$n_1 = N_1 \frac{\theta_1}{2\pi} \tag{8.5-1}$$

Similarly,

$$n_2 = N_2 \frac{\theta_2}{2\pi} \tag{8.5-2}$$

is the number of teeth moving past the contact point for gear 2 when its shaft rotates an angle θ_2. Because the two gears are matched, they must have the same tooth separation, which requires $n_1 = n_2$ for proper gear operation (Fig. 8.5-1). Thus, on equating (8.5-1) and (8.5-2), we get

$$\theta_2 = \theta_1 \frac{N_1}{N_2} \tag{8.5-3}$$

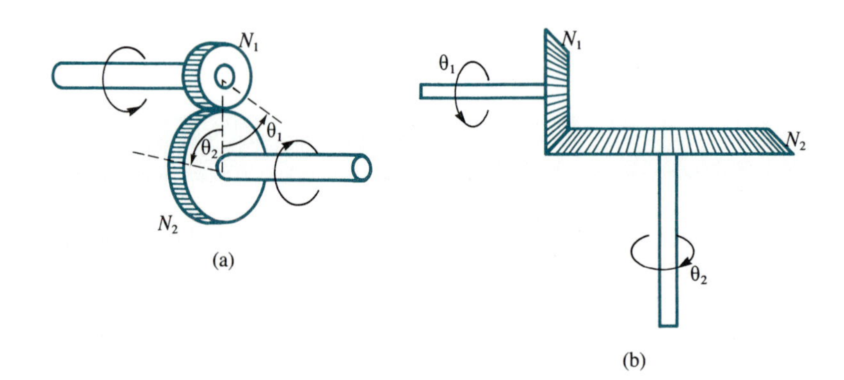

FIG. 8.5-1. Gears: (a) straight and (b) right-angled (beveled).

The quantity N_1/N_2 is called the *gear ratio.*

If θ_1 and θ_2 are varying with time, the rates of rotation of the shafts, denoted by ω_1 and ω_2, respectively, are the time derivatives of shaft angles. From (8.5-3) we have

$$\omega_2 = \omega_1 \frac{N_1}{N_2} \tag{8.5-4}$$

DC Motors

The two most important parts of a dc motor are its armature and its field. The *armature* is the rotating part to which the shaft is connected. A winding on the armature accepts dc current through appropriate sliding contacts from a source external to the motor. Armature current establishes part of the magnetic fields required inside the motor to cause rotation. The *field* refers to a winding on the stationary (housing) part of the motor called the *stator.* A dc field current establishes the remaining part of the magnetic fields needed to make the motor run. In some dc motors the effect of the field winding can be obtained from permanent magnets, and no actual windings are present.

Dc motors are basically devices where shaft speed (rotation rate) is a function of an input control voltage. Two approaches are possible. In the *field-controlled* dc motor a control voltage v_f is applied to the field winding while a constant current is maintained in the armature winding. For a constant value of v_f shaft speed would be constant and shaft angle θ_M would be the integral of shaft speed. In a control system v_f is not constant, and the transfer function of the motor is needed. If $\Theta_M(\omega)$ and $V_f(\omega)$ are the Fourier transforms of $\theta_M(t)$ and $v_f(t)$, respectively, the transfer function (control voltage to shaft angle) can be shown to be

$$\frac{\Theta_M(\omega)}{V_f(\omega)} = \frac{K_M}{j\omega[1 + j(\omega/\omega_f)][1 + j(\omega/\omega_L)]} \tag{8.5-5}$$

where

$$K_M = \frac{K_m}{\sigma_L R_f} \tag{8.5-6}$$

$$\omega_f = \frac{R_f}{L_f} \tag{8.5-7}$$

$$\omega_L = \frac{\sigma_L}{J} \tag{8.5-8}$$

Here R_f and L_f are the resistance and inductance of the field winding. The friction† and inertia of the motor's load are σ_L and J, respectively; K_m is the slope of the motor's torque-field current characteristic. The unit of K_M is the radian per

† The load's (viscous) friction represents the amount of resistive torque created at a given shaft speed. It has the unit of torque per radian per second, or newton-meter per radian per second.

second per volt. Constants ω_f and ω_L are called the *field* and *motor* (load) *break frequencies*, respectively. Often $\omega_f \gg \omega_L$, and the effect of ω_f can be neglected. When this constraint is true, (8.5-5) becomes

$$\frac{\Theta_M(\omega)}{V_f(\omega)} = \frac{K_M}{j\omega[1 + j(\omega/\omega_L)]} \tag{8.5-9}$$

We consider an illustrative example.

EXAMPLE 8.5-1

A particular field-controlled dc motor has negligible field inductance. The motor is connected through a gear train to an output shaft. A long time after 24 V is applied to the motor's field winding, the output shaft rotates at a constant rate of 4π rad/s. If it is known that $\omega_L = 0.8$ rad/s for the motor-gear combination, we find the transfer function of input field voltage to output shaft speed.

Because the gear train affects output shaft angle and speed by the same (gear ratio) factor, we treat the motor-gear combination as simply an equivalent motor. We note that shaft speed is given by $\omega_M(t) = d\theta_M(t)/dt$. If $\Omega_M(\omega)$ is the Fourier transform of $\omega_M(t)$, then $\Omega_M(\omega) = j\omega\Theta_M(\omega)$. From (8.5-5) with ω_f large we have the required transfer function:

$$H(\omega) = \frac{\Omega_M(\omega)}{V_f(\omega)} = \frac{K_M}{1 + j(\omega/\omega_L)} = \frac{K_M}{1 + j(\omega/0.8)}$$

By inverse Fourier transformation the impulse response is

$$h(t) = 0.8K_M u(t)e^{-0.8t}$$

For an input of $v_f(t) = 24u(t)$ the response, by convolution, is

$$\omega_M(t) = \int_{-\infty}^{\infty} v_f(t - \xi)h(\xi)\, d\xi = \int_{-\infty}^{\infty} 24u(t - \xi)[0.8K_M u(\xi)]e^{-0.8\xi}\, d\xi$$

$$= 24(0.8)K_M u(t) \int_0^t e^{-0.8\xi}\, d\xi = 24K_M u(t)(1 - e^{-0.8t})$$

For t large $\omega_M(\infty) = 24K_M$ must equal 4π rad/s, so $K_M = \pi/6$ rad/s·V, and

$$\frac{\Omega_M(\omega)}{V_f(\omega)} = \frac{\pi/6}{1 + j(\omega/0.8)}$$

In the other method of control of the dc motor, field current is maintained constant and speed is controlled by the applied armature voltage v_a. With the Fourier transform of $v_a(t)$ denoted by $V_a(\omega)$, the motor's transfer function can be shown to be

$$\frac{\Theta_M(\omega)}{V_a(\omega)} = \frac{K_m}{j\omega[-\omega^2 JL_a + j\omega(\sigma_L L_a + JR_a) + (\sigma_L R_a + K_m^2)]} \tag{8.5-10}$$

Here R_a and L_a are the armature winding's resistance and inductance, respectively, J and σ_L are defined as above for the field-controlled motor, and K_m is the slope of the motor's torque-armature current characteristic. In some motors the armature inductance is small enough to be neglected, and (8.5-10) reduces to

$$\frac{\Theta_M(\omega)}{V_a(\omega)} = \frac{K_M}{j\omega[1 + j(\omega/\omega_L)]} \tag{8.5-11}$$

where

$$K_M = \frac{K_m}{\sigma_L R_a + K_m^2} \tag{8.5-12}$$

$$\omega_L = \frac{\sigma_L R_a + K_m^2}{J R_a} \tag{8.5-13}$$

AC Motor

There are many forms of ac motors (Chap. 18). We shall here describe only the two-phase motor that is very useful in servomechanisms. It uses two windings; one is the reference to which an ac voltage (typically 60 or 400 Hz) is applied that has a constant peak amplitude. Motor characteristics are determined by an ac voltage with variable amplitude $v_c(t)$ applied to a control winding; its phase is either $+\pi/2$ or $-\pi/2$ relative to the reference winding's voltage.† This phase determines the direction of motor-shaft rotation.

The transfer function of the two-phase ac motor that relates shaft angle to control voltage can be shown to be

$$\frac{\Theta(\omega)}{V_c(\omega)} = \frac{K_M}{j\omega[1 + j(\omega/\omega_L)][1 + j(\omega/\omega_c)]} \tag{8.5-14}$$

where $V_c(\omega)$ is the Fourier transform of $v_c(t)$, and

$$K_M = \frac{K_T}{R_c(\sigma_L + \sigma_M)} \tag{8.5-15}$$

$$\omega_L = \frac{\sigma_L + \sigma_M}{J} \tag{8.5-16}$$

$$\omega_c = \frac{R_c}{L_c} \tag{8.5-17}$$

Here R_c and L_c are the resistance and inductance, respectively, of the control winding, J is the inertia of the load (includes the motor shaft), and σ_L is the viscous friction of the load. The constants K_T (unit is torque per ampere) and σ_M

† The shift in phase from $\pi/2$ to $-\pi/2$ can be the result of $v_c(t)$ changing sign from positive to negative, respectively.

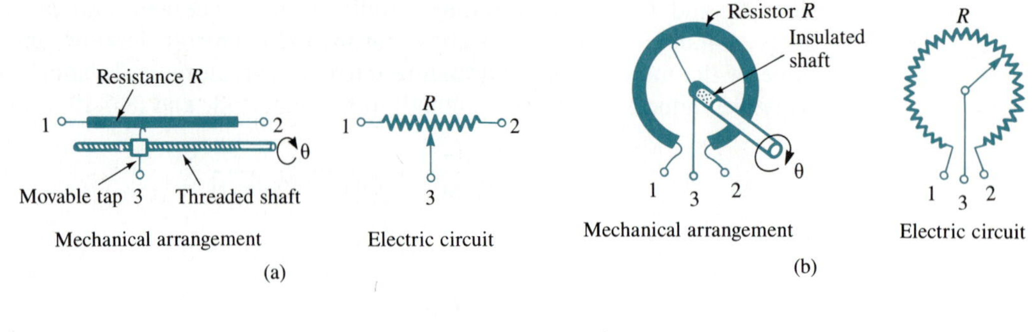

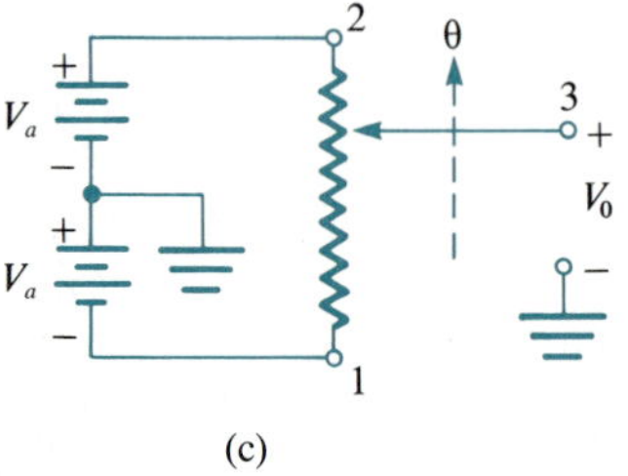

FIG. 8.5-2.
Potentiometers: (a) linear construction and (b) circular construction. (c) A possible electric circuit.

(unit is torque per radian per second) define the model for the motor's torque $T_M(t)$:

$$T_M(t) = K_T i_c(t) - \sigma_M \frac{d\theta_M(t)}{dt} \tag{8.5-18}$$

Thus, K_T is the stall torque per ampere of control current for the motor, and running torque decreases as a linear function of motor speed.

In many cases, $\omega_c \gg \omega_L$ in (8.5-14), and the transfer function is approximately

$$\frac{\Theta(\omega)}{V_c(\omega)} = \frac{K_M}{j\omega[1 + j(\omega/\omega_L)]} \tag{8.5-19}$$

Potentiometers

A potentiometer is a resistor having a resistance that, typically, is uniformly† spread either over a linear length or over all or a portion of a circle. A mechanically adjustable electrical tap is provided. Figure 8.5-2(a) and (b) show two possible potentiometers.

The most common application of the potentiometer is to produce a voltage that is proportional to the rotation angle of a mechancial shaft [Fig. 8.5-2(c)]. If

† Nonlinear resistance variation can also be obtained in potentiometers that follow prescribed laws such as logarithmic or cosine.

θ_{max} is the total possible angle of rotation and $\theta = 0$ corresponds to the center of this range, then

$$v_0 = \frac{2V_a}{\theta_{max}} \theta \qquad -\frac{\theta_{max}}{2} \leq \theta \leq \frac{\theta_{max}}{2} \tag{8.5-20}$$

for a linear potentiometer. We define $K_p = 2V_a/\theta_{max}$ as the transfer function of the potentiometer, so

$$v_0 = K_p \theta \qquad -\frac{\theta_{max}}{2} \leq \theta \leq \frac{\theta_{max}}{2} \tag{8.5-21}$$

Tachometer

A *tachometer* is a device that produces a dc voltage proportional to the rotational speed of a mechanical shaft,

$$v_0 = K_t \frac{d\theta}{dt} = K_t \omega \tag{8.5-22}$$

The tachometer can be considered as a small dc generator with break frequencies high enough to have negligible effect in most feedback loops.

Limiter

Not all control systems involve control of a mechanical quantity. Some involve control of electrical characteristics such as the frequency or phase of an electrical waveform. We shall next briefly describe several components useful to such systems.

The *limiter* is a device (see Chap. 9) that accepts a sinusoidal signal of varying amplitude and generates a sinusoidal signal with the same phase as the input but with constant peak amplitude. Its output has the form

$$v_0(t) = K_l \cos [\theta(t)] \tag{8.5-23}$$

where K_l is a constant with the unit of volts if $v_0(t)$ is a voltage.

Phase Detector

A phase detector is a two-input, one-output device. Its output is proportional to the difference in phase between two input sinusoidal signals whenever the phase difference is small. If the two input signals are $A_1 \sin \theta_1(t)$ and $A_2 \cos \theta_2(t)$, the phase detector's output is

$$v_0(t) = K_{ph}[\theta_1(t) - \theta_2(t)] \tag{8.5-24}$$

for small $\theta_1(t) - \theta_2(t)$. Here K_{ph} is a constant having the unit of volts per radian if $v_0(t)$ is a voltage.

Discriminator

A *frequency discriminator* is a device that produces a voltage proportional to the changes in the instantaneous angular frequency $\omega_i(t)$ of a sinuosidal signal from a nominal value. If the input signal is

$$v_i(t) = A_i \cos\left[\omega_0 t + \int_{-\infty}^{t} \omega_i(\xi)\, d\xi\right] \tag{8.5-25}$$

where ω_0 is the nominal angular frequency, the output is

$$v_0(t) = K_D \omega_i(t) \tag{8.5-26}$$

where K_D is the discriminator's constant (unit of volts per radian per second if v_0 is a voltage).† The constant ω_0 is called the *crossover frequency* of the discriminator; it is its center frequency. In an FM (frequency modulation) radio receiver, for example, a discriminator is often present with a crossover frequency of $\omega_0/2\pi = 10.7$ MHz.

Voltage-Controlled Oscillator

The *voltage-controlled oscillator*, or VCO, is a device that generates a sinusoidal signal having an instantaneous angular frequency that is a linear function of an input (control) voltage $v_i(t)$. Thus,

$$v_0(t) = A_0 \cos\left[\omega_0 t + \theta_0 + K_V \int_{-\infty}^{t} v_i(\xi)\, d\xi\right] \tag{8.5-27}$$

where A_0, ω_0, θ_0, and K_V are constants. The VCO's constant K_V has the unit of radians per second per volt if $v_i(t)$ is a voltage.

Mixer

A *mixer* is a device that accepts two input sinusoidal signals of the forms

$$v_1(t) = A_1 \cos[\omega_1 t + \phi_1 + \theta_1(t)] \tag{8.5-28}$$

$$v_2(t) = A_2 \cos[\omega_2 t + \phi_2 + \theta_2(t)] \tag{8.5-29}$$

and provides an output signal of the form

$$v_0(t) = V_0 \cos[(\omega_1 \pm \omega_2)t + (\phi_1 \pm \phi_2) + \theta_1(t) \pm \theta_2(t)] \tag{8.5-30}$$

Here ϕ_1 and ϕ_2 are constants and ω_1 and ω_2 are the average angular frequencies of

† An *ideal discriminator* behaves according to (8.5-26). In some practical discriminators the output amplitude is a function of the input's amplitude according to $v_0(t) = K_D A_i \omega_i(t)$, where K_D has the unit $(\text{rad/s})^{-1}$. Such a device is not usually desired since the output voltage is a function of input-signal amplitude *and* frequency instead of frequency alone. A limiter is then used to remove the dependence on A_i.

$v_1(t)$ and $v_2(t)$. For the case of the plus signs, the mixer is an *up-converter*, since the average frequency $\omega_1 + \omega_2$ of $v_0(t)$ is larger than either of the inputs. The minus signs correspond to a *down-converter*, where the average frequency of the output is the difference in the two input frequencies.

In general, V_0, A_1, and A_2 in (8.5-28) through (8.5-30) can be time-dependent. In some mixers V_0 is a constant independent of both A_1 and A_2. In the most common mixer, V_0 is proportional to *one* of the two inputs:

$$V_0 = K_{mix} A_i \qquad i = 1 \text{ or } 2 \tag{8.5-31}$$

where the constant K_{mix} has the unit of volts per volt if v_0 and the inputs are voltages. Finally, it is also possible for V_0 to be approximately equal to KA_1A_2, where K is a constant with the unit of V^{-1} when all signals are voltages.

8.6 SOME CONTROL-SYSTEM EXAMPLES

In this section some control systems are discussed that utilize the components defined in the preceding section.

Speed-Control Loop

Figure 8.6-1(a) sketches a control loop to stablize the speed of a dc motor that drives a cooling fan for a load. A gear takeoff is used to drive a tachometer. The tachometer's output voltage is subtracted from a reference voltage, which is used to establish motor speed. The voltage difference is performed by the differential amplifier of voltage gain A. We assume that the motor is a field-controlled machine defined by (8.5-9), so the amplifier's output is the motor's field voltage. Define $\Omega_M(\omega)$, $V_t(\omega)$, $V_{ref}(\omega)$, and $V_f(\omega)$ as the respective Fourier transforms of $\omega_M(t)$, the motor's speed, $v_t(t)$, $v_{ref}(t)$, and $v_f(t)$. Then

$$\Omega_M(\omega) = j\omega\Theta_M(\omega) = \frac{K_M V_f(\omega)}{1 + j(\omega/\omega_L)} = \frac{K_M A[V_{ref}(\omega) - V_t(\omega)]}{1 + j(\omega/\omega_L)} \tag{8.6-1}$$

Now since

$$V_t(\omega) = K_t \frac{N_1}{N_2} \Omega_M(\omega) \tag{8.6-2}$$

where N_1 and N_2 are the numbers of teeth on the two gears, (8.6-1) reduces to

$$\Omega_M(\omega) = \frac{K_s}{1 + j(\omega/\omega_s)} V_{ref}(\omega) \tag{8.6-3}$$

where

$$K_s = \frac{K_M A}{1 + (K_M A K_t N_1/N_2)} \tag{8.6-4}$$

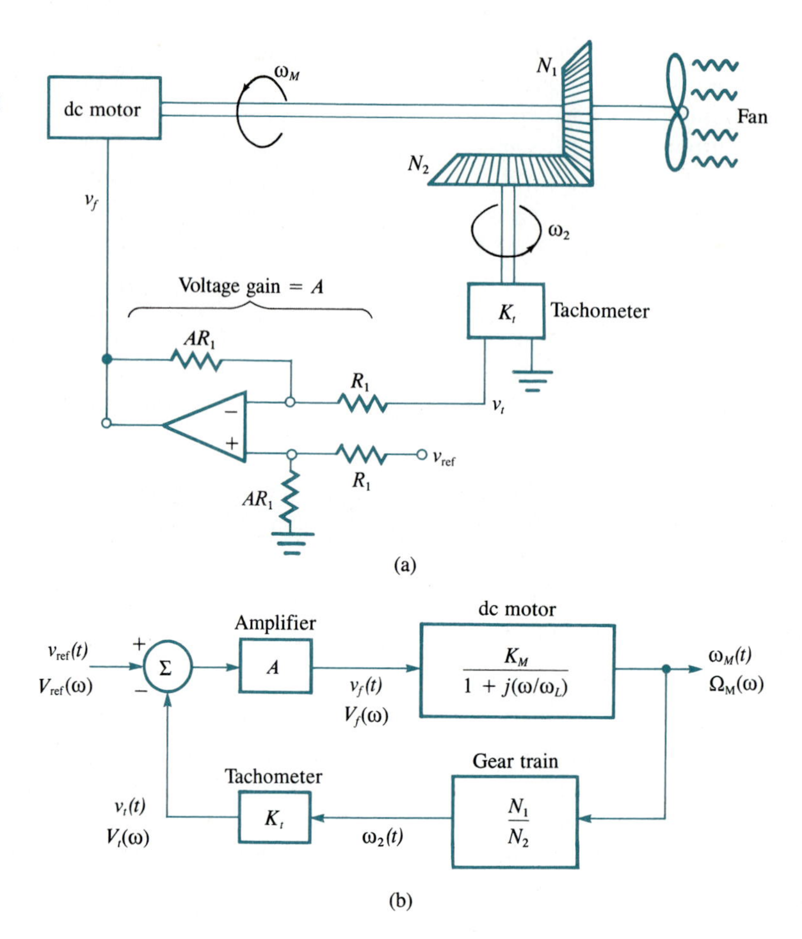

FIG. 8.6-1. (a) A feedback system to control the speed of a dc motor and (b) its block diagram.

$$\omega_s = \omega_L\left(1 + \frac{K_M A K_t N_1}{N_2}\right) \tag{8.6-5}$$

It is to be observed from (8.6-3) and (8.6-5) that the bandwidth of the closed-loop system, which is ω_s, is larger than the bandwidth of the motor alone, which is ω_L. This fact means that the closed-loop system has improved the motor's transient performance.

Transient behavior can be studied by inverse Fourier transformation of (8.6-3). The time function $\omega_M(t)$, which has a transform in product form, can be found as the convolution of the time functions having the transforms in the product, from (2.5-12). For $v_{ref}(t) = V_r u(t)$, where V_r is the applied voltage's amplitude, we have

$$\omega_M(t) = \mathcal{F}^{-1}\{\Omega_M(\omega)\} = \int_{-\infty}^{\infty} v_{ref}(t - \xi)[K_s\omega_s u(\xi)]e^{-\omega_s\xi}\, d\xi$$

$$= K_s\omega_s \int_0^\infty v_{\text{ref}}(t - \xi)e^{-\omega_s \xi}\, d\xi = K_s\omega_s V_r u(t) \int_0^t e^{-\omega_s \xi}\, d\xi$$

$$= K_s V_r u(t)(1 - e^{-\omega_s t}) \tag{8.6-6}$$

This function indicates that $\omega_M(t)$ rises toward its final value of

$$\omega_M(t)\big|_{t\to\infty} = K_s V_r = \frac{K_M A V_r}{1 + (K_M A K_t N_1/N_2)} \tag{8.6-7}$$

in an exponential manner.

EXAMPLE 8.6-1

We assume that a field-controlled dc motor is defined by

$$\frac{\Omega_M(\omega)}{V_f(\omega)} = \frac{8}{1 + j(\omega/0.2)}$$

It is to be used in the loop of Fig. 8.6-1 when $N_1/N_2 = 2$, $K_t = 15$ mV/(rad/s), and $A = 1000$. We find the closed-loop transfer function and the required constant voltage V_r that will cause a motor speed of 955 rpm (revolutions per minute).

From (8.6-4) and (8.6-5):

$$K_s = \frac{8(1000)}{1 + 8(1000)(0.015)(2)} = \frac{8000}{241} \approx 33.195 \text{ rad/(s·V)}$$

$$\omega_s = 0.2(241) = 48.2 \text{ rad/s}$$

From (8.6-3) the closed-loop transfer function is

$$\frac{\Omega_M(\omega)}{V_{\text{ref}}(\omega)} \approx \frac{33.195}{1 + j(\omega/48.2)}$$

The required constant value of $v_{\text{ref}}(t)$ is found from (8.6-7):

$$V_r = \frac{\omega_M(t)\big|_{t\to\infty}}{K_s} = \frac{955(2\pi)/60}{8000/241} \approx 3.013 \text{ V}$$

Position-Control Loop

Figure 8.6-2(a) sketches a control loop designed to position the pointing angle of a TV camera in response to a desired direction defined as the angle of rotation of a knob (on potentiometer 2). A field-controlled dc motor is assumed, but it could equally well be an armature-controlled device. The motor drives the camera through a gear train. The shaft of the camera's turntable is connected to potentiometer 1 to generate a feedback voltage v_p. Voltage v_p is subtracted from a reference v_{ref} generated by the independent reference potentiometer. Finally, the

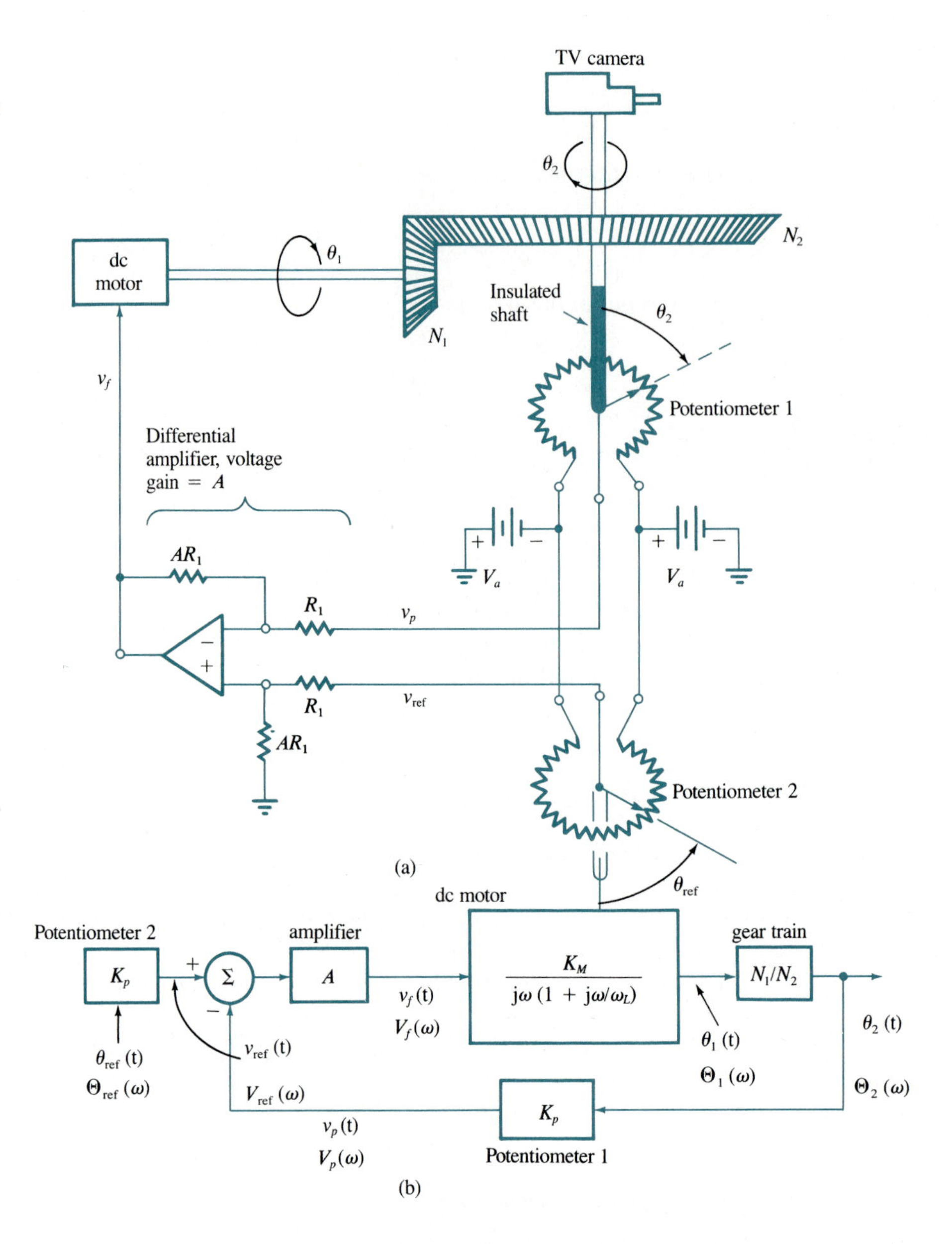

FIG. 8.6-2. (a) A control loop to position a TV camera and (b) its equivalent block diagram.

motor's control voltage v_f is the output of the (differential) amplifier with voltage gain A. The block diagram of the loop, based on a motor defined by (8.5-9), is shown in Fig. 8.6-2(b). Direct analysis gives

$$\frac{\Theta_2(\omega)}{\Theta_{ref}(\omega)} = \frac{1}{1 + j2\zeta(\omega/\omega_n) - (\omega/\omega_n)^2} \tag{8.6-8}$$

where

$$\zeta = \frac{1}{2}\sqrt{\frac{\omega_L}{(N_1/N_2)K_M A K_p}} \tag{8.6-9}$$

$$\omega_n = \sqrt{\omega_L\left(\frac{N_1}{N_2}\right)K_M A K_p} \tag{8.6-10}$$

It can be shown that the (transient) response of a system defined by (8.6-8) to a step command of reference angle will be a smooth function that asymptotically approaches the desired position if $\zeta \geq 1$. For $\zeta < 1$ the response is more rapid but overshoots the desired value and cycles about the value with decreasing-amplitude oscillations. The oscillations are less damped and more severe as ζ is made smaller. The condition $\zeta \geq 1$ places an upper limit on the allowable open-loop gain constant:

$$\left(\frac{N_1}{N_2}\right)K_M A K_p \leq \frac{\omega_L}{4} \tag{8.6-11}$$

It results, however, that this loop is readily compensated by using the network of Fig. 8.4-9. The reader can appreciate this fact by sketching the Bode plots of the open-loop gain. With compensation, loop gain can be increased up to the point where higher break frequencies (neglected in the motor of Fig. 8.6-2) become important.

Phase-Locked Loop

As an example of an electronic control loop, we consider the *phase-locked loop* of Fig. 8.6-3. This system is designed to force the phase of the voltage-controlled oscillator (VCO) to follow the phase of the reference signal. We have

$$\begin{aligned} v_1(t) &= K_{\text{ph}}[\omega_0 t + \theta_0 + \theta_{\text{ref}}(t) - \omega_0 t - \theta_0 - \theta_V(t)] \\ &= K_{\text{ph}}[\theta_{\text{ref}}(t) - \theta_V(t)] \end{aligned} \tag{8.6-12}$$

$$\theta_V(t) = K_V \int_{-\infty}^{t} v_2(\xi)\, d\xi \tag{8.6-13}$$

where K_{ph} and K_V are the phase detector and VCO constants, respectively. If

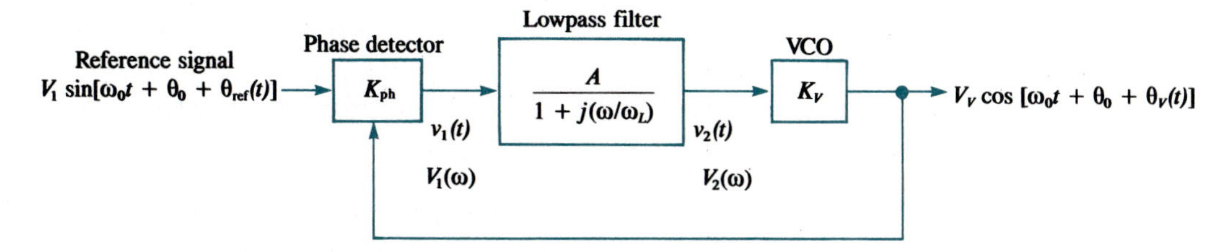

FIG. 8.6-3. A phase-locked loop.

$\Theta_V(\omega)$, $V_2(\omega)$, $V_1(\omega)$, and $V_{ref}(\omega)$ represent the Fourier transforms of $\theta_v(t)$, $v_2(t)$, $v_1(t)$, and $v_{ref}(t)$, we have

$$V_1(\omega) = K_{ph}[\Theta_{ref}(\omega) - \Theta_V(\omega)] \tag{8.6-14}$$

$$\Theta_V(\omega) = \frac{K_V V_2(\omega)}{j\omega} \tag{8.6-15}$$

$$V_2(\omega) = \frac{A}{1 + j(\omega/\omega_L)} V_1(\omega) \tag{8.6-16}$$

On combining these three expressions, we obtain

$$\frac{\Theta_V(\omega)}{\Theta_{ref}(\omega)} = \frac{1}{1 + j2\zeta(\omega/\omega_n) - (\omega/\omega_n)^2} \tag{8.6-17}$$

where

$$\zeta = \frac{1}{2}\sqrt{\frac{\omega_L}{K_V A K_{ph}}} \tag{8.6-18}$$

$$\omega_n = \sqrt{\omega_L K_V A K_{ph}} \tag{8.6-19}$$

Because (8.6-17) has the same *form* as (8.6-8), we conclude that the performance characteristics of the two loops are similar. Thus, the phase-locked loop must be given the same concern about transient performance and compensation as was given to the position-control loop.

Frequency-Control Loop

The control loop of Fig. 8.6-4 is designed to force the instantaneous angular frequency of a voltage-controlled oscillator (VCO) to follow the instantaneous angular frequency of the input signal $v_i(t)$. Loop analysis is straightforward. The

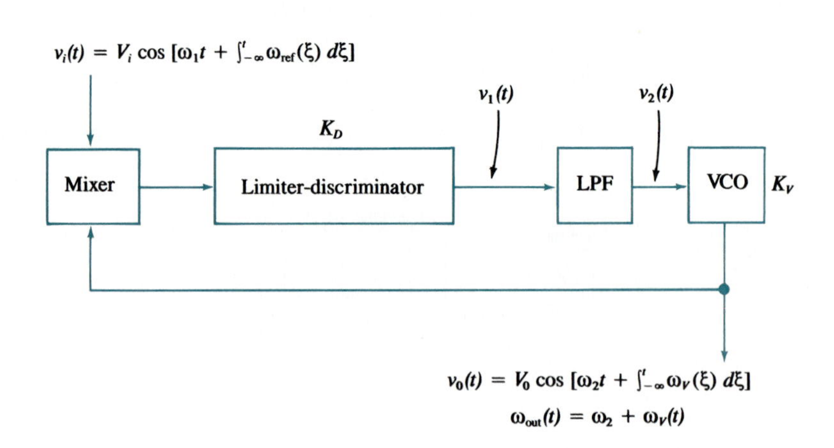

FIG. 8.6-4. A frequency-control loop.

VCO's instantaneous frequency is $\omega_2 + \omega_V(t)$, where

$$\omega_V(t) = K_V v_2(t) \tag{8.6-20}$$

The mixer's output instantaneous frequency is the difference between that of the input and that of the VCO; it is $\omega_1 - \omega_2 + \omega_{ref}(t) - \omega_V(t)$. The limiter-discriminator combination produces a voltage $v_1(t)$ that is proportional to the frequency changes from the nominal design frequency, selected to be $\omega_1 - \omega_2$. Thus,

$$v_1(t) = K_D[\omega_{ref}(t) - \omega_V(t)] \tag{8.6-21}$$

Finally, we Fourier-transform these last two expressions to get

$$\Omega_V(\omega) = K_V V_2(\omega) \tag{8.6-22}$$

$$V_1(\omega) = K_D[\Omega_{ref}(\omega) - \Omega_V(\omega)] \tag{8.6-23}$$

When these are combined with

$$\frac{V_2(\omega)}{V_1(\omega)} = \frac{A}{1 + j\omega/\omega_L} \tag{8.6-24}$$

which is the assumed form of the lowpass filter's (LPF's) transfer function, we have

$$\frac{\Omega_V(\omega)}{\Omega_{ref}(\omega)} = \left(\frac{K_D K_V A}{1 + K_D K_V A}\right) \frac{1}{\{1 + [j\omega/\omega_L(1 + K_D K_V A)]\}} \tag{8.6-25}$$

In steady state (8.6-25) indicates that $\omega_V(t)$ will be smaller than $\omega_{ref}(t)$ by a factor $K_D K_V A/(1 + K_D K_V A)$. However, for a large loop gain $K_D K_V A$, $\omega_V(t) \approx \omega_{ref}(t)$, as desired.

EXAMPLE 8.6-2

In the loop of Fig. 8.6-4, $K_V = 3\pi \times 10^4$ rad/(s·V), $K_D = 10^{-4}/2\pi$ V/(rad/s) for the discriminator, and $A = 20$ and $\omega_L = 5000\pi$ rad/s for the lowpass filter. We find the transient response for a step change in $\omega_{ref}(t)$ at $t = 0$. Here $K_D K_V A = (10^{-4}/2\pi)(3\pi \times 10^4)(20) = 30$, and $\omega_L(1 + K_D K_V A) = (155 \times 10^3)\pi$ rad/s. Thus,

$$\frac{\Omega_V(\omega)}{\Omega_{ref}(\omega)} = \frac{(150 \times 10^3)\pi}{(155 \times 10^3)\pi + j\omega} = H(\omega)$$

is the loop's transfer function. Its impulse response is easily found to be

$$h(t) = (150 \times 10^3)[\pi u(t)]e^{-(155 \times 10^3)\pi t}$$

For a step of change in the input frequency defined by

$$\omega_{ref}(t) = \Delta\omega u(t)$$

we have, by convolution,

$$\omega_V(t) = \int_{-\infty}^{\infty} \omega_{\text{ref}}(t - \xi)h(\xi)\, d\xi$$

$$= \int_{-\infty}^{\infty} \Delta\omega u(t - \xi)\,(150 \times 10^3)[\pi u(\xi)]e^{-(155\times 10^3)\pi\xi}\, d\xi$$

$$= \Delta\omega(150 \times 10^3)\pi u(t) \int_0^t e^{-(155\times 10^3)\pi\xi}\, d\xi$$

or

$$\frac{\omega_V(t)}{\Delta\omega} = \tfrac{30}{31}\, u(t)[1 - e^{-(155\times 10^3)\pi t}]$$

As $t \to \infty$ and the transient interval dies out, $\omega_V(t) \to \frac{30}{31}\Delta\omega \approx 0.968\,\Delta\omega$. An error remains in the response of $(1 - 0.968)\,\Delta\omega = 0.032\,\Delta\omega$ (or 3.2% of $\Delta\omega$).

PROBLEMS

8-1. A rather poor op amp has a finite gain of only $A_0 = 50$ but is otherwise ideal. If $R_2 = 22\ \text{k}\Omega$, what exact value of R_1 would be needed to give a gain of -15 for the circuit of Fig. 8.1-1? If the op amp's gain could be increased by 20% to 60, what new gain occurs when using the exact value of R_1?

8-2. Find the gain v_O/v_{in} for the circuit of Fig. P8-2. For the same values of R_1 and R_2, how does the gain compare with the same circuit when $R_3 = 0$ or $R_4 = \infty$?

FIG. P8-2.

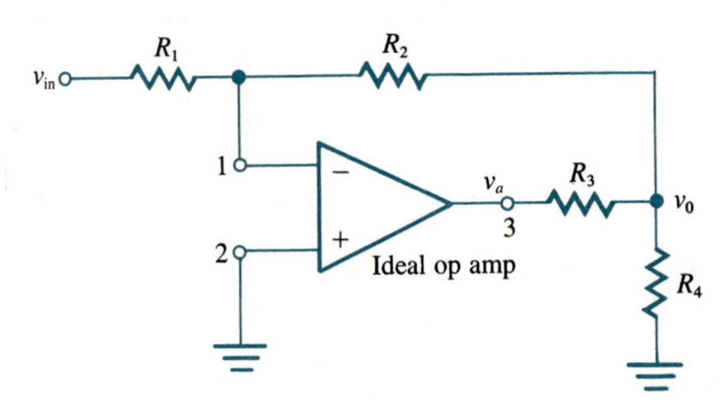

8-3. Determine v_O/v_{in} for the circuit of Fig. P8-3 if the op amps are ideal.

FIG. P8-3.

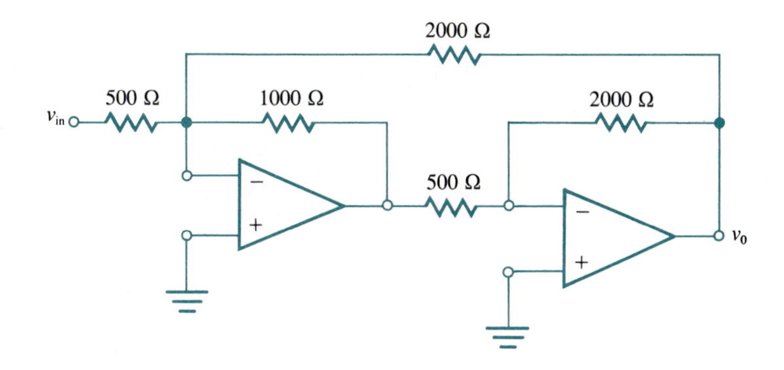

8-4. Let $R_1 = 1000\ \Omega$ and $v_{in} = 10$ V (dc) in the inverting amplifier of Fig. 8.1-1 and assume a voltmeter is connected between terminal 3 and ground to accurately measure voltages between 1 and 10 V (negative voltmeter terminal at terminal 3). Determine what voltage indication corresponds to R_2. Note that this circuit is an electronic *ohmmeter* capable of measuring the value of R_2.

8-5. In the amplifier of Fig. 8.1-2, $R_1 = 1000\ \Omega$ and $R_2 = 2200\ \Omega$. The op amp is ideal, except that its output cannot exceed ± 12 V at a current of ± 10 mA. (a) What is the minimum load resistor that can be added between terminal 3 and ground? (b) What is the largest allowable magnitude for v_{in} when using the minimum load resistance?

8-6. Find v_O/v_{in} for the circuit of Fig. P8-6 if the op amp is ideal.

FIG. P8-6.

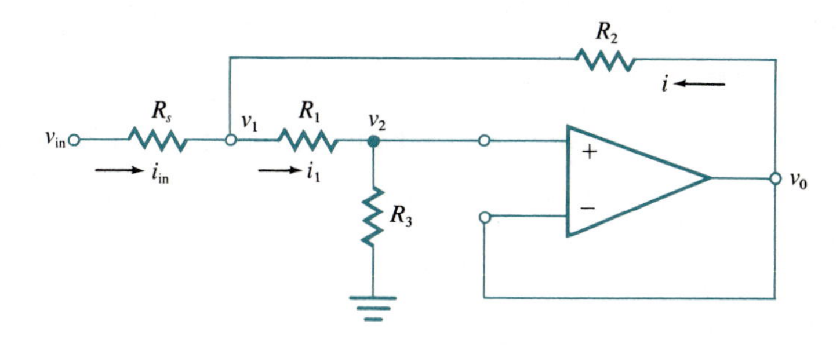

8-7. If $R_f = 10\ \text{k}\Omega$ in Fig. 8.2-1, design an ideal inverting summing amplifier for two inputs v_{b1} and v_{b2} so that $v_O = -12v_{b1} - 8v_{b2}$.

8-8. Let $R_f = 27\ \text{k}\Omega$ in Fig. 8.2-1, and find resistors R_n, $n = 1, 2, \ldots, N$, so that $v_O = -(v_{b1} + 2v_{b2} + 2^2v_{b3} + \cdots + 2^{N-1}v_{bN})$.

8-9. In a certain noninverting summing amplifier $R_m = (m + 0.5)50\ \Omega$, $m = 1, 2, 3, 4$, and 5. Also, $R_d = 470\ \Omega$ and $R_f = 2700\ \Omega$. Find v_O.

8-10. If $R_d = 820\ \Omega$ and $M = 6$ in the circuit of Fig. 8.2-2, find R_f so

$$v_O = \sum_{i=1}^{6} v_{am}$$

8-11. Prove that (8.2-3) is true for the amplifier of Fig. 8.2-2.

8-12. Find an expression for v_O in the summing amplifier of Fig. P8-12 if the op amp is ideal.

FIG. P8-12.

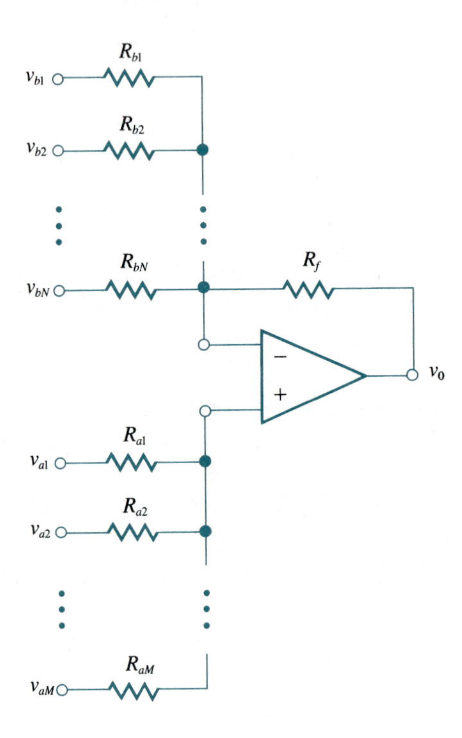

8-13. Determine v_O in terms of the input voltages for the circuit of Fig. P8-13.

FIG. P8-13.

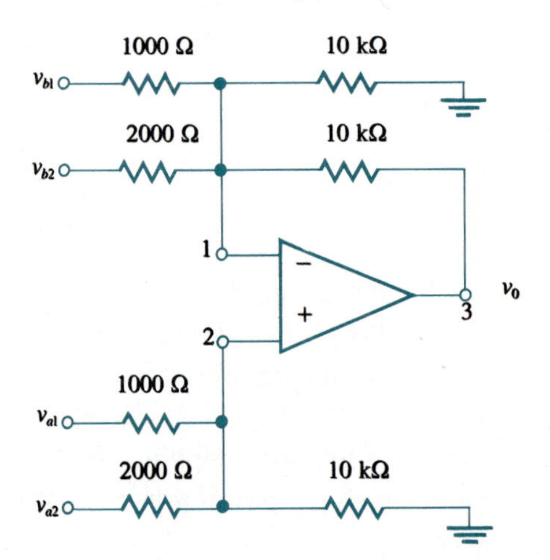

8-14. Find the input resistance between terminals a and b of the amplifier of Fig. 8.2-3 when $R_3 = R_1$ and $R_4 = R_2$.

8-15. Let $R_2 = R_1$ in the circuit of Fig. 8.2-3 and replace resistor R_3 by a capacitor C. If terminals a and b are connected together and a single input v_{in} is applied, find the transfer function of the network. If R_4 varies from 0 to ∞, what is the effect on the transfer function?

8-16. If the op amp in Fig. P8-16 is ideal, find v_0 in terms of v_a and v_b.

FIG. P8-16.

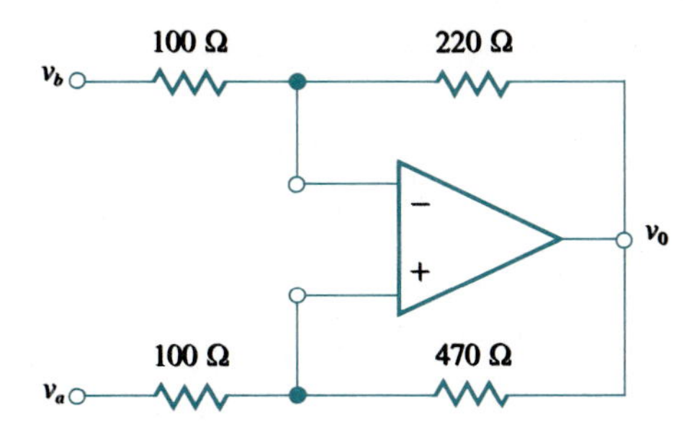

8-17. If a negative impedance converter is used as shown in Fig. P8-17, show that the load current i_L is

$$i_L = \frac{v_{in}}{R}$$

FIG. P8-17.

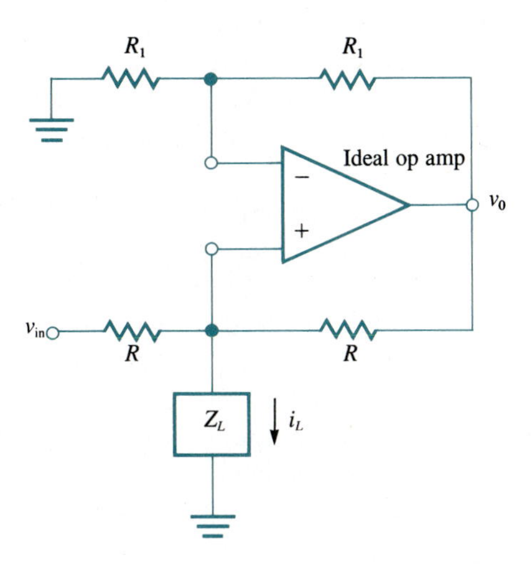

which is independent of Z_L. The network is a *voltage-to-current converter*, since the load sees a current source.

8-18. Determine how (8.2-12) is changed if the op amp in Fig. 8.2-5 has finite gain A_0 but is otherwise ideal.

★8-19. Find the input impedance Z_{in} for the *generalized impedence converter* of Fig. P8-19 if the op amps are ideal.

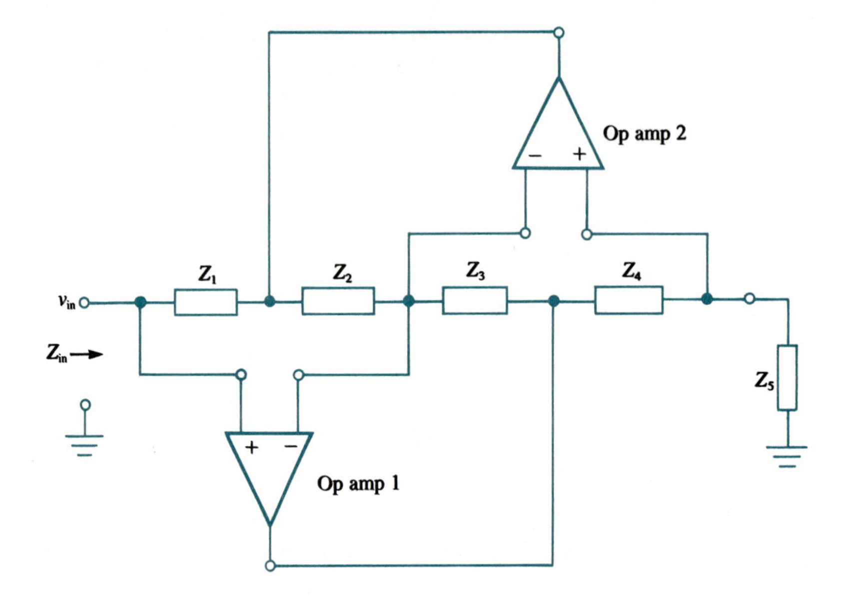

FIG. P8-19.

8-20. In the generalized amplifier of Fig. 8.2-9(a), let $Z_1(\omega)$ be comprised of a resistor R_A in parallel with a capacitor C_A. Similarly, let $Z_2(\omega)$ represent a resistor R_B in parallel with a capacitance C_B. Find R_A, R_B, C_A, and C_B so that the circuit has the same transfer function as (8.2-25).

8-21. For the circuit of Problem 8-20, let $R_B = 1000R_A$ and sketch 20 log $|H(\omega)|$ versus log (ω) for (a) $R_AC_A = 100R_BC_B$, (b) $R_AC_A = R_BC_B$, and (c) $R_AC_A = R_BC_B/100$.

8-22. An op amp has $A_0 = 2 \times 10^6$ and $\omega_u = \pi \times 10^7$ rad/s. What 3-dB bandwidth will an inverting amplifier have that uses the op amp if $R_1 =$ 50 Ω and (a) $R_2 = 5$ kΩ, (b) $R_2 = 500$ kΩ, and (c) $R_2 = 1$ MΩ?

8-23. In the offset-voltage-nulling circuit of Fig. 8.3-2(a), $R_1 = 1500$ Ω, $R_2 =$ 22 kΩ, $R_6 = 100$ kΩ, $R_5 = 470$ kΩ, $R_4 = 220$ Ω, and $V = 12$ V. (a) What range of input offset voltages can be generated at terminal 2 of the op amp? (b) If the op amp generates no output offset, what range of voltages will occur in the output as the potentiometer varies through its full range? (c) What resistance should R_3 have for a good design?

8-24. In an inverting amplifier $R_1 = 4.7$ kΩ and $R_2 = 68$ kΩ. The op amp has bias currents $I_{b1} = 50$ nA and $I_{b2} = 60$ nA but is otherwise ideal. (a) What value of R_3 in Fig. 8.3-3 should be used? (b) When the input signal is zero, what residual output offset voltage occurs?

8-25. In the closed-loop system of Fig. 8.4-2(a), $H_1(\omega) = 12/[1 + j(\omega/4)]$ and

$H_2(\omega) = 100/[1 + j(\omega/10)]$. Write the equation for the closed loop's transfer function.

8-26. The open-loop gain of a feedback system is

$$H_1(\omega)H_2(\omega) = \frac{50}{j\omega[1 + j(\omega/0.03)][1 + j(\omega/3)]}$$

Sketch the Nyquist plot for this function and show that the closed loop is unstable.

8-27. Make a Nyquist plot for the following open-loop-gain function:

$$H_1(\omega)H_2(\omega) = \frac{(2 \times 10^5)(70 + j\omega)}{j\omega(10 + j\omega)(700 + j\omega)}$$

Is the closed-loop system stable?

8-28. Sketch the idealized Bode plots of amplitude and phase for the open-loop-gain function

$$H_1(\omega)H_2(\omega) = \frac{12(0.7 + j\omega)}{(0.003 + j\omega)(0.04 + j\omega)(7 + j\omega)}$$

and use them to determine the approximate values of gain and phase margins of the closed-loop system.

8-29. Find the exact gain and phase margins for the system of Problem 8-28.

8-30. (a) Design a compensation network of the form in Fig. 8.4-9, with $R_1/R_2 = 10$, to compensate the open-loop-gain function

$$H_1(\omega)H_2(\omega) = \frac{100}{j\omega(1 + j2\omega)}$$

(b) Sketch the idealized Bode plots of magnitude and phase. (c) What are the approximate gain and phase margins?

8-31. The numbers of teeth on three gears in sequence are $N_1 = 20$ (input shaft's gear), $N_2 = 86$, and $N_3 = 250$ (output shaft's gear). If the output shaft rotates at 60 rpm, what is the speed of the input shaft? Discuss how the middle gear enters into this problem.

8-32. A relatively large field-controlled dc motor is defined by $R_f = 60\ \Omega$, $L_f = 25$ H, and $K_m = 75$ N·m/A. For the motor and load $J = 18$ N·m per rad/s^2, and $\sigma_L = 2$ N · m per rad/s. Write the motor's transfer function that defines the response of the motor's shaft angle to field voltage.

8-33. Armature inductance is negligible and $R_a = 1.2\ \Omega$ in an armature-controlled dc motor with load. The motor produces 0.01 N·m of torque per ampere of armature current. The transfer function of load-shaft speed to armature voltage of the motor-load combination is

$$\frac{\Omega_M(\omega)}{V_a(\omega)} = \frac{17.25}{1 + j(\omega/15)}$$

(a) What are J and σ_L for the motor-load combination? (b) What value of constant armature voltage will cause a steady-state local-shaft speed of 1000 rpm?

8-34. A control voltage of 6 V is suddenly applied to an armature-controlled dc motor that is initially at rest. Motor-shaft speed reaches 90% of its final constant speed of 200 rad/s after 0.55 s. Write the transfer function relating motor-shaft speed to control voltage. Assume armature inductance is negligible.

8-35. A two-phase ac servo motor with a gear train has no load attached. The motor-gear combination's torque constant is $K_T = 0.05$ N·m/A and $R_c = 60\ \Omega$. Full shaft speed (no attached load) is 200 rad/s when the motor's torque is 0.001 N·m. If the rotor's inertia is $J = 2 \times 10^{-5}$ N·m per rad/s^2, find σ_M, K_M, and ω_L for the unloaded motor-gear system. Assume $\omega_c \gg \omega_L$ and a stall torque of 0.05 N·m at full control voltage.

8-36. Find K_p (in volts per radian) for a potentiometer with a resistance of 10 kΩ spread linearly over a 330° circular arc and supplied by +5 V and −5 V at the two ends.

8-37. A given circuit produces the product $v_0(t) = K_{12}v_1(t)v_2(t)$ when $v_1(t)$ and $v_2(t)$ have the forms

$$v_1(t) = A_1 \sin [\omega_1 t + \Delta\theta_1(t)]$$

$$v_2(t) = A_2 \cos [\omega_2 t + \Delta\theta_2(t)]$$

and K_{12} has the unit of V^{-1}. Determine $v_0(t)$ and discuss how the circuit could be used as a phase detector. Does anything have to be added to make a phase detector? What is K_{ph}?

8-38. A frequency discriminator in an FM radio produces 1.5 V when the FM signal's angular frequency departs from the crossover value by $2\pi \times 10^5$ rad/s. What is K_D?

8-39. Assume (8.5-28) and (8.5-29) define input signals to a product device for which the output is

$$v_0(t) = K_{12}v_1(t)v_2(t)$$

with K_{12} having the unit of V^{-1}. (a) Discuss what needs to be done to use the device as an up-converting mixer. (b) Repeat part (a) for a down-converting mixer. (c) Define K_{mix} for parts (a) and (b).

8-40. Suppose the control voltage in the loop of Fig. 8.6-1(a) varies according to $v_{ref}(t) = A_{ref} \cos (\omega_r t)$, where ω_r is small. This reference will cause the motor's shaft speed to vary slowly in a sinusoidal way. Determine the actual speed of the fan (motor's shaft). Does it follow the command? Discuss.

8-41. Use the circuit of Fig. 8.4-9 with $G = 6.3$ (16 dB) to compensate the position-control loop of Fig. 8.6-2(a) for any value of ω_L that is less than $\omega_u/10$, where ω_u is the angular frequency at which the loop gain's magnitude is unity. The compensation network is placed between the amplifier and the motor. (a) What is the loop gain of the compensated loop? (b) Sketch the idealized Bode magnitude plot of the gain in part (a).

8-42. Let $\zeta = 1$ in (8.6-8), and show that the transfer function becomes the square of a simpler function.

8-43. Prove that (8.6-11) is true if $\zeta \geq 1$ in the loop defined by (8.6-8).

★ 8-44. Assume $\zeta \geq 1$, and show that (8.6-17) can be written as

$$\frac{\Theta_V(\omega)}{\Theta_{\text{ref}}(\omega)} = \frac{1}{(1 + j\omega a)(1 + j\omega b)}$$

where a and b are real constants. (*Hint:* Replace $j\omega/\omega_n$ by the variable s and find the roots of the denominator that make the expression true; then replace s by $j\omega/\omega_n$.)

8-45. Determine and give the Bode plots of (8.6-17) for $\zeta = 1.0$, 0.5, 0.2, and 0.05. Note the behavior that corresponds to "ringing" in the loop's step response for $\zeta < 1$.

8-46. Assume $K_D = 0.01$ V/(rad/s) $K_V = 10^3$ (rad/s)/V, $A = 100$, and $\omega_L/2\pi =$ 1000 Hz in the loop of Fig. 8.6-4. If $\omega_{\text{ref}}(t) = \Delta\omega_{\text{ref}} \cos \omega_r t$, what value of ω_r will cause the peak variation in $\omega_V(t)$ to be $\Delta\omega_{\text{ref}}/\sqrt{2}$?

DESIGN EXERCISES

D8-1. A *Butterworth* filter is one in which the transfer function $H(\omega)$ has a magnitude

$$|H(\omega)| = \frac{|A_0|}{\sqrt{1 + (\omega/\omega_3)^{2n}}} \qquad n = 1, 2, \ldots$$

where A_0 is the value at dc and ω_3 is the angular frequency at which $|H(\omega)|$ is 3 dB below its maximum value $|A_0|$. The index n is called the *order* of the filter. A second-order filter is realizable by the function

$$H(\omega) = \frac{A_0}{1 + \sqrt{2}(j\omega/\omega_3) + (j\omega/\omega_3)^2}$$

Design a second-order Butterworth filter for which $\omega_3/2\pi = 20$ kHz and A_0 is any necessary value. Use only one op amp, which can be assumed ideal in the design, but include an offset balancing circuit capable of giving ± 4 mV of input-voltage variation to account for practical effects ignored in the "ideal" assumption.

D8-2. A fourth-order Butterworth filter can be realized as a cascade of two second-order filters if the coefficient $\sqrt{2}$ in the second-order filter of Design Problem D8-1 is changed to 0.765 and 1.848, respectively, in the cascaded sections. Design a fourth-order filter to give any convenient overall gain A_0 and a ± 4 mV offset adjustment at the input to the first second-order network of the cascade. Assume $\omega_3/2\pi = 20$ kHz.

COMPUTER EXERCISES

Software required in solving the following computer exercises is available from the student's instructor. The software is provided free of charge, as part of the solutions manual, to the instructor, who is authorized by the publisher to freely make copies as needed for students' use.

C8-1. The speed-control loop of Example 8.6-1 is modified to add a filter in the feedback path. The filter is placed after the amplifier of gain $A = 1000$ and prior to the motor's field-control input point. The filter's transfer function is

$$H_F(\omega) = \frac{1 + j(\omega/2.4)}{[1 + j(\omega/0.5)][1 + j(\omega/30)]}$$

Write equations for (a) the loop-gain function and (b) the loop's transfer function $\Omega_M(\omega)/V_{\text{ref}}(\omega)$. (c) Determine the loop's approximate gain and phase margins from idealized Bode plots. Plot the exact loop-gain-magnitude and -phase functions as found from computer program FRESP. Is the loop stable?

C8-2. In the text the VCO of Fig. 8.6-4 was assumed to change frequency instantaneously in response to the input $v_2(t)$. That is, its transfer function was assumed to be

$$H_V(\omega) = \frac{\Omega_V(\omega)}{V_2(\omega)} = K_V$$

Assume a more practical VCO has a transfer function

$$H_V(\omega) = \frac{K_V[1 + j(\omega/\omega_2)]}{[1 + j(\omega/\omega_1)][1 + j(\omega/\omega_3)]}$$

Use computer program FRESP to plot the open-loop gain's magnitude and phase functions when the new VCO is assumed with $\omega_1 = 3 \times 10^4$ rad/s, $\omega_2 = 8.5 \times 10^4$ rad/s, and $\omega_3 = 5 \times 10^5$ rad/s. Assume other loop parameters are defined as in Example 8.6-2. What are the loop's gain and phase margins? Is the loop stable?

CHAPTER 9

Nonlinear Electric Circuits

9.0 INTRODUCTION

Most of the circuits developed in the preceding discussions of this book have been linear. Naturally, many of the practical applications of electrical engineering must deal with various forms of nonlinear circuits. In this chapter we introduce a few of the most important nonlinear circuits to serve as practical examples.

9.1 CHOPPER-STABILIZED AMPLIFIER

In instrumentation circuits the problem often arises as to how small dc voltages (say a few millivolts or less) should be amplified to obtain a more useful voltage level. The most obvious solution would be to use an op amp. However, recall that op amps suffer from practical effects (voltage and current offsets) that lead to unwanted output dc voltages that can interfere with, and even be mistaken for, the small desired dc voltage. Sometimes, a better approach is to first convert the small desired dc voltage to an *ac* voltage, use an ac amplifier that does not suffer from offsets, and then convert the amplified ac signal back to the equivalent amplified dc signal. The overall network that accomplishes these tasks is called a *chopper-stabilized amplifier*.

Figure 9.1-1(a) illustrates the *functions* needed in a well-designed chopper-

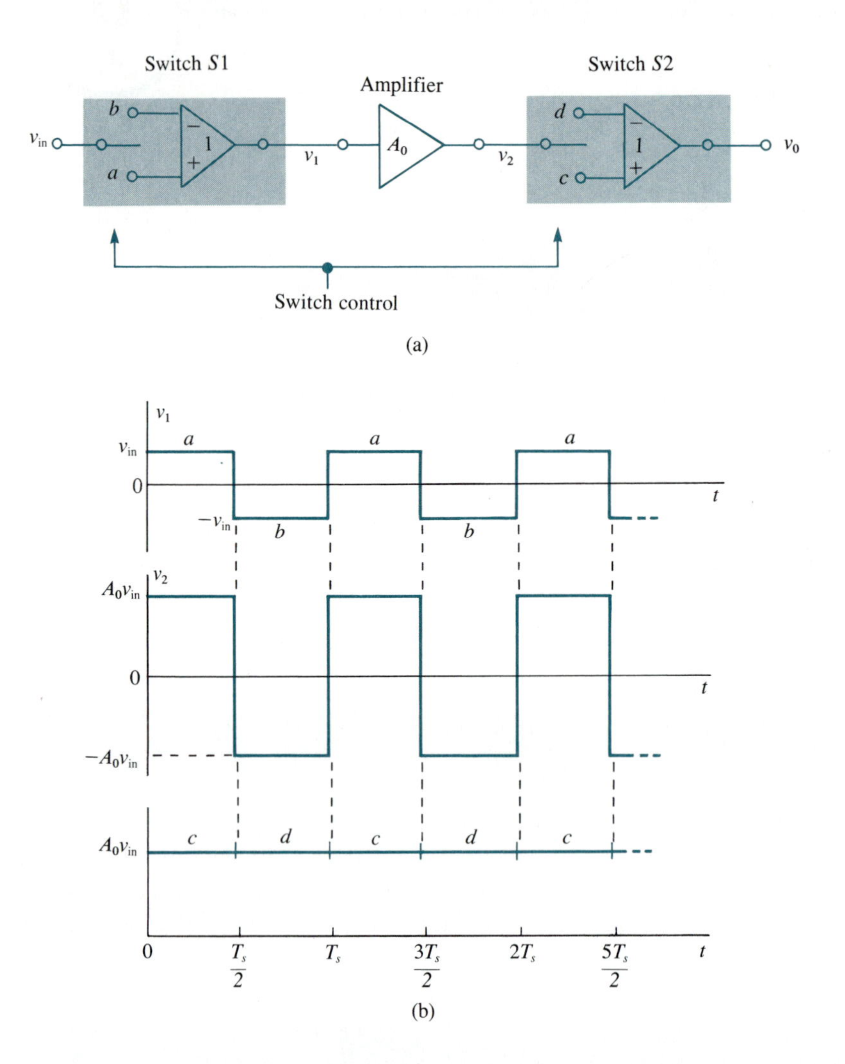

FIG. 9.1-1. (a) Functions of a chopper-stabilized amplifier and (b) the applicable waveforms.

stabilized amplifier. The input (small) signal v_{in} is passed through a switch $S1$ that first connects v_{in} to the amplifier for a time interval $T_s/2$ and then connects $-v_{in}$ to the amplifier for a second time interval $T_s/2$. Although $S1$ is shown as a mechanical single-pole double-throw switch followed by a (stable) ac amplifier with a gain of unity, a practical switch would be electronic. The square-wave signal v_1 is then amplified by the ac amplifier of gain A_0 to obtain v_2. The amplified square wave v_2 is then passed through a second switch $S2$ that is *synchronized* with $S1$. By examination of the applicable voltages in Fig. 9-1-1(b), it is clear that $S2$ converts the amplified ac square wave into the desired amplified dc voltage A_0v_{in}.

The careful reader may recognize that the operation of switch $S1$ is a *natural-sampling operation* of the type described in Sec. 2.7, as shown in Fig. 2.7-2(a). The sampling signal $s_p(t)$ in Fig. 2.7-2(a) is here a square-wave signal of amplitudes

± 1. The action of switch $S2$ is just a second sampling operation that conveniently recovers the original signal being sampled (v_{in} here). Because signals being sampled as in Sec. 2.7 could vary with time (do not have to be dc), the reader may suspect that the input to the chopper-stabilized amplifier can also be variable. Indeed, this suspicion is true. If v_{in} varies with time and has a spectral extent† of W_v (in radians per second), the main constraints on the network are that the sampling rate $2\pi/T_s = \omega_s$ must exceed $2W_v$ and the ac amplifier's gain must be constant at A_0 for all frequencies except those very near dc.

9.2 RECTIFIERS AND POWER SUPPLIES

Rectifiers were briefly introduced in Sec. 6.4, where diodes were the principal subject. In this section diode rectifiers are examined more closely in their role as a critical element in the conversion of an ac voltage to a dc voltage. For convenience, diode rectifiers will be divided into two categories: those that are mainly used in low-power instrumentation applications and those involving higher powers, such as in dc power supplies. The voltage that we wish to convert to a dc voltage is assumed to have the form

$$v_{in}(t) = V_{in} \sin \omega_0 t \tag{9.2-1}$$

where V_{in} is a constant peak amplitude (voltage) and ω_0 is a constant angular frequency having a period

$$T_0 = \frac{2\pi}{\omega_0} \tag{9.2-2}$$

Instrumentation Rectifiers

The simplest form of half-wave rectifier is sketched in Fig. 9.2-1(a) along with the appropriate waveforms. When $v_{in} > 0$, the diode is forward-biased, turns on, and acts like a switch that connects the input to the output. Thus, $v_0 = v_{in}$, ideally, when $v_{in} > 0$. When $v_{in} \leq 0$, the diode is back-biased and disconnects the output from the input. Thus, $v_0 = 0$, ideally, when $v_{in} \leq 0$.

The rectifier of Fig. 9.2-1(a) suffers from the nonlinearities of the diode at small signals, especially when the load current and the applied voltage are small. A better half-wave rectifier is shown in Fig. 9.2-1(b). It is called a *precison rectifier*. It masks the diode nonlinearities by using the high gain of an op amp and a feedback loop (Problem 9-2). When $v_{in} < 0$, we have $v_1 > 0$, so $D2$ conducts, $D1$

† Recall that spectral extent is the angular frequency above which the spectral content of a baseband signal $v_{in}(t)$ can be considered negligible for the application involved.

FIG. 9.2-1. Half-wave rectifiers: (a) simple and (b) precision.

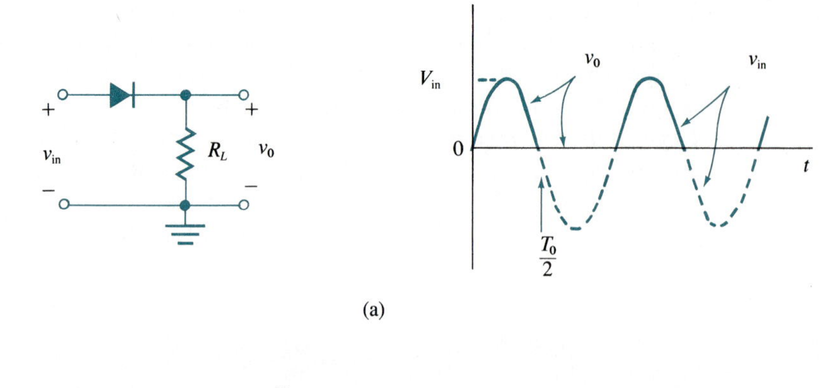

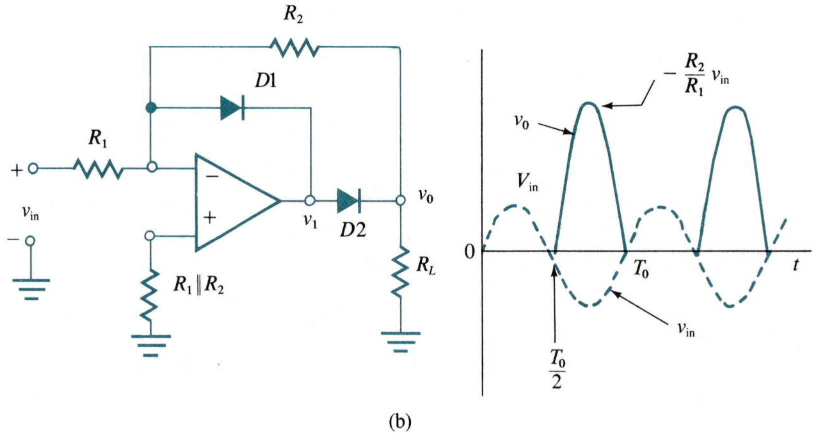

is cut off, and v_O is $-(R_2/R_1)v_{\text{in}}$. When $v_{\text{in}} \geq 0$, then $v_1 \leq 0$, $D2$ is cut off, and $D1$ conducts. Diode $D1$, which ideally has zero resistance when conducting, causes the gain of the inverting amplifier to be zero. Thus, $v_O \approx 0$ when $v_{\text{in}} \geq 0$, as shown in the applicable voltage waveforms.

If the directions of both $D1$ and $D2$ in Fig. 9.2-1(b) are reversed, the output-voltage waveform will consist of amplified negative half-cycles of the input that occur when $v_{\text{in}} \geq 0$. This behavior allows two such circuits to provide a full-wave rectification when used with a unity-gain differential amplifier, as illustrated in Fig. 9.2-2(a).

A disadvantage of the full-wave rectifier of Fig. 9.2-2(a) is the need for three op amps. A simpler full-wave circuit, called a *bridge rectifier* because of the diode arrangement, is shown in Fig. 9.2-2(b). Its output voltage, as taken across R_L as shown, is

$$v_O(t) = \left(1 + \frac{R_L}{R_1}\right)|v_{\text{in}}(t)| \tag{9.2-3}$$

since the circuit is just a noninverting amplifier.

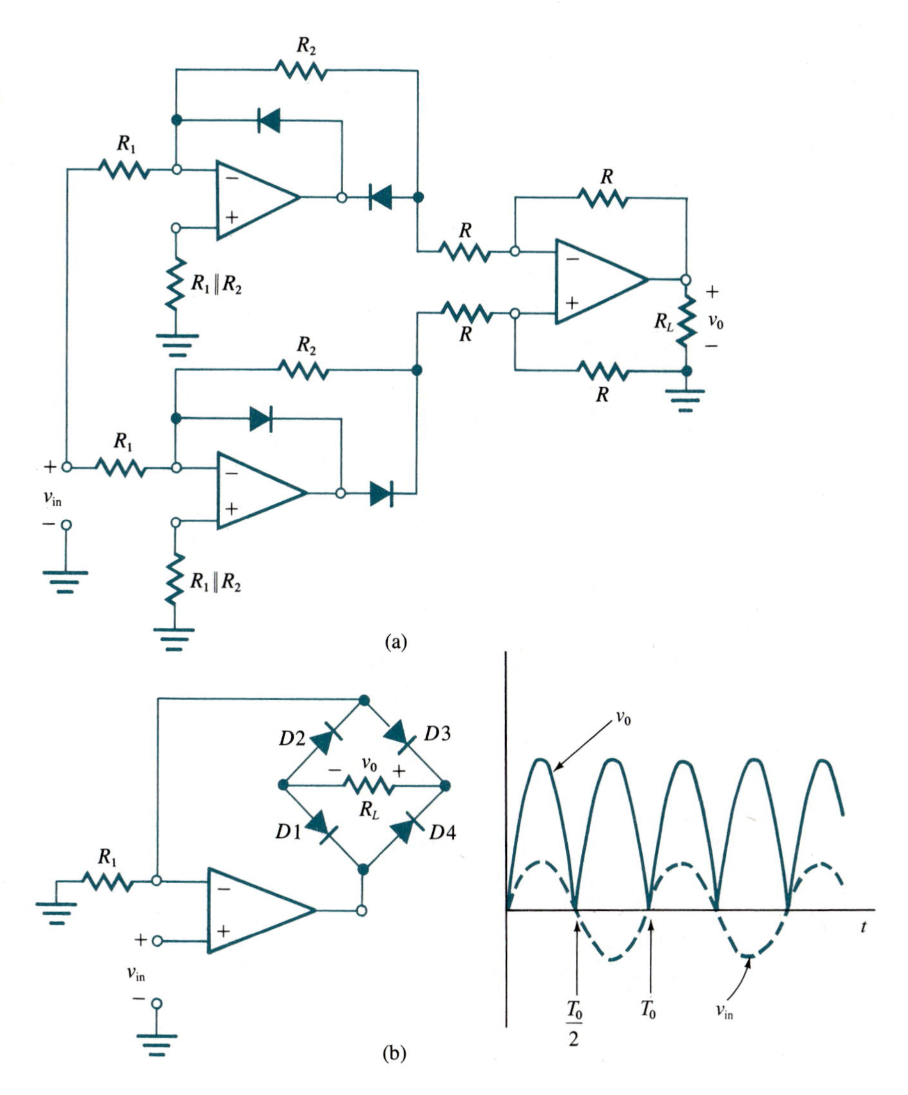

FIG. 9.2-2. Full-wave rectifiers that (a) combine two half-wave rectifiers of the types in Fig. 9.2-1(b) and (b) use a bridge circuit.

In both the half- and full-wave rectifiers discussed above the output voltages consist of periodic pulses of voltage that are proportional to half-cycles of the input voltage. The dc component of the outputs, from the use of the Fourier series, can be shown to be

$$V_{\mathrm{dc}} = \begin{cases} \dfrac{2V_{\mathrm{peak}}}{\pi} & \text{full-wave} \\ \dfrac{V_{\mathrm{peak}}}{\pi} & \text{half-wave} \end{cases} \tag{9.2-4}$$

where V_{peak} is the peak amplitude of the half-cycle pulses at the output.

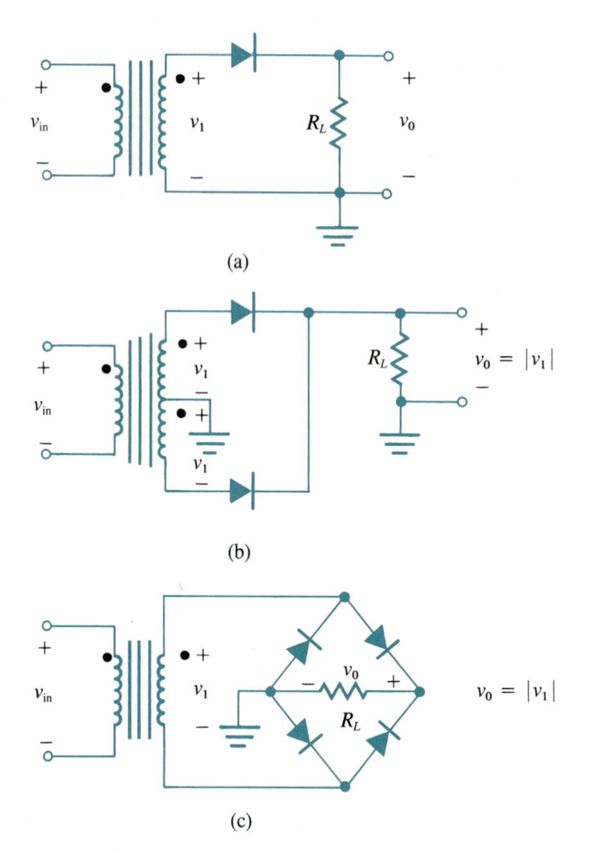

FIG. 9.2-3. Power rectifiers: (a) half-wave and (b) and (c) full-wave.

Because of power limitations on most op amps, instrumentation rectifiers are not normally found in power supplies. Other rectifier forms are used.

Power Rectifiers

By the use of high-current diodes and transformers, rectifiers can be made suitable for use in power supplies. Several forms of these *power rectifiers* are shown in Fig. 9.2-3. In these circuits v_{in} could represent the voltage from a 60-Hz, 115-V power line. In this case, the transformers, which give the advantage of isolation between the ac source and the dc circuit, would be iron-core, high-coupling units with either step-up or step-down windings. A full-wave rectifier to develop 5 V dc for a transistor supply, for instance, would use a step-down transformer to reduce the 115-V power line to about a 7.9 V peak amplitude for v_1, from (9.2-4). The advantage of the full-wave rectifier of Fig. 9.2-3(c) over that of (b) is that it can use a smaller transformer that requires no center tap.

FIG. 9.2-4. (a) A rectifier having a capacitor-input power-supply filter and (b) its applicable output waveform. (c) An alternative inductance-type filter for the circuit between terminals a and b.

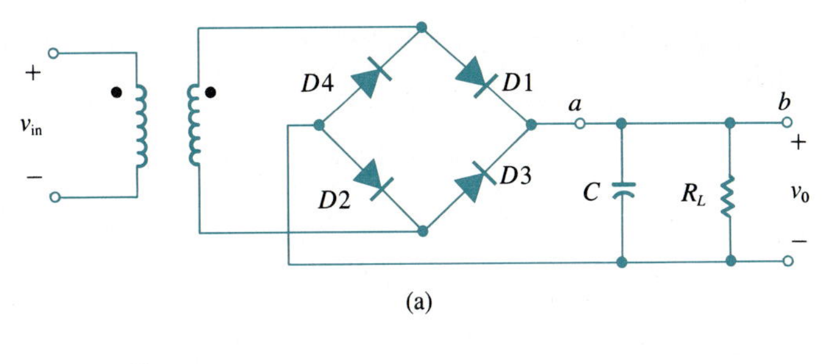

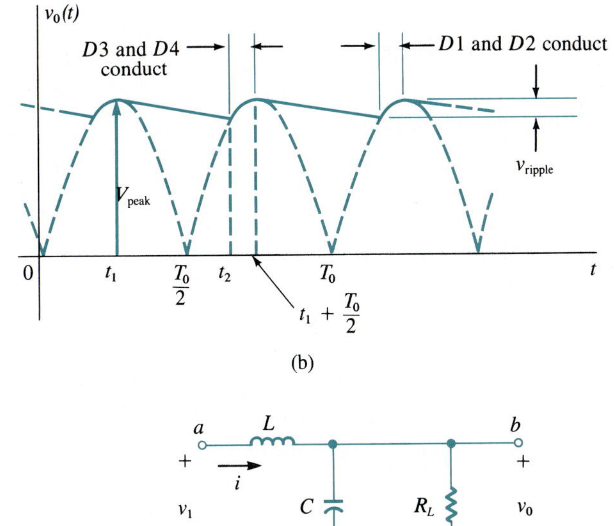

Filtering

Rectifiers alone would be of limited value in generation of dc voltage, because their outputs contain considerable voltage variation called *ripple*. Filters are added to remove the ripple. The simplest filter is a large capacitor connected across the rectifier's output in parallel with the load (resistor R_L). Figure 9.2-4(a) illustrates this addition to the rectifier of Fig. 9.2-3(c). The output waveform of Fig. 9.2-4(b) is the same as that of Fig. 6.4-5 discussed in Chapter 6. Analysis shows that in part of a typical half-period interval, say from t_1 to t_2, the output decays exponentially between times of diode conduction:

$$v_0(t) = V_{\text{peak}} e^{-(t-t_1)/R_L C} \qquad t_1 \leq t \leq t_2 \tag{9.2-5}$$

In a typical case $R_L C$ is large enough that the decay rate of $v_0(t)$ is small enough to assume that the conduction time $t_1 + (T_0/2) - t_2$ is small. This fact means that

$$t_2 - t_1 \approx \frac{T_0}{2} = \frac{\pi}{\omega_0} \tag{9.2-6}$$

and since

$$e^{-\epsilon} \approx 1 - \epsilon \qquad |\epsilon| \ll 1 \tag{9.2-7}$$

we have

$$v_0(t_2) = V_{\text{peak}} e^{-(t_2 - t_1)/R_L C} \approx V_{\text{peak}}\left(1 - \frac{\pi}{\omega_0 R_L C}\right) \tag{9.2-8}$$

The dc output voltage becomes

$$V_{\text{dc}} \approx \frac{V_{\text{peak}} + v_0(t_2)}{2} = V_{\text{peak}} - \frac{\pi V_{\text{peak}}}{2\omega_0 R_L C} \tag{9.2-9}$$

The ripple in the output is

$$v_{\text{ripple}} = V_{\text{peak}} - v_0(t_2) = \frac{\pi V_{\text{peak}}}{\omega_0 R_L C} \qquad \text{(peak-to-peak)} \tag{9.2-10}$$

For a half-wave rectifier with the product $R_L C$ large enough to give a small ripple, the dc voltage is given by (9.2-9) without the 2 in the second right-side term's denominator. The peak-to-peak ripple is twice that of (9.2-10).

EXAMPLE 9.2-1

A power supply constructed according to Fig. 9.2-4 has $V_{\text{peak}} = 15$ V, $\omega_0/2\pi = 60$ Hz, and $R_L = 100\ \Omega$. We find capacitance C that will make the load's ripple 5% of V_{peak}, and with this value of C we then find V_{dc} and load current.

From (9.2-10)

$$C = \frac{\pi V_{\text{peak}}}{\omega_0 R_L v_{\text{ripple}}} = \frac{1}{2(60)(100)(0.05)} \approx 1667\ \mu\text{F}$$

From (9.2-9)

$$V_{\text{dc}} \approx V_{\text{peak}}\left(1 - \frac{\pi}{2\omega_0 R_L C}\right) \approx 15\left[1 - \frac{10^6}{4(60)(100)(1667)}\right] \approx 14.63\ \text{V}$$

The dc load current is $I_{\text{dc}} \approx 14.63/100 = 146.3$ mA.

A disadvantage of the single-capacitor filter is that very large capacitors are needed to maintain small ripple, especially for high-current loads. By the use of an additional resistor and capacitor in the form of a lowpass filter (see Problem 9-5), both the ripple and the sizes of the capacitors can be reduced at the expense of a lower load voltage.

Another form of power-supply filter uses an inductor and capacitor, as shown in Fig. 9.2-4(c). Generally, the filter gives good regulation of load voltage with variations in load current. Analysis is difficult, however. For very large load resistance R_L the load's dc voltage approaches V_{peak}. As R_L decreases, the dc load voltage decreases to $2V_{\text{peak}}/\pi$ and then remains almost constant as R_L decreases further. The continued drop in the dc voltage is mainly due to the resistance that is present in a practical inductance.

The action of the inductance is to cause the impedance seen by the rectifier diodes to be large enough that they conduct current over most of their conducting half-cycle periods. It is roughly equivalent to the rectifiers' feeding a resistive load, although not the same. There is, however, a value of L, called the *critical inductance*, given by

$$L \geq \frac{R_L}{3\omega_0} \tag{9.2-11}$$

for which the constant-conduction-current model is a good approximation and analysis is straightforward. In this case the voltage $v_1(t)$ in Fig. 9.2-4(c) is the series of half-cycle pulses (dashed curve) of (b). A Fourier series expansion gives

$$v_1(t) = \frac{2V_{\text{peak}}}{\pi} - \frac{4V_{\text{peak}}}{\pi} \sum_{n=1}^{\infty} \frac{1}{(4n^2 - 1)} \cos 2n\omega_0 t \tag{9.2-12}$$

The constant term is the dc voltage. If the resistance of the inductance is negligible, this is also the dc voltage across the load R_L. The summation in (9.2-12) is the instantaneous ripple on $v_1(t)$.

To examine the load's ripple, we assume

$$\frac{1}{\omega_0 C} \ll R_L \tag{9.2-13}$$

so that the instantaneous ripple across C is

$$v_{\text{ripple}}(t) = -\frac{4V_{\text{peak}}}{\pi} \sum_{n=1}^{\infty} \frac{\cos 2n\omega_0 t}{(4n^2 - 1)(1 - 4n^2\omega_0^2 LC)} \tag{9.2-14}$$

For the usual case where

$$n\omega_0 L \gg \frac{1}{n\omega_0 C} \tag{9.2-15}$$

(9.2-14) becomes

$$v_{\text{ripple}}(t) \approx \frac{V_{\text{peak}}}{\pi\omega_0^2 LC} \sum_{n=1}^{\infty} \frac{1}{n^2(4n^2 - 1)} \cos 2n\omega_0 t \tag{9.2-16}$$

The peak amplitude of the component of (9.2-16) at frequency $4\omega_0$ is seen to be 20 times smaller than that of the lowest-frequency term at $2\omega_0$. Load peak-to-peak ripple is, therefore, approximately due to the lowest-frequency term and is

$$v_{\text{ripple}} \approx \frac{2V_{\text{peak}}}{3\pi\omega_0^2 LC} \qquad \text{(peak-to-peak at load)} \tag{9.2-17}$$

In some power supplies that use an inductance-input filter and where the load can open-circuit—or R_L can otherwise be large enough for (9.2-11) not to be true—a *bleeder resistance* can be added to the output in parallel with the load to make (9.2-11) true.

EXAMPLE 9.2-2

We design an inductance-input power supply for the same rectifier and load as in Example 9.2-1 and compare the output ripple of the two. Since $R_L = 100\ \Omega$, (9.2-13) requires

$$C \gg \frac{1}{\omega_0 R_L} = \frac{1}{120\pi(100)} \approx 26.53\ \mu\text{F}$$

Choose $C = 265.3\ \mu$F. From (9.2-15) for the worst case of $n = 1$:

$$L \gg \frac{1}{\omega_0^2 C} = \frac{1}{(120\pi)^2(265.3 \times 10^{-6})} = 0.0265\ \text{H}$$

Choose $L = 0.265$ H. From (9.2-17)

$$\frac{v_{\text{ripple}}(\text{peak-to-peak})}{V_{\text{peak}}} = \frac{2 \times 10^6}{3\pi(120\pi)^2(0.265)(265.3)} \approx 0.0212$$

Thus, for the same conditions the inductance-input filter has reduced the ripple to about 2.1% from 5% for the capacitor-input filter. The required capacitor is much smaller than before, but a rather bulky and heavy inductance is now necessary.

Inductance-input-filtered power supplies are used mainly in high-voltage applications. In many other modern applications the improved ripple reduction is achieved by using a capacitance filter in conjunction with an electronic regulation circuit.

Electronic Regulator

Ideally, a power supply should maintain a constant load voltage. Variations, whether due to load-resistance changes or ripple, should not occur. Both these effects can be reduced by use of an electronic regulator, as sketched in Fig. 9.2-5. To the left of terminal a the circuit is just a full-wave-rectifier power supply with a capacitance filter. The regulator consists of transistor Q, resistors R_1, R_2, and R_3, the zener diode Z, and the differential amplifier with voltage gain A_0. Func-

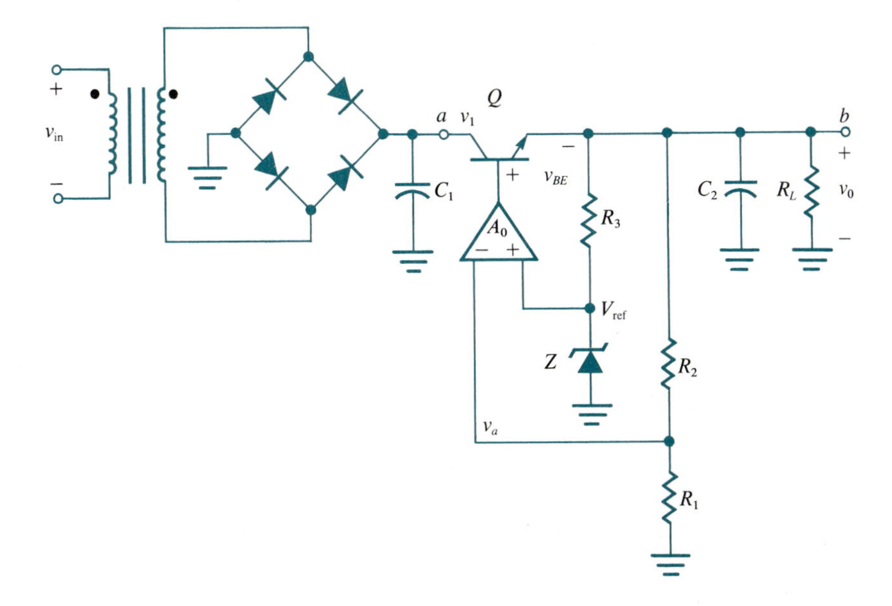

FIG. 9.2-5. A power supply with a series regulator and a full-wave rectifier.

tionally, Q acts like a variable resistance that compensates for variations in v_0. Capacitor C_2 filters high-frequency ripple components. The zener's voltage V_{ref} establishes the operating point of the system (dc level of v_0).

Suppose v_0 increases by a small amount, which causes v_a to increase and v_{BE} to decrease. A decrease in v_{BE} causes a decrease in emitter (load) current, which, in turn, reduces the original increase in load voltage, and a regulatory effect takes place. Careful study reveals that Q is an emitter-follower that drives the load R_L. The collector's power supply is v_1. The rest of the regulator provides a negative-feedback loop to control the emitter-follower's output. Because Q is in series with the load, the regulator is called *series*. *Shunt* regulators also are possible, where the controlling transistor is in parallel with the load.

Modern regulators can be purchased in IC form. They are available in many voltage values (v_0). Most require voltage v_1 to be at least 2 to 3 V larger than the required regulated output voltage. Many of these regulators incorporate protection circuits that allow the supply terminals to be shorted together or to be open-circuited without harming the supply. Ripple and load variations in voltage can be reduced to 0.01% or more by using these IC regulators.

9.3 CIRCUITS FOR SOME BASIC FUNCTIONS

In this section we discuss several circuits that perform basic functions. Some of these functions (limiters, phase detectors, and mixers) were used previously (Chap. 8) without showing how they could be implemented.

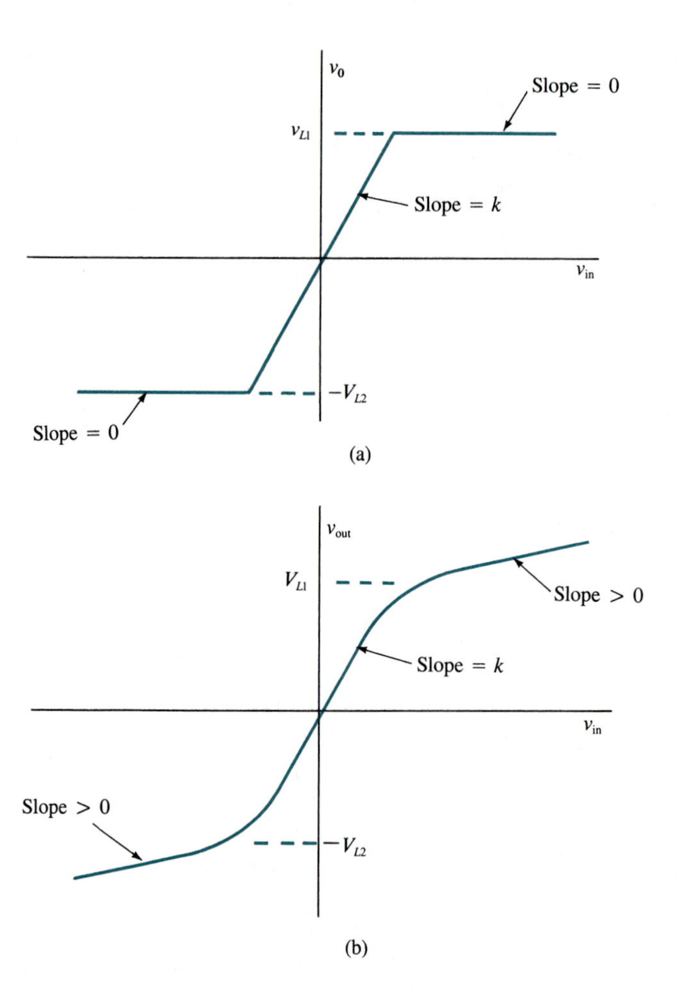

FIG. 9.3-1.
Limiters: (a) hard and (b) soft.

Limiters

A limiter is a nonlinear device that has an output-input voltage (or possibly current) characteristic, as shown in Fig. 9.3-1. As the input v_{in} increases in amplitude, a point is reached where the output "saturates." In the *hard limiter* of Fig. 9.3-1(a) the output becomes constant at a level V_{L1}; in the *soft limiter* of Fig. 9.3-1(b) the output only increases slowly above V_{L1}. A similar effect is seen for $v_{in} < 0$. These limiters offer limiting effect for both positive and negative values of v_{in} and can be called two-polarity limiters. Single-polarity limiters are also possible.

Limiters usually employ diodes (regular or zener).† Some common circuits

† Sometimes, limiting is obtained by driving "linear" circuits, such as amplifiers, into large-signal saturation.

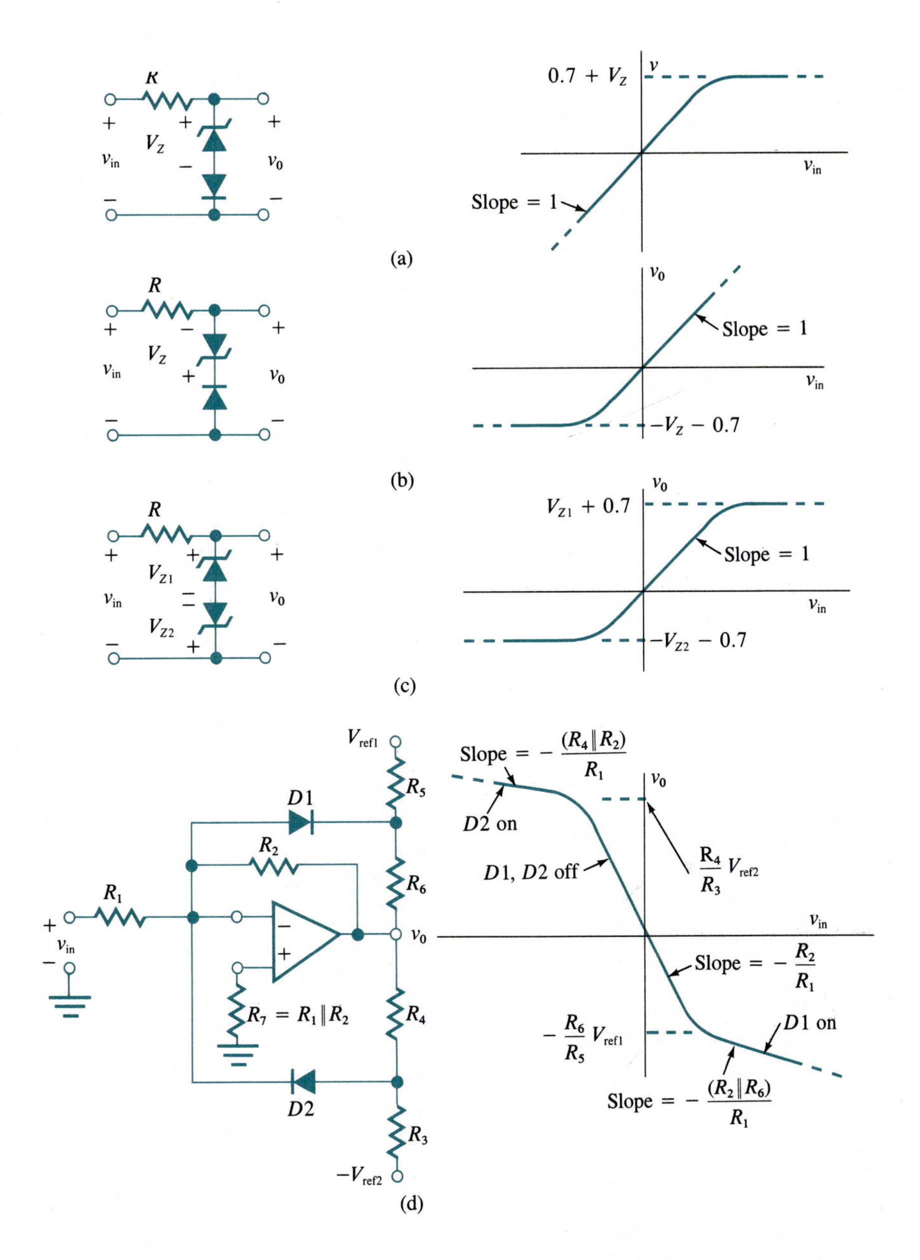

FIG. 9.3-2. Limiters: (a) and (b) single-polarity and (c) and (d) two-polarity.

are shown in Fig. 9.3-2. Single-polarity limiters are given in Fig. 9.3-2(a) and (b). The limit levels are controlled by choice of the zener diode's voltage and the cut-in voltage of the regular diode (assumed to be 0.7 V for a silicon diode). A two-polarity limiter is given in Fig. 9.3-2(c). Any load impedance connected across the output terminals is assumed to be large relative to R.

A very flexible two-polarity limiter is sketched in Fig. 9.3-2(d) along with its

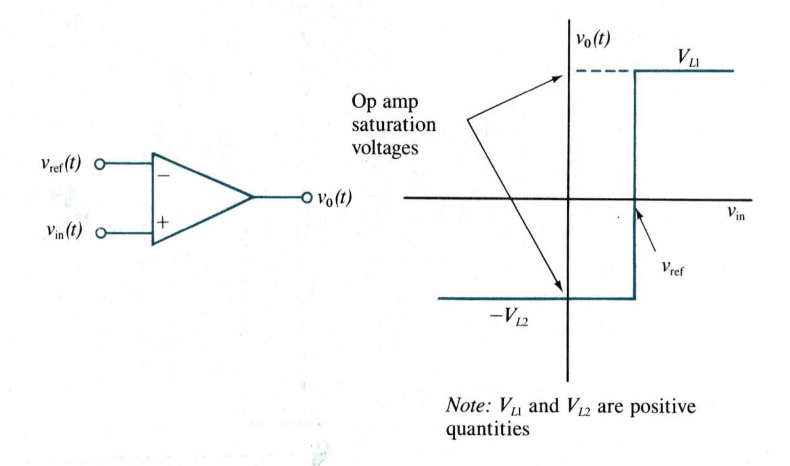

FIG. 9.3-3.
A comparator.

approximate output-input characteristic.† By choice of resistors and the positive reference voltages, V_{ref1} and V_{ref2}, both limit levels and the hard/soft-limiting effect can be controlled.

Comparator

A *comparator* is a two-input, single-output circuit. If the input voltages are $v_{in}(t)$ and $v_{ref}(t)$, the output voltage is ideally a constant level V_{L1} for $v_{in}(t) > v_{ref}(t)$ and a second constant level $-V_{L2}$ for $v_{in}(t) < v_{ref}(t)$. For $v_{in}(t) = v_{ref}(t)$ we ideally have $v_0 = 0$.

A simple comparator consists of an op amp without feedback, as shown in Fig. 9.3-3. Op amp saturation determines the output limit levels. By a reversal of the two inputs, the output-voltage polarity can be reversed.

The limiter of Fig. 9.3-2(d) can serve as a comparator by either removing R_2 or making it large. Minor modifications to provide a second input are made as depicted in Fig. 9.3-4.

EXAMPLE 9.3-1

We use the circuit of Fig. 9.3-4(b) to establish a comparator that has limit levels at the output of 0 V when $v_{in}(t) < +2$ V and -5 V when $v_{in}(t) > +2$ V. Thus, $v_{ref}(t) = -2$ V. We use $R_1 = 10$ kΩ and 24 V as the magnitude of any dc reference voltages required. Finally, we require the slope of the output-input characteristic to be small in the limiting regions.

Since $v_{ref}(t) = -2$ V, the input will be $+2$ V at the transition point between

† Analysis has assumed small diode forward resistances. That is, $R_f \ll R_5 || R_6$ and $R_f \ll R_3 || R_4$. Also assumed are $R_2 \gg R_5 || R_6$ and $R_2 \gg R_3 || R_4$.

FIG. 9.3-6. Phase detector structures that use (a) a single-throw switch or (b) a double-throw switch. (c) An implementation of the structure of (b).

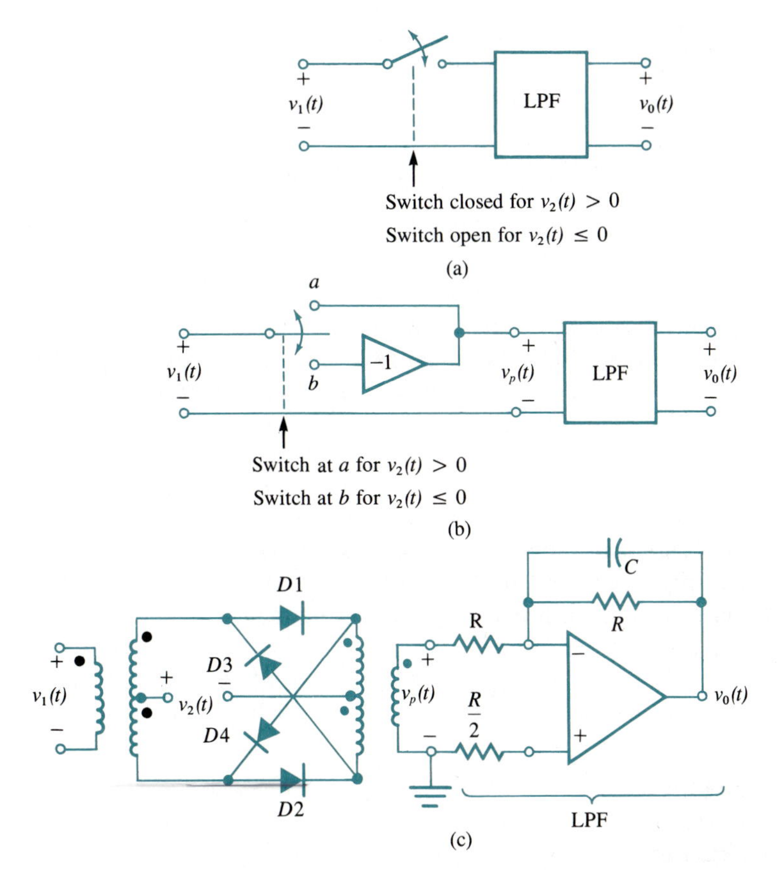

slowly varying, a Fourier series that is found by assuming $\theta_2(t)$ constant will closely represent $s_p(t)$. It is

$$s_p(t) = \frac{4}{\pi} \sum_{n=1}^{\infty} \frac{\sin n\pi/2}{n} \cos [n\omega_0 t + n\theta_2(t)] \tag{9.3-5}$$

On substituting (9.3-5) into (9.3-4) and reducing the algebra, we have

$$v_p(t) = \frac{2A_1}{\pi} \sin [\theta_1(t) - \theta_2(t)] + \{\text{terms at frequencies } 2n\omega_0,\ n = 1, 2, \ldots\} \tag{9.3-6}$$

The lowpass filter will remove all but the first term, so the phase detector's response is

$$v_0(t) = \frac{2A_1}{\pi} \sin [\theta_1(t) - \theta_2(t)] \tag{9.3-7}$$

Whenever $\theta_1(t) - \theta_2(t)$ is small,

$$v_0(t) \approx \frac{2A_1}{\pi}[\theta_1(t) - \theta_2(t)] = K_{\text{ph}}[\theta_1(t) - \theta_2(t)] \tag{9.3-8}$$

where the phase detector's constant K_{ph} is

$$K_{\text{ph}} = \frac{2A_1}{\pi} \tag{9.3-9}$$

The bandwidth of the lowpass filters in Fig. 9.3-6 need only be as large as the spectral extent of $\theta_1(t) - \theta_2(t)$. If this bandwidth is a significant fraction of ω_0, the filters may have to be high-order since they must remove the higher-frequency terms, the closest of which is at ω_0 minus the spectral extent.†

The phase detector of Fig. 9.3-6(c) is of the structure analyzed above. The output is given by (9.3-8) if the gain of the overall transformer pair is unity. It also uses a simple lowpass filter, which is adequate in most cases. Its action presumes $A_2 \gg A_1$ for proper operation. This constraint means that diode currents due to $v_1(t)$ are small relative to those due to $v_2(t)$. Thus, $v_2(t)$ only acts to turn on the proper diodes to connect the transformers together. The output $v_0(t)$ is then due to $v_1(t)$ passing through the transformers. The reader should trace the detailed operation as an exercise.

The constant K_{ph} for a phase detector having the structure of Fig. 9.3-6(a) will be half that of the phase detector of (b). It will also have a term at frequency ω_0 that must be filtered out by a lowpass filter. A practical realization follows that in Fig. 9.3-6(c) if diodes $D3$ and $D4$ are omitted.

Mixer

To implement a mixer (defined in Sec. 8.5), we need a product operation. Since the phase detector structures discussed above act as a product, they can also serve as mixers. Now, however, the two inputs are not at the same nominal frequency but have the forms

$$v_1(t) = A_1 \cos(\omega_1 t + \theta_1) \tag{9.3-10}$$

$$v_2(t) = A_2 \cos(\omega_2 t + \theta_2) \tag{9.3-11}$$

Here ω_1 and ω_2 are constants, but A_1, A_2, θ_1, and θ_2 may vary slowly. By retracing the analysis of Fig. 9.3-6(b), we find

$$\begin{aligned} v_p(t) = &\frac{2A_1}{\pi}\{\cos[(\omega_2 + \omega_1)t + \theta_2 + \theta_1] + \cos[(\omega_2 - \omega_1)t + \theta_2 - \theta_1]\} \\ &- \frac{2A_1}{3\pi}\{\cos[(3\omega_2 + \omega_1)t + 3\theta_2 + \theta_1] + \cos[(3\omega_2 - \omega_1)t + 3\theta_2 - \theta_1]\} \end{aligned}$$

† Ideally, the nearest term is at $2\omega_0$ minus the spectral extent, but practical effects usually create a "leakage" term at ω_0.

$$+ \{\text{terms at frequencies } (n\omega_2 \pm \omega_1),\ n = 5, 7, 9, \ldots\} \tag{9.3-12}$$

Many frequencies are contained in $v_p(t)$, but the two at frequencies $\omega_2 \pm \omega_1$ are the largest in amplitude and are the ones of most interest. By use of a bandpass filter at frequency $\omega_2 + \omega_1$ to remove all other components of $v_p(t)$, an up-converting mixer is achieved. Down-conversion follows use of a bandpass filter at frequency $\omega_2 - \omega_1$.

EXAMPLE 9.3-2

The input signals to a mixer have frequencies $\omega_1/2\pi = 4.5$ MHz and $\omega_2/2\pi = 9.7$ MHz. We list some of the frequencies contained in the product signal before filtering:

n	$\frac{n\omega_2 + \omega_1}{2\pi}$ (MHz)	$\frac{n\omega_2 - \omega_1}{2\pi}$ (MHz)	Relative Amplitude
1	14.2	5.2	1
3	33.6	24.6	$\frac{1}{3}$
5	53.0	44.0	$\frac{1}{5}$
7	72.4	63.4	$\frac{1}{7}$

Now suppose we desire an up-converter with an output frequency $(\omega_2 + \omega_1)/2\pi = 14.2$ MHz. This mixer requires a filter with a narrow enough bandwidth to remove the nearest component of the product signal, which is at 5.2 MHz (or 9.0 MHz away from the desired center frequency of 14.2 MHz). The next-nearest component to be removed is at 24.6 MHz (or 10.4 MHz away from 14.2 MHz). All other frequencies are farther away and of lower amplitudes, so they are more easily filtered.

9.4 POSITIVE FEEDBACK AND OSCILLATION

In many engineering applications it is necessary to generate periodic signals having a particular waveform shape, such as square, triangular, sawtooth, or sinusoidal. Such periodic repetition is called *oscillation*. In this section ways of generating sinusoidal waveforms are discussed. The following section then concentrates on other signal shapes.

For oscillations to occur, a feedback path around a device capable of gain (amplifier) is needed. Recall from the earlier discussion of the feedback loop of Fig. 8.4-2 that oscillations would occur at the angular frequency ω_0 whenever

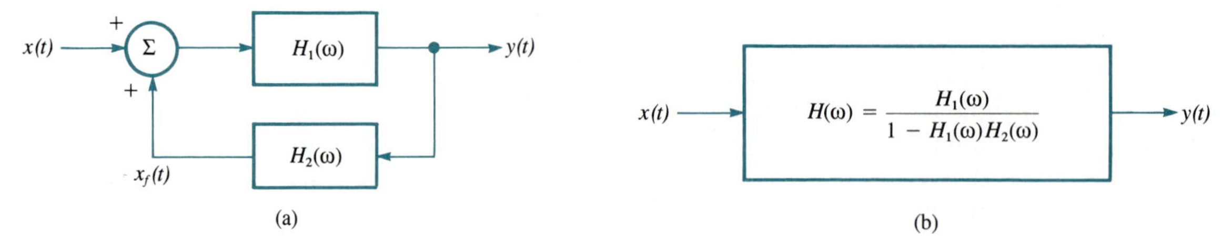

FIG. 9.4-1.
(a) A feedback loop drawn to illustrate positive feedback and (b) its equivalent transfer function.

$|H_1(\omega_0)H_2(\omega_0)| > 1$ and the phase angle of $H_1(\omega_0)H_2(\omega_0)$ was $-\pi$—in other words, when $H_1(\omega_0)H_2(\omega_0) < -1$. The loop is referred to as having *negative feedback* because of the negative sign at the summing junction in Fig. 8.4-2(a). If the loop is redrawn as sketched in Fig. 9.4-1(a), which is now referred to as having *positive feedback* because of the positive sign at the summing junction, the condition for oscillation becomes

$$H_1(\omega_0)H_2(\omega_0) > 1 \qquad (9.4\text{-}1)$$

This condition is known as the *Barkhausen criterion.*

Any loop with positive feedback for which (9.4-1) is true will start to oscillate, even with no externally applied signal. It only takes the smallest disturbance to initiate oscillations (such as from low-level noise in the circuit). It is important to note that the frequency of oscillation is determined by the *phase* of the loop gain $H_1(\omega)H_2(\omega)$; it is the angular frequency at which the phase is zero.

Because loop gain $H_1(\omega)H_2(\omega)$ is defined as the gain around the opened loop with small-signal linear operation, its phase will establish the oscillation frequency only when oscillations start and signal amplitudes in the closed loop are small. As signal levels grow, the amplifier (active device) eventually saturates and signal levels stabilize. Under the nonlinear condition the final frequency of oscillation may be different from the frequency at the start. The frequency could also be the same (stay constant) if saturation affects only the amplitudes and not the phases of voltages around the loop.

All the preceding considerations can be summarized by use of Fig. 9.4-2. An oscillator consists of a positive-feedback loop with no external input, as shown in Fig. 9.4-2(a). The equivalent form of Fig. 9.4-2(b) simply lumps all loop components together as one network with loop gain $L(\omega)$, where

$$L(\omega) = H_1(\omega)H_2(\omega) \qquad (9.4\text{-}2)$$

The behavior of $L(\omega)$ can be examined by breaking the loop at any convenient point (such as point X), inserting a signal source, such as v_{in} in Fig. 9.4-2(c) and (d), and calculating $L(\omega)$ as the ratio of the voltage at the output side of the break to the input voltage (of course, v_{in} must be in complex form for this ratio to be valid). At the output side of the break an impedance [Z_i in Fig. 9.4-2(c) and (d)] must be added that is equal to the input impedance seen by the source. Once $L(\omega)$

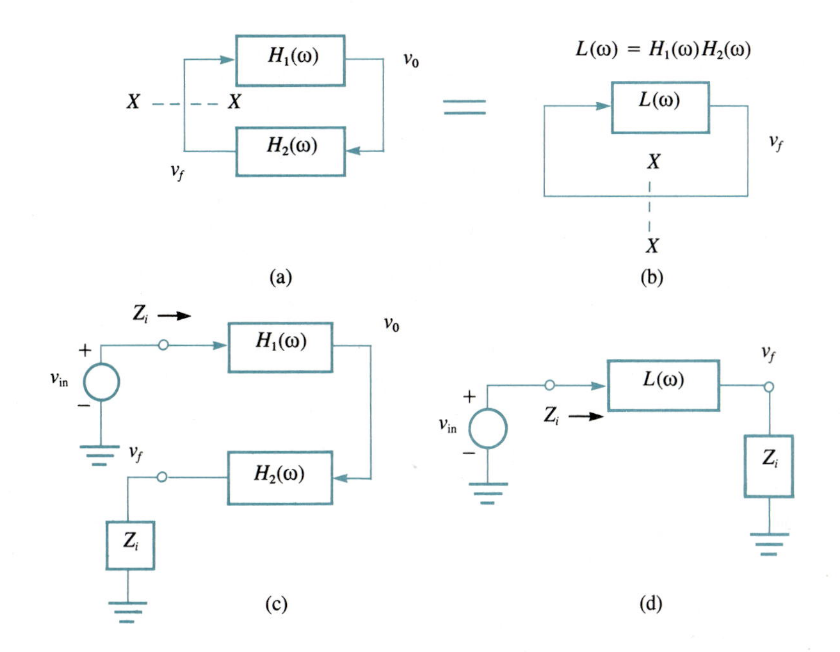

FIG. 9.4-2. (a) and (b) Oscillator feedback loops; (c) and (d) broken loops used to find the oscillation frequency.

is found, the frequency of oscillation is determined as that for which the phase of $L(\omega)$ is 0 (or -2π). An example will illustrate these facts.

EXAMPLE 9.4-1

A *phase-shift* oscillator uses a three-section *CR* highpass filter as a feedback path around an amplifier with sign inversion, as sketched in Fig. 9.4-3(a). Since one section of the filter will produce a phase shift of magnitude less than $\pi/2$ for finite frequencies, three sections are needed to guarantee a phase shift of $-\pi$ at a finite frequency. Thus, starting at point X and going around the loop, we have a phase shift of $-\pi$ through the amplifier because of its sign inversion and a phase shift of $-\pi$ through the filter. Total phase shift is -2π (or 0). Thus, the feedback signal will be in phase with the original signal at X, and oscillations are sustained.

To analyze the loop gain, we break the circuit at point X, as shown in Fig. 9.4-3(b). Analysis starts from equations derived at the three filter nodes, using Kirchhoff's current law:

$$\frac{v_3}{R} + \frac{v_3 - v_2}{1/j\omega C} = 0$$

$$\frac{v_2}{R} + \frac{v_2 - v_1}{1/j\omega C} + \frac{v_2 - v_3}{1/j\omega C} = 0$$

$$\frac{v_1}{R} + \frac{v_1 - v_0}{1/j\omega C} + \frac{v_1 - v_2}{1/j\omega C} = 0$$

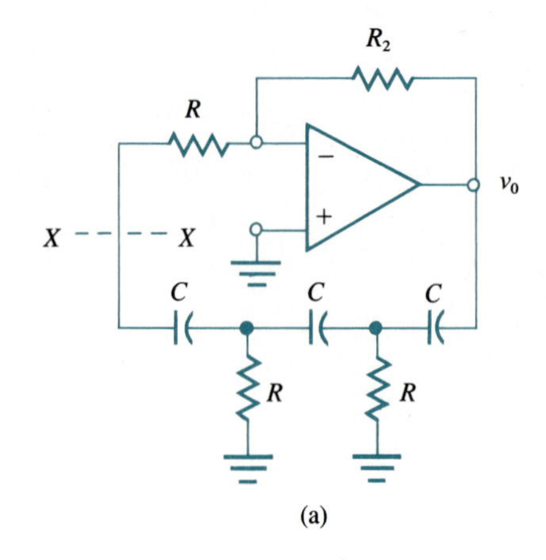

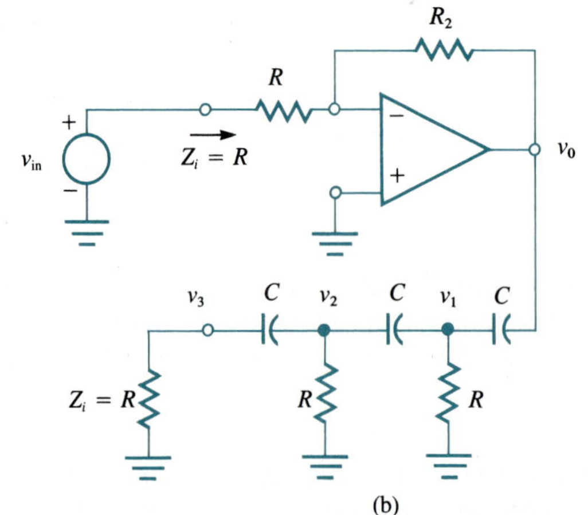

FIG. 9.4-3. (a) A phase-shift oscillator and (b) its loop broken for analysis of loop gain.

After straightforward algebraic reduction we obtain

$$\frac{v_3}{v_0} = \frac{-j\omega^3 R^3 C^3}{(1 - 6\omega^2 R^2 C^2) + j\omega RC(5 - \omega^2 R^2 C^2)}$$

The phase of this expression is

$$\text{Phase of } \frac{v_3}{v_0} = -\frac{\pi}{2} - \tan^{-1}\left[\frac{\omega RC(5 - \omega^2 R^2 C^2)}{1 - 6\omega^2 R^2 C^2}\right]$$

which is $-\pi$ when $\omega = \omega_0 = 1/(RC\sqrt{6})$. Thus,

$$\left.\frac{v_3}{v_0}\right|_{\omega_0 = 1/(RC\sqrt{6})} = -\frac{1}{29}$$

and

$$L(\omega_0) = -\frac{R_2}{R}\left(\left.\frac{v_3}{v_0}\right|_{\omega_0 = 1/(RC\sqrt{6})}\right) = \frac{R_2}{29R}$$

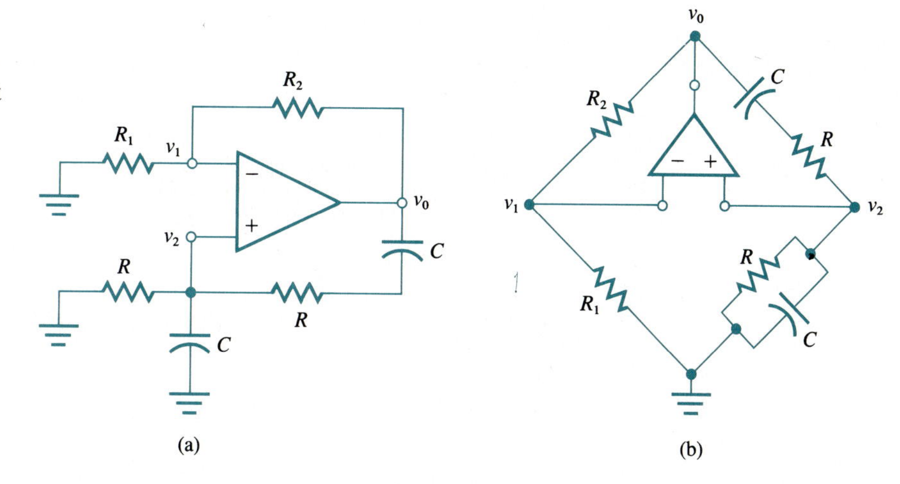

FIG. 9.4-4. (a) A Wein bridge oscillator and (b) its circuit drawn in bridge form.

To satisfy Barkhausen's criterion, we require $L(\omega_0) > 1$, which means $R_2 > 29R$ is necessary for oscillation. Equivalently, the magnitude of the op amp's gain must be at least 29.

Wien Bridge Oscillator

The *Wien bridge oscillator* of Fig. 9.4-4(a) is often implemented at lower frequencies with an op amp as shown. This sinusoidal-waveform oscillator's frequency is

$$\omega_0 = \frac{1}{RC} \tag{9.4-3}$$

The proper op amp gain for oscillation requires

$$R_2 > 2R_1 \tag{9.4-4}$$

Redrawing the circuit as in Fig. 9.4-4(b) shows why it is called a bridge. With $R_2 = 2R_1$ and $\omega = \omega_0 = 1/RC$, the oscillation frequency, it can be shown that $v_1 = v_2 = v_0/3$ and the bridge is said to be balanced.

Colpitts Oscillator

For frequencies above about 1 MHz, sinusoidal oscillators often use tuned circuits (called *tank circuits*, because they contain a large stored energy) consisting of an inductance and a capacitor in parallel that are resonant at the frequency of oscillation. The *Colpitts oscillator* of Fig. 9.4-5(a) is a good example. Analysis shows (Problem 9-22) that the oscillation frequency is

$$\omega_0 = \frac{1}{\sqrt{LC_1C_2/(C_1 + C_2)}} \tag{9.4-5}$$

which is the value at which L resonates with the series combination of C_1 and C_2.

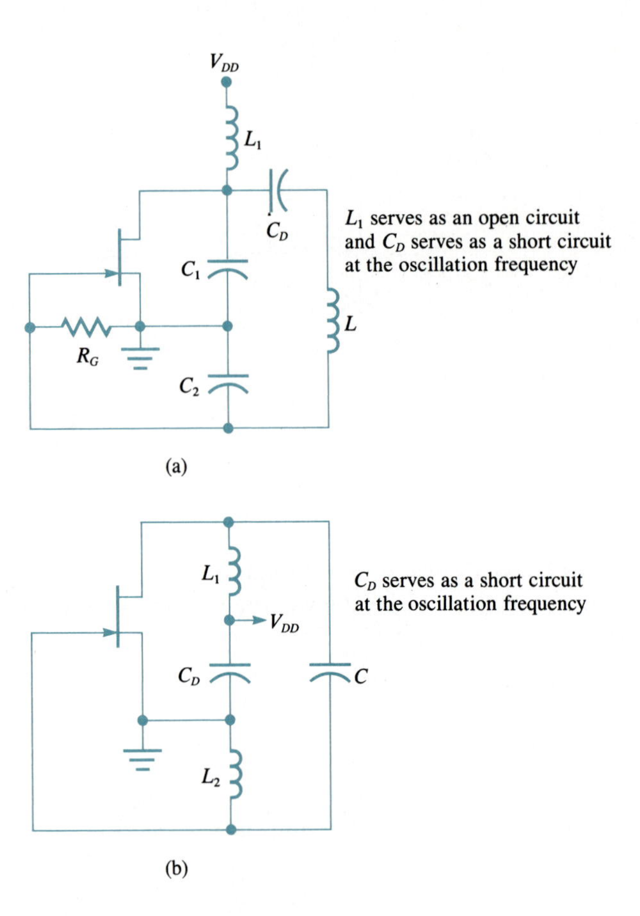

FIG. 9.4-5. Tuned-circuit oscillators: (a) Colpitts and (b) Hartley.

Oscillations are possible only if

$$g_m r_o > \frac{C_2}{C_1} \tag{9.4-6}$$

where g_m and r_o of the JFET are given by (6.6-12) and (6.6-13), respectively.

Hartley Oscillator

The *Hartley oscillator* of Fig. 9.4-5(b) is another popular tuned-circuit oscillator. Its oscillation frequency is

$$\omega_0 = \frac{1}{\sqrt{(L_1 + L_2)C}} \tag{9.4-7}$$

and oscillations require (Problem 9-25)

$$g_m r_o > \frac{L_1}{L_2} \tag{9.4-8}$$

Both the Hartley and Colpitts oscillators can be implemented with other active devices, such as the BJT or MOSFETs.

Crystal Oscillator

If the frequency of any oscillator is measured over time, it is found to vary. This fluctuation is due to many factors but generally decreases as the bandwidth of the tuned (bandpass) circuit is reduced. This means that very narrowband tuned circuits are desired when good frequency stability is required. With ordinary inductors the ratio of oscillation angular frequency ω_0 to bandwidth (in radians per second), called the *Q factor*, is limited to several hundred at most and usually is less than 100. For precise frequency control and high stability, tuned circuits with higher Q factors are needed.

Some natural crystals, such as quartz, have piezoelectric properties, where they mechanically vibrate at very precise rates when excited by an electric voltage applied between metal plates attached to the crystal. Electrically, the crystal itself behaves like a series RLC circuit. However, because there is capacitance between the metal plates, it can also behave as a parallel resonant circuit. The symbol for a piezoelectric crystal is shown in Fig. 9.4-6(a); its equivalent circuit is shown in (b). Capacitance C_p is due to the plates (a few picofarads), inductance L is usually very large (up to hundreds of henrys), and C is small (as small as about 0.001 pF). Resistance R determines the Q factor, which can be as large as about 10^6. If the effect of R is neglected, the impedance Z_X of the crystal is

$$Z_X = \frac{-j(1 - \omega^2 LC)}{\omega C_p\{[(C + C_p)/C_p] - \omega^2 LC\}} = jX \tag{9.4-9}$$

A plot of crystal reactance X is sketched in Fig. 9.4-6(c). The series and parallel resonance frequencies, respectively, are

$$\omega_s = \frac{1}{\sqrt{LC}} \tag{9.4-10}$$

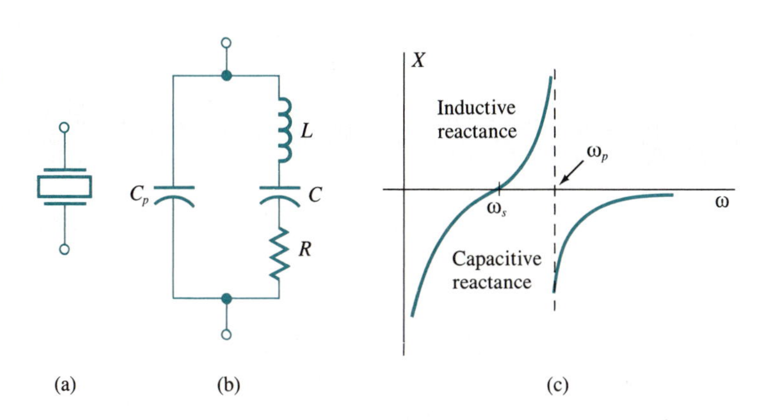

FIG. 9.4-6. (a) The electrical symbol for a piezoelectric crystal that has the equivalent circuit of (b). (c) The crystal reactance when R is neglected.

$$\omega_p = \sqrt{\frac{C + C_p}{LCC_p}} = \omega_s\sqrt{1 + \frac{C}{C_p}} \tag{9.4-11}$$

A region of frequencies from ω_s to ω_p exists where the crystal looks inductive. This fact allows the inductance in the Colpitts oscillator to be replaced by the crystal in order to realize a very stable crystal device called a *Pierce oscillator*.

Piezoelectric crystals may be cut for frequencies from a few kilohertz to about a hundred megahertz with frequency stabilities of better than 0.0001% (one part per million). Quartz-crystal oscillators are common in digital wristwatches, which accounts for their high accuracy.

EXAMPLE 9.4-2

A particular quartz crystal has $C_p = 5$ pF, $C = 0.001$ pF, $Q = 5 \times 10^4$, and $\omega_s/2\pi = 5 \times 10^6$ Hz. If Q is given by

$$Q = \frac{\omega_s L}{R}$$

we find L, R, and $\omega_p/2\pi$. From (9.4-10)

$$L = \frac{1}{\omega_s^2 C} = \frac{1}{(2\pi)^2(25 \times 10^{12})(10^{-15})} = \frac{10}{\pi^2} \approx 1.013 \text{ H}$$

From the expression for Q

$$R = \frac{\omega_s L}{Q} = \frac{2\pi(5 \times 10^6)(10)}{(5 \times 10^4)(\pi^2)} = \frac{2000}{\pi} \approx 637 \ \Omega$$

Finally, from (9.4-11)

$$\frac{\omega_p}{2\pi} = \frac{\omega_s}{2\pi}\sqrt{1 + \frac{C}{C_p}} = 5 \times 10^6\sqrt{1 + \frac{0.001}{5}} = (5 \times 10^6)\sqrt{1.0002} \approx 5.0005 \text{ MHz}$$

Note that ω_p is only 0.01% larger than ω_s for this crystal. It is typical for ω_s and ω_p to be very close together.

9.5 NONSINUSOIDAL WAVEFORM GENERATION

In this section several oscillators are developed that generate nonsinusoidal waveforms.

Pulse and Square-Wave Oscillators

A periodic sequence of pulses can be generated by the circuit of Fig. 9.5-1(a), which is called an *astable multivibrator*. The applicable waveforms are shown in

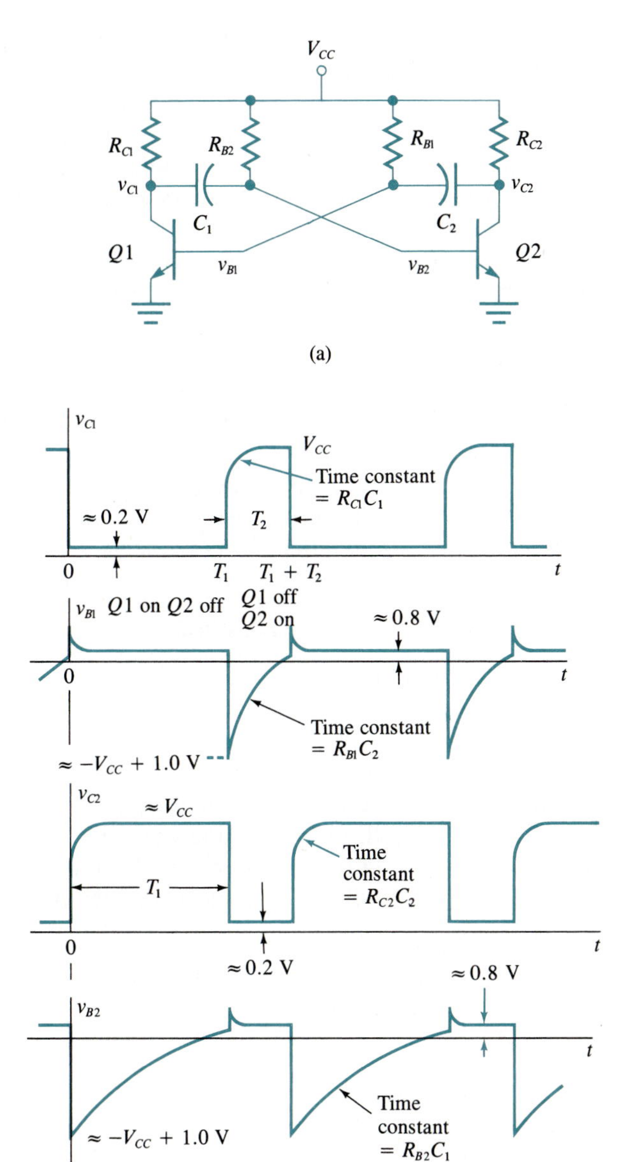

FIG. 9.5-1. (a) A multivibrator and (b) its waveforms.

Fig. 9.5-1(b). By proper choice of component values the circuit can produce square waves.

To understand operation, assume initially that transistor $Q1$ is on (in saturation) and $Q2$ is off (cut off). Just after $t = 0$ the collector voltage of $Q1$ will be at its saturation value (around 0.2 V) while its base voltage v_{B1} will be its saturation value (around 0.8 V). Since $Q2$ is off, its base voltage will be a large negative value (about $-V_{CC}$ plus the base and collector saturation voltages). This condition means that the voltage across C_1 is large and negative on the side that is the base of $Q2$. Because $Q2$ is off, it draws little current; so the supply feeds current into C_1 to charge it toward the level of the supply, which is V_{CC}. This charging (through R_{B2}) raises v_{B2} until it reaches the cut-in voltage of $Q2$ (about 0.6 to 0.7 V). When the cut-in voltage is reached, $Q2$ begins to conduct, and v_{C2} decreases, which decreases v_{B1} because the voltage across C_2 cannot change abruptly. These changes take place rapidly, causing v_{B1} to quickly decrease to a large negative value (about $-V_{CC}$ plus the base and collector saturation voltages), which cuts off $Q1$, while $Q2$ is left on. These changes occur at time T_1 in the waveforms of Fig. 9.5-1(b). The interval T_1 has the approximate duration

$$T_1 \approx R_{B2}C_1 \ln 2 \tag{9.5-1}$$

if V_{CC} is large relative to the collector and base saturation voltages and base cut-in voltage (both transistors are assumed identical).

For times from T_1 to $T_1 + T_2$ operation is the same as described in the preceding paragraph, except that the transistors are reversed ($Q2$ is now on and $Q1$ is off), and the charging interval that alters v_{B1} is now determined by R_{B1} and C_2:

$$T_2 \approx R_{B1}C_2 \ln 2 \tag{9.5-2}$$

The period T_0 of oscillation is $T_1 + T_2$, so

$$T_0 = T_1 + T_2 \approx (R_{B1}C_2 + R_{B2}C_1) \ln 2 \tag{9.5-3}$$

For generation of narrow pulses we may choose $T_2 < T_1$ and take v_{C1} as the output. For good performance the time constant $R_{C1}C_1$ that determines the rise time of the pulses must be small. For large-duration pulses (greater than $T_0/2$) the output can be taken as v_{C2}, and the time constant $R_{C2}C_2$ should be small.

For generation of square waves it is typical practice to choose $R_{C1} = R_{C2}$, $R_{B1} = R_{B2} = R_B$, $C_1 = C_2 = C$, so

$$T_0 = T_1 + T_2 \approx 2R_BC \ln 2 \tag{9.5-4}$$

EXAMPLE 9.5-1

In an astable multivibrator $R_{C1} = R_{C2} = 1\ \text{k}\Omega$, $R_{B1} = R_{B2} = R_B = 10\ \text{k}\Omega$, $C_1 = C_2 = C = 0.01\ \mu\text{F}$, and the period and time constants are to be found. Since this is a square-wave generator, (9.5-4) gives the period:

$$T_0 \approx (2 \times 10^4)(10^{-8}) \ln 2 \approx 1.386 \times 10^{-4}\ \text{s} = 138.6\ \mu\text{s}$$

which corresponds to a repetition frequency of $f_0 = 1/T_0 \approx 7213$ Hz. The time constant that determines the period is

$$R_BC = 10^4(10^{-8}) = 10^{-4}\ \text{s} = 100\ \mu\text{s}$$

The time constant that determines the rise time of the output waveform is

$$R_C C = 10^3(10^{-8}) = 10^{-5} = 10\ \mu\text{s}$$

This time constant is 10 μs/69.3 μs = 0.1443, or 14.43%, of the half-period of the square wave.

Square Waves from Sine Waves and Zero-Crossing Detection

Square waves may be readily generated from a sinusoidal signal. The sinusoid need only be passed through either a limiter or a comparator. By combining the operation with a highpass filter and rectifier, we obtain a zero-crossing detector. Figure 9.5-2 depicts the circuit and waveforms when a comparator is assumed. If the sine wave is the input v_{in}, the output v_0 of the comparator is the square wave.

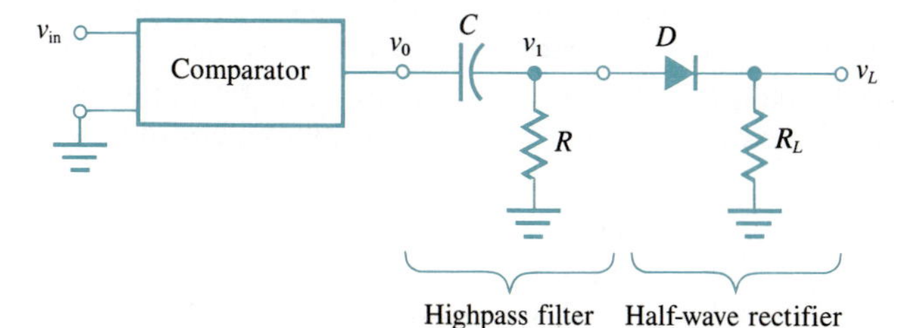

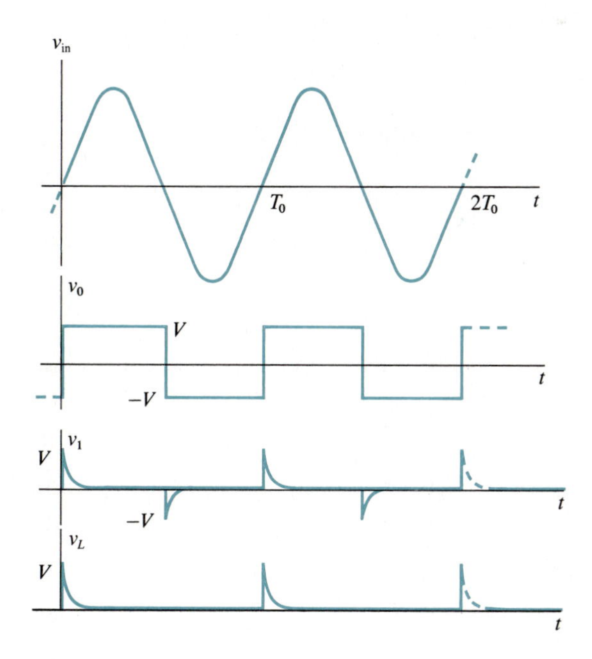

FIG. 9.5-2. A zero-crossing detector and its applicable waveforms.

If $RC \ll T_0$, the highpass filter's output is a sequence of positive and negative alternating-polarity spikes, as shown in the waveform sketches. By rectification, only the positive spikes are preserved. These spikes occur at the times of the sine wave's positive-going zero crossings, and a zero-crossing detector results. By reversal of the diode's direction a negative-going zero-crossing detector is possible.

Relaxation Oscillator

A circuit useful for square-wave generation at lower frequencies is shown in Fig. 9.5-3(a); it uses an op amp and is called a *relaxation oscillator*. The op amp acts as a comparator. To describe performance, assume first that $v_0 = V_0$, the saturation output level of the op amp, and v_1 is at some large negative level. Then $v_2 = V_0R_1/(R_1 + R_2)$ and $v_1 < v_2$, which sustains the output level at V_0. However, capacitor C will begin charging through R to try to reach a voltage V_0. When v_1 exceeds $v_2 = V_1 = V_0R_1/(R_1 + R_2)$, the op amp will abruptly change v_0 from V_0 to $-V_0$. After the time at which this change occurs, the capacitor's voltage will try to charge from V_1 to $-V_0$. However, when it reaches $-V_1$, it drops below v_2, and the output changes again to V_0, as shown in the applicable waveforms of Fig. 9.5-3(b). Thereafter, the entire process repeats periodically.

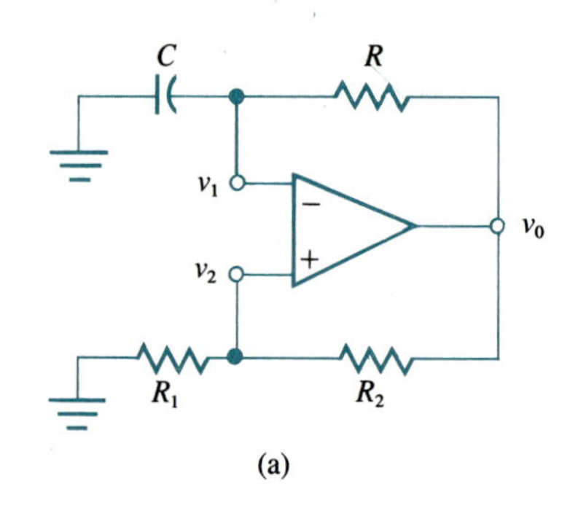

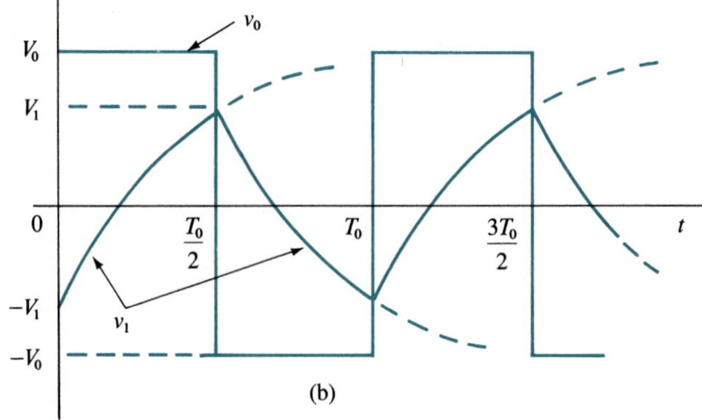

FIG. 9.5-3. (a) A relaxation oscillator and (b) its waveforms.

It can be shown that the period of the square wave for the circuit of Fig. 9.5-3(a) is

$$T_0 = 2RC \ln\left(1 + \frac{2R_1}{R_2}\right) \tag{9.5-5}$$

For frequencies $f_0 = 1/T_0$ above those for which the op amp behaves as a broadband comparator (typically a few tens of kilohertz), other implementations are required.

Triangular-Wave Generator

The signal v_1 in Fig. 9.5-3(a) is approximately triangular, as shown in (b). However, the waveform half-period segments have some curvature and are not very linear. Better linearity can be achieved by charging capacitor C through a constant-current source rather than through resistor R. In essence, we desire to replace the feedback path comprised of R and C by a more ideal integrator. If the noninverting integrator of Fig. 8.2-6 is used, the circuit of Fig. 9.5-4(a) results. An inverting integrator (Fig. 8.2-7) can also be used if its output is fed back to the positive input terminal of the relaxation oscillator's op amp, as shown in Fig. 9.5-4(b).

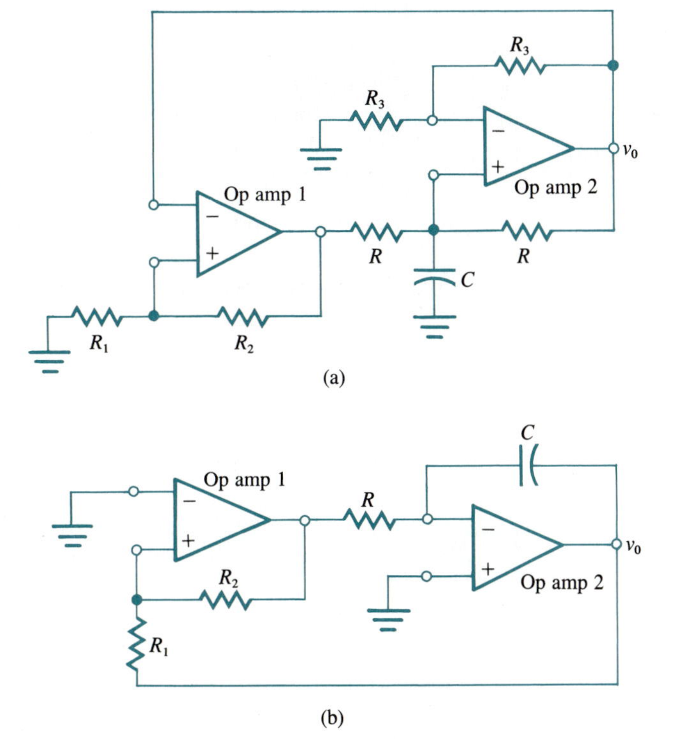

FIG. 9.5-4. A triangular-waveform generator that uses (a) a noninverting integrator and (b) an inverting integrator.

The period of oscillation for the generator of Fig. 9.5-4(b) can be shown to be (Problem 9-31)

$$T_0 = \frac{4RCR_1}{R_2} \tag{9.5-6}$$

The output waveform v_0 will vary triangularly between voltage extremes that are

$$v_{max} = V_0 \frac{R_1}{R_2} \tag{9.5-7}$$

$$v_{min} = -V_0 \frac{R_1}{R_2} \tag{9.5-8}$$

where V_0 is the saturation level of the relaxation oscillator's op amp 1 at its output.

EXAMPLE 9.5-2

An op amp has saturation levels of ± 9 V in its output. A triangular generator of the form of Fig. 9.5-4(b) is to be constructed for which $C = 0.1\ \mu\text{F}$, $V_{max} = 6$ V, $V_{min} = -6$ V, and $T_0 = 10^{-4}$ s. We find values for R, R_1, and R_2.

From (9.5-6) and (9.5-7)

$$T_0 = 10^{-4} = \frac{4RCR_1}{R_2} = \frac{(4 \times 10^{-7})RR_1}{R_2}$$

$$V_{max} = 6 = V_0 \frac{R_1}{R_2} = 9 \frac{R_1}{R_2}$$

Solution gives

$$R = 375\ \Omega \qquad \text{and} \qquad \frac{R_2}{R_1} = 1.5$$

Since R_1 or R_2 can be chosen arbitrarily, we select $R_1 = 10\ \text{k}\Omega$, so $R_2 = 15\ \text{k}\Omega$.

The triangular generator of Fig. 9.5-4(a) is considered further in Problem 9-30.

PROBLEMS

9-1. Allow v_{in} of Fig. 9.1-1(a) to vary slowly with time. Redraw the applicable voltages as in Fig. 9.1-1(b) to show that the output $v_0(t) = A_0 v_{in}(t)$.

9-2. During the time when $v_{in} < 0$, so that $D2$ in Fig. 9.2-1(b) is conducting and has a resistance R_f, $D1$ is open-circuited. Find an equation for v_0/v_{in} if the op amp is ideal. In what way does v_0 depend on R_f or R_L during this time?

9-3. Sketch v_0 for the circuit of Fig. P9-3, and discuss in what way the network is a voltage-doubling half-wave rectifier.

FIG. P9-3.

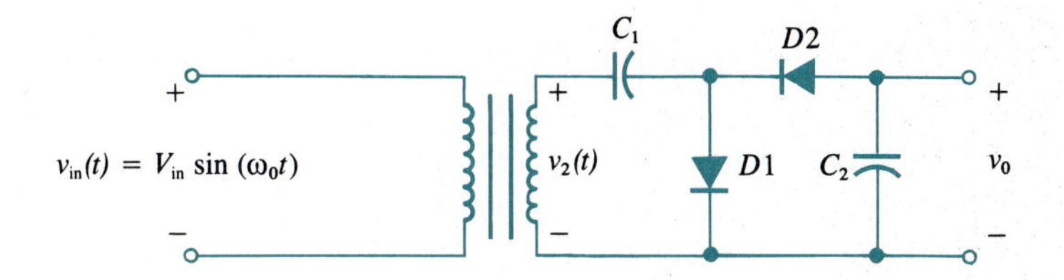

9-4. Work Problem 9-3 for the full-wave circuit of Fig. P9-4.

FIG. P9-4.

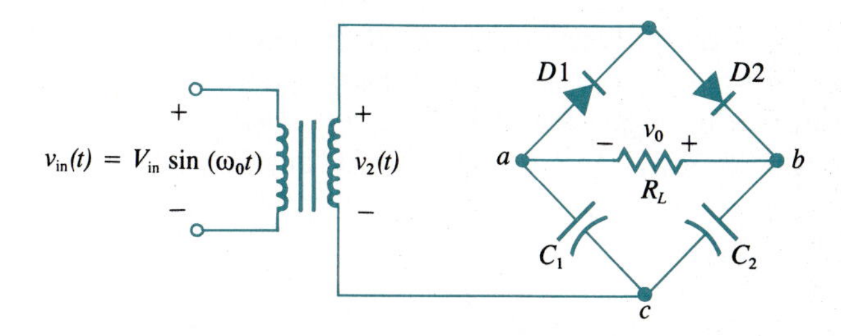

★9-5. For the power supply of Fig. P9-5, assume $R_L \gg 1/\omega_0 C_2$ and $R \gg 1/\omega_0 C_2$. Derive equations to approximate the dc voltage component of v_0 and the ac component at angular frequency $2\omega_0$. [*Hint*: First, prove that voltage v_1 is as shown in Fig. 9.2-4(b), except that the decay rate between diode conduction times is determined by the product RC_1. Next, analyze the remainder of the circuit, using a Fourier series for $v_1(t)$.]

FIG. P9-5.

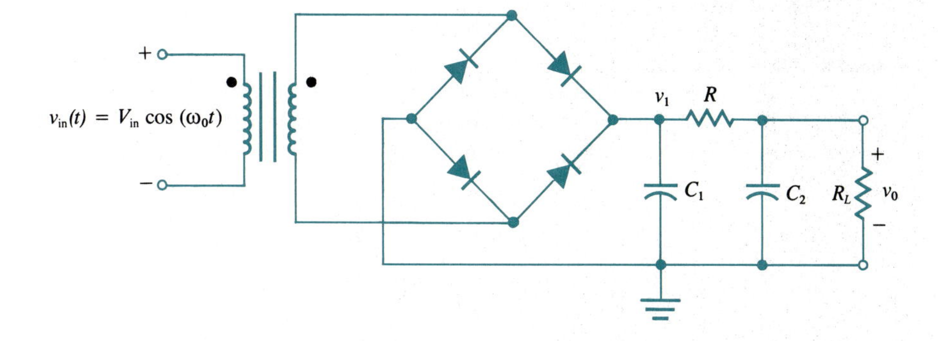

9-6. In the circuit of Fig. 9.2-2(b), assume that diodes $D1$ and $D3$ have resistances of 20 Ω when conducting, while $D2$ and $D4$ are 40 Ω. Let $R_1 = R_L = 480$ Ω. How does v_0 depend on these diode resistances?

9-7. The transformer in the power rectifier of Fig. 9.2-3(b) can be taken as ideal (lossless). The turns ratio is such that $v_1 = v_{in}/3$. The diodes have negligible forward resistance. If the average ac power being supplied by the input

$$v_{in}(t) = 115 \cos 120\pi t$$

is 18 W, what is R_L? What is the dc voltage across R_L?

9-8. In the power supply of Fig. 9.2-4(a) the transformer and diodes may be assumed ideal. If $R_L = 200\ \Omega$ and $C = 400\ \mu\text{F}$, determine what fraction of the period T_0 the diodes conduct. What is the peak-to-peak ripple voltage across R_L? Assume $\omega_0/2\pi = 60$ Hz.

9-9. A full-wave rectifier uses the filter of Fig. 9.2-4(c) with $L = 0.1$ H, $C = 100\ \mu\text{F}$, $R_L = 250\ \Omega$, and a source of frequency $\omega_0/2\pi = 400$ Hz. (a) Does this supply require a bleeder resistor? If so, determine its largest allowed resistance and add it to the output to give a new effective value for R_L. (b) What peak-to-peak ripple occurs?

9-10. Repeat the analysis procedure of the text that leads to (9.2-17), and obtain an expression for the ripple at the output of a full-wave rectifier having the filter of Fig. P9-10. Assume $1/\omega C_2 \ll R_L$, $1/\omega C_2 \ll \omega L_2$, $1/\omega C_1 \ll \omega L_1$, $1/\omega C_1 \ll \omega L_2$, all for any $\omega \geq \omega_0$. Use the expression with $L_1 = L_2 = L$, $C_1 = C_2 = C$, and find the minimum product LC that will give $|v_{\text{ripple}}/V_{\text{peak}}| < 0.01$ when $\omega_0/2\pi = 60$ Hz.

FIG. P9-10.

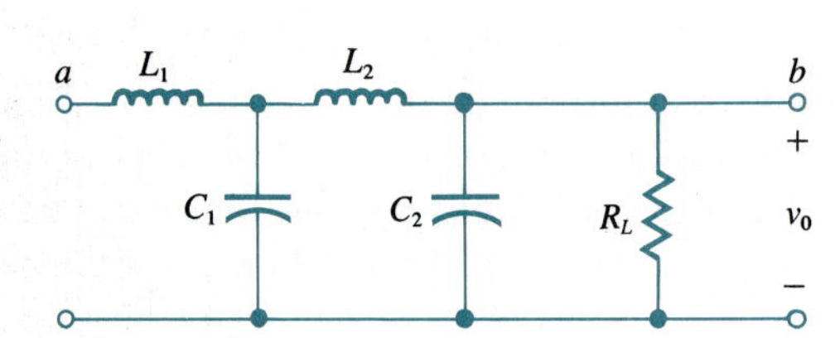

★9-11. Use the small-signal equivalent circuit of Fig. 6.5-6 for the transistor in Fig. 9.2-5, and analyze the circuit to show that the ratio of the ac component of v_0, denoted by Δv_0, to the ac component of v_1, denoted by Δv_1, is

$$\frac{\Delta v_0}{\Delta v_1} \approx \frac{r_\pi(R_1 + R_2)}{(\beta + 1)r_o R_1 A_0}$$

if $A_0 \gg 1$. Ignore the effect of C_2. Justify also that the dc component of v_0, denoted by V_0, is

$$V_0 = \left(1 + \frac{R_2}{R_1}\right)V_{\text{ref}}$$

9-12. Use results from Problem 9-11 to find the ripple reduction due to electronic regulation if $r_o = 1200\ \Omega$, $\beta = 40$, $r_\pi = 800\ \Omega$, $R_1 = 2200\ \Omega$, $R_2 = 8200\ \Omega$, and $A_0 = 50$ in a certain power supply.

9-13. Sketch the v_0-versus-v_{in} characteristic for the limiter of Fig. 9.3-2(d) if $R_4 = R_6 = 1000\ \Omega$, $R_3 = R_5 = 3300\ \Omega$, $V_{ref1} = V_{ref2} = 18$ V, $D1$ and $D2$ are ideal diodes, $R_2 = 7R_1$, and $R_1 = 1.2$ kΩ.

9-14. Work Problem 9-13 except assume that the diodes have a forward cut-in voltage of 0.7 V but otherwise behave as ideal.

9-15. In the comparator of Fig. 9.3-4(a), $R_1 = 15$ kΩ, $R_3 = 15$ kΩ, $R_4 = 3.3$ kΩ, $R_5 = 10$ kΩ, $R_6 = 4.7$ kΩ, $V_{ref1} = V_{ref2} = 24$ V, and the diodes are ideal. Determine v_0 for $v_{in} = -3, -1, 1$, and 5 V when v_{ref} is constant at -1.5 V.

★9-16. Follow the procedures given in the text and find the full equation for (9.3-6). Determine the amplitudes of the terms at frequencies $2\omega_0$, $4\omega_0$, and $6\omega_0$. If $s_p(t)$ is passed through the lowpass filter shown in Fig. 9.3-6(c) with $RC = 5/\omega_0$, what are the amplitudes of the same terms at the new output?

9-17. The two input signals to a down-converting mixer have frequencies of 10 and 7.1 MHz. The mixer's product signal $v_p(t)$ is passed through a bandpass filter having the transfer function that is approximated by

$$H(\omega) \approx \frac{W}{W + j(\omega - \omega_0)} \qquad \omega > 0$$

where $\omega_0/2\pi = 2.9$ MHz and $W/2\pi = 1.0$ MHz. What is the frequency of the nearest unwanted term in the output of the filter, and how large is it relative to the desired term?

9-18. Loop gain in a positive-feedback loop is

$$H_1(\omega)H_2(\omega) = \frac{j\omega RL/R_1}{R + j\omega L + (j\omega)^2 RLC}$$

Determine a condition on R_1 for the loop to oscillate, and find the frequency of oscillation.

9-19. The phase-shift oscillator of Fig. 9.4-3 must oscillate at a frequency of 10 kHz when $R = 5$ kΩ. Find C and R_2.

9-20. Work Problem 9-19 except assume the Wien bridge oscillator of Fig. 9.4-4 with $R_1 = 3.3$ kΩ.

9-21. Suppose resistor R_1 in the Wien bridge oscillator increases resistance as the current through it increases in magnitude. Discuss what effect it will have on the oscillator.

★9-22. Analyze the Colpitts oscillator of Fig. 9.4-5(a) and show that (9.4-5) and (9.4-6) are true.

9-23. A transistor for which $g_m = 0.013$ S and $r_o = 3.1$ kΩ is to be used in a

Colpitts oscillator with $C_1 = 100$ pF, $C_2 = 400$ pF, and $\omega_0/2\pi$ to be 10 MHz. (a) Will the circuit oscillate? (b) If so, what inductance L is required?

9-24. The transistor of Problem 9-23 is used in the Hartley oscillator of Fig. 9.4-5(b). Assume $\omega_0/2\pi = 12$ MHz, $L_1 = 3\ \mu$H, and $L_2 = 0.1\ \mu$H. (a) Will the circuit oscillate? (b) If so, what value of C is required?

★9-25. For the Hartley oscillator of Fig. 9.4-5(b), show that (9.4-7) and (9.4-8) are true.

9-26. Assume $C_p = 6$ pF, $C = 0.0005$ pF, $Q = \omega_s L/R = 4 \times 10^4$, and $\omega_s/2\pi = 10$ MHz for a quartz crystal. (a) What is the crystal's inductance? (b) If the crystal replaces the inductance in a Colpitts oscillator for which (9.4-5) applies with $C_1 = 100$ pF and $C_2 = 400$ pF, what is the exact frequency of oscillation?

9-27. An astable multivibrator is to produce positive pulses of 10.0 μs duration at a rate of 10,000 pulses per second. If $C_1 = 0.005\ \mu$F, $C_2 = 0.05\ \mu$F, $R_{C1}C_1 = 0.5\ \mu$s, and $R_{C2}C_2 = 5.0\ \mu$s are to be used, find R_{B1}, R_{B2}, R_{C1}, and R_{C2}. Where in the circuit of Fig. 9.5-1(a) should the output be taken?

9-28. An astable multivibrator has $R_{B1} = R_{B2} = R_B$, $R_{C1} = R_{C2} = R_C = 500\ \Omega$, and $C_1 = C_2 = C = 0.01\ \mu$F. It is used to generate square waves having a time constant that is not to exceed 2.5% of the period of oscillation. Find the largest allowable frequency of oscillation and the value of R_B necessary to produce this frequency.

9-29. In the relaxation oscillator of Fig. 9.5-3(a), $R = 1500\ \Omega$, $C = 0.01\ \mu$F, and R_1 and R_2 in parallel must equal R. If the oscillation frequency of the square waves is to be 40 kHz, find R_1 and R_2.

9-30. Show that the period of oscillations of the triangular-waveform generator of Fig. 9.5-4(a) is

$$T_0 = \frac{2RCR_1}{R_1 + R_2}$$

and its output voltage v_0 varies between extreme values of $\pm V_0 R_1/(R_1 + R_2)$, where $\pm V_0$ are the assumed output saturation levels of op amp 1.

9-31. For the triangular-waveform generator of Fig. 9.5-4(b), show that (9.5-6) through (9.5-8) are true.

9-32. A triangular-waveform generator as defined in Fig. 9.5-4(a) has an output waveform with a peak amplitude that is 70% of the op amp's output-saturation-voltage magnitude. The frequency of oscillation is 30 kHz when $C = 0.02\ \mu$F. What is R? (*Hint*: Use the results of Problem 9-30.)

DESIGN EXERCISES

D9-1. Design a dc power supply to provide 5V ± 0.3 V to a constant-resistance load of 100/3 Ω. A full-wave-rectifier design is to be made from available parts, which are (1) a 60-Hz power transformer with a 115-V-rms primary and a 16-V-rms secondary that has a center tap (8 V each side of the tap); the maximum power that can be extracted from the secondary is 2.2 W; (2) an inductance (choke) of 0.1 H with a series resistance of 50 Ω and a maximum current capacity of 250 mA; (3) two capacitors, each 2200 μF with a 16-V-dc voltage rating; (4) four power-rectifier diodes that can be approximated as ideal; and (5) resistors in standard values for 5% resistance tolerance and various power ratings. Determine the best design structure and then evaluate the supply for ripple, voltage, and current at the load.

D9-2. Design a practical Colpitts oscillator having the form of Fig. D9-2. The diode D with C_3, R_1, and C_S form a half-wave rectifier with a filter to provide negative feedback to the amplifier to limit the amplitude of oscillations. Resistor R_S establishes the desired operating point. Find the required component values that will cause the output voltage v_o to have a peak amplitude of 3 V into a high impedance and a frequency of 5 MHz. Assume the JFET has the characteristics of Fig. 6.6-3, with $V_A = 100$ V.

FIG. D9-2.

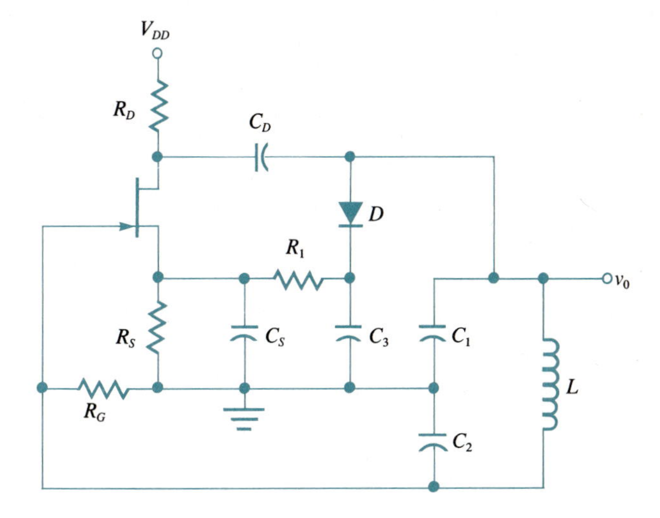

PART 3

Computers and Digital Devices

CHAPTER 10

Digital Fundamentals

10.0 INTRODUCTION

Attempts at making devices that could perform a programmed sequence of operations can be traced back to a few centuries ago. However, Charles Babbage, a brilliant English scientist and mathematician, is credited with the soundest early attempt around 1830. He established the basic fundamentals on which modern computers are built.

Digital computers hold an important role in our modern society. They have contributed to many commerical, industrial, and scientific developments. Scientists and engineers use mathematics as a language for modeling physical laws and properties. The exact algorithm for solving a specific problem has often been found, but the time required to manually perform the necessary steps and calculations was prohibitive [can you imagine manually computing a discrete Fourier transform (DFT) with $N = 1024$ points?]. Not only is the digital computer a valuable tool to perform these calculations at high speeds, but it can also compute with great accuracy. Computers are used in medical treatment, weather prediction, space exploration, and numerous other fields. Until quite recently, diverse activities such as manufacturing, business office operations, marketing, and home entertainment were conducted without the benefit of digital computers. This is no longer the case. Computer-controlled robots are becoming the work force in manufacturing. Typewriters have been replaced by word processors, and business decisions are based on computer analysis of data. Stereo receivers are digital, and television sets are remotely controlled.

The general-purpose digital computer is the best-known example of a digital system. Other examples include teletypewriters, dial-telephone switching exchanges, digital voltmeters, frequency counters, and other peripheral equipment. A characteristic of a digital system is its manipulation of finite, discrete elements of information. In digital systems, information is represented by signals (voltages or currents) that take on a limited number of discrete values and are processed by devices that normally function only in a limited number of discrete states. The lack of such practical devices capable of functioning *reliably* in more than two discrete states has resulted in the great majority of present digital devices being binary—that is, having signals and states limited to two values. Transistor circuitry that is either on or off has two possible signal values and can be constructed with extreme reliability. Transistors have revolutionized the computer industry.

The vast majority of present digital computers use the binary system, which has two binary digits (*bits*), 0 and 1. Internal representation of information in a digital computer is in groups of bits. Groups of bits can be used to represent numbers and symbols with special coding techniques. Hence, digital computers are used to manipulate and store numeric as well as symbolic data.

Figure 10.0-1 shows the interconnections of some digital modules that constitute the basic elements of a digital computer. The central processing unit (CPU), which consists of the arithmetic and logic unit (ALU) and the control unit, is the brain of the whole digital computer. It supervises the flow of information and the sequences of operations and performs the actual arithmetic and logic operations. The memory unit is used to store program codes, input/output, and intermediate data.

This chapter and the next three chapters are intended to present a basic foundation of digital systems that enables the reader to understand the operation of each module of a digital computer as illustrated in Fig. 10.0-1. The remainder of this chapter deals with the coding of data, basic gates, and the design of some of the components of the arithmetic logic unit.

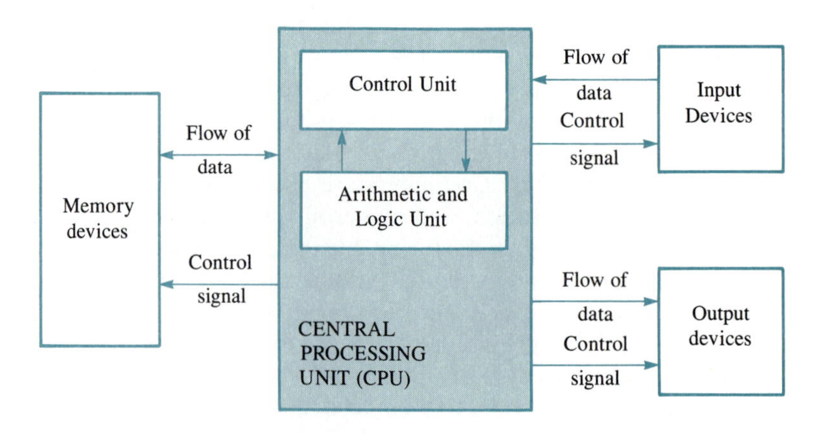

FIG. 10.0-1. The block diagram of a digital computer.

10.1 NUMBER SYSTEMS

In general, a number system is an ordered set of symbols (digits) with relationships defined for addition, subtraction, multiplication, and division. The base (radix) of the number system is the total number of digits in the system. For example, in our decimal system, the set of digits is {0, 1, 2, 3, 4, 5, 6, 7, 8, 9} and hence the base, or radix, is ten (10); in the binary system, the set of digits (bits) is {0, 1} and hence the base, or radix, is two (2).

There are two possible ways of writing a number in a given system: *positional notation* and the *polynomial representation.*

Positional Notation

A number N can be written in positional notation as follows:

$$N = (b_{n-1}b_{n-2} \cdots b_2b_1b_0 \cdot b_{-1}b_{-2} \cdots b_{-m})_r \tag{10.1-1}$$

where

$$\begin{aligned} r &= \text{base or radix of the system} \\ b_i &= i\text{th bit (digit)} \\ b_{n-1} &= \text{most significant bit (digit) MSB} \\ b_{-m} &= \text{least significant bit (digit) LSB} \\ n &= \text{number of integer bits (digits)} \\ m &= \text{number of fractional bits (digits)} \end{aligned}$$

and $0 \leq b_i \leq r - 1$ for all i, $-m \leq i \leq n - 1$.

Polynomial Representation

A number N can be written as a polynomial of the form

$$\begin{aligned} N &= b_{n-1}r^{n-1} + b_{n-2}r^{n-2} + \cdots + b_1r^1 + b_0r^0 + b_{-1}r^{-1} + \cdots + b_{-m}r^{-m} \\ &= \sum_{i=-m}^{n-1} b_ir^i \end{aligned} \tag{10.1-2}$$

where b_i, b_{-m}, b_{n-1}, m, n, and r are as defined before.

EXAMPLE 10.1-1

In positional and polynomial notation, the number $2536.47 is written as follows:

positional polynomial

$$(2536.47)_{10} = 2 \times 10^3 + 5 \times 10^2 + 3 \times 10^1 + 6 \times 10^0 + 4 \times 10^{-1} + 7 \times 10^{-2}$$

Hence, for the polynomial representation, $r = 10$, and

$$b_3 = 2 \qquad \text{(most significant digit)}$$

$$b_2 = 5 \qquad b_1 = 3 \qquad b_0 = 6 \qquad b_{-1} = 4$$

$$b_{-2} = 7 \qquad \text{(least significant digit)}$$

In the digital world, the most commonly used number systems are binary (base 2), octal (base 8), and hexadecimal (base 16).

Binary-Number System

The binary number is a base (radix) 2 system with two distinct digits (bits), 1 and 0. It is expressed as a string of 0s and 1s and a binary point, if a fraction exists. To convert from the binary to decimal system, express the binary number in the polynomial form and evaluate this polynomial by using decimal-system addition.

EXAMPLE 10.1-2

Convert the following binary numbers into their equivalent decimal numbers: (a) $(101101)_2$, (b) $(.101)_2$, and (c) $(101101.101)_2$.

Solution

(a)
$$(101101)_2 = 1 \times 2^5 + 0 \times 2^4 + 1 \times 2^3 + 1 \times 2^2 + 0 \times 2^1 + 1 \times 2^0$$
$$= 32 + 0 + 8 + 4 + 0 + 1$$
$$= (45)_{10}$$

(b)
$$(.101)_2 = 1 \times 2^{-1} + 0 \times 2^{-2} + 1 \times 2^{-3}$$
$$= .5 + 0 + .125$$
$$= (.625)_{10}$$

(c) By using the results obtained in parts (a) and (b), we get

$$(101101.101)_2 = (45.625)_{10}$$

It is often necessary to convert a decimal number into its equivalent binary number for further processing by the digital world. To accomplish this conversion, repeatedly *divide* the integer part of the decimal number by 2; the remainder after each division is used to form the equivalent binary number. This process is continued until a zero quotient is obtained. The binary number is formed by using the remainder after each division, with the first remainder being the least significant bit of the binary number to be formed.

EXAMPLE 10.1-3

Convert the decimal number $(75)_{10}$ into its binary equivalent.

Solution

	quotient	remainder
	$75 \div 2 = 37$	1 (LSB)
	$37 \div 2 = 18$	1
	$18 \div 2 = 9$	0
	$9 \div 2 = 4$	1
	$4 \div 2 = 2$	0
	$2 \div 2 = 1$	0
	$1 \div 2 = 0$	1 (MSB)

stop

Therefore, the binary number is

$$(1001011)_2$$

(MSB) (LSB)

The reader is encouraged to verify this answer by converting it to a decimal number, as previously outlined.

The conversion of a fractional decimal number into its equivalent binary number is obtained by repeatedly *multiplying* the decimal fraction by 2. If a 0 or 1 appears to the left of the decimal point of the product as a result of this multiplication, then a 0 or 1 is added to the binary fraction. This process is continued until the fractional part of the product is zero or the desired number of binary bits is reached.

EXAMPLE 10.1-4

Convert the fractional decimal number $(.4375)_{10}$ into its binary equivalent.

Solution

MSB of binary fraction

$$.4375 \times 2 = 0.8750$$
$$.8750 \times 2 = 1.7500$$
$$.7500 \times 2 = 1.5000$$
$$.5000 \times 2 = 1.0000$$

stop

LSB of binary fraction

Therefore,

$$(.4375)_{10} = (.0111)_2$$

Octal-Number System

Even though most computers and digital systems use the binary system, octal as well as hexadecimal (to be discussed next) numbers are useful tools for representing binary data. The octal-number system is a base 8 system and therefore has 8 distinct digits {0, 1, 2, 3, 4, 5, 6, 7}. It is expressed as a string of any combination of the 8 digits. The numbers $(6105)_8$, $(1010)_8$, $(347.6)_8$ are all valid octal numbers; however, the number $(648.2)_8$ is not a valid octal number. (Why?) To convert from octal to decimal, we follow the same procedure for converting from binary to decimal; that is, express the octal number in its polynomial form and evaluate this polynomial by using decimal-system addition.

EXAMPLE 10.1-5

Convert the following numbers into their equivalent decimal numbers: (a) $(367)_8$, (b) $(.240)_8$, and (c) $(367.240)_8$.

Solution

(a) $(367)_8 = 3 \times 8^2 + 6 \times 8^1 + 7 \times 8^0$

$= 192 + 48 + 7$

$= (247)_{10}$

(b) $(.240)_8 = 2 \times 8^{-1} + 4 \times 8^{-2} + 0 \times 8^{-3}$

$$= .250 + .0625 + 0$$

$$= (.3125)_{10}$$

(c) By using the results obtained in parts (a) and (b), we get

$$(367.240)_8 = (247.3125)_{10}$$

When conversion from decimal to octal is needed, one would apply the same procedure for conversion from decimal to binary; but instead of dividing by 2 for the integer part, we divide by 8 to obtain the octal equivalent. Also, instead of multiplying by 2 for the fractional part, we multiply by 8 to obtain the fractional octal equivalent of the decimal fraction. However, it is more common in the digital world to convert from binary to octal and vice versa.

The conversion from binary to octal is accomplished by grouping the binary number into groups of 3 bits each, starting from the binary point and proceeding to the right and to the left; we then replace each group by its octal equivalent.

EXAMPLE 10.1-6

Convert $(100101111011.01011)_2$ into its octal equivalent.

Solution

By grouping the bits into groups of 3 bits from the binary point, we obtain

$$100\ 101\ 111\ 011\ .\ 010\ 110$$

Observe that we needed to add a trailing 0 to complete the last group in the fractional part. Now, replacing each group by its octal "decimal" equivalent, we obtain

$$(100101111011.01011)_2 = (4573.26)_8$$

The conversion from octal to binary is accomplished by replacing each octal digit with its *3-bit* binary equivalent.

Hexadecimal-Number System

The internal structures of most present digital computers manipulate data in groups of 4-bit packets. Hence, a convenient method for representing binary data was developed by using the hexadecimal system. The hexadecimal-number system is a base 16 system and has 16 distinct symbols in the set: {0, 1, 2, 3, 4, 5, 6, 7, 8, 9, A, B, C, D, E, F}, where A is equivalent to decimal 10, B to 11, . . . , and F to 15. A hexadecimal number is expressed as a string of any combination of the 16 symbols. The numbers $(9A3.0B)_{16}$, $(1101.1)_{16}$, and $(0ABC.DEF)_{16}$ are all valid hexadecimal numbers. To convert from hexadecimal to decimal and vice versa, we follow the same procedure for conversion between decimal and octal, except that we now use 16 instead of 8.

EXAMPLE 10.1-7

Convert the following numbers into their equivalent decimal numbers: (a) $(2AB)_{16}$, (b) $(.F8)_{16}$, and (c) $(2AB.F8)_{16}$.

Solution

(a)
$$\begin{aligned}(2AB)_{16} &= 2 \times 16^2 + A \times 16^1 + B \times 16^0 \\ &= 2 \times 16^2 + 10 \times 16^1 + 11 \times 16^0 \\ &= (683)_{10}\end{aligned}$$

(b)
$$\begin{aligned}(.F8)_{16} &= F \times 16^{-1} + 8 \times 16^{-2} \\ &= 15 \times 16^{-1} + 8 \times 16^{-2} \\ &= (.96875)_{10}\end{aligned}$$

(c) By using the results obtained in parts (a) and (b), we get

$$(2AB.F8)_{16} = (683.96875)_{10}$$

The conversion from binary to hexadecimal is accomplished, as you might have expected, by grouping the binary number into groups of 4 bits each, starting from the binary point and proceeding to the right and to the left; we then replace each group by its hexadecimal equivalent.

EXAMPLE 10.1-8

Convert $(11101110100100.100111)_2$ into its equivalent hexadecimal number.

Solution

By grouping the bits into groups of 4 bits from the binary point and replacing each group with its hexadecimal equivalent, we obtain

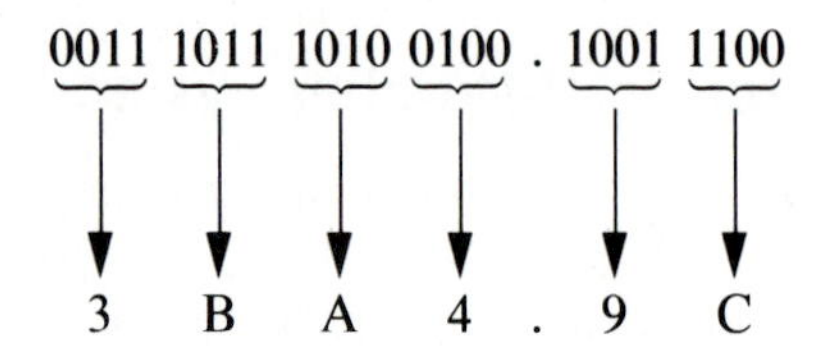

Hence,

$$(11101110100100.100111)_2 = (3BA4.9C)_{16}$$

The conversion from hexadecimal to binary is achieved by reversing the previous process.

Binary-Coded-Decimal (BCD) Numbers

Since people, who are comfortable with the decimal system, are the eventual users of digital computers and devices that manipulate data using the binary system, there is a need for decimal-to-binary conversion at the input of the digital device and from binary-to-decimal at the output of this device. The *binary-coded decimal* is a convenient code for resolving this interface problem. In a binary-coded decimal number each of the decimal digits is coded in binary, using 4 bits.

EXAMPLE 10.1-9

Convert $(163.25)^{10}$ into (a) binary and (b) BCD.

Solution

(a) $(163.25)_{10} = (10100011.01)_2$

(b) $(163.25)_{10} = (0001\ 0110\ 0011\ .\ 0010\ 0101)_{BCD}$

Table 10.1-1 lists the first 20 numbers in the decimal, binary, octal, hexadecimal, and BCD systems.

10.2 ARITHMETIC OPERATIONS IN A BINARY SYSTEM

Arithmetic operations in a binary system follow the same rules as in the decimal system, except that we use only the two allowable digits (0 and 1) and use base 2 for all computations.

Binary Addition

The basic rules for binary addition are as follows:

	(a)	(b)	(c)	(d)
Augend:	0	0	1	1
+ Addend:	+ 0	+ 1	+ 0	+ 1
= Sum:	= 0	= 1	= 1	= 10

carryover

In rule (d), we say that we have a sum of 0 and a carryover of 1.

TABLE 10.1-1. Number Representation in Different Systems

Base 10 (Decimal)	Base 2 (Binary)	Base 8 (Octal)	Base 16 (Hexadecimal)	BCD
00	00000	00	00	0000 0000
01	00001	01	01	0000 0001
02	00010	02	02	0000 0010
03	00011	03	03	0000 0011
04	00100	04	04	0000 0100
05	00101	05	05	0000 0101
06	00110	06	06	0000 0110
07	00111	07	07	0000 0111
08	01000	10	08	0000 1000
09	01001	11	09	0000 1001
10	01010	12	0A	0001 0000
11	01011	13	0B	0001 0001
12	01100	14	0C	0001 0010
13	01101	15	0D	0001 0011
14	01110	16	0E	0001 0100
15	01111	17	0F	0001 0101
16	10000	20	10	0001 0110
17	10001	21	11	0001 0111
18	10010	22	12	0001 1000
19	10011	23	13	0001 1001

EXAMPLE 10.2-1

Perform the following binary additions: (a) $(0101100)_2 + (0101010)_2$ and (b) $(0110.110)_2 + (0110.011)_2$.

Solution

(a)		(b)
1 0 1 0 0 0	carryover	1 1 0 1 1 0
0 1 0 1 1 0 0	augend	0 1 1 0 . 1 1 0
+ 0 1 0 1 0 1 0	addend	+ 0 1 1 0 . 0 1 1
= 1 0 1 0 1 1 0	sum	= 1 1 0 1 . 0 0 1

The reader might check these answers by converting from binary to decimal and performing a decimal addition.

Binary Subtraction

The basic rules for binary subtraction are as follows:

	(a)	(b)	(c)	(d)
Minuend:	0	0	1	1
− Subtrahend:	− 0	− 1	− 0	− 1
= Difference:	= 0	= 11	= 1	= 0
		↑ borrow-in		

Observe that in rule (b), a borrow occurs, because the subtrahend is greater than the minuend. As in decimal-system subtraction, the borrow-in for one stage must be subtracted from the minuend of the next significant stage.

EXAMPLE 10.2-2

Perform the following binary subtractions: (a) $(101011)_2 - (010101)_2$ and (b) $(11100.011)_2 - (10011.101)_2$.

Solution

(a)		(b)
0 1 0 1 0 0	borrow-in	0 0 0 1 1 1 0 0
1 0 1 0 1 1	minuend	1 1 1 0 0 . 0 1 1
− 0 1 0 1 0 1	subtrahend	− 1 0 0 1 1 . 1 0 1
= 0 1 0 1 1 0	difference	= 0 1 0 0 0 . 1 1 0

An alternative method to binary subtraction is the use of two's complement, as discussed at the end of this section.

Binary Multiplication

Multiplication in the binary system is very simple, since the digits of the multiplier are always 0 or 1. The basic rules for binary multiplication are as follows:

	(a)	(b)	(c)	(d)
Multiplicand:	0	0	1	1
× Multiplier:	× 0	× 1	× 0	× 1
= Product:	= 0	= 0	= 0	= 1

EXAMPLE 10.2-3

Perform the following binary multiplications: (a) $(100101)_2 \times (110011)_2$ and (b) $(101.011)_2 \times (110.001)_2$.

Solution

(a)

```
          100101    multiplicand
        × 110011    multiplier
        --------
          100101
         100101
        000000
       000000
      100101
     100101
  ------------
 = 11101011111      product
```

(b)

```
        101.011
      × 110.001
      ---------
        101 011
       0000 00
      00000 0
     000000
    101011
   101011
 ---------------
 = 100000.111011
         ↑ binary point
```

Observe that the positioning of the binary point is the same as that for the decimal point in the decimal system.

Binary Division

Binary division is even simpler than decimal division. Division by zero is not allowed, and the basic rules are as follows:

	(a)	(b)
Dividend:	0	1
÷ Divisor:	÷ 1	÷ 1
= Quotient:	= 0	1

The division procedure in the binary system is the same as that for the decimal system.

EXAMPLE 10.2-4

Perform the following binary divisions: (a) $(100111)_2 \div (11)_2$, (b) $(011001)_2 \div (100)_2$, and (c) $(11000.11)_2 \div (101.1)_2$.

Solution

```
(a)          quotient          1101
       Divisor | dividend   11 | 100111
                              -11
                               11
                              -11
                               011
                               -11
                                00
```

$\rightarrow (100111)_2 \div (11)_2 = (1101)_2$

```
(b)       0110.01       (c)          10 1.1
     100 | 011001            100x1 | 11000x1.1
         -100                      -1001
          0100                      001101
          -100                       -1001
           000100                     01001
             -100                     -1001
              000                      0000
```

Thus far, all of our discussions and illustrations have been with unsigned numbers, which might be thought of as positive numbers. However, in the real world, one deals with negative numbers as much as positive numbers. But since in a digital system, there are no + or − signs to represent positive or negative numbers, 0s and 1s are used, respectively, to represent these signs in the leftmost digit of the signed number. Signed binary numbers can be represented, as we shall see in the subsequent sections, in three different forms. But first, we need to introduce the *complementary form* of a given number.

Complements

The subtraction operation in digital computers can be simplified by using the complements of numbers; that is, we write the negative representation of the subtrahend and then *add* it to the minuend to obtain the difference.

Therefore, instead of performing $A - B$ by using a subtractor, we will perform $A + (-B)$ to obtain the same result by using an adder. This ability to perform a subtraction by using an addition of the complement of the subtrahend is an important characteristic of the digital computer. The available hardware (i.e., the adder) can then be used for subtraction as well as addition, which reduces the overall hardware and complexity of the central processing unit of the digital computer.

For each base r system, there are two types of complements, namely, the *radix complement*, also known as the *r's complement*, and the *diminished radix complement*, also known as the *(r − 1)'s complement*.

Radix Complement (r's Complement). The radix complement, denoted by $[N]_r$, for an n-digit and r-base number $(N)_r$ is defined as follows:

$$[N]_r = \begin{cases} (r^n)_r - (N)_r & \text{for } N \neq 0 \\ 0 & \text{for } N = 0 \end{cases} \tag{10.2-1}$$

EXAMPLE 10.2-5

Obtain (a) the 10's complement of the decimal number $(235)_{10}$ and (b) the 2's complement of the binary number $(011010)_2$.

Solution

(a) $$[235]_{10} = 10^3 - 235 = 1000 - 235 = 765$$

(b) $$[011010]_2 = 2^6 - (011010)_2 = (1000000)_2 - (011010)_2 = (100110)_2$$

The *two's complement* of a binary number can be obtained directly from the given number by copying each bit of the number, starting at the least significant bit and proceeding toward the most significant bit until the *first* 1 has been copied. After the first 1 has been copied, replace each of the remaining 0s and 1s by 1s and 0s, respectively.

EXAMPLE 10.2-6

Obtain the 2's complement of the following binary numbers: (a) $(101101000)_2$, (b) $(010101101)_2$, and (c) $(01010.00100)_2$.

Solution

(a) Since $(N)_2 = (101101000)_2$

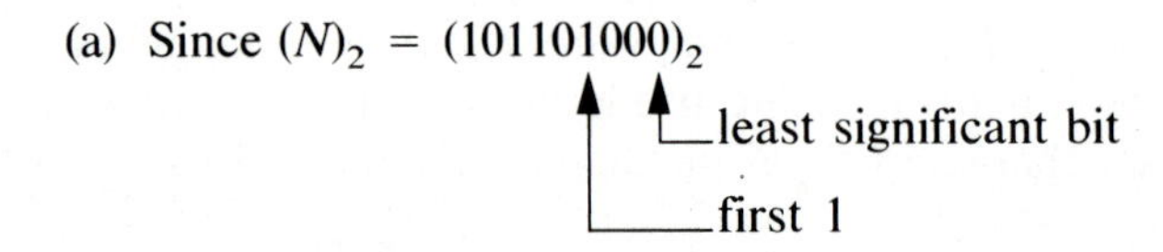

then

$$[N]_2 = (010011000)_2$$

(b) Since $(N)_2 = (010101101)_2$

↑ least significant bit and first 1

then

$$[N]_2 = (101010011)_2$$

(c) Since $(N)_2 = (01010.00100)_2$

↑ least significant bit

↑ first 1

then

$$[N]_2 = (10101.11100)_2$$

Diminished Radix Complement [(*r* − 1)'s Complement]. The diminished radix complement for an *n*-digit and *r*-base number $(N)_r$, denoted by $[N]_{r-1}$, is defined as

$$[N]_{r-1} = (r^n - 1)_r - (N)_r \qquad (10.2\text{-}2)$$

EXAMPLE 10.2-7

Obtain (a) the 9's complement of the decimal number $(235)_{10}$ and (b) the 1's complement of the binary number $(011010)_2$.

Solution

(a) $$\begin{aligned}[235]_9 &= (10^3 - 1) - 235 \\ &= 999 - 235 \\ &= 764\end{aligned}$$

(b) $$\begin{aligned}[011010]_1 &= (2^6 - 1) - (011010)_2 \\ &= (111111)_2 - (011010)_2 \\ &= (100101)_2\end{aligned}$$

Signed Binary Numbers

As mentioned earlier, the + and − of the signed binary numbers will be represented by a 0 and 1, respectively, in the leftmost bit of the signed number.

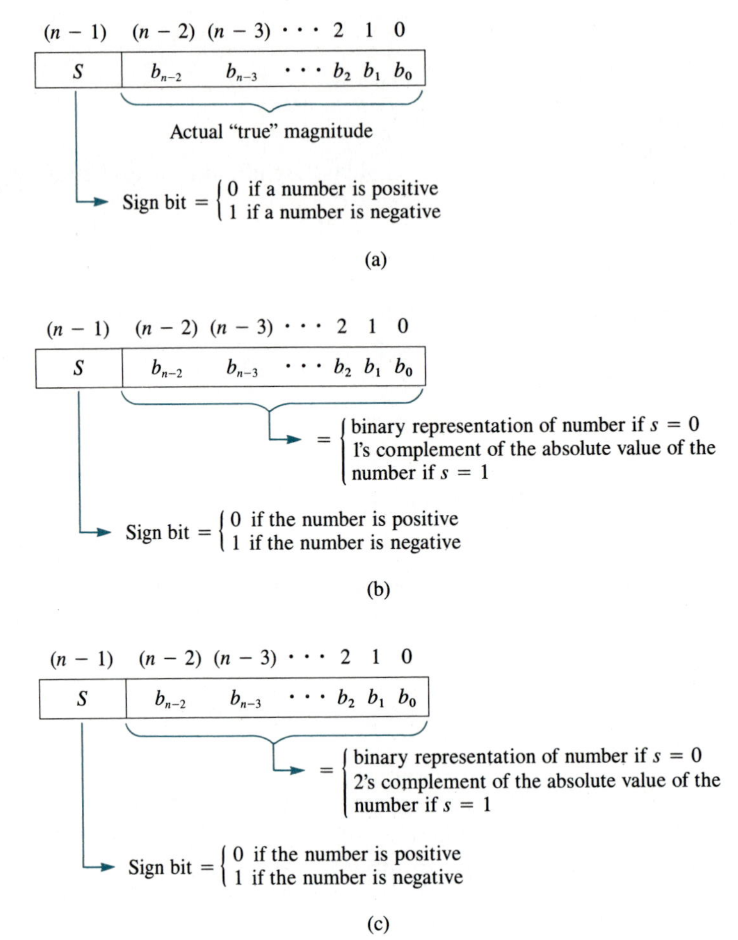

FIG. 10.2-1. Signed-binary-number formats: (a) sign magnitude, (b) one's complement, and (c) two's complement.

Observe that we did not say "the most significant bit" of the number. This leftmost bit will be called the sign bit. Hence, if we represent a signed binary number by using n bits, $(b_{n-1}b_{n-2}b_{n-3} \cdots b_2b_1b_0)$, then b_{n-1} will represent the sign of the number; and we are left with $(n - 1)$ bits to represent the actual magnitude of the signed number.

Three different methods of representing a signed number are introduced. Before we proceed to these discussions, it is important to realize that all numbers (signed and unsigned) are represented by a string of 0s and 1s in the computer. It is up to the user to treat the leftmost bit as a signed bit or as part of the magnitude.

Sign-Magnitude Format. In the sign-magnitude format, a signed binary number is represented by using n bits, as shown in Fig. 10.2-1(a).

EXAMPLE 10.2-8

Represent the following decimal numbers in their equivalent 8-bit sign-magnitude binary formats: (a) $(+47)_{10}$ and (b) $(-56)_{10}$.

Solution

(a) $(47)_{10} = (0101111)_2$ using 7 bits

so

$(+47)_{10} = 00101111$

↑ sign bit (positive)

(b) $(56)_{10} = (0111000)_2$ using 7 bits

so

$(-56)_{10} = 10111000$

↑ sign bit (negative)

One's-Complement Format. In 1's-complement format, an n-bit signed binary number is represented as shown in Fig. 10.2-1(b).

EXAMPLE 10.2-9

Repeat Example 10.2-8, but use 1's complement format.

Solution

(a) $(+47)_{10} = 00101111$

↑ true magnitude

↑ sign bit (positive)

(b) First, we obtain the 1's complement of the absolute value, as follows:

$$[0111000]_1 = (1000111)_2$$

Then we get

$(-56)_{10} = 11000111$

↑ 1's complement of the absolute value

↑ sign bit (negative)

Two's-Complement Format. In 2's-complement, an n-bit signed binary number is represented as shown in Fig. 10.2-1(c).

EXAMPLE 10.2-10

Repeat Example 10.2-8, but use 2's complement format.

Solution

(a) $(47)_{10} = 00101111$

- 0 — sign bit (positive)
- 0101111 — true magnitude

(b) First, we obtain the 2's complement of the absolute value:

$$[0111000]_2 = (1001000)_2$$

Then we get

$(-56)_{10} = 11001000$

- 1 — sign bit (negative)
- 1001000 — 2's complement of the absolute value

Table 10.2-1 lists the first ten positive and negative decimal integers in the three different forms, using 6-bit signed binary representation.

Signed-Binary-Number Arithmetic

The arithmetic operations of signed binary numbers in sign-magnitude format follow the same rules and procedures of decimal-system arithmetic. This procedure is not used in digital-computer arithmetic, since we must compare the signs of the two operands in order to determine the sign of the result. However, the 1's- and 2's-complement formats for signed-number representations offer a convenient method for basic signed-binary-number arithmetic. Most of today's digital computers use the 2's-complement system for integer arithmetic.

Two's-Complement Arithmetic

Addition. For two n-bit signed binary numbers in 2's-complement format, their addition is obtained by simply performing a binary addition on these two numbers, including the sign bits. If a carryover results from the leftmost bit, this carryover is discarded. The leftmost bit of the result will reflect the sign of the sum.

TABLE 10.2-1. Representation of Signed Binary Numbers Using 6 Bits

Signed Decimal Number	Format		
	Sign-Magnitude	1's Complement	2's Complement
+0	000000	000000	000000
−0	100000	111111	—
+1	000001	000001	000001
−1	100001	111110	111111
+2	000010	000010	000010
−2	100010	111101	111110
+3	000011	000011	000011
−3	100011	111100	111101
+4	000100	000100	000100
−4	100100	111011	111100
+5	000101	000101	000101
−5	100101	111010	111011
+6	000110	000110	000110
−6	100110	111001	111010
+7	000111	000111	000111
−7	100111	111000	111001
+8	001000	001000	001000
−8	101000	110111	111000
+9	001001	001001	001001
−9	101001	110110	110111

EXAMPLE 10.2-11

Perform the following additions by using 2's-complement arithmetic: (a) (13.75) + (−7.5) and (b) (−13.75) + (7.5).

Solution

(a) $(13.75)_{10} = (001101.11)_2$

$(7.5)_{10} = (000111.10)_2$

$$\begin{array}{rcr} (+13.75)_{10} & \rightarrow & (001101.11)_2 \\ +\ (-7.5)_{10} & \rightarrow & +(111000.10)_2 \\ \hline =\ (+6.25)_{10} & & =(1\ 000110.01)_2 \end{array} \rightarrow \text{Answer} = (000110.01)_2$$

(first 1 of the result: carryover, which must be dropped; following bit: sign bit)

(b)
$$\begin{array}{llll} & (-13.75)_{10} \rightarrow & (1\ 10010.01)_2 & \\ + & (+7.50)_{10} \rightarrow & +(0\ 00111.10)_2 & \\ \hline = & (-6.25)_{10} & =(1\ 11001.11)_2 \rightarrow & \text{Answer} = (111001.11)_2 \end{array}$$

↑ sign bit (negative)

Observe that to get the real magnitude of the final answer, we must take the 2's-complement of the direct-addition result if its sign bit is 1.

Subtraction. Two's-complement subtraction of two signed binary numbers can be achieved by simply performing an addition operation. This procedure arises from the fact that $A - B = A + (-B)$. Hence, subtraction in 2's-complement format of two signed numbers can be achieved by adding the 2's-complement of the subtrahend to the minuend. If a carryover results from the leftmost bit, this carryover is discarded. Also, the leftmost bit of the result will reflect the sign of the difference.

EXAMPLE 10.2-12

Perform the following subtractions by using 2's-complement arithmetic: (a) 52 − 39 and (b) 9.25 − 14.5.

Solution

(a)
$$\begin{array}{ll} (52)_{10} = (00110100)_2 & \rightarrow \\ (39)_{10} = (00100111)_2 & \end{array} \quad \begin{array}{r} (52)_{10} \\ -(39)_{10} \\ \hline =(13)_{10} \end{array} \rightarrow \quad \begin{array}{r} (00110100)_2 \\ +(11011001)_2 \\ \hline =(100001101)_2 \end{array}$$

↑ sign bit (positive)

↑ carryover

(b)
$$\begin{array}{ll} (9.25)_{10} = (001001.01)_2 & \rightarrow \\ (14.5)_{10} = (001110.10)_2 & \end{array} \quad \begin{array}{r} (9.25)_{10} \\ -\ (14.5)_{10} \\ \hline =(-5.25)_{10} \end{array} \rightarrow \quad \begin{array}{r} (001001.01)_2 \\ +(110001.10)_2 \\ \hline =(111010.11)_2 \end{array}$$

↑ sign bit (negative)

One's-complement arithmetic is very similar to 2's-complement arithmetic, except that we use the 1's-complement format for representing the signed numbers. But the carryover from the leftmost bit must be added to the rightmost bit of the result.

Overflow

Overflow is the condition under which the sum of two n-bit numbers requires $(n + 1)$ bits for its representation. This is a problem in digital computers, because the devices, known as *registers*, that are used to represent numbers have a finite number of digits. Hence, an n-bit device (register) cannot accommodate an $(n + 1)$-bit number. Therefore, most digital computers have a mechanism for raising a *flag*, indicating that a problem may exist, when an overflow is detected.

For an unsigned binary number where all the bits represent magnitude, the occurrence of an overflow after performing an addition is very easily detected by looking at the carryover of the most significant bit. The presence of a carryover indicates an overflow.

For signed binary numbers where the leftmost bit represents the sign of the number, overflow will not occur after performing an addition when the signs of the two numbers are different. (Why?) However, an overflow *may* occur when the two numbers have the same sign bits. For example, the addition of the two negative numbers $(10101111)_2$ and $(10001010)_2$ represented in 2's-complement format using 8 bits will produce an overflow; similarly, the positive numbers $(01011100)_2$ and $(01101111)_2$ will produce an overflow as a result of an addition operation.

Let us carry out the addition of these numbers and observe the carryin to the leftmost bit and the carryout of the leftmost bit, as shown below:

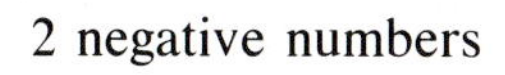

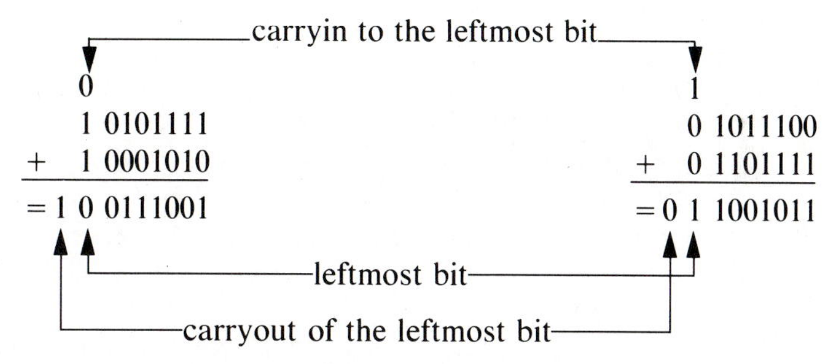

A quick observation will reveal that an overflow will occur when the carryin and carryout of the leftmost bit, the sign bit, are different. This condition can be easily tested by using a basic hardware device, as will be shown in Sec. 10.8, namely, an exclusive-OR gate.

10.3 ALPHANUMERIC CODES

When digital devices are to process information other than numerical data, alphanumeric codes are used. These codes are used to represent all letters, decimal digits, and special characters. A 7-bit binary code is the standard code used to

TABLE 10.3-1. ASCII (American Standard Code for Information Interchange)

	Most Significant Bits							
LSBs	**000**	**001**	**010**	**011**	**100**	**101**	**110**	**111**
0000	NULL	DLE	SP	0	@	P		p
0001	SOH	DC1	!	1	A	Q	a	q
0010	STX	DC2	"	2	B	R	b	r
0011	ETX	DC3	#	3	C	S	c	s
0100	EOT	DC4	$	4	D	T	d	t
0101	ENQ	NAK	%	5	E	U	e	u
0110	ACK	SYN	&	6	F	V	f	v
0111	BEL	ETB	'	7	G	W	g	w
1000	BS	CAN	(	8	H	X	h	x
1001	HT	EM	)	9	I	Y	i	y
1010	LF	SUB	*	:	J	Z	j	z
1011	VT	ESC	+	;	K	[	k	{
1100	FF	FS	,	<	L	\	l	|
1101	CR	GS	-	=	M	]	m	}
1110	SO	RS	.	>	N	∧	n	~
1111	SI	US	/	?	O	_	o	DEL

represent both upper- and lowercase letters, 10 decimal digits, and a number of special printable and nonprintable characters. This standard binary code for the alphanumeric characters is ASCII (American Standard Code for Information Interchange). Table 10.3-1 shows the 128 characters that can be coded by using ASCII. For example, the ASCII for the decimal digit 1 is 0110001 and for the letter T is 1010100.

A quick look at Table 10.3-1 reveals that the ASCII contains 34 nonprintable control characters. These control characters are represented in the ASCII table with their abbreviated names. They are used for formatting the printed text and for communication control when information is transmitted between remote terminals. Table 10.3-2 lists the abbreviated control characters and their functions. For example, the control character STX (start of text) is used to indicate the beginning of a text; ETX (end of text) is used to indicate the end of a text when it is transmitted over phone lines.

Another alphanumeric code used mostly by IBM in some of its early computer systems is the EBCDIC (Extended Binary-Coded-Decimal Interchange Code). In this code, 8 bits are used to represent the same character symbols as in ASCII; however, the bit assignment for these characters is different.

TABLE 10.3-2. Abbreviated Control Characters and Their Functional Name

NULL	Null	DLE	Data link escape
SOH	Start of heading	DCI	Device control 1
STX	Start of text	DC2	Device control 2
ETX	End of text	DC3	Device control 3
EOT	End of transmission	DC4	Device control 4
ENQ	Enquiry	NAK	Negative acknowledge
ACK	Acknowledge	SYN	Synchronous idle
BEL	Bell	ETB	End transmission block
BS	Backspace	CAN	Cancel
HT	Horizontal tab	EM	End of medium
LF	Line feed	SUB	Substitute
VT	Vertical tab	ESC	Escape
FF	Form feed	FS	File separator
CR	Carriage return	GS	Group separator
SO	Shift out	RS	Record separator
SI	Shift in	US	Unit separator
SP	Space	DEL	Delete

10.4 BOOLEAN ALGEBRA

Before introducing the different digital circuits of which a digital computer and other digital systems are constructed, we first present the basic mathematics required for the description of these circuits. This mathematical system is *boolean† algebra*. Boolean algebra is a mathematical system used to describe the different interconnections of digital circuits. The variables used in this algebra are known as boolean variables. They are binary variables that can assume only one of two distinct values (true or false).

It is often convenient to assign 1s and 0s to these variables. These variables are designated by any letter of the alphabet, such as A, B, C, etc., or any combination of the letters of the alphabet, like TEST, OUTPUT, etc. A boolean variable can also be a function of some other boolean variables. Boolean algebra has three fundamental logical operations: AND, OR, and NOT operations.

† Named for George Boole (1815–1864), a brilliant English mathematician whose formal education ended in the third grade.

Logical AND Operation

The AND operation is represented by the symbol · or by the absence of an operator. For example, $F = A \cdot B$ and $F = AB$ are two different ways of saying "F is equal to A AND B." The result of an AND operation is 1 if and only if the values of all its arguments are 1s. The possible binary values for the logical AND operation are as follows:

$$0 \cdot 0 = 0 \qquad 1 \cdot 0 = 0$$

$$0 \cdot 1 = 0 \qquad 1 \cdot 1 = 1$$

As an illustration of the AND operation, let us consider the circuit containing two switches A and B and a light bulb F that is connected to a power source, as in Fig. 10.4-1. Since the switches A and B can be only in one of two distinct positions (ON or OFF), A and B can be considered as boolean variables. Similarly, F is a boolean function since the light can be either ON or OFF. Now, if we assign 1s to the variables A and B when their corresponding switches are in the ON positions and a 1 to F when the light is ON, then it is clear that $F = 1$ (light is ON) if and only if $A = B = 1$ (switches are in the ON position). Also, if either A or B is 0 (switch is in the OFF position), then F is 0 (light is OFF), since both switches must be in the ON positions for the light to turn on.

The listing of all possible values of the arguments of a boolean function with its corresponding output value is known as the *truth table* of this function. Hence, the truth table of $F = A \cdot B$ is as follows:

Inputs		Output
A	B	$F = A \cdot B$
0	0	0
0	1	0
1	0	0
1	1	1

Since each argument (variable) of a boolean function can assume one of two distinct values (0 or 1), there are 2^n entries in the truth table of an n-variable function. The listing of all possible values of the arguments of a boolean function is easily obtained by first assigning all 0s to the first row of this table. Then,

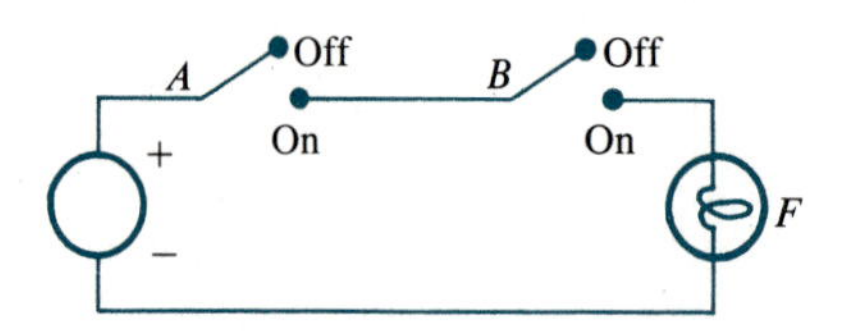

FIG. 10.4-1. An illustration of the AND operation.

alternate the 0s and 1s for the rightmost column of the arguments. In the next column, alternate the 0s and 1s every two rows, and alternate them every four rows for the third column; if there is a fourth column, the 0s and 1s would be alternated every eight rows; and so on. Once all the possible values of the arguments are listed, then the boolean function is evaluated for each possible entry of the table.

EXAMPLE 10.4-1. Prepare a truth table for $F = A \cdot B \cdot C$.

Solution

Since the number of arguments is 3 (A, B, and C), there are 2^3 possible entries.

Inputs	Output
A B C	$F = A \cdot B \cdot C$
0 0 0	0
0 0 1	0
0 1 0	0
0 1 1	0
1 0 0	0
1 0 1	0
1 1 0	0
1 1 1	1

Logical OR Operation

The OR operation is represented by the symbol +. For example, $F = A + B$ is read as "F is equal to A OR B or both." The result of an OR operation is 1 if the value of *any* of its arguments is 1. The truth table for the OR operation is as follows:

Inputs	Output
A B	$F = A + B$
0 0	0
0 1	1
1 0	1
1 1	1

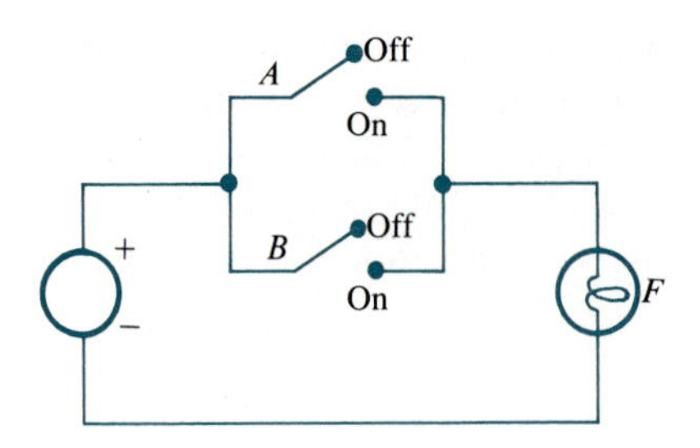

FIG. 10.4-2. An illustration of the OR operation.

Figure 10.4-2 is an illustration of the OR operation. Using the same analogy as for the circuit of Fig. 10.4-1, the reader should be able to see that $F = 1$ (light is ON) if either A or B (or both) is equal to 1 (switch is ON).

Logical NOT Operation

The NOT operation is represented by a bar over the variable. For example, $F = \overline{A}$ is read as "F equals NOT A." This operation is also known as the *complement* operation. The truth table for the NOT operation is as follows:

Input	Output
A	$F = \overline{A}$
0	1
1	0

We can take the complement of more than one variable. Thus the complement of $(A + B)$ is $\overline{(A + B)}$ and of $(A + B \cdot C)$ is $\overline{(A + B \cdot C)}$.

Unlike the OR and AND operations, which are binary operations that always require two operands, the NOT operation is a *unary* operation that requires only one operand. Observe that $\overline{\overline{0}} = \overline{1} = 0$; similarly, $\overline{\overline{1}} = \overline{0} = 1$. Therefore, a useful rule is that $\overline{\overline{A}} = A$.

The NOT operation has the highest priority, followed by the AND and the OR operations, in that order. For example, to evaluate the function $F = A \cdot \overline{B} + C$ when $A = 1$, $B = 1$, and $C = 0$, we first evaluate $\overline{B} = \overline{1} = 0$; second, we evaluate $A \cdot \overline{B} = 1 \cdot 0 = 0$; and finally, we evaluate $A \cdot \overline{B} + C = 0 + 0 = 0$. Therefore, F is equal to 0 when A equals 1, B equals 1, and C equals 0.

EXAMPLE 10.4-2

Evaluate $F = (\overline{A} \cdot B + \overline{C}) \cdot (A \cdot \overline{B} + C)$ for (a) $A = 0$, $B = 1$, and $C = 0$ and (b) $A = 1$, $B = 0$, and $C = 0$.

Solution

(a) $F = (\overline{0} \cdot 1 + \overline{0}) \cdot (0 \cdot \overline{1} + 0) = (1 \cdot 1 + 1) \cdot (0 \cdot 0 + 0)$
$= (1 + 1) \cdot (0 + 0) = (1) \cdot (0) = 0$

(b) $F = (\overline{1} \cdot 0 + \overline{0}) \cdot (1 \cdot \overline{0} + 0) = (0 \cdot 0 + 1) \cdot (1 \cdot 1 + 0)$
$= (0 + 1) \cdot (1 + 0) = (1) \cdot (1) = 1$

EXAMPLE 10.4-3

Prepare a truth table for $F = A \cdot \overline{B} + B \cdot C$.

Solution

After generating the eight (2^3) possible combinations for A, B, C, as shown in Table 10.4-1, we add a fourth column for $\overline{B}$, which corresponds to the complement of B. Two columns are now added listing the values for $A \cdot \overline{B}$ and $B \cdot C$ for each value of A, B, and C. Finally, we OR the values in the columns for $A \cdot \overline{B}$ and $B \cdot C$ to obtain the column that corresponds to $F = A \cdot \overline{B} + B \cdot C$. The result is shown in the last column.

Boolean Identities

George Boole's original algebra is a closed algebraic system that consists of a set of two elements, 0 and 1, an OR operation +, an AND operation ·, and a complement (NOT) operation $\overline{\ }$. Boolean algebra is used in digital design to reduce any logical function to its simplest form. Since boolean functions are eventually to be realized by using hardware devices, boolean algebra is a valuable tool for the reduction of the hardware.

TABLE 10.4-1. Truth Table for the Function $F = A \cdot \overline{B} + B \cdot C$

A	B	C	$\overline{B}$	$A \cdot \overline{B}$	$B \cdot C$	$F = A \cdot \overline{B} + B \cdot C$
0	0	0	1	0	0	0
0	0	1	1	0	0	0
0	1	0	0	0	0	0
0	1	1	0	0	1	1
1	0	0	1	1	0	1
1	0	1	1	1	0	1
1	1	0	0	0	0	0
1	1	1	0	0	1	1

TABLE 10.4-2. Basic Boolean Identities

Identity	Comments
1. $X + 0 = X$	
2. $X + 1 = 1$	
3. $X + X = X$	
4. $X + \overline{X} = 1$	
5. $X \cdot 0 = 0$	Identities 1–9 are basic to boolean algebra
6. $X \cdot 1 = X$	
7. $X \cdot X = X$	
8. $X \cdot \overline{X} = 0$	
9. $\overline{\overline{X}} = X$	
10. $X + Y = Y + X$	Commutative
11. $X \cdot Y = Y \cdot X$	Commutative
12. $X + (Y + Z) = (X + Y) + Z$	Associative
13. $X \cdot (Y \cdot Z) = (X \cdot Y) \cdot Z$	Associative
14. $X \cdot (Y + Z) = X \cdot Y + X \cdot Z$	Distributive
15. $X + Y \cdot Z = (X + Y) \cdot (X + Z)$	
16. $X + X \cdot Y = X$	Absorption
17. $X \cdot (X + Y) = X$	
18. $X \cdot Y + \overline{X} \cdot Z + Y \cdot Z = X \cdot Y + \overline{X} \cdot Z$	Consensus
19. $\overline{X + Y + Z} = \overline{X} \cdot \overline{Y} \cdot \overline{Z}$	De Morgan
20. $\overline{X \cdot Y \cdot Z} = \overline{X} + \overline{Y} + \overline{Z}$	De Morgan

Table 10.4-2 contains the most basic identities of boolean algebra. The first nine identities are the fundamental relations of boolean algebra and provide the basic foundation for the manipulation of boolean expressions. Identities 10 through 14 are very similar to the laws of ordinary algebra. Identities 10 and 11 are the commutative rules, 12 and 13 are the associative rules, and 14 is the distributive rule of boolean algebra. Identities 15 through 18 do not apply to ordinary algebra but are very useful in boolean algebra. Identity 16 is known as the *absorption identity*; identity 18 is known as the *consensus identity*. Finally, the last two identities are known as De Morgan's† rules.

Each of the 20 rules in Table 10.4-2 may be easily verified by substituting all possible values for the boolean variables and evaluating the left- and right-hand sides of each identity. This procedure is called a *proof by perfect induction*. To prove identity 16, for example, one would evaluate the left- and the right-hand

† Named for Augustus De Morgan (1806–1871), a brilliant English mathematician and logician.

TABLE 10.4-3. Proof by Induction for $X + X \cdot Y = X$

X	Y	$X \cdot Y$	Left-Hand Side $X + X \cdot Y$	Right-Hand Side X
0	0	0	0	0
0	1	0	0	0
1	0	0	1	1
1	1	1	1	1

sides of the identity for all possible values of x and y. This is tabulated in a truth table format, as shown in Table 10.4-3.

Since the consensus identity (identity 18) is the basis for a systematic minimization of a boolean function, we now prove this identity. We do this by creating a truth table for both the left- and right-hand sides of identity 18, as shown in Table 10.4-4.

De Morgan's rules (identities 19 and 20) govern the complement of an entire expression. Identity 19 may be stated as "the complement of the sum (OR) is equal to the product (AND) of the complements." Similarly, identity 20 may be stated as "the complement of the product (AND) is equal to the sum (OR) of the complements." The proofs of these two identities are left as exercises for the reader.

Simplification of Boolean Functions

Boolean functions are mathematical representations of interconnected logical devices. When these functions are implemented by using logical devices, each term in the function requires a logical device with inputs that correspond to each variable of the term. For example, the term $X \cdot Y \cdot Z$ will require a logical device with three inputs for the variables X, Y, and Z. Each term of a boolean expression

TABLE 10.4-4. Proof by Induction for $X \cdot Y + \overline{X} \cdot Z + Y \cdot Z = X \cdot Y + \overline{X} \cdot Z$

X	Y	Z	$\overline{X}$	$X \cdot Y$	$\overline{X} \cdot Z$	$Y \cdot Z$	Left-Hand Side $X \cdot Y + \overline{X} \cdot Z + Y \cdot Z$	Right-Hand Side $X \cdot Y + \overline{X} \cdot Z$
0	0	0	1	0	0	0	0	0
0	0	1	1	0	1	0	1	1
0	1	0	1	0	0	0	0	0
0	1	1	1	0	1	1	1	1
1	0	0	0	0	0	0	0	0
1	0	1	0	0	0	0	0	0
1	1	0	0	1	0	0	1	1
1	1	1	0	1	0	1	1	1

consists of *literal(s)*. A literal is a single variable within a term that may or may not be complemented. For example, the function $F = X \cdot Y + \overline{X} \cdot Z + Y \cdot Z$ has three terms and six literals; its equivalent function (see identity 18) $F = X \cdot Y + \overline{X} \cdot Z$ has only two terms and four literals. It should be clear that the smaller the number of terms and literals a boolean function has, the fewer logical devices are needed to implement this function. In general, one should minimize a boolean expression before realizing it by using physical devices. The simplification of boolean expression is achieved by applying the basic boolean identities to the expression for the purpose of reducing the number of terms and literals, if possible. Unfortunately, no systematic procedure exists that yields the best result. The only procedure available is the trial-and-error method, using the basic identities that become familiar with use. The following examples illustrate the simplification of boolean functions using the basic identities.

EXAMPLE 10.4-4

Using boolean identities, simplify $F = A + \overline{A} \cdot B$.

Solution

$$\begin{aligned} F &= A + \overline{A} \cdot B \\ &= A \cdot 1 + \overline{A} \cdot B && \text{by using identity 6} \\ &= A \cdot 1 + \overline{A} \cdot B + 1 \cdot B && \text{by using identity 18} \\ &= A + \overline{A} \cdot B + B && \text{by using identity 6} \\ &= A + B && \text{by using identity 16} \end{aligned}$$

EXAMPLE 10.4-5

Repeat Example 10.4-4 for $F = A \cdot B + \overline{B} \cdot C + A \cdot C \cdot D + A \cdot B \cdot D$.

Solution

$$\begin{aligned} F &= A \cdot B + \overline{B} \cdot C + A \cdot C \cdot D + A \cdot B \cdot D \\ &= A \cdot B + \overline{B} \cdot C + A \cdot C \cdot D && \text{by using identity 16} \\ &= A \cdot B + \overline{B} \cdot C + A \cdot C + A \cdot C \cdot D && \text{by using identity 18} \\ &= A \cdot B + \overline{B} \cdot C + A \cdot C && \text{by using identity 16} \\ &= A \cdot B + \overline{B} \cdot C && \text{by using identity 18} \end{aligned}$$

The above example is a good illustration of the use of the absorption and consensus identities (identities 16 and 18). First, we applied the absorption identity to eliminate the term $A \cdot B \cdot D$, using the term $A \cdot B$. Second, we used the consensus identity between the terms $A \cdot B$ and $\overline{B} \cdot C$ to add the term $A \cdot C$. Now, with the new term $A \cdot C$ present, we applied the absorption identity to eliminate the term $A \cdot C \cdot D$. Finally, we applied the consensus identity one more time to eliminate the term $A \cdot C$.

EXAMPLE 10.4-6

Using boolean identities, simplify

$$F = \overline{(\overline{A} + B + C) \cdot (\overline{A} + B + C) \cdot \overline{C}}$$

Solution

$$
\begin{aligned}
F &= \overline{(\overline{A} + B + C) \cdot (\overline{A} + B + C) \cdot \overline{C}} \\
&= \overline{(\overline{A} + B + C)} + \overline{(\overline{A} + B + C)} + \overline{\overline{C}} && \text{by using identity 20} \\
&= \overline{\overline{A}} \cdot \overline{B} \cdot \overline{C} + \overline{\overline{A}} \cdot \overline{B} \cdot \overline{C} + C && \text{by using identity 19} \\
&= A \cdot \overline{B} \cdot \overline{C} + A \cdot \overline{B} \cdot \overline{C} + C && \text{by using identity 9} \\
&= A \cdot \overline{B} \cdot \overline{C} + C && \text{by using identity 3} \\
&= A \cdot \overline{B} \cdot \overline{C} + C + A \cdot \overline{B} && \text{by using identities 6 and 18} \\
&= C + A \cdot \overline{B} && \text{by using identity 16}
\end{aligned}
$$

A boolean function may have more than one minimum representation. By definition, a minimum representation of a boolean function is equivalent to the original function and has no more terms or literals than any other representation of the same function.

EXAMPLE 10.4-7

Find the minimum representation for $F = B \cdot C + \overline{B} \cdot \overline{C} + A \cdot \overline{B} \cdot C$.

Solution

If we start the minimization by first applying the consensus identity between the second and the third terms, we obtain

$$
\begin{aligned}
F &= B \cdot C + \overline{B} \cdot \overline{C} + A \cdot \overline{B} \cdot C + A \cdot \overline{B} && \text{by using identity 18} \\
&= B \cdot C + \overline{B} \cdot \overline{C} + A \cdot \overline{B} && \text{by using identity 16}
\end{aligned}
$$

However, if we start the minimization by applying the consensus identity between the first and the last terms, we obtain

$$F = B \cdot C + \overline{B} \cdot \overline{C} + A \cdot \overline{B} \cdot C + A \cdot C \quad \text{by using identity 18}$$

$$= B \cdot C + \overline{B} \cdot \overline{C} + A \cdot C \quad \text{by using identity 16}$$

The reader is encouraged to verify that these two different expressions are equivalent.

Before presenting another method for the simplification of boolean expressions and different forms of their representations, we introduce the basic digital circuits of which a digital computer and other digital systems are constructed—the *logic gates*. Logic gates are used to obtain the realization of different terms of a boolean function.

10.5 FUNDAMENTAL LOGIC GATES

Logic gates are physical devices that perform the basic boolean operations. They are electronic circuits that operate on one or more input signals to produce an output signal. The input and output signals to these logic gates are the voltages or currents with the logic values of the boolean variables they represent. For example, a logical 1 may be represented by a positive voltage level, typically 5 V; a logical 0 is represented by a 0 voltage level.

Logical OR Gate

The OR gate is an electronic circuit realization of the boolean OR operation. The graphical representation of a two-input OR gate and its truth table are shown in Fig. 10.5-1. OR gates can have more than two inputs but only one output. Figure 10.5-2 shows three- and four-input OR gates. The output of an OR gate can be an

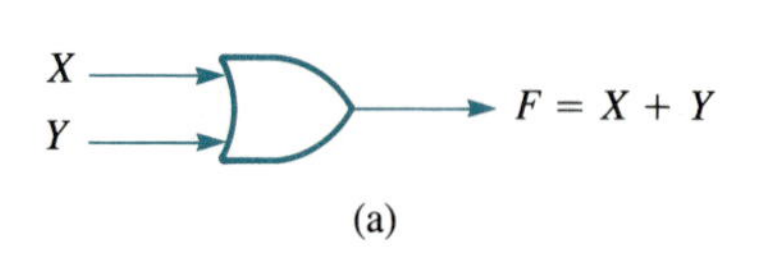

Inputs		Output
X	Y	F
0	0	0
0	1	1
1	0	1
1	1	1

(b)

FIG. 10.5-1. A two-input OR gate: (a) a physical representation and (b) the truth table.

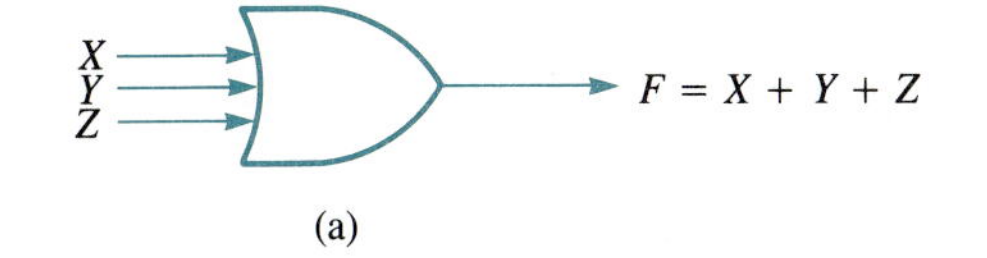

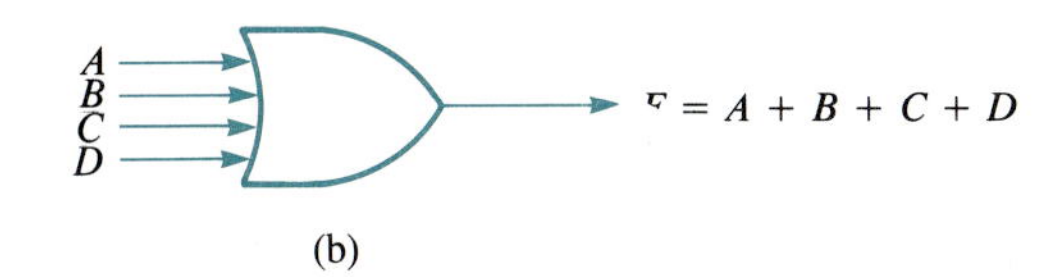

FIG. 10.5-2. (a) A three-input OR gate and (b) a four-input OR gate.

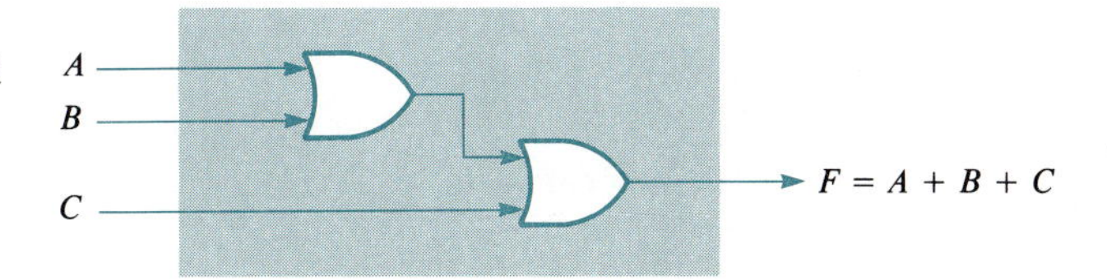

FIG. 10.5-3. A three-input OR gate using two-input OR gates.

input to another gate. Figure 10.5-3 shows how 2 two-input OR gates are connected to form a three-input equivalent OR gate.

Logical AND Gate

The AND gate is an electronic circuit realization of the boolean AND operation. The graphical representation of a two-input AND gate and its truth table are shown in Fig. 10.5-4. AND gates can have more than two inputs but only one output. Figure 10.5-5 shows three- and four-input AND gates. The output of an AND gate can be an input to another gate. Figure 10.5-6 shows the interconnections between 2 two-input AND gates and 1 two-input OR gate to produce the boolean expression $F = A \cdot B + C \cdot D$. Similarly, Fig. 10.5-7 shows the physical realization of the boolean expression $F = (W + X) \cdot (Y + Z)$.

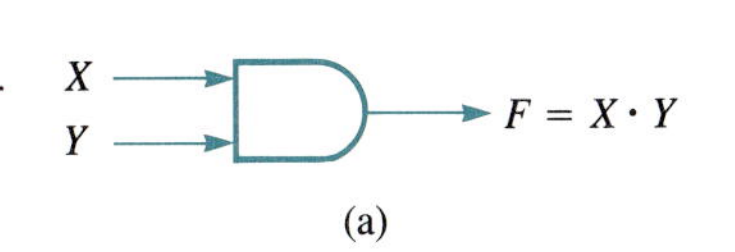

Inputs		Outputs
X	Y	F
0	0	0
0	1	0
1	0	0
1	1	1

(b)

FIG. 10.5-4. A two-input AND gate: (a) a physical representation and (b) the truth table.

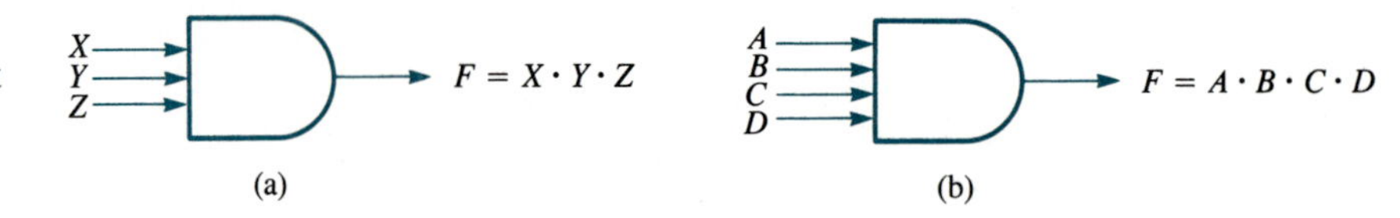

FIG. 10.5-5. (a) A three-input AND gate and (b) a four-input AND gate.

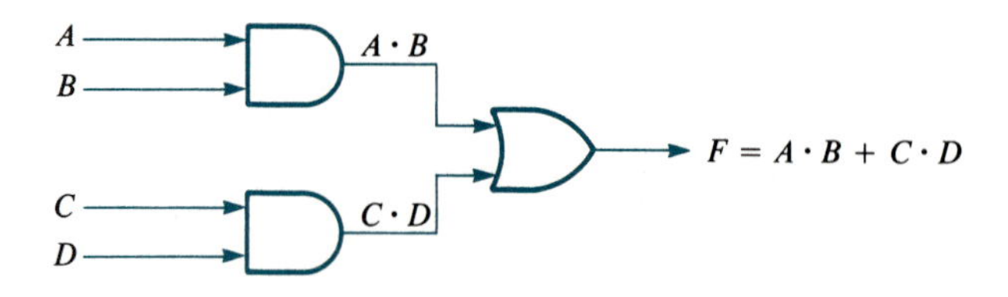

FIG. 10.5-6. A physical realization for the boolean function $F = A \cdot B + C \cdot D$.

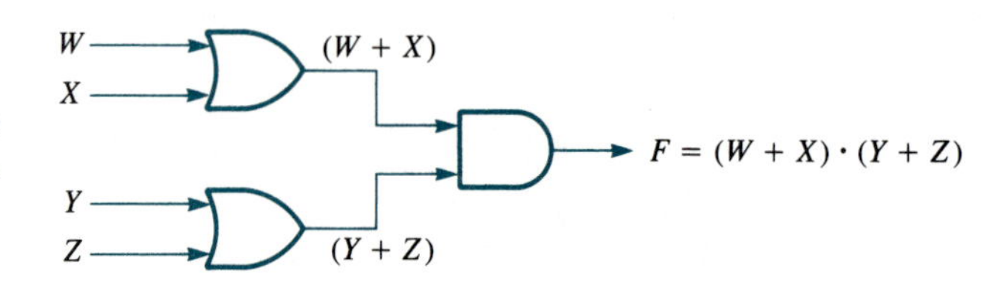

FIG. 10.5-7. A physical realization of the boolean function $F = (W + X) \cdot (Y + Z)$.

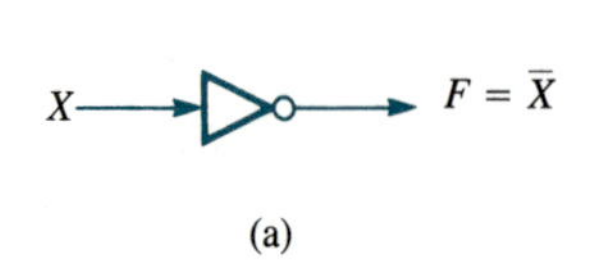

Input	Output
X	F
0	1
1	0

(b)

FIG. 10.5-8. A NOT gate: (a) a physical representation and (b) the truth table.

Logical NOT Gate

The NOT gate is an electronic circuit realization of the boolean NOT operation. Since the NOT operation is a unary operation, the NOT gate has *only* one input and one output. The graphical representation of the NOT gate and its truth table are shown in Fig. 10.5-8. The NOT gate is known as an *inverter*.

The interconnections of the basic logic gates (OR, AND, and NOT) form a *logic network*. This logic network is also known as a *combinational network*. In general, a combinational network is a logic network that does not contain memory devices. The boolean function describing a combinational network can be easily derived by systematically progressing from the input(s) to the output on the logic gates.

EXAMPLE 10.5-1

Derive the boolean function for the combinational network of Fig. 10.5-9(a).

Solution

Since the output ($\overline{X}$) of gate 1 is connected to one of the inputs of gate 2, then the

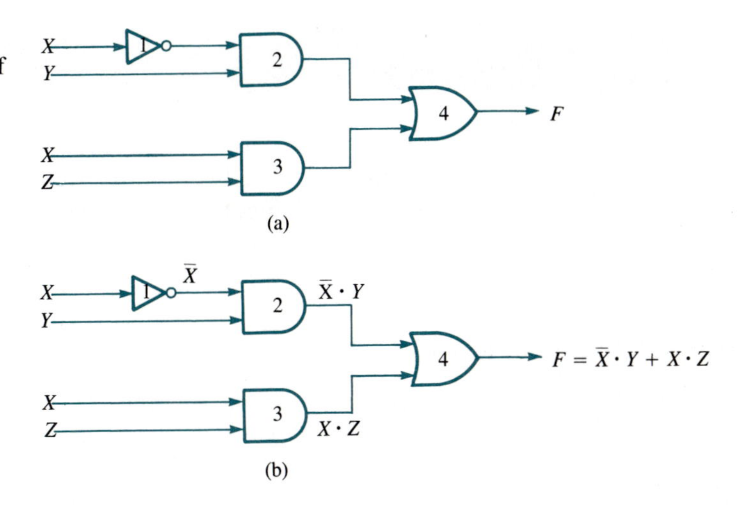

FIG. 10.5-9. The derivation of a boolean function for a combinational network.

output of gate 2 is $X \cdot Y$. Also, the output of gate 3 is $X \cdot Z$. Now since the inputs to gate 4 are the outputs of gates 2 and 3, the output (F) of gate 4 is equal to $\overline{X} \cdot Y + X \cdot Z$. This is shown graphically in Fig. 10.5-9(b).

Similarly, any boolean function can be transformed from an algebraic expression into a combinational network by using the basic logic gates. For example, the combinational network for $F = \overline{X} \cdot Y + X \cdot Z + Y \cdot Z$ is shown in Fig. 10.5-10.

A second look at the combinational networks shown in Figs. 10.5-9 and 10.5-10 reveals that these two networks are equivalent (see boolean identity 18). However, the network of Fig. 10.5-9 contains fewer logic gates. It is always desirable to implement a given boolean function by using the minimum number of components. This not only leads to an economical network but also will reduce the overall complexity, which will help in the troubleshooting of the network when one of the logic gates fails to operate properly.

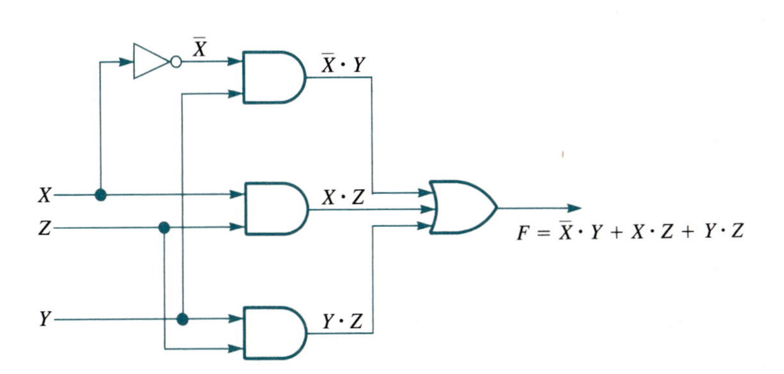

FIG. 10.5-10. A combinational network for $F = \overline{X} \cdot Y + X \cdot Z + Y \cdot Z$.

10.6 STANDARD FORMS

Since a boolean function can be written in different forms, there are different combinational networks describing the same function. However, certain forms of boolean functions lead to more desirable combinational networks. These forms are the *standard forms*. There are two standard forms: *sum-of-products* (SOP) and *product-of-sums* (POS).

Sum-of-Products Form

The sum-of-products form consists of the logical sum (OR) of *product terms*. A product term is the logical product (AND) of some variables. These variables may or may not be in complemented form. For example, the terms $A \cdot B \cdot \overline{C}$ and $\overline{A} \cdot \overline{B}$ are valid product terms, and the boolean function $F = \overline{A} \cdot \overline{B} + A \cdot B \cdot \overline{C}$ is in the sum-of-products form. When a product term contains each of the n variables of the function, it is called a *minterm*. For n variables, there are 2^n possible minterms. A minterm assumes the logical value 1 for only one combination of the n variables. The ith minterm of n variables is denoted by m_i, where the subscript i, $0 \leq i \leq 2^n - 1$, represents the decimal equivalent of the binary number obtained when replacing a variable in a minterm by 1 and its complement by 0. For example, the four minterms of the two variables A and B are $m_0 = \overline{A} \cdot \overline{B}$, $m_1 = \overline{A} \cdot B$, $m_2 = A \cdot \overline{B}$, and $m_3 = A \cdot B$.

Any boolean function can be expressed, algebraically, as a sum (OR) of minterms. An expression of this form is known as a *canonical sum of products*. Given a truth table for a logic function, its sum-of-products form can be obtained by taking the sum of the minterms that correspond to a 1 in the output column of the table. For example, the truth table given in Table 10.6-1 describes a boolean

TABLE 10.6-1. Truth Table for an Arbitrary Function *F*

m_i	A	B	C	F
m_0	0	0	0	0
m_1	0	0	1	1
m_2	0	1	0	0
m_3	0	1	1	1
m_4	1	0	0	0
m_5	1	0	1	0
m_6	1	1	0	1
m_7	1	1	1	1

function F that has an output of 1 that corresponds to minterms m_1, m_3, m_6, and m_7; therefore, F can be expressed as

$$F(A, B, C) = m_1 + m_3 + m_6 + m_7$$
$$= \overline{A} \cdot \overline{B} \cdot C + \overline{A} \cdot B \cdot C + A \cdot B \cdot \overline{C} + A \cdot B \cdot C$$

When expressing a boolean function as a sum of minterms, we usually represent it in a compact form by listing the values of i, the subscript of the ith minterm, in a summation form. For example, the above function is expressed as

$$F(A, B, C) = \Sigma\, m_i(1, 3, 6, 7)$$

where $\Sigma\, m_i(\ \ )$ means the sum of all the minterms whose subscript i is given inside the parentheses.

EXAMPLE 10.6-1

Given $F(A, B, C) = \Sigma\, m_i(2, 3, 4, 5)$, (a) prepare its truth table, (b) express F in a canonical sum-of-products form, and (c) minimize F in an SOP form.

Solution

(a) The truth table is shown in Table 10.6-2.

(b) $F(A, B, C) = m_2 + m_3 + m_4 + m_5$

$$= \overline{A} \cdot B \cdot \overline{C} + \overline{A} \cdot B \cdot C + A \cdot \overline{B} \cdot \overline{C} + A \cdot \overline{B} \cdot C$$

(c) Applying the consensus identity between the first and second terms and between the third and fourth terms of the function given in (b), we obtain

$$F(A, B, C) = \overline{A} \cdot B \cdot \overline{C} + \overline{A} \cdot B \cdot C + \overline{A} \cdot B + A \cdot \overline{B} \cdot \overline{C} + A \cdot \overline{B} \cdot C + A \cdot \overline{B}$$

Finally, using the absorption identity we obtain

$$F(A, B, C) = \overline{A} \cdot B + A \cdot \overline{B}$$

TABLE 10.6-2. Truth Table for $F = \Sigma\, m_i(2, 3, 4, 5)$

A	B	C	F
0	0	0	0
0	0	1	0
0	1	0	1
0	1	1	1
1	0	0	1
1	0	1	1
1	1	0	0
1	1	1	0

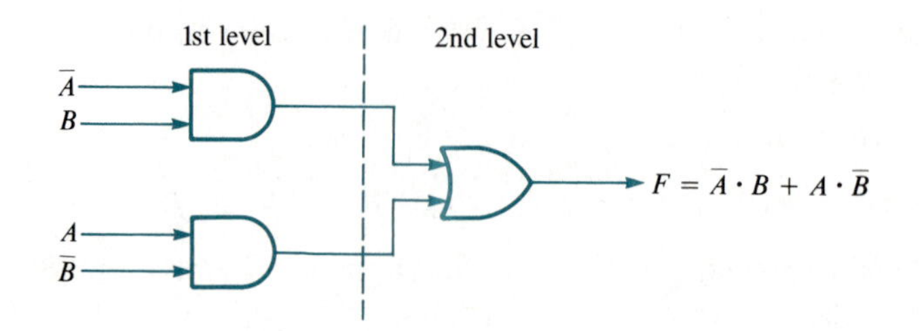

FIG. 10.6-1. A two-level realization for $F = \overline{A} \cdot B + A \cdot \overline{B}$.

Sum-of-products expressions can be implemented by using *two-level* (AND-OR) *networks*, where each product term requires an AND gate (except for a term with a single variable) and the logical sum of these terms is obtained by using an OR gate with inputs from the AND gates or the single variables. In the following discussion, it is assumed that the variable and its complement are always available, and therefore inverters are omitted from the circuit diagrams. Figure 10.6-1 shows a two-level implementation for the function of Example 10.6-1(c).

Product-of-Sums Form

The product-of-sums form consists of the logical product (AND) of *sum terms*. A sum term is the logical sum (OR) of some variables. These variables may or may not be in complemented form. For example, the terms $(A + B + C)$, $(A + \overline{B} + \overline{C})$, and $(\overline{A} + B)$ are valid sum terms; and the boolean function

$$F(A, B, C) = (A + B + C) \cdot (A + \overline{B} + \overline{C}) \cdot (\overline{A} + B)$$

is in the product-of-sums form. When a sum term contains each of the n variables of the function, it is called a *maxterm*. The ith maxterm of n variables, denoted by M_i, is the complement of the ith minterm m_i of the same n variables; that is, $M_i = \overline{m}_i$. Table 10.6-3 lists the eight possible maxterms of the variables A, B, and C and their corresponding minterms.

TABLE 10.6-3. Maxterms and Minterms for Three Variables

A	B	C	i	Maxterm (M_i)	Minterm (m_i)
0	0	0	0	$(A + B + C)$	$\overline{A} \cdot \overline{B} \cdot \overline{C}$
0	0	1	1	$(A + B + \overline{C})$	$\overline{A} \cdot \overline{B} \cdot C$
0	1	0	2	$(A + \overline{B} + C)$	$\overline{A} \cdot B \cdot \overline{C}$
0	1	1	3	$(A + \overline{B} + \overline{C})$	$\overline{A} \cdot B \cdot C$
1	0	0	4	$(\overline{A} + B + C)$	$A \cdot \overline{B} \cdot \overline{C}$
1	0	1	5	$(\overline{A} + B + \overline{C})$	$A \cdot \overline{B} \cdot C$
1	1	0	6	$(\overline{A} + \overline{B} + C)$	$A \cdot B \cdot \overline{C}$
1	1	1	7	$(\overline{A} + \overline{B} + \overline{C})$	$A \cdot B \cdot C$

Any boolean function can be expressed, algebraically, as a product of maxterms. An expression of this form is known as a *canonical product-of-sums*. Given a truth table for a logic function, its product-of-sums form can be obtained by taking the product (AND) of the maxterms that correspond to a 0 in the output column of the table. So for the function described in Table 10.6-1, there are four 0s in the output column; therefore, F can be expressed as

$$F = M_0 \cdot M_2 \cdot M_4 \cdot M_5$$
$$= (A + B + C) \cdot (A + \overline{B} + C) \cdot (\overline{A} + B + C) \cdot (\overline{A} + B + \overline{C})$$

Similarly, when expressing a boolean function as a product of maxterms, we usually represent it in a compact form by listing the values of i, the subscript of the ith maxterm, in a product form. For example, the above function is expressed as

$$F(A, B, C) = \Pi\ M_i(0, 2, 4, 5)$$

where $\Pi\ M_i(\quad)$ means the product of all the maxterms whose subscript i, is given inside the parentheses.

The product of sums can be implemented by using two-level (OR-AND) networks, where each term requires an OR gate (except for a term with a single variable) and the product of these terms is obtained by using an AND gate with inputs from the OR gates or the single variables.

EXAMPLE 10.6-2

Given $F(A, B, C) = \Pi\ M_i\ (0, 1, 6, 7)$, (a) prepare its truth table, (b) express F in a canonical product-of-sums form, (c) minimize F in a POS form, and (d) obtain a two-level realization for F of part (c).

Solution

(a) The truth table is shown in Table 10.6-4.

(b) $F(A, B, C) = (A + B + C) \cdot (A + B + \overline{C}) \cdot (\overline{A} + \overline{B} + C) \cdot (\overline{A} + \overline{B} + \overline{C})$

TABLE 10.6-4. Truth Table for $F = \Pi\ M_i(0, 1, 6, 7)$

A	B	C	F
0	0	0	0
0	0	1	0
0	1	0	1
0	1	1	1
1	0	0	1
1	0	1	1
1	1	0	0
1	1	1	0

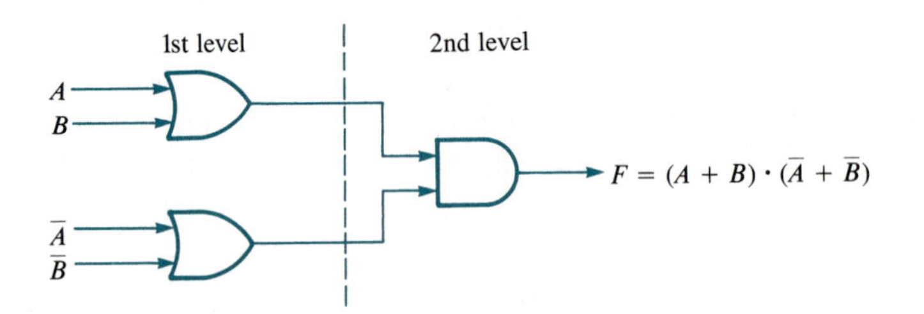

FIG. 10.6-2. A two-level realization for $F = (A + B)(\overline{A} + \overline{B})$.

(c) $F(A, B, C) = (A + B) \cdot (\overline{A} + \overline{B})$ from identity 15 in Table 10.4-2.
(d) The two-level realization for part (c) is shown in Fig. 10.6-2.

10.7 MAP SIMPLIFICATION

Even though the truth table uniquely represents a logic function, this same function may appear in different algebraic forms. Boolean identities were used for the simplification of a given algebraic form, but this procedure, as we discussed earlier, lacks the systematic process that guarantees the minimum form. The map method provides a convenient graphical procedure for obtaining a minimum SOP or POS of a boolean expression up to five variables. Maps for a larger number of variables are possible; however, they are very cumbersome to manipulate. The map method is known as the Karnaugh map, or simply the K map. This is a modified form of the truth table.

A K map is a diagram made up of cells (squares), one for each minterm of the function to be represented. Therefore, an n-variable K map, representing an n-variable function, has 2^n cells. So a 2-variable K map has 4 cells, a 3-variable K map has 8 cells, a 4-variable K map has 16 cells, etc.

Figure 10.7-1(a) shows a two-variable K map. This map is redrawn in Fig. 10.7-1(b), with the row headings corresponding to the binary values of B and the column headings corresponding to the binary values of A. Similarly, Figs. 10-7-2 and 10.7-3 show a three- and four-variable K map, respectively. Note the code used in listing the row and column headings when more than one variable is needed. The code assignment must be arranged such that any two physically adjacent cells differ by only one variable value; that is, cells that have a common

	$\overline{A}$	A
$\overline{B}$	$\overline{A} \cdot \overline{B}$	$A \cdot \overline{B}$
B	$\overline{A} \cdot B$	$A \cdot B$

(a)

$B \backslash A$	0	1
0	m_0	m_2
1	m_1	m_3

(b)

FIG. 10.7-1. Two-variable K maps.

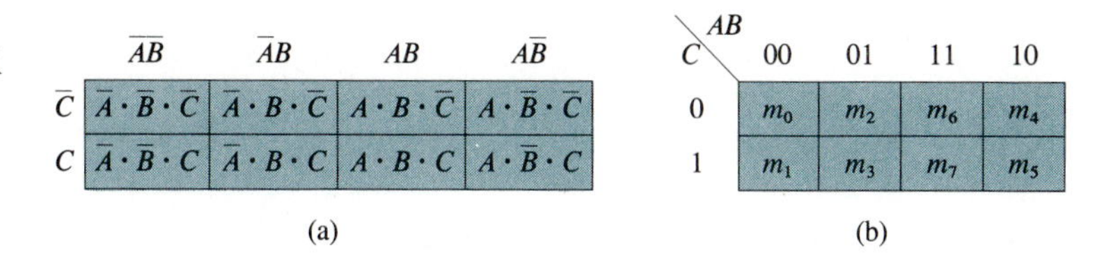

FIG. 10.7-2. Three-variable K maps.

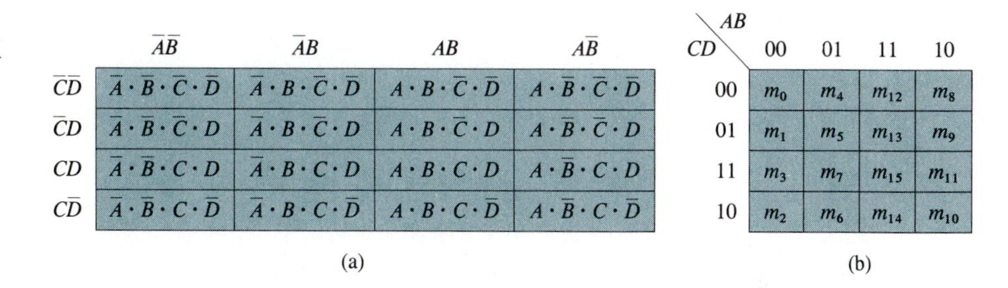

FIG. 10.7-3. Four-variable K maps.

side must correspond to assignments that differ by just one variable. For example, in Fig. 10.7-3, consider the cell corresponding to m_{15}, which has an assignment $A \cdot B \cdot C \cdot D = 1 \cdot 1 \cdot 1 \cdot 1$, and any adjacent cell. For the cell on the right, m_{11}, the assignment is $A \cdot \overline{B} \cdot C \cdot D = 1 \cdot 0 \cdot 1 \cdot 1$, which differs in the value of the variable B only. Similarly, the assignment for the cell on the left, which is $\overline{A} \cdot B \cdot C \cdot D = 0 \cdot 1 \cdot 1 \cdot 1$, differs only in the value of the variable A. Cells that differ in just one variable value are said to be adjacent. The reader is encouraged to verify that the cells in the top row of Fig. 10.7-3 are adjacent to the corresponding cells in the bottom row and that the cells in the rightmost column are adjacent to the corresponding cells in the leftmost column. Also, the cell corresponding to m_2 is adjacent to the cells corresponding to m_0 and m_{10}, in addition to its physical adjacency to the cells corresponding to m_3 and m_6.

A function can be represented in a K map by simply entering 1s in the cells that correspond to the minterms of the function.

EXAMPLE 10.7-1

Show the K-map representations of the following boolean function: (a) $F(A, B) = \Sigma\, m_i(0, 1, 2)$, (b) $F(A, B, C) = \Sigma\, m_i(0, 2, 3, 5, 7)$, and (c) $F(A, B, C, D) = \Sigma\, m_i(1, 3, 5, 6, 9, 10, 13, 14)$.

Solution

The K-map representations for the functions are shown in Fig. 10.7-4.

Since adjacent cells differ in only one variable, the product term, corresponding to these two adjacent cells for which the function has the value 1, is obtained

FIG. 10.7-4. The K map representation for Example 10.7-1.

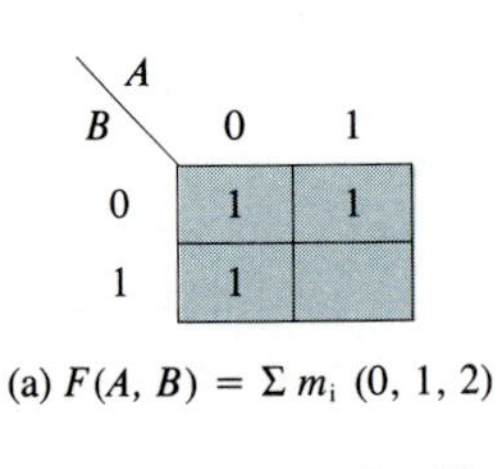

(a) $F(A, B) = \Sigma\, m_i\ (0, 1, 2)$

C \ AB	00	01	11	10
0	1	1		
1		1	1	1

(b) $F(A, B, C) = \Sigma\, m_i\ (0, 2, 3, 5, 7)$

CD \ AB	00	01	11	10
00				
01	1	1	1	1
11	1			
10		1	1	1

(c) $F(A, B, C, D) = \Sigma\, m_i\ (1, 3, 5, 6, 9, 10, 13, 14)$

by ANDing all those variables that are common to the two adjacent cells. For example, the product term that corresponds to the cells of m_0 and m_1 of Fig. 10.7-4(a) is $\overline{A}$, the product term that corresponds to the cells of m_0 and m_2 of Fig. 10.7-4(b) is $\overline{A} \cdot \overline{C}$, and the product term that corresponds to the cells of m_1 and m_3 of Fig. 10.7-4(c) is $\overline{A} \cdot \overline{B} \cdot D$.

Simplification of an n-variable boolean function by using a K map is achieved by grouping adjacent cells that contain 1s. The number of adjacent cells that may be grouped must always be equal to 2^m, where m, $0 < m < n$, is the number of cells that are adjacent to any cell in the group. The group of 2^m cells is called a *subcube* and is expressed by a product term using only $(n - m)$ literals.

Therefore, the larger the subcube is, the fewer literals are needed to express the product term. For example, in a four-variable K map, the possible subcubes that can be formed are as follows:

A 1-cell subcube covers 1 minterm and is expressed by a product term using 4 literals.

A 2-cell subcube covers 2 minterms and is expressed by a product term using 3 literals.

A 4-cell subcube covers 4 minterms and is expressed by a product term using 2 literals.

An 8-cell subcube covers 8 minterms and is expressed by a product term using 1 literal.

A 16-cell subcube covers 16 minterms and is expressed by the logical 1; that is, the function is always equal to 1.

For example, consider the boolean function F defined by the K map of Fig. 10.7-5. Grouping the cells that correspond to minterms m_1, m_3, m_5, and m_7, we obtain a four-cell subcube that can be expressed by the product term $\overline{A} \cdot D$.

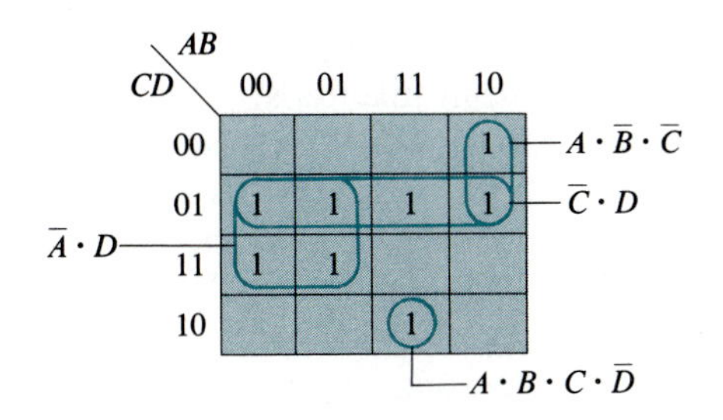

FIG. 10.7-5. A K map with subcubes for $F(A, B, C, D) = \Sigma\, m_i(1, 3, 5, 7, 8, 9, 13, 14)$.

Observe that the literals $\overline{A}$ and D are the only common literals to the four minterms m_1, m_3, m_5, and m_7. Similarly, grouping the cells that correspond to minterms m_1, m_5, m_{13}, and m_9, we obtain a four-cell subcube that can also be expressed by the product term $\overline{C} \cdot D$. Again, observe that the literals $\overline{C}$ and D are the only common literals to the four minterms m_1, m_5, m_{13}, and m_9. Now, grouping the two adjacent cells that correspond to minterms m_8 and m_9, we obtain a two-cell subcube that can be expressed by the product term $A \cdot \overline{B} \cdot \overline{C}$. Finally, the cell corresponding to m_{14} forms a one-cell subcube that can be expressed by the product term $A \cdot B \cdot C \cdot \overline{D}$.

The use of a cell for a given minterm in more than one subcube is allowed and justified by boolean identity 3 ($X + X = X$). Therefore, any cell may be included in as many subcubes as desired. Figure 10.7-6 shows some examples of cell

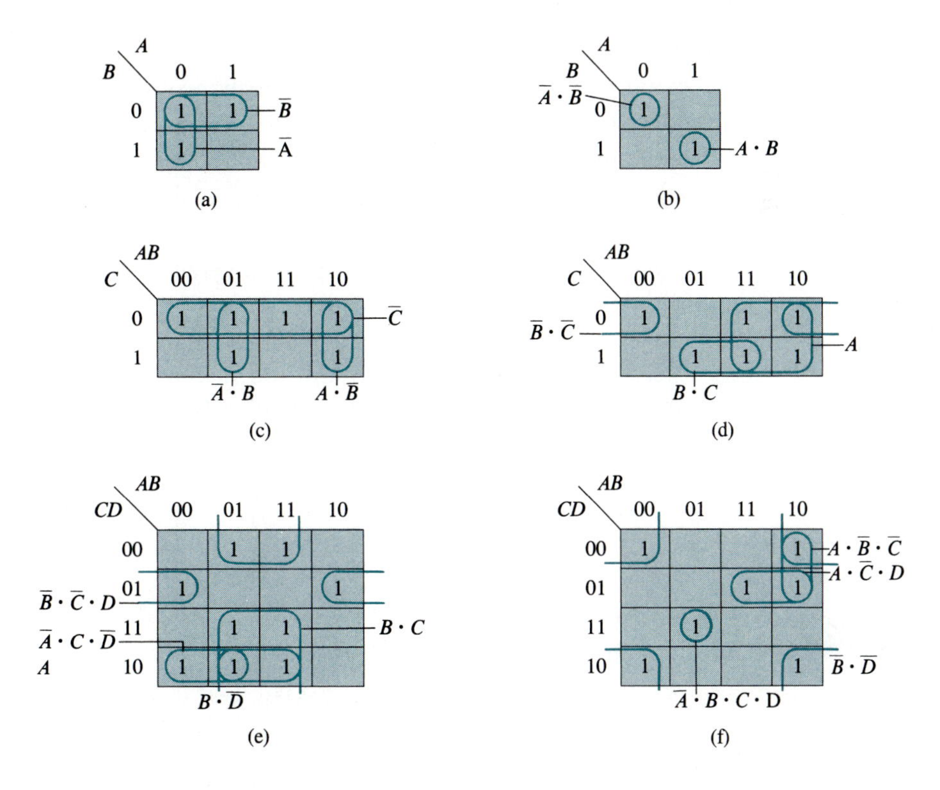

FIG. 10.7-6. Subcubes with one, two, and four cells.

groupings to form subcubes in two-, three-, and four-variable K maps representing two-, three-, and four-variable boolean functions.

Once a boolean function is plotted in a K map and its different subcubes are formed, this boolean function can be expressed as the logical sum of product terms that correspond to the minimum set of subcubes that covers all its 1 cells. Therefore, in forming a subcube, we should not select a subcube that is totally contained in another subcube. The product term representing the subcube containing the maximum possible number of adjacent 1 cells in the map is known as a *prime implicant*. If a prime implicant represents a subcube that contains at least one 1 cell that is not covered by any other subcube, then this prime implicant is known as an *essential prime implicant*. However, if *all* the 1 cells of a subcube of a prime implicant are covered by some other subcubes, then this prime implicant is known as an *optional prime implicant*. For example, all the product terms representing the subcubes in the maps of Fig. 10.7-6 are essential prime implicants. Also, all of these prime implicants are essential except for the prime implicant $A \cdot \overline{B} \cdot \overline{C}$ in Fig. 10.7-6(f), which is an optional prime implicant since all its 1 cells are covered by the two prime implicants $\overline{B} \cdot \overline{D}$ and $A \cdot \overline{C} \cdot D$. The minimized boolean expression is obtained by the logical sum (OR) of all the essential prime implicants and some other optional prime implicants that cover any remaining 1 cells not covered by the essential prime implicants.

The procedure for map simplification of an n-variable boolean function can be summarized in the following steps:

1. Plot the function in an n-variable K map.
2. Obtain all prime implicants that correspond to subcubes of the maximum adjacent 1 cells.
3. Determine the essential prime implicants.
4. Start the minimum expression with the logical sum of the essential prime implicants.
5. In the map, check the 1 cells that are covered by the subcubes expressed by the essential prime implicants.
6. If all the 1 cells of the map are checked, then no other terms are required; otherwise, add to your expression the minimum number of optional prime implicants that correspond to the subcubes that include the unchecked 1 cells.

The following example illustrates the application of this procedure.

EXAMPLE 10.7-2

Obtain a minimum boolean expression for

$$F(A, B, C, D) = \Sigma\, m_i(1, 3, 4, 5, 6, 7, 10, 12)$$

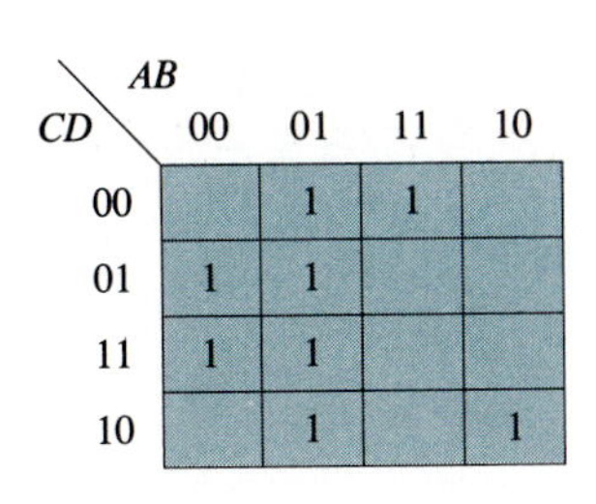

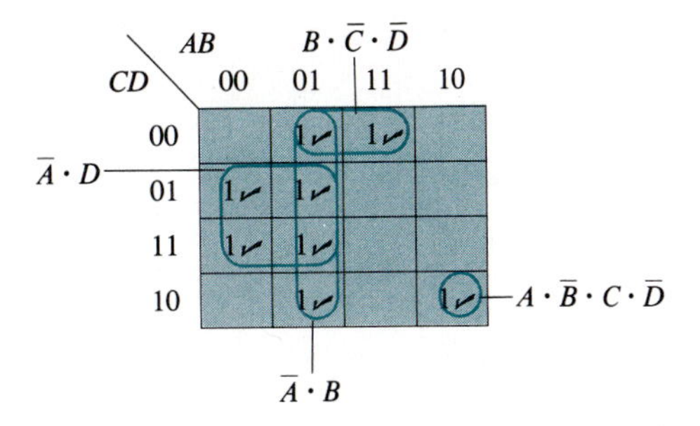

FIG. 10.7-7. An illustration of map simplification: (a) plot of $F(A, B, C, D) = \Sigma\, m_i(1, 3, 4, 5, 6, 7, 10, 12)$, and (b) extraction of Prime Implicants.

Solution

1. The plot of F is shown in Fig. 10.7-7(a).
2. The prime implicants, as shown in Fig. 10.7-7(b), are $\overline{A} \cdot D$, $\overline{A} \cdot B$, $B \cdot \overline{C} \cdot \overline{D}$, and $A \cdot \overline{B} \cdot C \cdot \overline{D}$.
3. Since each of the subcubes contains at least one 1 cell that is not covered by any other subcube, all prime implicants are essential.
4. $F(A, B, C, D) = \overline{A} \cdot D + \overline{A} \cdot B + B \cdot \overline{C} \cdot \overline{D} + A \cdot \overline{B} \cdot C \cdot \overline{D}$.
5. All the checked 1 cells are shown in Fig. 10.7-7(b).
6. Since all the 1 cells are checked,

$$F(A, B, C, D) = \overline{A} \cdot D + \overline{A} \cdot B + B \cdot \overline{C} \cdot \overline{D} + A \cdot \overline{B} \cdot C \cdot \overline{D}$$

Product-of-Sums Minimization

The minimum product-of-sums expression for a boolean function can be easily derived by using K maps in an almost identical procedure for the minimum sum-of-products expression. This is accomplished by using the 0s of the map and obtaining prime implicants by using the 0 cells. Since the 0 cells represent the complement of the function, complementing the minimum sum-of-products expression for the 0s of the function produces the minimum product-of-sums form of the function.

Don't-Care Conditions

In designing a digital system, we often encounter a situation where the output cannot be specified due to physical or logical constraints. For example, consider the logic circuitry represented by the black box of Fig. 10.7-8, where the input lines B and C are physically connected together. For this circuit, the input combination $A = 0$, $B = 1$, and $C = 0$ is not possible, and therefore, the output corresponding to this combination cannot be specified. Such an output is known

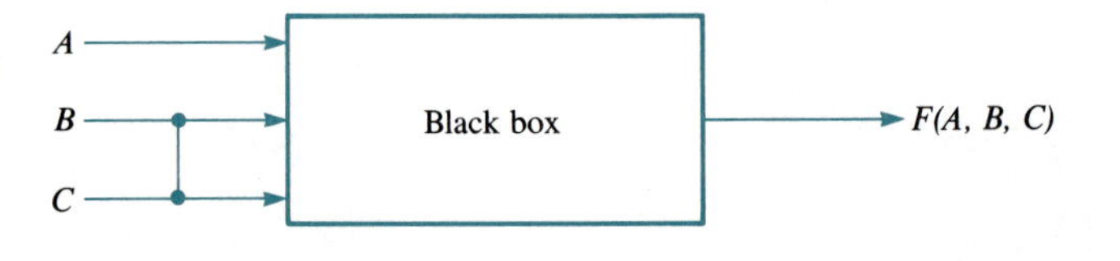

FIG. 10.7-8. An illustration of the physical constraint for a don't-care output.

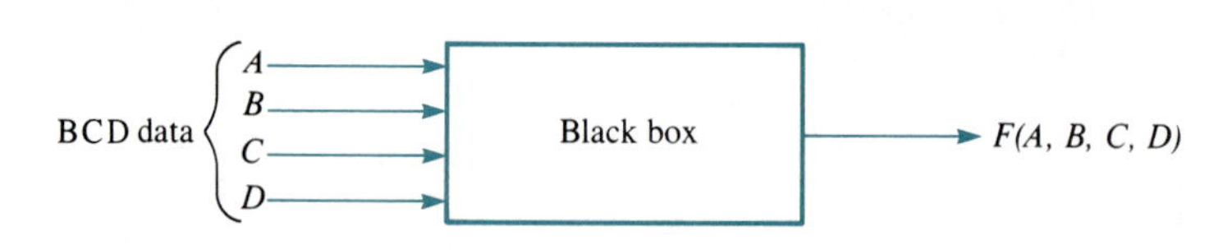

FIG. 10.7-9. An illustration of a logical constraint for a don't-care output.

as a *don't-care output*. As another example, consider the logic circuitry with four input lines representing BCD data, as shown in Fig. 10.7-9. For this circuit, six of the input combinations, namely, 1010, 1011, 1100, 1101, 1110, and 1111, are not valid BCDs, and therefore, the output of this circuitry cannot be specified. They are known as don't-care outputs. A don't-care output is represented by the letter d in the truth table and the K map.

Since d's represent don't-care outputs in a K map, they can be treated as 0s or 1s when we form subcubes. That is, when forming a subcube, we treat some or all of the d's as 1s if, by doing so, we obtain a larger subcube containing the maximum number of adjacent 1 and d cells. Similarly, we use some or all of the d's as 0s if, by doing so, we obtain a larger subcube containing the maximum number of adjacent 0s and d cells. However, we should not form a subcube that contains only d's. This procedure is illustrated by the following example.

EXAMPLE 10.7-3

For the function specified by the truth table in Fig. 10.7-10(a), obtain (a) a minimum sum-of-products expression and (b) a minimum product-of-sums expression.

Solution

We first plot the function F in a four-variable K map, as shown in Fig. 10.7-10(b).

(a) The extraction of prime implicants for F is shown in Fig. 10.7-10(c), and the minimum sum-of-products expression is

$$F = \overline{C} \cdot \overline{D} + B \cdot \overline{C}$$

(b) The extraction of prime implicants for the complement of F is shown in Fig. 10.7-10(d), and the minimum product-of-sums is obtained as follows:

$$\overline{F} = C + \overline{B} \cdot D$$

FIG. 10.7-10. Example of SOP and POS derivations: (a) truth table, (b) K map, (c) prime implicants for F, and (d) prime implicants for $\overline{F}$.

A	B	C	D	F
0	0	0	0	1
0	0	0	1	0
0	0	1	0	0
0	0	1	1	0
0	1	0	0	1
0	1	0	1	1
0	1	1	0	0
0	1	1	1	0
1	0	0	0	1
1	0	0	1	0
1	0	1	0	d
1	0	1	1	d
1	1	0	0	d
1	1	0	1	d
1	1	1	0	d
1	1	1	1	d

(a)

CD \ AB	00	01	11	10
00	1	1	*d*	1
01	0	1	*d*	0
11	0	0	*d*	d
10	0	0	*d*	d

(b)

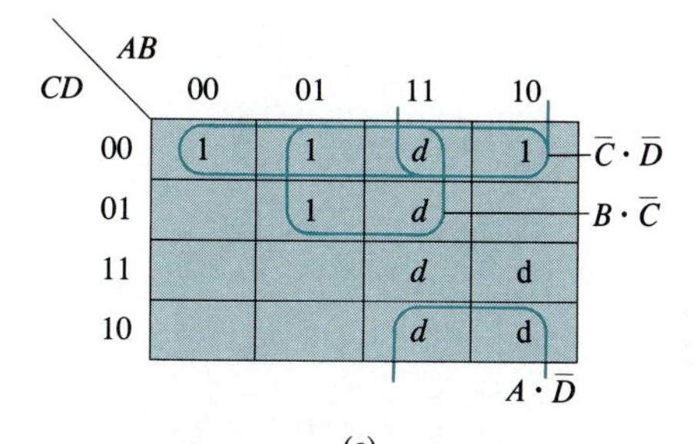

(c)

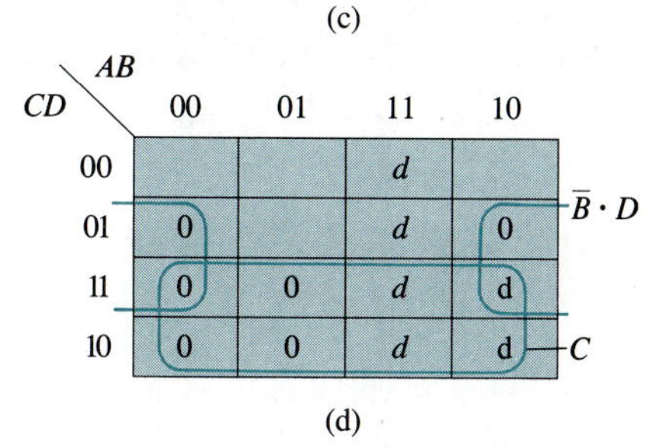

(d)

Complementing both sides of the above equation and using De Morgan's rules, we obtain

$$F = \overline{C} \cdot (B + \overline{D})$$

Notice that in Fig. 10.7-10(c) and (d) we did not form subcubes that cover *d*'s only.

10.8 MORE LOGIC GATES

Even though any boolean expression can be implemented by using the basic logic gates (namely, AND, OR, and NOT gates), four other types of logic gates—namely, NAND, NOR, exclusive-OR, and exclusive-NOR gates—are often used

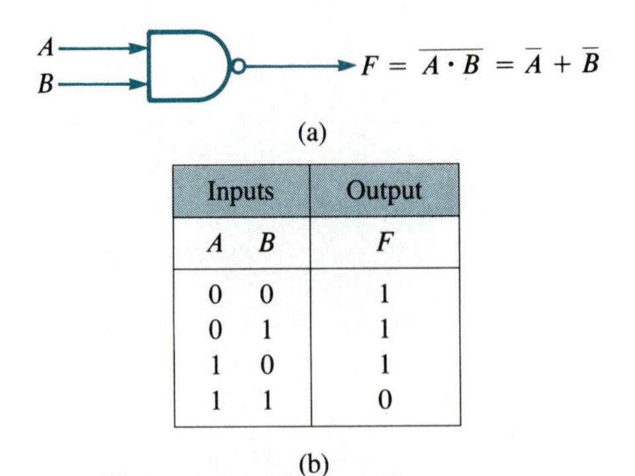

(a)

Inputs		Output
A	B	F
0	0	1
0	1	1
1	0	1
1	1	0

(b)

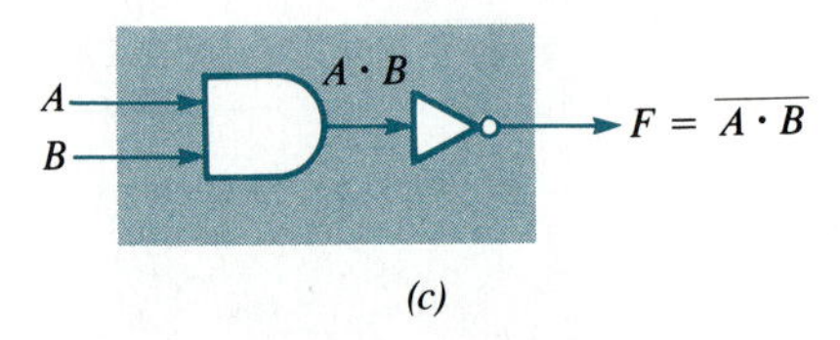

(c)

FIG. 10.8-1. A two-input NAND gate: (a) a graphical representation, (b) the truth table, and (c) an equivalent AND-NOT circuit.

in digital design. It is often desirable to realize a circuit by using only one type of basic gate. In such circuits, one of the following gates can be used.

NAND Gates

The graphical representation of a two-input NAND gate and its truth table are shown in Fig. 10.8-1(a) and (b), respectively. Figure 10.8-1(c) is the equivalent circuit of the NAND gate using an AND gate followed by an inverter.

The NAND gate is a functionally complete gate; that is, any of the basic operations of the AND, OR, and NOT gates can be obtained by using *only* NAND gate(s), as shown in Fig. 10.8-2. Therefore, any boolean function can be realized by using NAND gates *only*. The procedure for obtaining a minimum-NAND-gates realization for a boolean expression is as follows:

1. Obtain a minimum two-level AND-OR realization, where no inputs are directly connected to the OR gate; that is, every input must pass through an AND gate.
2. Redraw the above circuit by simply replacing all the AND and OR gates by NAND gates.

EXAMPLE 10.8-1

Using a minimum number of NAND gates, realize the following boolean function:

$$F(A, B, C) = \Sigma\, m_i(0, 3, 4, 5, 7)$$

Solution

1. From the K map shown in Fig. 10.8-3(a), the minimum sum-of-products form is

$$F = A \cdot \overline{B} + B \cdot C + \overline{B} \cdot \overline{C}$$

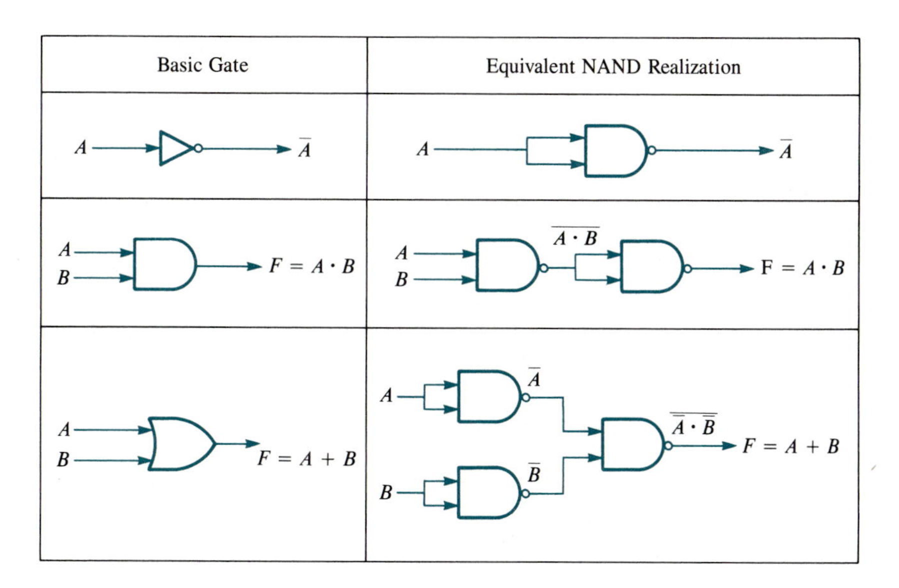

FIG. 10.8-2. NAND realizations of NOT, AND, and OR gates.

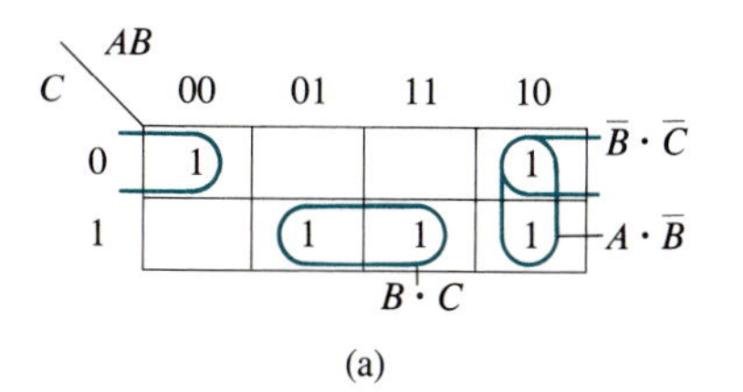

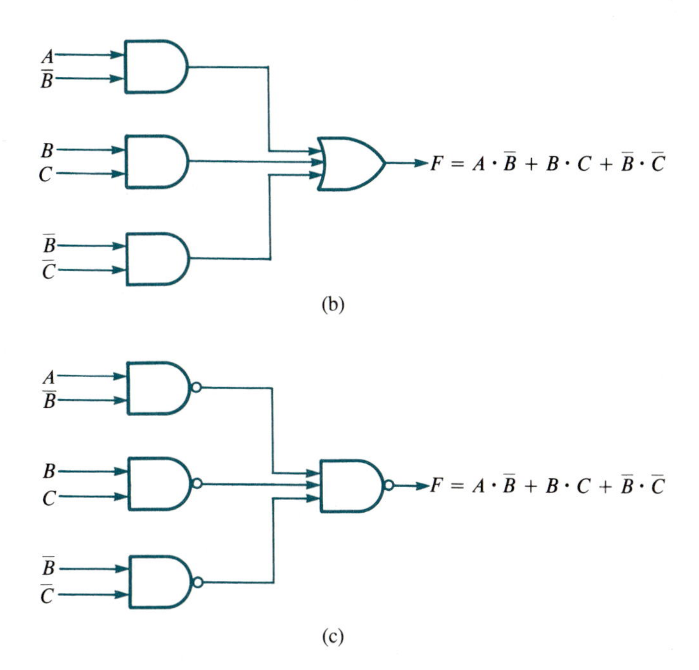

FIG. 10.8-3. (a) A K map for $F = \Sigma\, m_i(0, 3, 4, 5, 7)$; (b) an AND-OR realization of F; and (c) a NAND-NAND realization of F.

2. The two-level AND-OR realization is shown in Fig. 10.8-3(b).
3. The two-level NAND-NAND equivalent is shown in Fig. 10.8-3(c).

NOR Gate

The graphical representation of a two-input NOR gate and its truth table are shown in Fig. 10.8-4(a) and (b), respectively. Figure 10.8-4(c) depicts the equivalent circuitry of a NOR gate using an OR gate followed by an inverter.

Like the NAND gate, the NOR gate is functionally complete; that is, any of the basic operations of the AND, OR, and NOT gates can be obtained by using only NOR gate(s), as shown in Fig. 10.8-5. Therefore, any boolean function can be realized by using NOR gates *only*. The procedure for obtaining a minimum-NOR-gates realization for a boolean expression is similar to that for NAND-gates

FIG. 10.8-4. A two-input NOR GATE: (a) a graphical representation, (b) the truth table, and (c) an equivalent OR-NOT circuit.

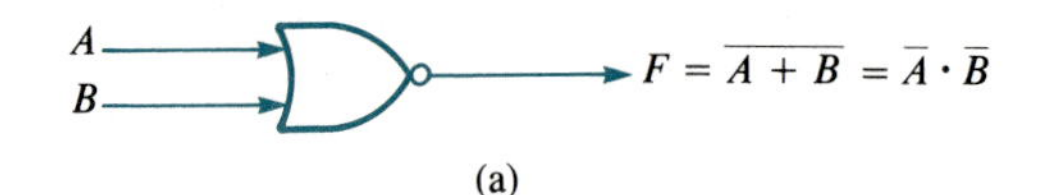

(a)

Inputs		Output
A	B	F
0	0	1
0	1	0
1	0	0
1	1	0

(b)

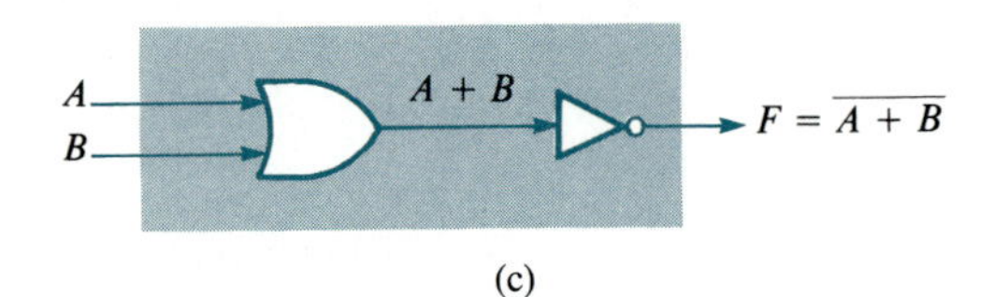

(c)

FIG. 10.8-5. NOR gate realizations of NOT, OR, and AND gates.

Basic Gate	Equivalent NOR Realization
$A \rightarrow \bar{A}$	$A \rightarrow \bar{A}$
$A, B \rightarrow A + B$	$A, B \rightarrow \overline{A + B} \rightarrow A + B$
$A, B \rightarrow A \cdot B$	$A \rightarrow \bar{A}$; $B \rightarrow \bar{B}$; $\bar{A}, \bar{B} \rightarrow A \cdot B$

FIG. 10.8-6. A two-input XOR gate: (a) a graphical representation and (b) the truth table.

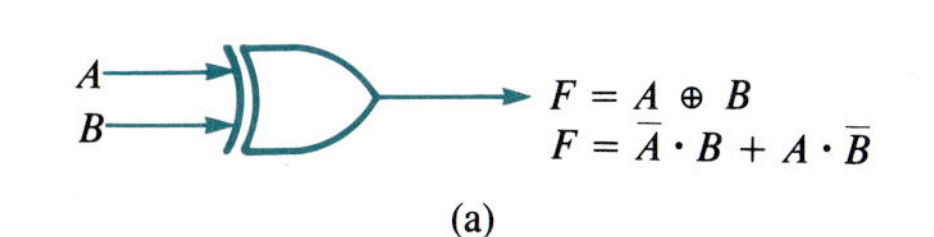

Inputs		Output
A	B	F
0	0	0
0	1	1
1	0	1
1	1	0

(b)

realization, except that we start with a minimum product-of-sums and an OR-AND realization.

Exclusive-OR Gate (XOR)

The graphical representation of a two-input exclusive-OR (XOR) gate and its truth table are shown in Fig. 10.8-6(a) and (b), respectively. The XOR operation is commonly denoted by $\oplus$. This gate is often referred to as a binary comparator. In general, the output of an XOR gate is a logical 1 if and only if there is an odd number of logical 1s at the input side.

Exclusive-NOR Gate (XNOR)

The graphical representation of a two-input exclusive-NOR (XNOR) gate and its truth table are shown in Fig. 10.8-7(a) and (b), respectively. This gate is also known as the EQUIVALENT gate, and its operation is commonly denoted by $\odot$. The XNOR gate is the complement of an XOR gate; that is, its output is a logical 1 if and only if there is an even number of logical 1s at its inputs.

10.9 INTEGRATED CIRCUITS (ICS)

The logic gates described so far, and some other logic devices to be described in the following chapters, are contained inside integrated circuits (ICs), also known as *chips*. An integrated circuit is a semiconductor material containing the elec-

FIG. 10.8-7. A two-input XNOR gate: (a) a graphical representation and (b) the truth table.

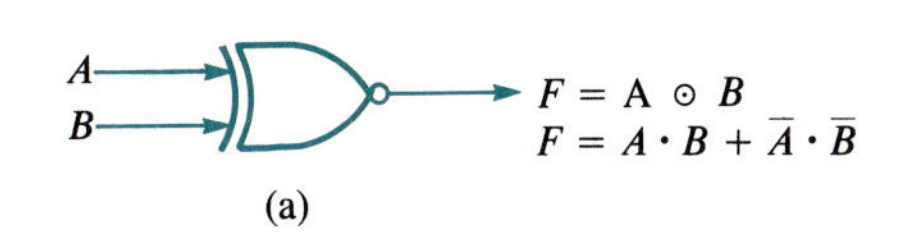

Inputs		Output
A	B	F
0	0	1
0	1	0
1	0	0
1	1	1

(b)

TABLE 10.9-1. Power and Speed Characteristics of TTL Families

YYY	Characteristics
L	Low-power version of standard TTL
LS	Low-power Schottky (lower power and higher speed than standard TTL)
S	Schottky (higher speed than standard TTL)
AS	Advanced Schottky (higher speed and lower power than S-TTL)
ALS	Advanced low-power Schottky (an improved LS-TTL)

tronic components for the digital devices. The integrated circuit is mounted in a plastic or ceramic container that has a set of pins that provides connections to the input(s) and the outputs(s) of the digital devices. Each pin is assigned a pin number that identifies its location on the chip. These pin numbers are usually not printed on the chip itself, but they always occur in a standard arrangement, with pin number 1 always marked. The number of pins may vary from 14 pins in a small IC to 114 or more in larger ICs.

Each IC has an identification number printed on the surface of the package. Integrated circuits are available in different logic families. The transistor-transistor logic (TTL) family is the most popular one. Each TTL IC has a standard 74YYYXXX identification number printed on the surface of the package. The last two or three digits (XXX) specify the function of the device. For example, 00 specifies an IC containing four NAND gates; 86 specifies an IC containing four XOR gates. The middle one, two, or three characters (YYY) specify the power consumption and the speed of the operation of the device. Table 10.9-1 lists the power and speed characteristics of different TTL families. For example, the 74LS00 and 74ALS00 are commercially available ICs that contain four NAND gates each in the low-power Schottky and advanced low-power Schottky series, respectively. Each manufacturer publishes a catalog (data book) that contains the exact logic functions, power consumption, and speed characteristics of its IC devices.

Figure 10.9-1(a) shows a 74LS08 integrated-circuit package with a notch that indicates the position of pin 1. Figure 10.9-1(b) is the pin diagram for the 74LS08 showing the gate layout inside the package and the corresponding pin numbers.

Since the corresponding logic functions of each of the TTL series are the same, we will adopt the apostrophe notation to replace YYY representing the actual TTL series. That is, we will use (′) rather than specifying L, LS, S, AS, or ALS, because in this chapter and the following three chapters we are concerned only with the logic functions of the ICs. For example, 74′00 will be the notation used for 74L00, 74LS00, 74AS00, or 74ALS00.

Integrated circuits are commonly divided into classes, based on the number of transistors per chip, as follows:

Small-scale integration (SSI) devices contain up to 100 transistors.

Medium-scale integration (MSI) devices contain from 100 to 1000 transistors.

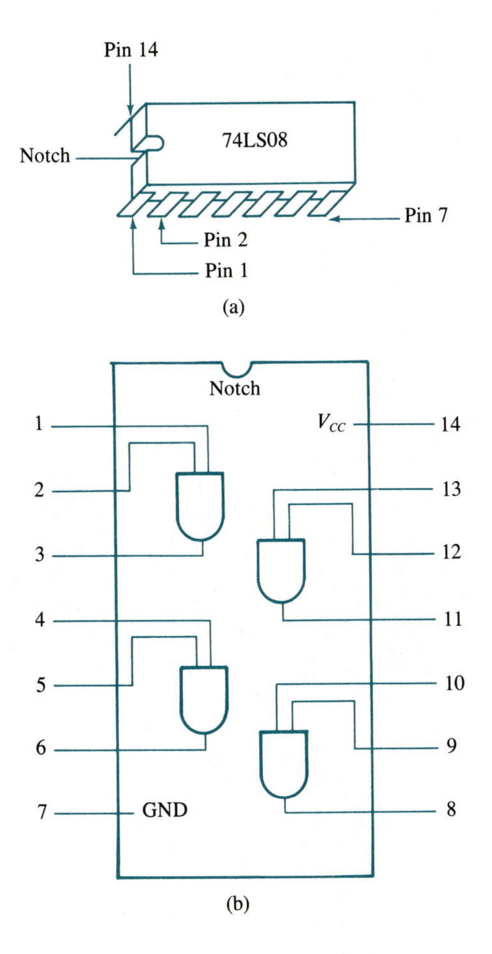

FIG. 10.9-1. A 74LS08 IC: (a) the IC package and (b) the pin-diagram.

Large-scale integration (LSI) devices contain from 1000 to 10,000 transistors. Very large-scale integration (VLSI) devices contain 10,000 or more transistors.

10.10 DESIGN OF COMBINATIONAL NETWORKS

With the background established in the previous sections of this chapter, we are now ready to consider some practical design problems. The design procedure can be outlined by the following steps:

1. Determine the number of inputs and outputs from the specification of a computer circuit and assign a symbolic name to each circuit.
2. Derive the truth table that describes the relationship between the inputs and output(s).

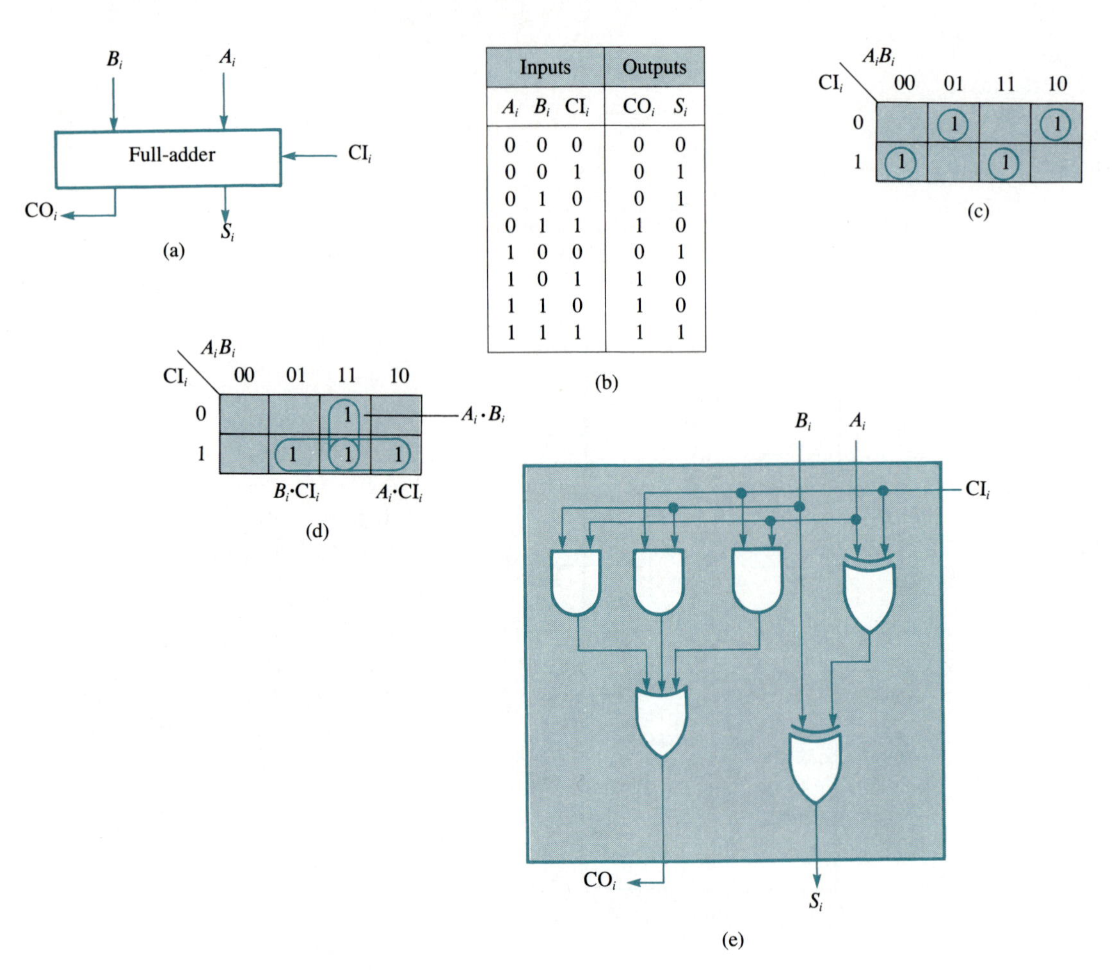

Inputs			Outputs	
A_i	B_i	CI_i	CO_i	S_i
0	0	0	0	0
0	0	1	0	1
0	1	0	0	1
0	1	1	1	0
1	0	0	0	1
1	0	1	1	0
1	1	0	1	0
1	1	1	1	1

FIG. 10.10-1.
The design of a full-adder: (a) a block diagram, (b) the truth table, (c) and (d) the K maps and (e) a logic diagram.

3. Derive a minimum boolean expression for each of the output functions.
4. Draw the logical circuit.

This procedure is best illustrated through the following practical design problems.

Full-Adder (FA)

A full-adder, as shown in Fig. 10.10-1(a), has three inputs and two outputs. It is a combinational network that performs the binary addition of the three inputs. The inputs A_i and B_i represent the two significant bits to be added, and CI_i represents the carryin from the previous significant bits. The outputs S_i and CO_i represent

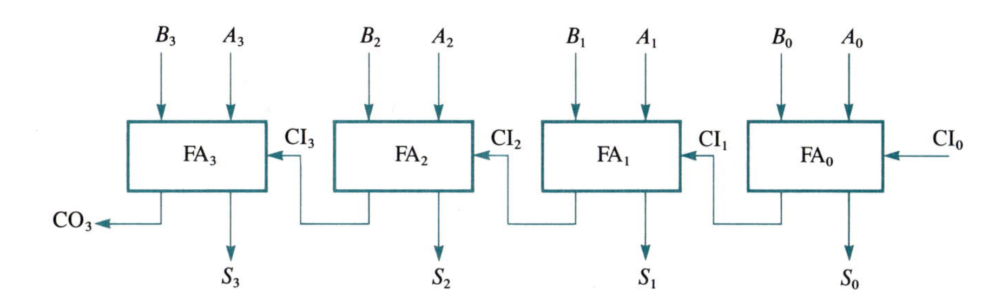

FIG. 10.10-2. A 4-bit binary adder.

the sum and carryout bits, respectively. For example, if $A_i = 1$, $B_i = 0$, and $CI_i = 1$, then $S_i = 0$ and $CO_i = 1$; that is, the addition of A_i, B_i, and CI_i $(1 + 0 + 1)$ will produce the binary number $(10)_2$, which corresponds to $S_i = 0$ and $CO_i = 1$. The truth table of the full-adder is shown in Fig. 10.10-1(b). The reader is encouraged to verify that the values of the outputs are determined from the addition of the three input bits. Using K maps, as shown in Fig. 10.10-1(c) and (d) for S_i and CO_i, respectively, we obtain the following boolean expressions for S_i and CO_i:

$$S_i = \overline{A}_i \cdot \overline{B}_i \cdot CI_i + \overline{A}_i \cdot B_i \cdot \overline{CI}_i + A_i \cdot \overline{B}_i \cdot \overline{CI}_i + A_i \cdot B_i \cdot CI_i$$

$$CO_i = A_i \cdot B_i + A_i \cdot CI_i + B_i \cdot CI_i$$

The above expressions can be realized by using any type of logic gates. However, a quick look at the S_i expression reveals that this is an XOR expression for the variables A_i, B_i, and CI_i. Therefore, the logic circuit for the full-adder can be simplified as shown in Fig. 10.10-1(e).

To obtain the binary addition of two n-bit binary numbers, we cascade n full-adder circuits together, with the carryin (CI_i) of a full-adder being connected to the carryout (CO_{i-1}) of the previous full-adder. The interconnectin of four full-adders to provide the addition of two 4-bit binary numbers is shown in Fig. 10.10-2.

MSI (medium-scale integrated-circuit) packages are available that contain 4- and 8-bit binary adders. The 74LS283, shown in Fig. 10.10-3(a), is an example of an MSI package containing a 4-bit binary adder. Figure 10.10-3(b) is a block diagram of an 8-bit binary adder using two 74LS283 devices.

Magnitude Comparator

A 2-bit magnitude comparator, as shown in Fig. 10.10-4(a), has two 2-bit binary numbers as inputs and three outputs. It is a combinational network that indicates whether the binary number $(A_1A_0)_2$ is equal to, less than, or greater than the binary number $(B_1B_0)_2$. Let us denote the output that corresponds to $(A_1A_0)_2 = (B_1B_0)_2$ by F_E, that corresponds to $(A_1A_0)_2 > (B_1B_0)_2$ by F_G, and that corresponds to $(A_1A_0)_2 < (B_1B_0)_2$ by F_S. The functional description of this magnitude comparator is given in Fig. 10.10-4(b) in the form of a truth table. Observe that $F_E = 1$ if and only if $(A_1A_0)_2 = (B_1B_0)_2$, $F_G = 1$ if and only if $(A_1A_0)_2 > (B_1B_0)_2$, and $F_S = 1$ if and only if $(A_1A_0)_2 < (B_1B_0)_2$, as shown in Fig. 10.10-4(b).

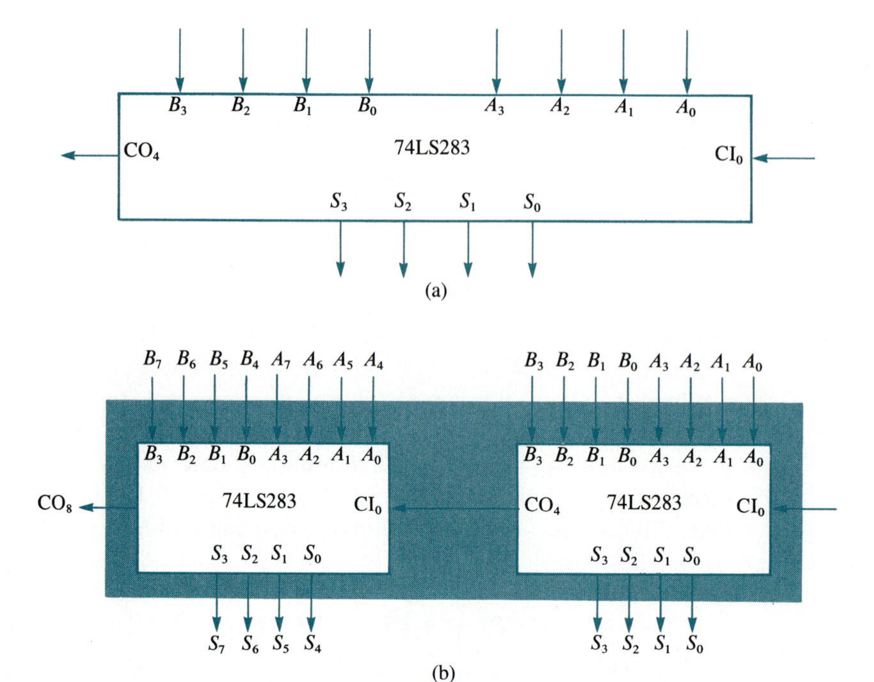

FIG. 10.10-3. A 74LS283 binary adder: (a) a functional block diagram and (b) an 8-bit binary adder.

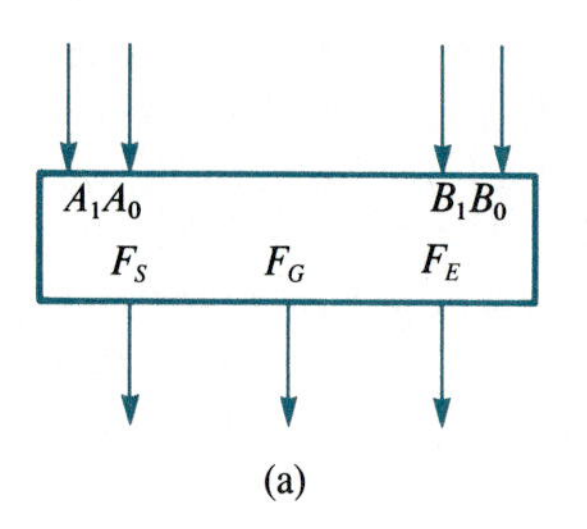

A_1	A_0	B_1	B_0	F_S	F_G	F_E
0	0	0	0	0	0	1
0	0	0	1	1	0	0
0	0	1	0	1	0	0
0	0	1	1	1	0	0
0	1	0	0	0	1	0
0	1	0	1	0	0	1
0	1	1	0	1	0	0
0	1	1	1	1	0	0
1	0	0	0	0	1	0
1	0	0	1	0	1	0
1	0	1	0	0	0	1
1	0	1	1	1	0	0
1	1	0	0	0	1	0
1	1	0	1	0	1	0
1	1	1	0	0	1	0
1	1	1	1	0	0	1

(b)

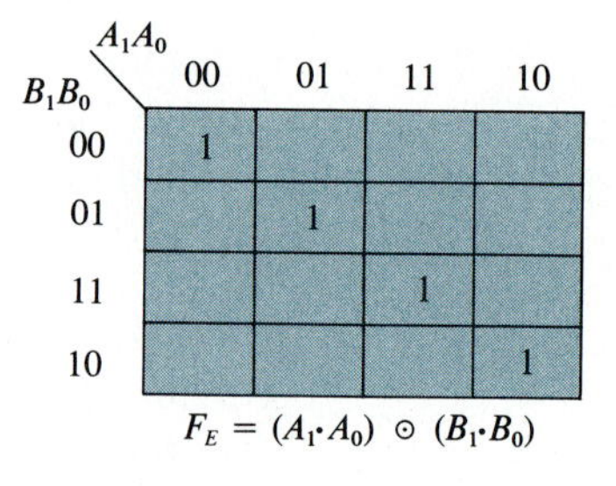

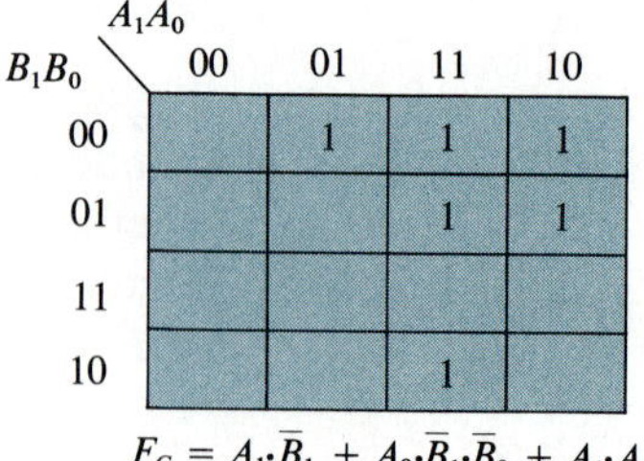

(c)

FIG. 10.10-4. The design of a 2-bit magnitude comparator: (a) a functional block diagram, (b) the truth table, (c) the K map, and (d) the logic diagram.

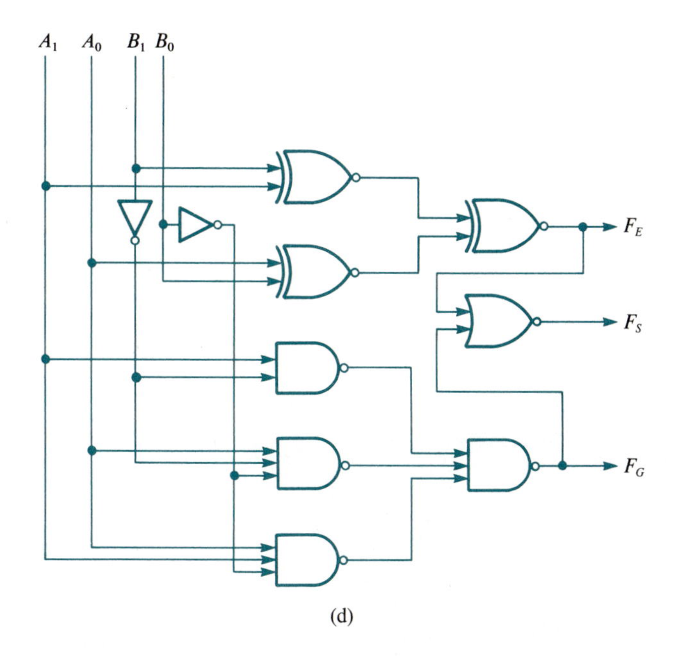

(d)

Now that the truth table is complete, we follow our design procedure and realize the logic circuitry for the 2-bit magnitude comparator, as shown in Fig. 10.10-4(c) and (d). Observe that we implemented F_E by using only 3 two-input XNOR gates. Also, we realized only two outputs, F_E and G_G, directly, since the third output F_S can be realized more economically by using the fact that if $F_E = 0$ and $F_G = 0$, then F_S has to be 1. Therefore, F_S can be expressed as $F_S = \overline{F_E} \cdot \overline{F_G}$ and can be realized by only 1 two-input NOR gate.

MSI packages that contain the logic circuitry for 4-bit magnitude comparators are commercially available. The 74LS85, shown in Fig. 10.10-5(a), is an example of such a device. In addition to the inputs and outputs we described above, the 74LS85 has three more inputs used for cascading such a device to obtain a $4n$-bit magnitude comparator, where $n = 1, 2, 3, \ldots$. Figure 10.10-5(b) is a block diagram of an 8-bit magnitude comparator using two 74LS85 devices.

10.11 DECODERS, ENCODERS, AND MULTIPLEXERS

Decoders

In the digital world, an n-bit binary code is capable of encoding up to 2^n distinct elements of information. For example, a 3-bit code can encode up to 8 distinct elements, and an 8-bit code can encode up to 256 distinct elements. A decoder is a

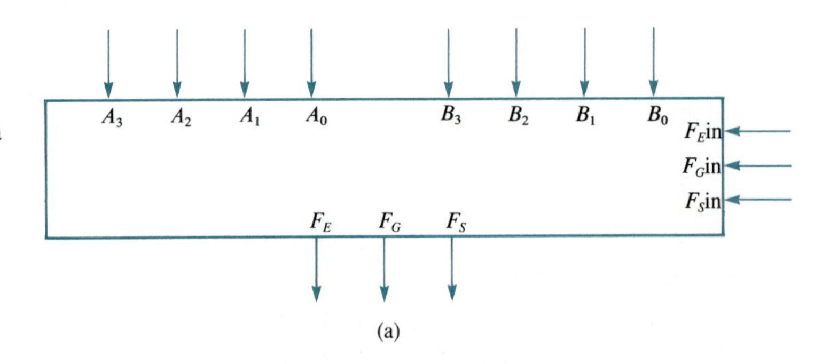

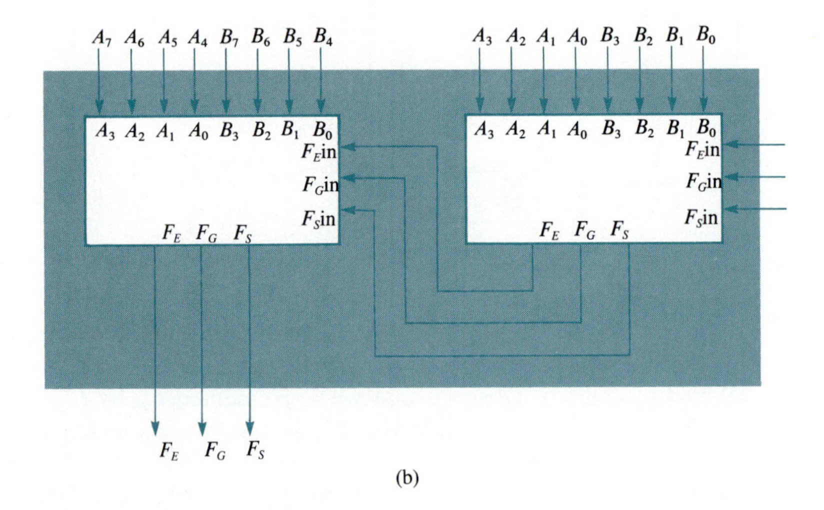

FIG. 10.10-5. A 74LS85 magnitude comparator: (a) a functional block diagram and (b) an 8-bit magnitude comparator.

combinational network that decodes (converts) the n-bit binary-coded input to m outputs, $m \leq 2^n$.

The block diagram of a 3-bit to 8-element decoder is shown in Fig. 10.11-1(a). The three inputs are decoded into eight outputs, one for each combination of the input variables. The functional description of the 3-to-8 decoder is shown in Fig. 10.11-1(b) in the form of a truth table. Observe that for each input combination, there is only one output that is equal to 1; that is, each combination selects only one of the eight outputs. The logic diagram of the 3-to-8 decoder is shown in Fig. 10.11-1(c).

Since decoding is so common in digital design, integrated circuits for decoders are available in different forms. Decoders are commercially available as MSI packages in the form of 2-to-4, 3-to-8, and 4-to-10 decoders. For example, the 74LS42 is a 4-to-10 decoder, and the 74LS138 is a 3-to-8 decoder.

Encoders

Encoding is the converse of the decoding operation. It is the process of forming an encoded representation of a set of inputs. An encoder is a combinational network

FIG. 10.11-1.
A 3-to-8 decoder: (a) a functional block diagram, (b) the truth table, and (c) a logic diagram.

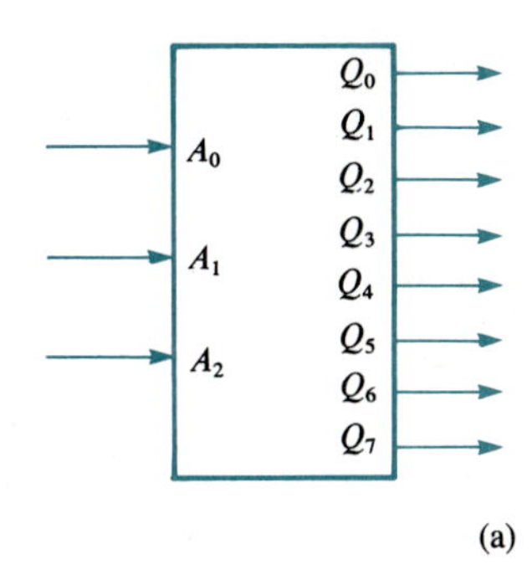

(a)

A_0	A_1	A_2	Q_7	Q_6	Q_5	Q_4	Q_3	Q_2	Q_1	Q_0
0	0	0	0	0	0	0	0	0	0	1
0	0	1	0	0	0	0	0	0	1	0
0	1	0	0	0	0	0	0	1	0	0
0	1	1	0	0	0	0	1	0	0	0
1	0	0	0	0	0	1	0	0	0	0
1	0	1	0	0	1	0	0	0	0	0
1	1	0	0	1	0	0	0	0	0	0
1	1	1	1	0	0	0	0	0	0	0

(b)

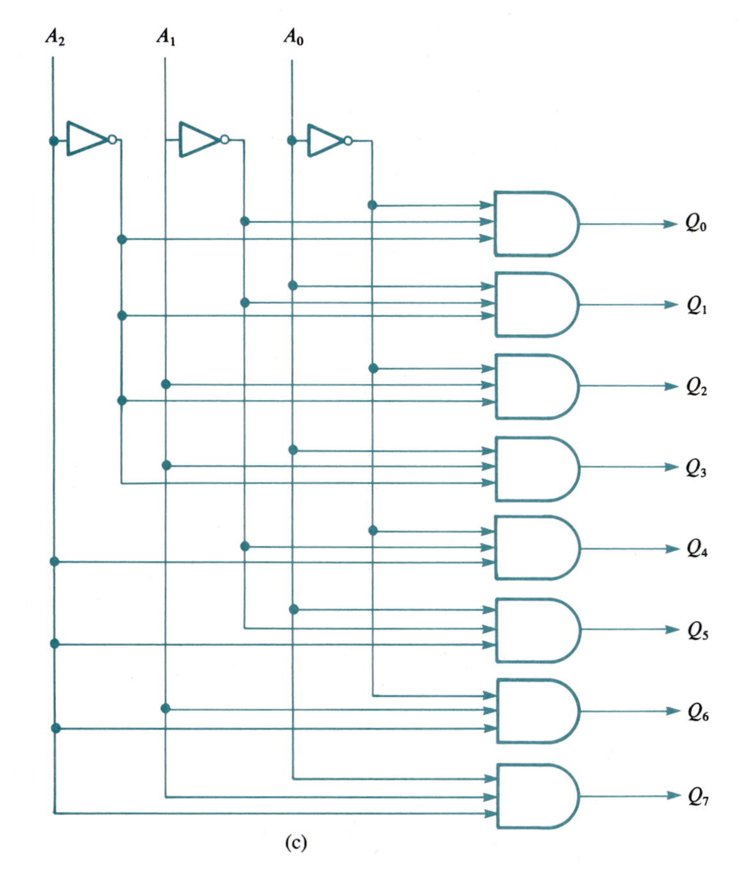

(c)

that generates an n-bit binary code that uniquely identifies the one out of m activated inputs, $0 \leq m \leq 2^n - 1$.

The block diagram of an 8-element to 3-bit encoder is shown in Fig. 10.11-2(a). The eight inputs, I_0 through I_7, are encoded by using a 3-bit binary code, $n_2n_1n_0$. The functional description of the 8-to-3 encoder is shown in Fig. 10.11-2(b) in the form of a truth table. Observe that for this definition of an

FIG. 10.11-2. An 8-to-3 encoder: (a) a functional block diagram, (b) the truth table, and (c) a logic diagram.

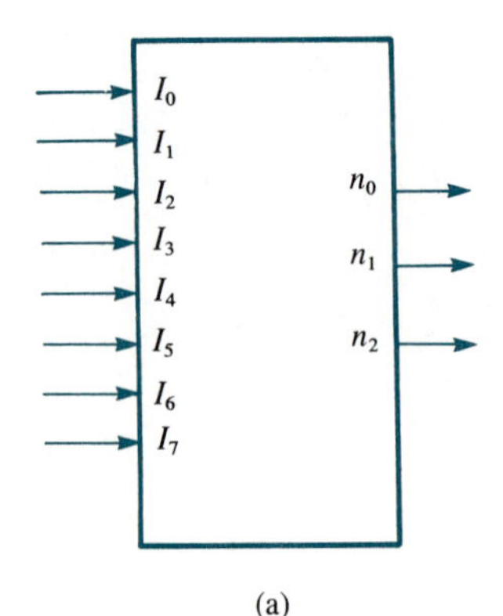

(a)

Inputs								Outputs		
I_7	I_6	I_5	I_4	I_3	I_2	I_1	I_0	n_2	n_1	n_0
0	0	0	0	0	0	0	1	0	0	0
0	0	0	0	0	0	1	0	0	0	1
0	0	0	0	0	1	0	0	0	1	0
0	0	0	0	1	0	0	0	0	1	1
0	0	0	1	0	0	0	0	1	0	0
0	0	1	0	0	0	0	0	1	0	1
0	1	0	0	0	0	0	0	1	1	0
1	0	0	0	0	0	0	0	1	1	1

(b)

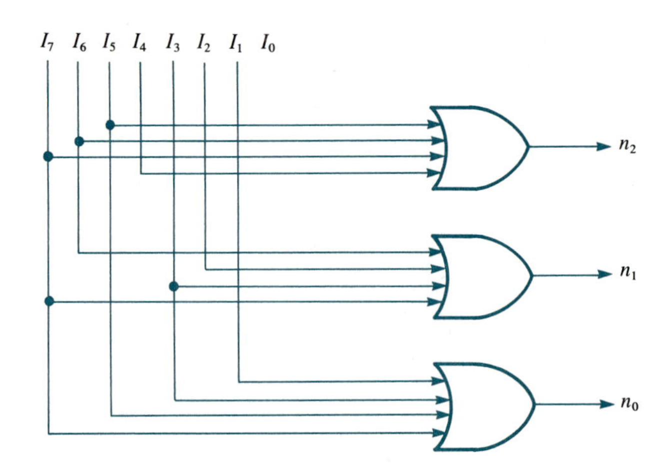

FIG. 10.11-3. A mechanical multiplex switch.

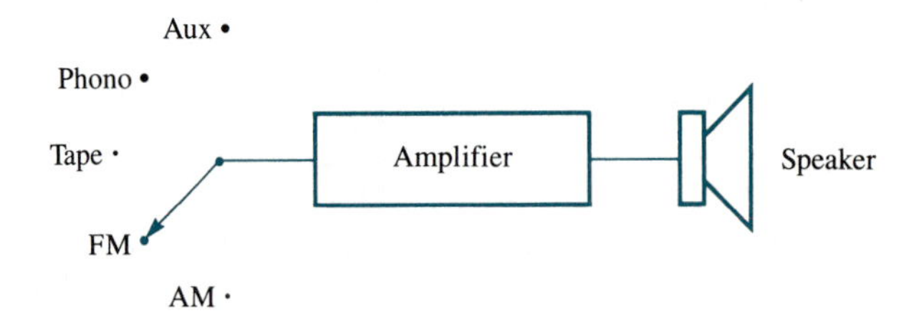

encoder, only one of the eight inputs is allowed to be activated at any given time. The logic diagram of the 8-to-3 encoder is shown in Fig. 10.11-2(c).

Multiplexers

A multiplexer is a combinational network that selects one of several possible input signals and directs that signal to a single output terminal. It is analogous to a mechanical switch, such as the selector switch of a stereo amplifier, as shown in Fig. 10.11-3. The selection of a particular input line is controlled by a set of

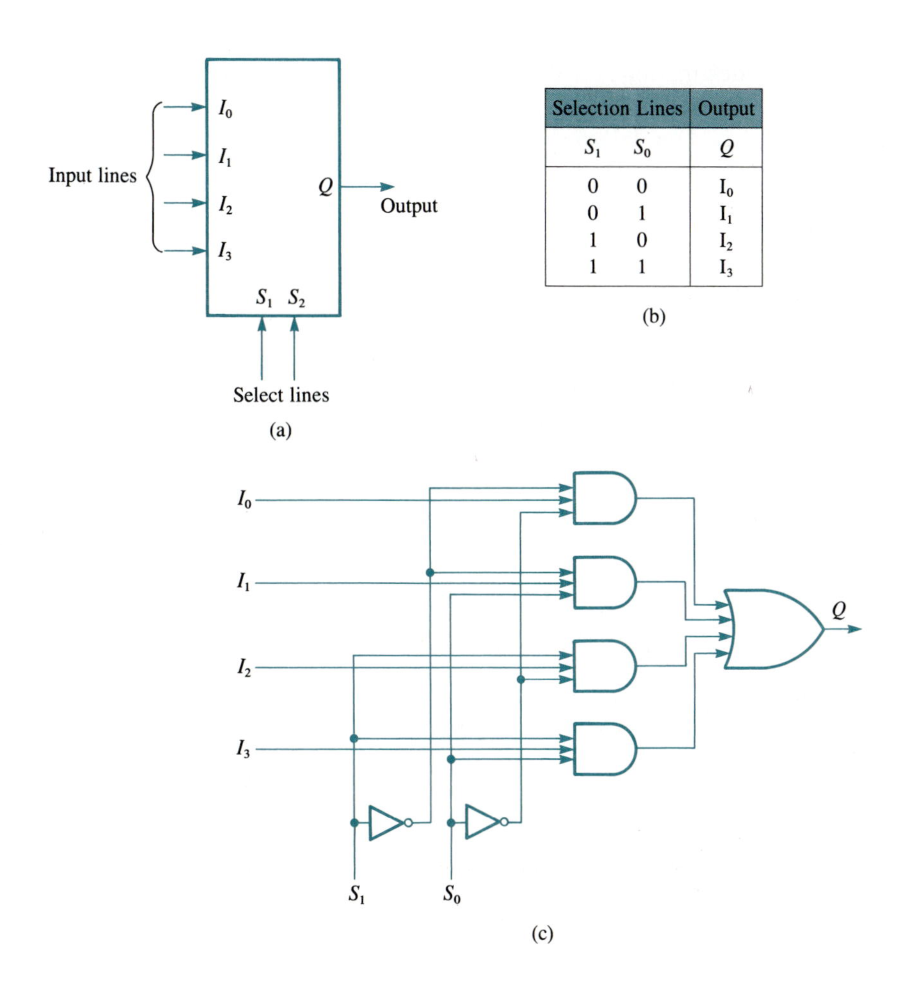

Selection Lines		Output
S_1	S_0	Q
0	0	I_0
0	1	I_1
1	0	I_2
1	1	I_3

FIG. 10.11-4. A 4-to-1 multiplexer: (a) a functional block diagram, (b) the truth table, and (c) a logic diagram.

selection variables. Usually, a multiplexer with n selection variables can select one out of 2^n input signals.

The block diagram of a 4-to-1 multiplexer is shown in Fig. 10.11-4(a). The functional description of the 4-to-1 multiplexer is shown in Fig. 10.11-4(b) in the form of a truth table. Each of the four inputs I_0 through I_3 is selected by S_1 and S_0 and directed to the output Q. The logic diagram of the 4-to-1 multiplexer is shown in Fig. 10.11-4(c). To demonstrate the circuit operation, consider the case where $(S_1S_0)_2 = (10)_2$, which corresponds $(S_1S_0)_2 = 2$ in decimal. Tracing the input signals I_0 through I_3, we obtain $Q = I_2$; that is, only the input whose *address* equals 2 is directed to the output. In general, only the input whose address is given by the select lines is directed to the output.

Two-, 4-, 8-, and 16-to-1 multiplexers are commercially available as MSI packages. For example, the 74LS151 is an 8-to-1 multiplexer, and the 74LS352 is a dual 4-to-1 multiplexer.

Finally, if we wish to send one input signal to one of n output lines, we would use a *demultiplexer*. A demultiplexer sends data from one source to one of several destinations. A demultiplexer is a data distributor, whereas the multiplexer is a data selector.

The design of a 1-to-4 demultiplexer will be left as a design problem for the reader.

PROBLEMS

10-1. Convert the following binary numbers into decimal: (a) 1011011, (b) .00101, and (c) 11101.101.

10-2. Convert the following decimal numbers into binary: (a) 239, (b) 0.375, and (c) 1259.00125.

10-3. Convert the following octal numbers into decimal: (a) 257, (b) .321, and (c) 2103.45.

10-4. Convert the following decimal numbers into octal: (a) 65,535, (b) 0.125, and (c) 379.25.

10-5. Convert the following octal numbers into binary: (a) 3651, (b) .214, and (c) 4125.016.

10-6. Convert the following binary numbers into octal: (a) 1011010, (b) .110101, and (c) 1110110111.1011.

10-7. Convert the following hexadecimal numbers into decimal: (a) 1ABC2, (b) .0E3, and (c) 256.72.

10-8. Convert the following decimal numbers into hexadecimal: (a) 1023, (b) .00125, and (c) 449.375.

10-9. Convert the following binary numbers into hexadecimal: (a) 1011011101, (b) .11101, and (c) 1101110001.11011110.

10-10. Convert the following hexadecimal numbers into binary: (a) 2ABF5, (b) 0.0DC5, and (c) 15CE.FB3.

10-11. Convert the following decimal numbers into BCD: (a) 1978, (b) 0.659, and (c) 2153.436.

10-12. Convert the following BCD numbers into decimal: (a) 010101100111, (b) .100110000100, and (c) 10010010.00000001.

10-13. Perform the following binary arithmetic operations: (a) $(1011010110)_2 + (0010111101)_2$, (b) $(0110110.1110)_2 + (1010011.1011)_2$, (c) $(110101101)_2 - (101101010)_2$, (d) $(1101.0101)_2 - (0101.1010)_2$, (e) $(10110111)_2 \times (01010110)_2$, (f) $(10111.010)_2 \times (01101.101)_2$, (g) $(10101100)_2 \div (1010)_2$, and (h) $(1110.1101)_2 \div (10.1)_2$.

10-14. Obtain the 1's complement of the following binary numbers: (a) 01110110, (b) 0.101110, and (c) 01101.0001.

10-15. Obtain the 2's complement of the following binary numbers: (a) 00110110, (b) 0.000100, and (c) 01101110.0110110.

10-16. Represent the following signed numbers in 1's-complement format, using 8 bits: (a) $(+127)_{10}$, (b) $(-127)_{10}$, (c) $(+\ 56)_{10}$, and (d) $(-74)_{10}$.

10-17. Repeat Problem 10-16 using 2's-complement format.

10-18. Perform the following arithmetic operations, using 2's-complement arithmetic: (a) 32 + 17, (b) 25 − 14, and (c) 45 − 64.

10-19. Repeat Problem 10-18 using 1's-complement arithmetic.

10-20. Prepare a truth table for each of the following boolean expressions: (a) $F(A, B) = A \cdot \overline{B} + \overline{A} \cdot B$, (b) $F(A, B, C) = A \cdot \overline{B} \cdot \overline{C} + \overline{A} \cdot \overline{B} \cdot C + A \cdot B \cdot C$, and (c) $F(A, B, C, D) = A \cdot B + \overline{C} \cdot \overline{D}$.

10-21. Without using K maps, simplify the following boolean expressions: (a) $F = A \cdot \overline{B} + A \cdot B$, (b) $F = A \cdot C + C \cdot D + B \cdot C \cdot D$, and (c) $F = A \cdot B \cdot \overline{C} + B \cdot C + A \cdot B \cdot D + B \cdot C \cdot D$.

10-22. Using K maps, simplify the functions given in Problem 10-21.

10-23. Using K maps, simplify the following boolean functions into their minimum (SOP) form: (a) $F(A, B) = \Sigma\, m_i(0, 1, 3)$, (b) $F(A, B, C) = \Sigma\, m_i(0, 2, 3, 4, 5, 6)$, and (c) $F(A, B, C, \text{D}) = \Sigma\, m_i(0, 4, 5, 6, 7, 12, 14)$.

10-24. Repeat Problem 10-23 for the following boolean functions: (a) $F(A, B) = \Pi\, M_i(0, 2)$, (b) $F(A, B, C) = \Pi\, M_i(0, 6)$, and (c) $F(A, B, C, D) = \Pi\, M_i(1, 3, 11, 14, 15)$.

10-25. Derive the boolean expressions for the logic circuits of Fig. P10-25.

FIG. P10-25.

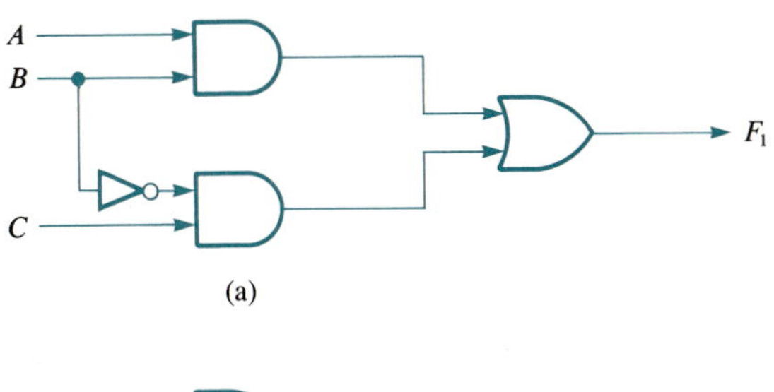

(a)

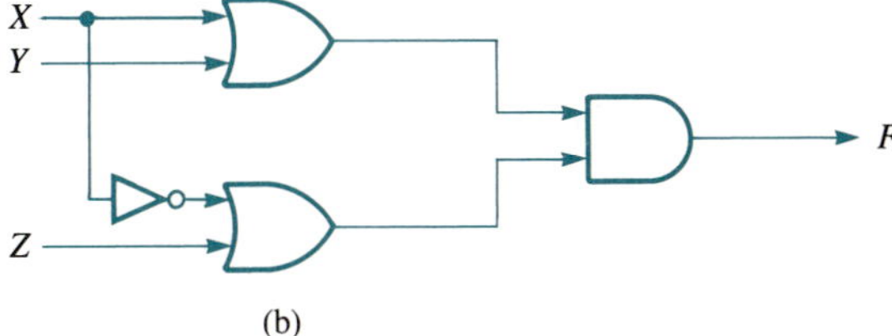

(b)

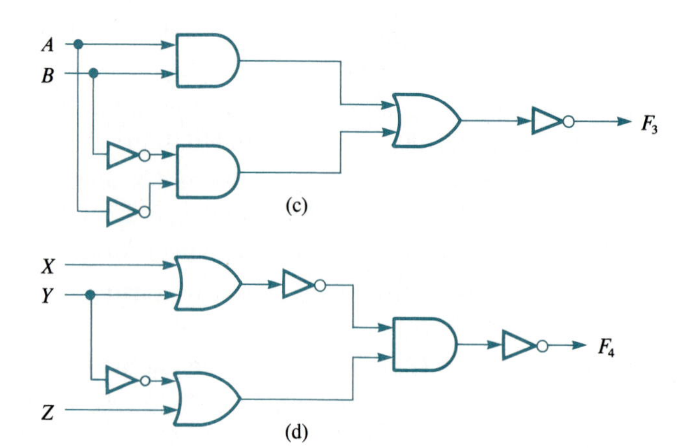

10-26. Draw the logic diagram for the following boolean expressions, without simplification: (a) $F_1 = A \cdot B + \overline{B} \cdot C + A \cdot B \cdot D + A \cdot C \cdot D$, (b) $F_2 = (X + Y) \cdot (\overline{X} + Z) \cdot (Y + \overline{Z})$, (c) $F_3 = \overline{\overline{A \cdot C} + B \cdot \overline{C} + A \cdot B \cdot C}$, and (d) $F_4 = \overline{(X + Y) \cdot \overline{(X + \overline{Z})} \cdot (Y + Z)}$.

10-27. Repeat Problem 10-26 with simplification.

10-28. Derive the boolean expressions for the logic circuits of Fig. P10-28.

FIG. P10-28.

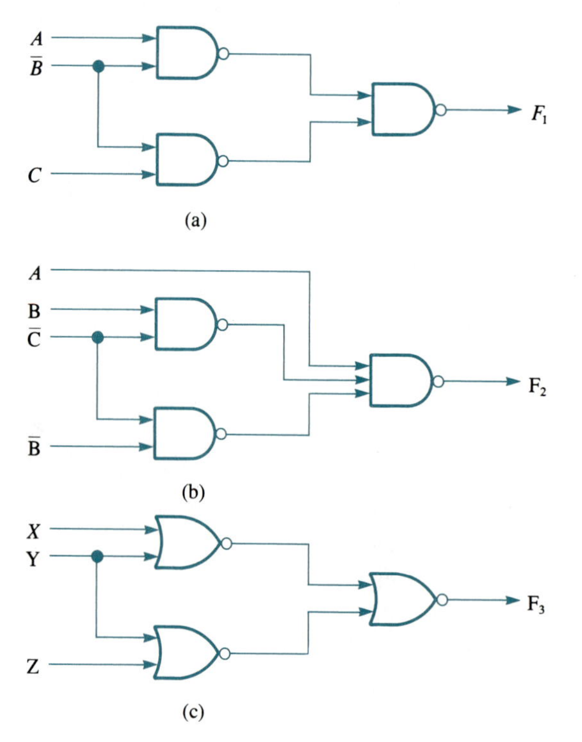

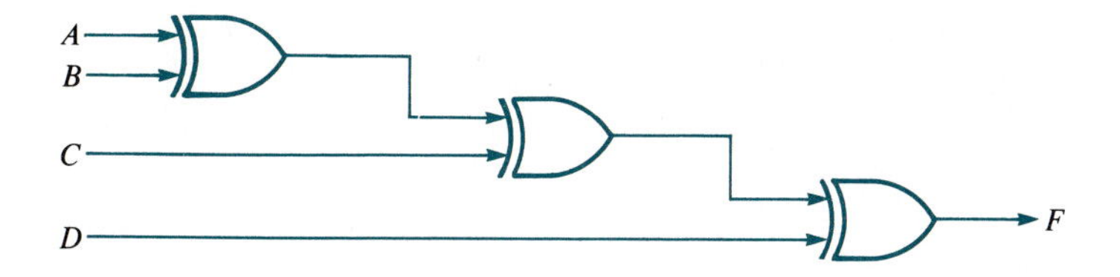

10-29. Obtain a minimum two-level NAND-NAND realization for the following expressions: (a) $F_1(A, B) = \Sigma\ m_i(0, 3) + d_1$, (b) $F_2(A, B, C) = \Sigma\ m_i(1, 6) + \Sigma\ d_i(2, 4, 5)$, and (c) $F_3(A, B, C, D) = \Sigma\ m_i(0, 4, 5, 7, 13) + \Sigma\ d_i(2, 6, 8, 10, 11)$, where $\Sigma\ d_i(\quad)$ means the sum of minterms that correspond to don't-care outputs.

10-30. Repeat Problem 10-29 using NOR gates only.

10-31. The logic circuit of Fig. P10-31 is known as a *parity checker* for a 4-bit binary number. (a) Derive its boolean expression. (b) Prepare its truth table. (c) Can you realize the same function by using fewer logic gates?

FIG. P10-31.

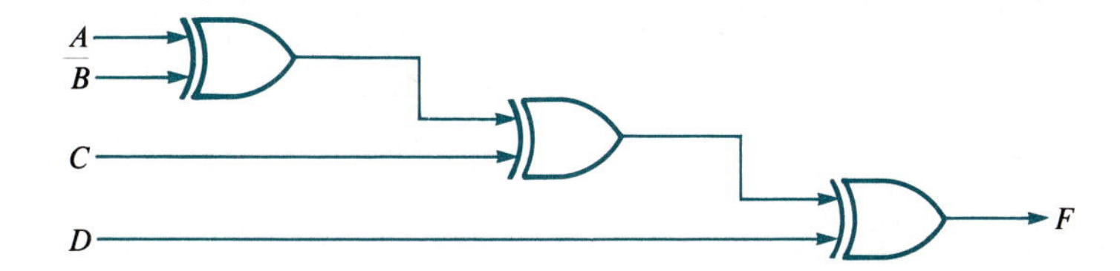

10-32. The logic circuit of Fig. P10-32 is one part of the magnitude comparator used to compare two binary numbers. (a) Prepare its truth table. (b) Can you describe this circuit?

FIG. P10-32.

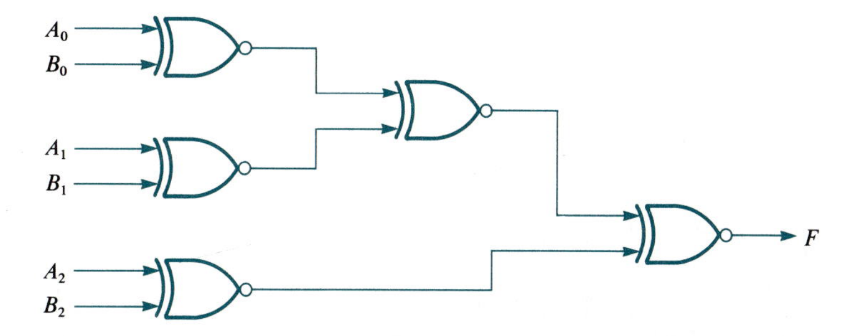

10-33. The pin diagrams for 74′00, 74′02, 74′08, 74′32, and 74′86 ICs are shown in Fig. P10-33. Now, show the pins' interconnection to realize (a) $F_1 = A \cdot B + C \cdot D$, using AND-OR; (b) $F_1 = A \cdot B + C \cdot D$, using NAND-NAND; (c) $F_2 = (x + y) \cdot (z + w)$, using OR-AND; (d) $F_2 = (x + y) \cdot (z + w)$, using NOR-NOR; and (e) $F_3 = A \oplus B \oplus C$.

FIG. P10-33.

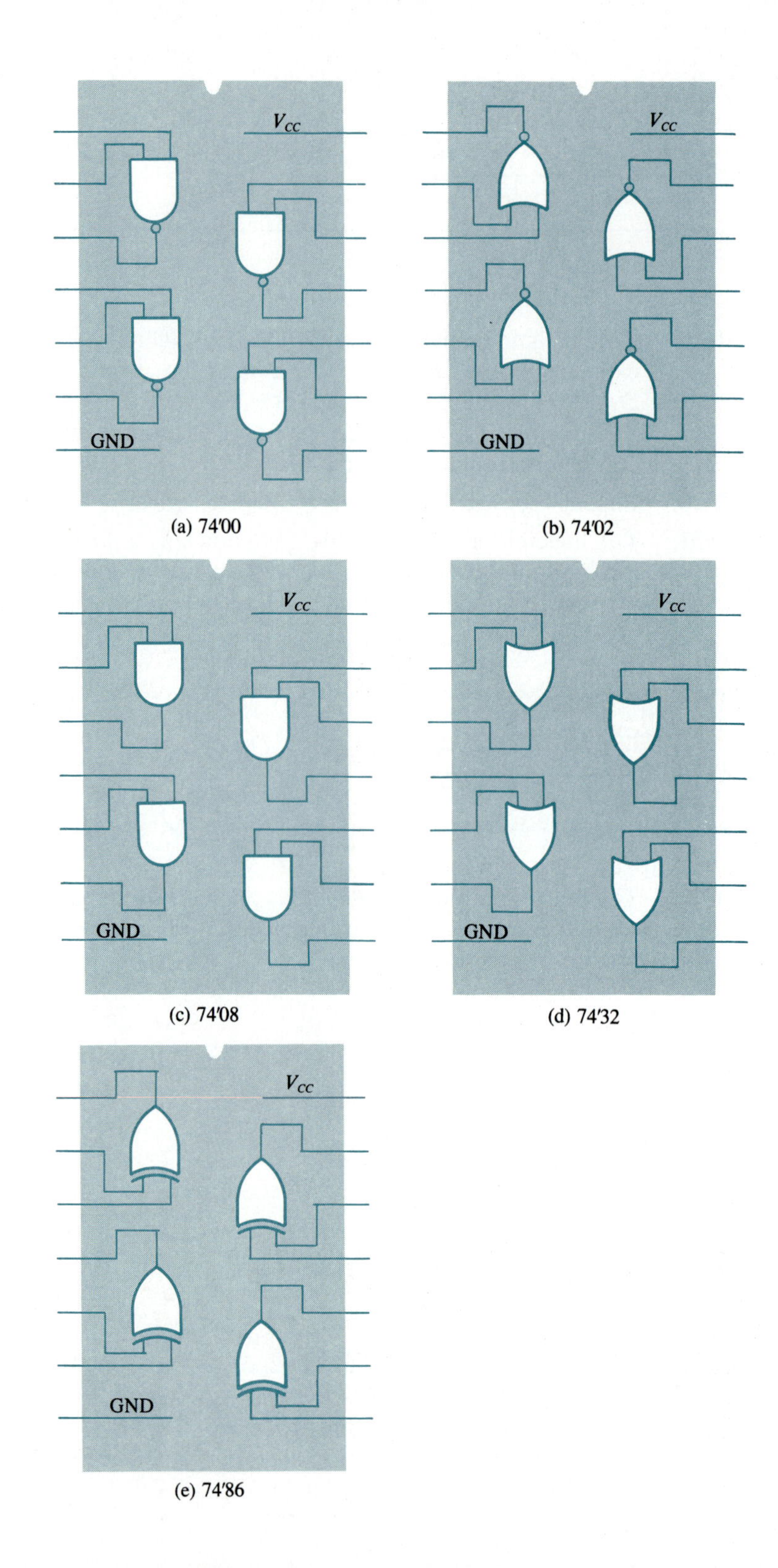

(a) 74'00

(b) 74'02

(c) 74'08

(d) 74'32

(e) 74'86

10-34. Implement the following boolean functions by using one 3-to-8 decoder and 3 three-input OR gates:

$$F_1(A, B, C) = \Sigma\, m_i(1, 2, 3)$$

$$F_2(A, B, C) = \Sigma\, m_i(2, 4, 6)$$

$$F_3(A, B, C) = \Sigma\, m_i(3, 5, 7)$$

10-35. Implement the following boolean functions by using 8-to-1 multiplexers:

$$F_1(A, B, C) = \Sigma\, m_i(0, 2, 4, 6)$$

$$F_2(A, B, C) = \Sigma\, m_i(1, 3, 7)$$

DESIGN EXERCISES

D10-1. Design a combinational network whose input is a 4-bit binary number and whose output is the 1's complement of the input number.

D10-2. Design a combinational network for a full-subtractor that performs the subtraction between 2 bits. This network should take into account that a 1 may have been borrowed by a previous stage.

D10-3. Redesign the full-subtractor, using full-adder circuitry and any other combinational gates.

D10-4. Using only four full-adders and four XOR gates, design a 4-bit adder/subtractor. That is, the same circuit should perform the addition or the subtraction of two 4-bit binary numbers. (*Hint*: $1 \oplus B = \overline{B}$ and $0 \oplus B = B$).

D10-5. A long hallway has three doors, one at each end and one in the middle. A switch is located at each door to control the hallway's light. Assuming that the light is off when switch variables have the values 0, 0, and 0, design a combinational network that controls the light.

D10-6. Design a combinational network that will be used to control an alarm bell. This alarm bell is to be installed in a room to protect it from unauthorized entry. Sensor devices provide the following logic signals:

$C = 1$	The control system is active.
$D = 1$	The room door is closed.
$M = 1$	There is a motion in the room.
$O = 1$	The room is open to public.

D10-7. A majority function is generated in a combinational network when the output is equal 1 if the input variables have more 1s than 0s. The output is 0 otherwise. Design a four-input majority function.

D10-8. Design an 8-to-1 multiplexer.

D10-9. Design a 16-to-1 multiplexer using two 8-to-1 multiplexers and one 2-to-1 multiplexer.

D10-10. Design a 1-to-4 demultiplexer. The block diagram and the functional description of this demultiplexer are as given in Fig. D10-10.

FIG. D10-10.

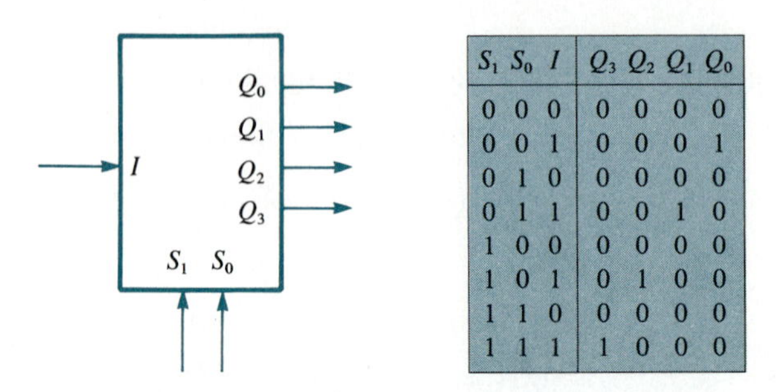

S_1	S_0	I	Q_3	Q_2	Q_1	Q_0
0	0	0	0	0	0	0
0	0	1	0	0	0	1
0	1	0	0	0	0	0
0	1	1	0	0	1	0
1	0	0	0	0	0	0
1	0	1	0	1	0	0
1	1	0	0	0	0	0
1	1	1	1	0	0	0

COMPUTER EXERCISES

The following problems are intended to be solved using the NUMCOM program provided in the software package. This software package is available to your instructor with the permission to make as many copies as needed for class use.

C10-1. Use the NUMCON computer program to convert the following decimal numbers into binary, octal, and hexadecimal: (a) 2576, (b) 0.75025, and (c) 39,732.00125.

C10-2. Use the NUMCON computer program to convert the following binary numbers into decimal, octal, and hexadecimal: (a) 1011011010, (b) .1011101, and (c) 111011011.111011.

C10-3. Use the NUMCON computer program to convert the following octal numbers into decimal, binary, and hexadecimal: (a) 124576, (b) .27312, and (c) 6456.0134.

C10-4. Use the NUMCON computer program to convert the following hexadecimal numbers into decimal, binary, and octal: (a) 67A3B, (b) .0BC2, and (c) 59F.ED0F.

CHAPTER 

Digital System Components

11.0 INTRODUCTION

The preceding chapter was primarily devoted to the analysis and design of combinational logic networks. The outputs of combinational networks are solely dependent on the input conditions at the instant of time when the outputs are being observed. That is, the outputs of combinational networks at any instant of time are a function of the inputs at the same instant of time. Combinational networks have no memory. Although every digital system is likely to have combinational networks, most systems encountered in practice also include storage elements. Computers store data for various periods of time, and the basic gate networks can be adapted for storage. The basic devices used for storage in digital computers are called *latches* and *flip-flops*. Latches and flip-flops provide the memory elements of a digital system.

This chapter begins with a description of latches and flip-flops and their characteristics. Discussion of the use of these devices and the basic combinational gates to perform different functions in a digital system follows the introduction to these devices. Different types of memory are also discussed in this chapter. We then proceed to a discussion of display devices and analog-to-digital (A/D) and digital-to-analog (D/A) converters.

11.1 LATCHES AND FLIP-FLOPS

Flip-flops are the basic devices for storing information in a digital system. They maintain their binary states indefinitely as long as power is delivered to the system. There are different types of flip-flops with different circuit implementations. However, all types of these devices have the same two characteristics: (1) All flip-flops have two outputs, with one of them being the complement of the other, and (2) all flip-flops are bistable devices, that is, devices with only two stable outputs. The most basic types of flip-flops respond to changes in input-signal levels and are known as *latches*. Flip-flops constructed from latches respond to input changes only at a momentary transition in a control input signal, usually known as the *clock*.

Latches

Figure 11.1-1(a) shows a block diagram of an *SR* latch. It has two inputs, designated by S and R, and two outputs, designated by Q and $\overline{Q}$. Figure 11.1-1(b) shows the actual implementation of the *SR* latch using NOR gates. The functional truth table of the *SR* latch is shown in Fig. 11.1-1(c). Observe that for the input combination $S = 0$ and $R = 0$, the outputs $Q = Q_p$ and $\overline{Q} = \overline{Q}_p$, where the subscript p denotes the previous values of Q and $\overline{Q}$. That is, for this input combination, the outputs do not change. So if the present outputs of the *SR* latch are $Q = 1$ and $\overline{Q} = 0$, then applying 0s at the S and R input means that Q and $\overline{Q}$ will still remain at 1 and 0, respectively. Now, if we apply a 0 and a 1 to S and R, respectively, then Q will be reset to 0 and $\overline{Q}$ will be set to 1, regardless of their previous values. Also, if we apply a 1 and a 0 to S and R, respectively, then Q will

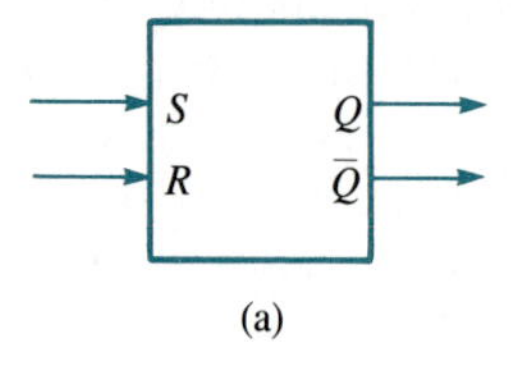

(a)

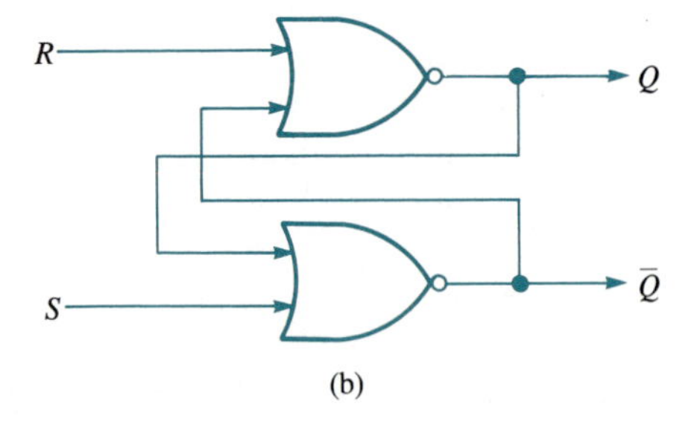

(b)

Inputs		Outputs	
S	R	Q	$\overline{Q}$
0	0	Q_p	$\overline{Q}_p$
0	1	0	1
1	0	1	0
1	1	Not allowed	

(c)

FIG. 11.1-1. An *SR* latch: (a) a block diagram, (b) logic implementation, and (c) the truth table.

FIG. 11.1-2. A D latch: (a) logic implementation and (b) the truth table.

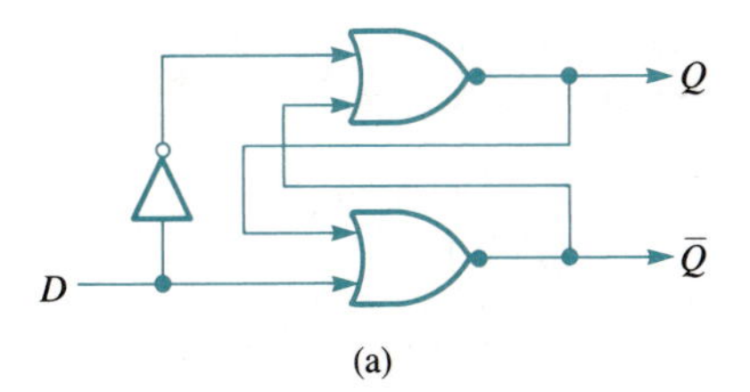

Input	Outputs	
D	Q	$\overline{Q}$
0	0	1
1	1	0

(b)

be set to 1 and $\overline{Q}$ will be reset to 0, regardless of their previous values. However, the input combination $S = 1$ and $R = 1$ is not allowed for an SR latch, since it resets both Q and $\overline{Q}$ to 0 and therefore violates the fact that Q and $\overline{Q}$ must always be the complement of each other.

The undesirable condition for $Q = \overline{Q} = 0$ can be easily avoided by ensuring that the S and R inputs are never set to 1 at the same time. This is easily accomplished, as shown in Fig. 11.1-2(a). This modified SR latch is known as the *D latch*. The functional truth table of the D latch is shown in Fig. 11.1-2(b). Observe that for this device the output Q will always follow the input D. This device is often referred to as a *delay element*.

Commercially available latches are usually provided with an additional control input signal that determines when the outputs of the latch can be changed. For example, Fig. 11.1-3(a) shows the block diagram of a D latch with an extra control line E. This control line acts as an enable signal; that is, the device will act as a D latch when $E = 1$, and the device will not change its state when $E = 0$, as shown by the functional truth table of Fig. 11.1-3(b). The 74′100 is a good example of such a device.

In analyzing digital networks containing memory devices, we often find it convenient to draw the graphical representation of the input and output signals in the time domain. This graphical representation is known as the *timing diagram*. The convention is to draw a positive signal, $+V$, for a logical 1 and no signal, or 0 V, for a logical 0. Figure 11.1-4 shows the timing diagrams for the SR latch of Fig. 11.1-1 and for the D latch of Fig. 11.1-3. Observe that, from the timing diagram of the SR latch, the outputs Q and $\overline{Q}$ will change whenever there exists an input condition on S and R that requires them to change according to its truth table. However, from the timing diagram of the D latch with the enable input E, Q will follow D if and only if the control signal E is a logical 1.

FIG. 11.1-3. An enabled D-latch: (a) a block diagram and (b) the truth table.

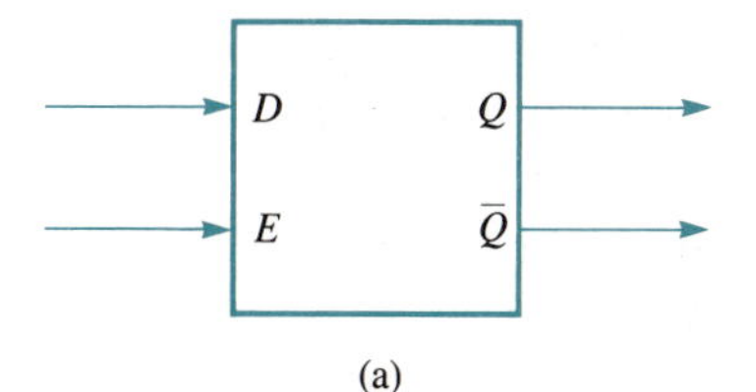

Inputs		Outputs	
E	D	Q	$\overline{Q}$
0	d	Q_p	$\overline{Q}_p$
1	0	0	1
1	1	1	0

(b)

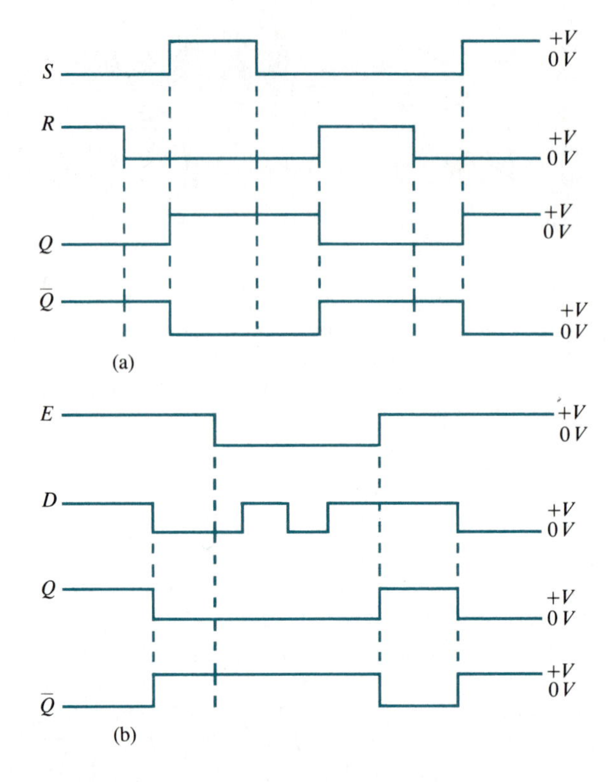

FIG. 11.1-4. Examples of timing diagrams: (a) for an *SR* latch and (b) for a *D* latch.

Flip-Flops

It is often desirable in a digital system to maintain a constant output for a circuit for one period of time regardless of the changes in the inputs. Since latches do not possess this property, flip-flops are introduced. Most flip-flops are constructed of basic latches and are controlled by an input signal known as the *clock*. The clock, denoted by CK, is a square-wave signal, as shown in Fig. 11.1-5(a). Flip-flops are designed to change states (outputs) either at the *rising edges* (also known as *positive edges*) of the clock pulses, as shown in Fig. 11.1-5(b), or at the *falling*

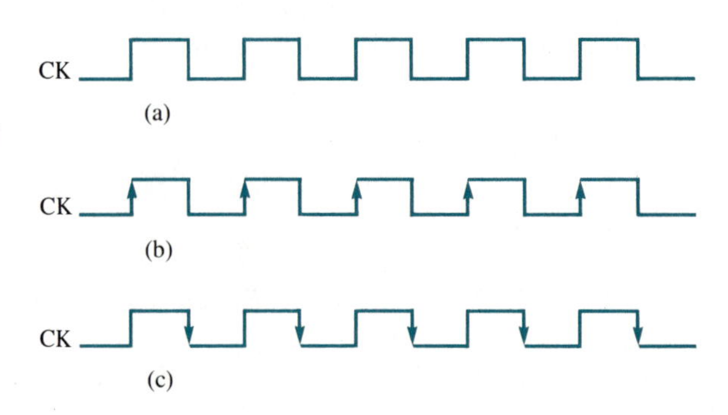

FIG. 11.1-5. (a) A clock signal, (b) positive edges of a clock signal, and (c) negative edges of a clock signal.

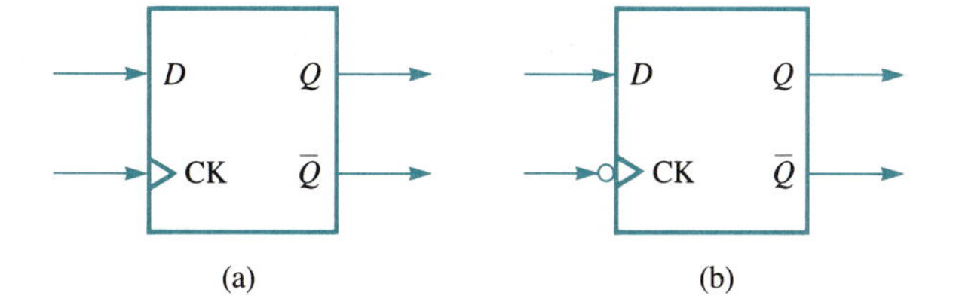

FIG. 11.1-6. A D flip-flop: (a) positive-edge-triggered and (b) negative-edge-triggered.

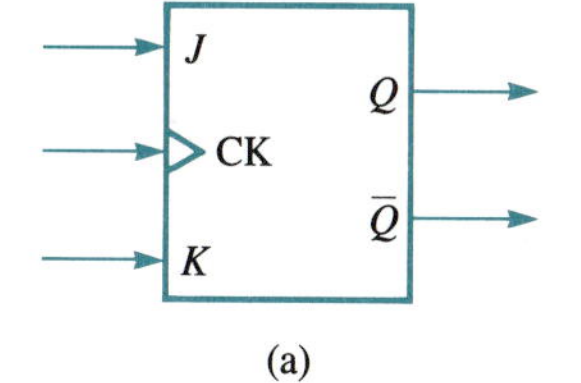

Inputs		Outputs	
J	K	Q	$\bar{Q}$
0	0	Q_P	$\bar{Q}_P$
0	1	0	1
1	0	1	0
1	1	$\bar{Q}_P$	Q_P

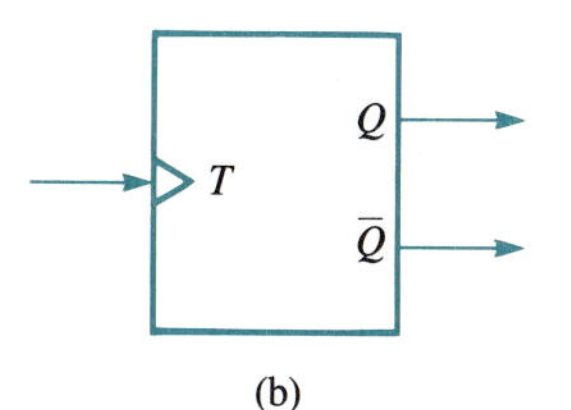

Inputs	Outputs	
T	Q	$\bar{Q}$
0	Q_P	$\bar{Q}_P$
1	$\bar{Q}_P$	Q_P

FIG. 11.1-7. Edge-triggered flip-flops: (a) JK and (b) T.

edges (also known as *negative edges*) of the clock pulses, as shown in (c).

When flip-flops respond to input changes only at the active edges of the clock, they are said to be *edge-triggered flip-flops*. Edge-triggered flip-flops sample their inputs only at the clock edge and change their outputs only as a result of the clock edge. Figure 11.1-6(a) shows the graphical representation of a positive-edge-triggered D flip-flop; the graphical representation of a negative-edge-triggered D flip-flop is shown in (b). The two most common flip-flops that are commercially available are the JK and T flip-flops, whose graphical representations and truth tables are shown in Fig. 11.1-7.

Most commercially available flip-flops, such as the 74′109, also include two extra control signals. They are the *preset* and *clear* control input signals. These input signals, when activated, will set ($Q = 1$) or clear ($Q = 0$) the flip-flop regardless of the other input signals. Figure 11.1-8(a) shows the block diagram of a positive-edge-triggered D flip-flop with preset and clear control signals. The functional truth table for this device is shown in Fig. 11.1-8(b), and its timing diagram for some arbitrary input signals is shown in (c).

It should be clear from the timing diagram of Fig. 11.1-8(c) that the clock is the synchronizing element in a digital system. It causes the outputs to change, if at all, at a precise predetermined instant of time. When clock signals are used with edge-triggered flip-flops and other basic combinational gates, some practical systems can be constructed. Unlike the combinational system, this system has memory

FIG. 11.1-8. A positive-edge-triggered flip-flop with preset and clear: (a) block diagram, (b) the truth table, and (c) a timing diagram.

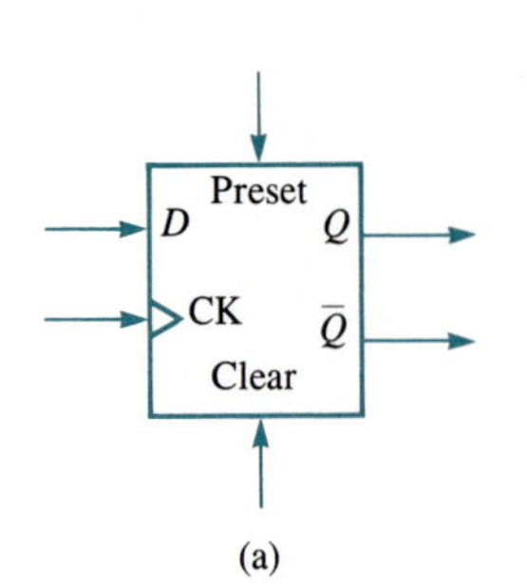

(a)

Inputs				Outputs	
Preset	Clear	CK	D	Q	$\bar{Q}$
1	0	d	d	1	0
0	1	d	d	0	1
0	0	↑	0	0	1
0	0	↑	1	1	0
1	1	d	d	Not allowed	

(b)

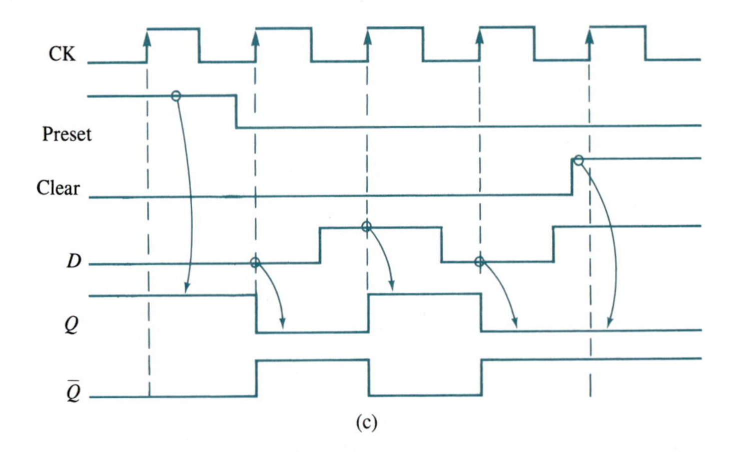

(c)

elements, and therefore, its output is a function of its present state and its present inputs. This type of system is known as a *sequential system*. Some examples of sequential systems are considered in the next section.

11.2 REGISTERS

A register is a collection of flip-flops, where each flip-flop is used to store one bit of information. In general, a register may contain some basic combinational gates, in addition to the flip-flops, to perform different binary arithmetic and logic operations. Figure 11.2-1 shows a simple register that is used to transfer the data available at D_3, D_2, D_1, and D_0 into Q_3, Q_2, Q_1, and Q_0 at the positive edge of the clock signal, which is connected to the four different flip-flops. This type of register can be used to sample and hold binary data. Such registers are commercially available and are classified as MSI functions. An example of an MSI package is the 74LS174, which contains a 6-bit register with a master clear input.

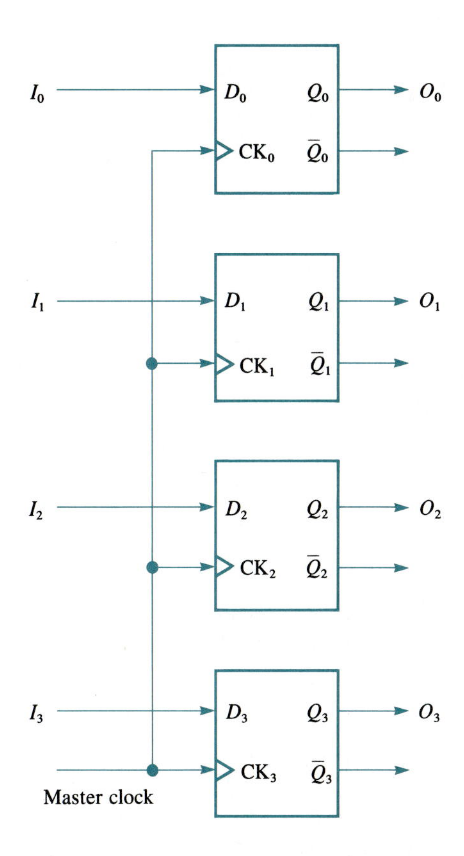

FIG. 11.2-1. A 4-bit register.

Serial-Input Shift Register

A serial shift register is a group of flip-flops in which binary information can be shifted in one of two directions. The binary information is transferred in 1 bit at a time by shifting the bits out of one flip-flop and into an adjacent one. Figure 11.2-2(a) shows a 4-bit serial shift-left register using *D* flip-flops. For this register, the rightmost flip-flop will receive the data in while transferring its output to D_1 at every clock cycle. Similarly, the second-from-the-right flip-flop will receive Q_0 while transferring its output to D_2, etc. This operation is best illustrated with a timing diagram, as shown in Fig. 11.2-2(b).

The shift register shown in Fig. 11.2-2 can easily be modified to perform a shift-right operation, and with the addition of some basic logic gates, the same register can be used to shift either left or right. The design of a shift-left/right register is left as an exercise for the reader.

Parallel-Input Shift Register

One good application for the serial-input shift register is the serial-to-parallel data conversion often needed in computer communications. However, when parallel

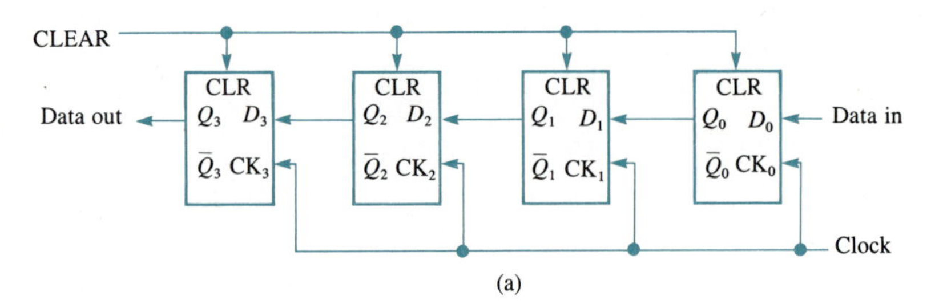

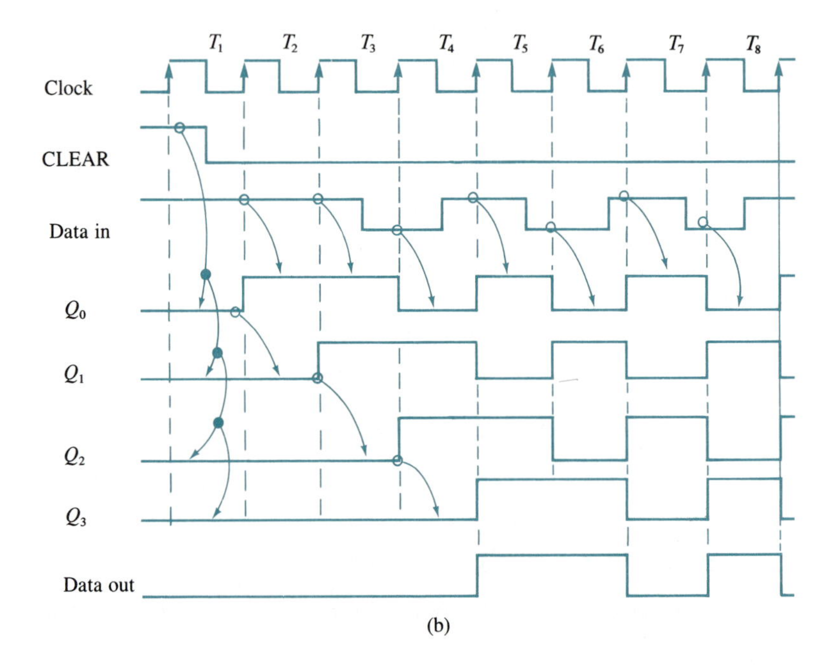

FIG. 11.2-2. A 4-bit shift-left register: (a) block diagram and (b) a timing diagram.

data are to be transmitted serially (bit-by-bit), a device for converting parallel-to-serial data is required. Such a device is called a parallel-input shift register. Figure 11.2-3 shows a block diagram of a 4-bit parallel-input shift-right register. The reader is encouraged to verify that when the control signal load/$\overline{\text{shift}}$ is at logic 1, then the values of the input lines I_0 through I_3 will be transferred to Q_0 through Q_3 at every clock pulse of the master clock. But when the control signal load/$\overline{\text{shift}}$ is at logic 0, then a shift-right operation on the data (next MSB, Q_3, Q_2, Q_1, and Q_0) will occur at every clock pulse of the master clock; that is, for every clock pulse, the value of next MSB will transfer to Q_3, the value of Q_3 to Q_2, the value of Q_2 to Q_1, and the value of Q_1 to Q_0.

Commercially available shift registers usually include the shift-left, shift-right, parallel-input, and no-change operations. These registers are known as *universal registers*. The 74LS178 is a good example of a universal 4-bit shift register.

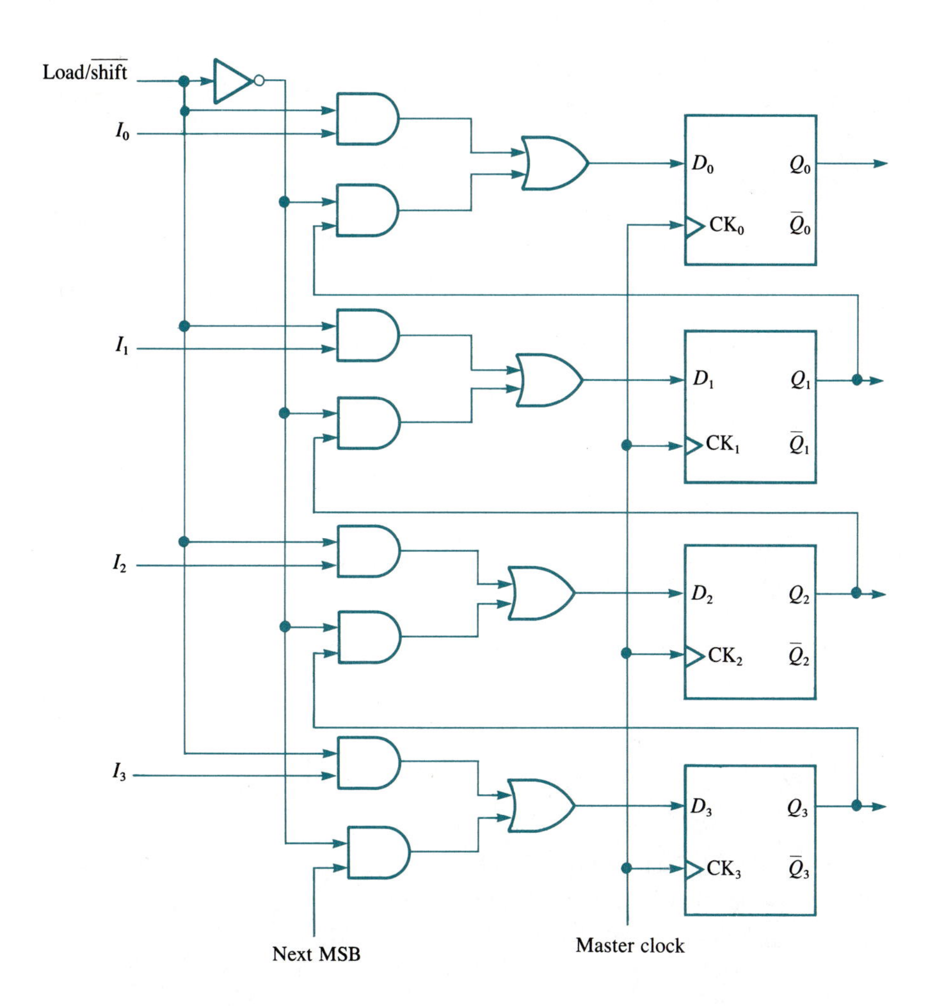

FIG. 11.2-3. A 4-bit, parallel-input, shift-right register.

11.3 COUNTERS

Timers, frequency meters, computers, and numerous other digital devices contain digital counters for counting events. A counter is a register that goes through a predetermined sequence of states when input pulses are received.

There are two types of counters, *ripple* (also known as *asynchronous*) and *synchronous counters*. In ripple counters, the output of each flip-flop activates the next flip-flop throughout the entire sequence of the counter's states. In a synchronous counter, all flip-flops are activated (triggered) simultaneously, since all the flip-flops' clock inputs are connected to a master clock.

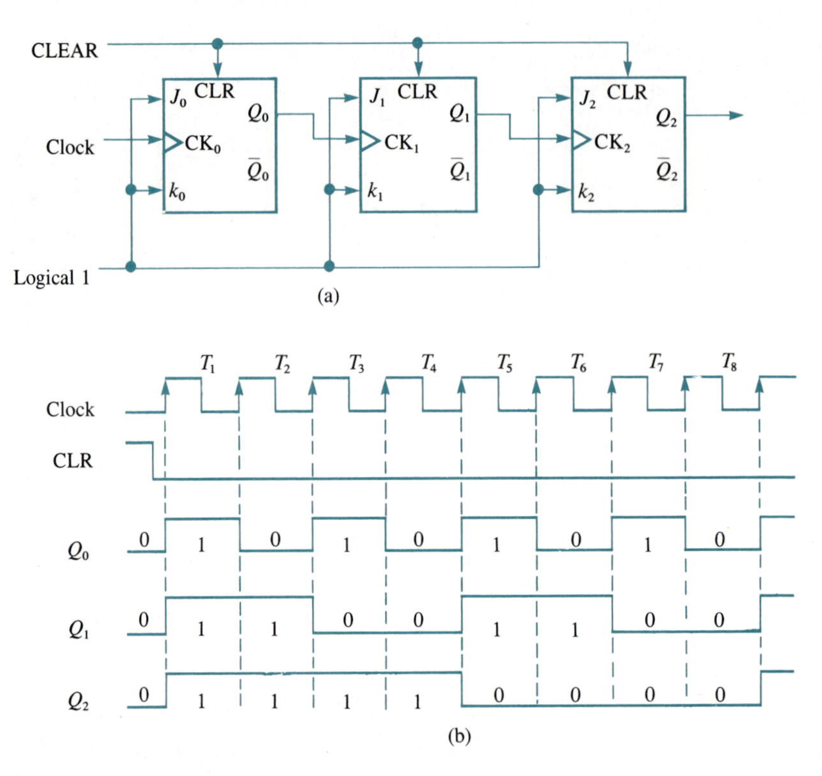

FIG. 11.3-1. A 3-bit ripple counter: (a) logic diagram and (b) a timing diagram.

Ripple Counters

Figure 11.3-1(a) shows the logic diagram of 3-bit ripple counter using *JK* flip-flops. Observe from the timing diagram of Fig. 11.3-1(b) that when the clear signal equals 0, the output of the leftmost flip-flop will change its state every clock pulse. The output of the second flip-flop, Q_1, is controlled by Q_0 and therefore will change its state every time Q_0 changes from 0 to 1. Similarly, the output of the second flip-flop will control the output of rightmost flip-flop.

Table 11.1-1 shows the outputs Q_0, Q_1 and Q_2 for the first 8 clock pulses. Observe that this counter is counting from 000 to 111. After the count reaches 111, counting begins again from 000. Therefore, a 3-bit counter will cycle through 8 states, 000 through 111. Similarly, a 4-bit counter will cycle through 16 states, 0000 through 1111. In general, an *n*-bit ripple counter will cycle through 2^n states, 0 through $2^n - 1$.

Counters that cycle through 2^n states, from 0 to $2^n - 1$, are known as *up-counters*. In some applications it is often desirable to count down from a preset number, and the circuits used are known as *down-counters*. It is interesting to notice that the circuit of Fig. 11.3-1(a) is considered as an up-counter if the outputs are taken from $Q_2\ Q_1\ Q_0$, and it is considered as a down-counter if the outputs are taken from $\bar{Q}_2\ \bar{Q}_1\ \bar{Q}_0$.

An *n*-bit ripple counter that cycles through the 2^n states is known as *divide-by-2^n counter*. They are also known as *modulo-2^n* binary counters.

TABLE 11.3-1. Outputs of a 3-Bit Ripple Counter

Clock	Q_2	Q_1	Q_0
	0	0	0
T_1	0	0	1
T_2	0	1	0
T_3	0	1	1
T_4	1	0	0
T_5	1	0	1
T_6	1	1	0
T_7	1	1	1
T_8	0	0	0

EXAMPLE 11.3-1

Design a modulo-5 binary ripple counter using *JK* flip-flops.

Solution

A modulo-5 counter counts up from 0 to 4 and then clears all flip-flops on the fifth pulse. Since we need to count up to 4, $(100)_2$, three flip-flops are required. The logical block diagram for this counter is shown in Fig. 11.3-2. The AND gate is used to detect the state 101 and clear all flip-flops.

Asynchronous ripple counters are commercially available as MSI packages. For example, the 74LS93 is a 4-bit ripple counter.

Synchronous Counters

One disadvantage of ripple counters is the slow speed of operation caused by the long time required for changes in state to ripple through the flip-flops. This

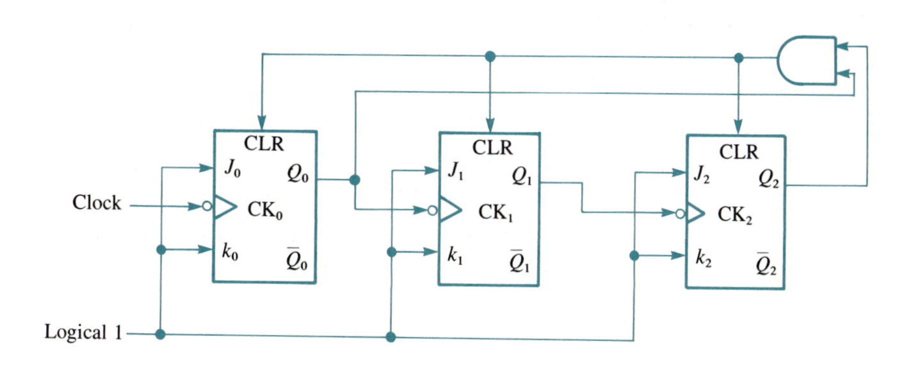

FIG. 11.3-2. A modulo-5 ripple counter.

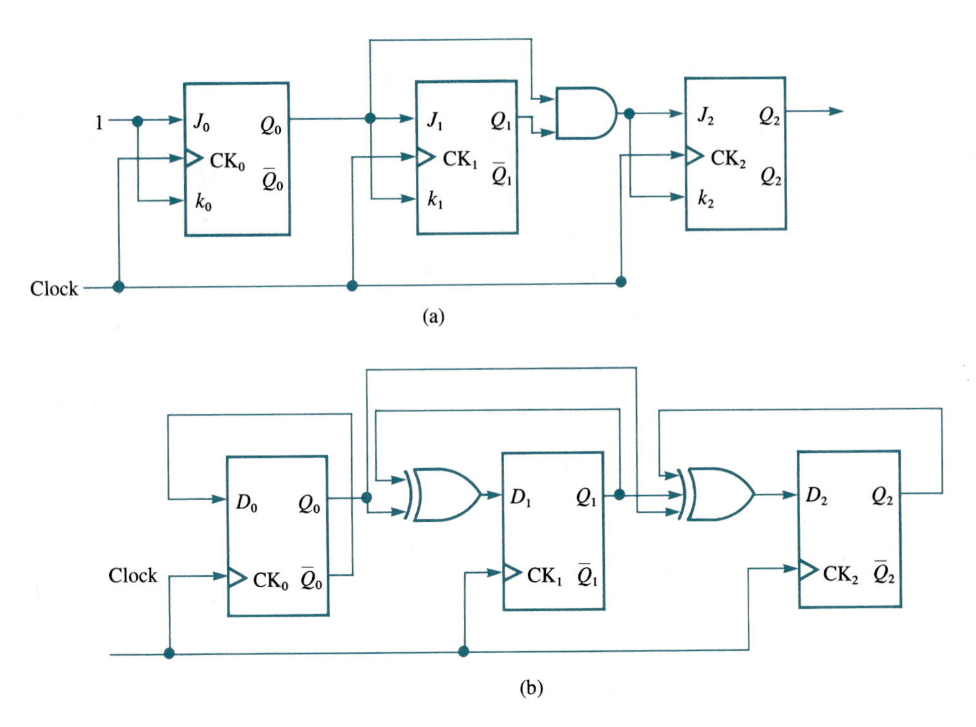

FIG. 11.3-3. A 3-bit synchronous counter: (a) *JK* flip-flops implementation and (b) *D* flip-flops implementation.

problem can be avoided by using *synchronous counters*. Counters in which the clock pulses are applied to trigger all flip-flops simultaneously are called synchronous counters. Because flip-flops are triggered simultaneously, additional control logic is required to determine which flip-flops, if any, must change state.

Figure 11.3-3(a) shows the logic diagram of a 3-bit binary counter using *JK* flip-flops. The implementation of the same synchronous counter using *D* flip-flops is shown in Fig. 11.3-3(b).

Synchronous counters can be designed to cycle through any sequence of states. The following procedure outlines the steps of the design of a synchronous counter using *D* flip-flops.

1. Construct a *counter table* that lists the states (sequence) the counter will cycle through.
2. Construct a *counter design table*. This is the counter's truth table, with its input side being the flip-flops' outputs, as given in the counter table, and its output side being the flip-flops' inputs, which are to be determined.
3. Fill in the values for the flip-flops' inputs by using the following rule:

   For a given row in the counter design table and a specific flip-flop, enter the next row's value in the input column of the table to its corresponding flip-flop's input in the output column.

4. Obtain minimum boolean expressions for each of the inputs of the flip-flops listed in the output side of the counter design table.
5. Draw the block diagram for the synchronous counter by first connecting all the flip-flops' clock inputs to a single clock input and connecting the inputs of the flip-flops using the minimal boolean expressions derived in step 4.

The above procedure is best illustrated through an example.

EXAMPLE 11.3-2

Design a synchronous down-counter that will cycle through the states 111, 110, 101, . . . , 010, 001, 000, 111, etc.

Solution

The counter table is as shown in Fig. 11.3-4(a). The counter design table as described by steps 2 and 3 of the above procedure is shown in Fig. 11.3-4(b). Observe that the value for D_2 in the first row of Fig. 11.3-4(b) is obtained by simply entering the value of Q_2 in the second row. Similarly, the value of Q_1 in the third row is entered for the value of D_1 in the second row. Also, the value of Q_0 in the fourth row is entered for the value of D_0 in the third row, etc.

The minimum boolean expressions for each of the inputs of the flip-flops are derived by using K maps, as shown in Fig. 11.3-4(c). Finally, the logic diagram for this counter is shown in Fig. 11.3-4(d).

The above procedure can be slightly modified to design any counter using different types of flip-flops. Designs using *JK* flip-flops usually lead to the most economical realization of counters. The design of a counter using *JK* flip-flops will be left as an exercise for the reader.

Various *n*-bit synchronous counters are commercially available as MSI packages. For example, the 74′163 is a universal 4-bit synchronous counter; that is, the 74′163 is capable of cycling through 16 states, 0000 through 1111.

The 74′163 has a synchronous master clear input, CLR, which, when activated, will reset the counter to zero at the next clock pulse. Also, this counter has one synchronous load input, LD, which, when activated, will initialize the counter to the 4-bit data appearing at the input lines of the counter. Therefore, this counter is considered to be a programmable counter.

Ring Counters

Figure 11.3-5(a) shows a 4-bit ring counter using *D* flip-flops. As in a synchronous counter, all flip-flops of the ring counter are triggered simultaneously; however, the output of each flip-flop drives only an adjacent flip-flop. In a ring counter, a

FIG. 11.3-4. The design of a synchronous down-counter: (a) the counter table, (b) the counter design table, (c) the equations for the flip-flop inputs, and (d) a logic diagram.

Q_2	Q_1	Q_0
1	1	1
1	1	0
1	0	1
1	0	0
0	1	1
0	1	0
0	0	1
0	0	0
1	1	1

(a)

States			Flip-Flops' Inputs		
Q_2	Q_1	Q_0	D_2	D_1	D_0
1	1	1	1	1	0
1	1	0	1	0	1
1	0	1	1	0	0
1	0	0	0	1	1
0	1	1	0	1	0
0	1	0	0	0	1
0	0	1	0	0	0
0	0	0	1	1	1

(b)

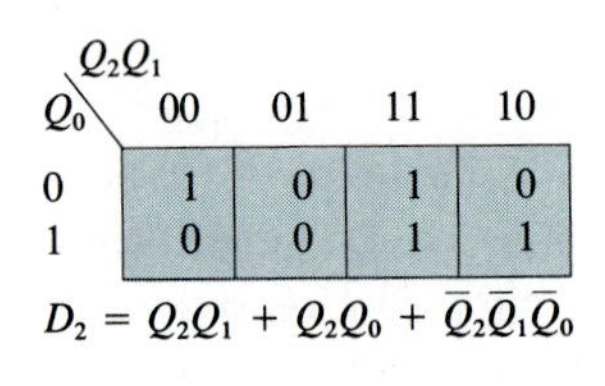

$D_2 = Q_2Q_1 + Q_2Q_0 + \bar{Q}_2\bar{Q}_1\bar{Q}_0$

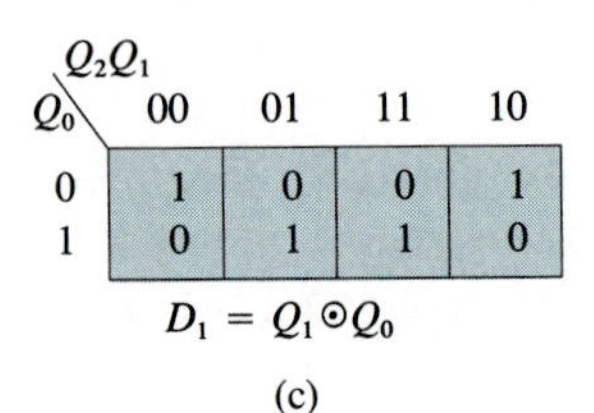

$D_1 = Q_1 \odot Q_0$

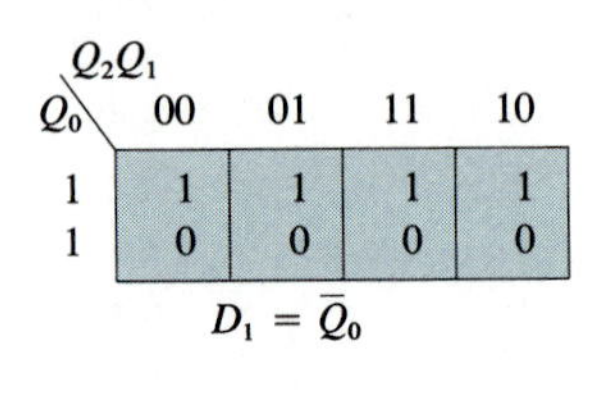

$D_1 = \bar{Q}_0$

(c)

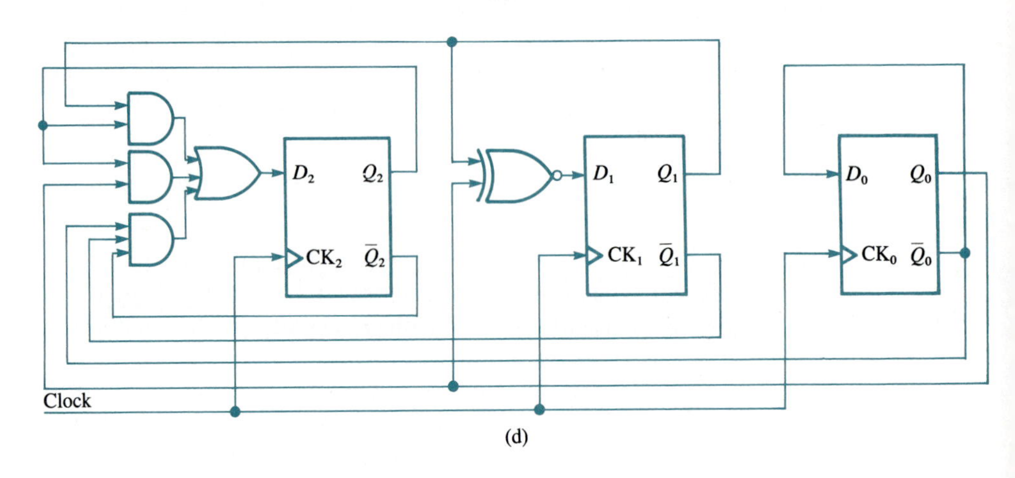

(d)

FIG. 11.3-5. A 4-bit ring counter: (a) logic diagram using D flip-flops and (b) a timing diagram.

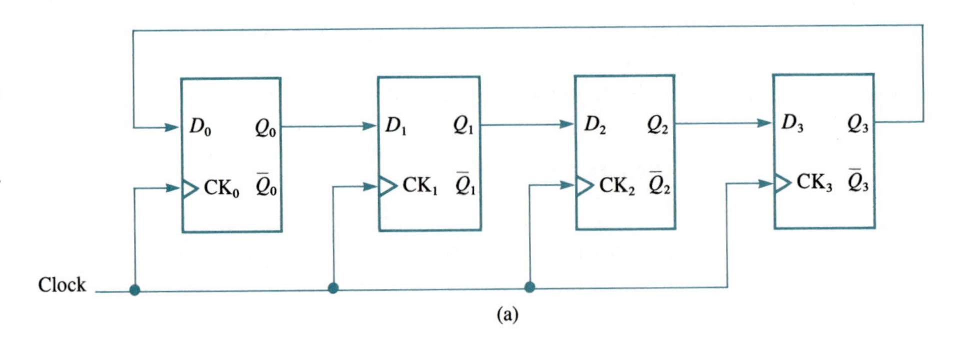

(a)

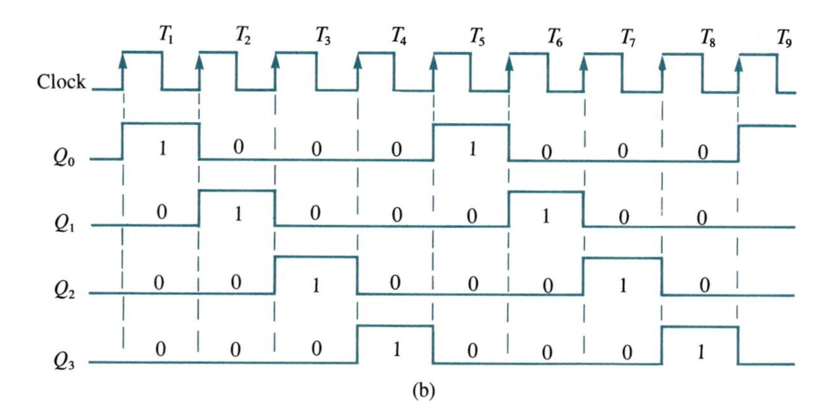

(b)

single pulse propagates through the ring, while all remaining flip-flops are at the zero state.

A modulo-N ring counter requires N flip-flops and no other gates. Figure 11.3-5(b) shows the timing diagram of a modulo-4 ring counter. Even though modulo-N ripple and synchronous counters require only $\log_2 N$ flip-flops, they may actually require more components than a ring counter.

11.4 MEMORY

Computers use memory to store both programs and data that determine its operation. For a digital computer, memory can be divided into three different types: *random-access memory*, *mass storage*, and *archival storage*.

In random-access memory, each memory location can be accessed in a time equivalent to any other memory location. Random-access memory includes *read-and-write memory* (RAM), *read-only memory* (ROM), programmable read-only memory (PROM), and erasable programmable read-only memory (EPROM). Mass-storage memory refers to a relatively large amount of memory. The time required to access data in a mass-storage device is relative to its location in the device. Mass storage has a relatively large storage capacity and is typically lower in cost per bit than random-access memory. Magnetic disk memory is a good example of a mass-storage media. Archival storage is long-term storage and may require user intervention for access by the system. Its access time is very slow, but it has a very large capacity. Magnetic tape is the most common archival medium.

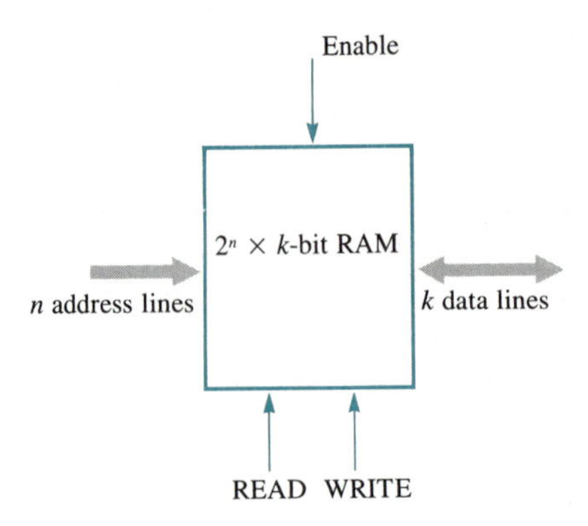

FIG. 11.4-1. A block diagram of a RAM device.

Read-and-Write Memory (RAM)

In read-and-write memory, the computer can write ("store") data into memory and read ("retrieve") them back later. RAM is *volatile*; that is, its contents are retained only as long as power is present.

Figure 11.4-1 shows a typical block diagram of a RAM device. Conceptually, it is a collection of 2^n addressable storage locations, each of which contains k bits. That is, the RAM of Fig. 11.4-1 includes a $2^n \times k$ matrix of memory cells. Each cell may be a flip-flop or a capacitor. The n address lines are decoded to select k cells. To read data from a selected set of k cells requires the enable and read control signals to be asserted. To write data to a selected set of k cells requires the enable and write control signals to be asserted.

There are two types of RAM, *static* and *dynamic*. A static RAM is the read-and-write memory that retains its data as long as power is applied and without any further action from the computer. Each cell of a static RAM is a flip-flop. Dynamic RAM is the read-and-write memory that requires continuous actions from the computer to maintain its contents. Each cell of a dynamic RAM is a capacitor, which leaks charges and therefore requires continuous refreshing to maintain its value. Static RAM is used in microprocessor-based systems that require small memory; dynamic RAM is used in large memory systems because of its lower cost and greater density.

RAM is commerically available in different sizes. The most common static-RAM sizes are 2K $\times$ 8, 8K $\times$ 8, and 32K $\times$ 8 (K = 1024). Although different part numbers are used by different manufacturers, the common part numbers are 6116, 6264, and 62256, respectively. Dynamic-RAM packages are also commercially available in 16K-bit, 64K-bit, 256K-bit, and 1M-bit (M = mega = 2^{20}) sizes.

Read-Only Memory (ROM)

In read-only memory, the computer can only read data from memory. ROM is nonvolatile, and data are physically and permanently stored in it. ROM is used to store data and programs that do not change during the operation of the system.

ROM is a truly nonvolatile device and maintains its contents even when its power is shut off.

There are three types of read-only memory: *mask-programmed*, *programmable*, and *erasable programmable* ROMs. The mask-programmed ROMs are read-only devices that are programmed during the manufacturing of the chip itself and are generally the less expensive devices for mass production. One of the principal application areas for mask-programmed ROMs is for data storage, such as character-font memory for laser printers.

Programmable read-only memory (PROM) is a field-programmable memory. That is, a PROM device fabricated by the manufacturer contains all 0s, and the user programs this device by electrically changing appropriate 0s to 1s. This is accomplished by destroying the fuses in the cells where 1s are desired; this is an irreversible process. PROMs are economical in small quantities.

Erasable programmable read-only memories (EPROMs) are widely used for program storage in microprocessor systems. They are nonvolatile and can be erased and reprogrammed when changes are necessary. They are very useful during product development, where they can be *repeatedly* erased and reprogrammed with new software versions. They are used for low-to-moderate-volume production. EPROMs are programmed by an instrument called a *PROM programmer*. The contents of EPROMs can be erased by shining an ultraviolet light into the window in the top of the device. The EPROM can then be reprogrammed, and this cycle can be repeated many times.

For remote-area applications where removing an EPROM device for alteration is not practical, *EEPROMs* (electrically erasable programmable ROMs) are used. The contents of EEPROMs may be altered without removing them from the circuits in which they are used. Special pins are provided in these devices such that when they are electrically activated, they permit the rewriting of selected memory locations.

Commercially available EPROMs and EEPROMs are 8-bits wide, but their sizes are different. For example, the Intel 2716 is a 2K × 8-bit EPROM; the 27512 is a 64K × 8-bit EPROM. The Intel 2816 is a 2K × 8-bit EEPROM.

Magnetic Disk Memory

Magnetic disk memory is a nonvolatile memory that provides large storage capabilities with moderate access time. Magnetic disk memories store data on one or more rigid, aluminum circular platters coated with iron oxide. The circular platters are called *disks*. The disks, which are usually grouped together into a *disk pack*, are separated by small air spaces to allow access for the *read/write head*. Figure 11.4-2 shows both the single disk and the disk pack. The surface of the disk is divided into concentric rings of data called *tracks*. Although tracks may resemble the grooves on an audio record, they do not spiral inward as on a record. Tracks are circular, and the start touches the end of each track. Even though outer tracks have larger circumferences than inner tracks, it is more convenient to store an equal amount of data on each track. The most common disks have 11-in

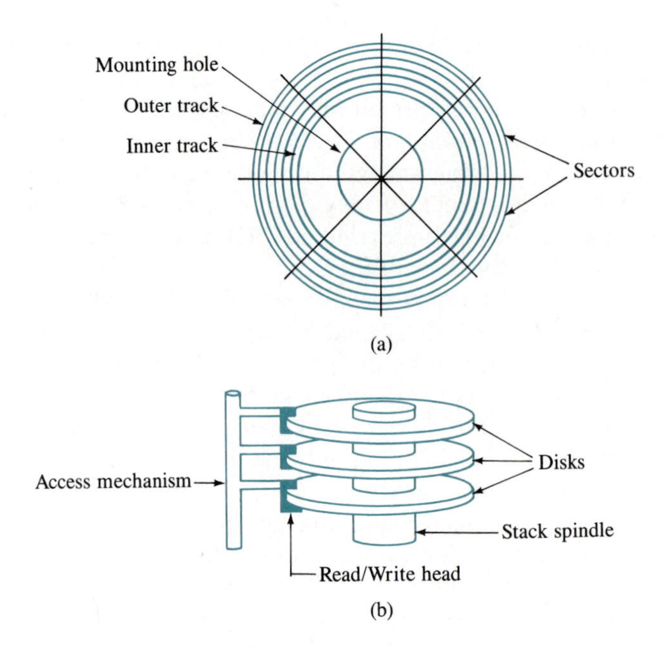

FIG. 11.4-2. A magnetic disk memory: (a) the tracks and sectors of a disk and (b) a disk pack.

diameters and provide 200 tracks per surface. These tracks are numbered from 0 to 199, starting with the outside perimeter of the disk.

Each track of a disk is treated as an array of dot positions, each of which is treated as a bit that will set to the magnetic equivalent of 0 or 1. Track packing density is usually 4000 bits per inch.

Disks are mounted on a common spindle, and all disks rotate at the same speed. A typical speed is 3600 rpm. With this speed, it will take at most $\frac{1}{60}$ of a second for any desired location of a track to swing into place. Tracks with the same number on all adjacent disks are referred to as a *cylinder* of the disk. Just as the surface of a disk is divided into tracks, so the circumference of a track is divided into sections called *sectors*. The type of the disk and its format determine the number of sectors. On a given disk, sectors are all of a fixed size. A typical disk has 17 sectors per track and 512 bytes (1 byte = 8 bits) of information per sector. By moving the read/write magnetic recording head across the surface of the disk and rotating the disk into position under the head, we can quickly access any desired sector. That is why magnetic disks are called *random-access* storage devices. In contrast to our use of a tape, we can access any sector directly without having to pass through the whole set of records sequentially.

Data are read from and stored into disks in blocks. The size of the block corresponds to the size of the sector. Even though data can be of any size, the actual disk read/write performed by the computer is only done in full, complete sectors.

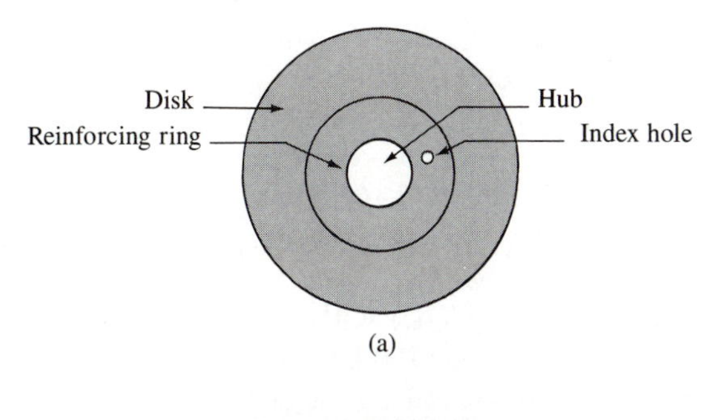

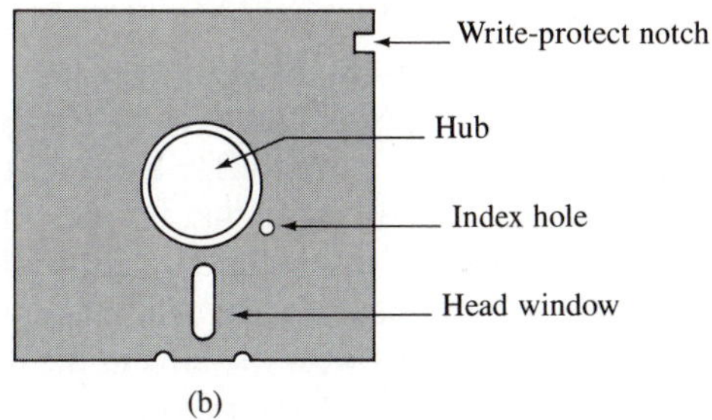

FIG. 11.4-3. (a) A mini-floppy disk and (b) a disk jacket.

Floppy-Disk Memory

The most popular form of magnetic storage devices in use today are *floppy disks*, which are also called *flexible disks*. They are low-cost, medium-capacity, and nonvolatile memory devices. They are made of a very soft flexible material, mylar plastic with magnetically sensitive iron oxide coating.

Floppy disks are currently found in three different sizes: (1) the $3\frac{1}{2}$-in micro-floppy disk, (2) the $5\frac{1}{4}$-in mini-floppy disk, and (3) the original 8-in standard floppy, which has fallen from popular use. Figure 11.4-3 shows a $5\frac{1}{4}$-in mini-floppy. A plastic jacket covers the mylar media for protection. The read/write head contacts the disk through the long slot in the jacket known as the *head window*. The jacket has a circular opening in the center, which is slightly larger than the circular hole, known as a *hub* of the disk; it exposes a ring of the disk. This ring allows the disk drive to clamp its drive spindle on it. The drive spindle, and thus the $5\frac{1}{4}$-in mini-floppy, rotates at 360-rpm speed.

A small index hole just to the outside of the center hub provides a rotational position reference that allows the disk drive to locate the beginning of the data. This index hole is in both the jacket and the disk, so that when the disk rotates, they align once per revolution. The alignment of the index holes produces an index mark, which is used by the disk drive to locate a given sector.

The write-protect notch is the small square opening in the jacket of the disk that prevents writing to the disk. It is covered to write-protect the disk.

The $5\frac{1}{4}$-in mini-floppy can be either single- or double-sided; that is, data can be stored and/or read from only one side or from both sides. The single-sided disk looks exactly like the double-sided one; however, data are recorded only on its bottom surface. On the double-sided disk, data are recorded on both surfaces. (Today, it is almost unheard of for a computer to have single-sided drives.)

The original mini-floppy disks were single-sided with 8 sectors per track and a total of 40 tracks. Called *single-sided/single-density* (SS/SD) floppies, they have all but disappeared. Today, mini-floppy disks are either *double-sided/double-density* (DS/DD) with 9 sectors per track and a total of 40 tracks per side or *double-sided/quad-density* (DS/QD) with either 9 or 15 sectors per track and a total of 80 tracks per side. About 720 kilobytes of data can be stored in a double-density mini-floppy disk, and about 1.2 megabytes of data can be stored in a quad-density mini-floppy disk.

The $3\frac{1}{2}$-in micro-floppy disk, also known as a *microdiskette*, is enclosed in a rigid protective case. Figure 11.4-4 shows a typical microdiskette. Inside, the microdiskette is the same very soft plastic material with magnetically sensitive iron oxide coating. Outside, the jacket is rigid and thus provides extra protection. The diskette is reinforced at its center with a metal disk, known as a *drive hub*, that allows the drive to attach to the disk and spin accurately. The disk surrounding the *chucking hole* aligns with the pins in the drive and provides a better linkage to the drive motor than the clamping technique used with larger disks. A sliding door, known as the *sliding shutter*, covers and protects the disk surface from outside contamination. The write protection is signaled by a plastic tab that slides into the write-protect notch. Microdiskettes are recorded in quad-density format, and 2-megabyte microdiskettes are commercially available. At the time of the writing of this book, several manufacturers were introducing higher-capacity and smaller-size diskettes. Although 4- to 16-megabyte $3\frac{1}{2}$-in diskettes are in the experimental stages, 2-in diskettes have been introduced and used in new electronic cameras and portable personal computers.

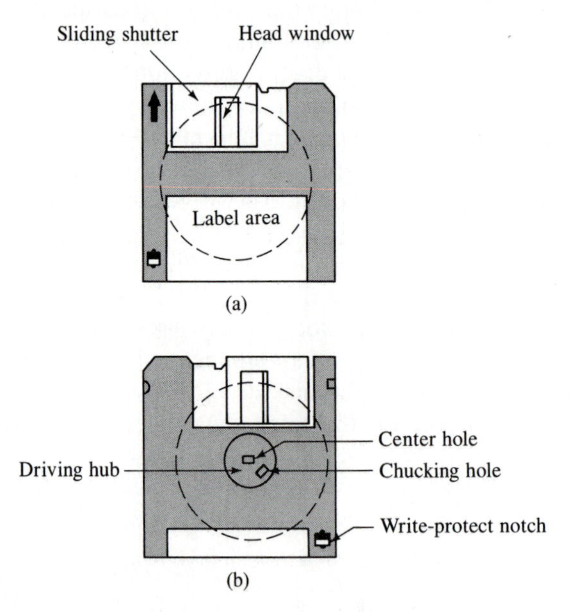

FIG. 11.4-4. A micro-floppy disk: (a) front view and (b) back view.

Magnetic Tape

Magnetic tapes are a particularly popular form of mass-storage device. Modern technology has made the cost of magnetic tapes very low, and thus they are the ideal devices for storing vast quantities of information inexpensively.

Magnetic tape is a flexible plastic tape coated with a thin film of some ferromagnetic material. The width of the tape varies from $\frac{1}{4}$ of an inch to 3 in. However, the most commonly used tapes are $\frac{1}{2}$ in wide, 2400 or 3600 ft long, and contained in a $10\frac{1}{2}$-in reel. Although tape densities vary, 200, 556, 800, 1600, 6250, and 12,500 bits per inch (BPI) are standard.

Even though magnetic tapes are inexpensive media for storing huge quantities of data, they lack the small access times required for fast operations. When information is retrieved from or updated on it, the entire tape must be read sequentially, which results in a slow access time.

Other Types of Memory

In addition to the rigid and floppy magnetic disks and magnetic tape, two newer types of secondary storage are increasingly used: the *Winchester disks* and the *videodisks*.

Unlike the magnetic disks previously discussed, Winchester disks are sealed modules that contain both the disk and the read/write mechanism. Since the package is sealed, little maintenance is required compared with the removable disks, which require an extremely clean environment. The Winchester's read/write heads have a very low mass, and only a small force is needed to keep them in position. Combined with the sealed, clean environment, these features allow the heads to "fly" closer to the disk, which in turn allows higher-density recording.

Videodisk, also known as *optical-disk memory*, is the latest entry into mass-storage devices. Currently available is the write-once optical-disk drive. The read-and-write drive is in its experimental stage at the time of the writing of this book.

The optical disk is constructed from a lightweight plastic material that contains a photosensitive layer for data storage. The photosensitive layer is covered with a thin plastic coating that protects it from contamination. Data are stored on the optical disk by energizing a high-intensity laser beam that causes the photosensitive layer to change its ability to reflect light. Currently, this process is irreversible; that is, data cannot be erased. Optical disks have high density. A 14-in optical disk can store up to 1 gigabyte of data (equivalent to almost 400,000 typewritten pages of information).

Data are read from the disk with a low-intensity laser beam that is focused through the plastic coating onto the photosensitive substrate. The change in reflected light, caused by the data originally stored on the disk, is sensed by a photosensor and converted back to a digital signal.

Unlike the magnetic disk, a 14-in optical disk has 40,000 tracks. Each track is divided into 25 sectors, with each sector holding up to 1 kilobyte (1024) of information. Optical disks are very reliable and durable; the disk can be destroyed only by breaking it.

Although magnetic and optical disks are effective mass-storage devices with a reasonable access time, they are not ideal for use in environments that are subject to vibration or mechanical stress. The ideal mass-storage device would have a capacity and cost per bit comparable to a magnetic or optical disk but would have no moving parts and use semiconductor technology for fabrication. Magnetic-*bubble-memory* technology has attempted this goal. Magnetic bubble memories are nonvolatile, solid-state memory devices. Bubble-memory chips are fabricated from orthoferrite material, a natural substance that is divided into alternating snakelike bands of magnetic fields. If a magnetic field is applied through a tiny metal loop held near the surface of an orthoferrite slab, the nearest snakelike region will be drawn into a thin magnetic cylinder called a *bubble*. The presence of a bubble denotes a binary 1, and its absence denotes a binary 0. Once a bubble has been formed in this way, it will remain on the surface of the chip indefinitely.

Four-megabit bubble-memory chips are commercially available. Bubble memory is compact, very reliable, and tolerant of dirty environments and vibration. The cost per bit for bubble memory remains relatively high, restricting its use to applications where its unique advantages justify the higher price.

11.5 DISPLAY DEVICES

Display devices can be categorized by the type and amount of information they display. All display devices fall into one of the following categories: on/off indicator, numeric, alphanumeric, or graphical display.

In addition to the above categories, display devices can be classified as *active* or *passive* devices. Active display devices emit light, as in *light-emitting diodes* (LEDs); passive display devices reflect or absorb light, as in *liquid-crystal displays* (LCDs).

Light-Emitting Diode (LED)

Light-emitting diodes are semiconductor display devices. Since there is no filament to burn out or glass to break, they are very reliable and rugged. They are inexpensive, are easily interfaced to digital logic, and do not require high voltage, as shown in Fig. 11.5-1. Most LEDs require approximately 10 mA of current flow to be illuminated completely. Since the forward voltage drop across an LED is approximately 1.70 V, the value of the current-limiting resistor, R, would be 330 Ω in this example.

LEDs are available in red, yellow, and green colors. Two-color LEDs are also available, which combine a red and a green LED in a single three-pin package. This allows a single LED to display red, green, or yellow. There are two common

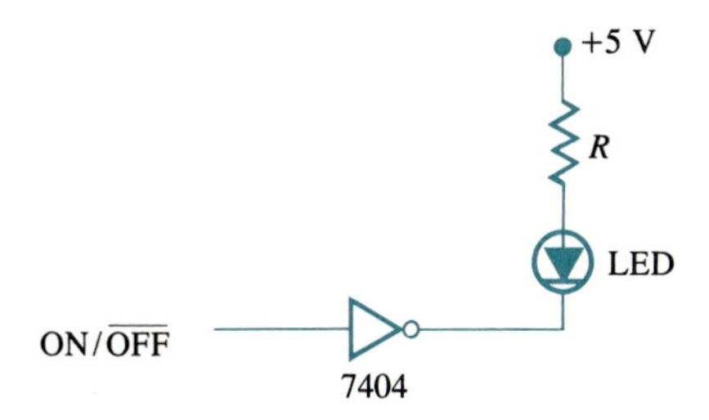

FIG. 11.5-1. A light-emitting diode.

sizes for LEDs, known as T-1 and T-1$\frac{3}{4}$, where the number after the T indicates the diameter of the lamp in units of eighths of an inch.

Liquid-Crystal Display (LCD)

Liquid-crystal display devices offer a good alternative to LEDs and their relatively high power consumption. A typical LCD display consumes only microwatts of power, over a thousand times less than an LED.

Liquid crystal is a substance that flows like a liquid but whose molecules orient themselves in the manner of a crystal. Liquid crystal is the heart of the LCD devices. When an electric field is applied to liquid-crystal material, the molecules are straightened out, and the display then absorbs the light and appears black. When no electric field is applied, the light is reflected, and the display appears as a silver mirror. LCDs can be more difficult to use in a design than LEDs, and the interface electronics is more complex. LCDs are most common in moderate-to-high-volume products, such as electronic wristwatches.

Seven-Segment Display

Seven-segment displays are the most commonly used numeric display devices. Figure 11.5-2(a) illustrates a common-cathode seven-segment LED display whose internal structure is nothing more than a single LED for each of the lettered segments as shown in Fig. 11.5-2(c). By an appropriate combination of segments, any digits can be displayed, as shown in Fig. 11.5-2(b).

Seven-segment displays are commercially available as individual digits or in groups of 4 to 12 digits in a single package. In addition to the single LED and 7-segment display devices, 10- and 16-segment display devices are also available. The 10-segment display device shown in Fig. 11.5-2(d) is a numeric display device that can display all the digits, allows the digit 1 to be centered, and can display the + sign. The 16-segment display shown in Fig. 11.5-2(e) is a widely used display device for alphanumeric data. It is also known as a *starburst* display device. This device is best-suited for displaying uppercase characters.

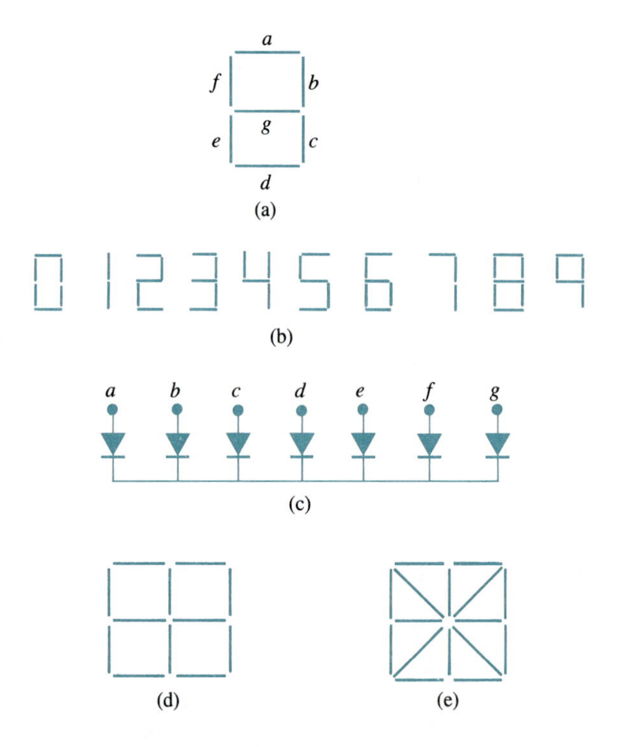

FIG. 11.5-2. Segment display devices: (a) a 7-segment display, (b) the digits displayed by a 7-segment display, (c) the internal structure of a 7-segment display, (d) a 10-segment display, and (e) a 16-segment display.

Cathode-Ray Tube (CRT)

The cathode-ray-tube (CRT) display is one of the oldest display technologies and remains one of the most popular. There are two basic types of CRT display devices in use today: the *raster-scanned* and the *graphics* displays. The raster-scanned display devices are very similar to commercial television sets, and the information is displayed in the same manner. Graphics display devices use different display techniques than commerical television and have extremely high resolution.

Figure 11.5-3(a) shows the main components of a typical CRT display device. When the filament is heated, the heat applied to the cathode causes electrons to become free. Once the electrons become free, they are attracted toward the screen of the CRT by the highly positive voltage applied to the screen. Grids shape the electrons into a narrow beam that strikes a small spot on the screen, which causes photons of light to be emitted from the phosphorous coating on the inside face of the picture tube. The type of phosphor determines whether the light that is emitted is white, green, or amber in a monochrome picture tube. A color picture tube contains three different filaments with their associated cathodes and has three different phosphors. These phosphors emit three different colors of light, red, green, and blue, which can be combined to generate any color.

If the electron beam's position remains fixed, only a single spot emits light; but if the beam is deflected, additional spots glow. When the beam is deflected

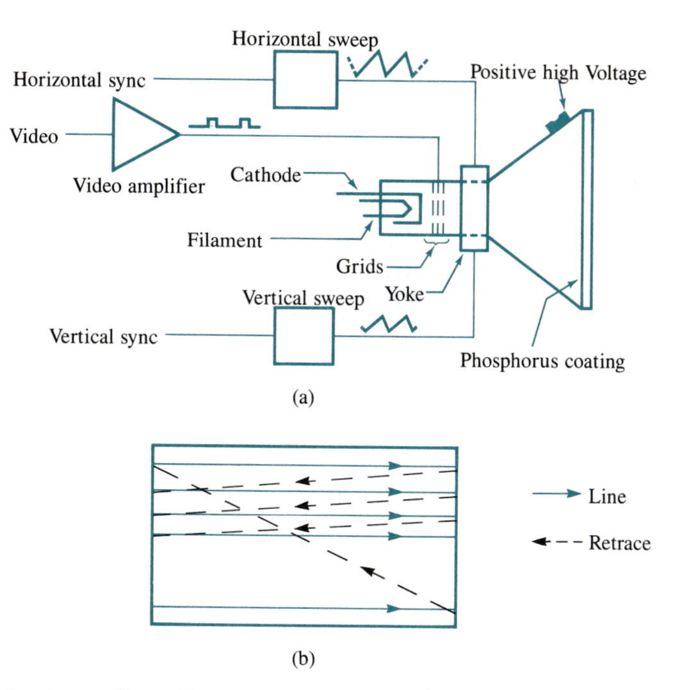

FIG. 11.5-3. (a) Construction of a CRT display and (b) raster lines.

horizontally, a line across the face of the CRT glows. When the beam is deflected either up or down, a vertical line glows. In a CRT, as in a television set, the beam is deflected by an external *yoke* placed around the neck of the picture tube. The yoke contains two coils for vertical electron beam deflection and two coils for horizontal deflection.

As the voltage across the horizontal-deflection coil increases, a magnetic field is created, which moves the beam across the screen from left to right. When the beam reaches the right-hand edge, a decreasing horizontal-deflection voltage returns the beam to the left. The retrace beam is cut off, so retrace is not visible. While the beam is returning to the left, a decreasing vertical-deflection voltage moves the beam slightly below the retrace line. The horizontal- and vertical-sweep voltages continue until the entire screen has been covered. The horizontal lines are known as *raster lines*. Figure 11.5-3(b) illustrates the raster found on a CRT display device.

The intensity of the electron beam can be varied as the horizontal- and vertical-deflection fields are swept. This variation in light intensity is the picture and is known as the *video*. A CRT video signal has only two levels: A 0 level causes a dark spot, and a 1 level causes a bright spot. The proper combination of 1s and 0s displays data on the screen of the CRT.

Each character displayed on the screen is produced by dot-matrix displays. Figure 11.5-4 shows two common dot-matrix sizes. The characters produced by a 5 × 7 dot matrix exist in a 7 × 8 matrix, and the characters produced by a 7 × 9 dot matrix exist in a 9 × 10 matrix, resulting in one line between character lines and two spaces between characters in a line.

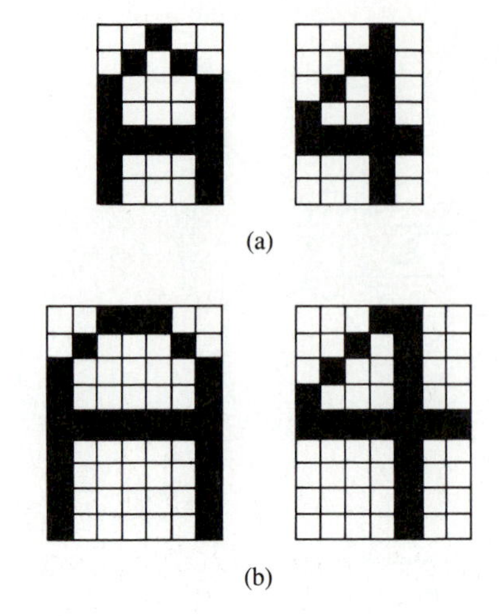

FIG. 11.5-4. Dot-matrix display fonts: (a) a 5 × 7 display and (b) a 7 × 9 display.

Graphics display terminals do not use scanning circuitry to generate lines on the CRT screen. Instead, the electron beam is positioned to any point on the screen using a vertical- and a horizontal-position signal. This allows the electron beam to be moved to any position on the screen to draw any required pattern.

11.6 DIGITAL-TO-ANALOG (D/A) CONVERTERS

The conversion of digital values to proportional analog values is a necessary task so that results of digital computations can be used in the analog world. A digital-to-analog converter (D/A) accepts an n-bit parallel digital code as input and provides an analog current or voltage as output. Two schemes for D/A conversion are commonly used: (1) a summing operational amplifier with binary-weighted input resistance, and (2) a constant-resistance ladder network with branches switched in and out, depending on the digital value to be converted.

Figure 11.6-1 shows the block diagram of a typical (D/A) converter. For an ideal D/A converter, the analog output for an n-bit binary-input code is

$$V_0 = -V_{ref}(b_0 + b_1 \times 2^{-1} + b_2 \times 2^{-2} + \cdots + b_{n-1} \times 2^{-n+1}) \qquad (11.6\text{-}1)$$

where

V_0 = analog output voltage

V_{ref} = reference analog input voltage

b_0 = most significant bit of binary-input code

b_{n-1} = least significant bit of binary-input code

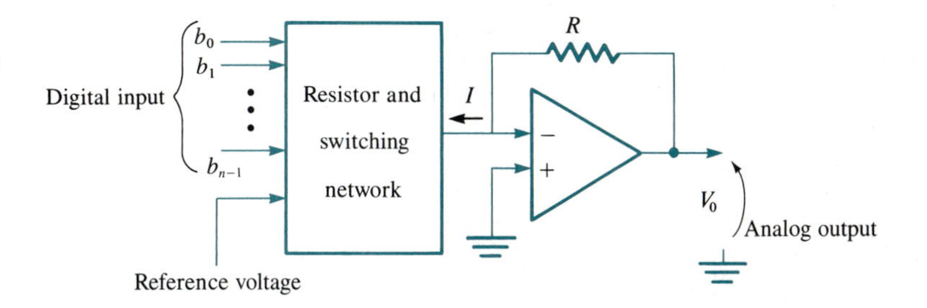

FIG. 11.6-1. A block diagram of a D/A converter.

The operational amplifier (op amp) is used at the output to provide current to voltage conversion and/or buffering. In some high-speed applications where a limited output-voltage range is acceptable, a resistor, instead of an op amp, provides the current-to-voltage conversion, thus eliminating the delay associated with the op amp.

Weighted-Resistor D/A Converter

The weighted-resistor 4-bit D/A converter, shown in Fig. 11.6-2, includes a reference-voltage source, a set of four electronically controlled switches, a set of four binary-weighted precision resistors, and an operational amplifier. Each binary bit of digital-input code controls its own switch. If the bit value is a 1, the switch closes. If it is a binary 0, the switch stays open. The resistor connected to the MSB, b_0, has a value of R. Each lower-order bit is connected to a resistor that is higher by a factor of 2; that is, b_1 is connected to $2R$, b_2 to $4R$, and b_3 to $8R$.

For the 4-bit D/A converter, the binary-input range is from 0000 to 1111. The maximum analog output voltage of $-1.875 \times V_{ref}$ will result when the input binary code is 1111. Proportionate analog outputs will result for smaller binary-input codes.

EXAMPLE 11.6-1

For the 4-bit D/A converter of Fig. 11.6-2 with $V_{ref} = -5$ V, compute the analog output voltage when (a) $b_0b_1b_2b_3 = 1111$, (b) $b_0b_1b_2b_3 = 1100$, and (c) $b_0b_1b_2b_3 = 0001$.

Solution

(a) From (11.6-1), we have

$$V_0 = 5 \times (1 + 1 \times 2^{-1} + 1 \times 2^{-2} + 1 \times 2^{-3})$$
$$= 5 \times 1.875 = 9.375 \text{ V}$$

(b) $$V_0 = 5 \times (1 + 1 \times 2^{-1} + 0 \times 2^{-2} + 0 \times 2^{-3})$$
$$= 5 \times 1.5 = 7.5 \text{ V}$$

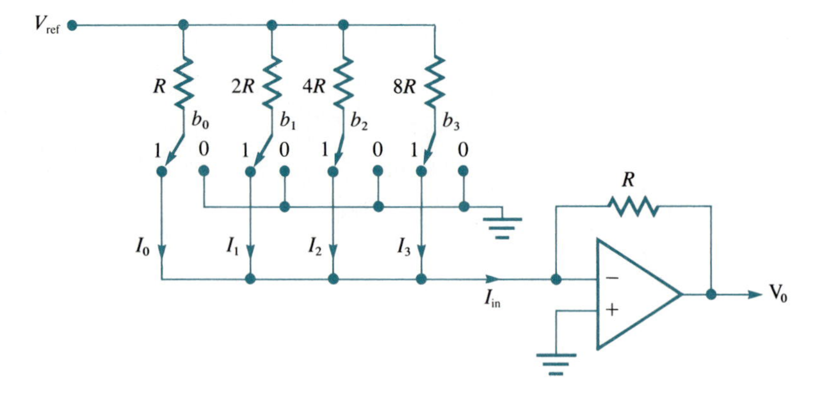

FIG. 11.6-2. A 4-bit weighted-resistor D/A converter.

$$\text{(c) } V_0 = 5 \times (0 + 0 \times 2^{-1} + 0 \times 2^{-2} + 1 \times 2^{-3})$$
$$= 5 \times 0.125 = 0.625 \text{ V}$$

An important design parameter of a D/A converter is ΔV, the smallest output-voltage change. This is known as the *resolution* of a D/A converter. For an n-bit D/A converter,

$$\Delta V = \frac{V_{ref}}{2^n - 1} \qquad (11.6\text{-}2)$$

One major disadvantage of the weighted-resistor D/A converter is that when the number of bits increases, the range of the weighted resistors becomes prohibitively large for accurate implementation as an integrated package.

R-2*R* Ladder D/A Converter

The 4-bit R-2R ladder D/A converter, shown in Fig. 11.6-3, contains a reference-voltage source, V_{ref}, a set of four switches, an op amp, and two resistors per bit. One resistor is in series with the bit switch, and the other, which is half the series resistor, is in the summing line.

Each switch contributes its appropriately weighted component to output voltage, resulting in a net output voltage that is proportional to the input binary code. For example, consider the case where only $b_0 = 1$. Then the MSB series resistor, 2R, is shunted by the equivalent resistance from point X of the ladder network to ground. The equivalent resistance, readily determined by combining parallel and series elements starting with the LSB, is 2R. Thus the input current, I_{in}, is $V_{ref}/2R$, and the output voltage, V_0, is $-V_{ref}/2$. When this analysis is continued, the analog output voltage can be expressed as

$$V_{out} = -V_{ref} \times (b_0 \times 2^{-1} + b_1 \times 2^{-2} + \cdots + b_{n-1} \times 2^{-n}) \qquad (11.6\text{-}3)$$

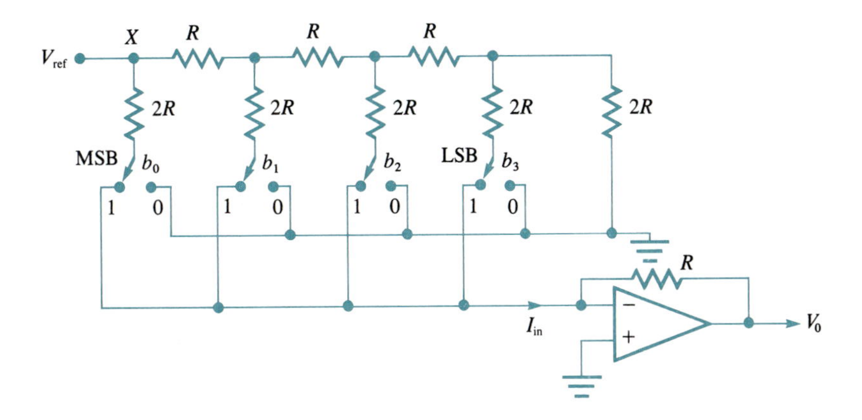

FIG. 11.6-3. A 4-bit R-2R ladder D/A converter.

The R-2R ladder D/A converter networks are relatively simple to manufacture because only two resistor values, R and 2R, are used. They are fast, practical, and reliable. One example of a commercially available D/A converter is the AD558, which is an 8-bit R-2R D/A converter.

2^nR D/A Converter

An n-bit 2^nR D/A converter requires 2^n resistors of equal value R and $2^{n+1} - 2$ analog switches. A 3-bit 2^nR D/A converter is shown in Fig. 11.6-4. The eight (2^3) resistors are connected in series to form a voltage divider, which provides eight analog voltage levels. The analog switches are controlled by the digital-input code such that each code creates a single path from the voltage divider to the converter output. A unity-gain amplifier is connected to the output to prevent loading of the voltage divider. Although this design requires a large number of components, they are economically manufactured as LSI packages.

EXAMPLE 11.6-2

Find the analog output voltage when the input to the 2^nR D/A converter of Fig. 11.6-4 is (a) 100 and (b) 010.

Solution

(a) For the binary input 100, the switches controlled by B_0, $\overline{B}_1$, and $\overline{B}_2$ are closed, and therefore, a path is created between the output V_0 and point 4, where the voltage is equal to $V_{ref}/2$. Hence, the analog output voltage is $V_{ref}/2$.
(b) For the binary input 010, the switches controlled by $\overline{B}_0$, B_1, and $\overline{B}_2$ are closed. A path between V_0 and point 6, where the voltage is equal to $V_{ref}/4$, is created. Therefore, the analog output voltage is $V_{ref}/4$.

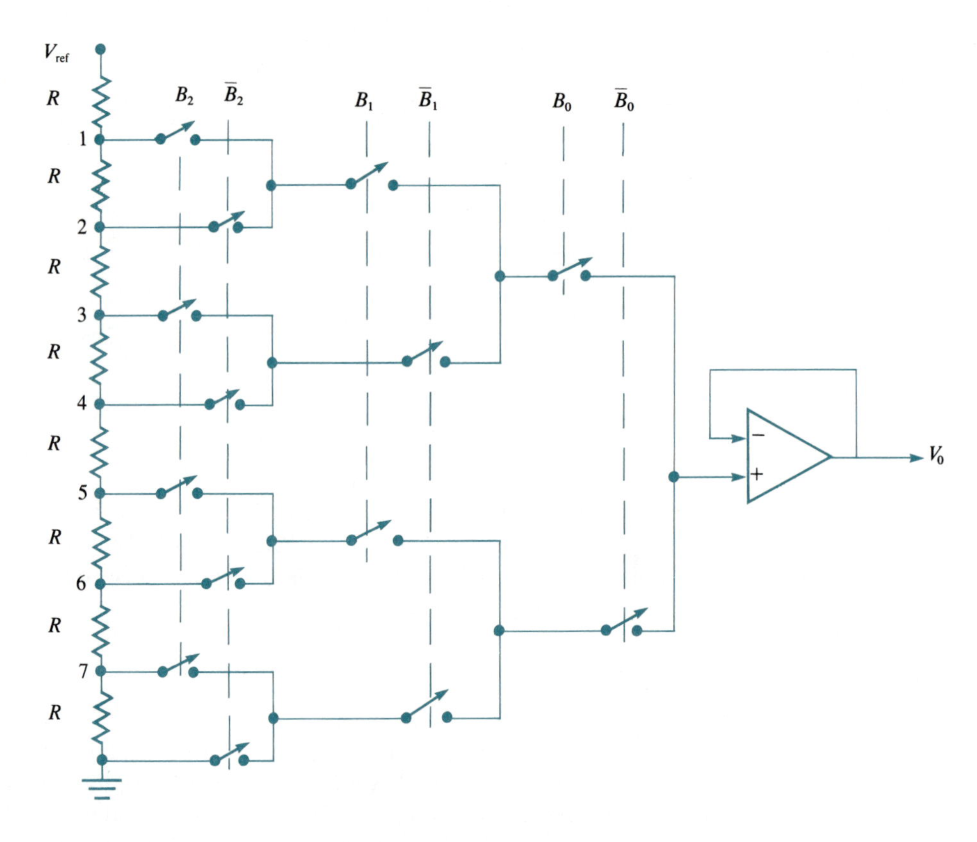

FIG. 11.6-4. A 3-bit 2^nR D/A converter.

11.7 ANALOG-TO-DIGITAL (A/D) CONVERTERS

The analog-to-digital (A/D) converter is a key part of many industrial, commercial, and military systems, because it is the interface between analog systems and digital systems. A/D converters have expanded the practical uses of digital equipment in many areas, such as process control, aircraft control, and telemetry. An A/D converter converts analog-input signals into digital-output data. Several different types of A/D converters exist, including *counter-controlled*, *successive-approximation*, and *dual-ramp* (dual-slope) *converters*.

Analog comparators are the basis of A/D conversion. The block diagram of an analog comparator is shown in Fig. 11.7-1. The input signals to the comparator are the analog voltages V_1 and V_2, and its output is the discrete voltage V_0. This comparator produces a high-level (logical 1) output signal if $V_1 \geq V_2$, and a low-level (logical 0) if $V_1 < V_2$. Analog comparators are commercially available. The LM311 is an analog comparator that is widely used by designers.

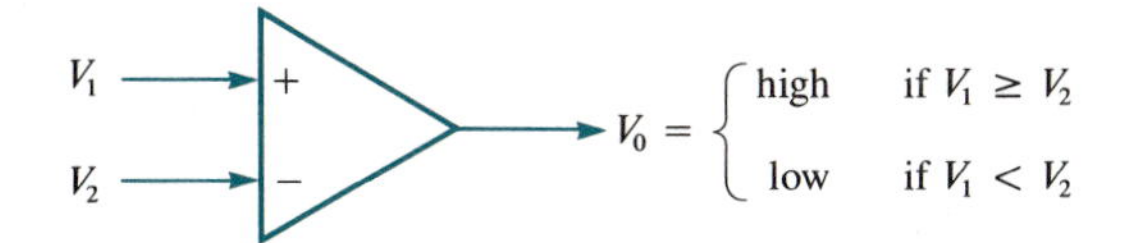

FIG. 11.7-1. An analog comparator.

Counter-Controlled A/D Converter

The block diagram of a counter-controlled A/D converter is shown in Fig. 11.7-2. The conversion from analog to digital is initiated by resetting the binary counter to zero, which produces a D/A output voltage $V_2 = 0$. If the analog input V_1 is larger than the D/A output voltage, the output of the comparator will be high, enabling the AND gate and allowing the counter to be incremented. As the counter is incremented, V_2 is increased. When V_2 gets slightly greater than the analog-input signal, the comparator signal becomes low, which causes the AND gate to stop the counter. The counter output at this point is the digital representation of the analog-input signal.

The major disadvantage of this method is the relatively long conversion time required to encode the analog-input signal. In the worst case, it will take 2^n clock periods to complete the A/D conversion, where n is the number of bits required to encode the analog signal.

Successive-Approximation A/D Converter

The successive-approximation A/D converter is much faster. It also uses a D/A converter, but the binary counter is replaced by a successive-approximation register (SAR), as shown in Fig. 11.7-3.

After a start-of-conversion pulse, the SAR sets the MSB to 1 and all other bits to 0. If the comparator indicates that the D/A converter's output is larger than the

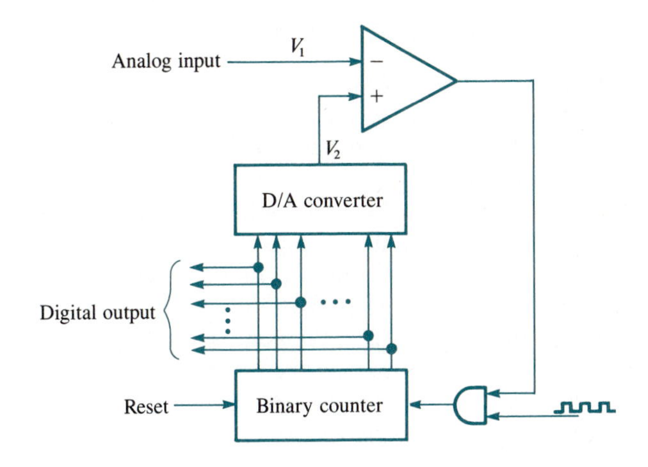

FIG. 11.7-2. A block diagram of a counter-controlled A/D converter.

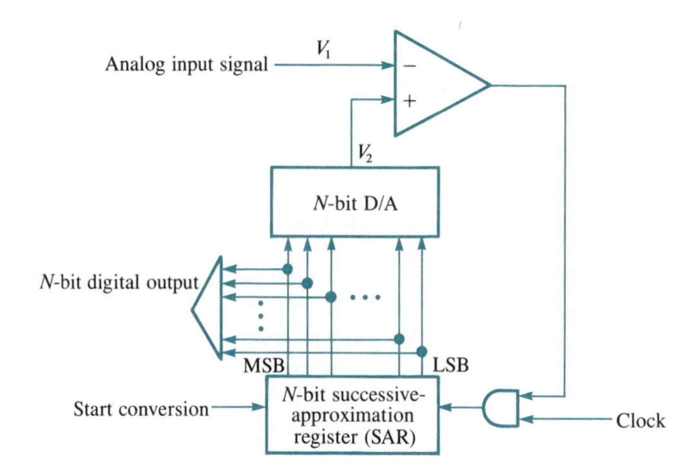

FIG. 11.7-3. A successive-approximation A/D converter.

signal to be converted, then the MSB is reset to 0, and the next bit is tried as the MSB. However, if the signal to be converted is larger than the D/A converter's output, then the MSB will remain 1.

This procedure is repeated for each bit until, at the end, the binary equivalent of the input analog signal is obtained. This procedure requires only n clock periods, rather than the 2^n clock periods needed by the counter-controlled A/D converter.

A popular 8-bit A/D converter that is commercially available is the National ADC0844. It is based on the successive-approximation technique.

Dual-Ramp (Slope) A/D Converter

The block diagram of a *dual-ramp*, also known as *dual-slope*, A/D converter is shown in Fig. 11.7-4(a). After a start-of-conversion pulse, the counter is cleared and the analog input, V_{in}, is selected as input to the integrator. When the output of the integrator, V_0, reaches 0, the counter starts to count. After a fixed amount of time T, the counter will overflow. The output of the integrator at the end of the fixed time T is proportional to the analog-input signal, as shown in Fig. 11.7-4(b).

At the end of T, that is, when the counter overflows, the reference voltage, V_{ref}, is selected and the integrator begins outputting a ramp with a positive slope. As V_0 increases, the counter is incremented until V_0 reaches the comparator threshold voltage (0 V), which stops the counter from being incremented again. The value of the counter is the binary code for the analog voltage V_{in}. Dual-ramp A/D converters can provide high accuracy at low cost.

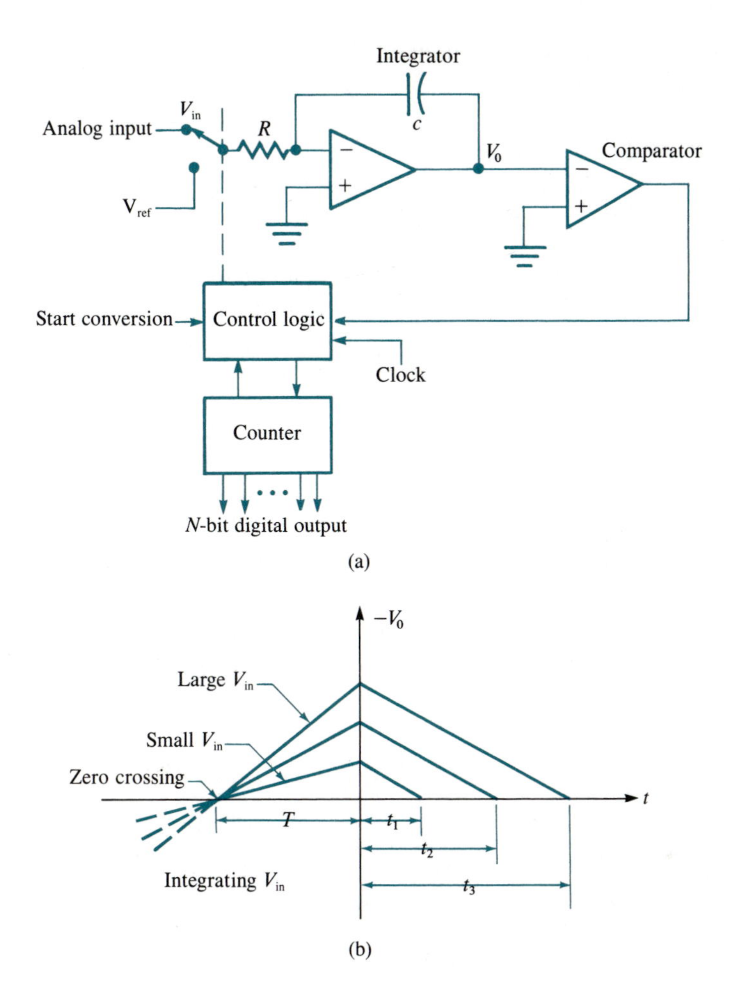

FIG. 11.7-4. A dual-ramp (slope) A/D converter: (a) a block diagram and (b) the timing sequence.

PROBLEMS

11-1. Show the logic diagram of an *SR* latch using only NAND gates.

11-2. Prepare the truth table for the NAND gate implementation of the *SR* latch.

11-3. Show the logic diagram of the enabled *D* latch using only NAND gates.

11-4. Complete the following timing diagram of an SR latch.

FIG. P11-4.

11-5. Complete the following timing diagram for the D latch of Fig. 11.1-3(a).

FIG. P11-5.

11-6. Modify the JK flip-flop of Fig. 11.1-7(a) to operate like the D flip-flop of Fig. 11.1-6(a).

11-7. Modify the JK flip-flop of Fig. 11.1-7(a) to operate like the T flip-flop of Fig. 11.1-7(b).

11-8. When the J and K inputs of a JK flip-flop are tied to logical 1, this device is known as a divide-by-2 counter. Complete the timing diagram for this counter.

FIG. 11-8.

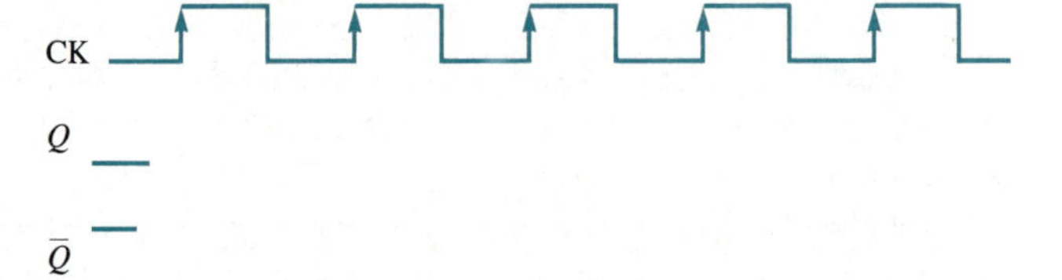

11-9. Redo the timing diagram of Fig. 11.2-2(b) for a 4-bit shift-right register.

11-10. If the content of the register of Fig. 11.2-2(b) is initially 0111, show the content of this register after six clock pulses, with data in being 101101.

11-11. Explain how a shift register can be used as a binary (a) divide-by-2 and (b) multiply-by-2 counter.

11-12. Draw the timing diagram for the modulo-5 ripple counter of Fig. 11.3-2.

11-13. A digital clock uses a quartz crystal of 32,768 Hz. How many flip-flops must a binary counter have to obtain 1 pulse per second?

11-14. Draw a timing diagram for the synchronous counter of Fig. 11.3-3(a).

11-15. Repeat Problem 11-14 for the counter of Fig. 11.3-3(b).

11-16. How many flip-flops must ring counter have to obtain a modulo-16 counter?

11-17. A typewritten page contains about 2500 computer characters (bytes). How many 500-page books can be stored on 2400 ft of 1600-BPI magnetic tape?

11-18. How many bits are required if a D/A converter must detect a 1-V change when $V_{ref} = 15$ V?

11-19. For the 4-bit D/A converter of Fig. 11.6-2, find (a) the maximum analog output voltage, (b) the minimum analog output voltage, and (c) the smallest detectable analog output voltage when $V_{ref} = -10$ V.

11-20. A 4-bit *R*-2*R* ladder D/A converter uses a −10-V reference-voltage source. Find (a) the analog output voltage when the binary-input code is 1100, and (b) the reference voltage to obtain the corresponding decimal-output voltage.

11-21. An 8-bit A/D converter is driven by a 1-MHz clock. Determine the maximum conversion time if (a) it is a counter-controlled D/A converter, and (b) it is a successive-approximation D/A converter.

11-22. Use the dual-slope A/D converter of Fig. 11.7-4. (a) Determine the total charge on the integrator due to the input voltage V_{in} during the signal integration time *T*. (b) If the discharge time is t_d, find an expression for t_d in terms of V_{in}, V_{ref}, and *T*.

11-23. A 10-bit *R*-2*R* ladder network D/A converter has an MSB resistor value of 10 Ω. What is the value of the LSB resistor?

11-24. Describe the basic difference between (a) weighted-resistor and *R*-2*R* ladder D/A converters, and (b) counter-controlled and successive-approximation A/D converters.

DESIGN EXERCISES

D11-1. Design a 4-bit shift-right register using *JK* flop-flops.

D11-2. Design a 4-bit shift-left/right register using *D* flip-flops.

D11-3. Design a 4-bit universal shift register.

D11-4. Design a 3-bit ripple up/down-counter that will count up if the control line $U/D = 1$ and will count down if the control line $U/D = 0$. For example, if the counter is in state 011 and $U/D = 1$, then the next state should be 100; however, if the present state is 011 and $U/D = 0$, then the next state should be 010.

D11-5. Design a modulo-6 ripple counter using *D* flop-flops.

D11-6. Design a synchronous binary counter that will cycle through 00, 10, 11, and 01 states. Use *D* flip-flops.

D11-7. Design a synchronous counter that will cycle through states 000, 101, 011, 010, 100, and 101. The rest of the unspecified states should be considered as don't-care.

D11-8. Design a modulo-5 ring counter using *JK* flip-flops.

D11-9. Digital watches display time by turning on a certain combination of the 7-segment display device. For example, to display the digit 5, the segments a, c, d, f, and g are turned on. Internally, the watch represents a decimal digit using BCD code. For example, the digit 5 is represented by 0101. Now, design a BCD-to-7-segment decoder, as shown in Fig. D11-9. (a) Develop a truth table for turning on the segments. The truth table should have inputs A, B, C, and D and outputs a, b, c, d, e, f, and g. Binary codes above 1001 will never occur in BCD, and therefore the outputs for these codes should be don't-care. (b) Use K maps for the simplification of the output functions and then realize them by using combinational gates.

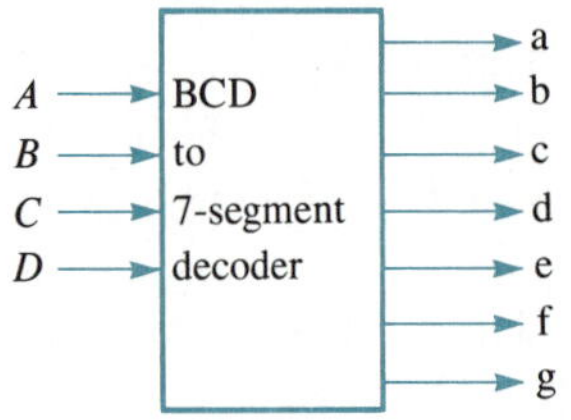

FIG. D11-9.
A 7-segment display.

A, B, C, D → BCD to 7-segment decoder → a, b, c, d, e, f, g

CHAPTER 12

Computer Systems

12.0 INTRODUCTION

The evolution of computers has profoundly influenced our way of life and revolutionized the design of electronics systems. Being surrounded by this ever-changing high technology, engineers and scientists must be familiar with the basic functions and organizations of computer systems in order to utilize their hardware and software capabilities effectively.

The aim of this chapter is to introduce the terminology and basic functions of the hardware and software components of computer systems. The chapter begins with the classification of different types of computer systems. The hardware and software organizations of computer systems follow. The tools for programming and managing a computer system, such as programming languages and operating systems, are discussed. The architecture and organization of a microcomputer system are introduced, followed by a discussion of the fundamental concepts of microprocessors' architecture and operations. The instruction set of a popular microprocessor is described, and the applications of different instruction types are introduced through numerous assembly-language programs. Finally, the concepts of stacks and interrupts are introduced.

12.1 TYPES OF COMPUTER SYSTEMS

In general, computers are machines that accept data and instructions, perform some predefined operations on the data, and make the results available to their users in different forms. Today, there is a wide variety of computers in terms of performance, capacity, size, and price. Computers can be grouped into four main classes: *microcomputers*, *minicomputers*, *mainframes*, and *supercomputers*.

Microcomputer

A microcomputer is a complete digital computer, available in a variety of sizes from multiple circuit boards down to all circuitry on a single chip. A microcomputer has a *microprocessor* as its central processing unit (CPU). A microprocessor is a digital integrated circuit that contains the digital functions to process the information and control the operations of the microcomputer. Microcomputers have become a common part of everyday life. Home, school, and personal office computers have rapidly spread, bringing computing power and literacy to a wide spectrum of people. In addition, today's 16- and 32-bit microcomputer systems are designed for dedicated real-time applications in a distributed system. The cost of microcomputers ranges from a few hundred dollars to about $10,000.

Minicomputer

By the early 1960s, a key development in the computer world was the introduction of *minicomputers*. Minicomputers are designed primarily for real-time dedicated applications or as high-performance, general-purpose computers. The PDP-11 series, from Digital Equipment Corporation (DEC), are the most prominent 16-bit minicomputers. For example, the PDP-11/02 and PDP-11/04 16-bit minicomputers are primarily used as dedicated process controllers; the PDP-11/45 and PDP-11/70 minicomputers are high-performance, general-purpose multiuser computers, which can typically support 20 users.

The speed, instruction repertoire, and memory capacity of minicomputers have increased with advances in technology, so today's minicomputers can perform the functions that mainframes did in the 1960s. These advances in technology led to the introduction in the late 1970s of the 32-bit minicomputers, which are referred to as the *superminis*. The VAX 8600 from DEC is the most prominent one. The cost of minicomputers ranges from around $20,000 to about $350,000.

Mainframe

Mainframes are high-performance, general-purpose computers that range in price from $1 to $10 million. Mainframes are capable of executing in excess of 53 million instructions per second (MIPS), compared with about 5 MIPS for a supermini. IBM is the dominant supplier of mainframe computers, and its 3090/400 model has an instruction execution rate of about 53 MIPS and costs about $9½ million. The primary function of mainframes today is to support large databases, such as those found in government agencies, large businesses, and many universities. Mainframes have the processing power and storage capacity for extremely large database systems.

Supercomputer

Even with the power and performance of mainframes, there remain applications that are beyond the reach of the general-purpose mainframes. Solutions for some problems in aerodynamics, seismology, weather prediction, image processing, real-time voice recognition, and nuclear energy require high precision and a program that repetitively performs complex arithmetic operations on large arrays of numbers. To handle these types of problems, *supercomputers* have been developed. Supercomputers are very powerful, extremely high-performance computers that may cost more than $10 million. Supercomputers are capable of executing in excess of one billion floating-point operations per second (FLOPS). Since supercomputers are used in applications that usually require the processing of ordered data, such as matrices or vectors, a speed advantage is achieved by performing arithmetic operation in parallel. Supercomputers contain highly parallel processors. Cyber 205, Cray X-MP, and Control Data's STAR-100 are examples of supercomputers.

12.2 COMPUTER SYSTEM ORGANIZATION

All different types of digital computer systems include two principal components, *hardware* and *software*. Hardware refers to the physical components of a computer such as memory unit (MU), arithmetic and logic unit (ALU), control unit (CU), input/output (I/O) devices, and so on. Software refers to the collection of instructions (programs) that direct the operations of the digital computer's hardware.

Hardware

A digital computer is basically organized as shown in Fig. 12.2-1. The exact nature of the components making up the four basic sections of the computer may vary, and the different sections may overlap and share some components; however, the four functions associated with each section may be clearly identified. The memory unit (MU) is used to store both data and programs that are currently being processed and executed. The arithmetic and logic unit (ALU) processes the data taken from the MU and/or input devices and places the processed data back into the MU and/or output devices. The input/output (I/O) system may consist of a

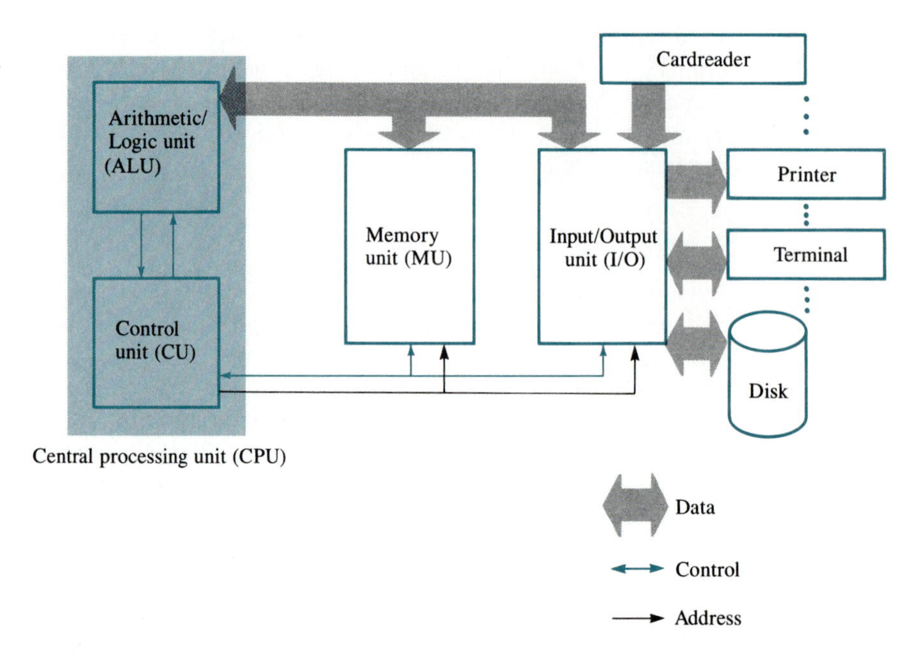

FIG. 12.2-1. Typical computer organization.

variety of devices for communicating with the external world and for storing large quantities of data. Keyboards, magnetic tape readers, and analog-to-digital (A/D) converters are examples of input devices; and line printers, plotters, and D/A converters are examples of output devices. Some devices, such as terminals and magnetic disk drives, provide both input and output capabilities. The control unit (CU) coordinates the activities of the MU, ALU, and I/O units. It retrieves instructions from programs resident in the MU, decodes these instructions, and directs the ALU to perform the corresponding processing steps. In addition, the CU oversees the I/O operations.

Software

The second major component of a computer system is the software. Software is generally divided into two categories: *system software* and *user software*. System software refers to the collection of programs that is provided by the computer system for the creation and execution of user programs. User software consists of those programs generated by the various users of the computer system for solving their specific problems.

A set of instructions and data specifying the solution of a particular problem is called a *program*. Since the basic unit of information in a digital computer is either 0 or 1, programs and data must therefore be expressed by using this binary system. Programs written by using 0s and 1s are known as *machine-language programs*. At this level of programming, instructions such as ADD and SUBTRACT must be represented by a unique pattern of 0s and 1s that the computer hardware can understand and execute directly. Writing programs in this form is tedious and error-prone. In addition, the programmer must have a detailed knowledge of the computer's structure.

The difficulty of machine-language programming is alleviated when programming symbols such as ADD and SUB, rather than 0s and 1s, are introduced. Programming at the symbolic level is called *assembly-language programming*. Assembly-language instructions match machine-language instructions on a more or less one-for-one basis but are written by using symbolic names, which are called *mnemonics*. An assembly-language programmer is also required to have a detailed knowledge of the computer's structure. To translate assembly-language programs into machine language, the computer uses an *assembler*. An assembler is a program that converts assembly-language programs into their equivalent machine-language programs. When an assembler is run in the computer for which it is to generate the machine code, this assembler is known as a *resident* or *self-assembler*. However, if the assembler is run in a computer to produce a machine code for a different computer, this assembler is known as a *cross-assembler*.

Even though assembly-language programming is an immense improvement over machine-language programming, it is still computer-oriented; that is, an assembly-language written for one computer may not run for a different computer. However, programs written by using *high-level languages* (HLL), such as FORTRAN, Pascal, BASIC, LISP, and C, could be run on virtually any computer. High-level languages are problem-oriented languages that permit us to write programs in forms as close as possible to the human-oriented languages. Each high-level-language statement may require many machine-language instructions.

There are two distinct methods for converting high-level-language programs into machine-language programs. In one method, each high-level-language statement is executed as soon as it is translated into its equivalent set of machine-language instructions. A system functioning in this way is known as an *interpreter*. BASIC is probably the most common interpretive language. Interpretive languages such as BASIC are very inefficient for programs with repetitive constructs, i.e., loops. The interpreter translates the high-level instructions in the loop structure on every pass through the loop, which is inefficient since the translation is the same on every pass. This inefficiency is corrected by *compilers*. A compiler translates the complete high-level language into machine language. Once the compilation (i.e., translation) of the whole program has been completed, execution of the program may be initiated. Once a program has been successfully compiled, it can be executed as many times as needed without recompilation. FORTRAN, PASCAL, and C are three examples of compiled languages.

Operations such as selecting a given compiler or assembler for translating a given program into machine language, starting and stopping the execution, and managing for different resources of the computer system are done by a set of system programs called the *operating system*. The operating system performs resource management and human-to-machine translation. The operating system can be thought of as layers of software on top of the hardware that let programmers concern themselves with their programs and data structure rather than with maintaining and manipulating the hardware. An operating system is designed to support a given computer architecture, and therefore, not every operating system is usable on every micro, mini, or mainframe computer. DOS, VMS, and UNIX are three common operating systems available for microcomputers, minicomputers, and mainframes.

12.3 ARCHITECTURE OF A MICROCOMPUTER SYSTEM

In the early 1970s, Intel Corporation introduced a large-scale digital integrated circuit that revolutionized the computer industry. It is the *microprocessor*. A microprocessor is an LSI device that combines the functions of the arithmetic/logic unit and control unit to produce a central processing unit on a single chip.

The microprocessor is the CPU of a microcomputer system. Replacing the ALU and CU of Fig. 12.2-1 by a microprocessor, one obtains the general layout of the major component of a microcomputer system. This system is known as a *stored-program computer* because instructions and data are stored in the same memory.

The major components of a microcomputer system are connected to each other by a *bus*. A bus is a set of wires carrying address, data, and control signals. The address lines are unidirectional signals that specify the address of a memory location or I/O device that will communicate with the microprocessor. The number of address lines varies from one microprocessor to another. This number is usually between 16 and 32. With a 24-bit address bus, like the one found in most of today's microcomputers, the microprocessor can access over 16 million (2^{24}) memory locations. Memory is usually organized in blocks of 8, 16, or 32 bits. The data bus, which is used to carry data between the CPU, memory, and I/O devices, is a bidirectional bus, and its size may vary from 8 to 32 bits. The signals provided on the control bus are used to synchronize the memory and I/O operations, select either memory or the I/O device, and request either the read or write operation from the selected device.

12.4 ARCHITECTURE OF A MICROPROCESSOR

CPU Architecture

A typical microprocessor (μp) consists of a set of dedicated registers, a set of general-purpose registers, an ALU, a CU, and address, data, and control buses, as shown in Fig. 12.4-1. These different componets of the μp are interconnected through an internal bus.

The set of dedicated registers usually consists of a *program counter* (PC), a *stack pointer* (SP), and a *program status word* (PSW).

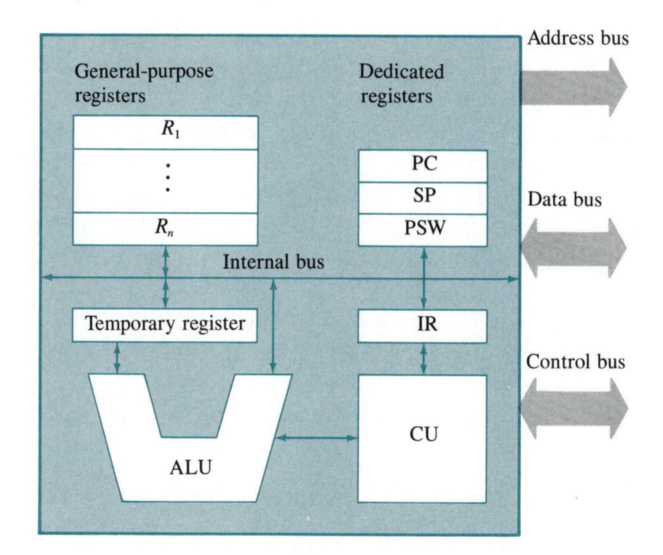

FIG. 12.4-1. Typical microprocessor architecture.

The program counter is a register used to hold the address of the memory location containing the next instruction to be executed. After each instruction is fetched (i.e., read) from the memory, the program counter is automatically updated to point to the next memory location from which the next instruction is to be fetched.

At times, the μp may have to be temporarily diverted from its main task and take up different operations, such as an interrupt or subroutine calls. When this occurs, the μp will save the status and the information related to the current position of the present task in a dedicated area of the random-access memory known as the *stack*. The stack pointer (SP) is used to point to the most recently saved data or the next available memory in the stack. The SP is used to save and/or retrieve data without explicitly specifying its address in memory.

When one must know the status of an operation, the program status word (PSW) is used. It is a collection of status bits, also known as *flags*, that are set to logical 1 or logical 0, depending on the result of the previous instruction. For example, if the addition operation of two 8-bit numbers produced a carryout of the MSB, then one bit of the PSW, known as the *carry flag* (CY), will be set. Conditional-jump (i.e., branch) instructions use the PSW's bits to make decisions and redirect the program flow.

Two dedicated registers used by the microprocessor cannot be accessed by the programmer (i.e., user): the *instruction* and the *temporary registers*. When the μp fetches an instruction from the memory location pointed to by the PC, it places it in the instruction register (IR), where it is decoded to determine what operation is to be performed. The temporary register is used by the ALU to temporarily hold some intermediate data for its arithmetic and/or logic operations.

The number and size of general-purpose registers vary from one microprocessor to another, depending on their intended performance. General-purpose registers are used to temporarily hold data while they are being processed and to maintain pointers to different sections of external memory. The principal task of the microprocessor is to fetch and decode instructions from memory and then generate the appropriate control signals needed to carry out these instructions. This task is performed by the control unit. The CU is the heart of the microprocessor, and it consists of the timing and data-routing circuits, such as multiplexers and decoders. The CU decodes the instruction being processed and properly establishes data paths among the various elements of the microprocessor.

Internal Operations

The internal operations of a microprocessor can be summarized by the following four steps:

1. The instruction to be executed is fetched from the memory location pointed to by the PC.
2. The instruction is decoded and the PC is updated to point to the next executable instruction.
3. The instruction is executed.
4. The above three steps are repeated.

In the fetch cycle, the microprocessor places the address, available in the program counter, in its address bus and issues a control signal requesting the memory to place the machine instruction in the data bus. The microprocessor then reads the instruction from the data bus and places it in the instruction register for decoding. Once the instruction is decoded, the control unit will generate the proper signals to perform the operations required by the instruction. When the execution of the instruction is completed, the program counter is updated, and this cycle is repeated.

However, if the instruction is determined to be a conditional branch instruction, the program status word (PSW) is first examined. If the test condition is met, the program counter is set to the branch address. If the test condition is not met, the program counter is updated to point to the instruction following the one just fetched. For unconditional branch instructions, the PC is set to the branch address from which the next instruction will be fetched.

The four steps described above are illustrated in the flowchart given in Fig. 12.4-2.

EXAMPLE 12.4-1

Suppose the PC register of the CPU of Fig. 12.4-1 contains $(0000\ 0000\ 0110\ 0100)_2$, $R_1 = (1101\ 0011)_2$ and $R_2 = (0011\ 1100)_2$. (a) What is the address of the next instruction to be fetched and executed? (b) If the accessed memory location contains a 2-byte code corresponding to the instruction "move the contents of R_1 to R_2," then what will be the contents of registers R_1 and R_2 and the PC?

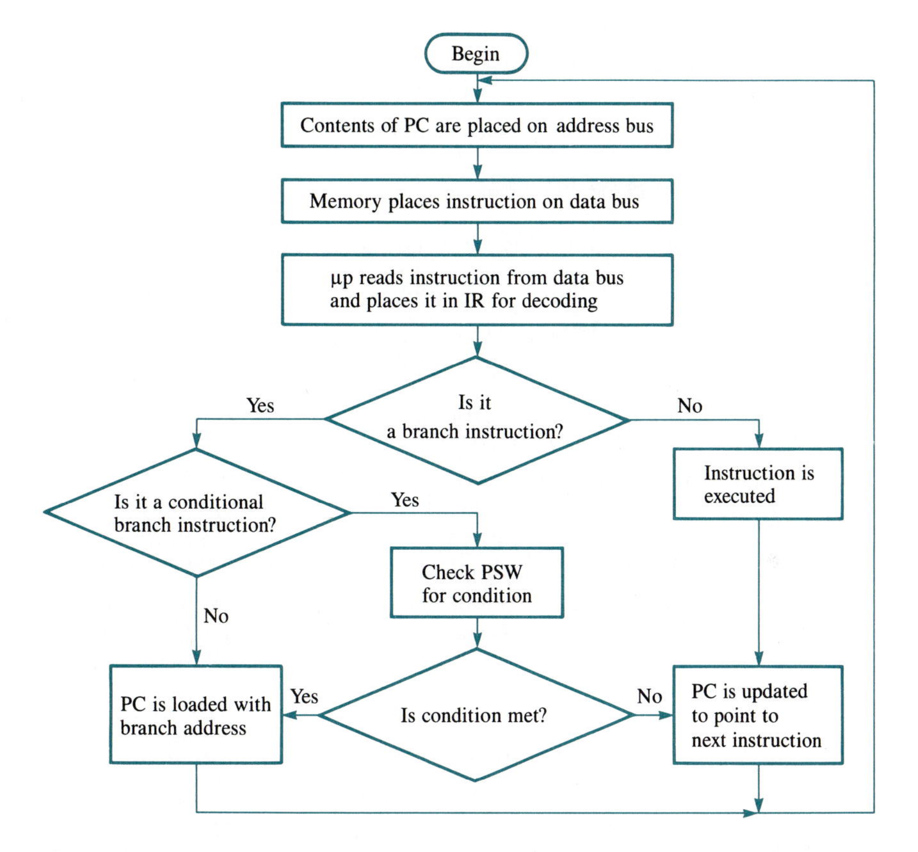

FIG. 12.4-2. The internal operations of a microprocessor.

Solution

(a) Since the content of the PC is the address of the next instruction to be fetched and executed, this instruction will be fetched from the memory location whose address is $(0000\ 0000\ 0110\ 0100)_2$ [i.e., $(100)_{10}$].

(b) After the execution of the 2-byte instruction from memory location $(0000\ 0000\ 0110\ 0100)_2$, the PC will be updated to point to the next instruction in memory. Therefore, the PC will contain $(0000\ 0000\ 0110\ 0110)_2$ which is the original value plus the length of the instruction just fetched. Since the move instruction does not modify its source, R_1 in this example, then R_1 and R_2 will both contain $(1101\ 0011)_2$.

Machine-Language Instructions

A machine-language instruction is a command to the microprocessor to manipulate data, and a sequence of these instructions constitutes a machine-language program.

Op code	Operand 1	...	Operand *n*

FIG. 12.4-3.
The general format of machine-language instruction.

Instruction Format

In general, a machine-language instruction has two components (or *fields*):

1. Op code field (or operation field).
2. Operand field(s).

The op code field is a string of 0s and 1s that specifies the type of operation the microprocessor is to perform. The operand field(s) specifies the information needed by the microprocessor to carry out the instruction. Figure 12.4-3 shows the general format of machine-language instruction.

While all instructions must have one op code field to specify how the data are to be manipulated, they may have zero, one, or more operand field(s) to specify the data, if any, to complete the operation described by the op code field. The operand field may contain the actual data, the address of the actual data, or a pointer to the address of the actual data. The following three instructions are examples of zero-, one-, and two-operand field instructions:

Zero-operand instruction	**HLT**	
One-operand instruction	**CLR**	R_1
Two-operand instruction	**ADD**	R_1, R_2

HLT is the mnemonic code for the HALT (stop) instruction, which instructs the microprocessor to stop the execution and to remain in an idle state. CLR is the mnemonic for the CLEAR instruction which, in this example, is instructing the microprocessor to clear the contents of R_1. For the third instruction, assuming that this microprocessor uses R_2 as the source and R_1 as the destination registers, then ADD R_1, R_2 instructs the microprocessor to add the contents of R_1 and R_2 and to place the result in R_1, that is, $R_1 = R_1 + R_2$.

The format and length of the machine-language instructions vary from one microprocessor to another. The size of the op code determines the number of distinct operations that can be specified. For example, op code fields 4 and 8 bits long can specify 16 and 256 distinct operations, respectively.

Even though microprocessors can only execute programs in machine language, assembly-language programming is used to simplify program writing and interpretation. Assembly-language programs must be converted by an assembler to machine code. In general, an assembly-language instruction may be comprised of four fields using the following format:

LABEL: OP_CODE OPERAND ; COMMENTS

The *LABEL* field is an optional field that specifies a symbolic name to be assigned to an instruction's location. The *OP_CODE* field contains the instruction mnemonic for the operation to be performed. The *OPERAND* field specifies the data, if any, that are required by the instruction. Operands may be expressed as binary, octal, hexadecimal, or decimal numbers; or they may be expressed as symbolic names with preassigned values. Finally, the *COMMENTS* field is an optional field that provides a description of the way the instruction relates to the purpose of the program. This field is necessary for the clarification and the overall documentation of the program.

Addressing Modes

The ways the operands can be specified in machine-language instructions are called *addressing modes*. A given addressing mode specifies the steps that must be taken by the CPU to obtain the proper operand; that is, the address mode of an operand specifies how to obtain the actual data to be manipulated. Different microprocessors support different addressing modes. The four typical addressing modes supported by popular microprocessors are examined here. (They are the basic addressing modes of the Intel 8085 microprocessor.)

1. *Immediate addressing*. In this addressing mode, the data to be manipulated are part of the instruction. For example, the instruction "**MVI A**, 195" loads register **A** with the integer 195. In this case, the operand "195" is the actual data to be loaded into register **A**, and it is explicitly specified in the instruction.
2. *Direct addressing*. In this addressing mode, the address of the location of the actual operand is specified. For example, the instruction "**LDA A**, 195" loads register **A** with the contents of the memory location whose address is 195. In this case, the operand "195" is the address of the actual data to be loaded into register **A**.
3. *Register addressing*. In this addressing mode, the actual data to be manipulated are in one of the microprocessor's internal registers and the operand specifies which of the registers to use to obtain the actual data. For example, the instruction "**MOV A, B**" moves the contents of register **B** into register **A**.
4. *Register indirect addressing*. In this addressing mode, the address of the actual data to be manipulated is contained in the register specified by the operand field. For example, the instruction "**MOV M**, 195" loads the integer 195 into memory location **M**. For the 8085 microprocessor, memory location M is pointed to by the register pair **H** and **L**. Therefore, this instruction will load "195" into the memory location whose address is given by the contents of registers **H** and **L**.

EXAMPLE 12.4-2

Assume that a program segment using 8085 mnemonic is as given below.

Memory Address	Instruction	; Comments
100	**MVI A**, 03	; load register **A** with 03
102	**MVI L**, 250	; load register **L** with 250
104	**MVI H**, 0	; load register **H** with 0

```
105      STA 200      ; store register A in memory
                      ; location 200
108      INR A        ; increment register A
                      ; A = A + 1
109      MOV B, A     ; move a copy of register A
                      ; to register B
110      MOV M, A     ; move the contents of
                      ; register A into the memory
111      HLT          ; stop the execution
.        .
.        .
.        .
200      XX           ; contents of memory
.        .            ; locations 200 and 250
.        .            ; are undefined
.        .            ;
250      XX           ;
```

(a) What are the addressing modes given in memory locations 100, 104, 105, and 110? (b) List the zero-, one-, and two-operand instructions. (c) Pretend that you are the CPU and your PC is initialized to 100. Now, what are the contents of registers **A**, **B**, **H**, and **L** and memory locations 200 and 250 after the execution of the above code?

Solution

(a)

Instruction	Addressing Mode
MVI A, 03	immediate
MOV H, L	register
STA 200	direct
MOV M, A	register indirect

(b)

Zero Operand	One Operand	Two Operands
HLT	**INR A**	**MVI A**, 03
	STA 200	**MVI L**, 250
		MVI H, 0
		MOV B, A
		MOV M, A

(c) The instruction at address 105, namely **STA** 200, directly affects the contents of memory location 200. This instruction stores the contents of register **A** in memory location 200. Since register **A** is initialized to 03 by the instruction given in memory location 100, then memory location 250 will contain 3.

The instructions at memory locations 104 and 102 initialize the **H** and **L** registers to 0 and 250, respectively. When **H** and **L** are combined to form a register pair **HL**, their combined value is 250. Now, the instruction at address 110 will load the memory location whose address is in **HL** with the contents of register **A**. Therefore, memory location 250 will contain 4, which is the value of register **A** after it was incremented by the instruction in location 108.

Therefore, we have the following contents in the registers and memory locations:

Register **A** will contain 4.

Register **B** will contain 4.

Register **H** will contain 0.

Register **L** will contain 250.

Memory location 200 will contain 3.

Memory location 250 will contain 4.

Instruction Set

A complete list of the op codes that a given microprocessor can recognize and execute is called an *instruction set*. Using any microprocessor to accomplish a given task depends on a detailed knowledge of its instruction set and addressing modes. A clear understanding of the instruction set is necessary to use the microprocessor effectively.

Currently, a bewildering variety of microprocessors are available. They are commonly characterized by the number of bits they can process in one instruction. The number of bits, also known as the *word length*, is generally determined by the width of the data path. Eight-, 16-, and 32-bit microprocessors are in widespread use. The Intel 8085, the Zilog Z80, and the Motorola 6800 are three popular 8-bit microprocessors. The Intel 8086 and 80286, the Motorola MC68000, and the Zilog Z8000 are 16-bit microprocessors that dominate the market. Finally, the Intel 80386, the Motorola MC68020, and the National NS32032 are truly powerful 32-bit microprocessors that dominate the very high end of the market.

Since different microprocessors support different instruction sets and addressing modes, and since we have limited space in this book, we will focus our attention on only one of these microprocessors, the Intel 8085.

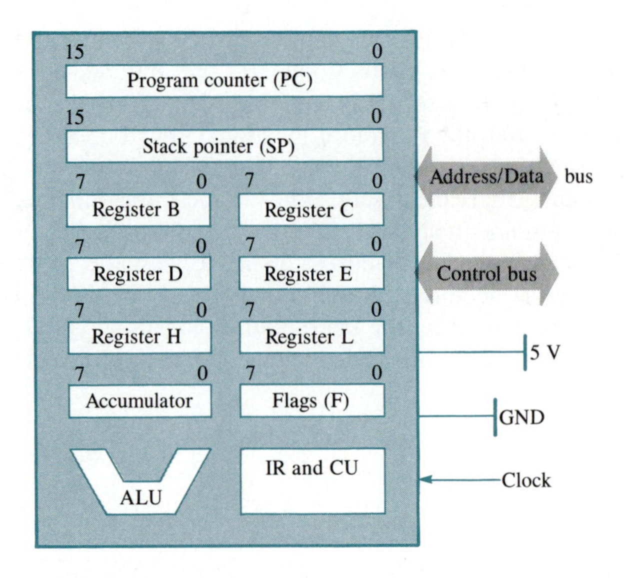

FIG. 12.5-1. The 8085's internal architecture.

12.5 THE 8085 MICROPROCESSOR

The Intel 8085 was designed as an enhanced version of the Intel's popular 8080. The 8080 was the first commercially successful 8-bit microprocessor, developed and designed by Intel Corporation in 1974. The 8085 μp is a 40-pin chip that operates from a single +5-V power supply. It executes the 8080 μp instruction set plus two minor additions to support the new hardware.

8085 Programming Model

The 8085 is a register-oriented microprocessor with a 16-bit **PC**, a 16-bit **SP**, an 8-bit PSW (flag), six 8-bit general-purpose registers (**B**, **C**, **D**, **E**, **H**, and **L**), and an 8-bit register known as the *accumulator* (**A**). Figure 12.5-1 shows the internal architecture of the 8085 μp.

The 16-bit **PC** and **SP** registers can provide an address range from 0 to $(2^{16} - 1)$, or a total of 65,536 memory locations. Hence, the 8085 is capable of directly addressing 64K (2^{16}) memory locations. The 8-bit flag register has only 5 of its 8 bits used as flags. As shown in Fig. 12.5-2, they are the *carry* (**CY**), *auxiliary carry* (**AC**), *parity* (**P**), *sign* (**S**), and *zero* (**Z**) flags. These bits are set or reset as a result of arithmetic and/or logic operations and are used to help in the decision making and branching under software control. The six general-purpose registers are used for data and address storage within the microprocessor. In addition, some instructions use these registers in pairs to perform 16-bit operations, and the **HL**, **BC**, and **DE** pairs can be used as address pointers. Finally,

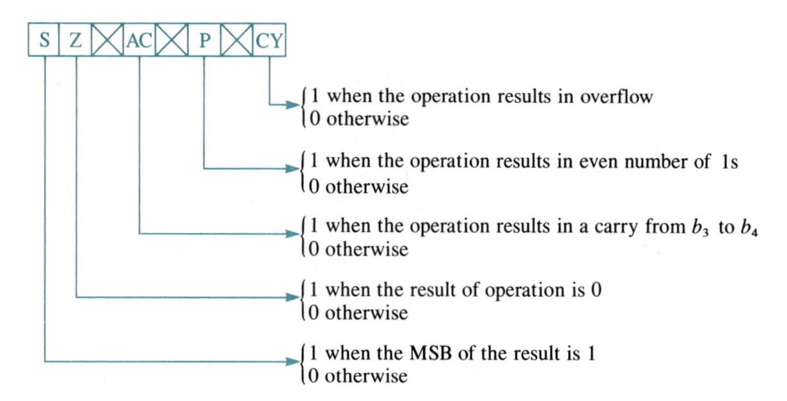

FIG. 12.5-2. The format of the 8085 flag register.

register **A**, which is known as the accumulator, is the most versatile register. Most of the arithmetic and logic operations require that one of their operands be in the accumulator, and the result will be placed in the accumulator. Also, when input/output operations are performed between the CPU and I/O devices, the accumulator is used to receive and transmit the data from and to the I/O devices.

EXAMPLE 12.5-1

Show the contents of all flags after the addition of registers **A** and **B** for (a) **A** = $(01101101)_2$ and **B** = $(11001001)_2$ and (b) **A** = $(01101000)_2$ and **B** = $(10000111)_2$.

Solution

(a)

		b_7	b_6	b_5	b_4	b_3	b_2	b_1	b_0
A →		0	1	1	0	1	1	0	1
+ **B** →	+	1	1	0	0	1	0	0	1
(Result in **A**)	1	0	0	1	1	0	1	1	0

CY = 1 since a carryout of the MSB is generated

P = 1 since the 8-bit result contains an even number of 1s

AC = 1 since a carry from b_3 to b_4 is generated

Z = 0 since the result does not equal zero

S = 0 since the MSB (i.e., b_7) of the result is zero

(b)

		b_7	b_6	b_5	b_4	b_3	b_2	b_1	b_0
A →		0	1	1	0	1	0	0	0
+ **B** →	+	1	0	0	0	0	1	1	1
(Result in **A**)		1	1	1	0	1	1	1	1

CY = 1 since a carryout of the MSB is generated

P = 0 since the 8-bit result contains an odd number of 1s

AC = 0 since no carry from b_3 to b_4 is generated

Z = 0 since the result does not equal zero

S = 1 since the MSB (i.e., b_7) of the result is 1

8085 Instruction Set

The 8085 instruction set consists of 74 distinct instructions, which, when combined with different addressing modes, can provide 246 distinct operations. The length of one instruction can vary from 1 to 3 bytes (1 byte = 8 bits). The first byte of the instruction is its op code. The second and third bytes, if any, are the operands. The mnemonics assigned to the instructions are designed to indicate the function of the instruction. The instruction set is divided into six categories, as described in the following sections. The symbols and abbreviations used in the description of the instruction set are defined in Table 12.5-1.

Data-Transfer Group. The data-transfer instructions move data between memory and registers and between registers themselves. Tables 12.5-2 describes the mnemonics used for this group. The 8085 does not allow for data transfer between two memory locations directly. Data-transfer instructions do not affect the flags.

TABLE 12.5-1. Symbols and Abbreviations

Symbols	Meaning
adrs	16-bit address value
data8	8-bit data value
data16	16-bit data value
rp	Register pairs **HL**, **DE**, or **BC**
$rp_{B,D}$	Register pairs **BC** or **DE**
r, r_1, r_2	One of the general-purpose register **A**, **B**, **C**, **D**, **E**, **H**, or **L**
M	8-bit memory location whose address is in register pair **HL**
( )	The contents of the memory location or registers enclosed in the parentheses
Port	8-bit address of an I/O device
$\longleftarrow$	Is transferred to

TABLE 12.5-2. 8085 Data-Transfer Instructions

Mnemonics	Description
MOV r_1, r_2	Move the contents of r_2 to r_1; $(r_1) = (r_2)$
MOV M, r	Move the contents of register **r** to memory
MOV r, M	Move the contents of memory to register **r**
MVI r, data8	Move immediate 8-bit data to **r**; (**r**) = **data8**
MVI M, data8	Move immediate 8-bit data to memory
LDA adrs	Load the accumulator with the contents of the memory location whose address is **adrs**
STA adrs	Store the contents of the accumulator into the memory location whose address is **adrs**
LHLD adrs	Load the **H** and **L** registers directly from the memory locations whose first address is **adrs**
SHLD adrs	Store the **H** and **L** registers directly in the memory locations whose first address is **adrs**
LXI rp, data16	Load the register pair r_p with immediate 16-bit data
LDAX $rp_{B,D}$	Load the accumulator from the memory location whose address is in register pair $rp_{B,D}$
STAX $rp_{B,D}$	Store the accumulator in the memory location whose address is in register pair **rp**
XCHG	Exchange the contents of registers **H** and **L** with the contents of registers **D** and **E**
XTHL	Exchange the top of the stack with registers **H** and **L**

EXAMPLE 12.5-2

Consider the following simple data-manipulation program.

Instruction #	Mnemonics	; Comments
1	**MVI A**, 123	; load register **A** with 123
2	**MVI C**, 100	; load register **C** with 100
3	**MVI B**, 0	; load register **B** with 0
4	**LXI H**, 200	; load register pair **HL** with 200
5	**STAX B**	; store a copy of register **A** into the memory ; location whose address is given by the ; contents of registers **B** and **C**
6	**MVI E**, 255	; load register **E** with 255
7	**MOV D, B**	; move a copy of register **B** to **D**
8	**XCHG**	; exchange the contents of **H** and **L** with ; **D** and **E**
9	**HLT**	; stop

(a) What are the contents of registers **A**, **B**, **C**, **D**, **E**, **H**, and **L** after the execution of the program? (b) What is the content of the memory location whose address is 100?

Solution

(a) Tracing the execution of the above program, we can tabulate the contents of each register after the execution of each instruction as follows:

After	Contents						
Instruction #	**A**	**B**	**C**	**D**	**E**	**H**	**L**
1	123						
2	123		100				
3	123	0	100				
4	123	0	100			0	200
5	123	0	100			0	200
6	123	0	100		255	0	200
7	123	0	100	0	255	0	200
8	123	0	100	0	200	0	255
9	123	0	100	0	200	0	255

A blank for a given entry denotes that the content of the register is unknown.
(b) The only reference to memory in the program is given in instruction #5, that is, **STAX B**, which stores the contents of register **A** (i.e., 123) in the memory location whose address is given by the register pair **BC**. Since **BC** contains 100, memory location 100 will contain the data 123.

Arithmetic Group. The arithmetic instructions are designed to add, subtract, increment, or decrement data in registers or memory. Table 12.5-3 describes the mnemonics used for this group.

EXAMPLE 12.5-3

Consider the following simple data-manipulation program.

Instruction #	Mnemonics	; Comments
1	**LXI H**, 76	; load register pair **HL** with 76
2	**LXI B**, 20	; load register pair **BC** with 20

TABLE 12.5-3. 8085 Arithmetic Instructions

Mnemonics	Description
ADD X*	Add the contents of **X** to the accumulator, **(A)** = **(A)** + **(X)**
ADC X	Add the contents of **X** and the **CY** flag to the accumulator, **(A)** = **(A)** + **(X)** + **(CY)**
ADI data8	Add the immediate 8-bit data to the accumulator, **(A)** = **(A)** + **data8**
ACI data8	Add the immediate 8-bit data and the **CY** flag to the accumulator, **(A)** = **(A)** + **data8** + **(CY)**
SUB X	Subtract the contents of **X** from the accumulator, **(A)** = **(A)** − **(X)**
SUI data8	Subtract the immediate 8-bit data from the accumulator, **(A)** = **(A)** − **data8**
SBB X	Subtract the contents of **X** from the accumulator, using the borrow (i.e., **CY**) flag, **(A)** = **(A)** − **(X)** − **(CY)**
SBI data8	Subtract the immediate 8-bit data from the accumulator, using the borrow (i.e., **CY**) flag, **(A)** = **(A)** − **data8** − **(CY)**
INR X	Increment the contents of **X** by 1, **(X)** = **(X)** + 1
DCR X	Decrement the contents of **X** by 1, **(X)** = **(X)** − 1
INX rp	Increment the contents of **rp** by 1, **(rp)** = **(rp)** + 1
DCX rp	Decrement the contents of **rp** by 1, **(rp)** = **(rp)** − 1
DAD rp	Add the contents of **rp** to **HL**, **(H)(L)** = **(H)(L)** + **(rp)**

*X denotes either **r** or **M**.

```
3    DAD B       ; add the contents of register pair
                 ; BC to the contents of register pair HL
4    MVI A, 38   ; load register A with 38
5    ADI 30      ; add 30 to the contents of register A
6    INR B       ; increment the contents of register B
7    DCX H       ; decrement the contents of register pair HL
8    MOV M, A    ; move the contents of register A to memory
9    HLT         ; stop
```

(a) What are the contents of registers **A**, **B**, **C**, **D**, **E**, **H**, and **L** after the execution of the program? (b) Which memory location is affected by the code segment and how?

Solution

(a) Tracing the execution of the code segment, we can tabulate the contents of each register after the execution of each instruction as follows:

After Instruction #	Contents						
	A	B	C	D	E	H	L
1						00	76
2		00	20			00	76
3		00	20			00	96
4	38	00	20			00	96
5	68	00	20			00	96
6	68	01	20			00	96
7	68	01	20			00	95
8	68	01	20			00	95
9	68	01	20			00	95

(b) Instruction #8 (i.e., **MOV M, A**) is the only instruction that references the memory. Since registers **H** and **L** contain 95, the contents of the accumulator at the time instruction #8 is being executed will be moved to memory location 95. Hence, memory location 95 will be loaded with 68.

Logical Group. The logic instructions are designed to perform logic (boolean) operations on data in registers, memory, and flags. Table 12.5-4 describes the mnemonics used for this group. The logical AND, OR, and exclusive-OR (XOR) instructions are used to clear, set, and complement, respectively, specific bits of the accumulator. The rotate instructions are used to shift the contents of the accumulator one position to the left or to the right. Two instructions are provided to set or complement the carry flag. Finally, two compare instructions are also provided to compare the contents of the accumulator with either another 8-bit register, the contents of an 8-bit memory location, or an 8-bit immediate data. The compare instructions do not affect the contents of the accumulator, register, or memory, but they do affect the condition flags. For example, if register **A** contains 55, then the instruction "**CPI** 55" will set the zero flag to 1 but leave register **A** unchanged.

EXAMPLE 12.5-4

Using some of the instructions discussed so far, write a sequence of instructions that load register **A** with the constants $(00001111)_2$, clear its b_0 and b_1, complement its b_3 and b_4, and set its b_6 and b_7, where b_0 and b_7 are the least and most significant bits, respectively, of register **A**.

TABLE 12.5-4. 8085 Logic Instructions

Mnemonics	Descriptions
ANA X*	Logical AND the contents of **X** with the accumulator, **(A)** = **(A)** · **(X)**
ANI data8	Logical AND the immediate **data8** with the accumulator, **(A)** = **(A)** · **data8**
ORA X	Logical OR the contents of **X** with the accumulator, **(A)** = **(A)** OR **(X)**
ORI data8	Logical OR the immediate **data8** with the accumulator, **(A)** = **(A)** OR **data8**
XOR X	Logical exclusive-OR the contents of **X** with the accumulator, **(A)** = **(A)** XOR **(X)**
XRI data8	Logical exclusive-OR the immediate **data8** with the accumulator, **(A)** = **(A)** XOR **data8**
RLC	Rotate the accumulator left, $(A_7) \rightarrow$ **(CY)**, $(A_7) \rightarrow (A_0)$, and $(A_i) \rightarrow (A_{i+1})$, $0 \le i \le 6$
RRC	Rotate the accumulator right, $(A_0) \rightarrow$ **(CY)**, $(A_0) \rightarrow (A_7)$, and $(A_i) \rightarrow (A_{i-1})$, $1 \le i \le 7$
RAL	Rotate the accumulator left through the **(CY)**, **(CY)** $\rightarrow (A_0)$, $(A_7) \rightarrow$ **(CY)**, and $(A_i) \rightarrow (A_{i+1})$, $0 \le i \le 6$
RAR	Rotate the accumulator right through the **(CY)**, **(CY)** $\rightarrow (A_7)$, $(A_0) \rightarrow$ **(CY)**, and $(A_i) \rightarrow (A_{i-1})$, $1 \le i \le 7$
CMA	Complement the accumulator
CMC	Complement the carry flag **(CY)**
STC	Set the carry flag, **(CY)** = 1
CMP X	Compare the contents of **X** with the accumulator, **(TEMP)** = **(A)** − **(X)**
CPI data8	Compare the accumulator with the immediate **data8**, **(TEMP)** = **(A)** − **data8**

*__X__ denotes either **r** or **M**.

Solution

Instruction #	Mnemonics	;	Comments
1	**MVI A**, 00001111B	;	**(A)** = 00001111
2	**ANI** 11111100B	;	**(A)** = 00001100
3	**XRI** 00011000B	;	**(A)** = 00010100
4	**ORI** 11000000B	;	**(A)** = 11010100

To load register **A** with $(00001111)_2$, instruction #1 is used. The AND operation is used to clear specific bits. After the execution of instruction #2, the accumulator is left with 0s corresponding to the 0s of the operand. The exclusive-OR operation is used to complement specific bits. After the execution of instruction #3, the accumulator is left with the complement of the bits corresponding to the 1s of the operand and with the original values for the bits corresponding to the 0s of the operand. Finally, the OR operation is used to set specific bits of the accumulator to 1. After the execution of instruction #4, the accumulator is left with 1s corresponding to the 1s of the operand and with the original values for the bits corresponding to the 0s of the operand.

Branch Group. Branch instructions transfer program control by changing the value of the program counter to the address of a nonconsecutive instruction. They alter the flow of normal sequential program execution, either conditionally or unconditionally.

There are three types of unconditional-branch instructions: *jump*, *call*, and *return*. The jump instruction, **JMP**, alters the normal sequential program flow by replacing the contents of the program counter with the address given in the instruction. The **JMP** instruction has the following format:

JMP **adrs**

The call instruction, **CALL**, alters the normal sequential program flow by first saving the present value of the **PC** in the stack and then loading the **PC** with the address given in the instruction. The **CALL** instruction is used to enter a procedure or a subroutine and has the following format:

CALL **adrs**

The return instruction, **RET**, alters the normal sequential program flow by replacing the contents of the **PC** with the contents of two consecutive memory locations whose address is the contents of the stack pointer (**SP**). The return instruction is a 1-byte instruction that does not require an operand. It is used to return the processor at the end of a procedure or subroutine to the main program.

In a conditional branch, transfer of program control occurs only if a specified condition is satisfied. The specific condition is usually that one of the flags be set or reset by the execution of a previous instruction. When the condition is not satisfied, the program execution continues with the instruction following the branch instruction. Conditional-branch instructions give the microprocessor the ability to make decisions about different actions on the basis of previous results.

The general formats for conditional jump, call, and return, respectively, are

J⟨condition⟩ **adrs**
C⟨condition⟩ **adrs**
R⟨condition⟩

where ⟨condition⟩ is one of the eight conditions given in Table 12.5-5. For example, the instruction "**JC adrs**" will cause the microprocessor to jump to the instruction whose address is "**adrs**" if and only if the carry flag is set to 1. Similarly, the instruction "**CNZ adrs**" will cause the microprocessor to save its program counter's present value and jump to the instruction whose address is "**adrs**" if and only if the zero flag is not set (**Z** = 0). Conditional- and unconditional-branch instructions are often used as decision elements in constructing basic logic structures.

TABLE 12.5-5. Conditions Used in a Conditional Branch

<Condition>	Meaning
C	Carry flag is set (**CY**) = 1
NC	Carry flag is not set (**CY**) = 0
Z	Zero flag is set (**Z**) = 1
NZ	Zero flag is not set (**Z**) = 0
M	Minus; sign flag is set (**S**) = 1
P	Plus; sign flag is not set (**S**) = 0
PE	Parity even; parity flag is set (**P**) = 1
PO	Parity odd; parity flag is not set (**P**) = 0

EXAMPLE 12.5-5

Write an assembly-language program segment that is equivalent to the following high-level program segment.

```
Start:     C = 0
           If (A = B)
                Then A = A + 1
                Else  A = A + 5
           Endif
Finish:    C = A
```

Solution

```
Start:     MVI C, 0        ; C = 0
           CMP B           ; compare A and B
           JZ Equal        ; if A = B, jump to Equal
           ADI 5           ; else, A = A + 5
           JMP Finish      ; unconditional branch to Finish
Equal:     INR A           ; then A = A + 1
Finish:    MOV C, A        ; C = A
```

The reader is encouraged to trace this code segment.

Input/Output Group. The input/output (I/O) instructions are designed to transfer data from an input device to the microprocessor in an input operation and from the microprocessor to an output device in an output operation. The 8085 μp has two I/O instructions: **IN** and **OUT**. The **IN** instruction inputs the data from an input device to the accumulator of the microprocessor and has the following format:

IN PORT

where **PORT** is an 8-bit address of an input device. The **OUT** instruction outputs the contents of the accumulator of the microprocessor to an output device and has the following format:

OUT PORT

where **PORT** is an 8-bit address of an output device. Since **PORT** is an 8-bit address, the 8085 can access up to 256 distinct I/O devices.

The **IN** and **OUT** instructions provide a mechanism that facilitates the interface of a microprocessor with I/O devices.

EXAMPLE 12.5-6

Write an assembly-language program segment that inputs the status of eight switches connected to input port 100 and determines whether it is equal to a predetermined value stored in register **B**. The result is indicated by a lighting of the LEDs connected to output port 200 if the input is equal to register **B**; otherwise, all the LEDs of port 200 are turned off.

Solution

Figure 12.5-3 depicts a block diagram for this example. From Fig. 12.5-3, it should be clear that to turn on the LEDs, the pattern 00000000 should be sent to port 200; the pattern 11111111 should be sent to port 200 to turn off the LEDs.

```
          IN 100               ; A = contents of port 100
          CMP B                ; compare registers A and B
          JZ Equal             ; if A = B, jump to Equal
          MVI A, 11111111B     ; set A with the pattern to
                               ; turn off the LEDs
          JMP Display          ;
Equal:    MVI A, 00000000B     ; set A with the pattern to turn on
                               ; the LEDs
Display:  OUT 200              ; send the proper code to port 200
```

Stack and Machine-Control Group. Table 12.5-6 lists the remaining instructions that the 8085 can recognize and execute. This group of instructions is used to manipulate the stack and stack pointer, to enable and to disable the interrupt system, to perform no operation, and to halt the processor. Stacks and interrupts are considered in the next section.

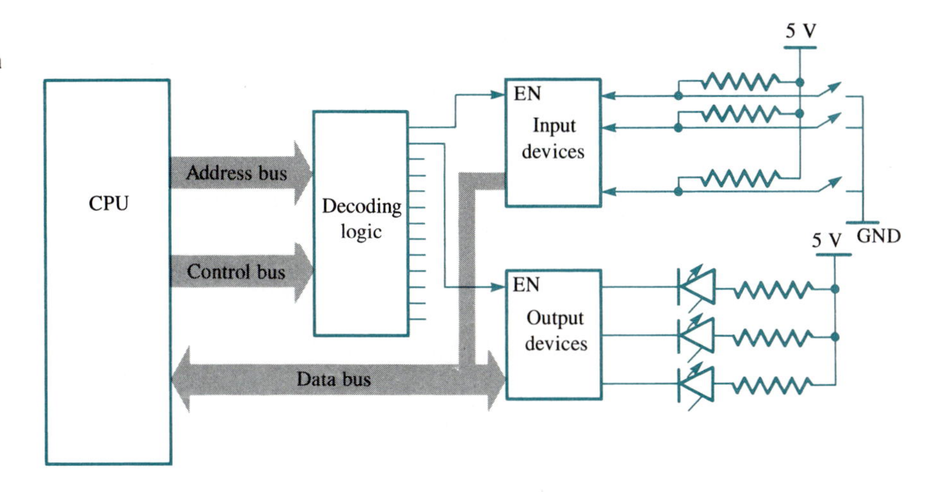

FIG. 12.5-3. A block diagram of the basic I/O structure.

12.6 STACKS AND INTERRUPTS

In order to understand the stack and its purpose, one must understand the concept of a subroutine. Assume that your assembly-language program requires the use of an operation that is not part of the instruction set, say multiplication, and that your program requires this operation several times in the course of program execution. You could write the sequence of instructions that simulate this operation each time that it is needed, but this would be redundant and could use a great

TABLE 12.5-6. Stack and Machine-Control Instructions

Mnemonics	Description
PUSH rp	Move the contents of **rp** onto the stack
PUSH PSW	Move the contents of the flags and **A** onto the stack
POP rp	Move the contents of the top of the stack into **rp**
POP PSW	Move the contents of the top of the stack into **PSW** and **A**
XTHL	Exchange the top of the stack with the contents of the **H** and **L** registers
SPHL	Move the contents of **H** and **L** into **SP**
EI	Enable the interrupt system
DI	Disable the interrupt system
HLT	The processor is stopped
NOP	No operation is performed
RIM	Move the serial input and interrupt mask into register **A** (read interrupt mask)
SIM	Move the contents of register **A** into the interrupt mask and serial output (set interrupt mask)
RST n	Transfer control to the instruction whose address is 8**n**, $0 \leq \mathbf{n} \leq 7$

deal of your memory. This situation calls for the use of a subroutine. A *subroutine* is a group of instructions that appear only once in the program code but can be executed from different points in the program. A subroutine has the following general format:

```
Sub_Name:       .
                .
                .
                Body_of_Subroutine
                .
                .
                .
                RET
```

where Body_of_Subroutine is the actual set of instructions that will be executed from different points in the program.

EXAMPLE 12.6-1

Write a subroutine called SUM that will compute the sum of 10 consecutive memory locations whose first location is pointed to by register pair **H** and **L**. The result is left in register **A**.

Solution

```
SUM:      MVI A, 0     ;  clear sum
          MVI C, 10    ;  initialize loop count to 10
AGAIN:    ADD M        ;  sum = sum + data_i
          INX H        ;  point to next data
          DCR C        ;  decrement loop count
          JNZ Again    ;  if loop count 0, repeat
          RET          ;  else return
```

The 8085 provides instruction for the call and for the return from a subroutine. When a **CALL** instruction is executed, the address of the next instruction (i.e., the present contents of the program counter) is pushed onto the stack. The contents of the program counter are then replaced by the address of the desired subroutine. When the subroutine is completed, a return instruction, **RET**, pops (i.e., restores) the previously stored address from the stack and puts it back into the program counter. Program execution then continues as though the subroutine had been coded inline.

The stack is simply an area of RAM that is addressed by the stack pointer (**SP**), which must be initialized by the program. The stack pointer must be set to point to the base of the stack. The base address of the stack is usually assigned to

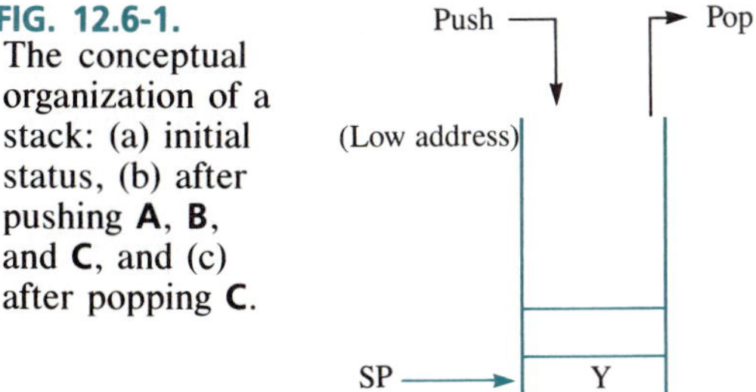

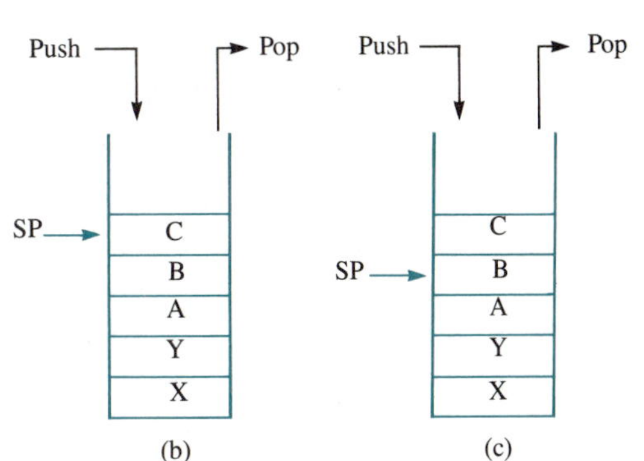

FIG. 12.6-1. The conceptual organization of a stack: (a) initial status, (b) after pushing **A**, **B**, and **C**, and (c) after popping **C**.

the highest available address in RAM. That is, when we push (insert) data into the stack, the stack expands into memory locations with lower addresses. As data are popped (removed), the stack decreases and its pointer (**SP**) is incremented back toward its base address. Stack operation is known as a last-in, first-out (LIFO) operation. The most recent data placed on the stack is called the *top of stack*. Figure 12.6-1 illustrates the conceptual organization of the stack.

The only limitation to the number of items that can be placed on the stack is the amount of RAM allocated for the stack. This allocated area of RAM is very important to the programmer, who must be certain that the stack does not expand into areas of RAM reserved for other data or tasks.

When external systems or devices want to gain the immediate attention of the μp, *interrupts* are used. Interrupts are external-hardware- or internal-software-initiated subroutine calls that interrupt the program currently being executed. External devices interrupt the μp through its interrupt pins, which, when activated, cause the μp to transfer control to a predefined routine known as the *interrupt service routine*.

There are two basic types of interrupt inputs: *maskable* and *nonmaskable* interrupt inputs. Maskable interrupt inputs can be enabled or disabled under program control; that is, maskable interrupt inputs may be programmed to enable or disable external devices from interrupting the μp. The nonmaskable interrupt is always enabled; that is, when the nonmaskable interrupt pin is activated by an external device, the μp is interrupted immediately. Nonmaskable interrupts are used to handle critical events such as power failure.

PROBLEMS

12-1. List the different types of computer systems and their characteristics.

12-2. What are the major components of a computer system?

12-3. List the functions of (a) CU, (b) ALU, (c) MU, and (d) I/O units.

12-4. List the characteristics of (a) machine-language programming, (b) assembly-language programming, and (c) high-level-language programming.

12-5. What is the difference between an interpreted and a compiled program?

12-6. What is an operating system?

12-7. What is the function of (a) an assembler, (b) a compiler, and (c) an interpreter?

12-8. What is a microprocessor?

12-9. What is a computer bus?

12-10. What determines the maximum size of memory that a microprocessor can access?

12-11. How wide should an address bus be to provide us with access up to 1 megabyte of memory?

12-12. What are the functions of signals provided on the control bus?

12-13. Show the general layout of the major components of a microcomputer system.

12-14. Show the internal structure of a typical microprocessor.

12-15. What are the functions of (a) a **PC**, (b) an **IR**, (c) an **SP**, and (d) a **PSW**?

12-16. Explain the functions of the control unit in the microprocessor.

12-17. If the **PC** of the CPU of Fig. 12.4-1 contains $(250)_{10}$, what will the **PC** contain after the execution of the instruction if (a) memory location $(250)_{10}$ contains a 2-byte instruction, and (b) memory location $(250)_{10}$ contains a 3-byte instruction?

12-18. What are the two components of a machine-language instruction?

12-19. What type of information may the operand field of an instruction contain?

12-20. What determines the distinct number of different operations that a μp can perform?

12-21. Show the general format of assembly-language instruction, and list the functions of its different fields.

12-22. Describe the basic addressing modes of the 8085 μp.

12-23. List three names of popular (a) 8-bit microprocessors, (b) 16-bit microprocessors, and (c) 32-bit microprocessors.

12-24. For the 8085 flag register, what are the values of **CY**, **Z**, **AC**, **P**, and **S** flags after the addition of the contents of registers **A** and **B** if (a) **A** = $(1011\ 1010)_2$ and **B** = $(1101\ 0001)_2$ and (b) **A** = $(0111\ 0010)_2$ and **B** = $(0000\ 1110)_2$?

12-25. Repeat Problem 12-24 for the operation AND instead of ADD.

12-26. Trace the execution of the following program segment which uses the 8085 instruction set to determine the contents of registers **A**, **B**, **C**, **D**, **E**, **H** and **L**. Which memory location is affected and how?

```
1    MVI A, 100
2    MOV B, A
3    STA 250
4    LXI H, 250
5    MOV E, M
6    MVI D, 66
7    XCHG
8    HLT
```

12-27. For the 8085 processor, write a sequence of instructions that (a) clears registers **A**, **B**, **C**, and **H**; (b) initializes memory locations 100 and 101 to 55H and 99H, respectively; and (c) swaps the contents of memory locations 200 and 201.

12-28. Repeat Problem 12-26 for the following code segment.

```
1    LXI H, 200
2    SUB A
3    MOV M, A
4    INX H
5    MOV M, A
6    ADI 99H
7    DAD H
8    MOV M, A
9    HLT
```

12-29. Write a sequence of instructions for the 8085 processor that will add the contents of four consecutive memory locations starting at address 100. Place the result in memory location 200.

12-30. Write a program segment for the 8085 processor that will add the contents of register pairs **HL** and **DE** and then place the result in register pair **BC**.

12-31. What are the effects of the following instructions?

```
1    ORI    10000001B
2    ANI    11100111
3    XRI    00000110
```

12-32. What will be the contents of register **A** after the execution of the three instructions given in Problem 12-31 if its original contents are 11110000B?

12-33. Write a sequence of instructions for the 8085 processor that replaces the four MSBs of memory location 100 with its four LSBs, clears the remaining bits, and places the result in memory location 200.

12-34. The 8085 instruction set does not provide a multiplication instruction. Using the fact that shifting a binary number to the left one position is equivalent to multiplying it by 2, write a sequence of instructions that provides a fast multiplication of the contents of register **A** by (a) 16 and (b) 17. Assume that the result can be contained in an 8-bit register.

12-35. Write a sequence of instructions for the 8085 processor that divides the contents of register **B** by 8.

12-36. Write an assembly-language program segment for the 8085 processor that is equivalent to the following high-level program statement.

B = ABS(A)

12-37. Write an assembly-language program segment for the 8085 processor that will increment register **C** by 1 if register **A** contains an even number of 1s; otherwise, it will decrement register **C** by 1.

12-38. Write an assembly-language program segment for the 8085 processor that will add the contents of 10 consecutive memory locations, starting at address 100. Assume that the sum will be an 8-bit number.

12-39. Write an assembly-language program segment for the 8085 processor that lights a set of eight LEDs connected to port 200 so that they show the contents of memory location 2000. An LED should be lit if the corresponding bit in location 2000 is 1. Draw a block diagram for the basic output structure.

12-40. Write an assembly-language program segment for the 8085 processor that inputs the values of eight switches connected to port 100, outputs these values to the LEDs connected to port 150, and rotates to the right the pattern on the LEDs eight times. Introduce a small delay between each rotation.

12-41. Discuss the advantages and disadvantages of a subroutine.

12-42. Write an assembly-language subroutine for the 8085 processor that clears an array of 100 consecutive bytes whose first address is in register pair **H** and **L**.

CHAPTER 13

Computer Networks

13.0 INTRODUCTION

Although advances in computers and communications technologies have had a tremendous impact on our society and our day-to-day lives, it has only been within the past 10 years or so that the power of these two technologies has merged. The declining cost and increased power of digital integrated circuits, digital computers, and peripheral devices have helped telecommunications evolve from voice-only switching networks to today's integrated voice, data, and video systems.

Telecommunications have made possible the transfer of computer resources over distance. *Computer communications networks* are the product of a combination of computers and telecommunications products. A computer network consists of interconnected *autonomous* computers that can exchange data. These computers need not be hard-wired (i.e., connected via copper wires); microwaves, communication satellites, and fiber optics can be used. The major aspects of computer communications networks are the subjects of this chapter.

This chapter begins by describing the concepts and characteristics of a network. Network architectures are discussed, and the open-systems interconnection (OSI) model is presented. The properties of most common transmission media are reviewed, and local-area networks (LANs) are examined through a discussion of today's most common LANs and their different protocols. We then proceed to a review of different methods of data transmission. Finally, modems and their modes of operations are discussed.

13.1 CONCEPT OF A NETWORK

In its most basic form, a computer communications network can be defined as an interconnected group of independent computers and peripheral devices that communicate with one another for the purpose of sharing software and hardware resources. The following are some specific goals of a computer communications network:

To provide access to special hardware devices and software packages that cannot be installed at each station because of their cost.

To transfer data from one station to another.

To provide a communications medium among widely separated users.

To rapidly disseminate data through the network.

To provide high reliability by having alternative sources of supply.

Computer networks can be classified into two categories: (1) *local-area networks* (LANs) and (2) *wide-area networks* (WANs). A LAN is a computer communications network where all of the resources are within the same building or campus. The stations, also known as *nodes*, within a LAN are physically linked with each other through twisted pairs of copper wires, coaxial cables, or fiber-optic cables. A WAN, also known as a *long-haul network*, is a computer communications network that connects geographically separated and dispersed comput-

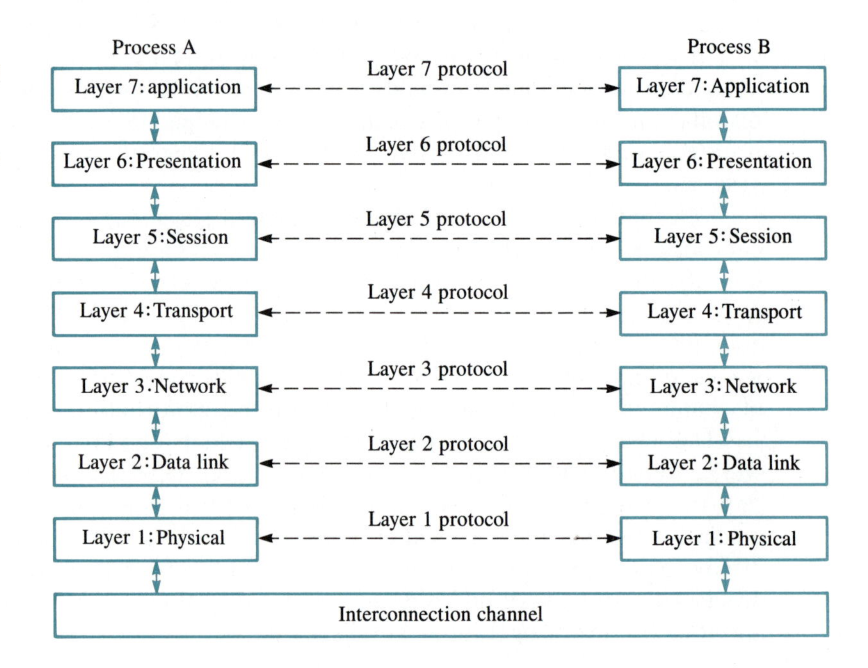

FIG. 13.2-1. The ISO model for an open-systems interconnection (OSI).

ers and their peripheral devices. Stations within a WAN communicate with each other through standard telephone lines, dedicated telephone lines, line-of-sight microwave systems, or fiber-optic links. A WAN can be classified into three subcategories: (1) *public-data networks*, (2) *private communications networks*, (3) *remote-access communications networks*.

The public-data networks are operated by a common carrier and provide services to a wide variety of subscribers who span a large geographical area. Dataphone Digital Services of AT&T, Tymnet, and Telnet are three public-data networks in the United States that provide digital-transmission services at reasonable cost. The subscribers usually own their connected devices. The private and the remote-access communications networks are either resource-sharing or mission-oriented networks for a specialized group of users. ARPANET (Advanced Research Projects Agency of the U.S. Department of Defense) and CYBERNET are two examples of private and remote-access, respectively, communications networks that span the United States, Western Europe, and some other countries to facilitate communication and to provide access to huge databases for their users.

13.2 NETWORK ARCHITECTURES

In a computer network, the different computers and devices communicate with each other through a well-established convention known as *protocols*. A protocol is a set of rules governing a time sequence of events that takes place between peer entities. Computer network architecture refers to the convention used to define how the different protocols of the system interact with each other to support the end users.

So that design complexity is reduced, most network architectures are organized as a series of *layers* or *levels* that provide the hierarchy of the network protocols. The most common network-architecture model is the open-systems interconnections (OSI).

OSI Model

The open-systems interconnection (OSI) model is a formal hierarchical identification of all computer-data-network functions that has been established by the International Standards Organization (ISO). This model, shown in Fig.13.2-1, is referred to as the *ISO model for open-systems interconnections*. This model of seven layers establishes the ground rules for the creation of a communications-network architecture that can be implemented and shared by anyone.

In Fig. 13.2-1, physical communications between layers are shown by solid lines and virtual communications by dotted lines; that is, no data are directly transferred from layer n, $1 \leq n \leq 7$, on process A to the same layer n on process

B. Instead, each layer passes and/or receives data and control information to and/or from the layer immediately below (or above) it, until the physical layer is reached and the actual communication occurs between processors through the interconnection channel.

Ideally, a computer communications network would require all the functions specified by the seven layers of the OSI standards. This network would provide communication between devices within a given network as well as communication between different networks. However, not all seven layers of the OSI standards need to be implemented to provide communication between devices in a network. In fact, only the lower three layers and the transmission channel are required to provide data communications between devices in a network. A brief discussion of the attributes and characteristics of OSI's seven layers depicted in Fig. 13.2-1 is given next.

OSI Layers

The physical layer provides the means for transferring data across the interconnection channel and controlling its use. It is concerned with the electrical and mechanical characteristics of the access to the interconnection channel. It provides the proper signals to regulate the synchronization of data flow and specifies the mechanical connectors such as sockets and plugs that are used in the interface.

The *data-link layer* is a medium (transmission) independent layer that is responsible for the transfer of data across the link. It is responsible for data-flow control by providing data-rate matching to prevent the receiver from being swamped by the transmitter. Also, it provides for the detection and correction of data-tansmission errors.

The *network layer* performs the networking functions, such as specifying the interface of the user into the network as well as the interface of two devices through the network. It establishes, maintains, and terminates connection across the network and can provide the network switching/routing and communications between different networks (internetworking).

The *transport layer* provides the end-to-end (i.e., source-to-destination) data integrity; that is, it provides the layers above it with a reliable transmission mechanism. It provides a transparent interface between the user and the network by isolating the upper three layers from the details of the lower three layers' operations.

The *session layer* is the user interface into the transport layer. This is the liaison between users who are exchanging data. It is responsible for managing, coordinating, and synchronizing the user data flow between applications. In addition, it provides a recovery system in case of failure.

The *presentation layer* provides the syntax (rules) of the representation of data between devices. It provides data, code, and file format conversions. It consists of many conversion tables (ASCII, EBCDIC, VIDEOTEXT, etc.), which are used to allow different types of equipment, using different formats, to properly communicate.

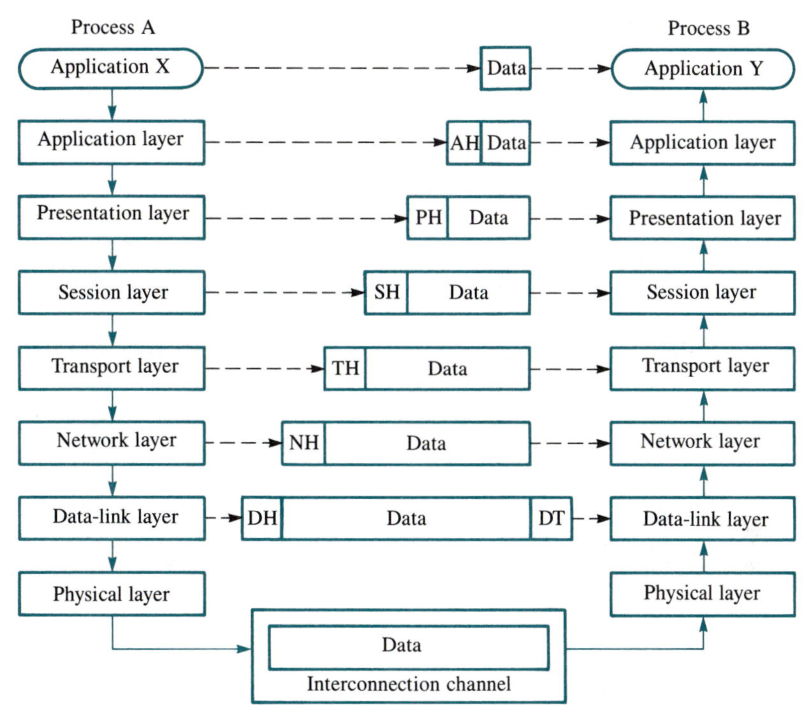

FIG. 13.2-2. An example of data flow in the OSI model.

The *application layer* provides support to process end users' applications, such as file management, programming languages, database management, and electronic mail. In addition to providing end users' service protocols, this layer provides the protocols for the network's services, such as network management functions.

Observe that the interconnection channel is not part of the OSI specification. For example, the interconnection channel can be copper or fiber-optic cables, as in a LAN, or it can be telephone lines or microwave, as in a WAN.

To illustrate the OSI model in operation, Fig. 13.2-2 shows an example of how an application X in process A can transmit data to application Y in process B. The application X gives the data to the application layer in process A, which then attaches an application header (AH) that contains the required protocol control information (PCI) for its peer layer 7 in process B. The resulting data and AH are passed as a new unit of data to the presentation layer in process A. The presentation layer may transform the new data, add a presentation header (PH) to the front, and transfer the result to the session layer. This process continues down through the data-link layer, where a header and a *trailer* are added to the data, which are passed to the physical layer as a *frame*. The physical layer transmits the frame to process B through the interconnection channel, bit by bit.

When the data are received by process B, the reverse process occurs. On this process, each layer strips off the outermost header, acts on the protocol-control information, and passes the remainder of the data up to next layer. This process continues up through the layers until it finally arrives at application Y in process B.

Observe that although the actual data are transmitted vertically down and up processes A and B, respectively, each layer is acting as though the data were transmitted horizontally. This shown in Fig. 13.2-2 by solid lines for up/down transfers and by dotted lines for horizontal transfers.

13.3 NETWORK TOPOLOGIES

In general, any computer communications network consists of a collection of *nodes* and *links* for users' applications. Nodes usually refer to the endpoints and consist of physical devices such as terminals, printers, PCs, and mainframes. Links usually refer to the transmission channels that connect these nodes. *Network topology* is defined as the geometrical arrangement of interconnected nodes. Figure 13.3-1 shows the most common network topologies.

The *bus topology*, also known as *multidrop topology*, is shown in Fig. 13.3-1(a). This topology is predominantly used in a LAN. It is made up of a

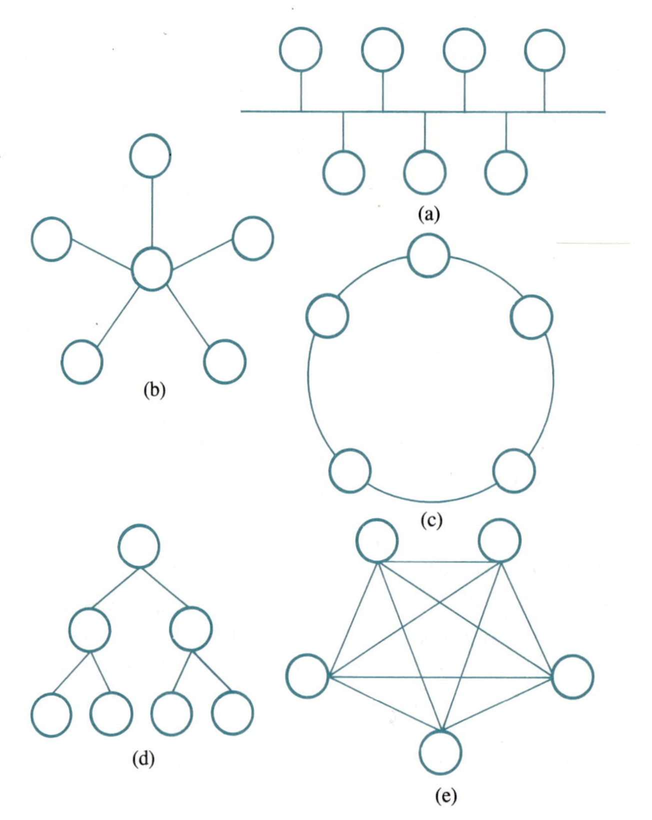

FIG. 13.3-1. Network topologies: (a) bus, (b) star, (c) ring, (d) tree, and (e) distributed.

common transmission medium (as the other topologies are) and a number of autonomous nodes connected to it. Data transmitted by one node can be received by all other nodes. Since the bus is shared, only one node can transmit at a time, and therefore, some form of bus-access control is required. Two common techniques (protocols) are used and will be discussed in the LAN section (Sec. 13.5).

The *star topology* is shown in Fig. 13.3-1(b). This topology consists of a central node that connects to all the subordinate nodes. Communication between any two or more nodes takes place through the central node, which provides the control of the network. PBX (private-branch exchange) systems are the most common star systems in use today.

The *ring topology* is shown in Fig. 13.3-1(c). It draws its name from the circular form of the connectivity created when each node is connected to the next adjacent node in a point-to-point fashion. Data transmitted from one node travels around the ring, passing through the intermediate nodes on its way to the destination node. Ring topology may employ centralized control, with one node designated as the controller, or decentralized control, with all nodes having equal status.

The *tree topology* is shown in Fig. 13.3-1(d). Tree topologies are hierarchical; that is, there is a single node that acts as the controlling node of the network. This is an ideal topology for data distribution from a large host computer. Some networks provide a great deal of intelligence in the intermediate nodes. This topology is used in most of the remote-access networks.

The *distributed topologies*, also known as *mesh*, unconstrained, or *hybrid networks*, have no definite configurations. They allow a large number of nodes to be connected together by point-to-point links. This topology provides an alternative route between nodes and is common in public and modern private communications networks. Figure 13.3-1(e) shows a *fully* distributed network, where every set of nodes may communicate directly with every other set through a single link.

13.4 TRANSMISSION MEDIA

In a computer communications network, data may be transmitted from one node to another through various transmission media (also known as physical channels). Transmission media can be either *bounded* or *unbounded*. In a bounded medium, the signals that represent the data are confined to the physical media. Examples of bounded media are twisted pairs of wires, coaxial cables, and optical-fiber cables. In an unbounded medium, such as the atmosphere, the ocean, and outer space, the transmission is wireless, using infrared radiation, lasers, microwave radiation, radio waves, and satellites. Since the bounded-type media are the media used in most LANs, we briefly discuss their different types.

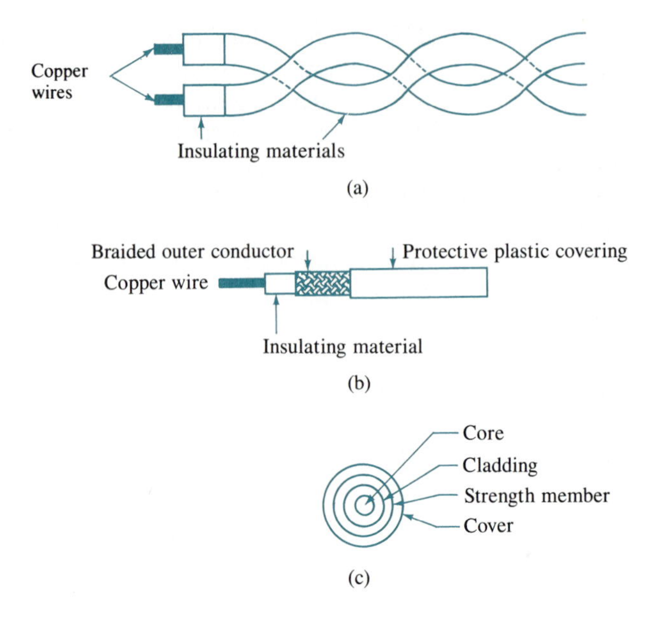

FIG. 13.4-1. Three types of bounded transmission media: (a) a twisted pair of wires, (b) a coaxial cable, and (c) a cross-sectional view of a fiber-optic cable.

Twisted Pairs

Twisted pairs of wires are comprised of two insulated copper wires that are twisted (wound) together so that each wire receives the same amount of exposure to interfering signals. These are the wires most commonly used in telephone systems. The size of the wires used varies from 22 AWG (American Wire Gauge) to 26 AWG. They have low attenuation at voice frequencies and high attenuation at high frequencies. They are used in low-performance and low-cost applications where a data rate of up to 1 megabit per second (Mbps) can be achieved for a transmission distance up to 1 km. Figure 13.4-1(a) shows a typical configuration for a twisted pair of wires.

Coaxial Cable

To achieve a higher transmission rate than the one achieved by twisted pairs of wires, coaxial cables are used. There are two kinds of coaxial cables. One kind is used for digital transmission and is known as baseband coaxial cable. The second kind is used for analog transmission (cable TV) and is known as *broadband coaxial cable*.

A coaxial cable consists of a central copper wire surrounded by an insulating material, as shown in Fig. 13.4-1(b). The insulating material is surrounded by either a braided or solid conducting material. It is called a coaxial cable because the two conductors are on the same "axis." Coaxial cables offer better immunity from interference than twisted pairs. Baseband coaxial cables are usually 50-Ω coaxial cables that can be used to transmit data at a rate of up to 10 Mbps over a distance of up to 2 km. Broadband coaxial cables are usually 75-Ω coaxial cables that can be used to transmit data at rates up to 500 Mbps over distances of up to 10 km.

Optical Fibers

Unlike twisted pairs and coaxial cables, optical-fiber cables are nonmetallic cables that use a beam of light to carry the data through *glass fibers*. The light, which is generated by either light-emitting diodes (LEDs) or lasers, is switched on and off to create (encode) the binary data. Since optical fibers carry light rays, the frequency of operation is that of light. Fiber-optic cables have very high immunity to interference and support data transmission of up to 1 Gbps (gigabits per second) over a distance of up to 100 km. They are smaller, lighter, and cheaper than metallic cables of the same capacity. A cross section of a typical optical fiber is shown in Fig. 13.4-1(c). The core material is usually glass, which is surrounded by either a plastic or glass cladding. The strength member is an elastic, abrasion-resistant material that provides additional strength to the glass fiber. In addition, the strength member and the cover isolate the fibers from the contamination of the environment.

13.5 LOCAL-AREA NETWORK

A local-area network (LAN) provides short-range communications systems that extend over a single compact facility such as a college campus or a military installation. A LAN transmits data between user stations and computers that are usually within a few hundred meters to a few thousand meters of each other. LANs operate using a variety of protocols. Due to the diversity of LAN industry, we will limit our discussion to the most common personal-computer-based LANs. The following discussion provides an overview of the most widely used LANs, the *Ethernet*,† *token-passing*, and *broadband* LANs.

Ethernet

Ethernet is the dominant *bus*-type LAN system and was invented by Xerox Corporation in 1973. After a few years of growth, Xerox, Digital Equipment, and Intel teamed together to establish an industry standard known as *DIX* Ethernet.

Figure 13.5-1(a) shows a typical Ethernet configuration. The transmission medium of the Ethernet is a baseband coaxial cable with a data-transmission rate of 10 Mbps. The protocol used in Ethernet is known as *carrier sense, multiple access with collision detection* (CSMA/CD). In this protocol, a station communi-

†Ethernet is a trademark of Xerox Corporation.

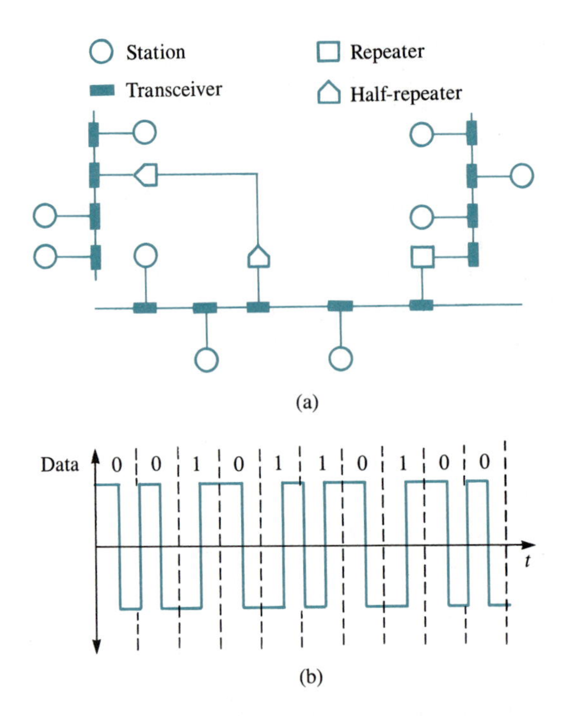

FIG. 13.5-1. Ethernet LAN: (a) a typical Ethernet configuration and (b) a Manchester code.

cates with another station by waiting until the bus (transmission medium) is idle, which is determined by carrier sensing. Once the bus is determined to be idle, the transmitting station sends a packet of data containing the destination address, the source address, the actual data, and some redundant check bits (check-sum) for transmission-error detection. The rest of the stations continuously monitor the data on the bus and accept those packets with their address and valid check-sum. Whenever a station receives valid data, this receiving station sends an acknowledgment to the source (i.e., sender). If the source does not receive this acknowledgment within a specified amount of time, it will retransmit the same data and assume that the original data were interfered with by either a noise in the bus or by a simultaneous transmission from another station (i.e., data collision).

Data are transmitted on the coaxial cable by using the *Manchester code*, as shown in Fig. 13.5-1(b). One advantage of the Manchester code is that it can be used as a synchronizing source, because the timing transition always occurs in the middle of every bit. Multiple access to the available data in the bus is provided by the passive taps (transceivers), which allow stations to be added or dropped without disrupting the system.

Table 13.5-1 lists the Ethernet's standard specifications.

Token Passing

In the CSMA/CD protocol of Ethernet, access to the bus—and hence to the network—requires a degree of competition between those stations trying to

TABLE 13.5-1. Ethernet Specification

Media	50 Ω (baseband) coaxial cable
Data rate	10 Mbps
Data-link protocol	CSMA/CD
Maximum number of nodes per network	1024
Maximum number of taps per segment	100
Maximum transceiver cable distance	50 m
Maximum separation between nodes	1500 m, using two repeaters
	2500 m, using half-repeaters
Maximum number of repeaters between two stations	2
Minimum transceiver separation	2.5 m

transmit data exactly at the same time. This competition, also known as *contention*, can be eliminated by providing a *token*; the access to the network is then determined by which station has this token. Token-passing networks use this access protocol and are implemented in two different topologies: (1) *token-bus network* and (2) *token-ring network*.

Token-Bus Network. A token-bus network is designed to support factory automation where real time and guaranteed response are needed. It is heavily supported by General Motors. As shown in Fig. 13.5-2, token-bus LANs use a bus topology. A token-bus-passing LAN eliminates the collisions found in the CSMA/CD systems by providing tables that contain the token-passing order for each station. The token-passing order as defined by tables is shown by the contour drawn as dashed lines in Fig. 13.5-2. Observe that if a station, like a printer or a plotter, never initiates communication, then it does not need to be in the polling sequence. However, if a station has a very high priority, then it will appear more than once in the table that defines the token-passing sequence.

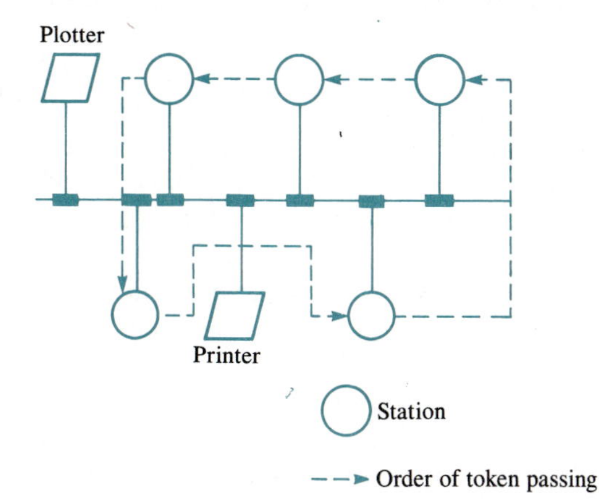

FIG. 13.5-2. A token-bus LAN.

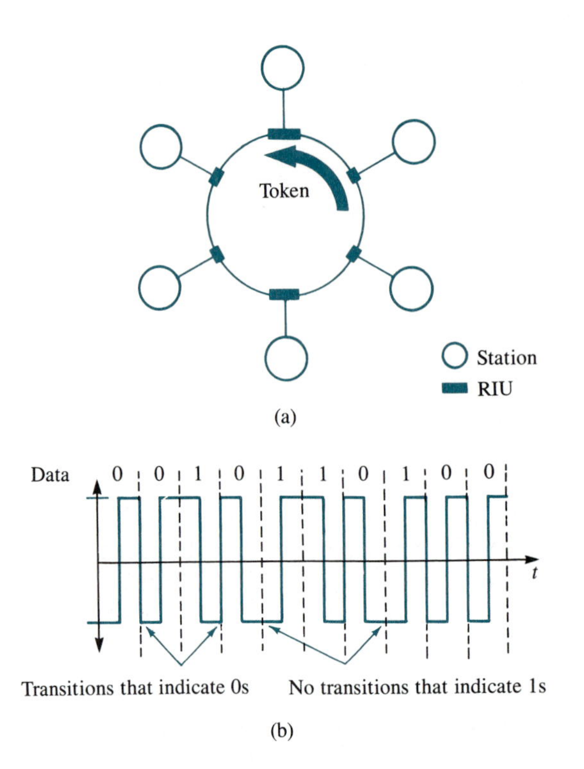

FIG. 13.5-3. A token-ring LAN: (a) a typical ring configuration and (b) the differential Manchester code.

Token-bus-passing LANs use a broadband coaxial cable (CATV) supporting data transmission rates of up to 20 Mbps for a given channel. MAP (manufacturing automation protocol) initiated by General Motors is an example of token-bus-passing LAN.

Token Ring. Token ring is the protocol used by IBM (International Business Machines) for its LANs. In a token-ring LAN, the closed-loop topology defines the order in which the token is circulated from one station to another. As shown in Fig. 13.5-3(a), a token-ring LAN consists of a collection of stations connected through *ring-interface units* (RIUs) by point-to-point lines.

When the network is idle, a token is passed around the ring sequentially from one node to the next adjacent node. When a station wishes to use the idle network, this station waits for the token to arrive and then removes it from the ring (i.e., marking the network as busy) and can transmit data. Upon acquiring the token, the transmitting station places the data, the destination address, a check-sum, and a busy token on the ring to be passed to the next adjacent station. As these new data are being circulated from one station to the next, each RIU regenerates the transmission and examines the destination address. If the destination address indicates that the data are destined for a given station in the ring, its RIU verifies their check-sum and copies them to this station.

Upon receiving the correct data, the receiving station appends an acknowledgment tag to the data as it regenerates them for transmission. As the data that have been circulated around the ring come back to the sending station, they are removed from the ring by the sender, which makes the token free, and the token passes to the next station. This method prevents any given station from monopolizing the network. However, if the free token passes around the network without being used, the same station can once again acquire the token and transmit data.

Some implementations of token-ring LAN employ priority schemes. With a priority scheme, a station with high priority may reserve the ring for the next transmission. In this scheme, the token contains a priority field. As the token is passed around the ring, only the station with the specified priority can use the token. However, stations with lower priority may reserve the token. Upon using the token, the station with high priority will change the priority field to the level reserved by lower-priority stations and release the token, making it available for other stations to acquire it.

Token-ring LANs use shielded twisted pairs of wires supporting data transmission rates of up to 4 Mbps. Data are transmitted by using *differential Manchester encoding*, as shown in Fig. 13.5-3(b). In this encoding technique, a 0 is represented by a transition in the voltage level of the signal at the start of a bit interval, and a 1 is represented by no transition in voltage level.

Broadband

Unlike Ethernet and token-passing networks, broadband LANs employ an analog-signaling technique. In this technique, an *RF modem* (radio-frequency modulator and demodulator) is used in each station in order to modulate and demodulate its transmitted and received data. Modems will be discussed in Sec. 13.7. Broadband-LAN architecture stems from the community-antenna television (CATV) technology. In a broadband LAN, it is possible to have multiple signals (users) on the same cable at the same time due to *frequency-division multiplexing* (FDM). FDM, discussed in the next section, allows multiple channels to be established within a system. In this system, the frequency spectrum of the cable is divided into independent sections of bandwidth. Each independent section forms a channel that can be used to transmit and/or receive independent data signals. Following the CATV standard and technology, the bandwidth of each channel is typically 6 MHz. Since the usable bandwidth of CATV is over 300 MHz, over 50 different channels can be used for different applications, such as data transmission, closed-circuit TV distribution, and point-to-point voice telephone and data circuits.

As shown in Fig. 13.5-4, a single-cable broadband LAN consists of an active base, known as the *headend*, and extends to different locations through a cable that splits from one into two or more cables by using *splitter* components. This single-cable broadband LAN must split the system bandwidth in order to separate the transmitting and receiving frequency bands. The system bandwidth is usually split in the middle, with all transmission signals on one side of the center frequency and all receiving signals on the other. The headend is responsible for transposing the frequencies of the transmitted signals into the receiving band. This

FIG. 13.5-4.
A broadband LAN.

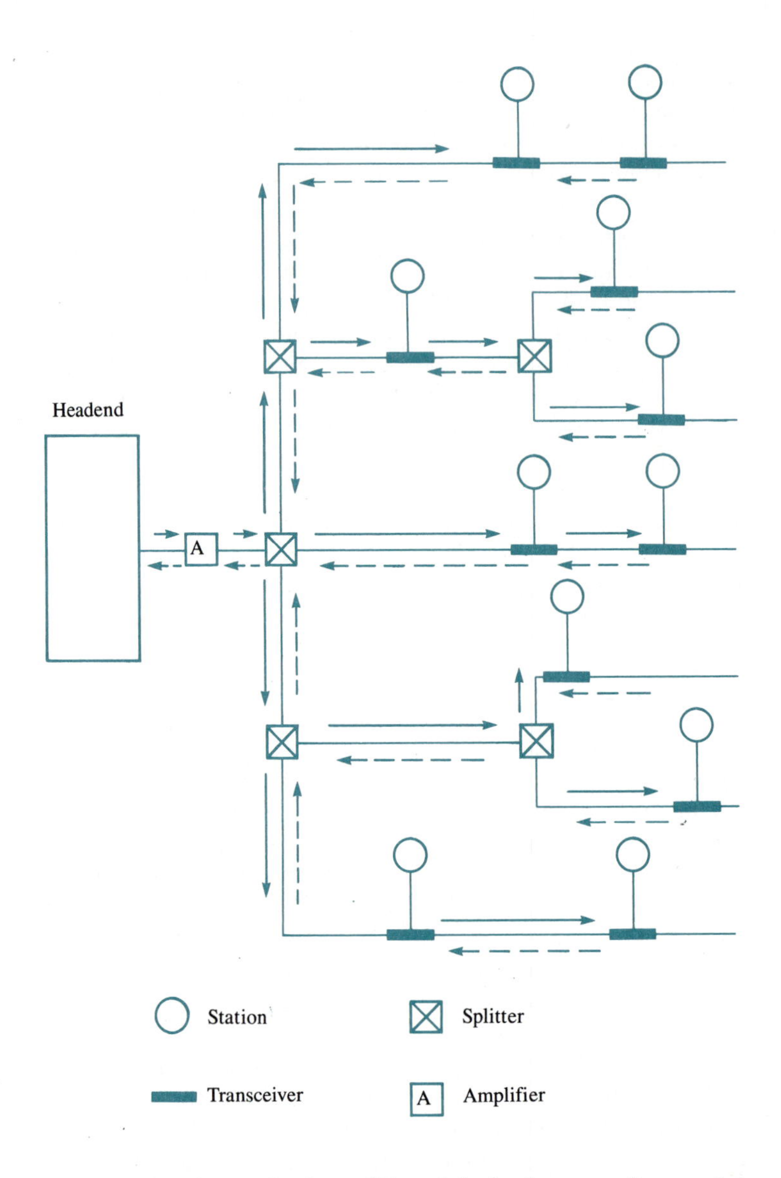

is shown in Fig. 13.5-4 by solid and dashed arrows for receiving and transmitting frequency bands, respectively.

Some commercially available broadband LANs use a two-cable system. One cable is used for data transmission and the other for data reception. This two-cable system allows most standard CATV equipment, which is designed for unidirectional transmission (i.e., from antenna to receiver), to be used directly.

Wangnet† is a broadband LAN that allows digital data, voice, and video to be

†Wangnet is a trademark of Wang Corporation, an office-automation-products company.

transmitted between stations. This LAN uses broadband transmission in a twin-coaxial-cable medium to provide connectivity between an unlimited number of stations.

13.6 DATA TRANSMISSION

Data transferred between two stations can be either *serial* or *parallel*. Serial data transfer refers to the transmission of data where the stream of bits moves one by one over a single line. However, if a group of bits moves over several lines at the same time, then the transmission is referred to as parallel data transfer.

Parallel Data Transmission

As shown in Fig. 13.6-1, when n bits are to be transmitted from one station to another, n separate lines (i.e., wires) must be provided. In this method, each bit of the character to be transmitted uses its own wire. In addition, some extra lines are used to provide controlling signals, which indicate to the receiving station when to sample the data lines for valid data. Figure 13.6-1 shows the parallel transmission of the letter A using ASCII code with even parity and providing a *strobe signal* that indicates available data to the receiver.

Parallel data transmission is used when the two stations are close to each other, usually within a few feet, as in a computer and a printer configuration. As the distance between the two stations increases, this parallel-interface technique becomes impractical, due to the cost of the wiring and the complexity of the line drivers and receivers required for transmitting and receiving electric signals on long cables.

Serial Data Transfer

The problems associated with parallel data transfer can be overcome by transmitting the digital data serially, bit by bit, over a single channel (i.e., wires). This

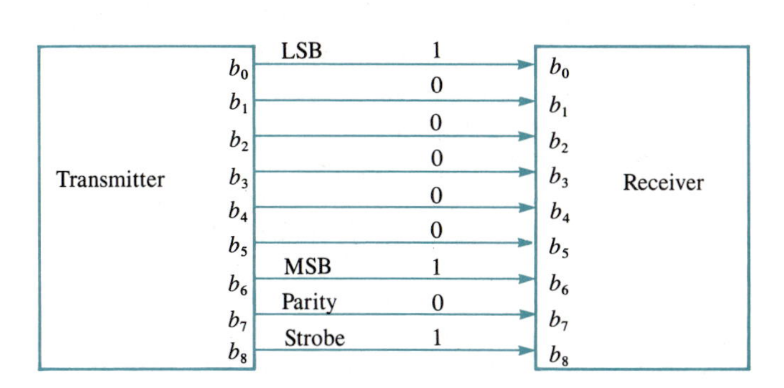

FIG. 13.6-1. Parallel data transmission.

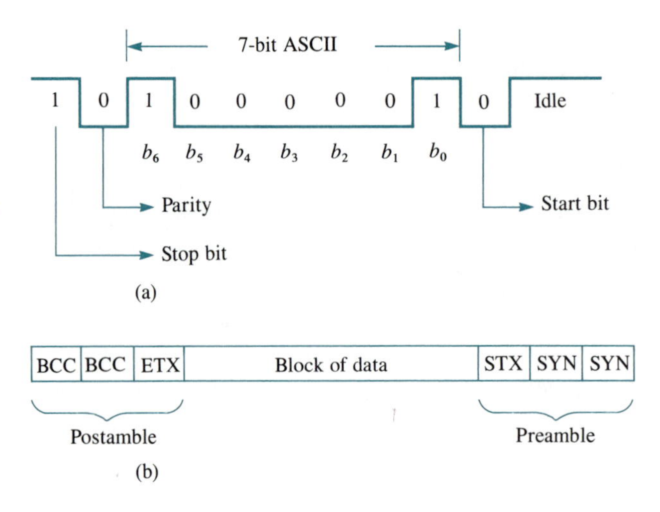

FIG. 13.6-2. Serial communications: (a) an asynchronous transmission and (b) the synchronous-transmission format.

technique allows data to be transmitted over a long distance. Since the stream of bits moves one by one over the single channel, some protocols must be established between the sender and the receiver to determine the beginning and the end of valid data. Serial data transfer can be either *asynchronous* or *synchronous*.

Asynchronous-Data-Transfer. In an asynchronous-data-transfer scheme, the bit streams of the data to be transmitted are divided into characters. Each character transmitted is then preceded by a *start bit* and followed by one or two *stop bits*. The start and stop bits are always of opposite value. The stop bit acts as the idle state of the transmission line; that is, when the stop bit is received by the receiver, it indicates the end of the bit stream of the character transmitted. Therefore, synchronization can be achieved by the receivers detecting the transition from the stop bit (i.e., idle state) to the start bit.

Figure 13.6-2(a) shows the format for asynchronous transmission of the letter A using 1 start bit, a 7-bit ASCII code, 1 even parity bit, and 1 stop bit.

LSI (large-scale integration) devices are commercially available for use in asynchronous-data-transfer. These devices are known as *UARTs* (universal asynchronous receivers/transmitters). They are used to accept parallel data (say from a computer), generate the proper start, stop, and parity bits, and output the stream of bits one by one at a programmable rate known as a *baud*† *rate* (number of bits per second). Also a UART can receive the serial data, strip off the start and stop bits, and output the equivalent parallel data.

Asynchronous-data-communication is most common in applications where low-speed terminals and small computers are used. However, when a higher rate of data transfer is desired, synchronous-data-communication is employed, because asynchronous communication requires at least 2 bits (start and stop) per 8

---

†Named in honor of J. M. E. Baudot, an early developer of the telegraph system.

bits (character) transmitted. This creates at least 20% overhead penalty; that is, 2 out of 10 bits transmitted are control bits.

Synchronous-Data-Transfer. Synchronous-data-communication is used for the transfer of large amounts of data at a high speed. In a synchronous-data-transfer scheme, the exact time for sending or receiving each bit of the data is predetermined before data are transmitted or received. In this method, data are transmitted in *blocks* of characters, with some special synchronizing clock signal either on a separate line or embedded with the data. In addition, each block of data is appended with a header and a trailer (control-information) known as *preamble* and *postamble*, respectively, to enable the receiver to identify the beginning and the end of each block of data.

The postamble also contains information for error checking of the entire block. Figure 13.6-2(b) shows the format used for synchronous-data-transfer, where the preamble block consists of the synchronizing characters (SYN) and the start-of-text (STX) character; the postamble consists of the end-of-text (ETX) character and the block-check characters (BCC).

USARTs (universal synchronous/asynchronous receivers/transmitters) are commercially available LSI devices. These devices provide the mechanism for synchronous data communication at programmable baud rates.

Frequency-Division Multiplexing (FDM)

As mentioned in the previous section, frequency-division multiplexing (FDM) is a technique for data transmission where the transmission frequency spectrum (i.e. bandwidth) is divided into smaller bands called *subchannels*, as illustrated in Fig. 13.6-3(a). This technique is widely used in telephone, radio, and cable TV systems. For example, the frequency spectrum of the AM-radio broadcasting system is about 1 MHz, and it is divided into smaller bands, each assigned to different radio stations.

Figure 13.6-3(b) illustrates a block diagram for the multiplexing process in FDM. In this process, each source signal S_i is used to modulate a different subcarrier frequency f_i. The resulting modulated subcarriers $S_{m1}, \ldots, S_{mi}, \ldots, S_{mn}$ are then combined by using the linear summing circuit to produce a total composite signal S_T. The composite signal S_T may then be transmitted directly over the channel or may be used to modulate a high-frequency carrier signal f_c, with the resulting signal S_C being transmitted. The subcarrier frequencies $f_1, \ldots, f_i, \ldots, f_n$ are usually chosen far enough apart that each signal spectrum is separated from all others.

The recovery (i.e., demultiplexing) process is illustrated by the FDM receiver shown in Fig. 13.6-3(c). First, the received signal S_C is demodulated, using the carrier demodulator to produce the composite multiplexed signal S_T. Then this signal S_T is applied to a bank of bandpass filters and the signals $S_{m1}, S_{m2}, \ldots, S_{mn}$ are recovered. Finally, subcarrier demodulators recover the source signals S_i from the signals S_{mi}.

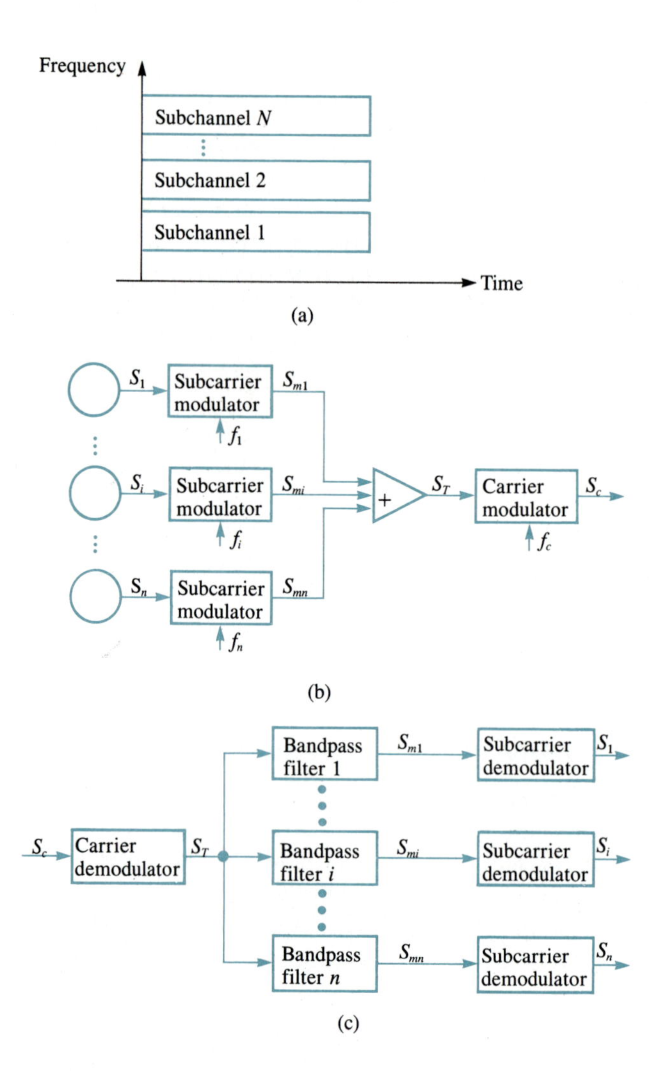

FIG. 13.6-3. Frequency-division multiplexing (FDM): (a) subchannels in FDM, (b) the multiplexing process, and (c) the demultiplexing process.

Transmission Modes

Data transmission between two stations may be handled in one of three modes: *simplex, half-duplex,* or *full-duplex* mode.

In a simplex mode, information travels only in one direction. This is the transmission mode used mainly in radio and TV broadcasts. This system is rarely used in data communications since the receiver cannot acknowledge the receipt of data.

In half-duplex mode, information may travel in both directions, but only one direction at a time. When the sender finishes the transmission and wants a reply from the receiver, the channel is turned around; that is, the transmitter becomes the receiver and vice versa. Radio communications systems are examples of half-duplex transmission modes.

In a full-duplex mode, information may travel in both directions simultaneously. This transmission mode uses two different carrier frequencies, as in telephone systems.

13.7 MODEMS

The most efficient way to interconnect computers and their peripheral devices is to use dedicated channels. But in some cases it is necessary to connect them through telephone lines, because telephone networks are extensive and the circuits from the network are available for use in data communication. However, since telephone networks are designed to carry human voices, which are analog signals with a narrow bandwidth (300 Hz to 3 kHz), they are not the best medium for direct digital-data transmission, which can have a wide frequency spectrum. When telephone lines are used to transmit digital data, it is necessary to convert the digital signals to an analog form with a relatively narrow bandwidth. A device that performs this function is called a *modem* (modulator/demodulator).

Basic Concepts

A modem is an electronic device that accepts digital data as a serial stream of bits and produces a modulated carrier signal as an output. This carrier signal is then transmitted over the telephone line to a similar modem at the receiving end, which converts (demodulates) the carrier signal back into its original serial stream of bits. Figure 13.7-1(a) depicts the input/output signals of a modem, and Fig. 13.7-1(b) shows its basic functional block diagram.

In the transmission process, the serial digital data to be transmitted are modulated, filtered, and amplified for analog transmission. At the receiving end of the circuit, the analog data received are amplified, filtered, and demodulated to produce serial digital signals. In addition to the serial input and output signals representing the original data, there are other control signals known as *handshake signals*. Handshake signals are used between modem and the digital devices and between transmitting and receiving modems to establish a proper communication session. Modems often include a ring detector for automatically answering a telephone call and dial-tone generators for originating a telephone call.

Transmission Modes

There are four different types of modems: half-duplex, full-duplex, synchronous, and asynchronous modems. When half-duplex modems are used, data can be transmitted in only one direction at a time and therefore can use all the bandwidth of the telephone lines. One of the disadvantages of the half-duplex operation is

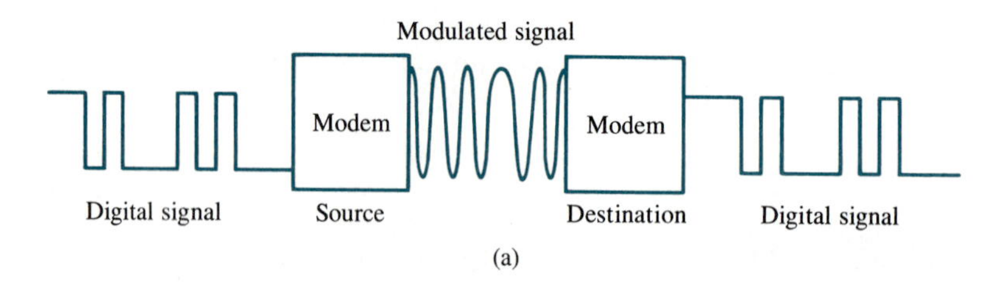

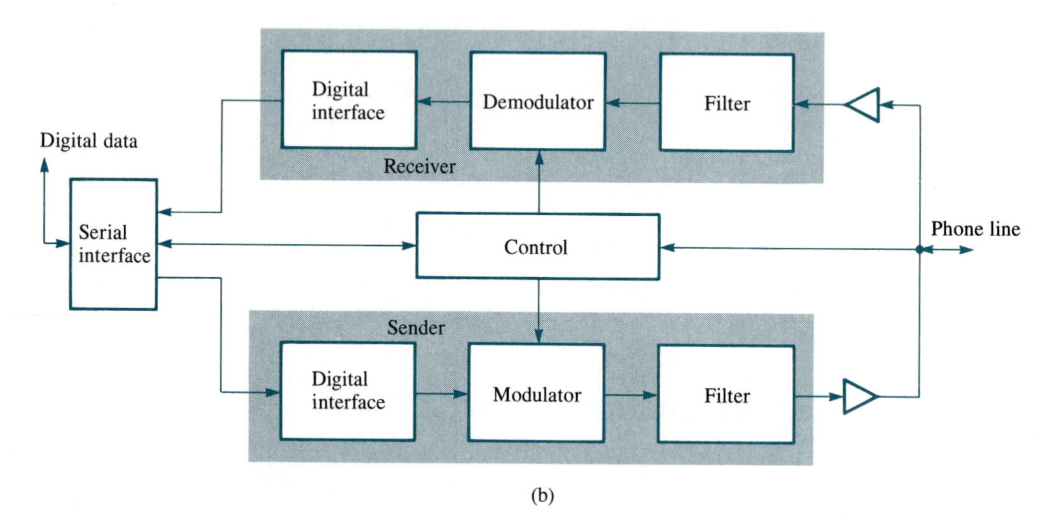

FIG. 13.7-1.
A modem: (a) input/output signals and (b) the basic functional block diagram.

that the receiving device must acknowledge the receipt of each error-free data or request retransmission of the data if an error is detected.

Full-duplex modems transmit data in both directions at the same time, and therefore, the bandwidth is shared, unless two different lines are used. Full-duplex modems transmit and receive data at different frequencies. When full-duplex modems are used, one modem must be designated as the *originating* modem and the other as the *answering* modem. This technique helps identify which modem should transmit using a given frequency. Full-duplex operation can avoid the turnaround delay associated with half-duplex by using a four-wire circuit.

Asynchronous modems are low-data-rate modems that can transmit serial data at a rate of up to 1800 bps (bits per second). As discussed in the previous section, asynchronous transmission has a built-in 20% overhead due to the added start-stop bits. Synchronous modems, on the other hand, are high-data-rate modems that can transmit serial data at a rate of up to 10,800 bps. Synchronous modems are more expensive than asynchronous modems, but they use the channels more efficiently than asynchronous modems.

Modulation

Low-data-rate modems (up to 1800 bps) use a simple form of modulation known as *frequency-shift keying* (FSK). In FSK modulation, one frequency represents 0 and another frequency represents 1. At higher data rates (2000–10,800 bps), another type of modulation is used: *phase-shift keying* (PSK) modulation. PSK, also known as *phase encoding*, uses only a single frequency, and data are represented by phase shifts of the carrier signal. FSK and PSK modulations are treated in more detail in Chapter 16.

Classification

Modems can be classified either as *voice-band* or *wideband* modems. Voice-band modems are low-to-high-speed (up to 10,800 bps) modems designed for use on dial-up, voice-grade standard telephone lines. A low-to-high speed modem can be either a stand-alone device or an integrated circuit board that connects directly to the communication channel (i.e., phone line). Some of these modems are known as *smart* modems; they are microprocessor-controlled. In addition to transmitting and receiving data, smart modems can respond to commands from the digital device and automatically connect, disconnect, and generate dialing tones on the telephone line. A good example of such a modem is the Hayes modem produced by Hayes Microcomputer Products.

A different type of voice-band modem is the *acoustic-coupler* device. The acoustic coupler is a low-speed (up to 600 bps) modem that connects acoustically to a standard telephone handset, using a microphone and a speaker in rubber cups. An acoustic coupler accepts a serial asynchrnous data stream from a digital device, modulates that data stream into the audio spectrum, and then transmits the audio tones over a dial-up telephone connection. When audible tones are picked up by the telephone earpiece, they are demodulated by the acoustic coupler into a serial data stream and transmitted to the digital device. Acoustic couplers have the advantage of portability.

Wideband modems are very high-speed (19,200 bps and up) modems designed for use with dedicated telephone lines that are not part of a dial-up telephone network. They use dedicated and permanent wideband communication channels. Dedicated lines are often four-wire lines, with one pair for each direction, and therefore, the bandwidth is not shared in the full-duplex operation. These can be economical when long, frequent communications between two stations are required. Presently, wideband modems are used mostly on private communications systems.

PROBLEMS

13-1. What are the objectives of a computer communications network?

13-2. What is the difference between a LAN and a WAN?

13-3. What are the three different classifications of long-haul networks?

13-4. What is a protocol?

13-5. Discuss the role of protocols in standardizing telecommunication techniques and procedures?

13-6. What do ISO and OSI stand for?

13-7. For the OSI model, name the layer providing (a) end-to-end data transfer through one link; (b) end-to-end data transfer through the network; (c) data transfer through the interconnection channel; (d) the syntax of data representation; (e) the protocols for the network services; (f) the recovery system for the network; and (g) the transparent interface between the user and the network.

13-8. Which of the seven layers of the OSI model are needed to provide communications between stations in a subnetwork?

13-9. What is a network topology?

13-10. Discuss the six common network topologies.

13-11. Give three examples of bounded transmission media.

13-12. Give three examples of unbounded transmission media.

13-13. What are the frequency and distance of operations for (a) twisted pairs of wires, (b) baseband coaxial cables, (c) broadband coaxial cables, and (d) fiber-optic cables?

13-14. Fiber-optic cables are ______________ cables that use ______________ to carry data through ______________. (Fill in the blanks.)

13-15. Discuss the protocol used by an Ethernet LAN.

13-16. What is the maximum number of nodes that can be supported in an Ethernet?

13-17. What is the transmission medium used in Ethernet?

13-18. What is the advantage of using the Manchester code for data transmission in Ethernet?

13-19. Using the Manchester data-encoding scheme, draw the pulse train for the data string 11010011010100.

13-20. Using the Manchester data-encoding scheme, find the data string represented by the pulse train shown in Fig. P13-20.

FIG. P13-20.

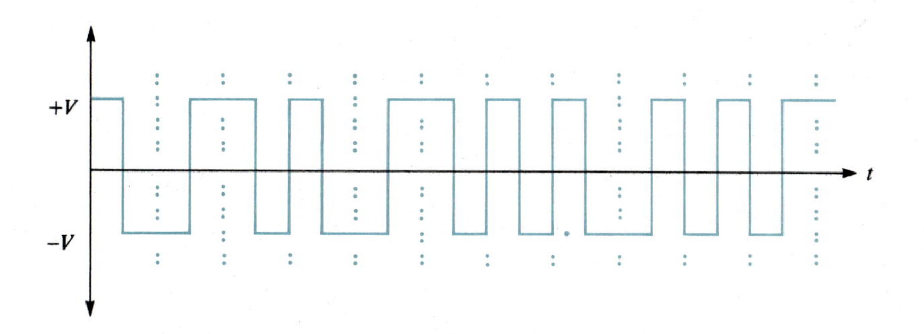

13-21. Repeat Problem 13-20 using Fig. P13-21.

FIG. P13-21.

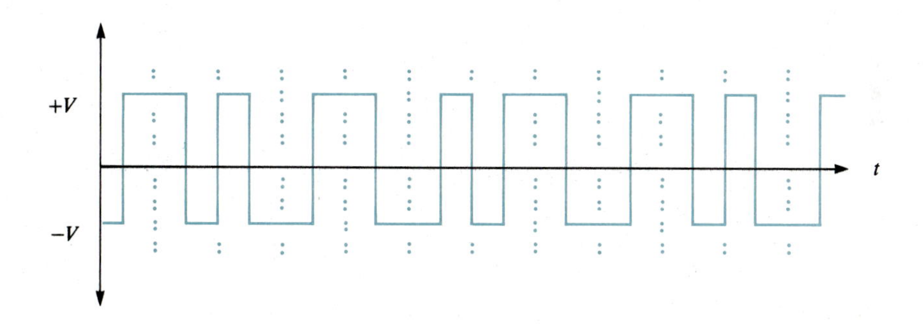

13-22. Repeat Problem 13-19 for the data string 00101100101011.

13-23. Using the differential Manchester data-encoding scheme, draw the pulse train for the data string 10000101111.

13-24. Repeat Problem 13-23 for the data given in Problem 13-19.

13-25. Repeat Problem 13-23 for the data given in Problem 13-22.

13-26. Explain the contention problem in Ethernet.

13-27. How does a token-bus network eliminate the contention problem of Ethernet?

13-28. Why is a token-bus network most suited for factory automation?

13-29. Which LAN is used by General Motors?

13-30. Which LAN was initiated by IBM?

13-31. What is the main difference between token-bus and token-ring LAN?

13-32. What type of signaling does broadband LAN use?

13-33. What can broadband LANs provide that Ethernet and token-passing LANs cannot? What makes this possible?

13-34. Complete the following table.

LAN	Media	Protocol	Speed	Signal	Example
Ethernet					
Token ring					
Token bus					
Broadband					

13-35. Show the code generated for the parallel transmission of the letters B and C using ASCII code with even parity and 1 strobe bit.

13-36. Draw the formats, as in Fig. 13.6-2(a), for asynchronous serial transmissions of the letters B and C.

13-37. What is a UART?

13-38. Using the format of Fig. 13.6-2(a) and a data rate of 110 bps, how many characters can a computer transmit to a serial printer in 1 min?

13-39. What is the overhead penalty in asynchronous serial communication when 1 start bit and 2 stop bits are used for each character (8 bits) transmitted?

13-40. What is the advantage of synchronous communication over asynchronous communication? Why?

13-41. Explain the basic principles of FDM.

13-42. A more accurate definition of baud rate is as follows: ''Baud rate is the number of signals per second.'' For this definition, (a) when does the baud rate equal the number of bits per second, and (b) when does the baud rate equal half the number of bits per second?

13-43. In full-duplex modems, why is it necessary to designate one modem as the originating modem and the other modem as the answering modem?

13-44. What type of modulation do low-data-rate modems use?

13-45. What are the main differences between a wideband modem and an acoustic coupler?

PART 4

Electrical Communications Systems

CHAPTER 14

Wave Propagation, Antennas, and Noise

14.0 INTRODUCTION

In this part of the book three chapters are dedicated to the description of electrical communications systems. One typical everyday communications system involves an announcer at a broadcast radio station who wishes to have his voice heard by you, the reader, through an ordinary radio receiver. His voice is first converted to an electrical (baseband) waveform by a microphone into which he speaks. The waveform is next processed by various electric circuits that change its form (shift it to a higher frequency) and raise its power level. The higher-power signal is then converted to an electromagnetic wave by an antenna that broadcasts the wave outward from the station in all directions. Although the electromagnetic wave is invisible to the eye, it conveys the station's transmitted power outward to all potential locations for receivers.

Next, a receiver at some distance from the station will have its own antenna that captures part of the transmitted wave and converts it back into an electric voltage (with a very small voltage amplitude, typically). This voltage is then amplified to a higher, more usable power level and converted in form by various signal-processing circuits to recover the original baseband waveform. Finally, the baseband waveform drives a speaker to create the audio sound that is heard.

The preceding description of a communications system would not be significantly different for other applications. For example, it would apply to television (TV) broadcasting by simply replacing the baseband waveforms by the television signal [containing both picture (video) and audio signals].

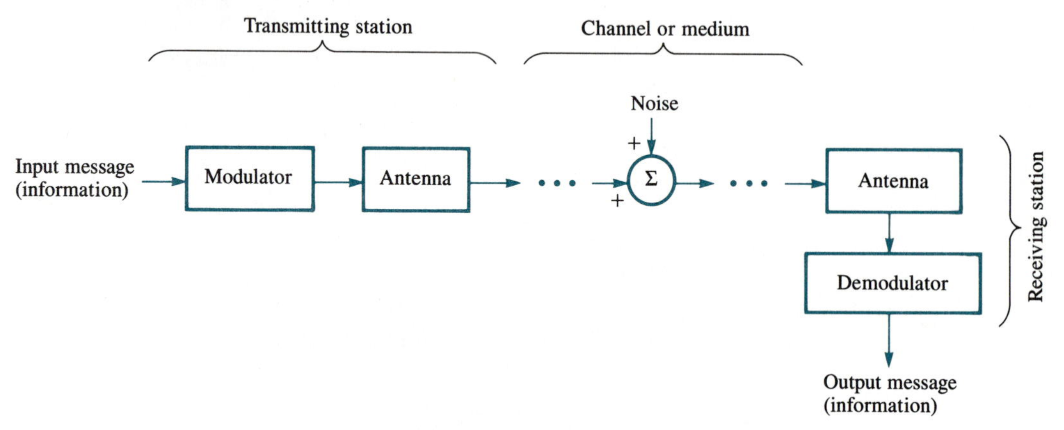

FIG. 14.0-1.
Components of a typical communications system.

To be more specific, a typical communications system contains the basic elements, or functions, illustrated in Fig. 14.0-1. The signal to be conveyed to the receiver is an electrical waveform, usually called the *message*, that represents the information source of interest. It might be a microphone's output, as in our earlier example, or it could be the bit stream of some digital computer that corresponds to financial data. The message is usually a baseband waveform.

The transmitting station uses a *modulator* to convert the message to a more useful form for transmission to the receiver. The conversion process is called *modulation*. The choice of type of modulation can affect system efficiency (the amount of received signal per dollar of system cost) or performance (the quality of the recovered message in the receiver). In the receiving station the *demodulator* must perform the inverse function of the modulator in order to recover the original message as closely as possible.

The receiver's output message is never exactly equal to the original message because of electrical noise in the system, practical component imperfections, and some theoretical limitations imposed by various modulators. The largest effects are usually due to noise. A system's electrical noise is actually generated by various noise sources throughout the system. However, for reasons that will subsequently be made clear, it is common to imagine (model) the noise as being generated by a single source located in the *channel*, or *medium*, that connects the transmitting and receiving stations. This fact accounts for the noise source shown in Fig. 14.0-1.

From the preceding discussions it is clear that a good understanding of a communication system having the form of Fig. 14.0-1 can be obtained by study of four fundamental subjects: electromagnetic waves and their propagation, antennas, noise, and modulators and demodulators. We shall reserve the following two chapters to discuss modulators and demodulators in detail. These operations are considered in Chapters 15 and 16 for analog and digital messages, respectively. In this chapter we discuss the remaining three fundamental subjects. Because of length and scope limitations of this book, our treatment of electromagnetic waves will be mainly limited to plausible arguments based on the reader's background in electric and magnetic fields derived from elementary physics. The wave concepts that apply to transmission lines are the most easily understood, so we develop these first.

14.1 WAVES AND TRANSMISSION LINES

A pebble thrown into a still pond creates a disturbance that causes ripples, or waves, to spread out (propagate) away from the disturbance. Clapping hands in a quiet room causes sound waves to propagate outward from the disturbance. A person moving one end of a long tight rope up and down cyclically, when the other end is tied to a fixed object, causes waves to propagate away from the person and down the rope. These are several examples of wave propagation. The speed of the waves is a characteristic of the medium supporting the waves. Sound waves obviously travel faster in air than do waves on a rope. Of special interest in every case is that a distant point sees no effect from the disturbance until a delay has occurred that equals the time it takes the wave to propagate from the disturbance to the point.

In an electrical network an electrical disturbance causes an electromagnetic wave† to propagate through the surrounding circuit. If the circuit distances of interest are small, such that propagation delays are less than $\frac{1}{8}$ of the period of the highest frequency of interest, for example, then delays are negligible and wave concepts may not need to be applied. This case corresponds to the ordinary circuit-analysis methods developed in preceding work.

However, if circuit dimensions are large, ordinary circuit analysis does not apply and wave concepts must be introduced. One common situation, applicable to all higher-frequency systems, concerns the coupling of signals between two distant parts of the system. One example is the connection of the high-power output of a transmitter located in a building to the transmitting antenna on a tower. Such connections are often made by special conductors called transmission lines, which usually consist of two or more parallel conductors‡ (to guide the waves) that are separated by insulating (dielectric) materials.

---

† An electric voltage and its resulting current may be considered to be a disturbance. The voltage corresponds to an associated electric field, while the current has an associated magnetic field. As the fields propagate, they carry associated voltages and currents to other parts of the circuit. The disturbance can therefore be taken as the electromagnetic fields.

‡ A single-conductor transmission line is also possible; it is called a *surface-wave transmission line*.

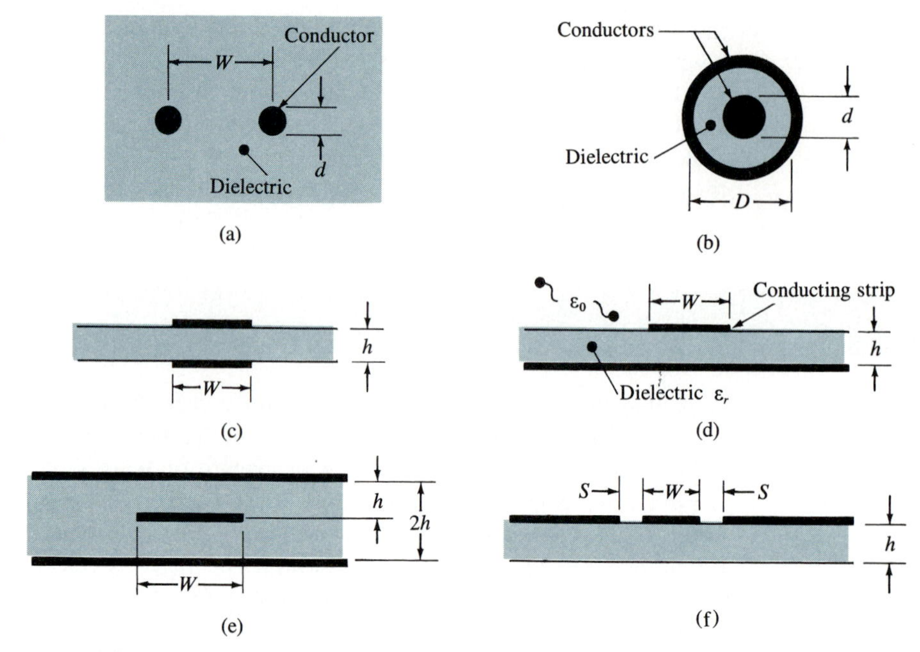

FIG. 14.1-1. Transmission lines: (a) a two-wire line, (b) a coaxial line, (c) a parallel stripline, (d) a microstrip line, (e) a stripline, and (f) a coplanar waveguide.

Geometry of Lines

Transmission lines are available in many forms. Figure 14.1-1 shows the cross sections of several popular types. A form of the two-wire line of Fig. 14.1-1(a) is used to connect some television antennas. The coaxial line of Fig. 14.1-1(b) is sometimes called a coaxial cable because of its appearance when implemented with flexible conductors. It is the most important and widely used of all the many possible cable-type transmission lines.

Other forms of transmission lines that are very useful in printed-circuit and integrated-circuit applications are sketched in Fig. 14.1-1(c)–(f). The microstrip line of Fig. 14.1-1(d) is probably the most widely used in both applications. The advantage of the stripline of Fig. 14.1-1(e) is the shielding it has from other circuits (the upper and lower plates are commonly at the same potential, and the center conductor is the "hot" lead). The coplanar geometry of Fig. 14.1-1(f) is especially useful in hybrid microwave integrated circuits; here the center conductor forms the "hot" lead and the two outer conductors are connected together electrically (ground). Its usefulness derives from its ability to connect both leads of lumped components (capacitors, resistors, etc.) to the circuit in the same plane [the similar connections in the geometries of Fig. 14.1-1(c), (d), and (e) involve connecting *through* the dielectrics to the lower ground plane].

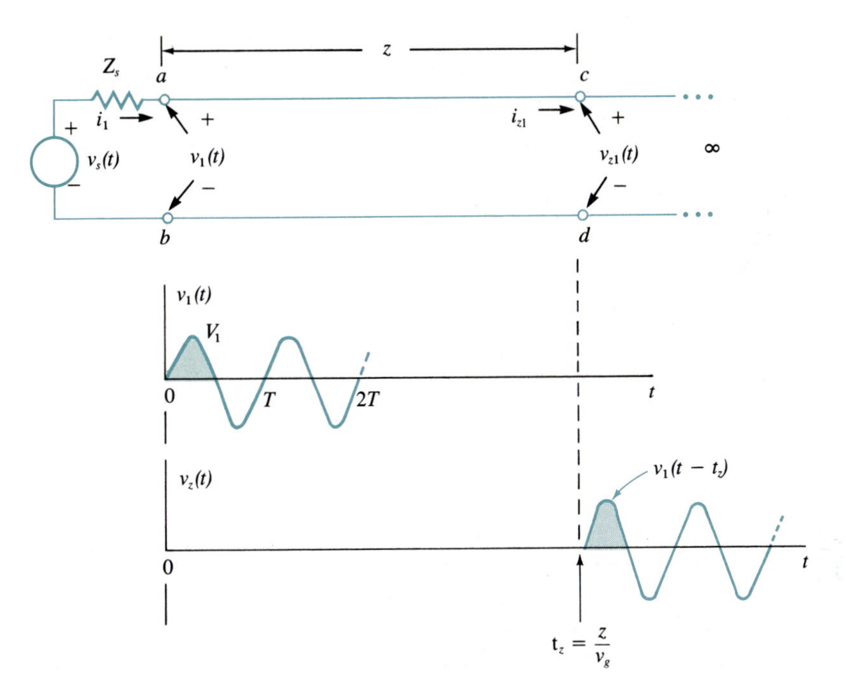

FIG. 14.1-2. A transmission line representation useful in illustrating wave propagation.

Waves on Transmission Lines

Some of the most important concepts involving waves propagating on transmission lines can be developed with the aid of Fig. 14.1-2, which represents any of the lines of Fig. 14.1-1. The representation is ideal in that the conductors and the dielectric are assumed to be lossless. A source of open-circuit voltage $v_s(t)$ and internal impedance Z_s drives the infinitely long line at terminals a, b. Assume initially that $v_s(t)$ is positive and increasing. Because of the line's insulation, no *conduction* current can flow in the line between points a and b. However, the line acts as a capacitance, so that positive charges build up at terminal a (an equal-magnitude negative charge is at b) in response to the positive voltage of the source. The charge flow constitutes a current $i_1(t)$ flowing into terminal a through Z_s. The current causes a voltage drop across Z_s, so that $v_1(t)$ is not equal to $v_s(t)$.

Voltage $v_1(t)$ gives rise to an electric field in the line between terminals a and b, while $i_1(t)$ gives rise to a magnetic field. Because the wires are conductors, the charges building up at the input terminals cannot stay fixed, and they tend to spread out (propagate) down the line. They do not move instantly, however, because the line exhibits an inductance effect† that retards charge migration (current). The charges, and therefore current, move down the wire at finite velocity v_g. As the charges propagate, the attendant electric and magnetic fields propagate, giving rise to an electromagnetic wave.

† Even a straight wire has some inductance.

Generally, the inductive and capacitive actions of the line compensate each other such that the line appears as a resistance,† denoted by Z_0, called the *characteristic impedance* of the line. Thus,

$$i_1(t) = \frac{v_1(t)}{Z_0} \tag{14.1-1}$$

$$v_1(t) = v_s(t) - i_1(t)Z_s \tag{14.1-2}$$

apply if we consider $v_s(t)$ to be a complex signal for analysis purposes. As $v_s(t)$ assumes values other than positive and increasing, as initially assumed, the results (14.1-1) and (14.1-2) remain valid. On solving these results simultaneously, we get

$$v_1(t) = \frac{Z_0}{Z_s + Z_0} v_s(t) \tag{14.1-3}$$

$$i_1(t) = \frac{1}{Z_s + Z_0} v_s(t) \tag{14.1-4}$$

As the electromagnetic fields associated with $v_1(t)$ and $i_1(t)$ at terminals a and b propagate down the line, they carry along the associated voltages and currents that are no different from $v_1(t)$ and $i_1(t)$ except for a delay in time. Thus, at terminals c and d in Fig. 14.1-2, which are a distance z from the input, the voltage and current are

$$v_{z1}(t) = v_1(t - t_z) \tag{14.1-5}$$

$$i_{z1}(t) = i_1(t - t_z) \tag{14.1-6}$$

where the delay time involved is

$$t_z = \frac{z}{v_g} \tag{14.1-7}$$

On substitution of (14.1-3) and (14.1-4), we get

$$v_{z1}(t) = \frac{Z_0}{Z_s + Z_0} v_s\left(t - \frac{z}{v_g}\right) \tag{14.1-8}$$

$$i_{z1}(t) = \frac{1}{Z_s + Z_0} v_s\left(t - \frac{z}{v_g}\right) \tag{14.1-9}$$

Because of the line's behavior, the impedance at terminals c, d, which is $v_{z1}(t)/i_{z1}(t)$ for the complex signals assumed here, is Z_0. In other words, the infinite line to the right of terminals c, d appears as an impedance Z_0 to the line left of terminals c, d. As a consequence, the line can be replaced by an impedance Z_0 at

† Although not very precise, a mental picture follows an analogy with a parallel resonant *RLC* circuit. At resonance (the frequency of interest) the inductive and capacitive reactances cancel, leaving only the resistance to account for power loss in the circuit. In the line the resistance (Z_0) accounts for the real power flowing down the line from the source.

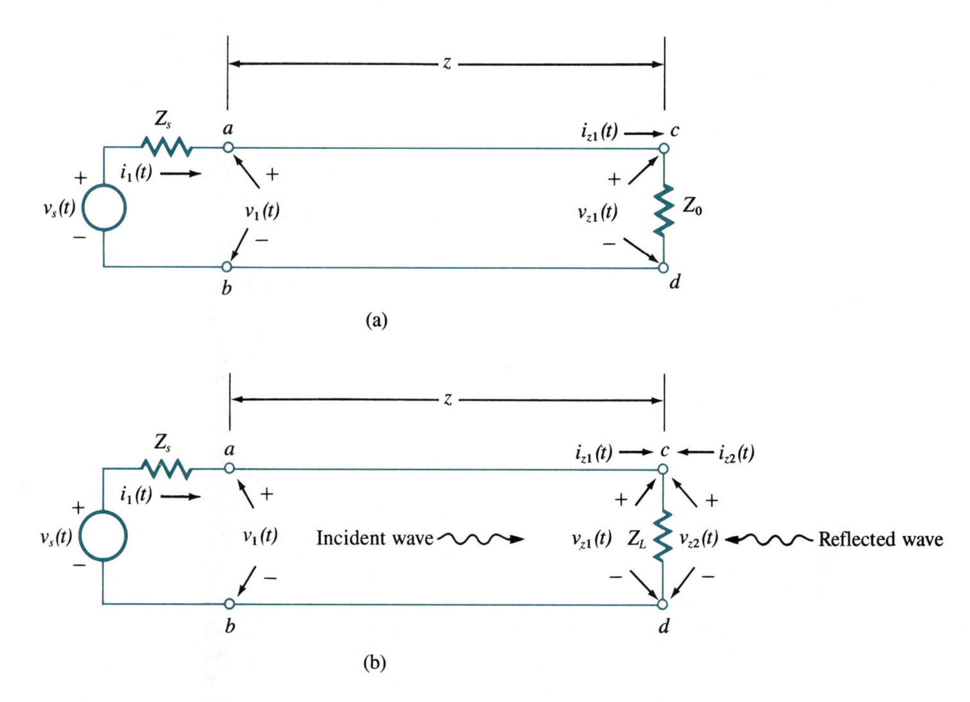

FIG. 14.1-3. A finite-length transmission line terminated (a) in its characteristic impedance and (b) in an arbitrary impedance.

terminals c, d, as shown in Fig. 14.1-3(a), with no effect on voltages and currents anywhere else on the line. This result is important, because it means that if a wave is launched on a real, finite-length transmission line, the wave will dissipate itself in the terminating load impedance if that impedance is equal to the line's characteristic impedance. We may think of this as a case where the load is *matched* to the line being used (the value of Z_0 will be different for different types of transmission lines).

If the load impedance, denoted by Z_L, as shown in Fig. 14.1-3(b), is *not matched* to the line, so that $Z_L \neq Z_0$, all the power in the incident wave defined by $v_{z1}(t)$ and $i_{z1}(t)$ is not dissipated in Z_L. This fact means that the load's voltage $v_L(t)$ and current $i_L(t)$ cannot be equal to $v_{z1}(t)$ and $i_{z1}(t)$, respectively. Physically, what happens is that a new (reflected) wave is generated that propagates back toward the source. The new wave will have a voltage $v_{z2}(t)$ and a current $i_{z2}(t)$ (positive toward the source) such that its power is the portion of the incident wave's power that is *not* absorbed in Z_L. At the load's terminals we must have

$$i_{z2}(t) = \frac{v_{z2}(t)}{Z_0} \tag{14.1-10}$$

$$v_L(t) = v_{z1}(t) + v_{z2}(t) \tag{14.1-11}$$

$$i_L(t) = i_{z1}(t) - i_{z2}(t) \tag{14.1-12}$$

$$i_L(t) = \frac{v_L(t)}{Z_L} \tag{14.1-13}$$

for the complex signals assumed. On solving these four relationships, we have

$$v_{z2}(t) = \Gamma_L v_{z1}(t) \tag{14.1-14}$$

$$i_{z2}(t) = \frac{\Gamma_L}{Z_0} v_{z1}(t) = \Gamma_L i_{z1}(t) \tag{14.1-15}$$

$$v_L(t) = (1 + \Gamma_L) v_{z1}(t) \tag{14.1-16}$$

$$i_L(t) = \frac{(1 - \Gamma_L)}{Z_0} v_{z1}(t) = (1 - \Gamma_L) i_{z1}(t) \tag{14.1-17}$$

where we define the *reflection coefficient* Γ_L of the load by

$$\Gamma_L = \frac{Z_L - Z_0}{Z_L + Z_0} \tag{14.1-18}$$

Note that if $Z_L = Z_0$, the reflection coefficient is zero and the reflected wave disappears from (14.1-14) and (14.1-15).

When the load in Fig. 14.1-3(b) causes a reflected wave, it travels back to the source, where it encounters the source's impedance Z_s acting as a "load" terminating the line. If $Z_s = Z_0$, there is no rereflection and the reflected wave is absorbed in the source as heat. If $Z_s \neq Z_0$, a rereflection can occur. Thus, if both the source and the load-end impedances are not equal to the line's characteristic impedance, multiple reflections occur. This effect is undesirable in most systems in practice, so the load or the source, or both, is usually designed to have impedance Z_0. An example will readily illustrate the importance of this choice.

EXAMPLE 14.1-1

A transmitter is connected to an antenna by a transmission line for which $Z_0 = 50\ \Omega$. The transmitter's source impedance is matched to the line, but the antenna is known to be unmatched and has a reflection coefficient $\Gamma_L = 0.32$. The transmitter creates a power of 15 kW in the incident wave to the antenna and will overheat to destruction if the reflected wave's power is more than 2 kW. We determine the antenna's radiated power and whether the transmitter will survive.

The reflection coefficient causes a reflected-wave amplitude of 0.32 times the incident wave's amplitude, from (14.1-14). The reflected power becomes $(0.32)^2(15)$ kW, or 1.536 kW. Since the transmitter is matched on the line, all the reflected power is dissipated in its output impedance. However, since it is less than 2 kW, the transmitter survives. The power radiated by the antenna becomes $15 - 1.536 = 13.464$ kW.

As a check on results, we use (14.1-18) to find

$$Z_L = \frac{Z_0(1 + \Gamma_L)}{1 - \Gamma_L} = \frac{50(1.32)}{0.68} \approx 97.06\ \Omega$$

From (14.1-16) the load's (antenna's) voltage is 1.32 times that of the incident wave. Since power is proportional to voltage squared, but inversely proportional to impedance, (14.1-16) gives

$$\text{Antenna power} = \frac{(1 + \Gamma_L)^2 Z_0}{Z_L} \cdot \text{incident power}$$

$$= \frac{(1.32)^2(50)(15)}{50(1.32)/0.68}\ \text{kW} = 13.464\ \text{kW}$$

as it should.

14.2 PRACTICAL TRANSMISSION LINES

In this section we briefly discuss two of the most widely used transmission lines, the coaxial line and the microstrip line.

The Coaxial Line

A coaxial transmission line is typically constructed as shown in Fig. 14.2-1. An inner conductor of outside radius a can be solid, stranded, or even hollow. The hollow conductor is mainly used in high-power applications, where a coolant circulates inside the conductor. The dielectric that separates the conductors is often polyethylene or Teflon (polytetraflouroethylene) in general-purpose, lower-power flexible lines. In higher-power rigid lines, air is often used. The outer conductor, having an inside radius b, usually consists of either single- or double-

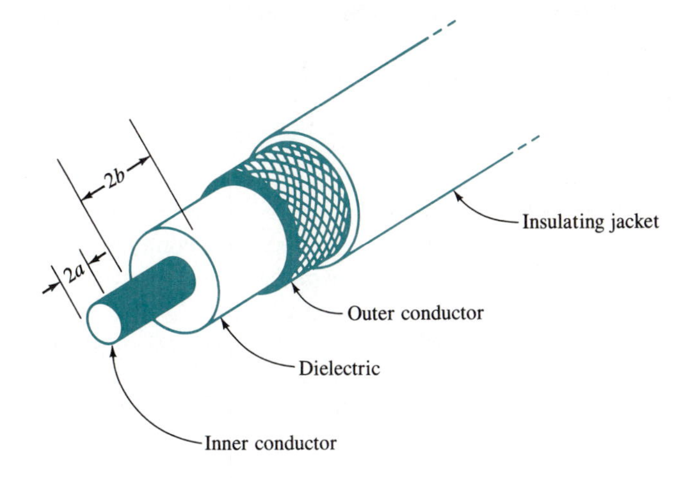

FIG. 14.2-1. Construction of a typical coaxial transmission line.

TABLE 14.2-1. Dielectric Constant and Loss Tangents for Several Dielectrics

		Loss Tangent, tan δ				
Material	ϵ_r	**1 kHz**	**1 MHz**	**100 MHz**	**3 GHz**	**25 GHz**
Dry air	1.00	—	—	—	—	—
Polyethylene	2.26	<0.0002	<0.0002	0.0002	0.00031	0.0006
Teflon	2.10	<0.0003	<0.0002	<0.0002	0.00015	0.0006

Source: Reference Data for Radio Engineers, 6th ed., Howard W. Sams, Indianapolis, 1975.

braided copper, often tinned or silvered. Double braid is preferred in applications requiring high isolation between the waves on the line and external electromagnetic fields. The outer conductor is solid copper in most rigid lines. The insulating jacket on flexible lines is often made by synthetic resin or Teflon.

Coaxial lines have been standardized and are specified by a radio guide, or RG, number. For example, RG-223/U stands for radio guide 223/universal; it is a small-sized, flexible, double-braided cable with an overall diameter of 5.5 mm (0.216 in), silvered-copper conductors, and characteristic impedance of 50 Ω. Most coaxial lines are either 50 or 75 Ω, but values of Z_0 from about 15 to 125 Ω can be found. Impedance Z_0 is related to the cable's parameters by

$$Z_0 = \frac{60}{\sqrt{\epsilon_r}} \ln\frac{b}{a} \tag{14.2-1}$$

where ϵ_r is the relative permittivity (dielectric constant) of the dielectric. Table 14.2-1 gives ϵ_r for some common dielectrics.

The velocity of propagation of waves in a coaxial line is

$$v_g = \frac{c}{\sqrt{\epsilon_r}} \tag{14.2-2}$$

where $c = 3 \times 10^8$ m/s is the speed of light. The delay τ of a cable of length L is therefore

$$\tau = \frac{L}{v_g} = \frac{L\sqrt{\epsilon_r}}{c} \tag{14.2-3}$$

EXAMPLE 14.2-1

A coaxial cable with a polyethylene dielectric connects an antenna to a receiver 28 m away. We determine the delay of the cable. From Table 14.2-1, $\epsilon_r = 2.26$; so

$$\text{Delay} = \tau = \frac{28\sqrt{2.26}}{3 \times 10^8} \approx 0.14\ \mu\text{s}$$

It is of interest to note that the wave propagates in the cable at $1/\sqrt{2.26} \approx 0.665$ times the speed of light.

Waves may propagate in transmission lines in various *modes*. A mode refers to the geometrical pattern of the electric and magnetic fields in the line. In a coaxial line the preferred mode is called TEM.† Propagation in this mode can occur at any frequency from zero (dc) up to a cutoff frequency f_c given approximately by

$$f_c = \frac{c}{\pi\sqrt{\epsilon_r}(a + b)} \quad \text{Hz} \qquad (14.2\text{-}4)$$

For $f \geq f_c$ undesired wave modes can occur. In practice $f < 0.95f_c$ is usually maintained. Cutoff frequency puts an upper bound of approximately 1 GHz on usable frequency for larger-diameter coaxial lines and around 20 GHz for smaller diameters.

Real coaxial lines attenuate waves owing to losses in the conductors and dielectric. The attenuation (decibels per meter) due to conductor losses is approximately

$$\text{Attenuation}\Big|_c = \frac{(1.373 \times 10^{-3})\sqrt{\rho f}}{Z_0}\left(\frac{1}{a} + \frac{1}{b}\right) \quad \text{dB/m} \qquad (14.2\text{-}5)$$

where ρ is the resistivity of the conductors (see Table 1.4-1). Attenuation due to dielectric losses is given by

$$\text{Attenuation}\Big|_d = (9.096 \times 10^{-8})\sqrt{\epsilon_r}\, f \tan\delta \quad \text{dB/m} \qquad (14.2\text{-}6)$$

where $\tan\delta$ is called the *loss tangent* of the dielectric. It is zero for a perfect dielectric and small for good dielectrics (see Table 14.2-1). The total line's attenuation is the sum of attenuations of conductors and dielectric.

EXAMPLE 14.2-2

We compute the losses at 100 MHz for RG-213/U, 50-Ω cable that has a = 0.445 mm (0.018 in), b = 1.473 mm (0.058 in), ϵ_r = 2.26 (polyethylene), and ρ = 1.63×10^{-8} Ω·m (silver):

$$\text{Attenuation}\Big|_c = \frac{1.373 \times 10^{-3}\sqrt{(1.63 \times 10^{-8})(10^8)}}{50}\left(\frac{1}{0.445 \times 10^{-3}} + \frac{1}{1.473 \times 10^{-3}}\right)$$

$$\approx 0.103 \text{ dB/m}$$

$$\text{Attenuation}\Big|_d = (9.096 \times 10^{-8})\sqrt{2.26}(10^8)(2)(10^{-4}) \approx 0.00273 \text{ dB/m}$$

Losses due to the conductors dominate. The total attenuation is 0.103 + 0.00273 ≈ 0.106 dB/m.

---

† TEM stands for transverse electromagnetic.

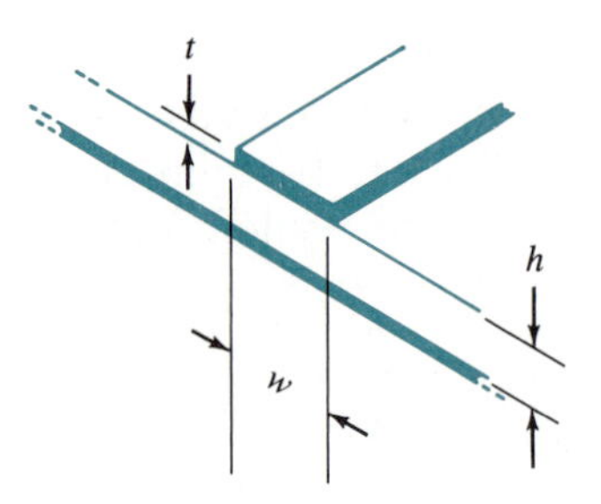

FIG. 14.2-2. Construction of a microstrip transmission line.

Finally, we note that the power-handling capacity of coaxial lines is up to several hundreds of kilowatts at lower frequencies with large-diameter rigid lines. Smaller, flexible lines can be found that can carry up to around 10 kW of average power at lower frequencies. Power capacity decreases approximately as $1/\sqrt{f}$ with frequency.

The Microstrip Line

The microstrip transmission line has the construction shown in Fig. 14.2-2. A conducting strip of width w and thickness t is separated from a conducting plane by a dielectric of thickness h and relative permittivity ϵ_r. The conductors are assumed to be of the same material with resistivity ρ. Waves propagate along the strip. Microstrip lines find their most important applications in modern microwave integrated circuits, where the dielectric is a ceramic-substrate material such as aluminum oxide.

Microstrip lines are usually found at frequencies above about 2 GHz and are useful to as high as about 60 GHz. However, it is important that the highest frequency to be used should not excite higher modes. Good practice† requires that the smaller of two upper-limit frequencies, denoted by $f_{c\mathrm{TM}}$ and $f_{c\mathrm{TR}}$, not be exceeded, where

$$f_{c\mathrm{TM}} = \frac{c \tan^{-1}(\epsilon_r)}{\pi h \sqrt{2(\epsilon_r - 1)}} \tag{14.2-7}$$

$$f_{c\mathrm{TR}} = \frac{c}{(2w + 0.8h)\sqrt{\epsilon_r}} \tag{14.2-8}$$

Other important parameters of microstrip lines, such as characteristic impedance and attenuation, are known but are somewhat complicated. We summarize the results here.†, ‡

† See T. C. Edwards, *Foundations for Microstrip Circuit Design*, Wiley, New York, 1981, p. 94.

‡ See K. C. Gupta, R. Garg, and I. J. Bahl, *Microstrip Lines and Slotlines*, Artech House, Norwood, Mass., 1979, p. 92.

$$Z_0 = \frac{84.78}{\sqrt{\epsilon_r + 1}} \left\{ \ln \left[\frac{4h}{w} + \sqrt{2 + 16\left(\frac{h}{w}\right)^2} \right] - \left(\frac{\epsilon_r - 1}{\epsilon_r + 1}\right)\left(0.226 + \frac{0.121}{\epsilon_r}\right) \right\} \quad \text{for } \frac{w}{h} < 3.3 \tag{14.2-9}$$

$$Z_0 = \frac{376.68}{\sqrt{\epsilon_r}} \left\{ \frac{w}{h} + 0.883 + 0.165\left(\frac{\epsilon_r - 1}{\epsilon_r}\right) + \left(\frac{\epsilon_r + 1}{\epsilon_r}\right)\left[0.462 + \frac{1}{\pi} \ln \left(0.94 + \frac{w}{2h}\right)\right] \right\}^{-1} \quad \text{for } \frac{w}{h} > 3.3 \tag{14.2-10}$$

$$\text{Attenuation}\Big|_c = \frac{(2.742 \times 10^{-3})A}{hZ_0} \sqrt{f\rho} \left[\frac{32 - (w_e/h)^2}{32 + (w_e/h)^2}\right] \quad \text{dB/m} \quad \text{for } \frac{w}{h} \le 1 \tag{14.2-11}$$

$$\text{Attenuation}\Big|_c = \frac{(12.12 \times 10^{-8})A\epsilon_{re}Z_0}{h} \sqrt{f\rho} \left[\frac{w_e}{h} + \frac{0.667w_e/h}{1.444 + (w_e/h)}\right] \quad \text{dB/m} \quad \text{for } \frac{w}{h} \ge 1 \tag{14.2-12}$$

In the above we define

$$A = \begin{cases} 1 + \dfrac{h}{w_e}\left(1 + \dfrac{1.25}{\pi} \ln \dfrac{4\pi w}{t}\right) & \dfrac{w}{h} \le \dfrac{1}{2\pi} \\ 1 + \dfrac{h}{w_e}\left(1 + \dfrac{1.25}{\pi} \ln \dfrac{2h}{t}\right) & \dfrac{w}{h} \ge \dfrac{1}{2\pi} \end{cases} \tag{14.2-13}$$

$$\frac{w_e}{h} = \begin{cases} \dfrac{w}{h} + \dfrac{1.25t}{\pi h}\left(1 + \ln \dfrac{4\pi w}{t}\right) & \dfrac{w}{h} \le \dfrac{1}{2\pi} \\ \dfrac{w}{h} + \dfrac{1.25t}{\pi h}\left(1 + \ln \dfrac{2h}{t}\right) & \dfrac{w}{h} \ge \dfrac{1}{2\pi} \end{cases} \tag{14.2-14}$$

$$\epsilon_{re} = \begin{cases} \dfrac{\epsilon_r + 1}{2} + \left(\dfrac{\epsilon_r - 1}{2}\right)\left(\dfrac{1}{\sqrt{1 + 12(h/w)}} - \dfrac{t/h}{2.3\sqrt{w/h}}\right) & \dfrac{w}{h} \ge 1 \\ \dfrac{\epsilon_r + 1}{2} + \left(\dfrac{\epsilon_r - 1}{2}\right)\left[\dfrac{1}{\sqrt{1 + 12(h/w)}} - \dfrac{t/h}{2.3\sqrt{w/h}} + 0.04\left(1 - \dfrac{w}{h}\right)^2\right] & \dfrac{w}{h} \le 1 \end{cases} \tag{14.2-15}$$

For the dielectric losses,

$$\text{Attenuation}\Big|_d = 9.096 \times 10^{-8} \frac{(\epsilon_{re} - 1)\sqrt{\epsilon_r}}{(\epsilon_r - 1)\sqrt{\epsilon_{re}}} f\sqrt{\epsilon_r} \tan \delta \tag{14.2-16}$$

To illustrate some numerical values, we use an example.

EXAMPLE 14.2-3

A microstrip line uses copper conductors ($\rho = 1.72 \times 10^{-8}\ \Omega\cdot\text{m}$), with $h = 1$ mm, $w/h = 0.7$ and $t/h = 0.07$, and has a dielectric for which $\epsilon_r = 8.79$ and $\tan \delta = 0.0002$ at 3 GHz. From (14.2-15), (14.2-14), and (14.2-13) we find $\epsilon_{re} = 5.682$, $w_e/h = 0.821$, and $A = 3.843$. From (14.2-9) we calculate $Z_0 = 61.24\ \Omega$. Finally, (14.2-16) and (14.2-11) give

$$\text{Attenuation}|_d = 0.121 \text{ dB/M}$$

$$\text{Attenuation}|_c = 1.185 \text{ dB/M}$$

Total loss for the line becomes $1.185 + 0.121 = 1.306$ dB/m. This is a relatively high loss per unit length. Fortunately, most microwave integrated circuits have short line lengths.

14.3 WAVEGUIDES

At higher frequencies, when power levels may be large or attenuation in transmission lines may be excessive, connections between system components are often made by using waveguides. A *waveguide* is usually a hollow, closed, rigid conductor (much like a pipe) through which waves propagate. The most common guides are either rectangular or circular in cross section, as shown in Fig. 14.3-1, but other shapes are possible. Most waveguides are rigid, but some flexible varieties are available. Because of the analogy with water pipes, waveguides are often called the "plumbing" in a system, and the engineers who work with waveguides are sometimes called "plumbers."

Modes and Operating Frequencies

As in transmission lines, waveguides propagate waves by modes. Unlike transmission lines, which operate at any frequency up to a cutoff value, waveguides have both upper *and lower* cutoff frequencies.

In the common rectangular guide the lower cutoff frequency above which operation must occur if operation is to be at the lowest, or dominant, mode is

$$f_c = \frac{1}{2a\sqrt{\mu\epsilon}} \tag{14.3-1}$$

where μ and ϵ apply to the dielectric filling the guide. For air, the usual case, (14.3-1) reduces to

$$f_c = \frac{c}{2a} \tag{14.3-2}$$

where c is the speed of light. The upper limit, above which undesired higher

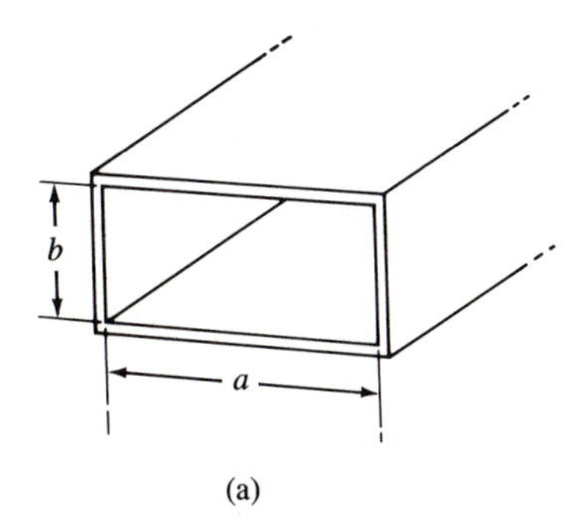

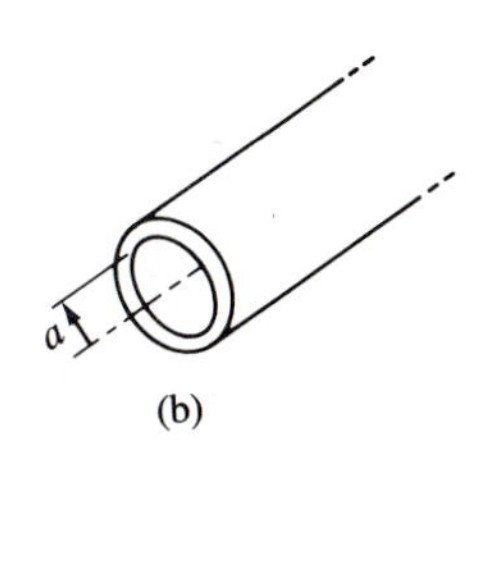

FIG. 14.3-1. Waveguides: (a) rectangular and (b) circular.

modes can propagate, cannot be larger than $2f_c$, which is realized only by choosing $b \leq a/2$. However, small attenuation requires b to be as large as possible compared with a. Standard waveguides available in practice have chosen $b \approx a/2$ to give the lowest attenuation possible while preserving the largest operating-frequency range. For several practical reasons the operating frequency should not come too close to the cutoff limits. A suggested range in practice is

$$1.25f_c \leq f \leq 1.90f_c \tag{14.3-3}$$

For a circular waveguide with inside radius a, the lower cutoff frequency for propagation by the dominant mode is

$$f_c = \frac{0.293}{a\sqrt{\mu\epsilon}} \tag{14.3-4}$$

which becomes

$$f_c = \frac{0.293c}{a} \tag{14.3-5}$$

for an air dielectric. The upper cutoff frequency is $1.307f_c$, so the maximum operating band to prevent higher modes is $f_c < f < 1.307f_c$.

Rectangular Waveguide

The characteristic impedance in waveguides is not constant with frequency, as it is in transmission lines. For a rectangular waveguide

$$Z_0 = \frac{\eta}{\sqrt{1 - (f_c/f)^2}} \tag{14.3-6}$$

where

$$\eta = \sqrt{\frac{\mu}{\epsilon}} \tag{14.3-7}$$

and for air

$$\eta = 120\pi \approx 377\ \Omega \tag{14.3-8}$$

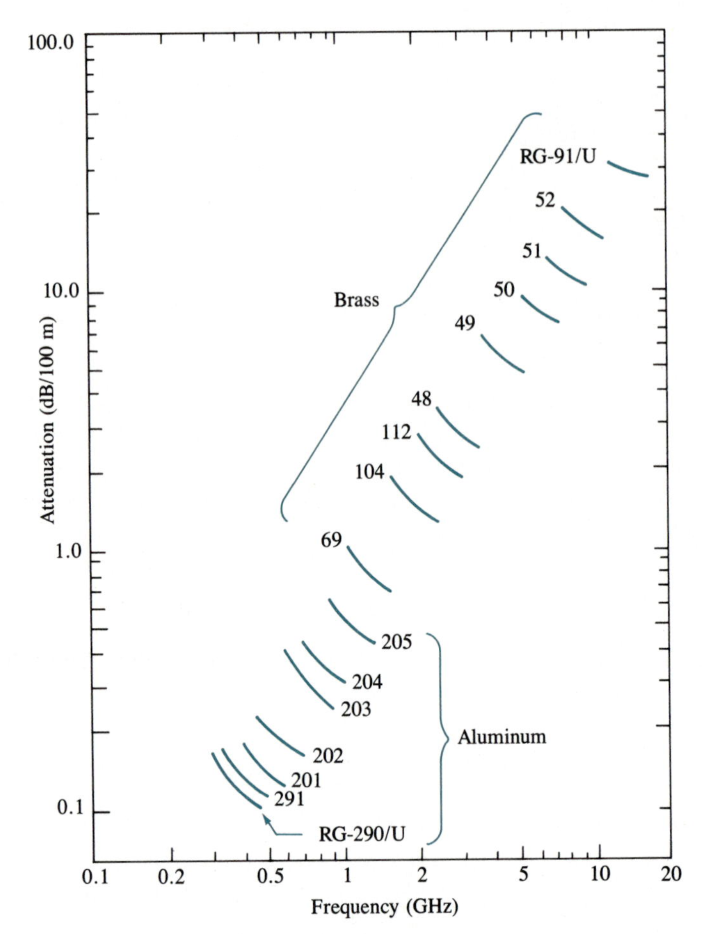

FIG. 14.3-2. Attenuation with frequency for various standard rectangular waveguides.

Attenuation in most waveguides is due mainly to conductor losses, since the dielectric is usually air. It is given by

$$\text{Attenuation}\bigg|_c = \frac{(0.458 \times 10^{-4})\sqrt{f\rho}[1 + (2b/a)(f_c/f)^2]}{b\sqrt{1 - (f_c/f)^2}} \quad \text{dB per unit length} \tag{14.3-9}$$

which is seen to vary with frequency. Standardized waveguides (where $a \approx 2b$) are available that cover frequencies from about 300 MHz to over 300 GHz. These guides use RG numbers similar to those for transmission lines. The attenuation and frequency range of some of these guides are shown in Fig. 14.3-2.

EXAMPLE 14.3-1

We use (14.3-9) to compute attenuation of the RG-49/U waveguide at its lowest frequency of operation, which is $f = 3.95$ GHz. For this guide $a = 4.755$ cm, $b = 2.215$ cm, and $\rho = 3.9 \times 10^{-8}$ Ω·m for the brass conductor. From (14.3-2)

$$f_c = \frac{3 \times 10^8}{2(4.755 \times 10^{-2})} \approx 3.155 \text{ GHz}$$

From (14.3-9)

$$\text{Attenuation}\bigg|_c = \frac{(0.458 \times 10^{-4})\sqrt{3.95(10)(3.9)}[1 + (4.43/4.755)(3.155/3.95)^2]}{(2.215 \times 10^{-2})\sqrt{1 - (3.155/3.95)^2}}$$

$$= 6.80 \text{ dB/100 m}$$

This value agrees well with the values shown in Fig. 14.3-2.

Equation (14.3-9) gives attenuation values in good agreement with experimental results for frequencies up to about 5 GHz. At higher frequencies, roughness of waveguide-conducting surfaces increases attenuation above theoretical values of (14.3-9).

Waveguide with a Circular Cross Section

A circular waveguide of radius a has a characteristic impedance given by (14.3-6), except that f_c is given by (14.3-4). Attenuation is

$$\text{Attenuation}\bigg|_c = 4.578 \times 10^{-5} \frac{\sqrt{f\rho\epsilon_r}[0.419 + (f_c/f)^2]}{a\sqrt{1 - (f_c/f)^2}} \quad \text{dB per unit length} \tag{14.3-10}$$

Because the circular guide's operating-frequency range, which does not exceed 31% of f_c, is less than that of the rectangular guide, which is less than 100% of f_c, it is less desirable for many applications. In some applications, however, its shape is an advantage. One example is in forming the rotary joint needed to feed power to a rotating antenna, which is used in some radar.

14.4 ANTENNA FUNDAMENTALS

In this section we shall discuss only the fundamental concepts needed to understand the role of an antenna as a power-coupling element of a system. Figure 14.4-1 defines some overall system terms and concepts of interest. A transmitter generates a power P_t; in some systems P_t is interpreted as peak power (power in a pulse at microwave frequency f, for example); in others P_t represents average power over a long time period. A loss L_t (which is a number larger than unity, the lossless case) represents the power reduction caused by the transmission line or waveguide that connects to the transmitting antenna, which couples energy into the medium.

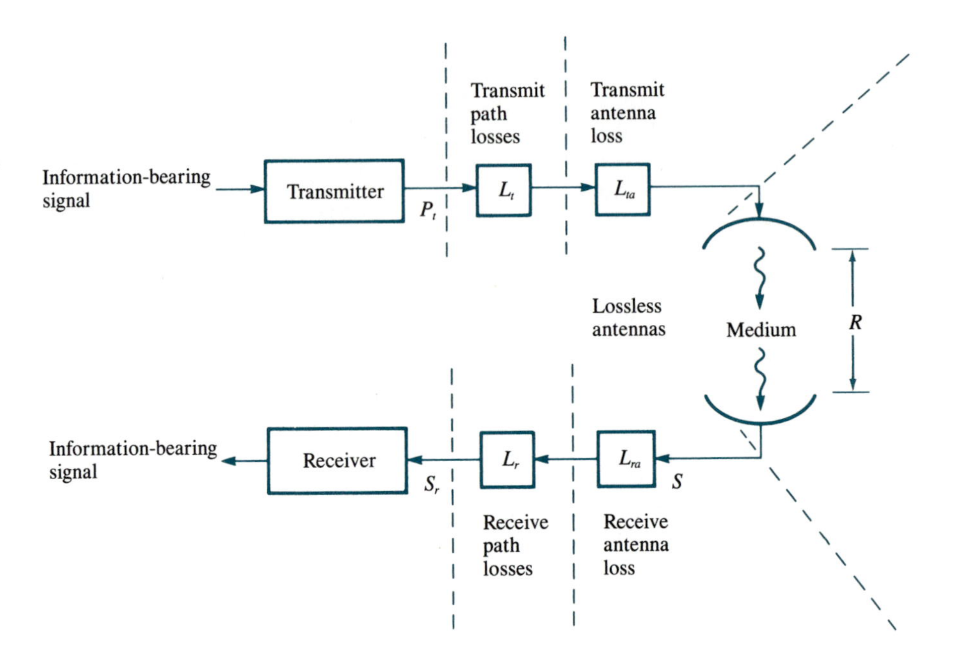

FIG. 14.4-1. Elements of a communications system that involves antennas.

Most antennas do not have much loss. There is always some due to such things as impedance mismatch, ohmic losses, and other effects related to detailed antenna theory. Whatever the total loss, as compared with an antenna without such practical effects, it is labeled as L_{ta}. Thus, we shall model a practical antenna that has loss as a loss in cascade with an ideal lossless antenna, which otherwise has all the properties of the actual antenna. Similar comments apply to the receiving antenna, where the loss is L_{ra}.

Finally, the antennas are separated by a distance R. The signal power available at the lossless-antenna output is S, and that at the receiver input is S_r, which is smaller than S due to the antenna loss L_{ra} and a receiving-path loss L_r.

Radiation-Intensity Pattern

Antennas do not radiate power equally in all directions in space. The *radiation-intensity pattern* describes the power intensity (which is power per unit solid angle, having the unit watts per steradian) in any spatial direction. For spherical coordinates centered on the antenna, as shown in Fig. 14.4-2(a), the radiation-intensity pattern might appear as shown in (b). The power intensity at any distant point R having angles θ and ϕ is plotted as a magnitude P from A. Clearly, the radiation pattern is a power-intensity function $P(\theta, \phi)$ of the angular directions θ and ϕ and appears as a surface, as illustrated in Fig. 14.2-2(b), when all possible values of θ and ϕ are considered.

In many communications problems it is desirable that the transmitting antenna concentrate most of its radiated power in one direction, that of the receiver.

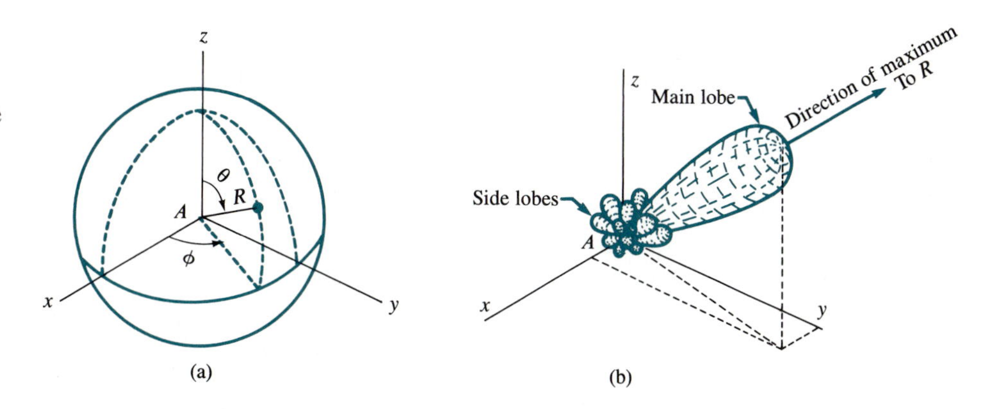

FIG. 14.4-2. (a) Spherical coordinates applicable to the radiation-intensity pattern of (b).

The radiation-intensity pattern will then contain one large *main lobe* in its surface. Other minor lobes, or *side lobes*, occur as well. The receiver will receive the maximum possible power if the transmitting antenna points its main-lobe maximum toward the receiver. Because antennas are *reciprocal elements*, the power output when used as a receiving antenna will also be determined by the radiation-intensity pattern, which now becomes a "receiving pattern." This means that received-power output will be maximum when the receiving antenna's main-beam maximum is directed toward the transmitting site.

If the radiation-intensity pattern is scaled such that the maximum intensity is unity, the result is commonly called the *radiation pattern*. In many problems where the radiation pattern has one dominant main lobe, it is not necessary to discuss the full three-dimensional pattern. Behavior may adequately be described in two orthogonal planes containing the maximum of the main lobe. Figure 14.4-3 illustrates a possible pattern in one of these planes for both polar (a) and linear (b) angle plots. The angular separation between points on the radiation pattern that are 3 dB down from the maximum is called the *beam width*. In the two orthogonal-plane patterns, called *principal-plane patterns*, we shall call these beam widths θ_B and ϕ_B. Their units are radians unless otherwise stated.

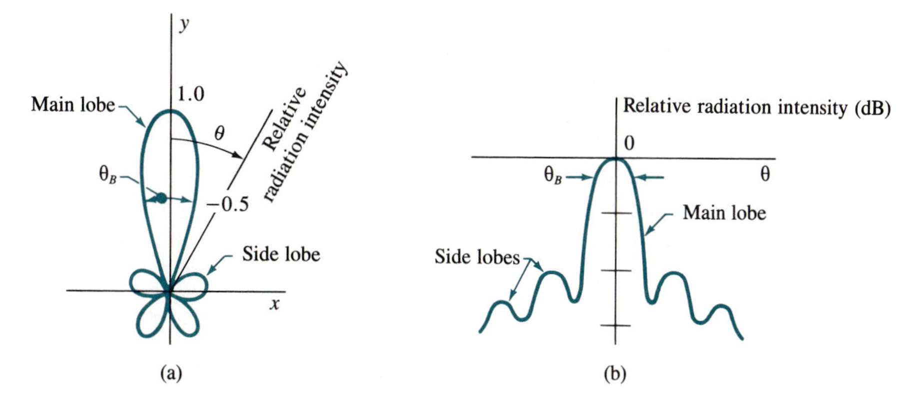

FIG. 14.4-3. Radiation patterns: (a) in polar coordinates and (b) in linear coordinates.

Directive Gain

Directive gain, also called *directivity*, is a measure of how concentrated an antenna radiates in the direction of its maximum. It is defined by the ratio

$$G_D = \frac{\text{maximum radiation intensity}}{\text{average radiation intensity}} \tag{14.4-1}$$

Gain G_D is based entirely on *radiated powers* and does not include the effect of any losses that may occur in the antenna. Since average radiation intensity is total radiated power, P_{rad}, divided by 4π steradians (sr), then

$$G_D = \frac{4\pi \text{ (maximum radiation intensity)}}{P_{\text{rad}}} \tag{14.4-2}$$

For most purposes a better gain definition is power gian, which includes the effects of antenna losses.

Power Gain

It is helpful to define an *isotropic antenna* as a lossless antenna that radiates its power uniformly in all directions.† If P is in the input power to such an antenna, the radiation intensity is a constant in any direction and is given by

$$\text{Radiation intensity} = \frac{P}{4\pi} \tag{14.4-3}$$

Power gain G is a measure of maximum radiation intensity of an antenna as compared with the intensity that would result from an isotropic antenna *with the same power input*. Hence G is defined by the ratio

$$G = \frac{\text{maximum radiation intensity}}{\text{radiation intensity of isotropic source with same power input}} \tag{14.4-4}$$

On using (14.4-3), we get

$$G = \frac{4\pi\text{(maximum radiation intensity)}}{P} \tag{14.4-5}$$

We observe that if the antenna is lossless, all the input power P is radiated power P_{rad}, so that directive and power gains are equal from (14.4-2):

$$G = G_D \qquad \text{lossless antenna} \tag{14.4-6}$$

When losses are present, as with any real antenna, $G < G_D$. The amount that G is less than G_D is accounted for by the *radiation-efficiency factor* ρ_r:

$$G = \rho_r G_D \tag{14.4-7}$$

† An isotropic antenna cannot be realized; it can only be approximated. It is, however, useful to imagine such an antenna and use it as a reference for comparison with real antennas.

where $0 < \rho_r \leq 1.0$

Antenna losses (transmitting and receiving) are related to ρ_r (ρ_{rt} for transmitting; ρ_{rr} for receiving) by

$$L_{ta} = \frac{1}{\rho_{rt}} \tag{14.4-8}$$

$$L_{ra} = \frac{1}{\rho_{rr}} \tag{14.4-9}$$

Because of our method of handling losses, we shall refer simply to antenna gain and use the symbol G, meaning either power or directive gain, since they are the same.

A useful, although approximate, expression for gain is

$$G = \frac{4\pi A_e}{\lambda^2} \tag{14.4-10}$$

where λ is the wavelength being transmitted (or received) and A_e is called the *effective aperture* or *effective area* of the antenna. For air as a medium λ is related to frequency f by

$$\lambda = \frac{c}{f} \tag{14.4-11}$$

where $c = 3 \times 10^8$ m/s is the speed of light.

Another useful result relates gain to the beam widths θ_B and ϕ_B of the pattern's main lobe. The result, which is only approximate, is

$$G = \frac{4\pi}{\theta_B \phi_B} \tag{14.4-12}$$

if beam widths are in radians, or

$$G = \frac{41.3 \times 10^3}{\theta_B \phi_B} \tag{14.4-13}$$

if units are degrees.

EXAMPLE 14.4-1

An antenna has beam widths of 3° and 10° in orthogonal planes and has a radiation-efficiency factor of 0.5. We find the maximum radiation intensity if 1000 W is applied to the antenna. From (14.4-13) gain is

$$G = \frac{41.3 \times 10^3}{3(10)} = 1.38 \times 10^3$$

or 31.4 dB.† If all power were radiated, then maximum intensity, from (14.4-2), would be $(1.38 \times 10^3)(10^3)/4\pi = 109.8$ kW/sr. However, only half is effectively transmitted according to (14.4-6), so the maximum intensity is 54.9 kW/sr.

---

† Gain in dB, denoted by G_{dB}, is related to gain as a numeric by $G_{dB} = 10 \log G$.

Aperture Efficiency

Some antennas have a physical aperture area A that can be identified (see Sec. 14.5). It is related to A_e by

$$A_e = \rho_a A \tag{14.4-14}$$

where ρ_a is called the *aperture efficiency* and $0 < \rho_a \leq 1.0$. One would expect intuitively that if all possible portions of the true physical surface were radiating power with equal intensity, an aperture with maximum possible efficiency would result ($\rho_a = 1.0$). Such is the case, and it is called *uniform illumination.*

If some portions of the antenna's radiating plane (aperture) radiate at a lower intensity level that the maximum, they may be considered as contributing to a less efficient antenna. This is referred to as *nonuniform illumination*, and $\rho_a < 1.0$. Thus, nonuniformly illuminated antennas are less efficient than uniformly illuminated ones, having less gain. From (14.4-10) and (14.4-14) G becomes

$$G = \frac{4\pi\rho_a A}{\lambda^2} \tag{14.4-15}$$

Although they give less gain, antennas with $\rho_a < 1$ may give lower side lobes than the more efficient uniformly illuminated antenna. The prices paid for the lower side lobes are broader main-lobe beam width and lower gain.

Signal Reception

Let a distant source produce a radiated *power density* $\mathcal{S}$ (unit is power per unit area) at a receiving antenna with effective area A_e (and real area A in the receiving antenna). The maximum power, denoted by S, that the antenna can produce during reception is

$$S = \mathcal{S}A_e = \mathcal{S}\rho_a A \tag{14.4-16}$$

The actual power depends on the receiver's impedance. As we subsequently show, it is the maximum power, called *available power*, that is important in system calculations.

Polarization

Most antennas transmit only one polarization of electromagnetic wave. That is, the electric field of the propagating wave is oriented relative to the antenna in only one direction. For example, the main lobe is usually† normal to the aperture, as illustrated in Fig. 14.4-4 for a circular aperture. If the local vertical at the antenna

† In most antennas the main lobe is in a direction normal to the plane of the aperture. However, for the important class of *phased-array antennas* the main lobe may be electronically steered to other angles away from the "broadside."

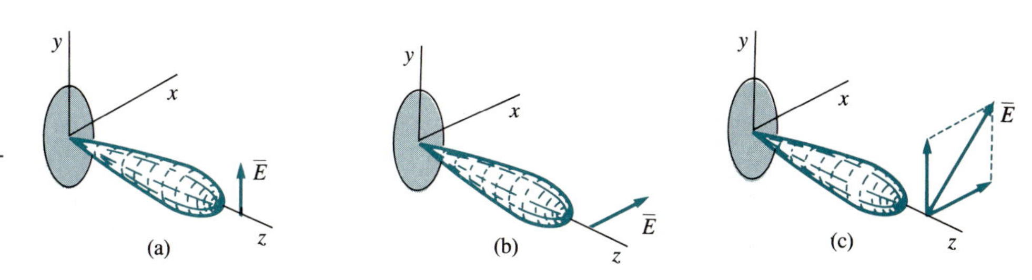

FIG. 14.4-4. Illustrations of electric field polarizations: (a) vertical, (b) horizontal, and (c) arbitrary linear.

is the y axis and the electric field lies in the vertical plane, as shown in Fig. 14.4-4(a), the radiation is said to be *vertically polarized*. If the electric field lies in the horizontal plane, as shown in Fig. 14.4-4(b), the radiation is said to be *horizontally polarized*. Both vertical and horizontal are special cases of *linear polarization*, as shown in Fig. 14.4-4(c). Here the electric field is still in a plane but has horizontal and vertical components that are in time phase. If the antenna main lobe points other than horizontally, vertical and horizontal have no meaning, and the above polarizations become simply linear polarizations.

Some systems are designed to transmit two linear orthogonal polarizations simultaneously that are not in time phase. *Circular polarization* is probably the most useful case, where horizontal and vertical electric fields are 90° out of time phase and have equal magnitude.† If the horizontal component of the radiation lags the vertical component by 90°, the resultant field appears to rotate counterclockwise in the xy plane with time, as one located at the antenna views the wave leaving the antenna, giving rise to what is called *left-hand* circular polarization. If the horizontal component leads the vertical component by 90°, the resultant field rotates clockwise with time in the xy plane and is called *right-hand* circular polarization. Circular is a special case of the more general *elliptical polarization*, where the two linear components have arbitrary relative amplitudes and arbitrary time phase. Elliptical polarization has seen little use in practice.

Antenna Impedance

The impedance, denoted by Z_a, looking into the feed-point terminals of an antenna is called the *antenna impedance*. Impedance Z_a will have both resistive and reactive components, in general. The resistive component has two parts, denoted by R_r and R_l, that, respectively, account for the power radiated by the antenna and for the power dissipated in the antenna itself as losses. Thus,

$$Z_a = R_r + R_l + jX_l \qquad (14.4\text{-}17)$$

Resistance R_r is called the *radiation resistance*, R_l is the *loss resistance*, and R_r +

† Circular polarization is useful in reducing undesired echoes from rain and other forms of precipitation in radar. It is also useful in space applications where the receiving-antenna orientation may change relative to that of the transmitter.

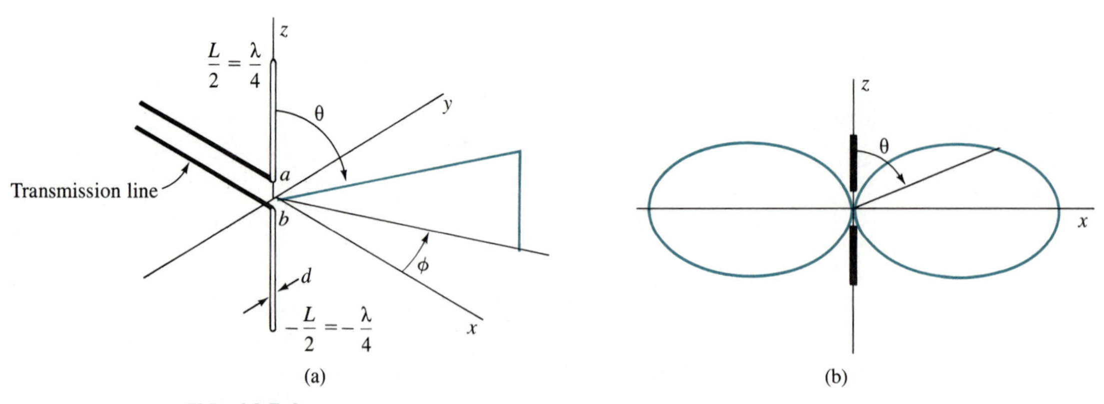

FIG. 14.5-1.
(a) The geometry of a half-wave-dipole antenna and (b) its radiation pattern in a principal plane.

R_l is the *antenna resistance*. Ideally, $R_r + R_l$ should equal the characteristic impedance of the feeding line or guide to prevent reflected power, and the *antenna reactance*, X_l, should be zero.

14.5 PRACTICAL ANTENNAS

The number and variety of antennas are almost endless. Thus, it is impractical to list and examine carefully the whole subject of real antennas, and no effort to do so will be attempted. Rather, we shall try to place a proper perspective on the whole subject and briefly discuss a few examples to illustrate some of the most important antennas.

For our purposes we may subdivide antennas into four types: wire, array, aperture, and lens. We shall briefly discuss three wire antennas, one of the array type, and four of the aperture type. The lens type, although useful in some radar and other applications, is the least important and is omitted.†

Half-Wavelength Dipole

The *half-wavelength-dipole antenna* is sketched in Fig. 14.5-1(a). It consists of a thin wire with a length of one half-wavelength ($\lambda/2$). The wire is broken in the center to allow excitation between terminals a, b by a transmission line. The half-wave dipole is an important practical antenna as a stand-alone antenna or as an

† For some details, see J. D. Kraus, *Antennas*, 2d ed., McGraw-Hill, New York, 1988.

element of a larger, more complicated antenna. The radiation (power) pattern is given by

$$P(\theta, \phi) = \frac{\cos^2[(\pi/2)\cos\theta]}{\sin^2\theta} \tag{14.5-1}$$

It is seen to be omnidirectional (a constant) in the ϕ coordinate for θ fixed, and it has a figure-eight shape in the θ coordinate for ϕ fixed, as shown in Fig. 14.5-1(b). The physical length L of the dipole (both halves), assuming air as the propagation medium, is related to the frequency of excitation by

$$L = \frac{\lambda}{2} = \frac{c}{2f} \tag{14.5-2}$$

where c is the speed of light.

The directive gain and effective area of the dipole are

$$G_D \approx 1.64 \tag{14.5-3}$$

$$A_e \approx 0.13\left(\frac{c}{f}\right)^2 = 0.13\lambda^2 \tag{14.5-4}$$

Other important parameters are the half-power beam width in the θ coordinate, which is $\theta_B = 78.08°$, and the antenna's input impedance at terminals a, b in Fig. 14.5-1(a), which is close to $73 + j42.5\ \Omega$. As a practical matter, the dipole can be driven by a transmission line of 75-Ω characteristic impedance. If the inductive component represents an unacceptable mismatch, the inductive reactance can be nulled by adding a 42.5-Ω capacitive reactance across terminals a, b to resonate the antenna.† An alternative would be to shorten the dipole slightly (a few percent below $\lambda/2$) until the reactance is zero; unfortunately, the radiation resistance also decreases slightly from 73 Ω.

The half-wave dipole is a relatively narrowband antenna. For a wire diameter d of about $\lambda/10^4$ (thin wire) the band of useful frequencies is about 3% of the design frequency found from (14.5-2) for a given length L. When the diameter is increased to about $\lambda/520$,‡ bandwidth increases to about 30%.

The wave radiated from the half-wave dipole is linearly polarized in the plane of the θ coordinate of Fig. 14.5-1(a).

Folded Half-Wave Dipole

The *folded-half-wave-dipole antenna* is a variation of the half-wave dipole that is used in TV, broadcast FM, and other applications. It is formed by adding a second conductor to "fold" the ends of the half-wave dipole together, as shown in Fig.

---

† Rather than use of a fixed capacitor, the reactance $-j42.5\ \Omega$ can be added in practice by a device called a *stub tuner*.

‡ Data from C. A. Balanis, *Antenna Theory, Analysis and Design*, Harper and Row, New York, 1982, p. 333.

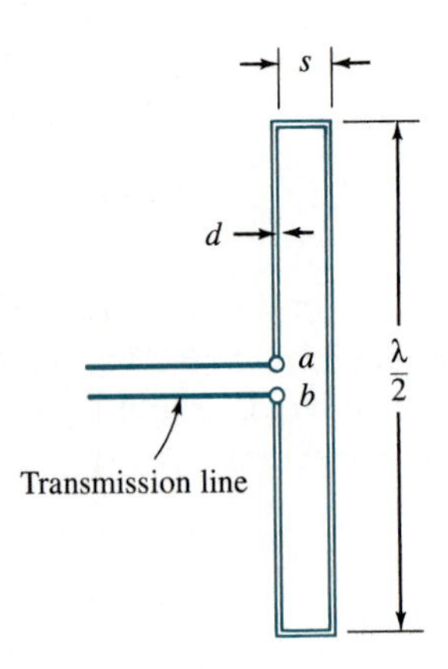

FIG. 14.5-2. A folded-dipole antenna.

14.5-2. The separation s between the long conductors is typically small ($s < \lambda/20$).

The folded dipole has the same radiation pattern, directive gain, and polarization as the half-wave dipole, but its input impedance at terminals a, b in Fig. 14.5-2 is 4 times as large. It also has improved bandwidth characteristics. Because its radiation resistance is $4(73) = 292\ \Omega$, the antenna is well-suited for use with 300-Ω TV cable.

Helical Antenna

Our third, and last, example of a wire antenna is the helical antenna that was discovered and extensively researched by J. D. Kraus. For the many variations of this interesting antenna the reader should see the specialized literature.† We shall discuss the form shown in Fig. 14.5-3. Here a round wire of diameter d is wound into a helix of axial length L. The turns separation S and the helix's (cylindrical) diameter D (circumference C) are both taken between conductor centers. The number of turns is $N = L/S$. The *pitch angle*, denoted by α, of the turns is defined by

$$\alpha = \tan^{-1}\frac{S}{C} = \tan^{-1}\frac{S}{\pi D} \qquad (14.5\text{-}5)$$

The helix attaches to the center conductor of a coaxial transmission line centered on the axis of the helix. (In some variations the line is offset to the periphery of the helix.) Attached to the outer conductor of the transmission line is a conducting sheet (ground plane) having a diameter of at least $\frac{3}{4}$ wavelength for the nominal design frequency of operation.

For proper antenna operation C, S, and N should satisfy

$$0.80 < \frac{C}{\lambda} < 1.15 \qquad (14.5\text{-}6)$$

† See Kraus, op. cit., chap. 7.

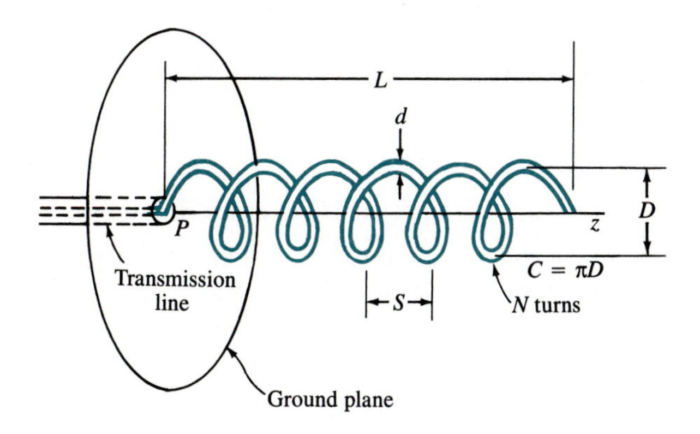

FIG. 14.5-3. The helical-beam antenna.

$$\frac{12\pi}{180} \text{ (or } 12°) < \alpha < \frac{14\pi}{180} \text{ (or } 14°) \tag{14.5-7}$$

$$N \geq 4 \tag{14.5-8}$$

With these conditions true, the radiation pattern has a dominant lobe with its maximum along the $+z$ axis of the helix, as given by

$$P(\theta, \phi) = \left[\sin\left(\frac{\pi}{2N}\right) \cos(\theta) \frac{\sin(N\psi/2)}{\sin(\psi/2)}\right]^2 \tag{14.5-9}$$

where

$$\psi = 2\pi\left[\frac{S}{\lambda}(1 - \cos\theta) + \frac{1}{2N}\right] \tag{14.5-10}$$

and θ is the angle from the $+z$ axis. Some small side lobes are also indicated by (14.5-9). Wave polarization in the $+z$ direction is nearly circular. Note that the pattern is symmetric about the $+z$ axis (called a *pencil-beam pattern*); its half-power beam width and directive gain, respectively, are†

$$\theta_B \approx \frac{52\lambda^{3/2}}{C\sqrt{NS}} \quad \text{degrees} \tag{14.5-11}$$

$$G_D \approx \frac{12NC^2S}{\lambda^3} \tag{14-5-12}$$

The input impedance seen by the transmission line at point P in Fig. 14.5-3 is almost purely resistive, as given by‡

$$Z_a \approx \frac{140C}{\lambda} \tag{14.5-13}$$

---

† Kraus, op. cit., pp. 281–284.

‡ Kraus, op cit., p. 278. For a peripheral feed $Z_a \approx 150\sqrt{\lambda/C}$.

Values range from 100 to 200 Ω in most applications.

We develop an example by assuming the near-optimum values of C and α, which are $C = \lambda$ and $\alpha = 14\pi/180$ (or 14°).

EXAMPLE 14.5-1

Antenna parameters will be found by assuming a 10-turn helix at $f = 500$ MHz, $C = \lambda$, and $\alpha = 14\pi/180$. Here $\lambda = c/f = 3 \times 10^8/5 \times 10^8 = 0.6$ m, so $C = 0.6$ m. From (14.5-5), $S = C \tan \alpha = 0.6 \tan (14\pi/180) \approx 0.15$ m, so $L = NS = 1.5$ m. From (14.5-11) through (14.5-13),

$$\theta_B \approx \frac{52(0.6)^{3/2}}{0.6\sqrt{1.5}} \approx 32.9°$$

$$G_D \approx \frac{12(10)(0.6)^2(0.15)}{(0.6)^3} \approx 30.0 \qquad \text{(or 14.8 dB)}$$

$$Z_a \approx \frac{140(0.6)}{0.6} = 140\ \Omega$$

Yagi-Uda Array

An antenna commonly used for television reception is the Yagi-Uda array of Fig. 14.5-4. It consists of an array of parallel dipoles all laying in the same plane. One, either a half-wave or folded-half-wave type, is the active element. The other conductors are called parasitic elements. One, the reflector, acts to reflect waves back toward the active element with a phase such as to enhance radiation in the $+z$ direction. The other elements, called directors, have lengths and spacings also chosen to enhance radiation in the $+z$ direction. The result is a radiation pattern with a principal lobe in the $+z$ direction. Polarization is that of the active element (linear, due to the dipole).

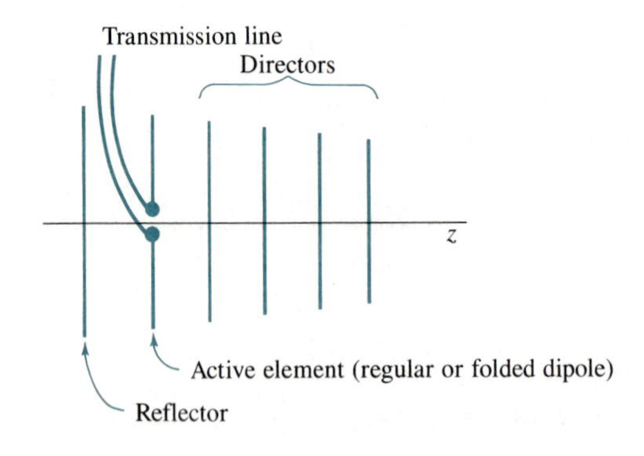

FIG. 14.5-4. A Yagi-Uda array antenna.

Parameters of gain, beam width, pattern, etc., are complicated functions of element lengths, spacings, and conductor diameters. In optimum designs none of these quantities are fixed. The active element is usually slightly shorter than $\lambda/2$ in length. The reflector (there is usually one, but sometimes two are used) is usually slightly longer than the active element and spaced from it by approximately $\lambda/4$. The lengths and spacings of the directors from the adjacent elements are often near 0.45λ and 0.35λ, respectively.

The Yagi-Uda array is usually seen with 3 to 12 elements, but units have been built with up to about 40 elements. Directive gain increases with the number of elements and is often in the range from 10 (10 dB) to 100 (20 dB). Bandwidth is usually small (up to about 10%, with 2% not unusual). Design frequencies from 100 to 1000 MHz are typical.

Pyramidal Horn

The pyramidal horn of Fig. 14.5-5 is formed by flaring the end of a rectangular waveguide. It is an aperture-type antenna because radiated energy emerges from the opening called an aperture. The radiation pattern contains a main lobe with its maximum in the direction of the z axis. Smaller side lobes also occur. If dimensions A and B are increased, for a fixed horn length L, directive gain first increases and then reaches a maximum before decreasing. For relatively long horns, where L is several wavelengths, the maximum directive gain is

$$G_D \approx \frac{2.05\pi AB}{\lambda^2} \tag{14.5-14}$$

This gain occurs when the aperture dimensions are†

$$A \approx \sqrt{3\lambda L} \tag{14.5-15}$$

$$B \approx 0.81A \tag{14.5-16}$$

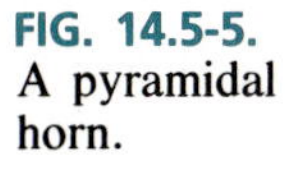

FIG. 14.5-5. A pyramidal horn.

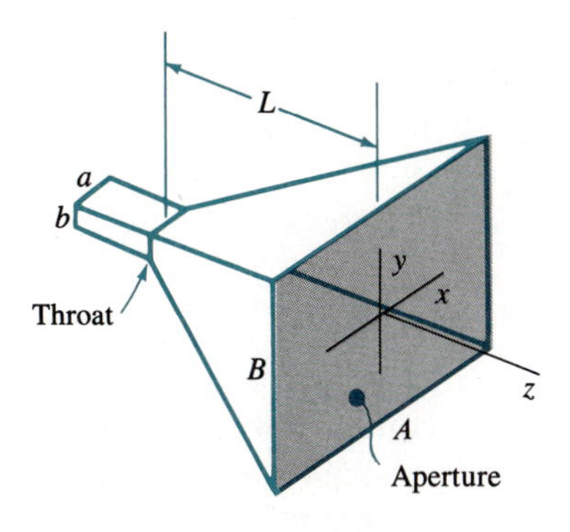

† O. P. Gandhi, *Microwave Engineering and Applications*, Pergamon Press, New York, 1981, p. 140.

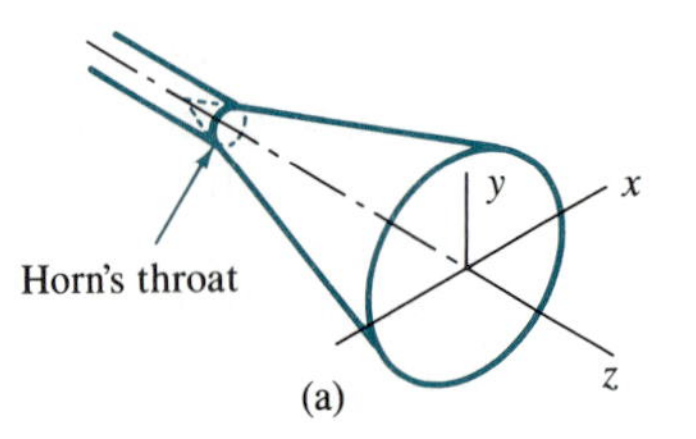

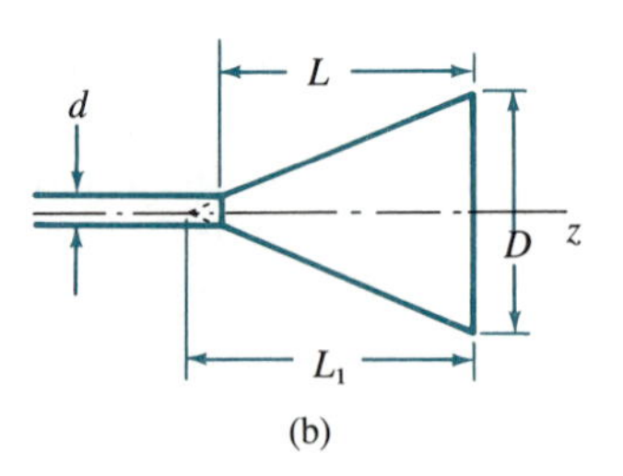

FIG. 14.5-6. (a) A conical horn and (b) its side view. All dimensions are inside dimensions.

The principal-plane beam widths for the optimum (maximum-gain) horn are known to be†

$$\theta_B \approx \frac{54\lambda}{B} \quad \text{(degrees: } yz \text{ plane)} \tag{14.5-17}$$

$$\phi_B \approx \frac{78\lambda}{A} \quad \text{(degrees: } xz \text{ plane)} \tag{14.5-18}$$

Conical Horn

The *conical-horn antenna* of Fig. 14.5-6 results when a cone is added to the end of a round waveguide. The general behavior, as the aperture of inside diameter D increases, is similar to the behavior of the pyramidal horn as its aperture dimensions increase. That is, for some fixed horn length L_1, there is a value of D given by‡

$$D \approx \sqrt{3.33\lambda L_1} \tag{14.5-19}$$

for which directive gain has its maximum value given by

$$G_D \approx 5.13\left(\frac{D}{\lambda}\right)^2 \tag{14.5-20}$$

The associated throat-to-aperture length L required by the optimum gain is given by

$$L \approx L_1\left(1 - \frac{d}{D}\right) \tag{14.5-21}$$

where d is the inside diameter of the waveguide. Beam widths for the resulting main beam, which is directed along the $+z$ axis, are

$$\theta_B \approx \frac{60\lambda}{D} \quad \text{(degrees: } yz \text{ plane)} \tag{14.5-22}$$

$$\phi_B \approx \frac{70\lambda}{D} \quad \text{(degrees: } xz \text{ plane)} \tag{14.5-23}$$

One of the principal uses of both conical-horn and pyramidal-horn antennas is as an illuminator for large-aperture paraboloidal antennas, which are capable of generating very narrow beam-width patterns (less than 1° in some cases).

---

† K. F. Lee, *Principles of Antenna Theory*, Wiley, New York, 1984, pp. 286, 289.

‡ The following five equations are adapted from A. P. King, "The Radiation Characteristics of Conical Horn Antennas," *Proc. I.R.E.*, **38**:249–251 (March 1950).

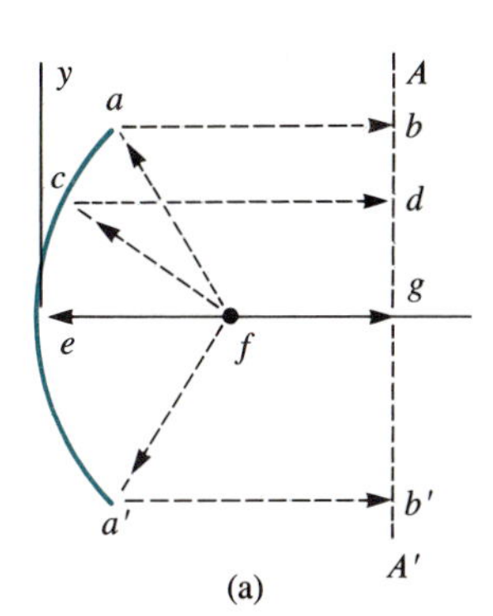

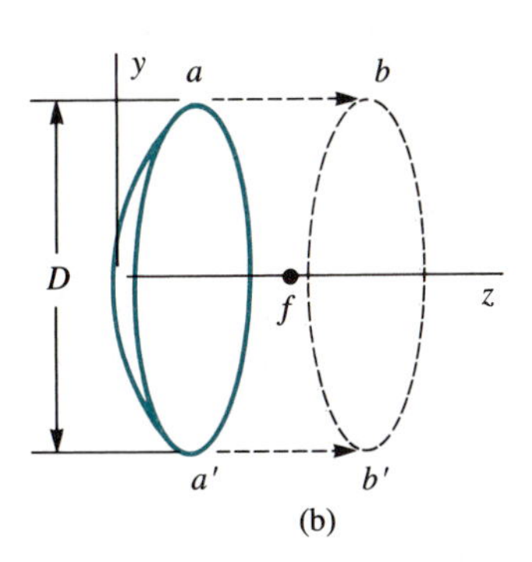

FIG. 14.5-7. (a) A parabola of focal length f. (b) A paraboloid of focal length f; it is the surface of revolution formed by the parabola of (a).

Paraboloidal Antenna

Consider a parabola with focal length f, as shown in Fig. 14.5-7(a). If an isotropic point source is located at f, a property of the parabola is that all waves reflected by the (conducting) parabola arrive at a line AA' with equal phase because the various path lengths, such as f-a-b, f-c-d, f-e-g, and f-a'-b', are equal. The line AA' is parallel to the line aa' defining the opening (aperture) of the finite-length parabola. The path lines shown may be thought of as *rays* representing elemental pieces of the circular wave front emanating from point f.

If the parabola of Fig. 14.5-7(a) is rotated about the z axis, a surface of revolution, called a paraboloid, is formed. It has the property that a spherical wave that emerges from the focus and reflects from the surface forms a plane wave (plane where all parts of the wave are equal in phase) in front of the paraboloid. This plane wave first forms at the aperture [plane aa' in Fig. 14.5-7(b)] but remains planar as it propagates (plane bb'). The action is, therefore, one of a collimator; it accepts broadcast (isotropic) radiation from the source and concentrates (collimates) part of it in one direction.† The radiation pattern of the *paraboloidal antenna*, as it is called, is mainly a dominant lobe in the z direction with smaller side lobes. The source at the focus is called the *feed*. Of course, radiation from the feed that does not intercept the paraboloid is unaffected and represents wasted energy, called *spillover*.

For minimization of spillover, feeds are not normally isotropic. They are typically small antennas themselves, such as pyramidal or conical horns, so that they have directivity and are able to concentrate most of their radiated energy toward the paraboloidal surface (the reflector). The radiation pattern of the feed is called the *primary pattern*. That of the overall antenna, the paraboloid with its feed, is the *secondary pattern*.

For a given focal length and feed primary pattern, antenna gain increases as D increases to reduce spillover and gather in more of the radiation from the feed.

---

† Automobile headlamps work in this fashion. Here the source at the focus is a small lamp, while the paraboloidal reflector concentrates the light to the front of the car. Flashlights behave similarly.

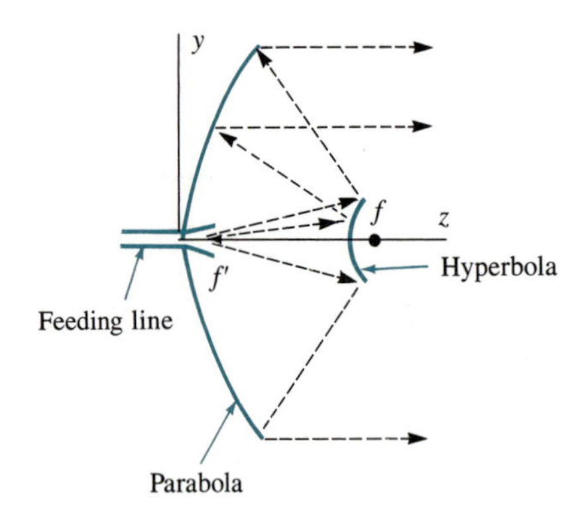

FIG. 14.5-8. A section of a Cassegrain antenna in a plane containing the z axis of symmetry.

However, as D becomes larger, a point is reached where directive gain is maximum. Any further increase in D creates an aperture area that is not effectively radiating additional energy, and aperture efficiency decreases; the result is lower directive gain. The ratio of f/D at which the maximum (optimum) occurs is a complicated function of many parameters. As a rough rule of thumb, the maximum tends to occur where D subtends an angle at the focus that is approximately equal to the -10-dB beam width of the primary pattern. With this condition the first minor lobe in the secondary pattern is about 22 to 25 dB below the main beam's maximum, and f/D ranges from 0.3 to about 0.5 for most paraboloidal antennas.†

Gain, beam width, and side-lobe levels depend heavily on the manner in which the intensity and phase vary over the aperture of the antenna. Most designs strive for equal phase over the aperture and an intensity that is symmetrical about the z axis. For such *illumination functions* the main lobe will be symmetric about the z axis. A constant aperture intensity (uniform illumination) gives the highest gain ($\rho_a = 1.0$), smallest 3-dB beam width ($58.4\lambda/D$ in degrees), and a peak side-lobe level of -17.6 dB from the main lobe's maximum. Real antennas have illuminations that decrease toward the aperture's edge. This behavior produces smaller gain (ρ_a from around 0.4 to about 0.8), larger half-power beam width (up to around $95\lambda/D$), and lower peak side-lobe levels (down to about -35 to -40 dB).

Paraboloidal antennas have found wide use as antennas for radar and communications. Because they are capable of forming very narrow (less than 1°) beam widths, they are especially useful in angle-tracking radars.

Cassegrain Antenna

The Cassegrain antenna, shown in Fig. 14.5-8, is a variation of the paraboloid that gives somewhat improved system performance. Here the feed is moved to the rear

---

† See M. I. Skolnik, *Introduction to Radar Systems*, 2d ed., McGraw-Hill, New York, 1980, pp. 237–239.

of the antenna, and it illuminates a conducting surface called a *subreflector* placed near the focus. If the subreflector is a hyperbola of revolution, or hyperboloid, it can be shown that a plane wave occurs in the aperture, as in the paraboloid. One advantage gained by the use of the subreflector is that it gives another parameter in design; its surface can be slightly modified to give some increase in aperture efficiency. Since the feed is moved to the rear of the antenna, waveguide length, and therefore losses, can be reduced.

14.6 RECEIVED SIGNAL POWER

Figure 14.4-1 can now be used to find the signal power S_r available to the receiver from its antenna when a power P_t/L_t is applied to the transmitting antenna. For a receiver located a distance R from the transmitter an *isotropic* transmitting antenna would cause a radiation power density† (power per unit area) of

$$\text{Power density} = \frac{P_t}{4\pi R^2 L_t} \tag{14.6-1}$$

For a real antenna that has power gain G_t and loss L_{ta} relative to an isotropic antenna,‡ (14.6-1) becomes

$$\text{Power density} = \frac{P_t G_t}{4\pi R^2 L_t L_{ta}} \tag{14.6-2}$$

If L_{ch} denotes any losses incurred by the wave in the channel and A_{re} is the effective area of the receiving antenna, the power that the receiving antenna is able to produce is

$$S = \frac{P_t G_t A_{re}}{4\pi R^2 L_t L_{ta} L_{ch}} \tag{14.6-3}$$

Finally, receiving-path losses L_{ra} (due to the antenna) and L_r are added, and a total system loss L is defined by

$$L = L_t L_{ta} L_{ch} L_{ra} L_r \tag{14.6-4}$$

so

$$S_r = \frac{P_t G_t A_{re}}{4\pi R^2 L} \tag{14.6-5}$$

† Radiation power density is the total radiated power divided by the area of a sphere of radius R for an isotropic antenna.

‡ We assume that the transmitting and receiving antennas point toward each other so that their gains are maximum.

If we use (14.4-10) with G_r representing the receiving antenna's gain, (14.6-5) becomes

$$S_r = \frac{P_t G_t G_r \lambda^2}{(4\pi)^2 R^2 L} \tag{14.6-6}$$

EXAMPLE 14.6-1

We calculate the available received power for two stations separated 48.28 km (30 mi), each having a circular-aperture antenna with a diameter of 2 m. Aperture efficiency is 0.5 at 4000 MHz. The total loss over the link is 7.94 (9 dB), and the transmitter generates 0.5 W. For wavelength, λ equals 300 divided by the frequency in megahertz, or 0.075 m. From (14.4-15), $G = G_t = G_r = 4\pi(0.5)\pi/(0.075)^2 \approx 3509$. From (14.6-5),

$$S_r = \frac{0.5(3509)^2(0.075)^2}{(4\pi)^2(48.28)^2(10^6)(7.94)} \approx 1.18 \times 10^{-8}\ \text{W}$$

All parameters required in (14.6-6) have been defined clearly except the channel loss L_{ch}. This loss is discussed in the following section.

14.7 NEAR-EARTH EFFECTS ON PROPAGATION

The computation of received-signal power in Sec. 14.6 assumes wave propagation directly between transmitting and receiving locations with no nonideal effects other than a loss to account for attenuation of the medium. This condition is approached only by two stations isolated from all other objects by distances large relative to the separation between stations (two satellites in space, for example). For most systems the presence of the earth and the earth's atmosphere cause a significant effect on propagation. We briefly discuss some of these effects in this section.

The atmosphere around the earth is made up of several layers, as illustrated in Fig. 14.7-1. The most significant layers are the troposphere, extending to an altitude of about 20 km, and the ionosphere, extending roughly from 50 to 600 km. A region of approximately free space separates the troposphere from the ionosphere.

The Troposphere

The troposphere contains the earth's air supply and all weather conditions. Air density, temperature, and humidity all decrease as a function of altitude. As a result, the refractive index of the air decreases with altitude, causing the velocity

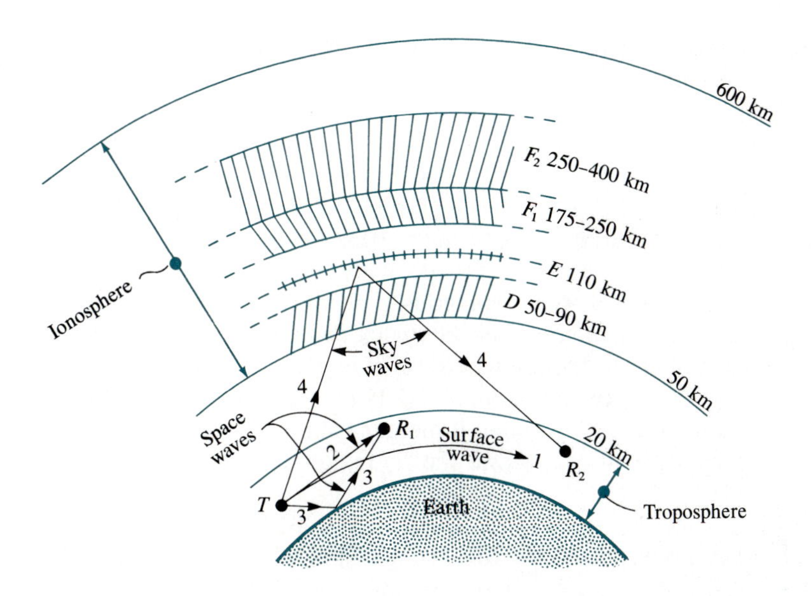

FIG. 14.7-1. The geometry of the earth's troposphere and ionosphere.

of wave propagation to increase. Thus, radio waves tend to bend downward or back toward the earth as they propagate through the troposphere. The amount of ray (wave) bending (called refraction) depends on how much of the troposphere the wave traverses and at what angle. Waves from directive antennas pointed near the horizon will experience a larger effect than if pointed vertically.

The Ionosphere

In the ionosphere, shown in Fig. 14.7-1, the rarified air becomes ionized, primarily due to ultraviolet sunlight. The most significant characteristics of the ionosphere affecting wave propagation are attenuation and reflection. The density of free electrons (ionization) reaches several maximum values versus altitude, giving rise to distinct "layers" or "regions" where waves may be reflected or attenuated. These layers are called D, E, F_1, and F_2 regions.

In general, the lower-frequency waves incident upon the earth side of the ionosphere tend to be reflected by the lower regions. As frequency increases, a wave may pass some lower regions, suffering mainly attenuation, and then be reflected back to earth by upper regions. At still higher frequencies (above about 30 MHz) waves may penetrate the entire ionosphere, with the only effect being attenuation, primarily. Thus, only frequencies above about 30 MHz may be used for space communications. This limit may approach 70 MHz in severe ionospheric activity.†

The D region, usually falling between 50 and 90 km in altitude, will reflect waves below about 300 kHz and attenuate higher-frequency waves. It gives strong

† P. David and J. Voge, *Propagation of Waves*, Pergamon Press, New York, 1969.

attenuation for 300 kHz $< f <$ 3 MHz, especially in the daytime, while only weakening waves for 3 MHz $< f <$ 30 MHz. The *D* region mostly disappears at night.

The *E* region has a relatively stable altitude of about 110 km. It is an important region for reflecting high frequencies, 3 MHz $< f <$ 30 MHz, during daytime and medium frequencies, 300 kHz $< f <$ 3 MHz, at night. Higher frequencies are reflected during daylight hours because ionization density increases during these times, due to the sun. Occasionally (up to about 50% of the time), localized and highly ionized areas analogous to clouds form, which give rise to what is called the *sporadic E* region. The effect of the sporadic *E* region is mainly to raise the maximum frequency where reflection is possible.

The F_1 region, from about 175 to 250 km, is distinct from the F_2 region only during the day. At night it merges with the F_2 region. Waves that penetrate the *E* region usually penetrate the F_1 region as well, with some attenuation being the main effect.

The F_2 region, from about 250 to 400 km or higher, provides the main means of long-distance high-frequency communication, 3 MHz $< f <$ 30 MHz, by wave reflection. Characteristics are more variable than those for the F_1 region and vary with day, season, and sunspot cycles.

Wave Types

Several different types of waves may individually or in combination be responsible for the received field at some point. As already mentioned, waves may be reflected from some region in the ionosphere. These are called *sky waves* and are illustrated by the rays marked 4 in Fig. 14.7-1. Sky waves account for nearly all very long-range communication. In the figure, the receiver R_2 is receiving a single reflection, but it is also possible to receive waves due to several bounces or "hops," with an earth reflection completing the hop cycle. Both single- and multiple-hop links may allow communication at distances beyond the transmitter's geometric horizon. Stations beyond the geometric horizon are said to be in the *shadow* or *diffraction zone*.

The sky wave applies only to frequencies below about 30 MHz. Above 30 MHz the ionosphere is almost always penetrated completely and no reflections occur. The sky wave is most important at the longest distances but may still be significant at intermediate distances. At closer distances its strength is usually exceeded by the ground wave (see below).

Space waves are waves made up of two components. They are shown as rays 2 and 3 in Fig. 14.7-1. One, ray 2, is called the *direct ray*; the other, ray 3, is the *indirect ray*. Thus, a space wave always involves one component reflected from the earth and applies to receivers above the horizon. Because rays 2 and 3 may, through interference, cause the space-wave field strength to alternately have maxima and minima with distance, the region above the geometric horizon is called the *interference zone*. In systems that use directive antennas, the effect of

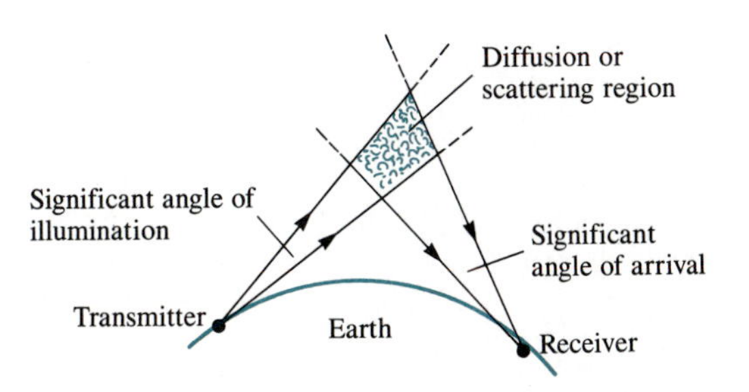

FIG. 14.7-2. The geometry relating to a tropospheric-scatter communication link.

the indirect ray is small if its path is defined by the side lobes of the antennas while the direct ray is through the main lobes.

A third wave type, the *surface wave*, hugs the earth's surface and bends with the earth's curvature by diffraction† as it progresses. Surface waves are most important at lower frequencies using vertical polarization and are the carriers of broadcast AM signals and noise signals caused by lightning. Distance for reliable communication at a given frequency is a complicated function of station heights, type of terrain, power, antenna type, etc.

At frequencies from about 100 to over 8000 MHz turbulence in the troposphere can cause irregularities in refractive index that can scatter energy several hundreds of kilometers over the geometric horizon. Systems based on these irregularities are called *tropospheric-scatter systems*, and the scattering can be considered due to a type of wave (*tropospheric wave*). A received signal power can be achieved that is thousands or even millions of times larger than with other methods. Tropospheric waves and sky waves are similar in that they are due to reflections (scattering). However, sky-wave reflections are mostly *specular*; that is, the wave arrives at the receiver from a single direction, as though it were reflected from a single point in the ionosphere. Tropospheric waves tend to be *diffuse*; that is, the received field is the result of waves arriving from many angles due to scattering from many points of turbulence, as seen in Fig. 14.7-2. The apparent gain of the antennas to be used is an average gain over the significant angles of illumination and arrival and are not necessarily the maximum gains.

Finally, we define a *ground wave*. It is not a distinct type of wave, such as a sky, space, surface, or tropospheric wave, but rather is a combination of the space and surface waves.

Effects of Clear Atmosphere

Waves traveling in a clear atmosphere undergo attenuation caused mainly by oxygen and water vapor absorption. The attenuation is very frequency-dependent, as shown in Fig. 14.7-3 for an atmosphere at 76 cm pressure and 7.5 g/m^3 of water content. For long paths that may span a considerable altitude, the effect of decreasing attenuation with altitude must be considered. Figure 14.7-4 gives the

† Diffraction is bending caused by the presence of a physical body (earth, buildings, etc.).

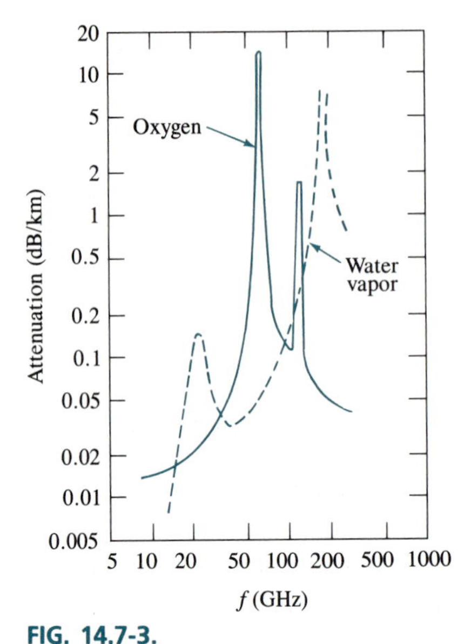

FIG. 14.7-3.
The theoretical attenuation due to oxygen at 76-cm pressure and water vapor at 7.5 g/m³ content. (Adapted from A. W. Straiton and C. W. Tolbert, "Anomalies in the Absorption of Radio Waves by Atmospheric Gases," © *Proceedings of the Institute of Radio Engineers*, vol. 48, May 1960, pp. 898–903, with permission.)

total attenuation for traversing the entire troposphere at different elevation angles from earth's horizontal.

A receiving antenna in the troposphere will produce noise. Parts of its output noise power are due to random waves from outer space (*cosmic noise*), from the atmosphere itself, from atmospheric disturbances such as thunderstorms, which produce lightning discharges, from human sources such as fluorescent lights and automobile ignitions, and from thermally induced radiations from objects like the earth, sun, and moon. The total noise available from an antenna is called *antenna noise*; it is discussed in the next section.

Effects of Unclear Atmosphere

Wave attenuation increases in an unclear atmosphere where rain, clouds, fog, snow, sleet, or hail may be present. Rain attenuation increases with increasing frequency and rainfall rate, as depicted in Fig. 14.7-5.

Attenuation due to clouds and fog at a temperature of 291 K and frequencies from 3 to 60 GHz can be found from equations given in the literature.† After some

† See D. E. Kerr, ed., *Propagation of Short Radio Waves*, vol. 13, M.I.T. Radiation Laboratory Series, Boston Technical Publishers, Lexington, Mass., 1964.

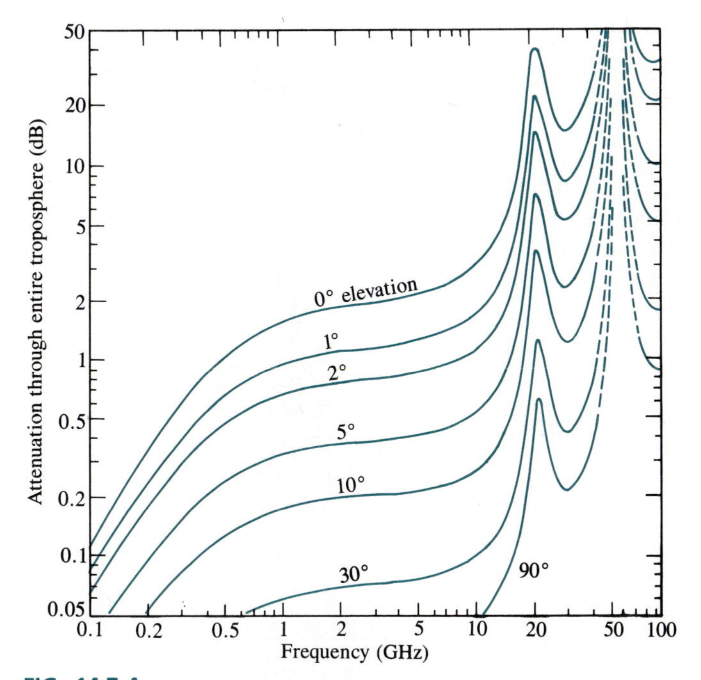

FIG. 14.7-4.
The total attenuation through the entire troposphere for various frequencies and elevation angles. (Adapted from L. V. Blake, "Tropospheric Absorption and Noise Temperature for a Standard Atmosphere," *Digest of the Institute of Electrical and Electronics Engineers, Professional Technical Group on Antennas and Propagation, 1963 International Symposium*, Boulder, Colorado, July 9–11, 1963. © I.E.E.E., 1963, with permission.)

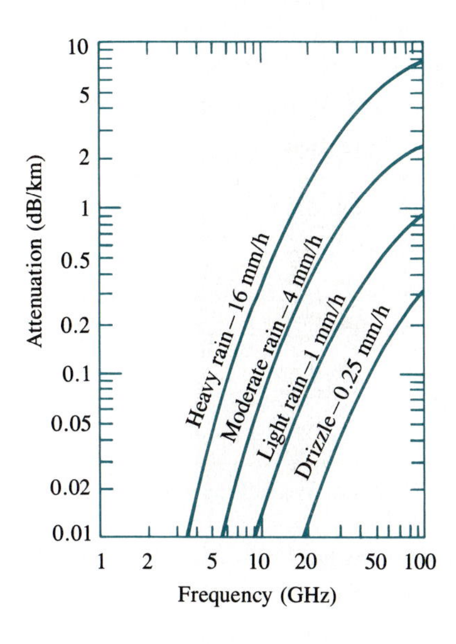

FIG. 14.7-5.
The attenuation due to rain at 291 K. (Adpated from D. E. Kerr, *Propagation of Short Radio Waves*, vol. 13, M. I. T. Radiation Laboratory Series, Boston Technical Publishers, Lexington, Mass. 1964.)

minor notation change, we have

$$\text{Attenuation (dB/km)} \approx \frac{0.148 f_{\text{GHz}}^2}{V_m^{1.43}} \tag{14.7-1}$$

where f_{GHz} is frequency in gigahertz and V_m is called *optical visibility* (meters). The International Visibility Code defines $V_m < 50$ m for dense fog, 50 m $\leq V_m <$ 200 m for thick fog, and 200 m $\leq V_m <$ 500 m for moderate fog. Only thick or dense fog produces significant attenuation rates for frequencies below 15 GHz.

Attenuation due to snow† at a temperature of 273 K (0°C) can be expressed as

$$\text{Attenuation (dB/km)} = (7.47 \times 10^{-5}) f_{\text{GHz}} I [1 + (5.77 \times 10^{-5}) f_{\text{GHz}}^3 I^{0.6}] \tag{14.7-2}$$

where I is snowfall intensity in millimeters of melted-water content per hour. For most values of f_{GHz} and I the second term is negligible. In fact, for frequencies below 15 GHz only moderate (4 mm/h) or heavier snowfall produces significant attenuation in most cases.

EXAMPLE 14.7-1

We find the attenuation of a wave traveling a distance of 8.05 km (5 mi) over a horizontal path when $f_{\text{GHz}} = 20$. From Fig. 14.7-3 the clear-atmosphere attenuation is $(0.13 + 0.02)(8.05) \approx 1.21$ dB. If the path were filled with heavy rain, the attenuation would be about $1.7(8.05) \approx 13.69$ dB of additional loss, from Fig. 14.7-5. If heavy snow was falling instead of rain,

$$\begin{aligned}\text{Attenuation} &\approx (7.47 \times 10^{-5})(20)(16)[1 + (5.77 \times 10^{-5})(8 \times 10^3)(16^{0.6})](8.05) \\ &\approx 0.66 \text{ dB}\end{aligned}$$

from (14.7-2). Finally, from (14.7-1) for dense fog, attenuation $\approx 0.148(20^2)(8.05)/50^{1.43} \approx 1.77$ dB. Clearly, rain contributes far more attenuation than snow or fog on this path.

Attenuation due to sleet and hail is difficult to evaluate but generally is said to be less than that of rain with an equivalent rate of precipitation.‡ It will not be discussed further here.

† K. L. S. Gunn and T. W. R. East, "The Microwave Properties of Precipitation Particles," *Q. J. R. Meteorolog. Soc.* **80**:522–545 (October 1954).

‡ M. I. Skolnik, *Introduction to Radar Systems*, 2d ed., McGraw-Hill, New York, 1980, p. 503.

14.8 NOISE IN RECEIVING SYSTEMS

Noise is a voltage having an amplitude that randomly fluctuates with time. In most receiving systems it is an undesirable quantity. Two broad categories of noise exist. One is noise that originates external to the system; the antenna generates this noise in response to random waves that arrive from cosmic sources, atmospheric effects, etc. The second category is internally generated noise; it is noise generated within all the circuits making up the receiver, including transmission lines, amplifiers, etc.

It is common to model internal noise as having been generated by an external source, the one representing the external noise. We develop this model in this section. Thus, since the receiving antenna's output is the "source" of external noise to the receiver, our model will show how the antenna can be imagined to generate extra noise of the correct amount to exactly account for the total noise emerging from the receiver's output.

Available Power Gain

Let dP_{as} be the incremental power (signal or noise) available from some source† in a small frequency interval df, and let any (noise-free) linear two-port network be driven at its input by the source. If dP_{ao} denotes the incremental power available at the network's output, with the input source connected, then *available power gain* of the network is defined by

$$G = \frac{dP_{ao}}{dP_{as}} \tag{14.8-1}$$

It is easy to show by repeated application of this definition that a cascade of N networks, for which the available power gain of network n is denoted by G_{an}, is

$$G_a = G_{a1}G_{a2} \cdots G_{aN} = \prod_{n=1}^{N} G_{an} \tag{14.8-2}$$

Resistive Noise Source

Any resistor generates a noise voltage across its terminals because of thermal agitation of electrons in the material. The noisy resistor of resistance R can be modeled by a Thévenin equivalent voltage source of noise in series with a *noise-free* resistance R. It can be shown that the available incremental noise power dN_a

† The actual power delivered to a load by the source depends on the load's impedance for a fixed source.

that can be extracted from the equivalent source in a small frequency band df (in hertz) is

$$dN_a = kT\,df \tag{14.8-3}$$

where $k = 1.38 \times 10^{-23}$ J/K is Boltzmann's constant and T is the resistor's physical temperature (in kelvin). Note that dN_a is independent of the load impedance and where in frequency the band df is located.

The behavior of a simple resistor as a source of (*thermal*) noise is the basis on which arbitrary noise sources are defined.

Arbitrary Noise Source

Let an arbitrary source of noise be capable of providing an incremental noise power dN_{as} in a small bandwidth df. We *assign* a temperature T_s to the source, called the *effective noise temperature*, so that when it is used in an equation of the form of (14.8-3), the correct available noise power results. Thus,

$$dN_{as} = kT_s\,df \tag{14.8-4}$$

so

$$T_s = \frac{dN_{as}}{k\,df} \tag{14.8-5}$$

It is important to realize that T_s is not necessarily the physical temperature T of a source. If the source is some passive, linear network with all components at physical temperature T, then $T_s = T$. For other sources $T_s \neq T$ (an antenna, for example). Temperature T_s can also depend on frequency, because dN_{as} may be frequency-dependent.

Effective Input Noise Temperature

Figure 14.8-1(a) depicts a real linear network that generates some internal noise that shows up at the output. The network is driven at its input by a source of noise temperature T_s and has an available power gain G_a. The output-available noise power due to the network, called *excess noise*, is imagined to have come from the driving source by assigning an increase in its temperature to account for the excess noise. The increase is called the *effective input noise temperature* of the network. The network is now considered noise-free, as shown in Fig. 14.8-1(b). Thus,

$$dN_{ao} = kT_s\,dfG_a + kT_e\,dfG_a \tag{14.8-6}$$

It is often convenient to represent the available noise powers as shown in Fig. 14.8-1(c) when cascades of networks are involved.

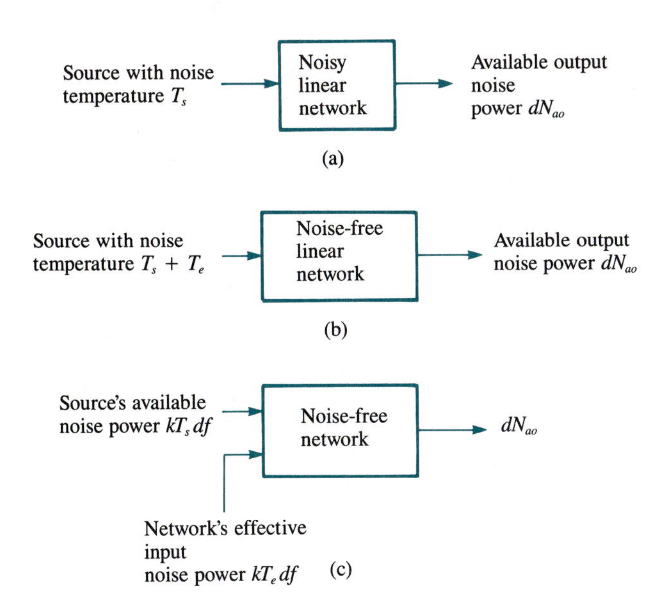

FIG. 14.8-1. (a) A noisy network driven by a noise source. (b) The network with its excess noise modeled as generated by the source. (c) An equivalent model.

Impedance-Matched Attenuator

As an example of a noisy real network, consider an impedance-matched attenuator. A waveguide or transmission line can be modeled as an attenuator when its losses are small. Let the attenuator have a physical temperature T_L and be driven by a *resistive* source with physical temperature T_L. Hence $T_s = T_L$. Since the attenuator's output is just another resistive source, $dN_{ao} = kT_L\,df$. For a loss $L_r = 1/G_a$, (14.8-6) gives

$$T_e = T_L(L_r - 1) \tag{14.8-7}$$

Now since T_e is the *increase* in the source's temperature to account for the attenuator's own noise, it does not depend on the source itself. Therefore, (14.8-7) can be used with sources for which $T_s \neq T_L$ if they are impedance-matched.

Real Networks—Noise Bandwidth

The preceding discussions allow the final developments necessary to define the noise performance of the typical receiving system shown in Fig. 14.8-2(a). The antenna is a source of noise with effective noise temperature T_a, called the *antenna temperature*. A loss between points A and B represents whatever receiving path components are present prior to the receiver's amplifier (these may be transmission lines, waveguides, rotary joints, isolators, filters, or even a mixer). The power loss is the reciprocal of gain, so $L_r \geq 1$. The loss is assumed to have a physical temperature T_L. The receiver is assumed to operate at a nominal (center) frequency f_0 and have available power gain $G_a(f)$. Its effective input noise

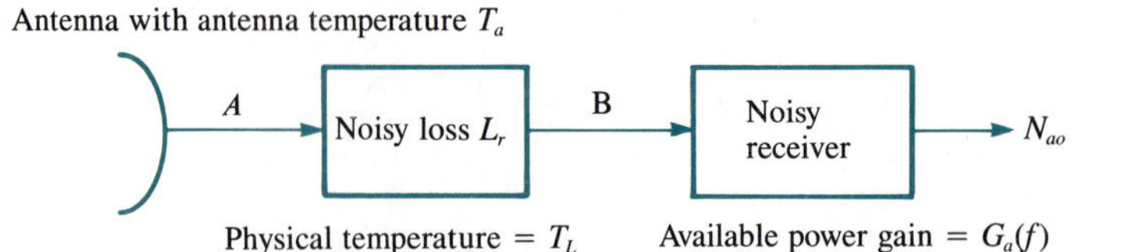

(a)

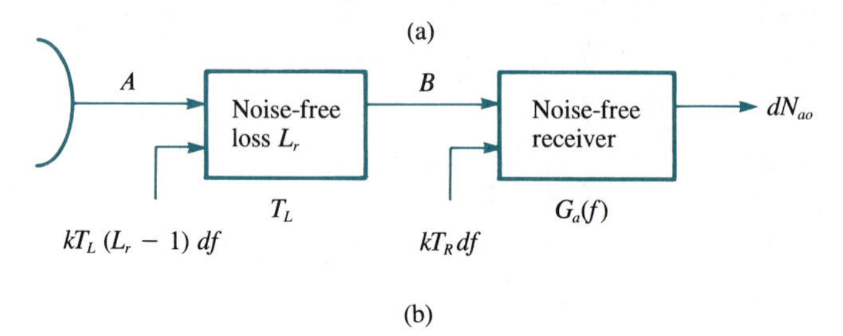

(b)

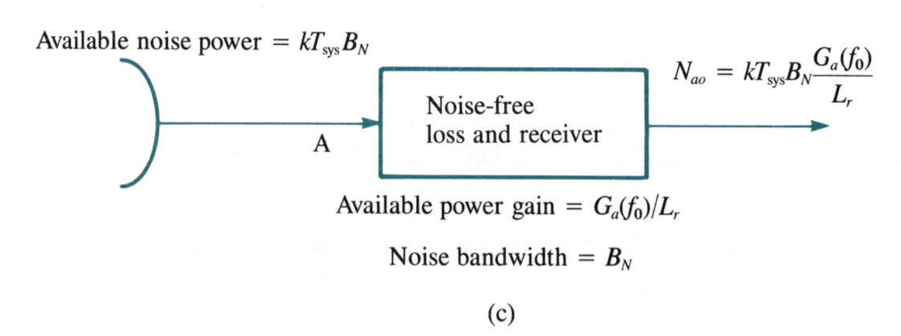

(c)

FIG. 14.8-2. (a) A receiving system and (b) and (c) two equivalent noise-free models.

temperature is $T_R(f)$. With these definitions the model of Fig. 14.8-2(b) applies. In a small frequency band df, dN_{ao} is

$$dN_{ao} = kT_a\,df\frac{G_a(f)}{L_r} + kT_L(L_r - 1)\,df\frac{G_a(f)}{L_r} + kT_R(f)\,dfG_a(f) \qquad (14.8\text{-}8)$$

The total output-available noise power N_{ao} is obtained by integrating (14.8-8). In most systems T_a, T_L, and L_r are nearly constant over the frequency range of interest in the receiver, so

$$N_{ao} = \int_0^\infty dN_{ao} = \frac{kT_a}{L_r}\int_0^\infty G_a(f)\,df + \frac{kT_L(L_r - 1)}{L_r}\int_0^\infty G_a(f)\,df + k\int_0^\infty T_R(f)G_a(f)\,df \qquad (14.8\text{-}9)$$

Now suppose that the actual receiver is replaced by an idealized one with the same nominal power gain $G_a(f_0)$, a rectangular passband of width B_N (in hertz) centered on f_0, and a *constant* effective input noise temperature $\overline{T}_R$. If the idealized system's output-available noise power must be the same as the real system is, then

$$N_{ao} = \frac{kT_aB_NG_a(f_0)}{L_r} + \frac{kT_L(L_r - 1)B_NG_a(f_0)}{L_r} + k\overline{T}_RB_NG_a(f_0) \qquad (14.8\text{-}10)$$

from (14.8-8). By equating (14.8-9) and (14.8-10) term by term, we find the two are equal when

$$\int_0^\infty G_a(f)\,df = B_N G_a(f_0) \tag{14.8-11}$$

$$\int_0^\infty T_R(f)G_a(f)\,df = \overline{T}_R B_N G_a(f_0) \tag{14.8-12}$$

or when

$$B_N = \frac{\int_0^\infty G_a(f)\,df}{G_a(f_0)} \tag{14.8-13}$$

$$\overline{T}_R = \frac{\int_0^\infty T_R(f)G_a(f)\,df}{\int_0^\infty G_a(f)\,df} \tag{14.8-14}$$

The term B_N is called the *noise bandwidth* of the receiver; $\overline{T}_R$ is called the *average input effective noise temperature.*

Data sheets on commercially available amplifiers often specify noise bandwidth. Temperature $\overline{T}_R$ may not be given but can be found from the *average standard noise figure*, denoted here by F_0, which is usually given. Noise figure F_0 is measured by assuming a resistive source at a *standard temperature* T_0 of 290 K; it is related to $\overline{T}_R$ by

$$\overline{T}_R = T_0(F_0 - 1) = 290(F_0 - 1) \tag{14.8-15}$$

Both F_0 and $\overline{T}_R$ are measures of the noisiness of an amplifier. In an ideal, noise-free unit, $F_0 = 1$ and $\overline{T}_R = 0$.

Finally, we may imagine that all the available output noise power is coming from the antenna, as sketched in Fig. 14.8-2(c). The equivalent noise temperature of the antenna, denoted T_{sys}, is now called the *system noise temperature*. Here

$$N_{ao} = \frac{kT_{sys}B_N G_a(f_0)}{L_r} \tag{14.8-16}$$

On equating (14.8-16) and (14.8-10), we find that

$$T_{sys} = T_a + T_L(L_r - 1) + \overline{T}_R L_r \tag{14-8-17}$$

The available system noise power at point A becomes

$$N_{aA} = kT_{sys}B_N \tag{14.8-18}$$

EXAMPLE 14.8-1

An engineer purchases an amplifier for which $F_0 = 3$ (4.77 dB), $f_0 = 4$ GHz, and $B_N = 14$ MHz. It is to be used with an antenna for which $T_a = 200$ K. The connecting-path loss is 1.45 (1.61 dB) when at physical temperature 250 K. We find the system noise power the engineer will have at the antenna's output. From (14.8-15) $\overline{T}_R = 290(3 - 1) = 580$ K. From (14.8-17)

$$T_{\text{sys}} = 200 + 250(1.45 - 1) + 580(1.45) = 1153.5 \text{ K}$$

From (14.8-18)

$$N_{aA} = (1.38 \times 10^{-23})(1153.5)(14 \times 10^6) \approx 2.23 \times 10^{-13} \text{ W}$$

14.9 SYSTEM PERFORMANCE WITH NOISE

An excellent measure of performance of many communication systems is the ratio of signal power to noise power at the system's output. This ratio is equal to the available signal power at point A in Fig. 14.8-2(a), given by

$$S_A = \frac{P_t G_t G_r \lambda^2}{(4\pi)^2 R^2 L_t L_{ta} L_{ch} L_{ra}} \tag{14.9-1}$$

from (14.6-6), divided by the available noise power given by (14.8-18). Hence,

$$\left(\frac{S}{N}\right)_A = \frac{P_t G_t G_r \lambda^2}{(4\pi)^2 R^2 L_t L_{ta} L_{ch} L_{ra} k T_{\text{sys}} B_N} \tag{14.9-2}$$

EXAMPLE 14.9-1

If the antennas of Example 14.6-1 are used with the receiver of Example 14.8-1, we calculate the signal-to-noise ratio of (14.9-2). Since $S_A = L_r S_r = 1.45(1.18 \times 10^{-8})$ W and $N_{aA} = 2.23 \times 10^{-13}$ W, we have

$$\left(\frac{S}{N}\right)_A = \frac{S_A}{N_{aA}} = \frac{1.45(1.18 \times 10^{-8})}{2.23 \times 10^{-13}} \approx 7.67 \times 10^4$$

or 48.8 dB.

PROBLEMS

A source for which $Z_s = 50\ \Omega$ launches a wave carrying 100 W on a 50-Ω transmission line that is lossless. A load of 50 Ω $\pm$ 10% terminates the line. (a) What is Γ_L for this load? (b) What is the largest power that can occur in the wave reflected from the load?

14-2. For the source and line of Problem 14-1, what load resistances will reflect a power of 10 W or less?

14-3. A source of impedance $Z_s = 100\ \Omega$ has an open-circuit voltage $v_s(t) = 12.5 \cos \omega_0 t$ and drives a 75-Ω transmission line terminated with a 75-Ω load. Find the current and voltage at the input terminals of the line.

★14-4. A transmission line of length z is terminated by a short circuit and driven by a matched source. Find an expression for the impedance of the line at terminals a, b in Fig. 14.1-3.

★14-5. Generalize Problem 14-4 by assuming any load impedance Z_L.

14-6. What ratio b/a is required if a coaxial cable using a polyethylene dielectric is to have $Z_0 = 50\ \Omega$?

14-7. A rigid 50-Ω coaxial transmission line has an air dielectric. If the inside radius to the outer conductor is 10 mm, what is the cutoff frequency f_c?

14-8. If the line in Problem 14-7 is copper, what maximum length can be used if losses are not to exceed 3 dB when $f = 3$ GHz?

14-9. A microstrip transmission line is to be designed with $w/h = 5$ and must use a dielectric with $\epsilon_r = 8.8$. What is the largest value of h that will allow operation at frequencies up to 20 GHz?

14-10. Compute Z_0 by (14.2-9) and (14.2-10) and compare the two when $w/h = 3.3$. Assume $\epsilon_r = 8.8$.

14-11. A microstrip transmission line is used at 3 GHz. Its parameters are $\epsilon_r = 2.1$ (Teflon), $w = 2$ mm, $h = 0.5$ mm, and $t = 0.2$ mm. What are the characteristic impedance and attenuation of the line if the conductors are copper?

14-12. Find the maximum usable frequency for the line of Problem 14-11.

14-13. Inside dimensions of an RG-290/U aluminum (where $\rho = 2.83 \times 10^{-8}$ Ω·m) waveguide are $a = 58.42$ cm and $b = 29.21$ cm. Find both the theoretical and practical frequency ranges of operation for this guide.

14-14. Work Problem 14-13 for an RG-139/U waveguide with $a = 0.8636$ mm and $b = 0.4318$ mm.

14-15. What diameter circular waveguide will have a lower cutoff frequency of 10 GHz? What is its largest usable frequency?

14-16. RG-52/U is a brass waveguide ($\rho = 3.9 \times 10^{-8}$ Ω·m) for which $a = 22.86$ mm and $b = 10.16$ mm. Determine Z_0 at the edges of the practical operating-frequency range of the guide.

14-17. Find the attenuations instead of Z_0 in Problem 14-16.

14-18. Find the attenuation of the waveguide in Problem 14-13 at frequencies of $1.25f_c$ and $1.90f_c$.

14-19. An antenna has an aperture area of 10 m², an aperture efficiency of 0.55, negligible losses, and is used at 5 GHz. (a) What is its power gain? (b)

What maximum power density can the antenna generate at a distance of 16 km if its input power is 1.5 kW?

14-20. The power gain of an antenna is 12,000. If its input power is 800 W, what maximum radiation intensity can it generate?

14-21. Determine the beam width of a half-wave-dipole antenna between the −10-dB points of its radiation pattern.

14-22. Find the effective area of a half-wave dipole at 2.5 GHz.

14-23. A helical antenna at 500 MHz is to have a 30° half-power beam width when optimally designed. (a) How long is the antenna? (b) How many turns does the helix have? (c) What is its directive gain?

14-24. A helical antenna has $C/\lambda = 1.0$, $L = 1.6$ m, and $N = 25$ turns. It is used at 1000 MHz. (a) Is this antenna optimum? (b) What is its beam width? (c) What is its directive gain?

14-25. A pyramidal horn has dimensions $A = 6\lambda$ and $B = 4.86\lambda$ at 6 GHz. Find (a) the directive gain and (b) the principal-plane beam widths. (c) Is this horn optimum?

14-26. A circular waveguide with a 2.5-cm inside diameter is expanded by adding a conical flare to produce an aperture with inside diameter of 5.771λ at 8 GHz. The cone's apex-to-aperture length is $L_1 = 10\lambda$. Find (a) the horn's length L, (b) directive gain, and (c) the principal-plane beam widths. (d) What is the total flare angle of the horn?

14-27. A paraboloidal antenna has an aperture efficiency $\rho_a = 0.5$ and diameter 100λ at 6 GHz. Illumination by the feed is such that beam widths of the principal-plane secondary patterns are equal. (a) What is the antenna's power gain? (b) What is the beam width?

14-28. Two stations communicate with each other from locations on mountain-tops 40 km apart. Each uses a paraboloidal antenna with 1° beam width (pencil beam), aperture efficiency $\rho_a = 0.8$, and radiation efficiency $\rho_r = 0.83$. If transmitting, a station has $P_t = 50$ W, and $L_t = 1.9$ at 8 GHz. When a station is receiving, $L_r = 1.5$. The link is designed for a channel loss of 2.51. (a) What is the diameter of the antennas? (b) What signal power is available at the receiver's input?

14-29. A short communications link of 10 km operates at 100 GHz. What clear-air loss L_{ch} is expected to occur?

14-30. A satellite has an elevation angle of 5° from horizontal as measured at a receiving site with a narrow-beam antenna at 12 GHz. What loss is encountered as the wave traverses the channel?

14-31. If moderate rain is falling over the entire communications link of Problem 14-29, what attenuation does it cause?

14-32. A thunderstorm of 5-km extent drops heavy rain as it passes between the stations of Problem 14-28. What attenuation does it cause in the received wave?

14-33. What maximum noise power can be extracted from a resistance R at temperature 290 K? Assume a 1-kHz frequency band.

14-34. An antenna has an available noise power of 1.66×10^{-15} W in a 1-MHz bandwidth. What is its antenna temperature?

14-35. A long piece of waveguide connects an antenna to a receiver. Its loss is 3.4 dB at 12 GHz when its physical temperature is 280 K. What is its effective input noise temperature?

14-36. An antenna has an effective noise temperature of 135 K. It couples through a waveguide that has a loss of 0.8 dB to a receiver. What effective noise temperature does the waveguide's output present to the receiver if the waveguide's physical temperature is 285 K?

14-37. The antenna and waveguide of Problem 14-36 feed a receiver for which $B_N = 10$ MHz, $G_a(f_0) = 10^{12}$, and $\overline{T}_R = 320$ K. (a) What is the system's noise temperature? (b) Find N_{ao} at the receiver's output.

14-38. Two stations in a communications link are separated by 30 km and use identical paraboloidal antennas having diameters of 50λ and aperture efficiencies of 0.6 at 35 GHz. Antenna losses are negligible. Antenna connection-path losses are $L_t = L_r = 1.33$ at physical temperature 285 K. Antenna temperature is 85 K. The receiver at either station has $G_a(f_0) = 10^7$, $\overline{T}_R = 250$ K, and $B_N = 12$ MHz. (a) What transmitter-output power is required to guarantee a system signal-to-noise power ratio of 45 dB when heavy rain falls over a distance of 6 km? (b) What available noise power occurs at the receiver's output? (c) What output-signal power occurs with and without rain when the transmitter power found in (a) is used?

CHAPTER 15

Analog Communications Systems

15.0 INTRODUCTION

In any communications system the way the transmitted waveform is structured has a large influence on the system's performance. Clearly, it must be some function of the message to be conveyed; otherwise, the receiver would have no way to decipher the message. Usually, the transmitted waveform is the result of varying either the amplitude, phase, or frequency of a basic signal called a *carrier*. Combinations of amplitude, phase, and frequency variations are also possible.

The carrier has the form $A \cos(\omega_0 t + \theta_0)$, where A, θ_0, and $\omega_0/2\pi$ are the carrier's amplitude, phase, and frequency, respectively. *Amplitude modulation* (AM) occurs when A is varied as a linear function of the message. In *phase modulation* (PM) a phase term that is a linear function of the message is added to the carrier. If the added phase is a linear function of the integral of the message, the result is called *frequency modulation* (FM). Since instantaneous angular frequency is the time derivative of instantaneous phase, the carrier's frequency in FM is a linear function of the message. Because FM and PM are closely related, we only discuss FM and AM in this chapter.

Every system will have a *modulator* at the transmitting station to structure the transmitted waveform. Correspondingly, there will be a *demodulator* in each receiver to recover the message from the received signal. Occasionally, we refer to the demodulator as a *detector*.

An *analog message* has one of a continuum of possible amplitudes at any given time. A *digital message* is formed when an analog signal is sampled, making it discrete in time, and quantized to make its sample amplitudes have discrete values. When the message to be sent over a communications system is analog, we refer to the system as analog. When the message is digital, the system is called digital. Digital systems are developed in the following chapter. In this chapter we discuss only analog systems.

15.1 STANDARD AMPLITUDE MODULATION (AM) SYSTEM

There are three basic broadcast systems in operation in the United States, two are radio and one is television. One of the radio systems uses a form of AM that we call *standard AM*. Each radio station is assigned a particular carrier frequency that is one of the values from 540 to 1600 kHz in 10-kHz increments. The *Federal Communications Commission* (FCC) regulates the assignments so as to minimize interference between nearby stations. Each station can occupy a channel bandwidth of only 10 kHz centered on its carrier. A typical receiver's frequency response is from 100 Hz to 5 kHz (within ±2 dB referred to 1 kHz). Distortion in the recovered message is often as low as 1% (maximum is 5% but can increase to 7.5% for large-amplitude messages). These characteristics are adequate for low-cost mass communication and general audio entertainment, but they clearly do not represent high-fidelity behavior.

Waveform and Spectrum

The standard AM signal has the form

$$s_{AM}(t) = [A_0 + f(t)] \cos (\omega_0 t + \theta_0) \tag{15.1-1}$$

where A_0, $\omega_0/2\pi$, and θ_0 are the unmodulated amplitude, frequency, and phase angle, respectively, of the carrier and $f(t)$ represents the message. The total amplitude of the carrier, which is $[A_0 + f(t)]$, is a linear function of the message. Figure 15.1-1 sketches the behavior of $s_{AM}(t)$ for a possible message.

If the maximum amplitude of $f(t)$ in (15.1-1) exceeds A_0, *overmodulation* is said to occur. Overmodulation is prevented in standard AM because it creates excessive distortion in the receiver. To prevent overmodulation, we require the maximum magnitude of $f(t)$ to satisfy†

$$|f(t)|_{max} \leq A_0 \tag{15.1-2}$$

† The actual FCC regulation only requires the negative peak of $f(t)$ to not exceed A_0, but (15.1-2) is more convenient for mathematical analysis and will be assumed here.

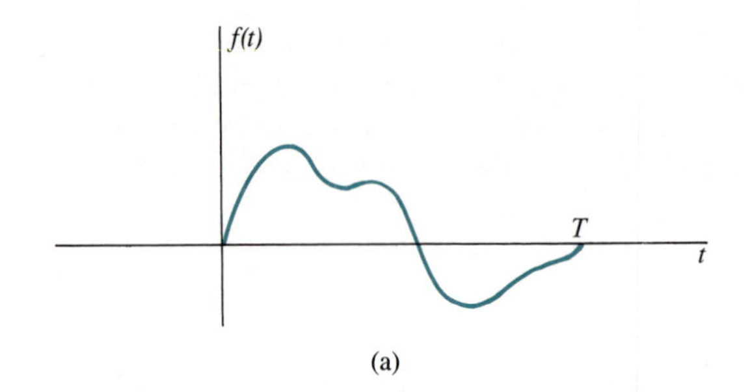

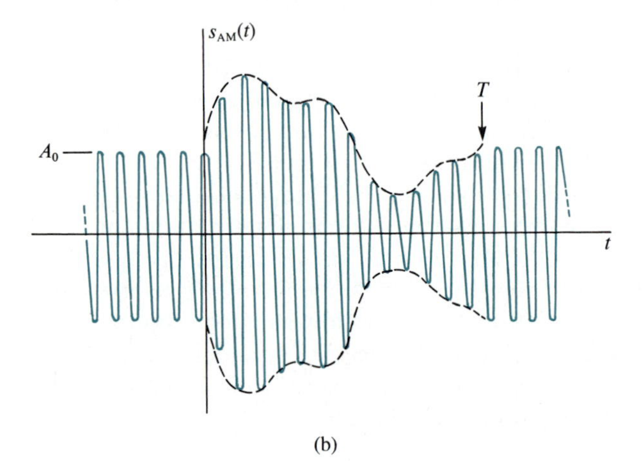

FIG. 15.1-1. (a) An information signal (message) and (b) its corresponding standard AM signal.

With this condition being true, the "envelope" of $s_{AM}(t)$ may become zero but never goes "negative" in Fig. 15.1-1.

To see the spectral behavior of $s_{AM}(t)$, we first assume that $f(t)$ has a Fourier transform $F(\omega)$,

$$f(t) \leftrightarrow F(\omega) \tag{15.1-3}$$

and then Fourier-transform (15.1-1). The spectrum of $s_{AM}(t)$, denoted by $S_{AM}(\omega)$, is

$$\begin{aligned} S_{AM}(\omega) &= [2\pi A_0 \delta(\omega - \omega_0) + F(\omega - \omega_0)] \frac{e^{j\theta_0}}{2} \\ &\quad + [2\pi A_0 \delta(\omega + \omega_0) + F(\omega + \omega_0)] \frac{e^{-j\theta_0}}{2} \end{aligned} \tag{15.1-4}$$

The spectrum $S_{AM}(\omega)$ is illustrated in Fig. 15.1-2(b) for the possible signal spectrum of (a); $\theta_0 = 0$ has been assumed in the sketches for simplicity. The spectral impulses are due to the carrier and are always present, even if the message disappears. The other two terms show that the effect of AM is to shift half-amplitude replicas of the signal's spectrum out to angular frequencies ω_0 and

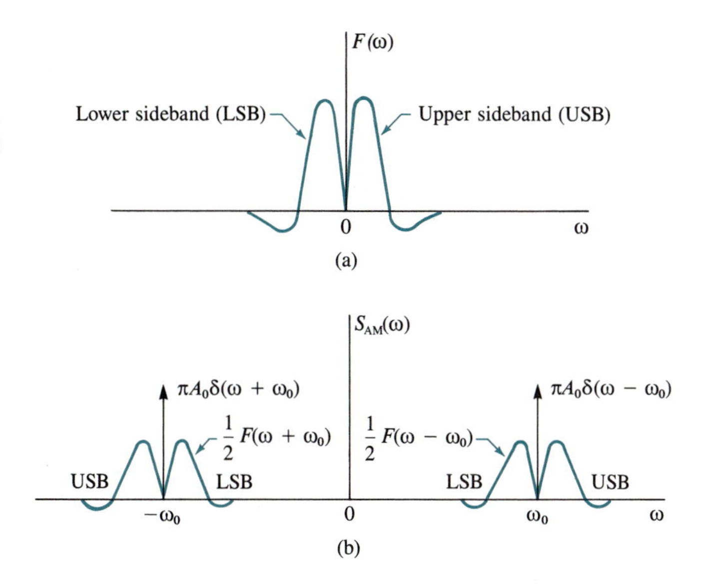

FIG. 15.1-2. (a) An information-signal spectrum and (b) its corresponding signal spectrum.

$-\omega_0$. If the maximum frequency extent of $f(t)$ is W_f (in radians per second), the frequency extent of the standard AM waveform is $2W_f$. The band of frequencies above ω_0 or below $-\omega_0$ is called the *upper sideband* (USB). That on the opposing side is called the *lower sideband* (LSB). Since AM transmits both USB and LSB, it is called *double-sideband modulation*.

Waveform Generation

The class C amplifier (see Sec. 7.8) serves as a common means of generating the high-power AM signal. Figure 15.1-3 illustrates a possible circuit. The active device may be a transistor for low-power stations (as low as 250 W) but is often a vacuum tube at the higher-power levels (up to 50 kW). Basically, when $f(t) = 0$, the circuit is an amplifier that creates the output carrier $A_0 \cos(\omega_0 t + \theta_0)$. When the message is present, it acts to increase or decrease the effective supply voltage seen by the amplifier. Thus, its gain increases and decreases with message amplitude, and AM results. Overmodulation can occur if the message's negative voltage at the secondary of transformer $T1$ exceeds the positive supply voltage; this condition reduces the effective supply voltage to zero, cutting off the active device.

Signal Power Distribution and Efficiency

For simplicity we shall assume in this subsection that $s_{AM}(t)$ is a voltage representing the wave that excites the transmitting station's antenna. The antenna is assumed to represent a real (resistive) impedance Z_0 to the transmitter that feeds it. The average power in any real voltage that exists across a resistance is

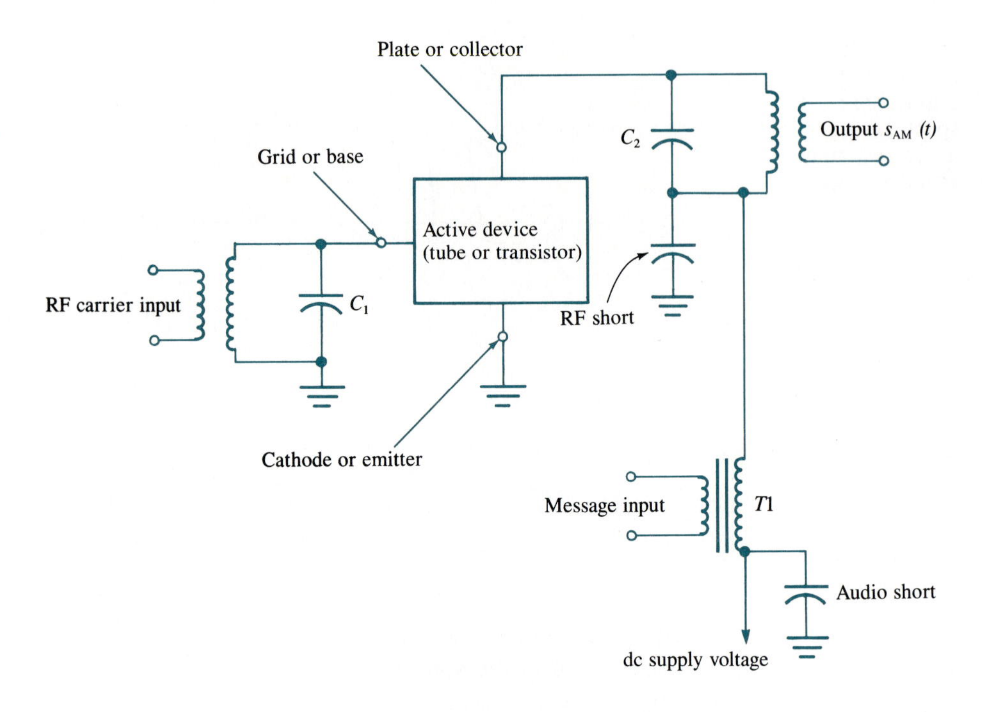

FIG. 15.1-3. A high-level modulator consisting of a class C amplifier for standard AM.

proportional to the average squared value of the voltage and is inversely proportional to the resistance. If we adopt the overbar to represent the time average, the average squared values of $s_{AM}(t)$ and $f(t)$ are $\overline{s_{AM}{}^2(t)}$ and $\overline{f^2(t)}$, respectively. With these definitions and assumptions the power in $s_{AM}(t)$ is found to be

$$P_{AM} = \frac{\overline{s_{AM}{}^2(t)}}{Z_0} = P_c + P_f = \frac{1}{2Z_0}[A_0^2 + \overline{f^2(t)}] \tag{15.1-5}$$

where $f(t)$ is assumed to have no dc content (the usual case),

$$P_c = \frac{A_0^2}{2Z_0} \tag{15.1-6}$$

is the power in the carrier, and

$$P_f = \frac{\overline{f^2(t)}}{2Z_0} \tag{15.1-7}$$

is the added power caused by modulation.

Only the power caused by the message contributes toward message quality in the receiver; it can be called useful power. The carrier's power, although important to the receiver's ability to recover the message with low-cost circuitry, is not useful power in the sense that it carries no information. Modulation *efficiency*,

η_{AM}, of the transmitted signal is defined as the ratio of useful power to total power:

$$\eta_{AM} = \frac{P_f}{P_c + P_f} = \frac{\overline{f^2(t)}}{A_0^2 + \overline{f^2(t)}} \tag{15.1-8}$$

The largest possible value of η_{AM} is 0.5 (or 50%), which occurs when $f(t)$ is a square wave of peak amplitude A_0. If $f(t)$ is a sinusoid, $\eta_{AM} \leq \frac{1}{3}$ (or 33.3%), as the following example demonstrates. For practical messages, such as voice and music audio, efficiency can be less than $\frac{1}{3}$.

EXAMPLE 15.1-1

If $f(t)$ is a sinusoid, $f(t) = A_m \cos \omega_m t$ with period $T_m = 2\pi/\omega_m$, its average squared value is

$$\overline{f^2(t)} = \frac{1}{T_m} \int_{-T_m/2}^{T_m/2} A_m^2 \cos^2 (\omega_m t)\, dt$$

$$= \frac{A_m^2}{8\pi} \int_{-2\pi}^{2\pi} (1 + \cos x)\, dx = \frac{A_m^2}{2}$$

Efficiency becomes

$$\eta_{AM} = \frac{A_m^2}{2A_0^2 + A_m^2} = \frac{(A_m/A_0)^2}{2 + (A_m/A_0)^2}$$

For no overmodulation $A_m \leq A_0$, so $\eta_{AM} \leq \frac{1}{3}$.

Message Demodulation

When the signal (15.1-1) arrives at the receiver, its demodulator (detector) must recover the message $f(t)$. The envelope detector of Fig. 15.1-4(a) is almost universally used (see also Sec. 6.4). Performance is good providing that the product RC is chosen properly. As we shall see in Sec. 15.3, the received AM signal is typically shifted in its "carrier" frequency to a lower value $\omega_{IF}/2\pi$ when it is applied to the detector. Its amplitude is, of course, proportional to the transmitted signal. Thus, the response $s_d(t)$ is proportional to the envelope of $s_{AM}(t)$, as shown by the dashed curve in Fig. 15.1-4(b) atop the half-cycles of the carrier's sinusoid at frequency $\omega_{IF}/2\pi$. The product RC is chosen to satisfy

$$\frac{\pi}{\omega_{IF}} \ll RC < \frac{1}{\omega_{m,\,\max}} \tag{15.1-9}$$

where $\omega_{m,\,\max}/2\pi = 5$ kHz in standard AM.

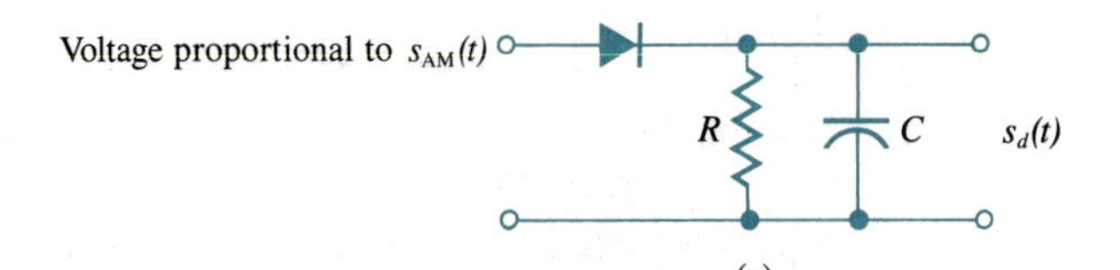

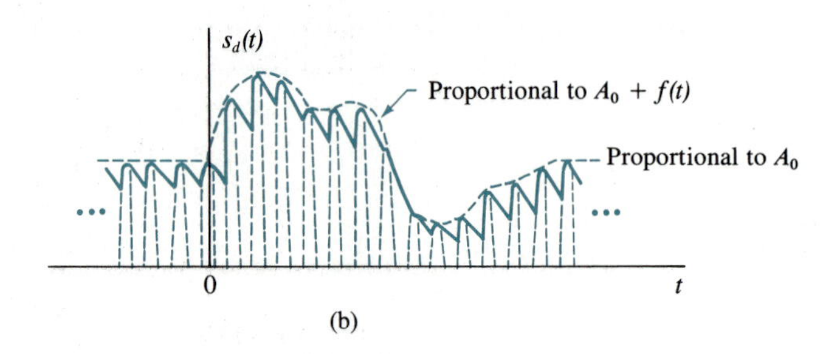

FIG. 15.1-4. (a) A standard AM envelope detector and (b) its response.

15.2 NOISE PERFORMANCE OF STANDARD AM

The actual signal-to-noise power ratio available at the antenna's output in a communications system was found in (14.9-2). We now write this ratio for the standard AM system as $(S_i/N_i)_{AM}$. It can be shown that the available signal-to-noise power ratio at the receiver's output, denoted by $(S_o/N_o)_{AM}$, is

$$\left(\frac{S_o}{N_o}\right)_{AM} = 2\eta_{AM}\left(\frac{S_i}{N_i}\right)_{AM} \tag{15.2-1}$$

In these definitions, the total power in the received AM signal has been used in $(S_i/N_i)_{AM}$, but only the useful power due to the message has been included in $(S_o/N_o)_{AM}$. Equation (15.2-1) indicates that the AM system should use the largest message for modulation that will not overmodulate so that efficiency η_{AM} is as large as possible.

EXAMPLE 15.2-1

Assume that the unmodulated carrier's power is 850 W in an AM transmitter that transmits a total power of 1000 W. We find the required value of $(S_i/N_i)_{AM}$ at a receiver if $(S_o/N_o)_{AM}$ must be 10^3 for good performance. Here $P_c = 850$ W and $P_c + P_f = 1000$ W, so $P_f = 150$ W and (15.1-8) gives $\eta_{AM} = 0.15$. From (15.2-1), $(S_i/N_i)_{AM} = 10^3/0.30 \approx 3333$ (or 35.2 dB).

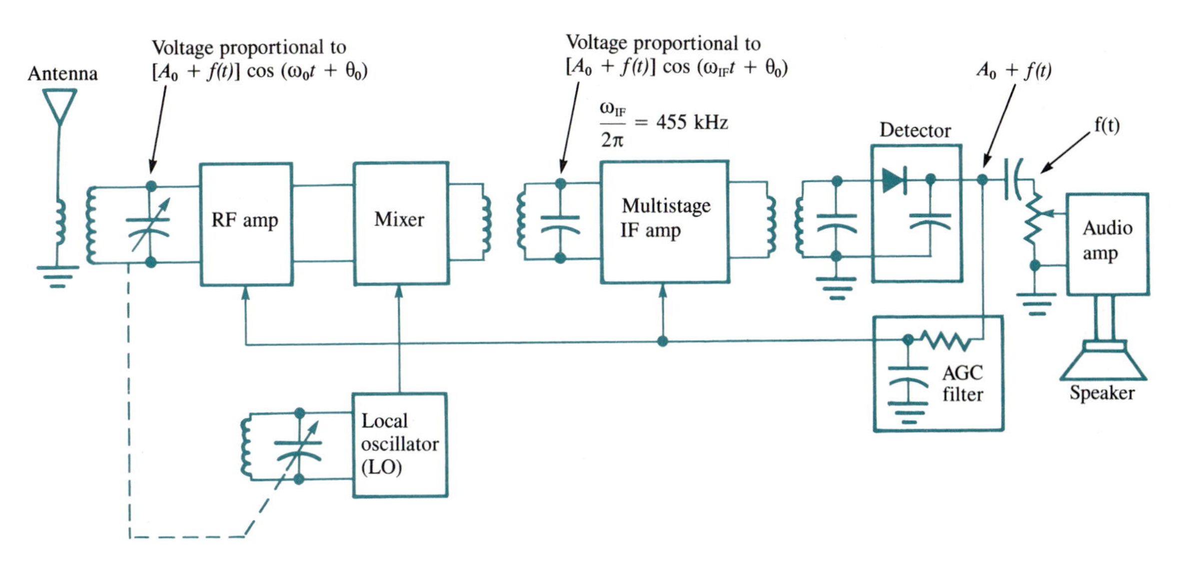

FIG. 15.3-1.
A superheterodyne radio receiver for standard AM signal reception.

15.3 AN AM RADIO RECEIVER

Modern standard AM radio receivers have the structure shown in Fig. 15.3-1. It is called a *superheterodyne receiver* because it uses a mixer, which is a device that is able to shift the nominal (center) frequency of the signal at its input to a new value in its output signal. To make this shift, it must have a second input called a *local-oscillator* (LO) *signal* with a frequency equal to the amount of shift desired.

The key to receiver operation is mainly in the operation of the mixer. The radio-frequency (RF) amplifier is tuned to the required station frequency $f_0 = \omega_0/2\pi$ by mechanical variation of its tuning capacitor. The local oscillator's frequency $f_{LO} = \omega_{LO}/2\pi$ is simultaneously shifted (via ganged capacitors)† such that the difference $|f_0 - f_{LO}|$ equals a constant $f_{IF} = \omega_{IF}/2\pi = 455$ kHz, called the *intermediate frequency* (IF). It is possible to design f_{LO} to be either higher or lower than f_0.

If f_{LO} is higher than f_0, then $f_{LO} = f_0 + f_{IF}$ and the mixer's response to a station at frequency f_0 will be at the desired IF frequency f_{IF}. The mixer will also respond to a signal at frequency $f_0 + 2f_{IF}$, called the *image frequency*, as sketched in Fig. 15.3-2(a). It is necessary to suppress this image response by using the

† Inexpensive receivers often omit the RF amplifier. More expensive sets are often electronically tuned by using frequency synthesizers.

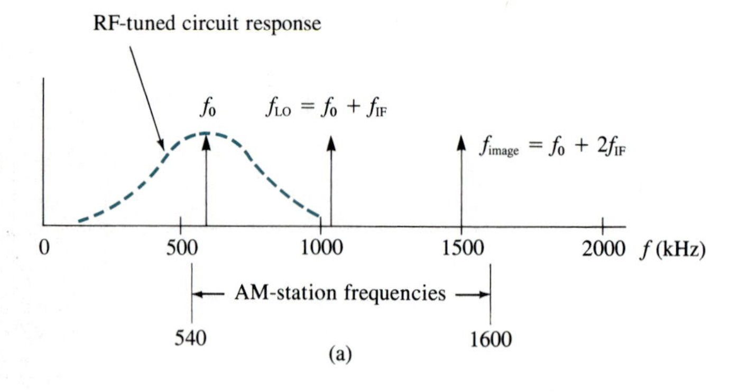

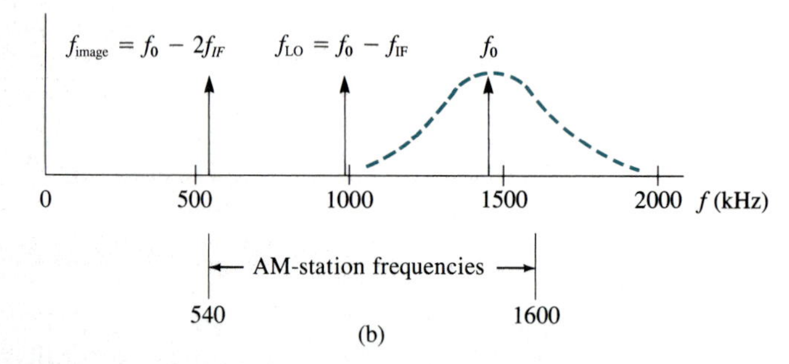

FIG. 15.3-2. AM station and image frequencies for (a) a high-side local oscillator and (b) a low-side local oscillator.

antenna and other RF-tuned circuits as a filter. A similar behavior occurs when f_{LO} is lower than f_0, as shown in Fig. 15.3-2(b). The RF circuits do not need to provide isolation from adjacent radio stations; they need only have sufficiently narrow bandwidth to remove the image a distance $2f_{IF}$ away from f_0.

Amplifiers in the IF circuits provide most of the gain needed to raise the small antenna signal to a level sufficient to drive the envelope detector. The detector's output contains a dc component proportional to A_0 and a component proportional to the audio message $f(t)$. The message is used to drive the speaker. The dc component is used in an *automatic-gain-control* (AGC)† loop to control the gain of RF and IF amplifiers (by controlling their operating bias points and, therefore, their gains). Loop action is to maintain approximately a constant IF level at the detector's input, even for large variations in antenna voltage.

The bandwidth of the IF circuit is about 10 kHz. It is the IF circuit that provides the principal rejection of adjacent channel interference.

15.4 OTHER FORMS OF AM

There are several variations of amplitude modulation in use. We describe three.

† Sometimes called *automatic volume control* (AVC).

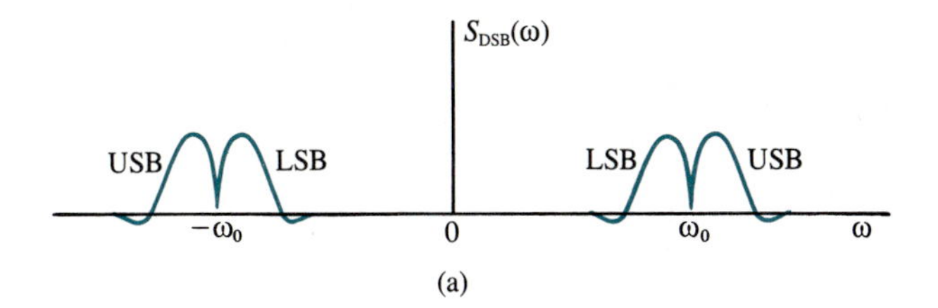

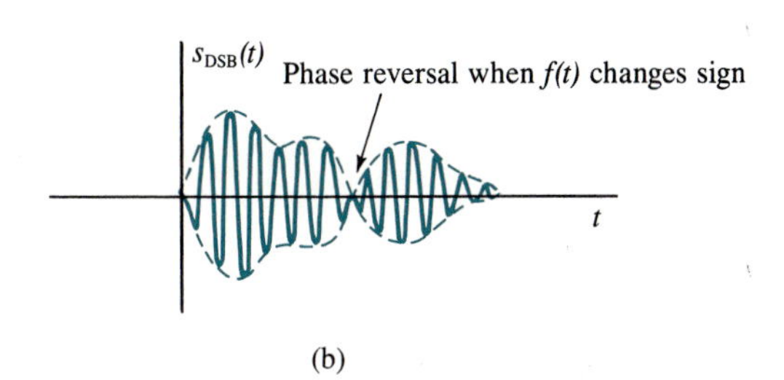

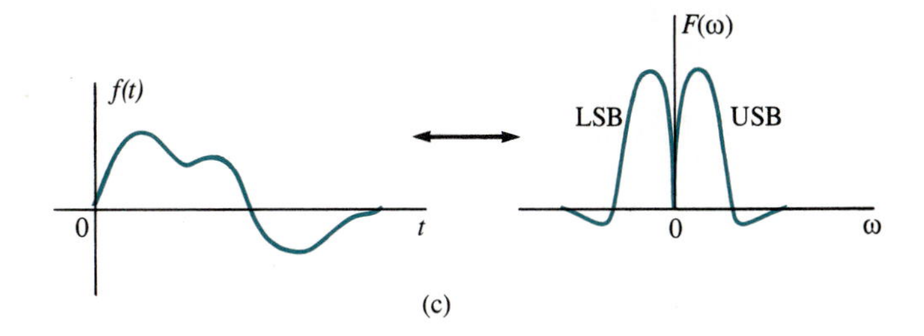

FIG. 15.4-1. (a) The spectrum of the (b) DSB suppressed-carrier signal corresponding to the message and spectrum of (c).

Suppressed-Carrier AM

Standard AM can be modified to improve efficiency by simply eliminating the carrier term A_0 in (15.1-1). The resulting waveform corresponds to *suppressed-carrier AM*. It is also called *double-sideband* (DSB) suppressed-carrier AM. The DSB waveform is

$$s_{DSB}(t) = f(t) \cos (\omega_0 t + \theta_0) \tag{15.4-1}$$

An example waveform is sketched in Fig. 15.4-1(b) for the message of (c). The spectrum of $s_{DSB}(t)$ is illustrated in Fig. 15.4-1(a).

From (15.1-8) with $A_0 = 0$, the efficiency of DSB is 1.0 (or 100%). The price paid for the increased power efficiency is added complexity, especially in the demodulator (because the simple envelope detector can no longer be used without additional circuitry).

The generation of $s_{DSB}(t)$ requires a device to produce the product of $f(t)$ and $\cos (\omega_0 t + \theta_0)$. One of the many types of product device is shown in Fig. 15.4-2(a), with $\theta_0 = 0$ assumed for simplicity. It is called a *balanced modulator*. Following the balanced modulator with a bandpass filter (or tuning the transformers to resonate at f_0) makes the sharp edges of the waveform in Fig. 15.4-2(b) smooth, as shown in Fig. 15.4-1(b). The reader is encouraged to think through the behavior of the balanced modulator as an exercise (Problem 15-13).

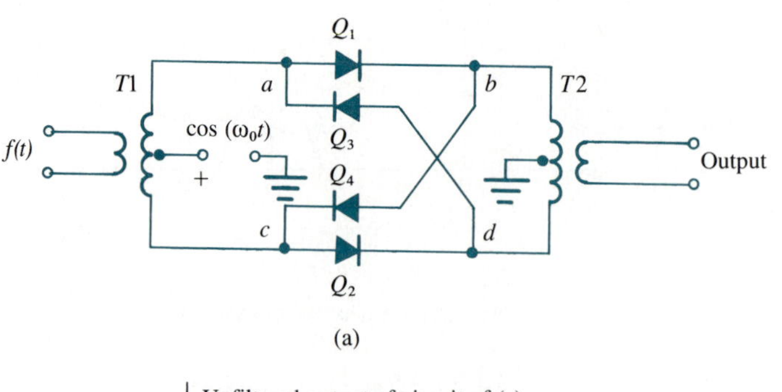

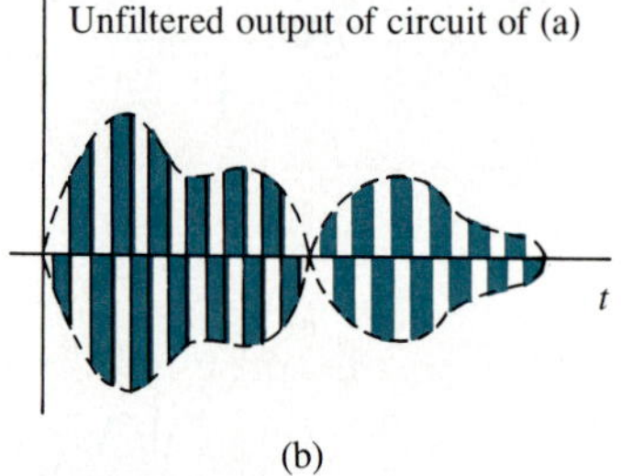

FIG. 15.4-2. (a) A four-diode balanced modulator and (b) its unfiltered-output waveform.

A balanced modulator followed by a lowpass filter can also be used for demodulating the DSB signal to recover $f(t)$. This form of demodulator is called a *coherent* or *synchronous detector*. One of its inputs is $s_{\text{DSB}}(t)$, and the other is a local carrier that must be phase-coherent (synchronous) with the transmitted "carrier"; that is, it must be proportional to $\cos(\omega_0 t + \theta_0)$. The need for the local carrier is a chief disadvantage of the DSB scheme. Sometimes, a *small* amount of *pilot carrier* is added to the DSB transmitted signal to aid the receiver in constructing the local carrier.

Single-Sideband AM

By study of the Fourier transform $F(\omega)$ of a *real* message $f(t)$, it can be shown that

$$F(-\omega) = F^*(\omega) \tag{15.4-2}$$

In other words, knowledge of $F(\omega)$ for positive (or negative) values of ω means $F(\omega)$ is also known for negative (or positive) ω from (15.4-2). This fact means that only one message sideband in the DSB spectrum of Fig. 15.4-1(a) needs to be transmitted (either USB or LSB). Filtering to remove one sideband of the DSB signal, as shown in Fig. 15.4-3, creates a *single-sideband* (SSB) signal. The principal advantage of the SSB signal is the 2:1 bandwidth savings over DSB or standard AM.

SSB is popular with amateur-radio operators because of its high efficiency ($\eta_{\text{SSB}} = 1$) and bandwidth savings. It can be demodulated, as in DSB, with a

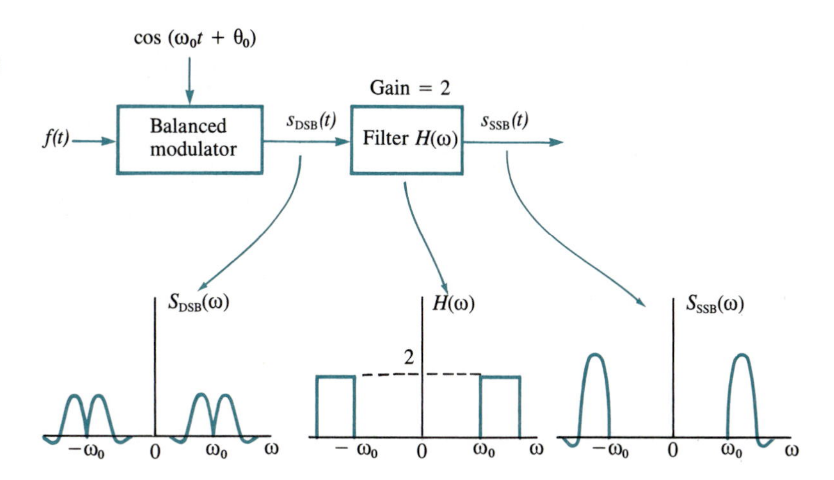

FIG. 15.4-3. A block diagram of the filter method for generating SSB.

synchronous detector, but all the problems associated with the need for a coherent local carrier are also present.

Vestigial-Sideband AM

Vestigial-sideband (VSB) AM is a variation of SSB where a small portion (or vestige) of the filtered sideband is allowed to remain. VSB is mainly used in the broadcast-television system. Generation of VSB is similar to generation of SSB, except that the sideband-removal filter has a slightly different transfer function. Signal recovery in the receiver uses a synchronous detector.†

Noise Performances

It can be shown that, to a good approximation,

$$\left(\frac{S_o}{N_o}\right)_{\text{DSB}} = 2\left(\frac{S_i}{N_i}\right)_{\text{DSB}} \tag{15.4-3}$$

$$\left(\frac{S_o}{N_o}\right)_{\text{SSB or VSB}} = \left(\frac{S_i}{N_i}\right)_{\text{SSB or VSB}} \tag{15.4-4}$$

These results would seem to imply that DSB outperforms SSB or VSB by a factor of 2. In reality, for the same messages and transmitted powers, all three systems give the same performance. The reason, of course, is that the bandwidth of the DSB system is twice that of the SSB or VSB systems, so its input noise power is twice as large.

† In a TV system the carrier is not fully suppressed, so an envelope detector can still be used.

15.5 FREQUENCY MODULATION (FM) SYSTEM

Radio stations that broadcast *frequency modulation* (FM) in the United States are each assigned one of 100 possible carrier (station) frequencies from 88.1 to 107.9 MHz. Each station broadcasts in a channel bandwidth of 200 kHz centered on the carrier. Thus, the FM band extends from 88.0 to 108.0 MHz. Frequency assignments are controlled by the FCC to prevent interference between stations. The FM system is able to present higher-quality audio to the user than AM because the FCC allows a larger audio band extending from 50 Hz to 15 kHz. Maximum distortion is below 2.5% in the midband but may be as large as 3.5% out of midband (100 Hz–7.5 kHz), although distortion at most stations is below 1%.

Although our discussions to follow will emphasize broadcast applications, FM finds many uses, such as in satellite links, aircraft altimetry, radars, amateur radio, and various two-way-radio applications. For the same message and transmitted power, an FM system can exhibit more freedom from interference and better performance in noise than in all the AM systems. The price paid for this performance is the need for larger bandwidth, up to 20 times more than in AM.

FM Waveform and Bandwidth

A sinusoid is said to be frequency-modulated if its instantaneous angular frequency $\omega_{FM}(t)$ is a linear function of the message. Thus,

$$\omega_{FM}(t) = \omega_0 + k_{FM} f(t) \tag{15.5-1}$$

where k_{FM} is a constant [the unit is radians per second per volt when $f(t)$ is a voltage] and ω_0 is the carrier's nominal angular frequency. Since instantaneous phase is the integral of instantaneous angular frequency, the FM signal becomes

$$s_{FM}(t) = A_0 \cos\left[\omega_0 t + \theta_0 + k_{FM} \int f(t)\, dt\right] \tag{15.5-2}$$

where A_0 is a constant amplitude and θ_0 is an arbitrary constant phase angle.

The maximum amount that $\omega_{FM}(t)$ departs from its nominal value in (15.5-1) is called *peak frequency deviation*, denoted by $\Delta\omega$; thus;

$$\Delta\omega = k_{FM} |f(t)|_{max} \tag{15.5-3}$$

Although the FM signal's bandwidth, denoted by W_{FM} (radians per second), is difficult to compute exactly, except for a few message forms, practical experience indicates that it is approximately

$$W_{FM} \approx 2(\Delta\omega + W_f) \tag{15.5-4}$$

where W_f is the spectral extent of $f(t)$. Equation (15.5-4) is known as *Carson's rule*.† In *narrowband FM*, defined by the condition $\Delta\omega < W_f$, (15.5-4) gives results

† After John R. Carson.

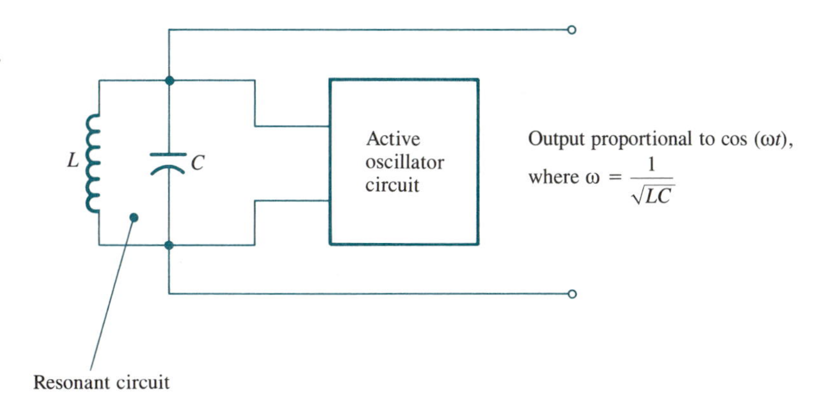

FIG. 15.5-1. Basic elements of an oscillator for use in a voltage-controlled oscillator (VCO).

in good agreement with experiment. For wideband FM, defined by $\Delta\omega \gg W_f$, (15.5-4) becomes more accurate if W_f is replaced by $2W_f$.

The performance of narrowband FM with noise is roughly equivalent to that of AM systems. Only wideband FM exhibits the marked improvement discussed earlier. For this reason, only the wideband system will be discussed further.

Generation of Wideband FM

Many techniques exist for generating FM waveforms. We discuss only the most common and, conceptually, the most simple. It is called the *direct method* and it uses a *voltage-controlled oscillator* (VCO) as a modulator. A VCO is an oscillator with an oscillation frequency equal to the resonant frequency of a tuned circuit, as shown in Fig. 15.5-1. If either the inductance or the capacitance is made voltage-sensitive to the message $f(t)$, frequency can be varied.

Several methods exist to obtain a voltage-variable reactance. One very simple method is to use the junction capacitance of a reverse-biased diode, which depends on the amount of bias. When total bias is composed of a dc voltage to establish an operating point and the message $f(t)$, capacitance, and therefore oscillator frequency, can be made to vary with $f(t)$. These diodes are usually called *varactors*. Varactor-controlled VCOs can have a quite linear frequency-voltage characteristic but often can give only small frequency deviations (small $\Delta\omega$).

Carrier-Frequency Stabilization

Any VCO is inherently unstable; otherwise, it would not be easy to vary its frequency to produce FM. In many applications it is necessary to stabilize the carrier's frequency. A station in the broadcast FM system, for example, must hold its carrier to within ±2 kHz of its assigned frequency. Figure 15.5-2 depicts a modulation method whereby the carrier's angular frequency ω_0 can be forced to a stable value.† The modulating signal is passed through a highpass filter (HPF),

† See L. W. Couch II, *Digital and Analog Communication Systems*, 3d ed., Macmillan, New York, 1990, p. 310.

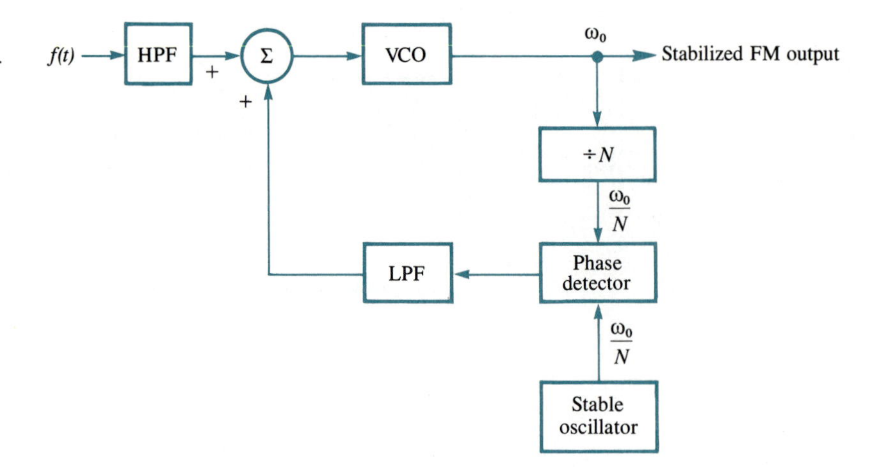

FIG. 15.5-2. An FM modulation method having carrier stabilization.

which only ensures that $f(t)$, as applied to the VCO, has no dc or very low-frequency content. The FM output, which may be subject to drift in ω_0, is applied to a frequency divider. The divider reduces both the center (carrier's) frequency and the FM signal's bandwidth by a factor of N. The bandwidth reduction helps to ensure a carrier component in the narrowband FM signal applied to the phase detector. The output voltage of the phase detector, when averaged by the loop's lowpass filter (LPF), is approximately proportional to the difference in the phases of the unstable carrier and the stable oscillator. This voltage is fed back to the VCO in a way that forces the frequency of the carrier to be N times that of the stable oscillator. The error in the stabilized carrier's angular frequency will be N times the frequency error of the stable oscillator.

The output voltage of the phase detector in Fig. 15.5-2 will also have a voltage component that fluctuates according to the highpass-filtered version of $f(t)$ that is applied to the VCO. However, since the filtered form of $f(t)$ has no low-frequency components, it does not pass through the loop's lowpass filter. It therefore does not significantly affect the carrier-stabilization characteristics of the loop.

Message Demodulation

Demodulators for FM broadly fall into two classes, *frequency discriminators* and *locked-loop demodulators*. The discriminator produces an output voltage proportional to the frequency variations that occur in the FM signal at its input. It was extensively used in older FM receivers but is being displaced in recent times by locked-loop demodulators. The locked-loop demodulators are conceptually more complicated than a discriminator but are more cost-competitive when implemented in integrated-circuit form. Most locked-loop demodulators provide better performance than a discriminator when signal levels are low and noise is a problem. For relatively high signal levels both give the same performance.

Many forms of frequency discriminator exist. One of the most useful is called the *ratio detector*. The basic circuit is sketched in Fig. 15.5-3(a). In this circuit C_0

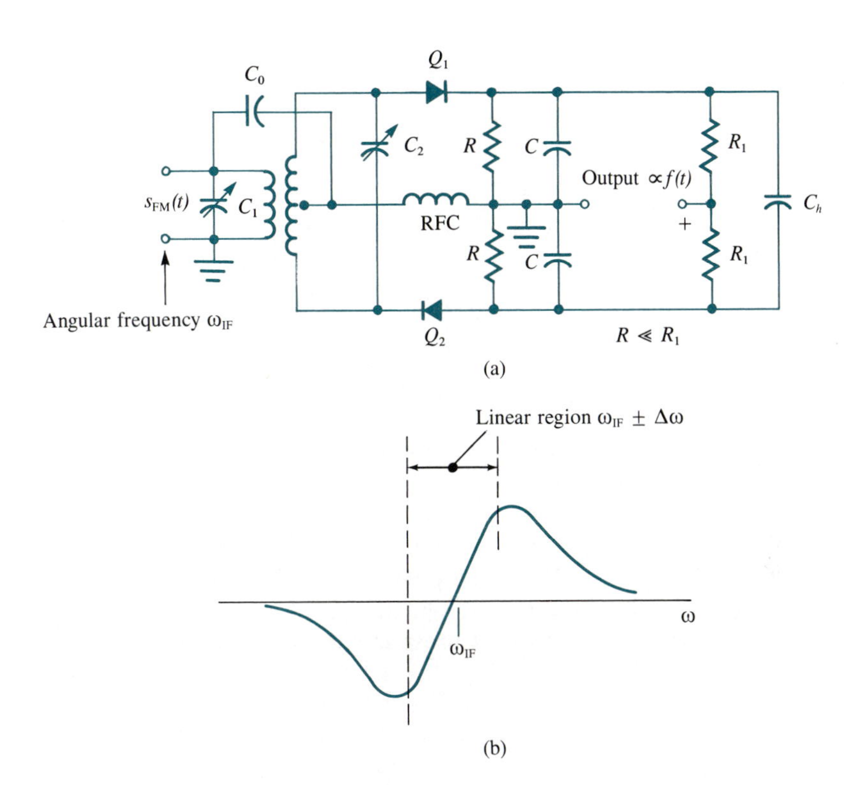

FIG. 15.5-3. (a) The ratio detector and (b) its output-input characteristic.

is a short circuit at frequency $\omega_{IF}/2\pi$, and the choke (RFC) is approximately an open circuit. Capacitors C_1 and C_2 are adjusted to resonate the transformer at ω_{IF}. Resistors R and capacitors C form envelope detectors with their respective diodes Q_1 or Q_2. Resistors R_1 are chosen large relative to R. The overall output-voltage variation with frequency variation in the input FM waveform is shown in Fig. 15.5-3(b). The behavior in the linear range is related to the phasing of the voltages driving the envelope detectors in a complicated way that we shall not describe. Behavior in the regions outside the response peaks is related to the bandwidth of the tuned transformer.

In all FM demodulators it is necessary that the output voltage be a function of only frequency changes and not of *amplitude* variations in the FM signal. Many forms of discriminator require a limiter at its input to remove amplitude variations. An AGC loop is commonly used (as in the AM receiver) to remove slow amplitude changes, but rapid changes are unaffected. In the ratio detector fast amplitude changes can be absorbed by the capacitor C_h; it, therefore, has the major advantage of not needing a separate fast limiter.

Several forms of locked-loop demodulators exist. We describe the most important one. It is the *phase-locked-loop* (PLL) demodulator of Fig. 15.5-4. Here the phase of a VCO's output signal is forced to follow (lock to) the phase of the input FM waveform with small error. Assume initially that only carrier with phase $\theta(t) = \omega_0 t + \theta_0$ is present at the input. If the VCO's output phase $\hat{\theta}(t)$ is the same

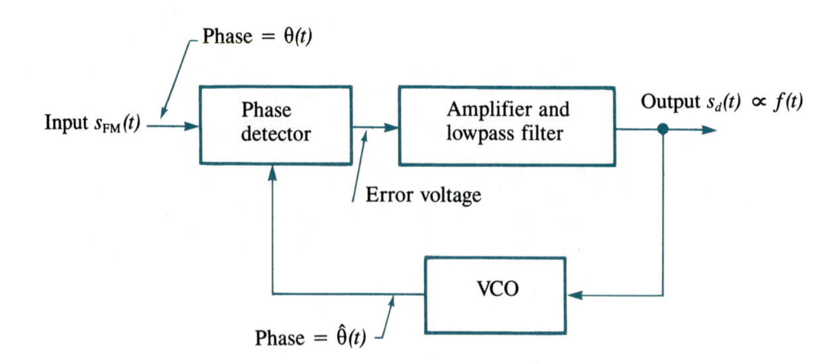

FIG. 15.5-4. A phase-locked loop for the demodulation of FM.

value, the output of the phase detector will be zero, and the loop is said to be locked. If $\hat{\theta}(t)$ tries to drift by a small amount $\delta\theta$, the phase detector will produce an error voltage proportional to $-\delta\theta$. When amplified and applied to the VCO input, the error will act to move the VCO output phase to reduce $\delta\theta$ to nearly zero. Thus, without FM, the loop will stay locked to the exact frequency and approximate phase of the carrier signal.

When modulation is present, the loop's action is similar. The VCO phase $\hat{\theta}$ will now follow the *instantaneous* phase $\theta(t)$ of the input signal, with only a small phase error. The amplified error voltage appearing at the VCO input will be proportional to $f(t)$, since the VCO acts as an integrator. This fact may also be reasoned as follows: Since FM phase variations in $\theta(t)$ are proportional to $\int f(t)\,dt$ and $\hat{\theta}(t) \approx \theta(t)$, then $\hat{\theta}(t)$ will have phase variations approximately proportional to $\int f(t)\,dt$. However, for a VCO, output-*frequency* variations are proportional to input voltage; so that if the required *phase* variations in $\hat{\theta}(t)$ are to be produced, the VCO input voltage must be proportional to $f(t)$, since phase is the integral of frequency.

The filter is selected to give the desired closed-loop bandwidth, which is often chosen as narrow as possible to reject noise but broad enough to give demodulation with minor distortion of $f(t)$.

15.6 NOISE PERFORMANCE OF FM

As for an AM receiver, we define the FM receiver's input signal-to-noise power ratio, denoted by $(S_i/N_i)_{FM}$, at the receiving antenna's output, as given by (14.9-2). Performance at the receiver's output, where the recovered message is available, is measured by the output-power signal-to-noise ratio, denoted by $(S_o/N_o)_{FM}$. Good performance corresponds to large values of $(S_o/N_o)_{FM}$. What performance is achievable depends on the type of demodulator.

Performance with a Discriminator

When the demodulator is a discriminator, it can be shown that

$$\left(\frac{S_o}{N_o}\right)_{\text{FM}} = 6\left(\frac{\Delta\omega}{W_f}\right)^3\left[\frac{\overline{f^2(t)}}{|f(t)|^2_{\text{max}}}\right]\left(\frac{S_i}{N_i}\right)_{\text{FM}} \tag{15.6-1}$$

where $|f(t)|_{\text{max}}$ is the maximum amplitude of $f(t)$, W_f is the spectral extent of $f(t)$, and $\Delta\omega$ is given by (15.5-3). *Crest factor*, denoted by K_{cr}, is a message-related constant defined by

$$K^2_{\text{cr}} = \frac{|f(t)|^2_{\text{max}}}{\overline{f^2(t)}} \tag{15.6-2}$$

It is a measure of how large the maximum amplitude of a message is in relation to its rms value. In terms of K_{cr} (15.6-1) becomes

$$\left(\frac{S_o}{N_o}\right)_{\text{FM}} = \frac{6}{K^2_{\text{cr}}}\left(\frac{\Delta\omega}{W_f}\right)^3\left(\frac{S_i}{N_i}\right)_{\text{FM}} \tag{15.6-3}$$

Equation (15.6-3) indicates that performance is proportional to $(S_i/N_i)_{\text{FM}}$, as one might expect. It also implies that $(S_o/N_o)_{\text{FM}}$ improves as the cube of the FM signal's bandwidth (which is $2\Delta\omega$). In reality, it only increases as the square, because $(S_i/N_i)_{\text{FM}}$ decreases inversely as $\Delta\omega$ increases (for the broadband input noise assumed). The decrease occurs because the receiver's input noise power must increase as bandwidth increases. The net increase in $(S_o/N_o)_{\text{FM}}$ by the square of $\Delta\omega$ is the most important advantage of FM. Simply increasing the bandwidths of $s_{\text{FM}}(t)$ and the system, all else being constant, increases system performance. There is a limit, unfortunately, to the possible increase. The limit, called the *threshold*, occurs approximately where $(S_i/N_i)_{\text{FM}}$ decreases to about 10. At this point the assumptions under which (15.6-3) was derived are no longer valid. Equation (15.6-3) also indicates a performance decrease as the square of the message's crest factor.

A more careful analysis, including the threshold region, for a sinusoidal message of angular frequency $\omega_m = W_f$, gives†

$$\left(\frac{S_o}{N_o}\right)_{\text{FM}} = \frac{3\beta^3_{\text{FM}}(S_i/N_i)_{\text{FM}}}{1 + (24\beta^2_{\text{FM}}/\pi)(S_i/N_i)_{\text{FM}} \exp\left[-(S_i/N_i)_{\text{FM}}\right]} \tag{15.6-4}$$

Here

$$\beta_{\text{FM}} = \frac{\Delta\omega}{\omega_m} \tag{15.6-5}$$

is called the *modulation index*. A plot of (15.6-4) is given in Fig. 15.6-1. On

† P. Z. Peebles, Jr., *Communication System Principles*, Addison-Wesley, Reading, Mass., 1976, p. 279.

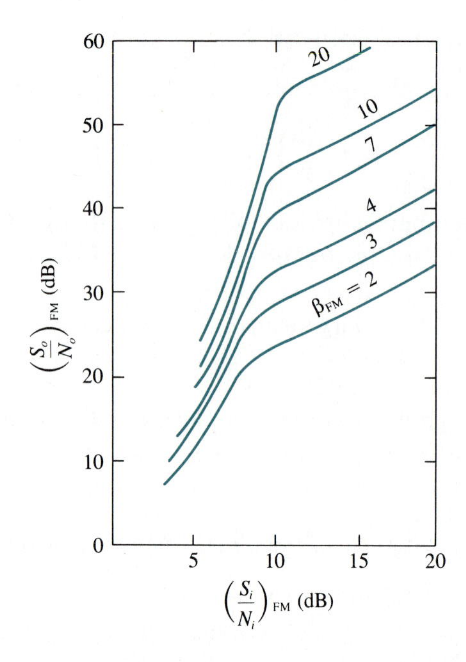

FIG. 15.6-1. FM output versus input signal-to-noise ratios when the modulation is due to a sinusoidal message.

recognizing that $K_{cr}^2 = 2$ for a sinusoid, we see that (15.6-3) equals the numerator of (15.6-4) and applies only in the linear part of the curves in Fig. 15.6-1.

EXAMPLE 15.6-1

Suppose a sinusoidal signal is the message in an FM system. In the receiver, *threshold signal-to-noise ratio* is defined as the value of $(S_i/N_i)_{FM}$ at which $(S_o/N_o)_{FM}$ falls to 0.794 (−1 dB) of the value given by (15.6-3). This value occurs when the denominator of (15.6-4) is $1/0.794 \approx 1.259$. We find the largest value of β_{FM} that will guarantee $(S_i/N_i)_{FM} = 10$ to be at or above the threshold value. From (15.6-4) we require

$$1 + \frac{24\beta_{FM}^2}{\pi}(10)e^{-10} \le 1.259$$

or $\beta_{FM} \le 8.642$. The value of $(S_o/N_o)_{FM}$ at threshold is $0.794(3)(8.642)^3(10) \approx 1.537 \times 10^4$ (or 41.9 dB).

Performance with Locked Loops

The performance of the phase-locked- or the frequency-locked-loop demodulator is the same as for the discriminator [given by (15.6-3)], provided that $(S_i/N_i)_{FM}$ is above the threshold value. However, the thresholds for locked loops are lower than for the discriminator, which makes it possible for loop-type receivers to operate at smaller signal-power levels. Such receivers are often found where transmitter power is at a premium, as in space communications applications.

Emphasis in FM

A careful study of the spectral properties of the noise at the output of the discriminator in an FM receiver shows that higher-frequency components of output noise power are accentuated proportional to ω^2. If a lowpass filter, called a *de-emphasis filter*, is added, the large-amplitude noise can be greatly reduced and $(S_o/N_o)_{FM}$ increased. However, the filter also acts on the message, causing distortion if nothing else were done. Distortion is prevented by passing the message at the transmitter through a compensating filter, called a *preemphasis filter*, before modulation occurs. It accentuates the higher frequencies in the message so as to exactly compensate for the effect of the de-emphasis filter. There is no overall effect on the message.

A common de-emphasis filter used in broadcast FM is shown in Fig. 15.6-2(d). Its transfer function $H_d(\omega)$ is

$$H_d(\omega) = \frac{1}{1 + j(\omega/W_1)} \tag{15.6-6}$$

where $W_1/2\pi = 2.12$ kHz (time constant $R_1C = 75\ \mu$s). The function $|H_d(\omega)|$ is sketched in Fig. 15.6-2(e). For perfect message recovery the preemphasis filter's transfer function must be

$$H_p(\omega) = \frac{1}{H_d(\omega)} = 1 + j\left(\frac{\omega}{W_1}\right) \tag{15.6-7}$$

over all important frequencies (out to about 15–20 kHz). The system's performance improvement with these filters in place can be shown to be

$$R_{FM} = \frac{(S_o/N_o)_{FM}\ (\text{emphasis})}{(S_o/N_o)_{FM}\ (\text{no emphasis})} = \frac{(W_f/W_1)^3}{3[(W_f/W_1) - \tan^{-1}(W_f/W_1)]} \tag{15.6-8}$$

where W_f is the spectral extent of $f(t)$, as usual.

A practical approximation to the required function of (15.6-7) is an amplifier of voltage gain $K = (R_1 + R_2)/R_2$ in cascade with the network of Fig. 15.6-2(b). The network's transfer function (less amplifier) is sketched in Fig. 15.6-2(c). Resistor ratio R_1/R_2 is selected sufficiently large to make $W_2/2\pi$ at least 15 kHz in broadcast FM.

For voice-type messages the improvement factor of (15.6-8) is realized. However, for broader-band messages such as quality music audio, part of the improvement R_{FM} is actually due to an increase in FM-signal bandwidth. Since channel bandwidth in broadcast FM is limited to 200 kHz, such signals require a reduction in the gain K. There is a consequential reduction in message power in the receiver and a reduction in the improvement factor R_{FM}. Factor R_{FM} with the bandwidth

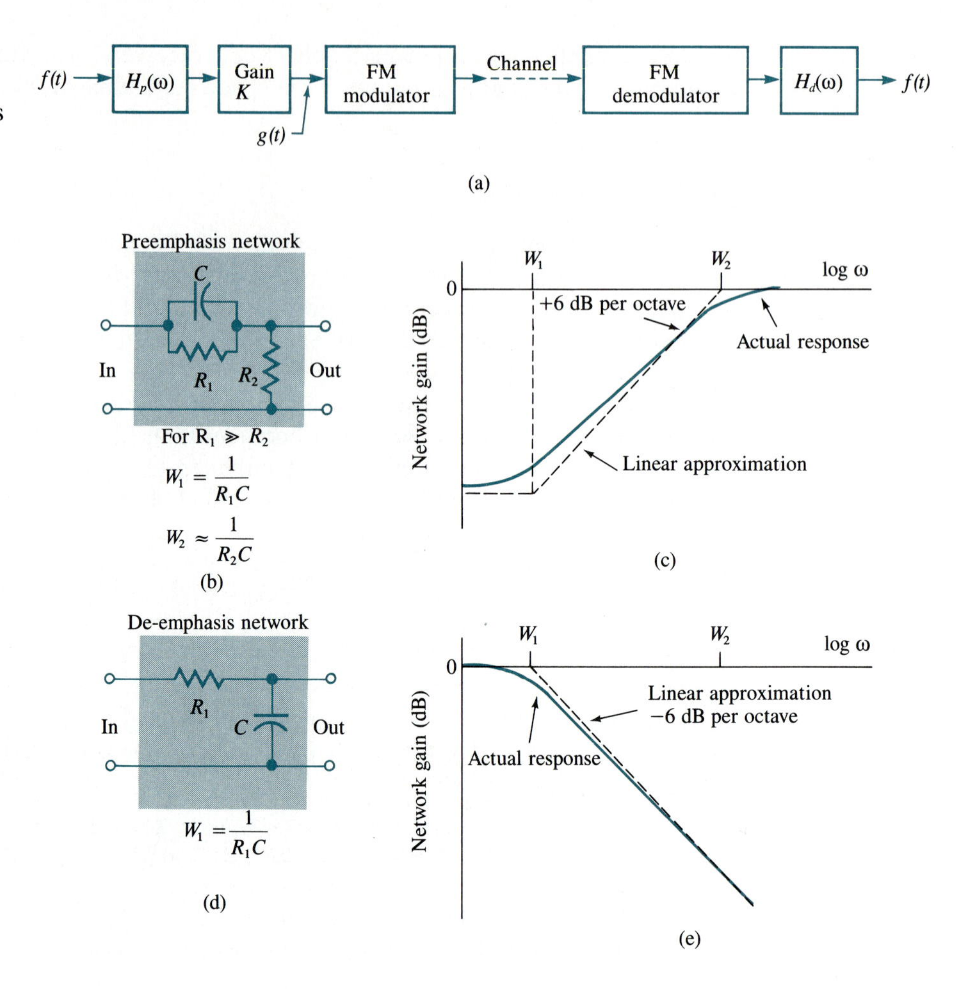

FIG. 15.6-2. (a) The placement of emphasis filters in an FM system. (b) A typical preemphasis network and (c) its response for broadcast FM. (d) A corresponding deemphasis network and (e) its response.

limitation is related to R_{FM} of (15.6-8) without any limitation by

$$\frac{R_{\text{FM}}\ \text{(with bandwidth limitation)}}{R_{\text{FM}}\ \text{(no bandwidth limitation)}} = \frac{1}{1 + (W_{\text{rms}}/W_f)^2(W_f/W_1)^2} \tag{15.6-9}$$

where W_{rms} is the rms bandwidth (in radians per seconds) of $f(t)$ defined by

$$(W_{\text{rms}})^2 = \frac{\displaystyle\int_{-\infty}^{\infty} \omega^2|F(\omega)|^2\, d\omega}{\displaystyle\int_{-\infty}^{\infty} |F(\omega)|^2\, d\omega} \tag{15.6-10}$$

The placement of the various emphasis components in an FM system is shown in Fig. 15.6-2(a).

EXAMPLE 15.6-2

For a broadcast FM system, assume $W_1/2\pi = 2.12$ kHz, $W_f/2\pi = 15$ kHz, and $W_{rms}/2\pi = 4.25$ kHz. We determine R_{FM} with and without bandwidth constraints. From (15.6-8)

$$R_{FM} = \frac{(15/2.12)^3}{3[(15/2.12) - \tan^{-1}(15/2.12)]} \approx 20.92 \qquad \text{(or 13.2 dB)}$$

for no bandwidth constraint. From (15.6-9)

$$R_{FM} \text{ (with bandwidth limitation)} = \frac{20.92}{1 + (4.25/15)^2(15/2.12)^2} \approx 4.17$$

(or 6.20 dB). The bandwidth constraint has resulted in a loss in emphasis improvement of 13.2 − 6.2 = 7.0 dB.

15.7 FM STEREO

In the early days of broadcast FM all stations were monaural. Currently, almost all have stereo capability; that is, they are able to transmit two independent audio messages. One, which we call the "left" message, usually represents the left side of the sound source (microphone on the left side of a symphonic orchestra, for example). The other corresponds to the right side of the source and is labeled the "right" message. Independent reception of these messages gives the listener a degree of "presence" that is unavailable in monaural, single-message reception.

The block diagram of an FM-stereo transmitter is shown in Fig. 15.7-1(a). Left and right messages, denoted by $f_L(t)$ and $f_R(t)$, respectively, undergo preemphasis and are then added, to create $f_1(t)$, and differenced, to give $f_d(t)$.† Signal $f_1(t)$ is the same as the message $f(t)$ would be in a monaural system. Both $f_1(t)$ and $f_d(t)$ have a spectrum spanning 50 Hz to 15 kHz. Signal $f_d(t)$ DSB-modulates (with carrier suppressed) a 38-kHz *subcarrier* to give $f_2(t)$. Both $f_1(t)$ and $f_2(t)$ are added to $f_3(t)$, a low-level pilot carrier at 19 kHz that is included to aid in the receiver's demodulation process; their sum is the final composite message $f_s(t)$.

In some stations $f_s(t)$ contains a fourth component, SCA, which stands for *subsidiary communications authorization*. It is a narrowband FM waveform on a 67-kHz subcarrier with a total bandwidth of 16 kHz. The spectrum of $f_s(t)$, including SCA, is sketched in Fig. 15.7-1(b). SCA is a monaural signal that cannot be recovered by the typical home receiver. It is a special signal available to fee-paying subscribers such as physicians and dentists for their waiting rooms, of-

† Preemphasis could be applied after the sum and difference operations, instead. The effects are the same.

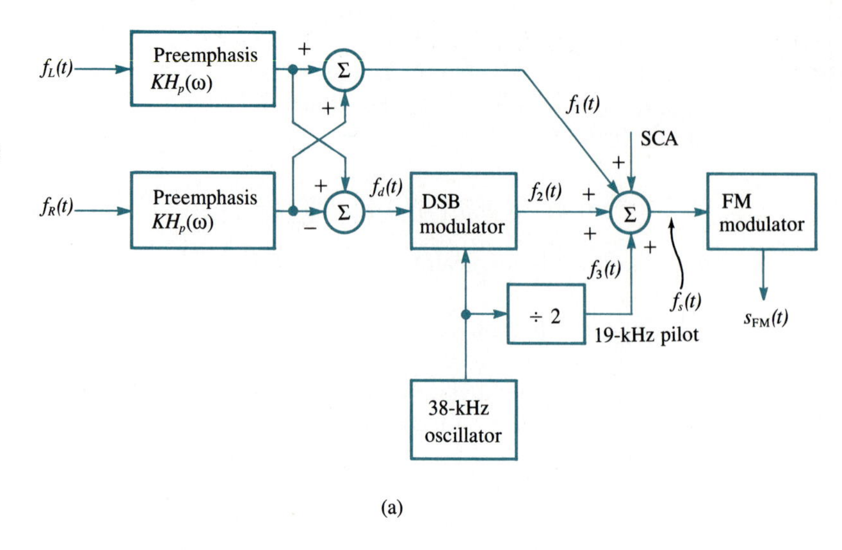

(a)

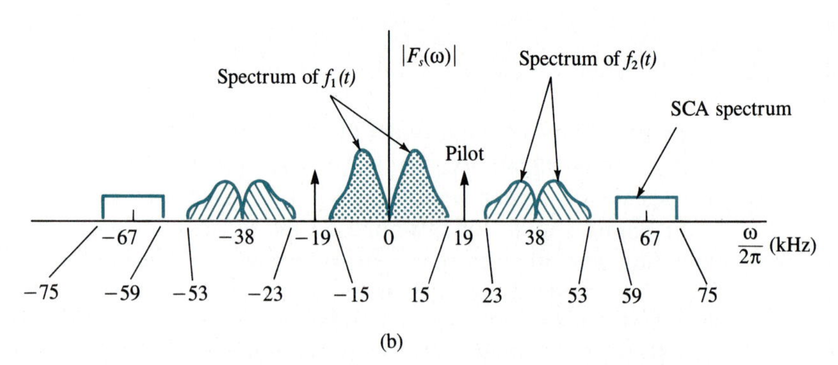

(b)

FIG. 15.7-1. (a) A block diagram for an FM stereo transmitter and (b) the spectrum magnitude of the composite signal $f_s(t)$.

fices, stores, and others that desire background music free of commercials. SCA can also be used for some nonaudio purposes (paging, digital data, some limited video information, etc.)

When SCA is present, it is allowed only 10% of the total peak-frequency deviation; the remaining 90% is allocated 10% for the pilot carrier and 80% for the sum of $f_1(t)$ and $f_2(t)$. When SCA is not present, 10% of the peak deviation is for the pilot carrier and 90% is for the sum of $f_1(t)$ and $f_2(t)$.†

Figure 15.7-2 illustrates a stereo demodulator. An FM demodulator, typically of the discriminator type, first recovers the composite message $f_s(t)$ from the main FM signal. Appropriate filters then select the spectrum portions that correspond to $f_1(t)$, $f_2(t)$, and $f_3(t)$. If the pilot carrier is passed through a frequency doubler, a coherent 38-kHz local-oscillator signal is generated so that a synchronous detec-

† F. G. Stremler, *Introduction to Communication Systems*, 2d ed., Addison-Wesley, Reading, Mass., 1982, p. 300.

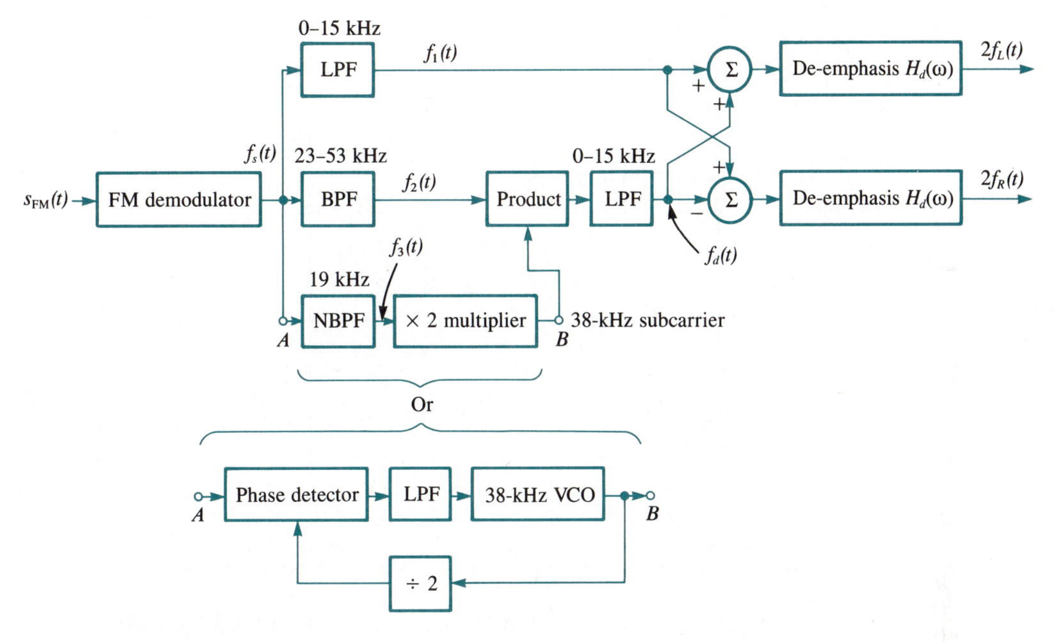

FIG. 15.7-2.
A block diagram of an FM stereo receiver.

tor (product device and lowpass filter) can recover $f_d(t)$ from the DSB waveform $f_2(t)$. Finally, addition and subtraction of $f_1(t)$ and $f_d(t)$ followed by de-emphasis filters lead to recovery of $f_L(t)$ and $f_R(t)$.

In some receivers the narrowband bandpass filter (NBPF) and multiplier in Fig. 15.7-2 can be replaced by the phase-locked loop, shown as an alternative method of local-carrier recovery.

The output signal-to-noise power ratio in stereo FM is smaller than in a monaural system with the same transmitted power, messages, and other parameters. The loss is roughly 22 dB.† The high loss can be tolerated because of the high power being transmitted by many FM stations.

15.8 AN FM RADIO RECEIVER

A block diagram showing the most fundamental parts of a modern FM receiver is drawn in Fig. 15.8-1. The "front-end," which consists of the antenna, radio-frequency amplifier, mixer, and local oscillator, functions in a manner similar to

† H. Taub and D. L. Schilling, *Principles of Communication Systems*, 2d ed. McGraw-Hill, New York, 1986, p. 386.

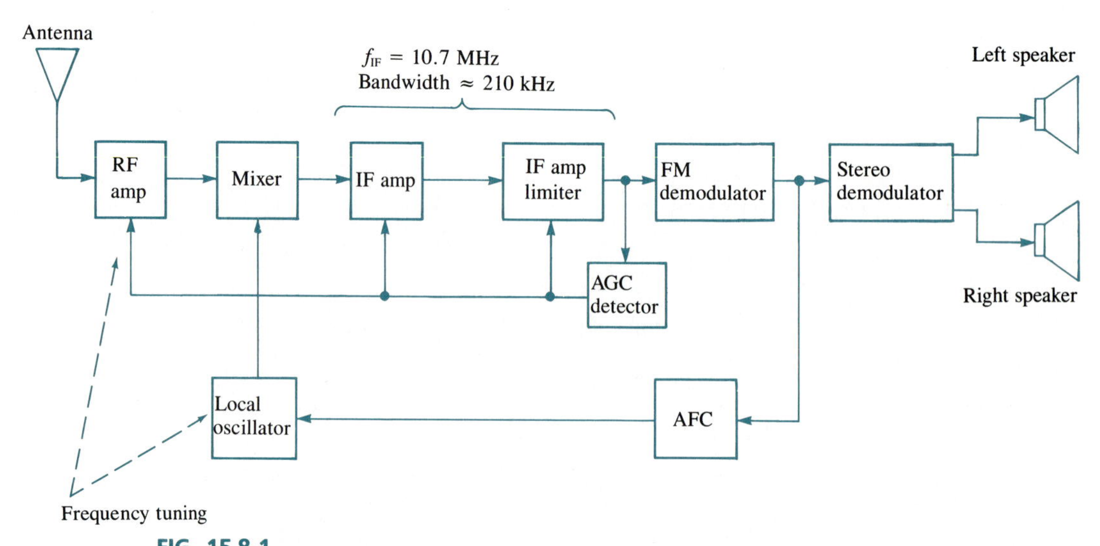

FIG. 15.8-1.
A block diagram of the elements of a typical FM broadcast receiver.

an AM receiver, except that the frequencies involved are different. In particular, the RF amplifier must eliminate the image-frequency band $2f_{IF}$ away from the station to which the receiver is tuned. The IF frequency is 10.7 MHz in FM, so the image is 21.4 MHz from the carrier's frequency f_0.

The IF amplifier is often divided into two parts. The highest-level stage is designed to limit at a proper level to drive the demodulator. In some inexpensive receivers this limiter is the only control over signal amplitude; very large received-signal levels result in saturation (limiting) of the lower-level amplifier stages. More expensive FM receivers may have automatic gain control added to reduce the gains of the RF and IF amplifier stages (as shown in Fig. 15.8-1).

The limited IF-amplifier output drives the FM demodulator. A heavily filtered output from the demodulator is often used to provide an automatic-frequency-control (AFC) loop through the local oscillator, which can be implemented to have electronic tuning by use of a varactor. The AFC loop locks the receiver to the selected station after manual tuning has been accomplished. In the more expensive receivers a meter may be present to show when the AFC loop is operating optimally.

Finally, the demodulator's response feeds the stereo demodulator, which is implemented as shown in Fig. 15.7-2.

15.9 TELEVISION (TV) SYSTEMS

Television is a method of using electric signals to convey aural and fixed or moving visual images from one point to another. Television is in use all over the world and in space. Because many variations exist in the systems implemented,

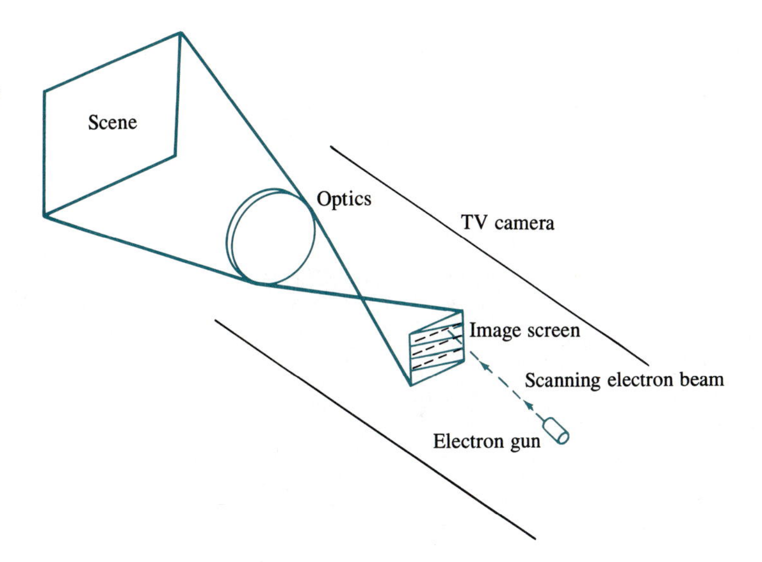

FIG. 15.9-1. The functions involved in a TV camera.

we shall discuss in detail only the standardized system used in the United States. It is known as the NTSC color-TV system (the National Television System Committee's work was adopted in 1954). In the early days of TV,† stations broadcast only monochrome (black and white) pictures. Today nearly all stations broadcast in color and nearly all TV receivers are implemented to receive color. In fact, the NTSC color system is *compatible* with monochrome receivers, so that even though the transmitted signal may carry color, the older monochrome receivers still function as though they were receiving black-and-white images. For these reasons we shall discuss only color TV.

The operation of television is best understood by first considering the basic concept on which it depends for the conversion of visual images to electric signals.

Raster Scanning

Visual images may be converted to electric signals for TV purposes through *raster scanning*. Figure 15.9-1 illustrates the *ideas* involved (actual circuits behave in a more complicated way, but we shall ignore these details for simplicity of discussion). A visual scene is focused at an image plane in a TV camera, somewhat as in an ordinary camera where the image occurs on the film. The image screen is photosensitive, however, so that if it is struck by a narrow electron beam, a voltage can be produced by the TV camera that is proportional to the intensity of the image at the point where the beam strikes. In some tubes the voltage is

† The first public broadcasts of TV were in 1927 in England and in 1930 in the United States. Television evolved from research conducted in the early 1920s.

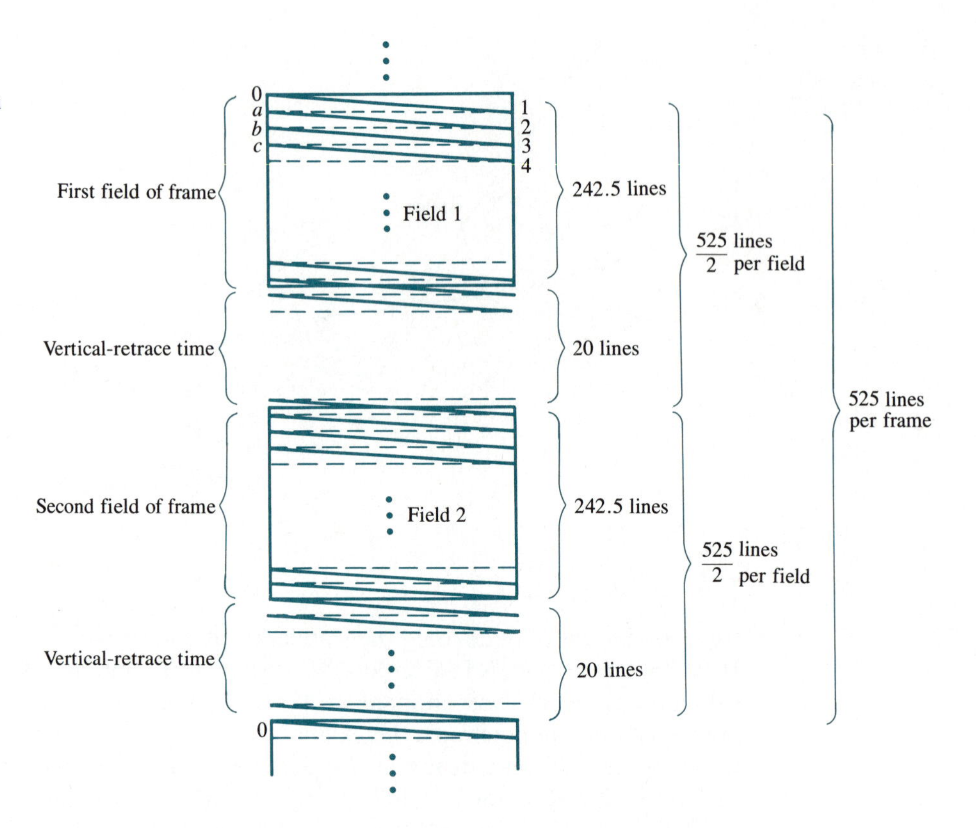

FIG. 15.9-2. The geometry of raster scanning in an NTSC TV system.

generated by photoemission from the target image (the most popular case); others are based on photoconduction.

When the image is scanned left to right and top to bottom in a system of closely spaced parallel lines, as shown, the camera's voltage will vary with time, according to where in the scene the electron beam is scanning. The timing of the scanning is tightly controlled. The receiving system also uses an electron beam to scan a photosensitive phosphor-coated surface in synchronism with that of the TV camera. However, its beam intensity is made to vary with the camera's output voltage, so that the brightness of the phosphor's glow matches the original image.

The TV image area (raster) has a standardized aspect ratio of four units of width for each three units of height. Because images change with time, the rate of scanning and the number of lines must both be large enough to provide quality image reproduction in the receiver. Good performance is achieved when the raster is scanned with 525 lines at a rate of 29.97 *frames* per second.

The exact manner of scanning is illustrated in Fig. 15.9-2. Assume, at time zero, that scanning starts at the upper left corner of the image. The first line traces to point 1 and then quickly retraces to point a, where line 1 is now complete. Line 2 starts at point a, traces to point 2, and quickly retraces to point b. Other lines progress in the same manner at the rate of 15,734.264 lines per second (63.556 μs

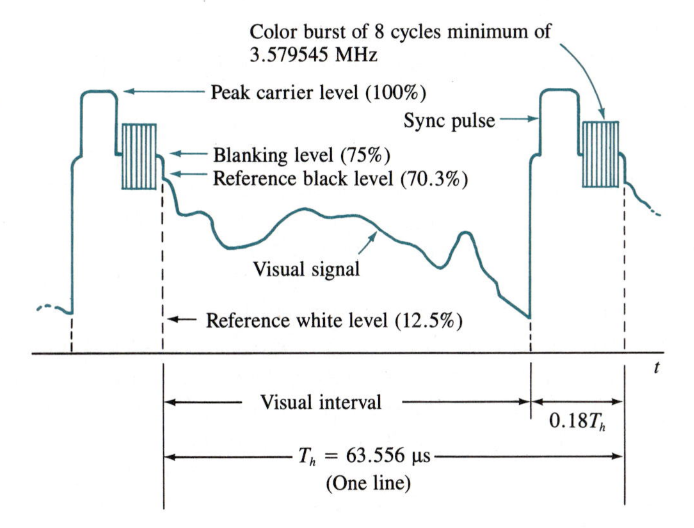

FIG. 15.9-3. A waveform representing one horizontal line of a raster scan in TV.

per line). When 242.5 lines are completed, the raster's visual area has been scanned once; this scan is called a *field*. The next 20 lines† are not used for visual information; during their time special signals are inserted for various reasons (testing, closed captions, etc.), and the beam is retraced vertically to begin a new field. The new field is shown as field 2.

The first line in field 2 starts in the top center of the raster; it is the last half of line 262. Lines in the scanning of field 2 proceed normally but are interlaced vertically between the lines produced in field 1. This *interlacing* helps remove flicker in some moving images. At the end of two fields, which make one frame, 525 lines have occurred (485 in the visual area), and the whole scanning sequence begins anew, with new frames generated at the rate of 59.94/2 = 29.97 frames per second.

The TV waveform representing one scan line is sketched in Fig. 15.9-3. To the visual voltage generated by the camera is added a blanking pulse with a duration of 18% of the horizontal-sweep period. The blanking pulse turns off the electron beam in the receiver's picture tube during the horizontal-retrace time. A synchronization (sync) pulse is added to the top; it is used in the receiver to synchronize its horizontal-scanning rate with that of the transmitter. A burst of at least 8 cycles of 3.579545 MHz, called the *color burst*, is added to the "*back porch*" of the blanking pulse for synchronizing the receiver's color circuits. Finally, the visual information fluctuates according to the image between the "black level" and the "white level," which are set at 70.3% and 12.5%, respectively, of the peak amplitude.

During the 20-line interval (Fig. 15.9-2) at the end of a frame, the horizontal

---

† The number 20 is a nominal value; it may be as small as 19 (for 243.5 visual lines) or as large as 21 (for 241.5 visual lines) in some systems.

blanking pulses are expanded until they merge to form a continuous blanking pulse. An array of various sync pulses are added on top for both horizontal- and vertical-synchronization purposes.

So far nothing has been said about how color is generated. The camera voltage discussed above was only related to the intensity of the image. If a filter is added to the camera optics of Fig. 15.9-1 so that only the color red passes through, the camera's voltage becomes proportional to the intensity of the amount of red in the image. Three such cameras, all synchronized and viewing the same image, are used in color TV to decompose the image into its color components of red, green, and blue. These are the three primary additive colors, which means that they are all that is needed to reconstruct the original scene at the receiver.

The color receiver uses a picture tube with three electron beams and a phosphor having red, green, and blue components. Each beam excites one color of phosphor. At any spot in the image the three colors separately glow with the proper intensities in response to the three transmitted color signals. The eye effectively adds the three colors together to reproduce the original scene in color.

The preceding discussion has revealed the fundamental concepts needed in color TV. It only remains to discuss the details of the transmitter and receiver to see how signals are processed.

Color-TV Transmitter

A block diagram of the most important functions in a TV transmitting station is given in Fig. 15.9-4. The red, green, and blue output signals of the TV camera have the full visual-signal bandwidth (about 4.2 MHz). They are not usually transmitted directly. Rather, three new waveforms are generated according to

$$m_Y(t) = 0.30m_{\mathrm{R}}(t) + 0.59m_{\mathrm{G}}(t) + 0.11m_{\mathrm{B}}(t) \tag{15.9-1}$$

$$m_I(t) = 0.60m_{\mathrm{R}}(t) - 0.28m_{\mathrm{G}}(t) - 0.32m_{\mathrm{B}}(t) \tag{15.9-2}$$

$$m_Q(t) = 0.21m_{\mathrm{R}}(t) - 0.52m_{\mathrm{G}}(t) + 0.31m_{\mathrm{B}}(t) \tag{15.9-3}$$

by the matrix circuit. Part of the reason these signals are used is to make color-TV signals compatible with monochrome receivers. Such receivers respond only to $m_Y(t)$, which is called the *luminance signal*. It has the full visual bandwidth and defines the brightness (whiteness or gray level) of the image.

Signals $m_I(t)$ and $m_Q(t)$, called *chrominance signals*, relate only to the color content of the image. In fact, the quantity $[m_I^2(t) + m_Q^2(t)]^{1/2}$ is called *saturation* (or color intensity). A very deep red is saturated; red diluted with white to give a light pink is nearly unsaturated. The quantity $\tan^{-1}[m_Q(t)/m_I(t)]$ is called *hue* (or *tint*). Experiments have shown that the chrominance signals do not have to have full visual bandwidth. In fact, the bandwidths of $m_I(t)$ and $m_Q(t)$ can be reduced to about 1.6 and 0.6 MHz, respectively, by lowpass filters and the eye can distinguish little effect. These filtered signals are $s_I(t)$ and $s_Q(t)$ in Fig. 15.9-4.

The filtered chrominance signal $s_I(t)$ modulates a *color subcarrier* at fre-

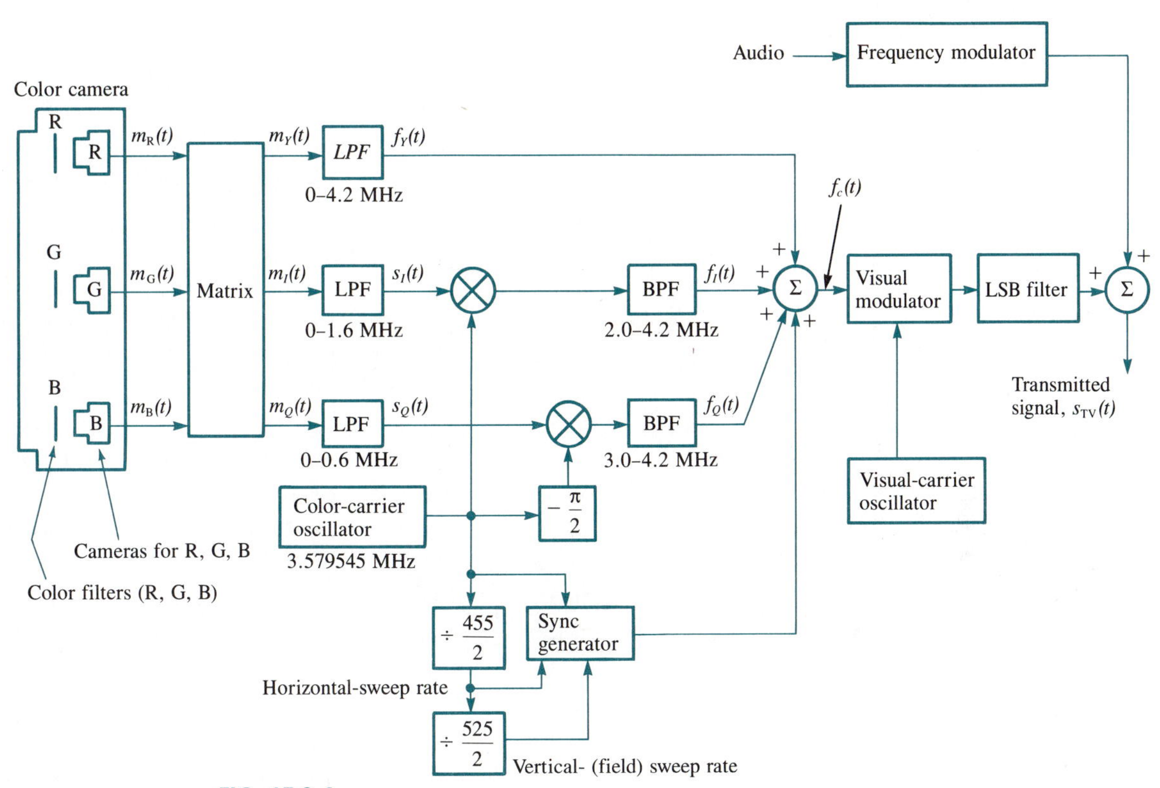

FIG. 15.9-4.
A TV transmitting station showing the most important functions.

quency 3.579545 MHz ± 10 Hz via DSB. The DSB signal is then filtered by the bandpass filter (BPF) of passband 2.0–4.2 MHz. Its effect is to remove part of the USB in the DSB signal so that $f_I(t)$ is approximately a VSB signal. The other chrominance signal modulates a quadrature-phase version of the color subcarrier to produce DSB. This DSB signal passes directly through the BPF with passband 3.0–4.2 MHz, ideally without change.

Next, the transmitter forms a composite "baseband" waveform $f_c(t)$ by adding $f_Y(t)$, $f_I(t)$, $f_Q(t)$, and sync pulses, as shown in Fig. 15.9-3. The composite signal has a bandwidth of about 4.2 MHz and modulates a visual carrier by standard AM. The standard AM signal is then filtered to remove part of the lower sideband. The resulting VSB signal and the audio-modulated aural carrier† are added to form the final transmitted signal $s_{TV}(t)$.

Broadcast television in the United States is regulated by the FCC. Each station is assigned a broadcast channel 6 MHz wide that is designated by a channel

---

† Audio modulates an aural carrier by FM having a peak frequency deviation of 25 kHz. The audio signal's frequency range is 50 Hz to 15 kHz. The audio system uses preemphasis with a 75-μs time constant.

TABLE 15.9-1 TV Channels in Use in the United States

	Channel Number	Band (MHz)	Channel Number	Band (MHz)	Channel Number	Band (MHz)
VHF channels	2	54–60	24	530–536	47	668–674
	3	60–66	25	536–542	48	674–680
	4	66–72	26	542–548	49	680–686
	5	76–82	27	548–554	50	686–692
	6	82–88	28	554–560	51	692–698
	7	174–180	29	560–566	52	698–704
	8	180–186	30	566–572	53	704–710
	9	186–192	31	572–578	54	710–716
	10	192–198	32	578–584	55	716–722
	11	198–204	33	584–590	56	722–728
	12	204–210	34	590–596	57	728–734
	13	210–216	35	596–602	58	734–740
			36	602–608	59	740–746
UHF channels	14	470–476	37	608-614	60	746-752
	15	476–482	38	614–620	61	752–758
	16	482–488	39	620–626	62	758–764
	17	488–494	40	626–632	63	764–770
	18	494–500	41	632–638	64	770–776
	19	500–506	42	638–644	65	776–782
	20	506–512	43	644–650	66	782-788
	21	512–518	44	650–656	67	788–794
	22	518–524	45	656–662	68	794–800
	23	524–530	46	662–668	69	800–806

number. There are 68 channels, as defined in Table 15.9-1† The way the spectrum of $s_{TV}(t)$ is distributed within the 6-MHz channel is shown in Fig. 15.9-5.

EXAMPLE 15.9-1

A TV station operates on channel 20. We find the station's visual carrier's frequency. From Fig. 15.9-5 the visual carrier is 1.25 MHz above the lower edge of the channel, which extends from 506 to 512 MHz from Table 15.9-1. Thus, the station's visual-carrier frequency is 507.25 MHz.

† The frequency bands assigned to channel numbers are different in the cable-TV (CATV) system. See L. W. Couch II, *Digital and Analog Communication Systems*, 3d ed., Macmillan, New York, 1990, p. 433.

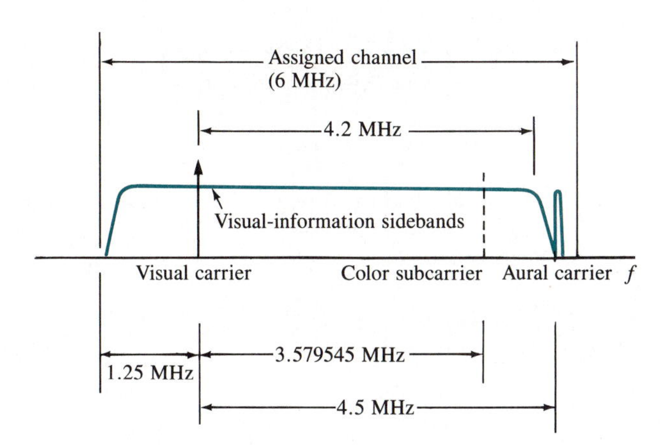

FIG. 15.9-5. A TV signal's spectrum.

Color-TV Receiver

The basic functions required in a color-TV receiver are given in Fig. 15.9-6. The early circuits form a straightforward superheterodyne receiver, except that the frequency-tuning local oscillator is typically a push-button-controlled frequency synthesizer in newer sets. Another difference between TV and radio receivers is that the IF circuitry in TV is especially tuned to give a filter characteristic required in vestigial-sideband modulation. The filter shapes the IF signal's spectrum so that envelope detection is possible.

The envelope detector's output contains the composite visual signal $f_c(t)$ and the frequency-modulated aural carrier at 4.5 MHz. The latter is processed in an FM demodulator to recover the audio information for the speaker. The former is acted on by appropriate filters to separate out signals $f_Y(t)$ and $[f_I(t) + f_Q(t)]$. Further processing of $[f_I(t) + f_Q(t)]$ by two synchronous detectors in quadrature provides the detection to recover $s_I(t)$ and $s_Q(t)$. An appropriate matrix forms linear combinations of $f_Y(t)$, $s_I(t)$, and $s_Q(t)$ that are close approximations of $m_R(t)$, $m_G(t)$, and $m_B(t)$ that originated in the transmitter.† These three signals control the three electron beams in the picture tube.

The envelope detector's output is also applied to circuits that separate the synch signals needed to lock in the horizontal- and vertical-sweep circuits of the receiver. These circuits also isolate the bursts of color carrier so that a phase-locked loop can lock to the phase of the color carrier and provide the reference signals for the chrominance synchronous detectors.

Not shown in Fig. 15.9-6 are many practical circuits needed in a real receiver, such as gates, clamps, dc-restoring circuits, sweep circuits, phase shifters, and others.

† Except for the effects of filtering and noise and practical effects, the equivalence would be exact.

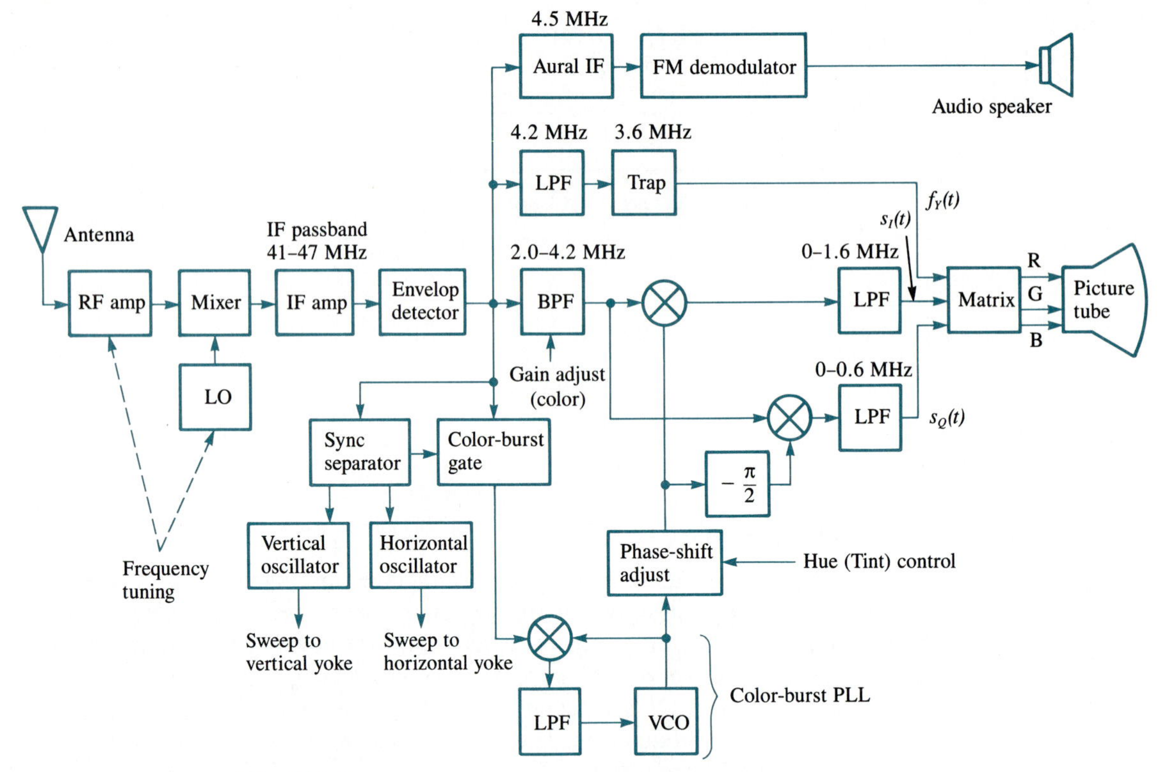

FIG. 15.9-6.
The basic functions in a color-TV receiver.

PROBLEMS

15-1. How many station frequencies are possible in the AM broadcast system in the United States?

15-2. In a standard AM waveform the message is $f(t) = \beta_{AM} A_0 \cos \omega_m t$, where $\beta_{AM} \leq 1$ is a constant called the *modulation index*. Find the amplitude of the sideband frequencies $\omega_0 \pm \omega_m$ relative to the carrier's amplitude in the spectrum of the AM waveform.

15-3. A message has a spectrum

$$F(\omega) = \begin{cases} K \cos \dfrac{\pi\omega}{2W_f} & -W_f \leq \omega \leq W_f \\ 0 & \text{elsewhere} \end{cases}$$

where K and W_f are positive constants. Sketch the spectrum of a standard AM signal that uses the message. What physical significance do K and W_f have in the modulation process?

15-4. Assume a standard AM waveform of 1-kW average power is being transmitted by an antenna with an input resistance of 75 Ω. If efficiency is 0.12, find $\overline{f^2(t)}$. What powers are in the carrier and sidebands?

15-5. In a standard AM system $A_0 = 220$ V, $[\overline{f^2(t)}]^{1/2} = 75$ V, and $Z_0 = 50\ \Omega$. Find (a) η_{AM} and (b) P_c, P_f, and P_{AM}.

15-6. If $R = 5600\ \Omega$ in the envelope detector of a standard AM receiver, what value should its capacitor have?

15-7. At the transmitter in a standard AM system, $P_f = 30$ W. In the receiver, $(S_o/N_o)_{AM} = 250$ when $(S_i/N_i)_{AM} = 3125$. What is the transmitter's unmodulated carrier power? What is the total transmitted power?

15-8. In a standard AM system $\eta_{AM} = 0.18$. The carrier component of the total received-signal power at the receiving antenna's output is 10^{-6} W. The available noise power at the same point is 1.1×10^{-10} W. What is the value of $(S_o/N_o)_{AM}$?

15-9. An AM radio receiver is tuned to a station at 890 kHz and has a high-side local oscillator. What is the image frequency?

★ **15-10.** Assume that an AM broadcast receiver is designed with a low-side local oscillator. Find the values of f_{LO} and f_{image} as f_0 is varied from 1600 to 540 kHz. Give reasons why the low side is not as good a choice as a high-side local-oscillator frequency.

15-11. Assume that the tuned circuits at frequency f_0 in a standard AM receiver attenuate frequencies at a rate of -40 dB per decade for frequencies outside the 3-dB bandwidth $2W_3$; that is, the relative voltage response of these circuits is $1/\{1 + j[(\omega - \omega_0)/W_3]\}^2$. What value of W_3 is needed to give 50 dB of attenuation to the nearest edge of a 10-kHz band centered on the image frequency? Assume a high-side local-oscillator frequency.

15-12. Sketch the transmitted signal for a DSB system for which $f(t) = A_m \cos \omega_m t$, $\omega_0 = 5\omega_m$, and $\theta_0 = 0$.

15-13. Discuss how the circuit of Fig. 15.4-2(a) works.

★ **15-14.** It can be shown that an SSB signal can be expressed exactly by

$$s_{SSB}(t) = f(t) \cos(\omega_0 t + \theta_0) \mp \hat{f}(t) \sin(\omega_0 t + \theta_0)$$

where upper and lower signs correspond to an SSB signal with USB and LSB sidebands not removed, respectively. Here $\hat{f}(t)$ is a waveform having the Fourier transform

$$\hat{F}(\omega) = -j[2u(\omega) - 1]F(\omega)$$

where $u(\omega)$ is the unit step function and $F(\omega)$ is the spectrum of $f(t)$. Find an expression for $\hat{f}(t)$ in terms of $f(t)$.

15-15. Interpret the second equation in Problem 15-14 as $F(\omega)$ acted on by a filter to produce $\hat{F}(\omega)$. Then use the interpretation to implement the first equation by a circuit to generate $s_{SSB}(t)$.

15-16. A receiver is designed to receive a DSB signal and produces $(S_o/N_o)_{DSB} = 5 \times 10^4$ with a particular message and transmitted power. If the transmitter is converted to SSB with the same power and message, the unchanged receiver can still demodulate the message. What value of (S_o/N_o) will the DSB receiver produce when receiving the SSB signal?

15-17. A commercial FM station broadcasts a signal with 180-kHz bandwidth when $|f(t)|_{max} = 2$ V. What is k_{FM} for the modulator?

15-18. An FM signal, for which $f(t) = 1.8 \cos \omega_m t$, is given by

$$s_{FM}(t) = 53 \cos \left\{183\pi(10^6)t + \frac{\pi}{3} + 6 \sin [\pi(10^4)t]\right\}$$

Find (a) A_0, (b) ω_0, (c) θ_0, (d) ω_m, (e) k_{FM}, and (f) β_{FM}.

15-19. An FM station's modulator has a sensitivity defined by $k_{FM} = 5\pi \times 10^4$ (rad/s)/V. A receiver uses a discriminator that has a gain constant of $10^{-5}/\pi$ V/(rad/s). Determine the signal at the receiver's output. Neglect noise.

15-20. In an FM system without emphasis $\Delta\omega/W_f = 5$, $K_{cr} = 4$, and $(S_o/N_o)_{FM}$ must be at least 4×10^4 (or 46.0 dB). (a) What is the smallest that $(S_i/N_i)_{FM}$ can be? (b) Is the system operating above threshold?

15-21. An FM system without emphasis produces $(S_i/N_i)_{FM} = 50$ and $(S_o/N_o)_{FM} = 324$. With all else held fixed, the transmitter increases $\Delta\omega$ by a factor of 4 to improve performance. The receiver's bandwidth is also increased by a factor of 4 to accommodate the wider-bandwidth signal. Find $(S_i/N_i)_{FM}$ and $(S_o/N_o)_{FM}$ in the modified receiver. Is the system operating above threshold?

15-22. If $\Delta\omega$ and receiver bandwidth in Problem 15-21 had been increased by a factor of 8 instead of 4, discuss the effect.

15-23. In the unmodified system of Probem 15-21, $\Delta\omega/W_f = 3$. What is the message's crest factor?

15-24. A sinusoidal message is conveyed over an FM system in which $\beta_{FM} = 7$ and no emphasis is present. In the receiver $(S_i/N_i)_{FM} = 11$. Find $(S_o/N_o)_{FM}$. Is operation above threshold?

15-25. A voice message for which $W_f = 2\pi(3.3 \times 10^3)$ rad/s and $W_{rms} = 2\pi(0.8 \times 10^3)$ rad/s is transmitted over an FM broadcast system with standard emphasis. Compare the improvements due to emphasis as given by both (15.6-8) and (15.6-9).

15-26. Assume that the DSB modulator of Fig. 15.7-1 is a simple product device and its oscillator input is $\cos \omega_{sc}t$, where $\omega_{sc}/2\pi = 38 \times 10^3$ Hz. In Fig. 15.7-2, assume that the 0–15-kHz LPF and the 23–53-kHz BPF both have a nominal voltage gain of unity. If the oscillator input to the product device is $\cos \omega_{sc}t$, what gain must the 0–15-kHz LPF have to produce the outputs indicated? For simplicity, assume no emphasis; i.e., the emphasis and deemphasis filters are absent.

15-27. Demonstrate that the image frequency for an FM system does not fall in the range 88.1–107.9 MHz regardless of choice of high- or low-side local oscillator.

★ **15-28.** In a typical color-TV receiver there are 485 lines in the image part of the raster scan. The image part of each line lasts 52.12 μs. If the horizontal line is divided into resolution cells (say alternating black and white squares) so as to have the same spacial size as the lines give vertically, estimate the bandwidth of the TV's video signal.

15-29. In the signals of (15.9-1)–(15.9-3), assume that the red, green, and blue signals have a maximum amplitude of unity (are normalized). If $m_I(t)$ and $m_Q(t)$ are the real and imaginary components of a color vector defined by a magnitude $[m_I^2(t) + m_Q^2(t)]^{1/2}$ and phase $\tan^{-1}[-m_Q(t)/m_I]$, plot points corresponding to fully saturated red, green, and blue colors. A popular TV instrument called a *vectorscope* plots colors as vectors (its axes are shifted 33° clockwise from those defined here, however).

CHAPTER 16

Digital Communications Systems

16.0 INTRODUCTION

An analog signal has been defined as having any one of a continuum of possible amplitudes at any given time. An analog system is one that processes the entire analog signal (or message) to convey information to a user (the receiver). We now define a *digital signal* as having any one of a finite number of discrete amplitudes at any given time. In general, the signal could be a voltage, a current, or just a number, such as 0 or 1.† Sometimes, we may call a digital signal a digital *message*. A communications system designed to process only digital signals to convey information is called digital.

Many modern communications systems are now designed to be digital instead of analog. In fact, the recent trend is to make as much of the system digital as possible. Digital systems are preferred over analog for many reasons: Discrete data are efficiently processed, while analog messages can also be transmitted by using suitable signal-processing methods to convert them to digital form; digital systems interface well with computers; they are very reliable and give high performance with low cost; they are flexible and able to accommodate a variety of messages with ease; security techniques are readily available to provide message privacy to users; they may be structured so that advanced signal-processing

† A signal for which only two amplitudes are possible is called a *binary digital signal*. This type of waveform is commonly used in computers and most digital communications systems.

techniques can be added after initial construction. These advantages come at the expense of some disadvantages too. The most serious of these are the added complexity required for system synchronization and the need for larger bandwidth than in an equivalent analog system.

Because a digital system sends information only in discrete amounts, its interface with a source having only discrete messages is direct. Characters (such as letters, numbers, punctuation symbols) needed to send a report from one company to another represent a source of discrete messages. Similarly, the numbers contained in a financial data sheet to be conveyed electronically from one bank to another represent a digital (discrete) source.

Digital systems can also convey analog messages if they are first converted to digital form. In Sec. 16.2 we shall discuss how the conversion is accomplished. Thus, a digital communications system, with suitable conversion of the form of message, can transmit any type of message. Systems currently exist that can simultaneously transmit audio, television, and digital data over the same channel.

16.1 OVERALL DIGITAL COMMUNICATIONS SYSTEM

Before the development of details it is helpful to place the main functions of an overall system in perspective. Figure 16.1-1 sketches these functions. For each function in the transmitting station there is an inverse operation in the receiver, so our main emphasis in discussions is on the transmitter. If an analog message is to

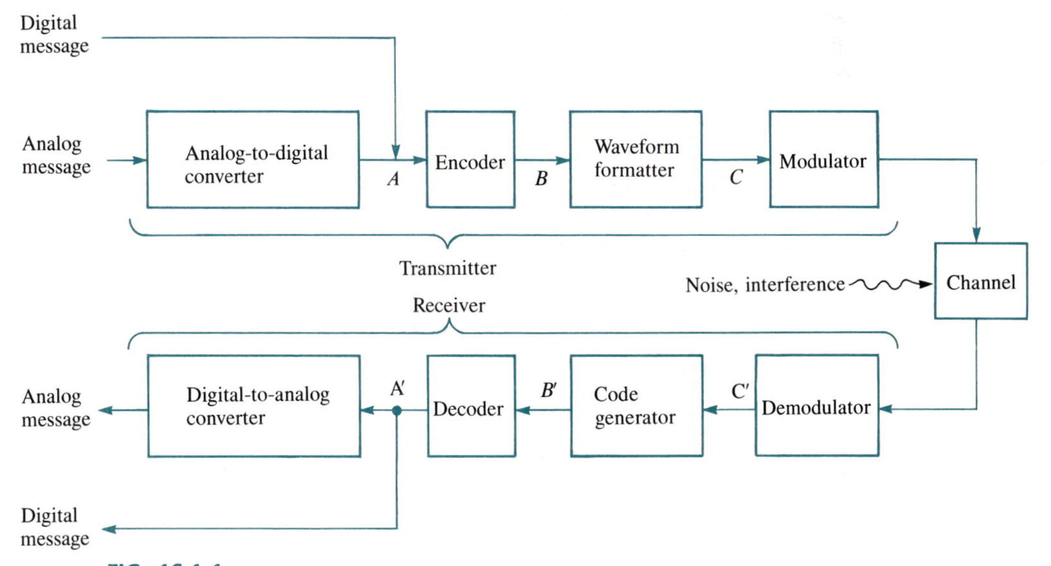

FIG. 16.1-1.
The main functions in a digital communications system.

be conveyed to the receiver, it must first be converted to a digital signal by an *analog-to-digital* (A/D) *converter*. The result is a signal at point A with one of L possible levels at any given time. If no analog message is involved, an equivalent L-level digital signal can be the input at point A directly.

Encoding is a critical function in all digital systems. It has a direct bearing on equipment complexity and performance in noise. Generally, the encoder operation consists of both *source-encoding* and *channel-encoding* functions. In source encoding each of the possible L levels at point A is converted to a sequence of the two levels 0 and 1.† The particular sequence of 0s and 1s that corresponds to a particular one of the L levels is called a *code word*. The set of all possible code words is known as a *binary code*, and the L-level to binary-level conversion is referred to as *coding*. The process of source encoding is not unique, and many codes are possible. We later discuss a simple code that leads to a simple system implementation. Other codes are more complicated but are more optimum in the sense that fewer binary digits (on the average) are needed to adequately represent the L-level source.

The purpose of channel encoding is to convert the binary code generated by the source encoder to another binary code that will lead to fewer errors in the receiver due to noise when demodulation takes place. The number of bits in a code word at the channel encoder's output is typically larger than at its input. The extra bits are added in special ways to allow for error identification and correction in the receiver. Channel encoding is such an advanced topic that we shall not consider it further in this book.

The waveform formatter in Fig. 16.1-1 accepts the sequence of binary digits from the encoder and converts each into a suitable waveform. It might, for example, assign a 2-μs pulse with 5-V amplitude to each digit 1 and a 2-μs interval of 0 V to every digit 0. Other waveform formats are possible, as we subsequently discuss.

The modulator (Fig. 16.1-1) serves a similar function to that of an analog system. It is the point where the digital waveform (message) modulates the amplitude, phase, or frequency of a carrier for transmission over the channel. However, in the digital system where the message has only two amplitudes, the modulation process is referred to as *keying*. In *amplitude-shift keying* (ASK) a carrier's amplitude is shifted (keyed) between two levels. *Phase-shift keying* (PSK) involves keying between two phase angles of the carrier, and shifting a carrier's frequency between two values is called *frequency-shift keying* (FSK). Many other forms of modulation are also possible. However, because of our limitation in scope, we limit our discussions to only ASK, PSK, one variation of PSK, and FSK.

The receiver functions in Fig. 16.6-1 are the inverses of those in the transmitter. Thus, except for errors due to noise, interference, and practical system

---

† We consider only the two-level, or binary, encoder. Each digit in the sequence is called a binary digit (or *bit*, for short). Thus, the sequence 011010 is a 6-bit sequence.

imperfections, the signals at points A', B', and C' would be identical to those at points A, B, and C, respectively. The digital-to-analog (D/A) converter reconstructs an analog message that is a close approximation to the original message. As we shall later find, this reconstructed message cannot equal the original waveform exactly (even in theory) because of the nature of the A/D and D/A operations.

In developing discussions of digital systems, we find it helpful to begin with the methods by which analog messages are converted to digital form.

16.2 ANALOG-TO-DIGITAL CONVERSION OF MESSAGES

A digital communications system transmits only discrete amounts of information at discrete times. Analog signals may be converted to the necessary form for transmission over the digital system by three operations: sampling, quantization, and coding.

Sampling

Sampling of an analog signal makes it discrete in time. We already found in Sec. 2.7 that a bandlimited signal can be recovered exactly from its samples taken periodically in time at a rate at least equal to twice the signal's bandwidth. Since any real signal has some frequency above which its spectral content can be taken as negligible, we may consider it bandlimited. Thus, sampling is the means by which an analog message is made discrete in time for use with digital systems. If a message $f(t)$ has a spectral extent W_f (in radians per second), the sampling rate f_s (samples per second) must satisfy

$$f_s \geq \frac{W_f}{\pi} \tag{16.2-1}$$

from the sampling theorem. The minimum rate W_f/π (samples per second) is called the Nyquist rate.

If the digital system could transmit the *exact* message samples, the receiver could reconstruct $f(t)$ exactly (no error) at all times. Unfortunately, the exact samples cannot be conveyed and must be converted to discrete samples in a process of quantization.

Quantization

Quantization consists of rounding exact sample values to the nearest of a set of discrete amplitudes called *quantum levels*. Figure 16.2-1(a) illustrates the quantization of samples of a message that is known to have values between 0 and 7 V. The quantizer is assumed to have eight quantum levels, 0, 1, 2, . . . , 7 V. As

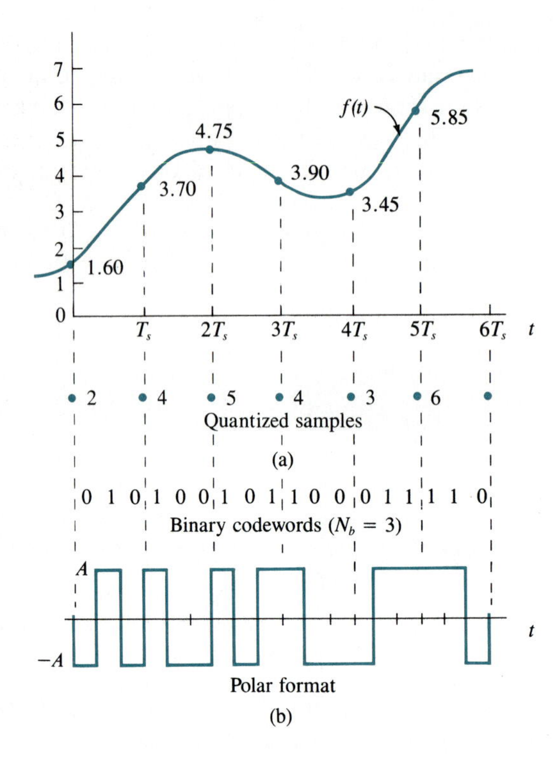

FIG. 16.2-1. (a) An analog waveform showing quantized samples and (b) the coding and waveform formatting of quantized samples.

shown, a sequence of exact samples is 1.60, 3.70, 4.75, 3.90, 3.45, and 5.85 V. Quantization results in a sequence of *quantized samples* (quantum levels) of 2, 4, 5, 4, 3, and 6 V.

The quantizer of Fig. 16.2-1(a) is called *uniform* because the separation between any two adjacent quantum levels is a constant called the *step size*, which we denote by δv (volts). Clearly, a uniform quantizer is not limited to messages with only nonnegative voltages. Figure 16.2-2 depicts the output- (quantum levels) versus input-voltage characteristic of an L-level quantizer to operate with messages having both positive and negative amplitudes of the same maximum magnitude. The characteristic is stairstep in shape, and two forms are possible. The form of Fig. 16.2-2(a) applies when L is an even integer and is called *midriser*. The *midtread* quantizer of Fig. 16.2-2(b) applies when L is odd. Both forms of quantizer are said to be saturated (or overloaded) when†

$$|f(t)| > \left(\frac{L-2}{2}\right)\delta v + \delta v = \frac{L}{2}\,\delta v \qquad \text{(overload)} \qquad (16.2\text{-}2)$$

Quantizers can also be designed to have *nonuniform* separation of quantum

† Our quantizers are assumed to have a "gain" of unity, so tread widths and riser heights all equal δv. Other gains are also possible (see Problem 16-7).

FIG. 16.2-2. Quantizer characteristics: (a) midriser when L is even and (b) midtread when L is odd.

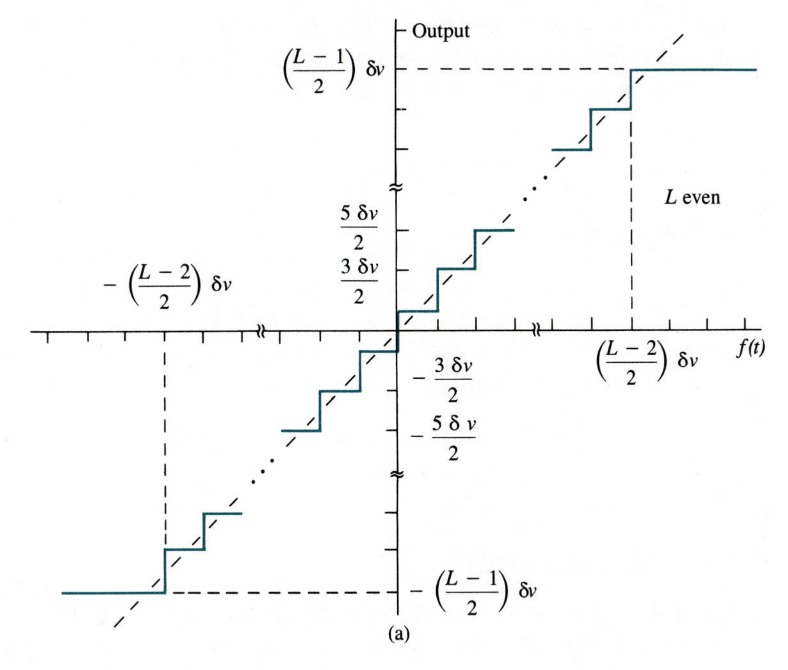

(a)

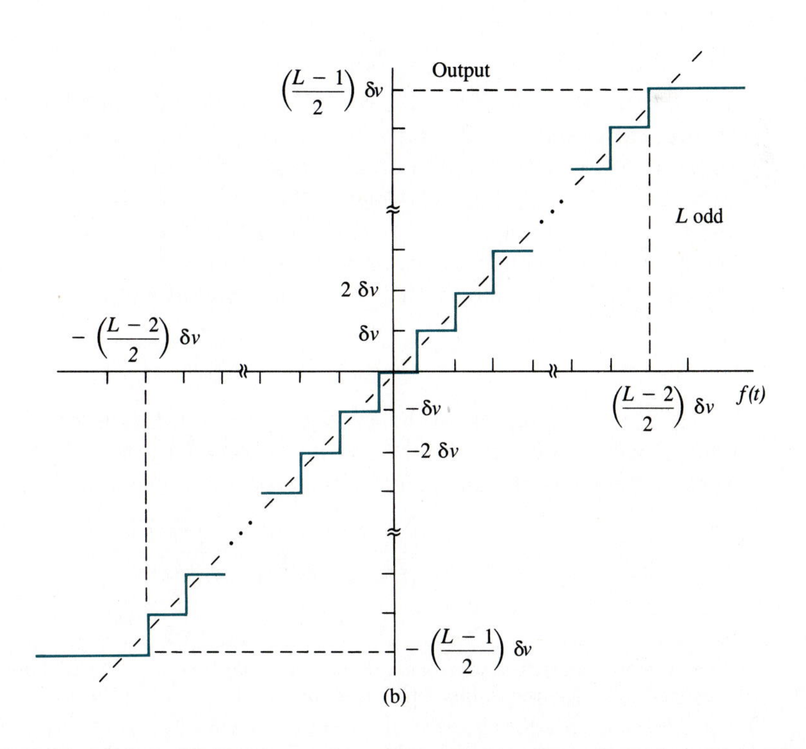

(b)

levels. These quantizers can lead to improved system performance over the uniform devices. Analysis is complicated, however, and we shall restrict our discussion to only the uniform quantizer.†

EXAMPLE 6.2-1

A uniform quantizer is to be designed to have 16 levels. The extreme (saturation) levels are to correspond to extreme values of the message of 1.4 V $\le f(t) \le$ 12.6 V. We find the quantum levels. A sketch of the quantizer's characteristic shows 16 "treads" of $f(t)$ between 1.4 V and 12.6 V. A tread width, denoted by δv, is $(12.6 - 1.4)/16 = 0.7$ V. The first quantum level is therefore at $1.4 + (\delta v/2) =$ 1.75 V. Other quantum levels, denoted by ℓ_i, are

$$\ell_i = 1.75 + (i - 1)\,\delta v = 1.75 + (i - 1)(0.7) \qquad i = 1, 2, \ldots, 16$$

Quantization Error

Careful study shows that sampling followed by quantization is equivalent to quantization followed by sampling. By taking the latter viewpoint, we may readily expose the system limitations caused by quantization. Consider Fig. 16.2-3, which shows a message $f(t)$ and its quantized version, denoted by $f_q(t)$. It's clear that $f_q(t)$ is equivalent to

$$f_q(t) = f(t) + \epsilon_q(t) \tag{16.2-3}$$

where $\epsilon_q(t)$ is a *quantization error*. Now a digital communications system can theoretically convey the sampled quantum levels [samples of $f_q(t)$] without error to the receiver (assuming, of course, that effects of noise and practical imperfections are neglected). This fact means that, theoretically, $f_q(t)$ can be recovered in the receiver without error. From (16.2-3) the recovery of $f_q(t)$ is viewed as the recovery of $f(t)$ with an error, or noise, $\epsilon_q(t)$ present. For a large number of levels (small δv) it can be shown that the mean-squared value of $\epsilon_q(t)$ is‡

$$\overline{\epsilon_q^2(t)} = \frac{(\delta v)^2}{12} \tag{16.2-4}$$

If S_o and N_q represent the average powers in $f(t)$ and $\epsilon_q(t)$, respectively, the best signal-to-noise ratio that can be achieved by any digital communications system that transmits an analog signal processed by a uniform quantizer becomes

$$\left(\frac{S_o}{N_q}\right) = \frac{\overline{f^2(t)}}{\overline{\epsilon_q^2(t)}} = \frac{12\overline{f^2(t)}}{(\delta v)^2} \tag{16.2-5}$$

† For a more complete discussion, see Peyton Z. Peebles, Jr., *Digital Communication Systems*, Prentice-Hall, Englewood Cliffs, N.J., 1987, pp. 58–78.

‡ Recall that the overbar represents the average or mean value, as adopted in Sec. 15.1.

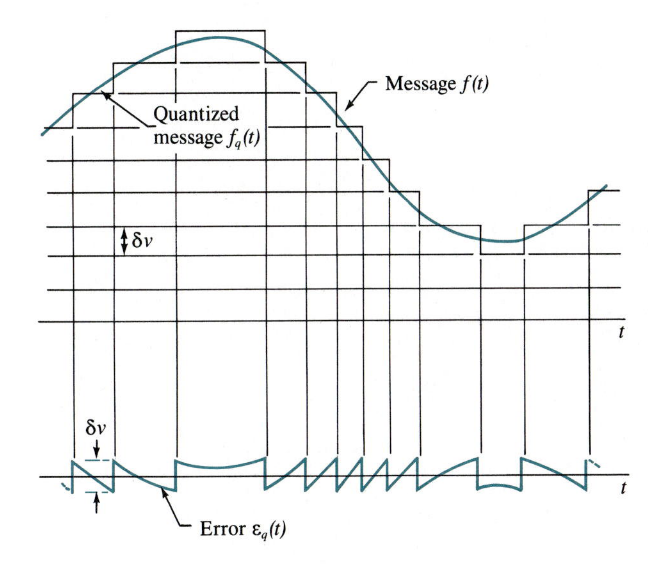

FIG. 16.2-3. A message, its quantized version, and the quantization error.

For the most common message, $f(t)$ fluctuates symmetrically between equal-magnitude extremes. That is, $-|f(t)|_{\max} \leq f(t) \leq |f(t)|_{\max}$. For this message, if a large number L of levels span its variations,

$$\delta v = \frac{2|f(t)|_{\max}}{L} \tag{16.2-6}$$

and

$$\left(\frac{S_o}{N_q}\right) = \frac{3L^2\overline{f^2(t)}}{|f(t)|^2_{\max}} \tag{16.2-7}$$

The ratio of peak-amplitude to message-rms value has already been defined as a message's crest factor K_{cr}:

$$K^2_{\text{cr}} = \frac{|f(t)|^2_{\max}}{\overline{f^2(t)}} \tag{16.2-8}$$

In terms of the crest factor (16.2-7) can be expressed as

$$\left(\frac{S_o}{N_q}\right) = \frac{3L^2}{K^2_{\text{cr}}} \tag{16.2-9}$$

We see that messages with large crest factors lead to poor performance.

Companding

A process called *companding* can be used to lower a waveform's crest factor to produce better performance. The procedure is to pass the message through a nonlinear network, called a *compressor*, that progressively compresses the message's larger amplitudes. The inverse operation in the receiver is called an *expan-*

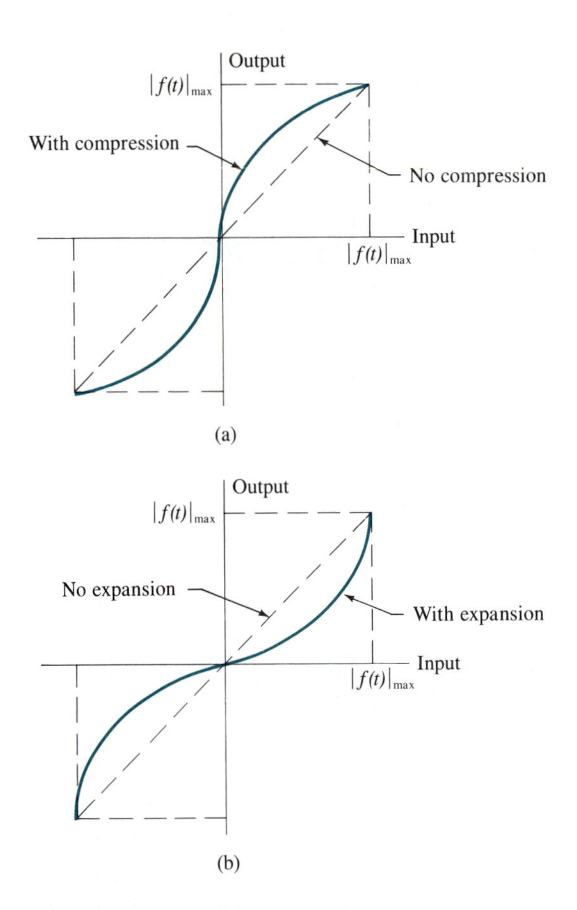

FIG. 16.2-4. Compandor characteristics: (a) compressor and (b) expandor.

dor and it restores the original message. Taken together, a compressor and expandor form a *compandor*.

The action of the compressor is to increase the rms-signal value for a given peak magnitude. It can be implemented in several ways. Perhaps the easiest to visualize is a nonlinear network acting on the message prior to quantization by a quantizer with uniformly separated quantum levels. A typical set of input-output characteristics for this form of compandor is sketched in Fig. 16.2-4. However, in practice, the equivalent effect can be achieved by a quantizer having nonuniform quantization.

Source Encoding

After quantization of message samples, the digital system must next code each quantized sample into a sequence of binary digits (bits) 0 and 1. One simple approach is to use the *natural binary code* of Table 16.2-1. For a code with N_b bits, integers N from 0 to $2^{N_b} - 1$ are represented by a sequence of digits, b_{N_b}, $b_{N_b-1}, \ldots, b_2, b_1$, according to

$$N = b_{N_b}(2^{N_b-1}) + \cdots + b_2(2^1) + b_1(2^0) \tag{16.2-10}$$

TABLE 16.2-1 Natural Binary Code for the First 21 Nonnegative Integers

Decimal Integer	Natural Binary Code Words
0	00000
1	00001
2	00010
3	00011
4	00100
5	00101
6	00110
7	00111
8	01000
9	01001
10	01010
11	01011
12	01100
13	01101
14	01110
15	01111
16	10000
17	10001
18	10010
19	10011
20	10100

Thus, the decimal level 12 can be described by a 4-bit code, and the code word for 12 is 1100, since $N = 12 = 1(2^{4-1}) + 1(2^{4-2}) + 0(2^{4-3}) + 0(2^{4-4}) = 2^3 + 2^2 = 12$. Here b_1 is called the *least significant bit* (LSB), and b_{N_b} is known as the *most significant bit* (MSB).

Generally, a natural binary code of N_b bits can encode $L_b = 2^{N_b}$ levels. If L levels span the message's variations, then we require

$$L \leq L_b = 2^{N_b} \tag{16.2-11}$$

Figure 16.2-1(b) shows the natural binary encoding of the message sketched in (a).

EXAMPLE 16.2-2

An encoder uses an 8-bit natural binary code to encode 256 voltage levels from -7.65 V to $+7.65$ V in steps of $\delta v = 0.06$ V. A symmetrically fluctuating message for which $|f(t)|_{\max} = 6.27$ V and $K_{cr} = 3$ is encoded. We find $L, \overline{f^2(t)}$, and (S_o/N_q) from (6.2-9). From (6.2-6), $L = 2(6.27)/0.06 = 209$. From (6.2.8), $\overline{f^2(t)} = (6.27)^2/9 = 4.368\ \text{V}^2$. From (6.2-9), $(S_o/N_q) = 3(209)^2/9 \approx 14{,}560$ (or 41.6 dB).

16.3 FORMATTING OF DIGITAL SIGNALS

Once the message's samples have been quantized and coded into a sequence of bits (binary 0s and 1s), a suitable waveform must be chosen to represent the bits. This waveform can then be transmitted directly over the channel, if no carrier modulation is involved, or used for carrier modulation. We refer to the process of waveform selection as formatting of the digital sequence.

Waveforms for Formatting

The *unipolar waveform* simply assigns a pulse to a code 1 and no pulse to a 0. If binary digits occur each T_b seconds (the bit interval's duration), the duration of a typical pulse can be less than or equal to T_b. The equality is usually selected because it produces the best system performance.

A *polar waveform* consists of a pulse of duration T_b for a binary 1 and a negative pulse of the same magnitude and duration for a 0. For the same magnitude of pulses the polar waveform produces better system performance in noise than the unipolar signal.

The *Manchester waveform* transmits a pulse of duration $T_b/2$ followed by an equal-magnitude, but negative, pulse of duration $T_b/2$ for each binary 1. A binary 0 causes the negative of this two-pulse sequence to occur. A main advantage of the Manchester signal is that it never contains a dc component, even when a long string of 0s or 1s might occur in the digital sequence.

Figure 16.3-1 illustrates unipolar, polar, and Manchester waveforms for a possible sequence of binary digits.

When processing polar or Manchester waveforms, a digital system must not lose track of polarity. An unknown sign inversion leads to the complement (1s and 0s replaced by 0s and 1s, respectively) of the correct digital sequence in the

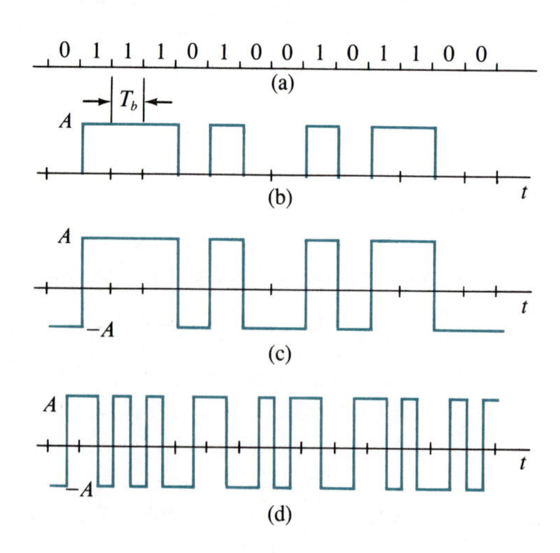

FIG. 16.3-1. (a) A binary digital sequence $\{b_k\}$ and its formats: (b) unipolar, (c) polar, and (d) Manchester.

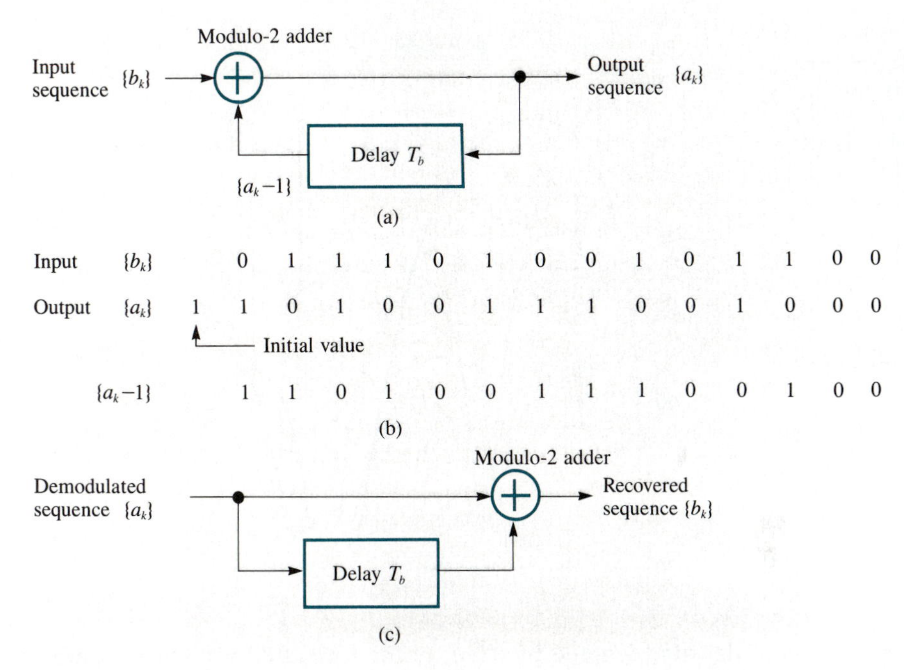

FIG. 16.3-2. (a) A differential encoder, (b) some representative sequences, and (c) the decoder.

receiver. An ingenuous technique, called *differential encoding*, removes the need to maintain polarity. The scheme converts the original sequence of digits, denoted by $\{b_k\}$, to a new sequence of digits, denoted by $\{a_k\}$, by using the differential encoder of Fig. 16.3-2(a). The output digit in the kth interval is

$$a_k = a_{k-1} \oplus b_k \tag{16.3-1}$$

where $\oplus$ represents modulo-2 additon ($0 \oplus 0 = 0$, $0 \oplus 1 = 1$, $1 \oplus 0 = 1$, and $1 \oplus 1 = 0$). The new sequence is used in waveform formatting. Some example sequences are given in Fig. 16.3-2(b), and the required decoder in the receiver is shown in (c).

Spectra of Formatted Waveforms

Examination of typical digital sequences indicates that digits fluctuate randomly between the two digits 0 and 1 with time. The formatted waveform then is a randomly fluctuating set of pulses defined by the chosen format. With random waveforms it is the *power spectral density*,† and not the Fourier transform, that is important in defining spectral content. If binary digits occur at the rate of one every T_b seconds and pulses in the format have magnitude A, the power spectral

† Power spectral density has the unit volts squared per hertz; if the random voltage exists across a resistance R, then the waveform's *power spectrum* (in watts per hertz) equals the power spectral density divided by R. Similarly, a random current has a power spectrum that equals R times the power spectral density (with unit of amperes squared per hertz).

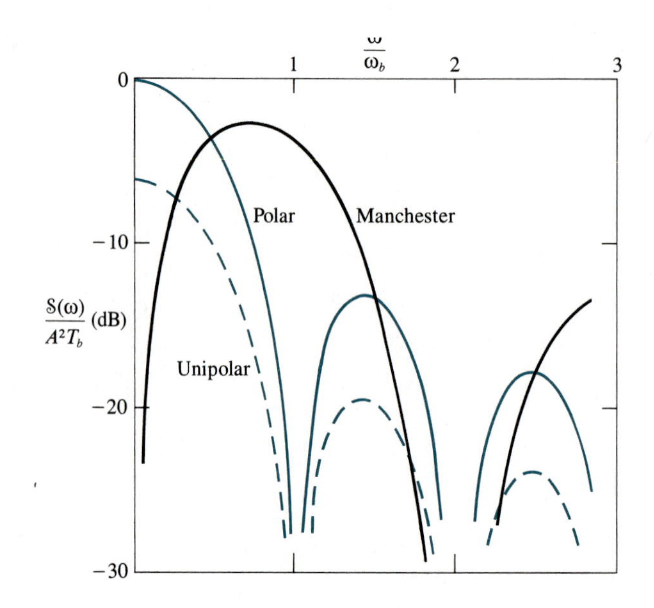

FIG. 16.3-3. Power spectral density for unipolar, polar, and Manchester waveforms.

density, denoted by $\mathcal{S}(\omega)$, of the formatted waveform can be shown to be†

$$\mathcal{S}(\omega) = \begin{cases} \frac{A^2\pi}{2}\,\delta(\omega) + \frac{A^2T_b}{4} Sa^2\left(\frac{\omega T_b}{2}\right) & \text{unipolar} \\ A^2T_b Sa^2\left(\frac{\omega T_b}{2}\right) & \text{polar} \\ A^2T_b \sin^2\left(\frac{\omega T_b}{4}\right) Sa^2\left(\frac{\omega T_b}{4}\right) & \text{Manchester} \end{cases} \tag{16.3-2}$$

all applicable for $-\infty < \omega < \infty$.

For comparison, the continuous portions of the functions of (16.3-2) are plotted in Fig. 16.3-3. The impulse term in the unipolar expression corresponds to a dc component having half the total power in the waveform. The unipolar and polar formats both have the same bandwidth and relative side-lobe level. The Manchester waveform has no spectral component at dc but requires twice the bandwidth of the other signals.

To get a feel for some numerical values of bandwidth, we show an illustrative example.

EXAMPLE 16.3-1

Assume an audio message is bandlimited to 15 kHz, sampled at twice the Nyquist rate, and encoded by a 12-bit natural binary code that corresponds to $L_b = 2^{12} = 4096$ levels, all of which are assumed to span the message's variations. We find the

† Recall that $Sa^2(x) = [\sin(x)/x]^2$.

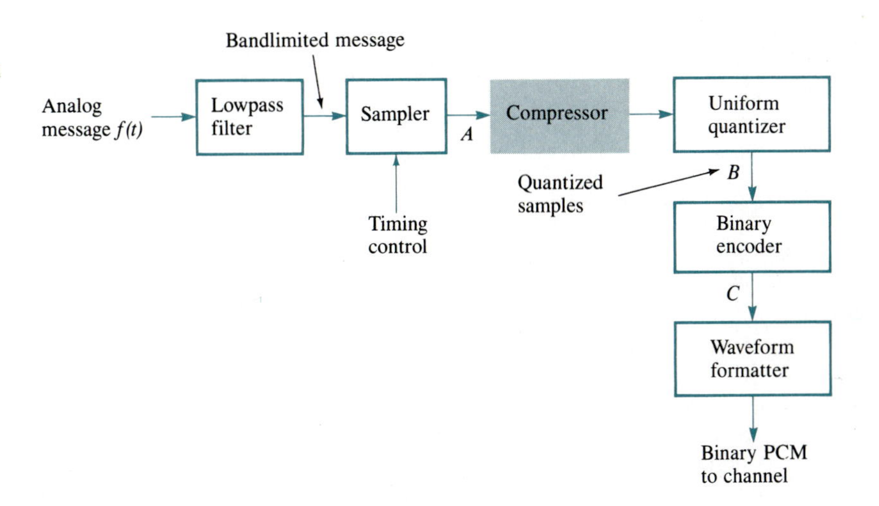

FIG. 16.4-1. A block diagram of the functions required to generate PCM.

first-null bandwidth required to support a polar-waveform format. We also find the best possible performance from (6.2-9) if $K_{cr} = 4$.

Since sampling is twice the Nyquist rate, it is 4 times the spectral extent, or $4(15 \times 10^3) = 60 \times 10^3$ samples per second. The period between samples is $1/(60 \times 10^3) = 100/6$ μs. This period must contain 12 pulses, one for each bit in the 12-bit code; so $T_b = 100/6(12)$ μs. Since the first-null bandwidth is $\omega_b/2\pi = 1/T_b$ Hz, it is $\omega_b/2\pi = 0.72$ MHz. From (6.2-9), $(S_o/N_q) = 3(4096)^2/16 \approx 3.146 \times 10^6$, or 65.0 dB.

16.4 PULSE-CODE MODULATION (PCM)

The processing of an analog signal by sampling, quantizing, and binary encoding is called *pulse-code modulation* (PCM). In this section we shall define a PCM system and determine its performance when noise is present.

PCM System

Because many different ways exist to implement a system, we shall define the PCM system through its necessary functions. Figure 16.4-1 gives a block diagram of the functions required to generate PCM. The lowpass filter is present only to guarantee that the message is bandlimited to the spectral extent for which the system is designed. The compressor is shown dashed because it could be present to increase performance but is not an essential function. In this section we shall assume that the PCM signal is transmitted directly over the (baseband) channel. In

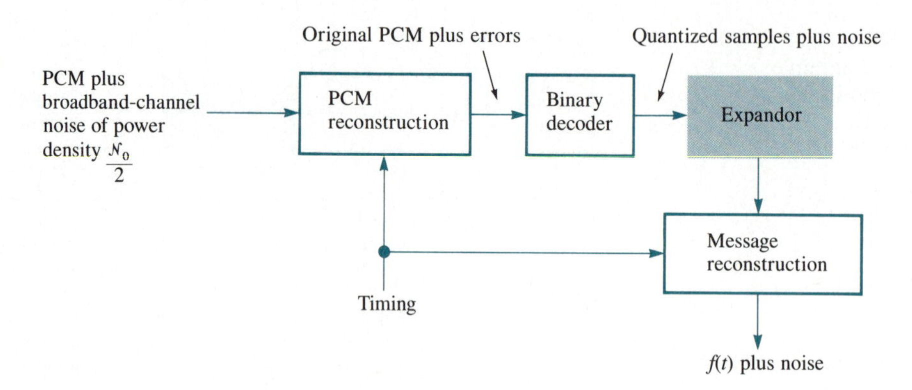

FIG. 16.4-2. A block diagram of the functions required to receive PCM.

Sec. 16.6 we shall discuss systems in which a carrier is modulated by the PCM signal or other digital signal.

The PCM signal that arrives at the receiver from the channel is corrupted by the noise generated within the receiver. In concert with the ways developed in Chapter 14 to handle receiver noise, we imagine this noise as coming from a source at the receiver's input (to the PCM reconstruction function). Because the noise is broadband, we shall imagine it to originate from a broadband source of power density $\mathcal{N}_0/2$ (watts per hertz). This level is related to (14.8-18) by $\mathcal{N}_0 = kT_{sys}$.

The receiving functions necessary to recover the message from the PCM signal (plus channel noise) are basically the inverses of those in the transmitter. Figure 16.4-2 illustrates the receiver based on the generator of Fig. 16.4-1. The first operation is to reconstruct the originally transmitted PCM signal as nearly as possible from the noise-contaminated received waveform. The reconstructed waveform will no longer have noise, but the *effects* of noise are present in the form of occasional incorrect pulses (a correct positive pulse in polar PCM may be demodulated as a negative pulse, for example). The binary decoder converts the code words represented by the reconstructed PCM signal into the corresponding quantum levels. If companding is used, the expandor maps the levels to the equivalent levels corresponding to the original quantized message samples. Finally, the message's reconstruction uses the theory of the sampling theorem to recover the original analog message plus the quantization error and a noise term. Message reconstruction is typically a lowpass filter or sample-hold device, as described in Sec. 2.7.

The reader will no doubt have already concluded that the most critical receiver operation is PCM reconstruction. This conclusion is correct, since it is at this point that the effect of noise is minimized through careful choice of circuit implementation. Some thought will reveal that the only knowledge required of the receiver to reconstruct the original PCM signal is whether the various transmitted bits are 0s or 1s.† This requirement reduces to a determination of which of two

† Strictly, the receiver must also know the *timing* of the incoming stream of pulses. We shall assume such timing exists such that the receiver is *synchronized* with the transmitter.

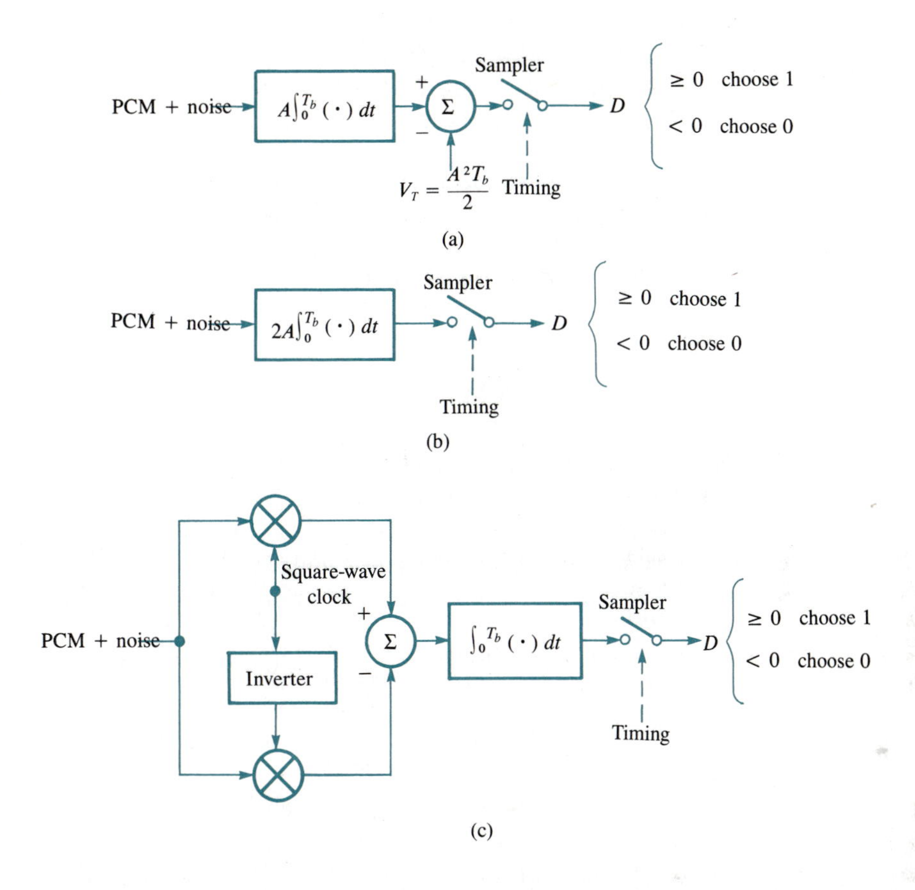

FIG. 16.4-3. PCM-reconstruction circuits for (a) unipolar, (b) polar, and (c) Manchester waveforms.

voltage levels was transmitted for any given pulse of a codeword. For example, with a unipolar stream of pulses of amplitude A, the two levels are A V and 0 V. For polar pulses of amplitudes $\pm A$, the two levels are A V and $-A$ V.

For each transmitted pulse, level determination reduces to a voltage measurement at some time instant. The measured voltage will equal the level caused by the pulse plus a noise-voltage part. The ability of the receiver to make the correct level determination is optimized if the ratio of the pulse-caused voltage to the noise-rms voltage is the largest possible at the measurement time. It can be shown that the circuits of Fig. 16.4-3 provide optimum PCM reconstruction when bits in the code words are equally probable (the usual case). At the end of each bit interval (duration T_b) the difference D between the integrator's output and a preset threshold V_T (which is zero for polar and Manchester PCM) is measured by sampling. If $D \geq 0$, it is decided that a binary 1 existed in the interval; if $D < 0$, a 0 is declared. In the unipolar system, V_T is equal to half the signal component of the integrator's output level at the sampling time when the input has a pulse present (binary 1). After the sample is taken, the integrator is discharged to 0 V in preparation for integration over the next bit interval.

The signal responses of the integrators in Fig. 16.4-3 for unipolar and polar PCM are straightforward to compute and the reader is encouraged to find these as an exercise. The Manchester circuit of Fig. 16.4-3(c) deserves additional comments. The periodic square-wave clock generates a voltage A for $0 < t < T_b/2$ and $-A$ for $T_b/2 < t < T_b$. Its fundamental frequency is, therefore, $1/T_b$ (hertz). The product of the clock and the incoming Manchester PCM waveform becomes a polar-PCM signal. The product with the clock inverted is the negative of a polar signal. After differencing in the summing junction, the response is a double-amplitude polar-PCM signal. The remainder of the circuit is just a reconstruction circuit for polar PCM, as in Fig. 16.4-3(b).

Bit-Error Probability

The presence of noise means that the PCM-reconstruction circuits of Fig. 16.4-3 may occasionally make a mistake in deciding what input pulse was received in a given bit interval. How often an error is made is determined by the probability of bit error, denoted by P_e. If $P_e = 10^{-3}$, we would expect 1 bit out of every 1000 bits to be in error, on the average. Similarly, $P_e = 10^{-2}/4$ would correspond to 1 bit error in every 400 bits, and so on. Detailed analysis of PCM shows that

$$P_e = \begin{cases} \frac{1}{2}\,\text{erfc}\left[\sqrt{\dfrac{A^2 T_b}{4\mathcal{N}_0}}\right] & \text{unipolar PCM} \\ \frac{1}{2}\,\text{erfc}\left[\sqrt{\dfrac{A^2 T_b}{\mathcal{N}_0}}\right] & \text{polar and Manchester PCM} \end{cases} \tag{16.4-1}$$

where A is the amplitude of the PCM pulses at the receiver's input. Here, also,

$$\text{erfc}\,(x) = 1 - \frac{2}{\sqrt{\pi}} \int_0^x e^{-\xi^2}\, d\xi \tag{16.4-2}$$

is called the *complementary error function*; it equals 1 at $x = 0$ and decreases to 0 as $x \to \infty$. For x larger than 2 the approximation

$$\text{erfc}\,(x) \approx \frac{e^{-x^2}}{x\sqrt{\pi}} \qquad x > 2 \tag{16.4-3}$$

is accurate to about 10.5% or better. Accuracy improves as x increases. For negative x

$$\text{erfc}\,(-x) = 2 - \text{erfc}\,(x) \tag{16.4-4}$$

Other forms of (16.4-1) are useful. It is customary in communications systems to assume that impedance levels are resistive and unity. This assumption is equivalent to normalizing voltages and currents under the constraint that impedances are real. Which viewpoint one wishes to use is unimportant, however, when ratios are involved, as in (16.4-1), because the impedance constant is the same for both signal and noise and cancels in the ratio. With the customary approach, the energy received for a binary digit 1 is A^2T_b for the unipolar, polar, and Manchester formats; energy for a binary 0 is also A^2T_b for polar and Manchester formats but 0 for the unipolar format. Since 1s and 0s each occur with probability $\frac{1}{2}$, the *average* energy received per bit is

$$E_b = \begin{cases} A^2T_b(\frac{1}{2}) + A^2T_b(\frac{1}{2}) = A^2T_b & \text{polar and Manchester PCM} \\ A^2T_b(\frac{1}{2}) + 0 = \dfrac{A^2T_b}{2} & \text{unipolar PCM} \end{cases} \tag{16.4-5}$$

In terms of E_b (16.4-1) becomes

$$P_e = \begin{cases} \frac{1}{2}\,\text{erfc}\left[\sqrt{\dfrac{E_b}{2\mathcal{N}_0}}\right] & \text{unipolar PCM} \\ \frac{1}{2}\,\text{erfc}\left[\sqrt{\dfrac{E_b}{\mathcal{N}_0}}\right] & \text{polar and Manchester PCM} \end{cases} \tag{16.4-6}$$

This form is useful in comparing performances of systems.

A final form for (16.4-1) can also be shown to be

$$P_e = \begin{cases} \frac{1}{2}\,\text{erfc}\left[\sqrt{\dfrac{1}{8}\left(\dfrac{\hat{S}_i}{N_i}\right)_{\text{PCM}}}\right] & \text{unipolar PCM} \\ \frac{1}{2}\,\text{erfc}\left[\sqrt{\dfrac{1}{2}\left(\dfrac{\hat{S}_i}{N_i}\right)_{\text{PCM}}}\right] & \text{polar and Manchester PCM} \end{cases} \tag{16.4-7}$$

where $(\hat{S}_i/N_i)_{\text{PCM}}$ is defined as the ratio of the power in the signal's peak amplitude, at the integrator's output at the sample time, to the average noise power at the same point.

EXAMPLE 16.4-1

At the sampler in a PCM system $(\hat{S}_i/N_i)_{\text{PCM}} = 32$. We find P_e by assuming a unipolar system and a polar system. Here

$$\sqrt{\frac{1}{8}\left(\frac{\hat{S}_i}{N_i}\right)_{\text{PCM}}} = \sqrt{\frac{32}{8}} = 2.0 \qquad \text{unipolar PCM}$$

$$\sqrt{\frac{1}{2}\left(\frac{\hat{S}_i}{N_i}\right)_{\text{PCM}}} = \sqrt{\frac{32}{2}} = 4.0 \qquad \text{polar PCM}$$

Since both values are 2 or more, we use (16.4-7) to compute

$$P_e \approx \frac{1}{2}\,\text{erfc}\,(2.0) \approx \frac{e^{-2^2}}{2(2)\sqrt{\pi}} \approx 2.58 \times 10^{-3}$$

$$P_e \approx \frac{1}{2}\,\text{erfc}\,(4.0) \approx \frac{e^{-4^2}}{2(4)\sqrt{\pi}} \approx 7.94 \times 10^{-9}$$

These results show that bit errors in a unipolar system occur much more frequently than in a polar system having the same signal-to-noise ratio.

Noise Performance of PCM

It remains to determine the effect of bit errors due to noise on the recovery of an analog message at the receiver's output. When receiver noise is negligible, only quantization error is present and (16.2-3) applies. When receiver noise is not negligible, we can write the recovered signal as $\hat{f}_q(t)$:

$$\hat{f}_q(t) = f_q(t) + \epsilon_n(t) = f(t) + \epsilon_q(t) + \epsilon_n(t) \tag{16.4-8}$$

In other words, there is now an error $\epsilon_n(t)$ in the reconstructed message $\hat{f}_q(t)$ due to noise. Because the quantization and receiver errors arise from different mechanisms, they may be taken as statistically independent; and the total output-waveform power is the sum of individual powers. The ratio of desired output-signal power to total output-noise power becomes

$$\left(\frac{S_o}{N_o}\right)_{\text{PCM}} = \frac{\left(\dfrac{S_o}{N_q}\right)}{1 + [\overline{\epsilon_n^2(t)}/\overline{\epsilon_q^2(t)}]} \tag{16.4-9}$$

where (S_o/N_q) is given by (16.2-5), or by (16.5-7) when (16.2-6) applies. The presence of $\epsilon_n(t)$ gives rise to a *threshold effect* in PCM, where noise causes *no* effect on performance when $\overline{\epsilon_n^2(t)} \ll \overline{\epsilon_q^2(t)}$ and great effect when $\overline{\epsilon_n^2(t)}$ is not negligible compared with $\overline{\epsilon_q^2(t)}$.

Our basic problem is to find $\overline{\epsilon_n^2(t)}$. To this end, we observe that a quantum level generated by the decoder block in Fig. 16.4-2 (no companding) will be exact only if all bits of a code word are recovered correctly by the PCM-reconstruction stage. Thus, $\hat{f}_q(t)$ will possess a noise error if a code word is in error, and a code word is in error if any one or more bit errors are made. Most PCM systems operate such that it is highly unlikely that more than 1 bit is in error in any one word. We assume this to be true and examine the effects of bit errors on a bit-by-bit basis for an N_b-bit word.

If an error occurs in the least significant bit, an error in $\hat{f}_q(t)$ of δv occurs. If such an error occurs in m words out of M possible words, the average- (mean-) squared error is $(\delta v)^2 m/M$. For a large number of words m/M is interpreted as the probability P_e of the least significant bit's being in error. Average-squared error in $\hat{f}_q(t)$ is then $(\delta v)^2 P_e$.

For the next least significant bit, a bit error causes an error in $\hat{f}_q(t)$ of $2(\delta v)$. The mean-squared error becomes $(2\,\delta v)^2 P_e$. Continuing the logic to the ith next least significant bit, the *mean*-squared error in $\hat{f}_q(t)$ is $(2^i\,\delta v)^2 P_e$. Now, recognizing that the total mean-squared error $\overline{\epsilon_n^2(t)}$ in $\hat{f}_q(t)$ is the sum of the contributions from each bit in the code word, we have

$$\overline{\epsilon_n^2(t)} = (\delta v)^2 P_e \sum_{i=0}^{N_b-1} (2^2)^i = (\delta v)^2 P_e \frac{2^{2N_b} - 1}{3} \tag{16.4-10}$$

On using the approximation $2^{2N_b} \gg 1$, true for most practical systems, and

(16.2-4), we have

$$\frac{\overline{\epsilon_n^2(t)}}{\overline{\epsilon_q^2(t)}} \approx 2^{2N_b+2}P_e \tag{16.4-11}$$

Finally, after use of (16.4-1), (16.4-11) allows (16.4-9) to be put in the useful form

$$\left(\frac{S_o}{N_o}\right)_{\text{PCM}} = \begin{cases} \dfrac{\left(\dfrac{S_o}{N_q}\right)}{1 + 2^{2N_b+1}\,\text{erfc}\left[\sqrt{\dfrac{1}{8}\left(\dfrac{\hat{S}_i}{N_i}\right)_{\text{PCM}}}\right]} & \text{unipolar PCM} \\[2ex] \dfrac{\left(\dfrac{S_o}{N_q}\right)}{1 + 2^{2N_b+1}\,\text{erfc}\left[\sqrt{\dfrac{1}{2}\left(\dfrac{\hat{S}_i}{N_i}\right)_{\text{PCM}}}\right]} & \text{polar or Manchester PCM} \end{cases} \tag{16.4-12}$$

Normalized plots of (16.4-12) for polar PCM, assuming all levels span the signal's variations ($L = L_b = 2^{N_b}$), are shown in Fig. 16.4-4. The threshold effect is clearly evident and occurs typically where $(\hat{S}_i/N_i)_{\text{PCM}}$ is in the range of 8 to 15 dB for N_b from 2 to 10. For a sinusoidal message, where $K_{\text{cr}}^2 = 2$, the curves are direct plots of $(S_o/N_o)_{\text{PCM}}$.

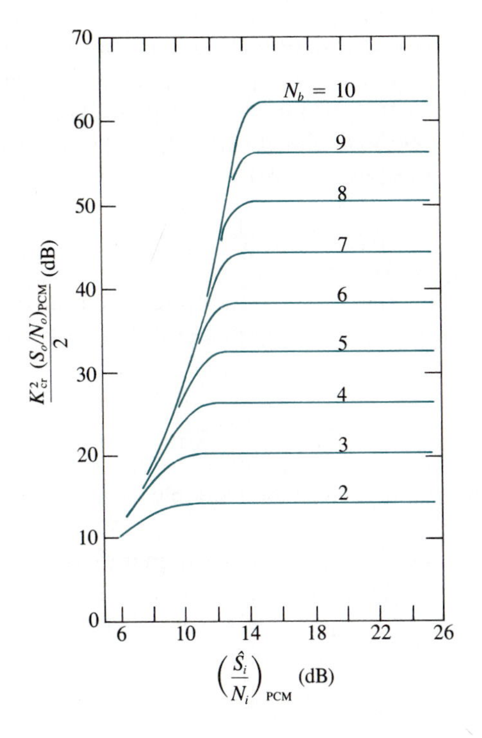

FIG. 16.4-4. Performance curves for a PCM system using a polar waveform.

EXAMPLE 16.4-2

For the polar-PCM system of Example 16.4-1, we assume $N_b = 8$ bits and determine whether the system's noise has much effect on performance. Here $(\hat{S}_i/N_i)_{\text{PCM}} = 32$, or 15.05 dB. From Fig. 16.4-4 with $N_b = 8$, operation is in the flat part of the curve, so receiver-noise effects are negligible. Performance is limited by quantization error.

16.5 TIME MULTIPLEXING OF DIGITAL SIGNALS

A powerful advantage that accrues to a digital communications system is the ability to combine many messages in time for transmission over the same channel on a single communications link. The ability derives from the fact that data from any single source are transmitted only at discrete times. When data from many sources in time are interlaced, a single link can handle all sources. The interlacing of data is called *time multiplexing*. Our discussions of multiplexing will center on examples to illustrate the concepts involved.

Multiplexing of Analog Signals

When we convert an analog signal to digital form, sampling, quantization, and coding are all involved. Time multiplexing can be done at any stage after sampling. Figure 16.5-1 demonstrates time multiplexing immediately after sampling for N similar messages. The trains of samples (shown as natural samples) can be added, as shown in Fig. 16.5-1(d) for the signal at point A, if the sampling pulses are properly interleaved. For N similar messages of spectral extent W_f (radians per second), the sampling interval T_s must satisfy

$$T_s \leq \frac{\pi}{W_f} \tag{16.5-1}$$

from the sampling theorem. The time per sampling interval that is allowed per message is called a *time slot*. For sampling pulses of duration τ and separation τ_g, called the *guard time*, we have

$$\tau + \tau_g = \frac{T_s}{N} \leq \frac{\pi}{NW_f} \tag{16.5-2}$$

The time required to gather in at least one sample of each message (T_s here) is called a *frame*.

The remainder of the system of Fig. 16.5-1(a) operates on the waveform of (d) as though it were from a single (composite) source. Thus, for N_b-bit encoding, each time slot in the output-PCM signal will have N_b bits of duration

$$T_b = \frac{T_s}{NN_b} \tag{16.5-3}$$

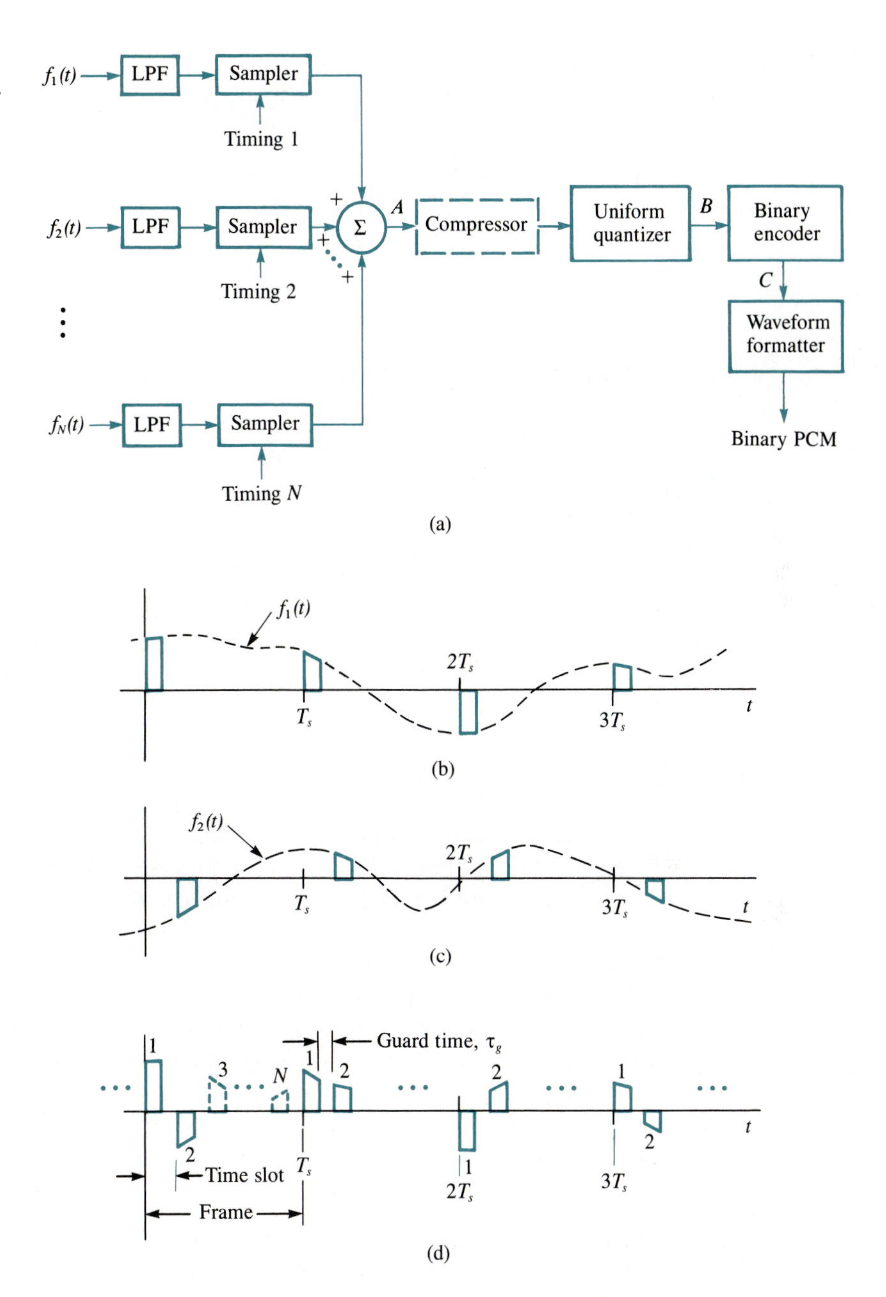

FIG. 16.5-1. (a) A time multiplexer for N similar analog signals and waveforms at point A for (b) message 1, (c) message 2, and (d) the full waveform.

It is clear that the system requires all sample trains to be at the same frequency; in other words, they must be derived from the same timing source, called a *clock*.

For simplicity, the preceding discussions have assumed that all frame time is used for messages. In practice, some time must be allocated for synchronization, because the receiver must know the start times of frames. Synchronization can be accomplished in many ways. One or more time slots can be dedicated to this use.

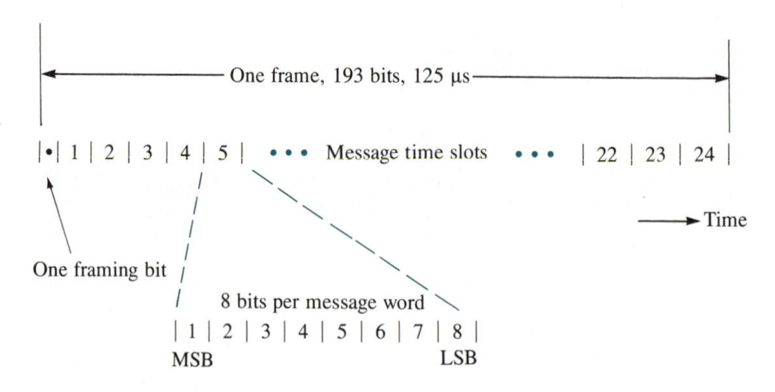

FIG. 16.5-2. The frame structure for a D3 channel bank.

In other cases a specific time interval is used that can be less than one time slot. The following example demonstrates this latter situation.

EXAMPLE 16.5-1

The American Telephone and Telegraph (AT&T) Company uses a device known as a D3 *channel bank* to multiplex 24 telephone messages, each having an 8-kHz sampling rate ($T_s = 1/(8 \times 10^3)$ s $= 125$ μs). Each sample uses 8-bit encoding, so there are $8(24) = 192$ message bits during each 125 μs. One extra bit, called a *framing bit*, is allowed for frame synchronization, giving a total of 193 bits per frame. The total bit rate is $193(8000) = 1.544$ megabits per second (Mbps).

Frame structure is sketched in Fig. 16.5-2. Frame synchronization uses 12 bits (12 frames). Odd-numbered frames are identified by the framing bit sequence 101010. Even-numbered frames have the framing bit sequence 001110, so that frames 6 and 12 are identified by the transition 01 and 10, respectively. After 12 frames the sequences repeat.

During any given message time slot the full 8 bits are used for message information during 5 out of every 6 frames. During 1 of every 6 frames the eighth (least significant) bit is used for *signaling*.† The signaling rate for each of the 24 messages is therefore 1 bit/6 frames at 8000 frames per second, or about 1.333 kilobits per second (kbps). The occasional "borrowing" of a message bit for purposes other than message information is called *bit robbing* or *bit stealing*.

The channel bank described in the foregoing example is a *synchronous* multiplexer, which is one where the digital structure of each input message is determined by a single master clock. When different clocks are involved, the multiplexer is called *asynchronous*.

† Signaling refers to conveying information about telephone number dialed, dial tone, busy signal, ringing, etc.

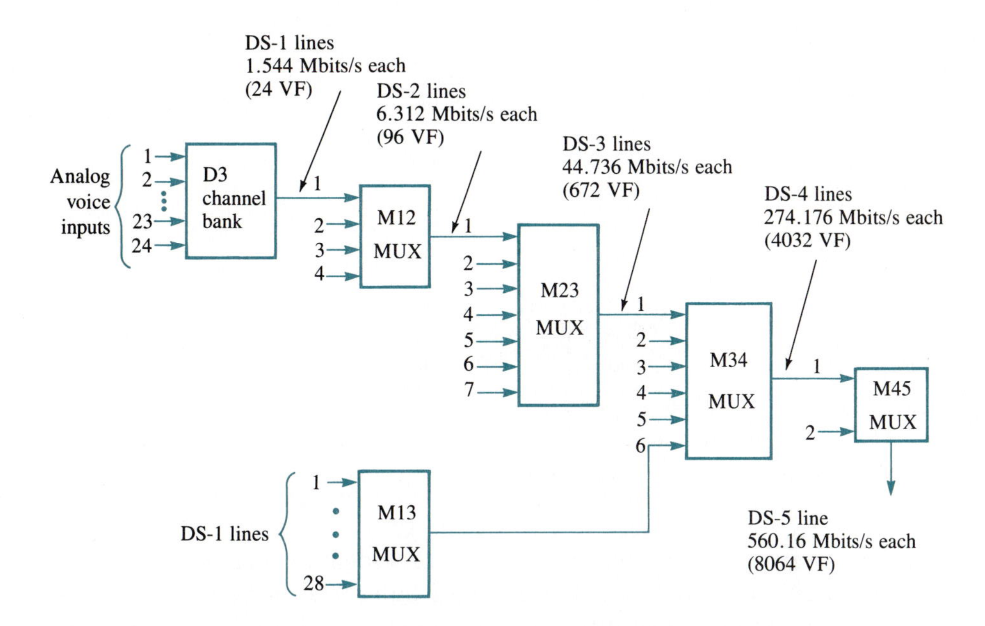

FIG. 16.5-3. The hierarchies of digital multiplexing in the United States and Canada.

Asynchronous Multiplexing

Much of today's traffic in the telephone industry is now handled by digital means. A hierarchy of digital multiplexing has been established, as depicted in Fig. 16.5-3. The output from the channel bank is a digital signal (DS) on a line said to carry level 1 multiplexing; hence, a DS-1 line carries a digital signal at a nominal bit rate of 1.544 Mbps. Four input DS-1 lines can be multiplexed to generate a level 2 signal on a DS-2 line by an M12 multiplexer (MUX); its output bit rate is nominally 6.312 Mbps. Other multiplexers are used to attain levels up to 5, as shown. All multiplexers (except the channel bank) are asynchronous.

The output bit rate of an asynchronous MUX is larger than the sum of the maximum bit rates of all its inputs (called *tributaries*). The output can be word-interleaved (as in the channel bank) or bit-interleaved (the M12 MUX is like this). Operation of a typical MUX is complicated to describe and not necessary here. Generally, however, bits are arriving at the MUX's input at a slower rate than are generated at its output. Occasionally, extra bits, called *stuff bits*, are added to maintain the output bit rate. Control bits are also added for synchronization and for defining where stuff bits were added so that the demultiplexer can properly recover the data bits. In the MUXs of Fig. 16.5-3 the output bit rates are about 1% to 2% larger than the sum of the nominal input rates to account for variations in these rates.

Sometimes, the MUXs described above are called *quasi-synchronous* because the clocks associated with tributaries may be very stable but not precisely synchronized. The word *asynchronous* then applies to sources where blocks of data are generated randomly in time but bits within a block have precise timing. A keyboard terminal is a good example. It generates one character at a time that is

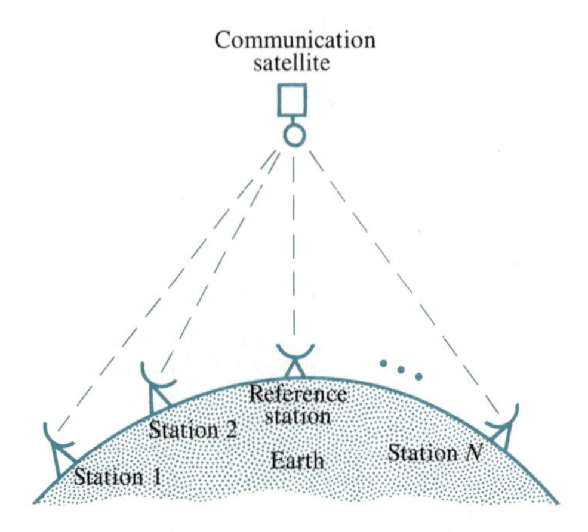

FIG. 16.5-4. Communications links through a satellite.

typically encoded to have one start bit, 7 bits to represent the character, which are encoded according to the American Standard Code for Information Interchange (ASCII),† one *parity bit* for error identification, and one stop bit. Multiplexers for this type of start-stop information tend to be either of the *fixed-assignment* or *demand-assignment* type.

A fixed-assignment multiplexer periodically provides access to the channel for each source independent of its actual need. Buffers are used to store incoming characters until selected for channel transmission. Character interleaving and character stuffing are typical, somewhat as in quasi-synchronous multiplexing. When an input source (terminal) is idle, dummy characters are transmitted and efficiency of channel use can be low.

Better efficiency can be achieved by the demand-assignment multiplexer. Here idle inputs are skipped and larger buffers are used to help average the input data rate. Output data rate is related more to the average input data rate, while the demand-assignment device must serve the total input *peak* data rate. Demand-assignment multiplexing has also been called *statistical multiplexing*.

Time-Division Multiple Access

The geometry of several (N) earth stations that communicate with each other through a communications satellite is illustrated in Fig. 16.5-4. *Time-division multiple access* (TDMA) is one important means by which each station timeshares the satellite and broadcasts to all other stations during its assigned time. All stations use the same up-link frequency, and all recieve a single down-link

† A list of standard code words for each keyboard character is available in many sources. For example, see L. W. Couch II, *Digital and Analog Communication Systems*, 3d ed., Macmillan, New York, 1990, p. 706.

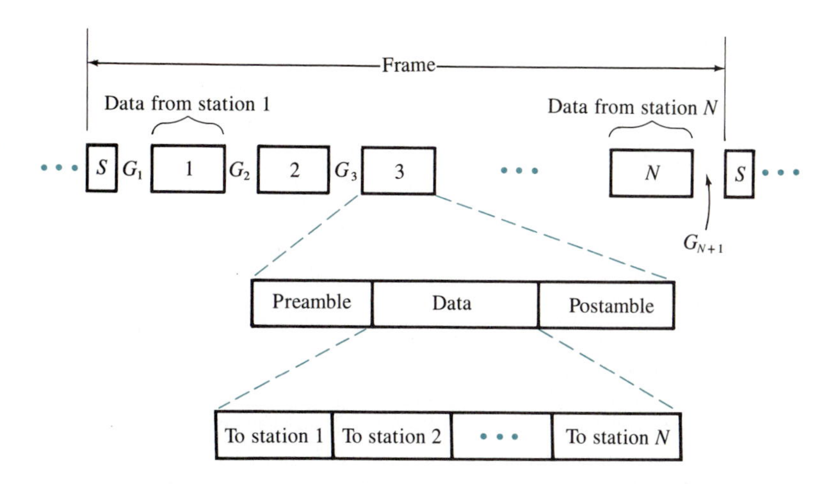

FIG. 16.5-5. The frame structure of the transmission from the satellite in a TDMA system.

frequency from the satellite. Frequencies are typically on the order of 6 GHz up and 4 GHz down, but the trend is to higher frequencies (14 GHz up, 12 GHz down) as the lower-frequency systems become traffic-saturated.

One station in the system of Fig. 16.5-4 is designated as the primary reference station. Its responsibility is to maintain the system synchronous according to the timing format given in Fig. 16.5-5. It transmits synchronization signals (S) at the start of each frame for use by all receiving sites. Next, each station, in turn, transmits a block (or burst) of its data through the satellite. The start of a block contains a *preamble* that is used by receiving stations for carrier and bit synchronization as well as a few other special functions. Following the preamble is a sequence of data blocks to the other stations. A *postamble* is sometimes included at the end of a station's transmission to initialize decoders in preparation for the burst from the next stations. Guard intervals $G_1, G_2, \ldots, G_{N+1}$ are present to absorb small errors in timing so that overlap of transmissions is avoided.

16.6 CARRIER MODULATION BY DIGITAL SIGNALS

Most digital communications systems use carriers modulated by the digital signal. Many modulation methods exist, and it is relatively easy to list 10 to 20. However, most of these reduce to variations of three fundamental methods called *amplitude-shift keying* (ASK), *phase-shift keying* (PSK), and *frequency-shift keying* (FSK). Therefore, we shall limit our discussions to only ASK, PSK, FSK, and one especially important variation of PSK, called *differential phase-shift keying* (DPSK). In all cases we assume only binary modulation.

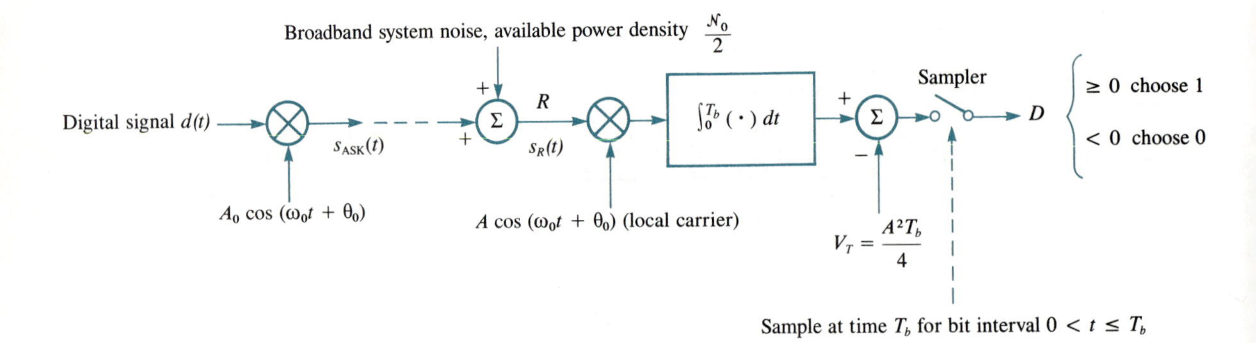

FIG. 16.6-1.
Optimum coherent-ASK communications system's functions.

Amplitude-Shift Keying (ASK)

In binary ASK, a carrier's amplitude is keyed between two levels,† usually off and on, which is the case examined here. For purposes of discussion, we assume rectangular pulses, the on and off conditions correspond to binary 1 and 0, respectively, and the probabilities are equal (0.5) that any binary digit will be a 0 to 1 in the digital bit stream.

Both coherent- and noncoherent-system implementations of ASK are possible, with the latter being the simpler. The necessary functions for an *optimum* coherent system are given in Fig. 16.6-1. We shall describe operation as it pertains to a bit interval assumed to start at $t = 0$ and extend to $t = T_b$. Operation in any other interval will be similar. The desired ASK signal is a pulse of carrier when a binary digit 1 occurs and no pulse when a 0 occurs. Hence, it is

$$s_{ASK}(t) = \begin{cases} A_0 \cos(\omega_0 t + \theta_0) & 0 < t < T_b, \text{ for } 1 \\ 0 & 0 < t < T_b, \text{ for } 0 \end{cases} \tag{16.6-1}$$

where A_0, ω_0, and θ_0 are, respectively, the peak amplitude, angular frequency, and phase angle of the modulated carrier. The signal $s_{ASK}(t)$ is equivalent to a carrier $A_0 \cos(\omega_0 t + \theta_0)$ modulated by a digital signal $d(t)$ that is 0 or 1 in a given bit interval, according to whether the bit is 0 or 1, respectively. Thus, the digital signal and product device are equivalent to the waveform formatter and modulator of Fig. 16.1-1.

---

† ASK systems are not as widely used as other systems for various reasons. However, most optical communications systems use intensity modulation of a light source, so there is a direct analogy between carrier amplitude and light level. Thus, the ASK concept remains quite important and modern for the optical applications.

The channel is modeled as a direct path from the modulator to the receiver's input, although in practice it could contain several components, such as a power amplifier, transmitting antenna, real channel, and a receiving antenna. It is convenient to think of point R in Fig. 16.6-1 as the output of the receiving antenna, if any. System noise is modeled as a broadband source† at point R with power density $\mathcal{N}_0/2$. In terms of the system's available noise power, as given by (14.8-18), we have

$$\mathcal{N}_0 = kT_{\text{sys}} \tag{16.6-2}$$

The received signal $s_R(t)$ will not be the same as $s_{\text{ASK}}(t)$ because of the channel's gain (or loss). We shall assume the two differ only in amplitude,‡ so

$$s_R(t) = \begin{cases} A \cos(\omega_0 t + \theta_0) & 0 < t < T_b, \text{ for } 1 \\ 0 & 0 < t < T_b, \text{ for } 0 \end{cases} \tag{16.6-3}$$

Only the receiver's demodulator and code generator are shown in Fig. 16.6-1 because the other receiving functions (Fig. 16.1-1) do not affect performance and are straightforward. The product device is equivalent to a synchronous detector that removes the input carrier. The required local carrier must be phase-coherent with the input signal, a major disadvantage of coherent ASK. The input to the integrator is a unipolar-PCM signal, so the remainder of the circuit is just a PCM receiver for this format [Fig. 16.4-3(a)].

The most commonly accepted measure of performance in a digital system is the probability that a bit error will occur. It can be shown that this probability, denoted by P_e, is

$$P_e = \tfrac{1}{2}\,\text{erfc}\left[\sqrt{\frac{A^2T_b}{8\mathcal{N}_0}}\right] \tag{16.6-4}$$

Two other useful forms of (16.6-4) can be established. In terms of the average received energy per bit interval, which is

$$E_b = \frac{A^2T_b}{2}\left(\frac{1}{2}\right) + 0 = \frac{A^2T_b}{4} \tag{16.6-5}$$

we have

$$P_e = \tfrac{1}{2}\,\text{erfc}\left[\sqrt{\frac{E_b}{2\mathcal{N}_0}}\right] \tag{16.6-6}$$

It can be shown that the ratio of the power in the signal *at the sampler at the sample time* (denoted by $\hat{S}_i$) to the average noise power (denoted by N_i) is

† This is often called a *white-noise* source.

‡ In addition to an amplitude change, a real channel can cause time delay and, if bandwidth limitations are present, can distort the received signal. Unstable channels can also cause these effects to change with time.

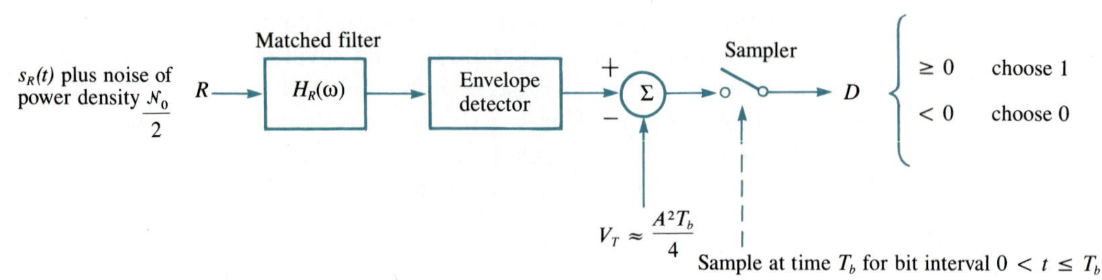

FIG. 16.6-2.
The optimum demodulator and code generator for a noncoherent-ASK system.

$$\left(\frac{\hat{S}_i}{N_i}\right)_{\text{ASK}} = \frac{A^2T_b}{4\mathcal{N}_0} \tag{16.6-7}$$

Hence

$$P_e = \tfrac{1}{2}\,\text{erfc}\left[\sqrt{\tfrac{1}{2}\left(\frac{\hat{S}_i}{N_i}\right)_{\text{ASK}}}\right] \tag{16.6-8}$$

The need for a coherent local oscillator is a serious disadvantage of the coherent-ASK system. The noncoherent system eliminates this need. The only changes required to realize the noncoherent system are replacing the synchronous detector and integrator in Fig. 16.6-1 with the filter and enveloped detector as given in Fig. 16.6-2. The filter's transfer function should ideally be

$$H_R(\omega) = \tfrac{1}{2}[G^*(\omega - \omega_0)e^{-j(\omega-\omega_0)T_b} + G^*(\omega + \omega_0)e^{-j(\omega+\omega_0)T_b}] \tag{16.6-9}$$

where * represents the complex conjugate and

$$G(\omega) = AT_be^{-j\omega T_b/2}\,Sa\left(\frac{\omega T_b}{2}\right) \tag{16.6-10}$$

The function $H_R(\omega)$ is difficult to realize exactly but can be closely approximated in practice. The filter is called a *matched filter* because its impulse response is matched to (has the same form as) the carrier pulse at its input.† This point is proved by an example.

EXAMPLE 16.6-1

From Fourier transform theory, if

$$h_R(t) \leftrightarrow H_R(\omega)$$

forms a Fourier transform pair, then

$$h_R^*(-t) \leftrightarrow H_R^*(\omega)$$

† A filter matched to a signal $s(t)$ in broadband noise has an impulse response $h(t) = Ks^*(t_0 - t)$, in general, where t_0 and K are arbitrary real constants.

forms another pair (*conjugation property*). From the frequency-shifting and time-shifting properties, we can also write

$$g(t) \leftrightarrow G(\omega)$$

$$g(t + T_b) \leftrightarrow G(\omega)e^{j\omega T_b}$$

$$g(t + T_b)e^{\pm j\omega_0 t} \leftrightarrow G(\omega \mp \omega_0)e^{j(\omega \mp \omega_0)T_b}$$

Using all these results allows (16.6-9) to be inverse-Fourier-transformed:

$$h_R^*(-t) = \mathcal{F}^{-1}\{H_R^*(\omega)\} = \tfrac{1}{2}g(t + T_b)e^{j\omega_0 t} + \tfrac{1}{2}g(t + T_b)e^{-j\omega_0 t}$$
$$= g(t + T_b)\cos\omega_0 t$$

From the inverse transform of (16.6-10), which is

$$g(t) = A \operatorname{rect}\left[\frac{t - (T_b/2)}{T_b}\right]$$

we have

$$h_R^*(-t) = A \operatorname{rect}\left[\frac{t - (T_b/2)}{T_b}\right] \cos\omega_0 t$$

Thus, because rect $(-x)$ = rect (x),

$$h_R(t) = A \operatorname{rect}\left[\frac{t - (T_b/2)}{T_b}\right] \cos\omega_0 t$$

This impulse response has the form of $s_{\text{ASK}}(t)$ when a carrier pulse is transmitted.

Bit-error probability in the noncoherent system is approximately given by

$$P_e \approx \frac{1}{2}\left[1 + \sqrt{\frac{1}{2\pi(E_b/\mathcal{N}_0)}}\right] \exp\left[-0.5\left(\frac{E_b}{\mathcal{N}_0}\right)\right] \qquad (16.6\text{-}11)$$

where E_b is given by (16.6-5). For $P_e < 10^{-4}$, the usual case, the noncoherent system requires approximately 1 dB more in transmitted-signal power to achieve the same value of P_e as in the coherent system. For most practical systems the 1-dB penalty is a small price to pay to remove the extra complexity of the coherent system.

Phase-Shift Keying (PSK)

The phase angle of a carrier is keyed between two values in *phase-shift keying* (PSK). The values are usually separated by π radians. We consider only this form of PSK, sometimes called *phase-reversal keying* (PRK). In the bit interval that we

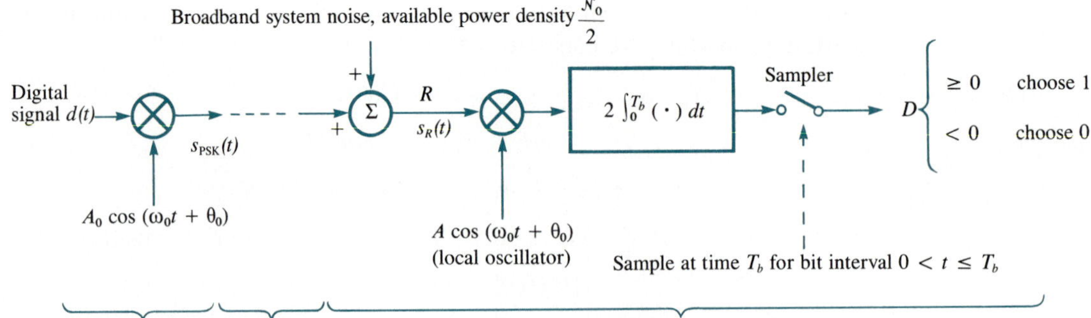

FIG. 16.6-3.
The functions of an optimum coherent-PSK communications system.

have chosen to discuss as typical (from $t = 0$ to $t = T_b$), the PSK waveform becomes

$$s_{PSK}(t) = \begin{cases} A_0 \cos(\omega_0 t + \theta_0) & 0 < t < T_b, \text{ for } 1 \\ -A_0 \cos(\omega_0 t + \theta_0) & 0 < t < T_b, \text{ for } 0 \end{cases} \tag{16.6-12}$$

It is equivalent to a carrier $A_0 \cos(\omega_0 t + \theta_0)$ being modulated in amplitude by a digital signal $d(t)$ with amplitudes -1 or 1 when the binary digit in the bit interval is 0 or 1, respectively. Figure 16.6-3 illustrates these points as well as the rest of the PSK system. The system is quite similar to the ASK system in Fig. 16.6-1. In fact, the only differences are (1) the received signal, which is now

$$s_R(t) = \begin{cases} A \cos(\omega_0 t + \theta_0) & 0 < t < T_b, \text{ for } 1 \\ -A \cos(\omega_0 t + \theta_0) & 0 < t < T_b, \text{ for } 0 \end{cases} \tag{16.6-13}$$

(2) the ''gain'' of 2 assigned to the integrator, and (3) the absence of a threshold when binary digits 0 and 1 occur with equal probability, as we assume. Similarly, receiver operation follows the same discussion given for ASK.

The probability of a bit error occurring in any given bit interval in coherent PSK is

$$P_e = \tfrac{1}{2} \operatorname{erfc}\left[\sqrt{\frac{A^2 T_b}{2\mathcal{N}_0}}\right] \tag{16.6-14}$$

The alternative form involving the average received energy per bit interval, which is

$$E_b = \frac{A^2 T_b}{2} \tag{16.6-15}$$

is

$$P_e = \tfrac{1}{2} \operatorname{erfc}\left[\sqrt{\frac{E_b}{\mathcal{N}_0}}\right] \tag{16.6-16}$$

EXAMPLE 16.6-2

Coherent-ASK and coherent-PSK systems both transmit the same average energy per bit interval and operate on the same channel such that $E_b/\mathcal{N}_0 = 18$. We determine P_e for the two systems. From (16.6-6), $P_e = 0.5 \text{ erfc } (\sqrt{\tfrac{18}{2}}) = 0.5 \text{ erfc } (3)$, or

$$P_e \approx \frac{1}{2}\frac{e^{-9}}{(3)\sqrt{\pi}} \approx 1.16 \times 10^{-5} \qquad \text{coherent ASK}$$

after use of (16.4-3). For PSK

$$P_e = \tfrac{1}{2} \text{ erfc } (\sqrt{18}) \approx \tfrac{1}{2} \text{ erfc } (4.243) \approx \frac{e^{-18}}{2(4.243)(\sqrt{\pi})} \approx 1.01 \times 10^{-9}$$

Clearly, the PSK system is superior to the ASK system, based on equal values of $E_b/\mathcal{N}_0$, which is a fair basis of comparison.

Differential Phase-Shift Keying (DPSK)

The coherent-PSK system suffers the same disadvantage as the coherent-ASK system; it must generate the coherent local carrier. The noncoherent-ASK system overcame this disadvantage. However, no truly noncoherent version of PSK is possible. The reason, of course, is that the PSK signal carries its information in the carrier's phase, but a noncoherent system purposely disregards phase and operates only on signal amplitude. PSK can still avoid the need of a local carrier by use of an ingenious technique called *differential phase-shift keying* (DPSK); it is sort of a pseudo-coherent system, where the receiver uses the received signal to act as its own "carrier."

The functions needed in a DPSK system are illustrated in Fig. 16.6-4. The leftmost product operation and the 1-bit delay constitute an analog implementation of the differential encoder (Fig. 16.3-2). The digital signal $d(t)$ is a polar waveform of levels $+1$ and -1 corresponding to binary digits 1 and 0, respectively. The output signal $a(t)$ from the differential encoder PSK-modulates a carrier to create the DPSK signal $s_{\text{DPSK}}(t)$. The applicable waveforms corresponding to an example sequence of message binary digits are illustrated in Fig. 16.6-5(a).

In the receiver the key to operation is the coherent detector, which is the product device followed by the wideband lowpass filter (WBLPF).† *Both* inputs to the product are derived from the output of a filter matched to the input pulse in a

† The WBLPF is needed only to remove the components in the output of the product device that are not at baseband; it mainly filters out a component at frequency $2\omega_0$, so it does not need to be narrowband.

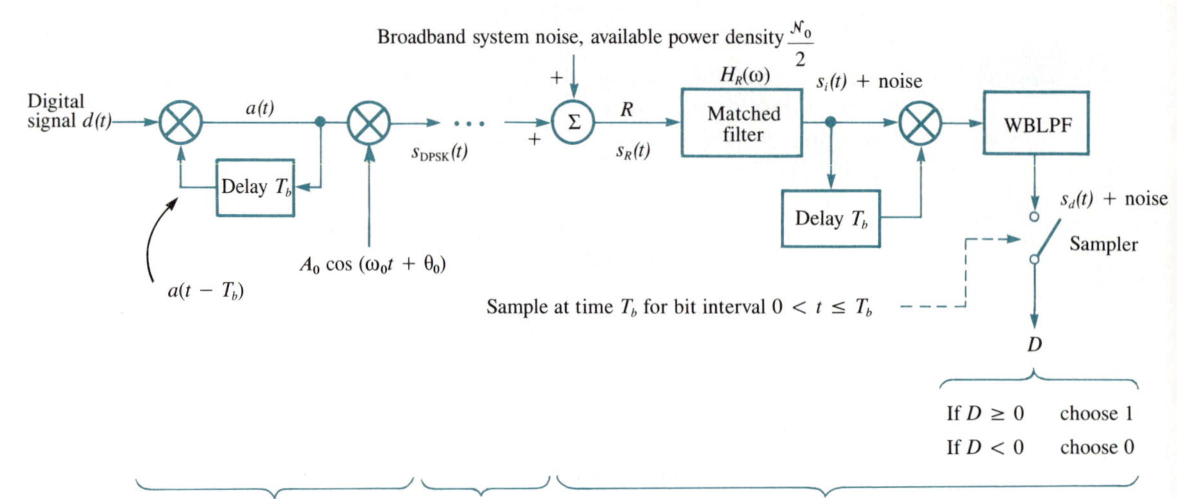

FIG. 16.6-4.
The functions required in a DPSK system.

single bit interval. Ideally, the filter's transfer function can be taken as

$$H_R(\omega) = \tfrac{1}{2}e^{-j(\omega-\omega_0)T_b/2} Sa\left[\frac{(\omega - \omega_0)T_b}{2}\right] + \tfrac{1}{2}e^{-j(\omega+\omega_0)T_b/2} Sa\left[\frac{(\omega + \omega_0)T_b}{2}\right] \tag{16.6-17}$$

The 1-bit delayed input to the product serves as the "local" oscillator for the coherent detector. In effect, the DPSK waveform in a given bit interval serves as its own local-oscillator signal in the following bit interval.

Let $s_i(t)$ be the signal component at the matched filter's output, and let $s_d(t)$ represent the signal component at the detector's output. The detector's action is such that $s_d(t)$ is a positive voltage if the phases of both $s_i(t)$ and $s_i(t - T_b)$ are the same, that is, have like phases. If the phases of $s_i(t)$ and $s_i(t - T_b)$ differ by π radians, $s_d(t)$ will be a negative voltage. At the sample time at the end of the bit interval these voltages will have maximum amplitudes. Thus, the *sign* of the voltage at the sampler depends on the phase relationship between $s_i(t)$ and its delayed replica. Figure 16.6-5(b) illustrates applicable phase and polarity relationships. It is apparent that the sign of $s_d(t)$ is of the same form as $d(t)$. Consequently, sampling to determine the sign of the detector's output can be used to determine the original digital bit sequence.

The noise performance of the DPSK system is determined by its probability of making a bit error, which can be shown to be

$$P_e = \tfrac{1}{2}e^{-A^2T_b/(2\mathcal{N}_0)} = \tfrac{1}{2}e^{-E_b/\mathcal{N}_0} \tag{16.6-18}$$

Here A is the peak amplitude of $s_R(t)$ in Fig. 16.6-4, and

$$E_b = \frac{A^2T_b}{2} \tag{16.6-19}$$

is the average energy per bit interval in $s_R(t)$.

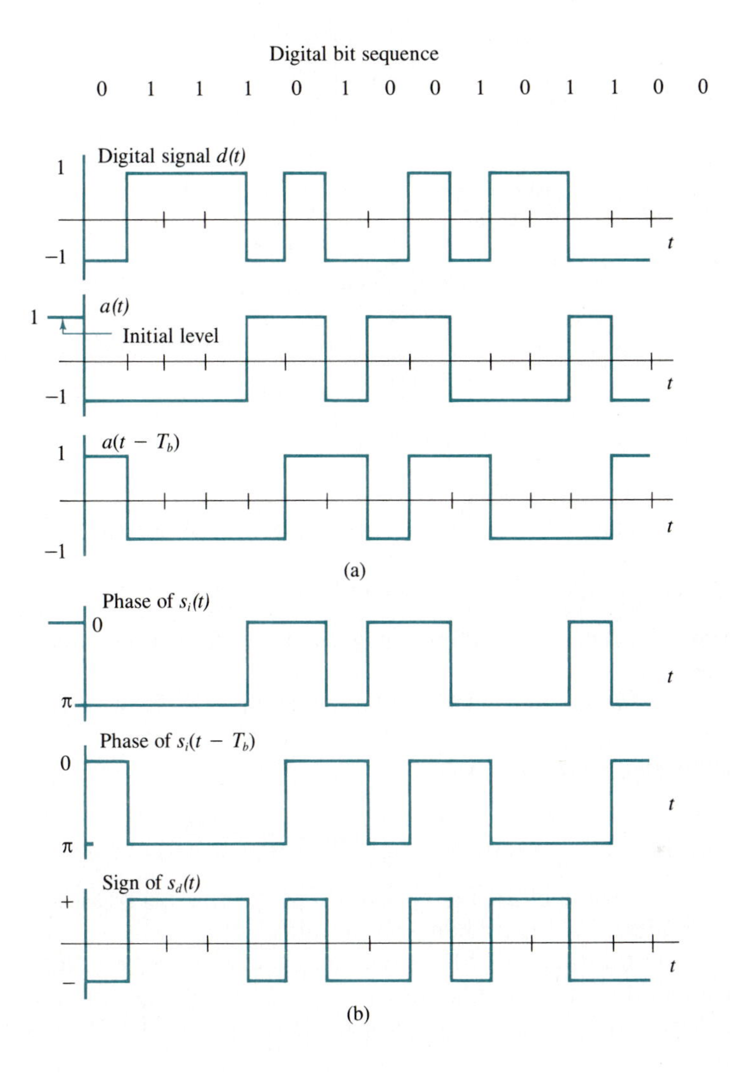

FIG. 16.6-5. (a) A digital-message sequence and the applicable modulator's waveforms in DPSK and (b) the waveform phases and polarities applicable to DPSK message recovery.

Frequency-Shift Keying (FSK)

As a last example of carrier-modulated digital communications systems, we discuss *frequency-shift keying* (FSK). FSK has both coherent and noncoherent implementations. The functional block diagram of a coherent system is sketched in Fig. 16.6-6. The transmitted signal $s_{\text{FSK}}(t)$ is generated by frequency modulation of a voltage-controlled oscillator (VCO). The digital signal $d(t)$ is assumed to have a polar format with amplitudes 1 and -1 corresponding to binary digits 1 and 0, respectively. Modulation keys the VCO's angular frequency between two values, according to

$$\omega_2 = \omega_0 + \Delta\omega \qquad d(t) = 1 \tag{16.6-20}$$

$$\omega_1 = \omega_0 - \Delta\omega \qquad d(t) = -1 \tag{16.6-21}$$

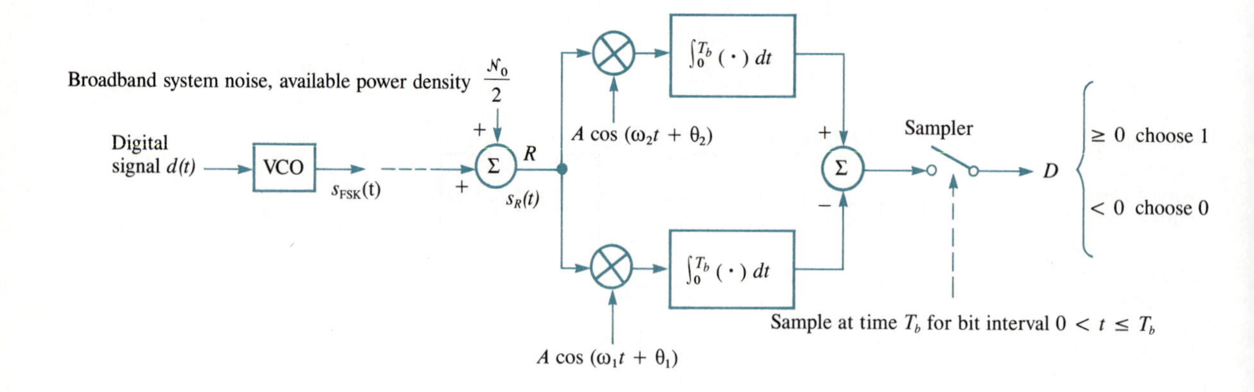

FIG. 16.6-6.
The functions required in a coherent-FSK system.

Here $\Delta\omega$ represents a frequency deviation from a nominal, or carrier, angular frequency ω_0.

In general, $\Delta\omega$ could have any value in FSK. However, so that bandwidth in the signal $s_{FSK}(t)$ is conserved, $\Delta\omega$ is usually not much larger than the minimum allowable value, which is†

$$\Delta\omega = \frac{\pi}{T_b} = \frac{\omega_b}{2} \tag{16.6-22}$$

With this value of $\Delta\omega$ the bandwidth of the channel needed to support $s_{FSK}(t)$ is approximately $2\omega_b$ (radians per second).

The transmitted FSK signal can be viewed as two unipolar-ASK signals in parallel (one at carrier frequency ω_2 and one at ω_1). The receiver becomes two coherent-ASK receivers in parallel, as shown in Fig. 16.6-6 (compare with Fig. 16.6-1).‡

The probability of a bit error occurring in any given bit interval, for the system of Fig. 16.6-6, is

$$P_e = \tfrac{1}{2} \operatorname{erfc}\left[\sqrt{\frac{A^2 T_b}{4\mathcal{N}_0}}\right] \tag{16.6-23}$$

where A is the peak amplitude of $s_R(t)$. In terms of the average received energy per bit (E_b), (16.6-23) becomes

† A type of FSK called *minimum-shift keying* (MSK) uses $\Delta\omega = \omega_b/4$. It is more complicated than coherent FSK and is not discussed here. Its noise performance can be superior to FSK's, however, and can equal that of coherent PSK.

‡ The reader should think about this model and justify why the difference, rather than the sum, occurs in Fig. 16.6-6.

$$P_e = \tfrac{1}{2}\,\text{erfc}\left[\sqrt{\frac{E_b}{2\mathcal{N}_0}}\right] \tag{16.6-24}$$

The functional block diagram of the noncoherent-ASK system is exactly the same as Fig. 16.6-6 if each product device and following integrator in the receiver is replaced by a corresponding matched filter followed by an envelope detector. The matched filters are defined by

$$H_{R_i}(\omega) = \frac{AT_b}{2}\left\{Sa\left[\frac{(\omega - \omega_i)T_b}{2}\right]e^{-j(\omega-\omega_i)T_b/2} + Sa\left[\frac{(\omega + \omega_i)T_b}{2}\right]e^{-j(\omega+\omega_i)T_b/2}\right\} \tag{16.6-25}$$

where ω_i, $i = 1, 2$, are defined in (16.6-20) and (16.6-21). The noncoherent system's bit-error probability can be shown to be

$$P_e = \tfrac{1}{2}\exp\left[-\frac{A^2T_b}{4\mathcal{N}_0}\right] = \tfrac{1}{2}\exp\left[-\frac{E_b}{2\mathcal{N}_0}\right] \tag{16.6-26}$$

where E_b is given by (16.6-19).

16.7 PERFORMANCE COMPARISONS OF CARRIER-MODULATED SYSTEMS

The noise performances of various digital communications systems can be compared on the basis of their bit-error probabilities P_e. The fair comparison requires that each system receive the same average signal energy per bit (E_b) and that each system operate with the same noise level at its input ($\mathcal{N}_0/2$). Thus, we compare systems when the ratio $E_b/\mathcal{N}_0$ is the independent parameter. We gather the various equations for reference:

$$P_e = \begin{cases} \dfrac{1}{2}\left[1 + \sqrt{\dfrac{1}{2\pi(E_b/\mathcal{N}_0)}}\right]\exp\left(\dfrac{-E_b}{2\mathcal{N}_0}\right) & \text{noncoherent ASK} \\[2ex] \tfrac{1}{2}\exp\left(\dfrac{-E_b}{2\mathcal{N}_0}\right) & \text{noncoherent FSK} \\[2ex] \tfrac{1}{2}\,\text{erfc}\left[\sqrt{\dfrac{E_b}{2\mathcal{N}_0}}\right] & \begin{array}{l}\text{coherent ASK}\\ \text{coherent FSK}\\ \text{unipolar PCM}\end{array} \\[2ex] \tfrac{1}{2}\exp\left(\dfrac{-E_b}{\mathcal{N}_0}\right) & \text{DPSK} \\[2ex] \tfrac{1}{2}\,\text{erfc}\left[\sqrt{\dfrac{E_b}{\mathcal{N}_0}}\right] & \begin{array}{l}\text{PSK}\\ \text{polar PCM}\\ \text{Manchester PCM}\end{array} \end{cases} \tag{16.7-1}$$

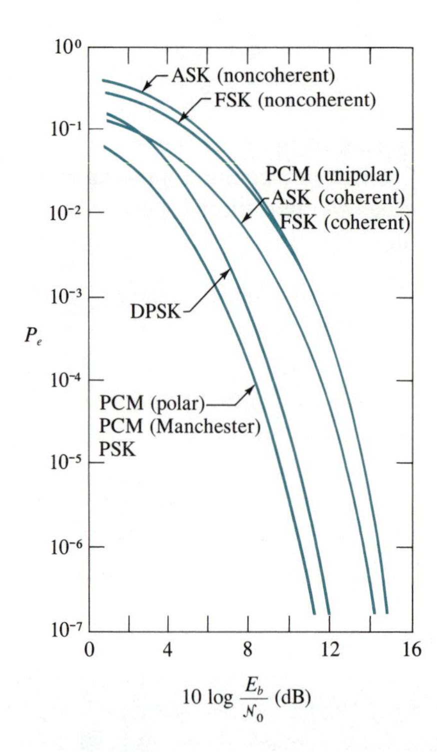

FIG. 16.7-1. Plots of P_e versus $E_b/\mathcal{N}_0$ (in decibels) for various digital communications systems.

Figure 16.7-1 shows plots of the equations of (16.7-1). Of the noncarrier (PCM) systems the polar- and Manchester-formatted systems are superior to the unipolar system. For the same value of P_e the unipolar system requires about twice (3 dB) as much transmitted energy per bit. This result is a consequence of not transmitting *any* energy during half the bits, on the average, in the unipolar system.

Of the carrier-modulated systems the coherent systems (PSK, FSK, ASK) perform better than their noncoherent versions (DPSK, noncoherent FSK, and noncoherent ASK, respectively), but only about 1 dB better (in $E_b/\mathcal{N}_0$ for the same P_e). In most cases in practice, it is worth the expense of 1 dB more in transmitter power to avoid having to generate a local carrier in the receiver. The PSK system is superior by about 3 dB (again in $E_b/\mathcal{N}_0$ for a given P_e) to both FSK and ASK, which have nearly the same performances.

PROBLEMS

16-1. An audio message has a spectral extent of 3 kHz. What minimum rate of sampling must be used to convert the message to digital form?

16-2. Work Problem 16-1 except assume a television signal with a spectral extent of 4.5 MHz.

16-3. An analog message has a spectral extent of 15 kHz. It is sampled at three times the Nyquist rate. What is its sampling rate?

16-4. A quantizer's quantum levels are separated by the step size $\delta v = 0.2$ V. The lowest and highest levels are -3.1 V and 3.1 V. A sequence of samples of a message is -2.13 V, -0.52 V, 0.96 V, 0.18 V, and -0.77 V. What is the corresponding sequence of quantized samples?

16-5. Is the quantizer of Problem 16-4 midriser or midtread?

16-6. What is the maximum amplitude that a message can have if the quantizer in Problem 16-4 is to not be saturated?

16-7. The quantizers of Fig. 16.2-2 can be said to have a "gain" of unity. Discuss the consequences of a quantizer with a "gain" of $K_q > 1$.

16-8. A symmetrically fluctuating message, for which $\overline{f^2(t)} = 1.96$ V^2 and $K_{cr} = 3.2$, is to be quantized so that $(S_o/N_q) = 2100$. The quantizer is to use the smallest number of levels that is a power of 2. How many levels does it have? What is δv? What are the extreme quantum levels?

16-9. A quantizer for a symmetrically fluctuating message has $\delta v = 0.04$ V and 128 quantum levels that span the extremes of the message exactly. (a) What is $|f(t)|_{max}$? (b) What is the largest crest factor the message can have if (S_o/N_q) must be at least 6600?

16-10. A compressor in a system is able to change a message's crest factor from 3.4 to 1.8 while keeping its peak amplitude constant. How many decibels of improvement in (S_o/N_q) can be expected in the system?

16-11. How many bits must a natural binary encoder have to work with the quantizers of Problems 16-4 and 16-9?

16-12. Make a table of binary code words where one digit (leftmost) represents the sign of an analog signal's quantized samples (0 for negative, 1 for positive) and the next three digits are natural binary code words for the magnitude of the quantized samples. The result is called a *folded binary code* (of 4 bits).

16-13. Let the digit sequence $b_4 b_3 b_2 b_1$ represent a 4-bit natural binary code for 16 levels labeled 0 through 15. Define a 4-bit *Gray code* by the digit sequence $g_4 g_3 g_2 g_1$, where

$$g_k = \begin{cases} b_4 & k = 4 \\ b_{k+1} \oplus b_k & k = 1, 2, 3 \end{cases}$$

Construct a table of natural binary code words and their corresponding Gray code words. What unique characteristic do you observe about the Gray code?

16-14. Sketch polar and unipolar waveforms for the digit sequence 100111001011011000. What would be a good guess about the probability that a binary 1 will occur in the next digit interval?

16-15. If the sequence of Problem 16-14 is labeled $\{b_k\}$ and is the input to the differential encoder of Fig. 16.3-2(a), find (a) $\{a_k\}$ when the output's initial value is 1, and (b) $\{a_k\}$ for an initial value of 0. (c) Put both sequences $\{a_k\}$ found in (a) and (b) through the differential decoder, and determine whether the original sequence $\{b_k\}$ is recovered in both cases.

16-16. Sketch the Manchester-formatted waveform for the first 10 digits of the sequence of Problem 16-14.

16-17. A Manchester waveform for which $T_b = 1$ μs is passed over a baseband channel having a bandwidth of 4 MHz. Explain the effect of the channel on the spectrum of the waveform at the receiver.

16-18. What noise level $\mathcal{N}_0/2$ at the input to a polar-PCM receiver corresponds to $P_e = 10^{-5}$ when $A = 6$ V and $T_b = 0.5$ μs?

16-19. (a) What minimum value of $(\hat{S}_i/N_i)_{\text{PCM}}$ is needed in a Manchester PCM system to realize $P_e = 10^{-4}$? (b) With $P_e = 10^{-4}$, $\mathcal{N}_0/2 = 4 \times 10^{-7}$ V^2/Hz, and $T_b = 0.3$ μs, what minimum input-pulse amplitude A is required?

16-20. In a polar-PCM system $N_b = 8$ and all levels span the variations of an analog message. If $(\hat{S}_i/N_i)_{\text{PCM}} = 20$, determine $(S_o/N_o)_{\text{PCM}}$ for the message if its crest factor is 2.7. Is the system operating above threshold?

16-21. Devise a sampling scheme (timing of samples) so that four voice messages each with 3-kHz bandwidth can be sampled at the Nyquist rate and time-multiplexed with samples taken at twice the Nyquist rate from six analog "monitoring" signals, each with 500-Hz bandwidth. What is the frame's duration?

16-22. A large radar site transmits 85 similar analog signals to a control room by a simple time multiplexing of natural samples over a single line. The signals are used to monitor various radar functions, and each has a 100-Hz bandwidth. If one time slot is used for synchronization, $\tau_g = 2\tau$, and sampling is at twice the Nyquist rate, what is τ?

16-23. Assume that $\omega_0 = 3\omega_b = 6\pi/T_b$, and sketch an ASK signal for the binary sequence 1011001001.

16-24. Pulses arriving at the receiver in a coherent-ASK system have a peak amplitude of 2.2 V. If the bit duration is 2.6 μs and the noise level is $\mathcal{N}_0/2 = 10^{-7}$ V^2/Hz, find P_e. For a 1-Ω impedance, what is E_b?

★ **16-25.** When a pulse of carrier of amplitude A during $0 < t \le T_b$ arrives at point R in Figs. 16.6-1 and 16.6-2, show that D at time T_b in both systems is the same if noise is neglected and $\omega_0 T_b \gg 1$.

16-26. In a coherent-ASK system $E_b/\mathcal{N}_0 = 20$. What value of $E_b/\mathcal{N}_0$ is required in a noncoherent-ASK system to give the same value of P_e as the coherent system?

16-27. Show that $E_b/\mathcal{N}_0$ in a coherent-ASK system must be precisely twice that in a coherent-PSK system if both systems have the same values of P_e.

16-28. In a DPSK system $P_e = 3 \times 10^{-4}$ when the received pulses are 1.6 V in amplitude. If the pulse amplitude increases such that $P_e = 2 \times 10^{-6}$, what is the new amplitude?

16-29. When the coherent-FSK receiver of Fig. 16.6-6 is viewed as two polar-ASK receivers in parallel, why does the summing junction produce a difference rather than a sum?

16-30. The input FSK pulse at point R in Fig. 16.6-6 is

$$s_R(t) = A \cos(\omega_2 t + \theta_2) \qquad 0 < t < T_b$$

and 0 elsewhere in t, when the bit interval corresponds to a 1. (a) Find the outputs from the two integrators, and show that the outputs are approximately $A^2T_b/2$ (upper integrator) and 0 (lower integrator) at the sample time $t = T_b$. (b) Show that the two outputs of (a) are reversed if $s_R(t)$ changes in frequency and phase to ω_1 and θ_1, respectively. Assume $\omega_0 T_b \gg 1$ in every case.

16-31. For $E_b/\mathcal{N}_0 = 12$, solve all the equations of (16.7-1) and observe the marked difference in performance of the various digital communications systems.

PART 5

Power Systems and Machinery

CHAPTER

Power System Fundamentals

17.0 INTRODUCTION

In modern society electric power is indispensable. The very fabric of our existence depends on electric power. It drives the machinery that processes our food and manufactures our conveniences (electric shavers, toasters, microwave ovens, clocks, etc.) and necessities (clothing). It drives the devices with which we communicate (radio, telegraph, telephone, etc.), that provide us security (home-security systems, police and fire radios, etc.), that brighten our nights (lights), and entertain us (movie houses, radios, TVs, videocassette recorders, and compact disks). We have become so accustomed to reliable, accessible electric power that we take it for granted. Our use and demand for electric power grow annually at the rate of 2 to 3% and our need doubles about every 24 to 35 years.

Every engineer, not just electrical engineers, should be aware of the ways in which electric power is generated and makes its way to the ultimate user. In this chapter we shall discuss power generation and distribution, touching on main topics and giving some, but avoiding excessive, detail.

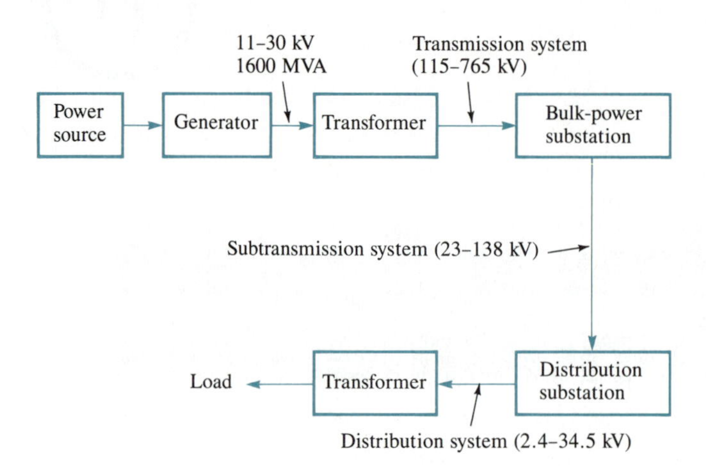

FIG. 17.1-1. The principal elements of a power system.

17.1 ELECTRIC POWER SYSTEMS IN THE UNITED STATES

Power is generated in the United States by many generating stations interconnected in an overall network called the *power grid* that spans the entire continental United States. Part of the network is federally owned and part is municipally owned, but the bulk is privately owned. Some stations in a given geographical area operate as "power pools" for economic and reliability advantages.

The structure of the parts of the power grid (the structure in a local area, for example) varies greatly. The principal elements of a power system are shown in Fig. 17.1-1. Except for a few direct current (dc) cases, by far the bulk of all generating stations are ac (alternating current) at 60-Hz frequency. The principal source of energy comes from the burning of fossil fuels (coal, oil) to generate steam that drives steam turbines. Turbines, in turn, drive electric generators (Sec. 17.2). Other important power sources are water (to turn the electric generator, in a waterwheel fashion) and nuclear energy, where nuclear reactions generate heat to drive the steam-turbine-generator chain. Although less widely used, other sources of energy are geothermal sources, wind, sun, and ocean currents and tides.

The power source of Fig. 17.1-1 drives an ac electric generator that is equivalent to a voltage source with a sinusoidal waveform at a 60-Hz frequency. It is capable of providing large amounts of power (up to around 1600 MW) to its load. An ac generator actually produces three-phase (abbreviated 3-ϕ) power. That is, *three* voltages are produced at 60 Hz with relative phase angles of 0, $-2\pi/3$, and $-4\pi/3$. Each of the three voltage sources provides part of the total power, ideally one-third in a "balanced load" (see Sec. 17.2 for more detail). Voltages of ac generators range from about 11 to 30 kV (rms),† as measured between a pair of the

† Recall that the rms (root-mean-squared) value of a sinusoidal voltage is $1/\sqrt{2}$ times its peak value.

three output terminals (called line-to-line voltage, or simply the *line voltage*).

So that the real power losses in the transmission of power to the user, or load, are minimized, the high-current medium-voltage generator outputs are raised to much higher voltages (with lower currents) by transformers for transmission over the power grid. Grid voltages usually range from 115 to 765 kV. Some nominal values in use are 115, 138, 230, 345, 500 and 765 kV. Several generators may be operated in parallel in the power grid to provide the total power needed. They are combined at a common ''point'' called a *bus*. A power bus can actually occupy acres of ground and is not just a single point. The network of transmission lines that forms the power grid is called the *transmission system* (Fig. 17.1-1).

The purpose of the transmission system is to combine the generated power from many sources to provide the power needs of many users. Many transmission lines exist around the country to form the nation's grid. A given transmission line typically feeds one or more *bulk-power substations* that use transformers to reduce the voltage to a lower level for retransmission over a *subtransmission network*. The transmission lines of the subtransmission network usually operate at voltages from about 23 to 138 kV, with 23, 34.5, and 69 kV being some popular values.

The subtransmission network's lines typically feed power to *distribution substations* that again reduce the voltage to a value typically in the range of 2.4 to 34.5 kV, with 4.16, 12.47, 24.9, and 34.5 kV being some representative values. Transmission lines of the distribution system commonly provide power to small industries, commercial establishments, and homes. A transformer is commonly used to set the voltage at the final user's location. For example, the power for a typical home is derived from a transformer that reduces the distribution line's voltage, sometimes called the *primary* or *feeder voltage*, to 120 and 240 V (the *secondary* or *consumer voltage*) using a three-wire line.

The reader should understand that the overall power network is very complicated. Generators can provide power at buses at the subtransmission or even the distribution system points and are not limited to buses at the transmission system's level. In fact, the trend in modern systems has been to use higher transmission voltages as the technology evolves. Some voltages considered to be transmission level a few years ago are now considered to be subtransmission level.

A load is any point where a user extracts power from the system. The power can be real, reactive, or some combination of both. Loads can be placed at any points in the system. Generally, however, it would be a very large user, such as a town or city, that would tie on to the system at the transmission level. A large consumer, such as a railroad or a large industry, might represent a load at the subtransmission level. Smaller industrial concerns, such as areas within a city or a rural area, would represent medium loads at the distribution level.

The preceding discussion has ignored many details to present a plausible explanation of the overall power system. In the rest of this chapter the most important system concepts and components required to make a practical system function are described.

17.2 POWER GENERATION

Three-Phase Generators

The generator in Fig. 17.1-1 is an electromechanical device whereby power supplied to rotate a mechanical shaft is converted to an electric voltage source capable of providing electric power. Mechanically, a generator consists of a *rotor* and a *stator*. The rotor is the rotating part of the machine driven by the shaft. A winding is placed on the rotor, which is excited by a dc current to establish a magnetic field that emanates radially from the rotor. The magnetic field is roughly analogous to that of a permanent magnet. The field is fixed in relation to the rotor but rotates in space as the rotor revolves. In high-speed machines the rotor generates only one north and one south pole and is called a two-pole generator. Slower machines (such as in a hydroelectric plant) can have many poles. If p is the number of poles (always an even positive integer), the angular rate of the shaft, denoted by ω_s, is related to the angular frequency ω of the electric voltage according to

$$\omega_s = \frac{2\omega}{p} \tag{17.2-1}$$

Thus, for $\omega/2\pi = 60$ Hz, a two-pole machine has a shaft speed of $\omega_s/2\pi = 60$ Hz $= 3600$ rpm. Generators in which shaft speed is locked or synchronized to the electric frequency, as in (17.2-1), are called *synchronous*.

The stator is the stationary part of the generator inside which the rotor revolves. Three windings, called *phase* or *load* windings, are placed on the stator; they are physically spaced in increments of 120° around the periphery. As the magnetic field of the rotor revolves, it cuts the conductors of the three windings and induces a voltage in each by Faraday's law (Sec. 1.4). If the magnetic field encounters the phase windings labeled *abc* in sequence, the generated phase voltages are†

$$v_a = E_s \exp j\omega t \tag{17.2-2a}$$

$$v_b = E_s \exp\left[j\omega t - j\left(\frac{2\pi}{3}\right)\right] \tag{17.2-2b}$$

$$v_c = E_s \exp\left[j\omega t - j\left(\frac{4\pi}{3}\right)\right] \tag{17.2-2c}$$

Here we assume the desired case where each voltage has the same peak amplitude E_s. The rms value of a phase voltage is $E_s/\sqrt{2}$. It is the rms voltage rather than peak voltage that is quoted when values are given in power systems.

† We use complex signals to represent the real waveforms. Recall that the real signal is the real part of the complex waveform. (See sec. 4.1.)

A generator for which (17.2-2) is true is called *three-phase* with sequence *abc*. Only one other sequence, *acb*, is possible; it occurs when any two phases in the *abc* sequence are interchanged. We shall restrict our discussions henceforth to only the *abc* sequence, also called a *positive sequence*.

Two methods are used to provide the direct current required to excite the rotor's field winding. In smaller machines the winding is accessible through terminals having electric contacts, called brushes, that slide on *slip rings*. There is one slip ring for each end of the winding. Another method, used on larger machines, relies on rectification of the output of a small 3-ϕ ac generator mounted directly on the shaft of the synchronous generator. It is called a *brushless exciter*. The exciter requires a (small) dc field current (to its stator). This current can either be externally provided or derived from rectifying the output of a small ac generator, called a *pilot exciter*, that is also on the rotor's shaft. The pilot exciter is small enough that it can use permanent magnets on its rotor to provide its necessary magnetic field. The combination of brushless exciter and synchronous generator produces a machine with lower maintenance requirements, because no mechanical contacts are necessary between the rotor and the stator. When the pilot exciter is added, the generator becomes independent of any external electric power sources.

The power capacity of large generators increases with increasing rotor and stator size. Achieving high steam-turbine efficiencies requires high shaft speeds. Thus, steam-driven turbine generators usually operate at 3600 rpm (two poles) or 1800 rpm (four poles) for 60-Hz ac. Such high speeds limit rotor diameters to about 1.1 m (3.5 ft) and 2.1 m (7 ft), respectively, for two- and four-pole machines, because of centrifugal forces. Large size is achieved by extending rotor lengths to around 5 to 6 times the diameters.† All these modern high-speed machines are of the horizontal-shaft form and have capacities up to about 1600 MVA for 1800-rpm machines and 1150 MVA in 3600-rpm machines.

On the opposite end of the speed scale are the slow-speed, multipole generators found in hydroelectric power plants. A typical machine runs at 360 rpm and has $p = 20$ poles for a 60-Hz ac frequency. Power capacity ranges up to about 120 MVA. Horizontal-shaft machines are usually used when the height (head) of the water source is large. In low-head cases vertical-shaft machines are common.‡

Choice of Three-Phase Voltage for Power Generation

The reader may wonder why three-phase voltages are generated for the power system instead of the much simpler system using a single-phase voltage. There are several compelling reasons. In a 1-ϕ system, motors and generators would involve instantaneous power flow that pulsates (see Fig. 4.5-1, for example). This condi-

† A. R. Bergen, *Power Systems Analysis,* Prentice-Hall, Englewood Cliffs, N.J. 1986, p. 6.

‡ J. Weisman and R. Eckart, *Modern Power Plant Engineering,* Prentice-Hall, Englewood Cliffs, N.J. 1985, p. 421 and sec. 8.8.

tion leads to nonuniform torque and vibration problems in high-power applications. Instantaneous power in a 3-ϕ system is constant (for a balanced system). In low-power applications (consumers and small motors), where the pulsating power is acceptable and 1-ϕ applications are encountered, the 3-ϕ system can still service all these needs. Finally, we note that for the same power requirements, 3-ϕ equipment for distribution and use of power can be physically smaller and less expensive than in a 1-ϕ system.

Wye and Delta Connections of Generators

A generator's three-phase voltages can be represented as three separate sources, as shown in Fig. 17.2-1(a). When connected as shown, the generator is equivalent to three 1-ϕ sources operating simultaneously and requiring six wires to connect the three loads. However, it can be shown (Problem 17-7) that the net current $i_n = i_a + i_b + i_c$ is zero when the loads are equal. If $i_n = 0$, there is no need for three of the six wires and they can be eliminated, as shown in Fig. 17.2-1(b). With this connection, called a *wye* (Y) *connection*, the system becomes a 3-ϕ *balanced*

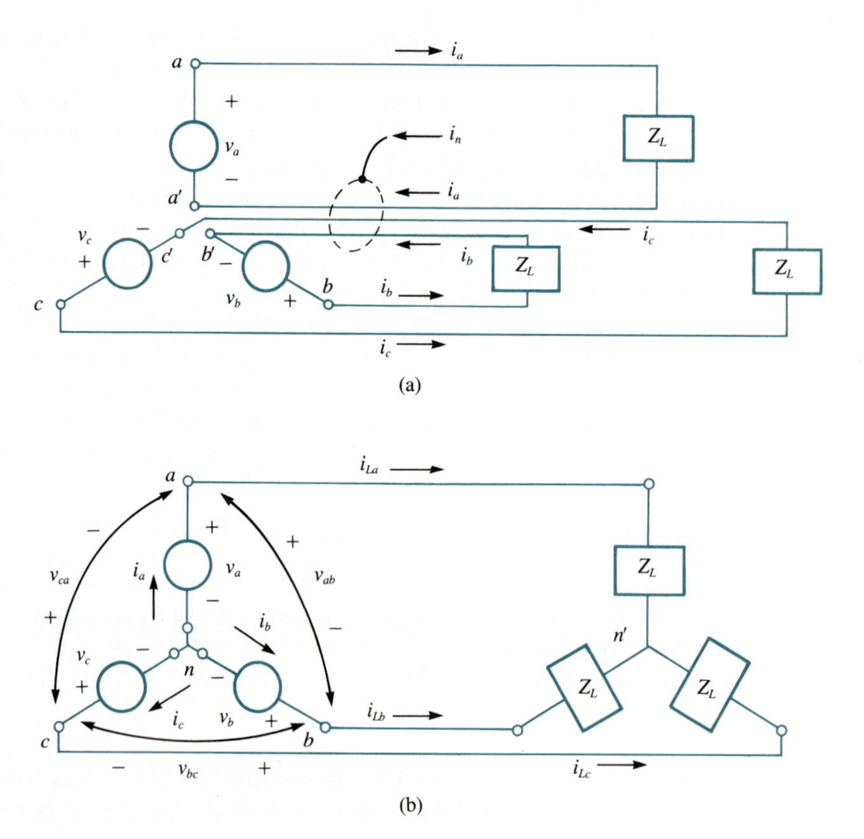

FIG. 17.2-1. (a) Phase voltages of a three-phase generator represented as three separate voltage sources and (b) an equivalent wye connection.

system. In a real system with slight unbalance, a fourth wire, called the *neutral*, is brought out from point n and connected to n'. The result is a four-wire, 3-ϕ connection. The amount of material needed in the (low-current) neutral wire is small in relation to the "hot" leads in a nearly balanced system. Sometimes, the neutral n can be grounded such that the earth serves as the neutral line.

For a wye-connected generator Fig. 17.2-1(b) indicates that *line currents* (i_{La}, i_{Lb}, i_{Lc}) are equal to their respective *phase currents* (i_a, i_b, i_c). Thus, for the balanced system

$$i_{La} = i_a = \frac{v_a}{Z_L} \qquad (17.2\text{-}3a)$$

$$i_{Lb} = i_b = \frac{v_b}{Z_L} \qquad (17.2\text{-}3b)$$

$$i_{Lc} = i_c = \frac{v_c}{Z_L} \qquad (17.2\text{-}3c)$$

The relationships between *line-to-line* (or simply *line*) *voltages* (v_{ab}, v_{bc}, v_{ca}) and the phase voltages are not as obvious. On reduction of the algebra, the results are

$$v_{ab} = v_a - v_b = \sqrt{3}\, e^{j\pi/6} v_a \qquad (17.2\text{-}4a)$$

$$v_{bc} = v_b - v_c = \sqrt{3}\, e^{j\pi/6} v_b \qquad (17.2\text{-}4b)$$

$$v_{ca} = v_c - v_a = \sqrt{3}\, e^{j\pi/6} v_c \qquad (17.2\text{-}4c)$$

which show that line voltages have amplitudes $\sqrt{3}$ times those of the phase voltages and phase shifts of $\pi/6$ (30°) relative to the phases of the phase voltages. We use an example to show that (17.2-4a) is true; (17.24b) and (17.2-4c) are proved in a similar way.

EXAMPLE 17.2-1

We prove (17.2-4a):

$$\begin{aligned} v_{ab} &= v_a - v_b = E_s e^{j\omega t} - E_s e^{j\omega t - j(2\pi/3)} \\ &= E_s e^{j\omega t - j(\pi/3)}(e^{j\pi/3} - e^{-j\pi/3}) \\ &= v_a e^{-j\pi/3} 2j \sin\frac{\pi}{3} = \sqrt{3}\, e^{j\pi/6} v_a \end{aligned}$$

Synchronous generators in most power systems are wye-connected with a neutral ground. However, a second connection method is also possible. It is called a *delta* (Δ) *connection* and is illustrated in Fig. 17.2-2. For a balanced system the

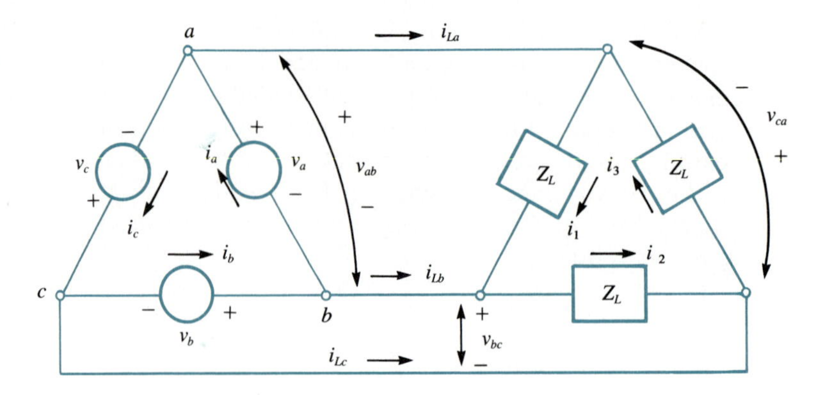

FIG. 17.2-2. Phase voltages in a three-phase generator having a delta connection.

load impedances are all equal, as shown. Clearly, line voltages equal phase voltages:

$$v_{ab} = v_a \tag{17.2-5a}$$

$$v_{bc} = v_b \tag{17.2-5b}$$

$$v_{ca} = v_c \tag{17.2-5c}$$

From analysis of the circuit it is found that

$$i_{La} = \sqrt{3}\, e^{-j\pi/6} i_a \tag{17.2-6a}$$

$$i_{Lb} = \sqrt{3}\, e^{-j\pi/6} i_b \tag{17.2-6b}$$

$$i_{Lc} = \sqrt{3}\, e^{-j\pi/6} i_c \tag{17.2-6c}$$

where the phase currents are

$$i_a = \frac{v_a}{Z_L} \tag{17.2-7a}$$

$$i_b = \frac{v_b}{Z_L} \tag{17.2-7b}$$

$$i_c = \frac{v_c}{Z_L} \tag{17.2-7c}$$

Equation (17.2-6) shows that the magnitude of any line current is $\sqrt{3}$ times the magnitude of any phase current. The phase of a given line current is shifted by $-\pi/6$ ($-30°$) relative to the phase of its respective phase current.

Wye and Delta Load Connections

In discussing the wye-connected generator (Fig. 17.2-1) we assumed a wye-connected load. Similarly, a delta-connected load was assumed in Fig. 17.2-2 for a delta-connected generator. Both load assumptions were for convenience, because

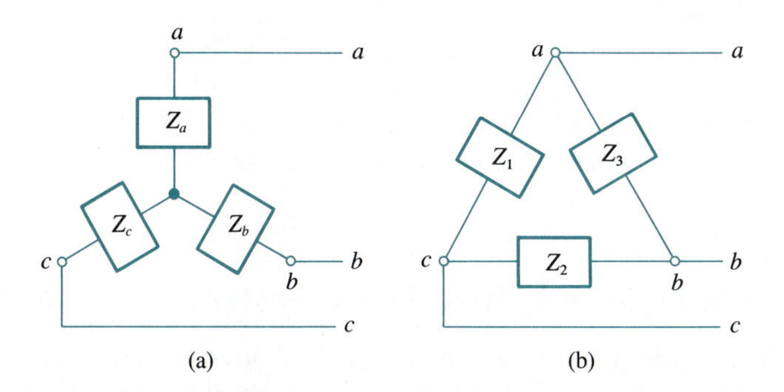

FIG. 17.2-3. Three-phase generator loads, (a) wye-connected and (b) delta-connected, that have equivalent terminal behavior when (17.2-8) is true.

any load in delta form can be converted to a wye form having the equivalent terminal characteristics (line voltages and currents) and vice versa.

To require 3-ϕ delta- and wye-connected loads, as shown in Fig. 17.2-3, to have the same terminal characteristics, we need only have the impedance between any pair of delta lines be equal to that between a corresponding pair of wye lines (with the third lines open-circuited in both). We readily find the requirements:

$$Z_b + Z_c = \frac{Z_2(Z_1 + Z_3)}{Z_1 + Z_2 + Z_3} \qquad \text{line } a \text{ open} \tag{17.2-8a}$$

$$Z_a + Z_c = \frac{Z_1(Z_2 + Z_3)}{Z_1 + Z_2 + Z_3} \qquad \text{line } b \text{ open} \tag{17.2-8b}$$

$$Z_a + Z_b = \frac{Z_3(Z_1 + Z_2)}{Z_1 + Z_2 + Z_3} \qquad \text{line } c \text{ open} \tag{17.2-8c}$$

If these three equations are solved for Z_a, Z_b, and Z_c in terms of Z_1, Z_2, and Z_3 (Problem 17-9), we have

$$Z_a = \frac{Z_1 Z_3}{Z_1 + Z_2 + Z_3} \tag{17.2-9a}$$

$$Z_b = \frac{Z_2 Z_3}{Z_1 + Z_2 + Z_3} \tag{17.2-9b}$$

$$Z_c = \frac{Z_1 Z_2}{Z_1 + Z_2 + Z_3} \tag{17.2-9c}$$

Thus, if we are given a delta-type load, an equivalent wye-type load can be constructed with impedances given by (17.2-9).

Alternatively, (17.2-8) or (17.2-9) can be solved for the delta-type load that is equivalent to a given wye-type load (Problem 17-10):

$$Z_1 = Z_a + Z_c + \frac{Z_a Z_c}{Z_b} \tag{17.2-10a}$$

$$Z_2 = Z_b + Z_c + \frac{Z_b Z_c}{Z_a} \tag{17.2-10b}$$

$$Z_3 = Z_a + Z_b + \frac{Z_a Z_b}{Z_c} \tag{17.2-10c}$$

In the case of a balanced 3-ϕ load, where $Z_1 = Z_2 = Z_3 = Z_\Delta$ and $Z_a = Z_b = Z_c = Z_Y$, an especially simple form results for both (17.2-9) and (17.2-10):

$$Z_\Delta = 3Z_Y \tag{17.2-11}$$

Power in Balanced Three-Phase Systems

Two of the advantages of a balanced 3-ϕ power system compared with three identical single-phase systems are that (1) the same total average power is provided with fewer wires, and (2) the power provided is constant—that is, it does not pulsate. These facts are briefly demonstrated below.

Let P represent the average power provided by one of three identical single-phase systems defined by Fig. 17.2-1(a) and (17.2-2). Then from (4.5-5), for system a,

$$P = \text{Re}\left[\frac{v_a i_a^*}{2}\right] = \text{Re}\left[\frac{E_s^2}{2Z_L^*}\right] = \frac{E_s^2 \cos\theta_L}{2|Z_L|} \tag{17.2-12}$$

where the load impedance is represented by $Z_L = |Z_L| \exp j\theta_L$. For systems b and c the same result is achieved:

$$\text{Re}\left[\frac{v_b i_b^*}{2}\right] = \text{Re}\left[\frac{E_s^2}{2Z_L^*}\right] = P \tag{17.2-13}$$

$$\text{Re}\left[\frac{v_c i_c^*}{2}\right] = \text{Re}\left[\frac{E_s^2}{2Z_L^*}\right] = P \tag{17.2-14}$$

Overall average power from the three identical 1-ϕ systems is therefore $3P$, as expected. Let $P_{3\text{-}\phi}$ represent the average power from the 3-ϕ system of Fig. 17.2-1(b); it is the sum of the powers generated in the phases. Thus,

$$P_{3\text{-}\phi} = \text{Re}\left[\frac{v_a i_a^*}{2}\right] + \text{Re}\left[\frac{v_b i_b^*}{2}\right] + \text{Re}\left[\frac{v_c i_c^*}{2}\right] = 3P \tag{17.2-15}$$

is the same as in the three 1-ϕ systems.

The instantaneous real power in the 3-ϕ system, denoted by $p_{3\text{-}\phi}$, is needed to prove that it is constant in time. It is the sum of the instantaneous powers generated in the three phases:

$$\begin{aligned}
p_{3\text{-}\phi} &= \text{Re}(v_a)\,\text{Re}(i_a) + \text{Re}(v_b)\,\text{Re}(i_b) + \text{Re}(v_c)\,\text{Re}(i_c) \\
&= E_s \cos(\omega t) \frac{E_s}{|Z_L|} \cos(\omega t - \theta_L) \\
&\quad + E_s \cos\left(\omega t - \frac{2\pi}{3}\right) \frac{E_s}{|Z_L|} \cos\left(\omega t - \frac{2\pi}{3} - \theta_L\right) \\
&\quad + E_s \cos\left(\omega t - \frac{4\pi}{3}\right) \frac{E_s}{|Z_L|} \cos\left(\omega t - \frac{4\pi}{3} - \theta_L\right)
\end{aligned}$$

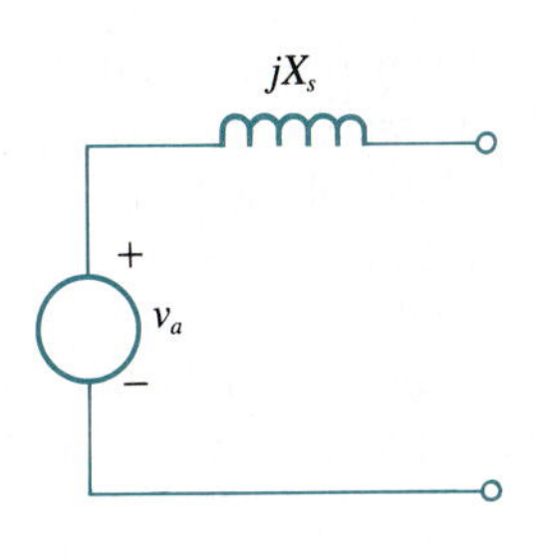

FIG. 17.2-4. An equivalent circuit for phase *a* of a three-phase generator.

$$= \frac{E_s^2}{2|Z_L|}\left[\cos\theta_L + \cos(2\omega t - \theta_L) + \cos\theta_L + \cos\left(2\omega t - \frac{4\pi}{3} - \theta_L\right)\right.$$
$$\left. + \cos\theta_L + \cos\left(2\omega t - \frac{8\pi}{3} - \theta_L\right)\right]$$
$$= \frac{3E_s^2\cos\theta_L}{2|Z_L|} = 3P \qquad (17.2\text{-}16)$$

which is a constant. To gain the advantage of power constancy, loads (motors, for instance) must be 3-ϕ, that is, use all three phases. A 1-ϕ motor operating on only one phase of a 3-ϕ system will have pulsating power. In higher-power applications where pulsation is important, motors are always 3-ϕ.

Because all power systems try to operate in a balanced condition, and since all of the three phases behave similarly, analysis is typically done on a *per-phase* basis. In a similar manner, when diagrams are made of power systems, they are typically *one-line diagrams*, meaning that a single line is used to represent all three phases, since the phases are similar. One-line diagrams are used extensively to model power transmission in the system.

Equivalent Circuit

A synchronous generator has an approximate equivalent circuit per phase, as shown in Fig. 17.2-4. The internal reactance X_s is called the *synchronous reactance*. More generally, there is also a resistance in series with X_s, but its effect is usually negligible.

17.3 POWER TRANSFORMERS

Transformers are critical elements in any power system. They allow the relatively low voltages from generators to be raised to very high levels for efficient power transmission. At the user end of the system, transformers reduce the voltage down to values most appropriate for the types of loads being serviced. In both types of use we refer to the *primary* of the transformer as the side (winding) connected to the source and the *secondary* as the output, or load, side. Since the power system is three-phase (3-ϕ), transformers must be 3-ϕ.

A 3-ϕ transformer can be assembled from three ordinary single-phase (1-ϕ) transformers or built by placing all windings on a single core. Each approach has its own advantages and disadvantages. The following discussions will mainly center on use of three separate transformers, because concepts are more easily visualized. Later, we briefly discuss the single 3-ϕ unit.

There are four ways in which 3-ϕ transformers may be connected: wye-wye (Y-Y), delta-delta (Δ-Δ), wye-delta (Y-Δ), and delta-wye (Δ-Y).

Wye-Wye Transformer

The wye-wye form of connection is illustrated in Fig. 17.3-1. The physical connection is shown in Fig. 17.3-1(a), but the second representation of (b) is often used. Symbols for one-line diagrams are shown in Fig. 17.3-1(c). In real transformers losses are usually small enough that they can be approximated as ideal. Thus, we may assume

$$v_a = \frac{v_A}{n} \qquad i_a = ni_A \tag{17.3-1a}$$

$$v_b = \frac{v_B}{n} \qquad i_b = ni_B \tag{17.3-1b}$$

$$v_c = \frac{v_C}{n} \qquad i_c = ni_C \tag{17.3-1c}$$

where n is the *transformer ratio* (same as the turns ratio of Sec. 1.4). For a balanced load, analysis of the circuit gives

$$i_{La} = ni_{LA} \tag{17.3-2a}$$

$$i_{Lb} = ni_{LB} \tag{17.3-2b}$$

$$i_{Lc} = ni_{LC} \tag{17.3-2c}$$

for line current relationships, and

$$v_{ab} = v_a - v_b = \frac{v_A - v_B}{n} = \frac{v_{AB}}{n} \tag{17.3-3a}$$

$$v_{bc} = v_b - v_c = \frac{v_B - v_C}{n} = \frac{v_{BC}}{n} \tag{17.3-3b}$$

$$v_{ca} = v_c - v_a = \frac{v_C - v_A}{n} = \frac{v_{CA}}{n} \tag{17.3-3c}$$

for line voltages. These equations show that all secondary voltages and currents are in phase with their respective primary quantities.

Advantages of the Y-Y transformer are that neutral points are available for connections or grounding, winding insulation must withstand only the phase voltages and not the higher line voltages, and two choices of load voltage are available if neutral lines are used. The last two advantages lead to some use in

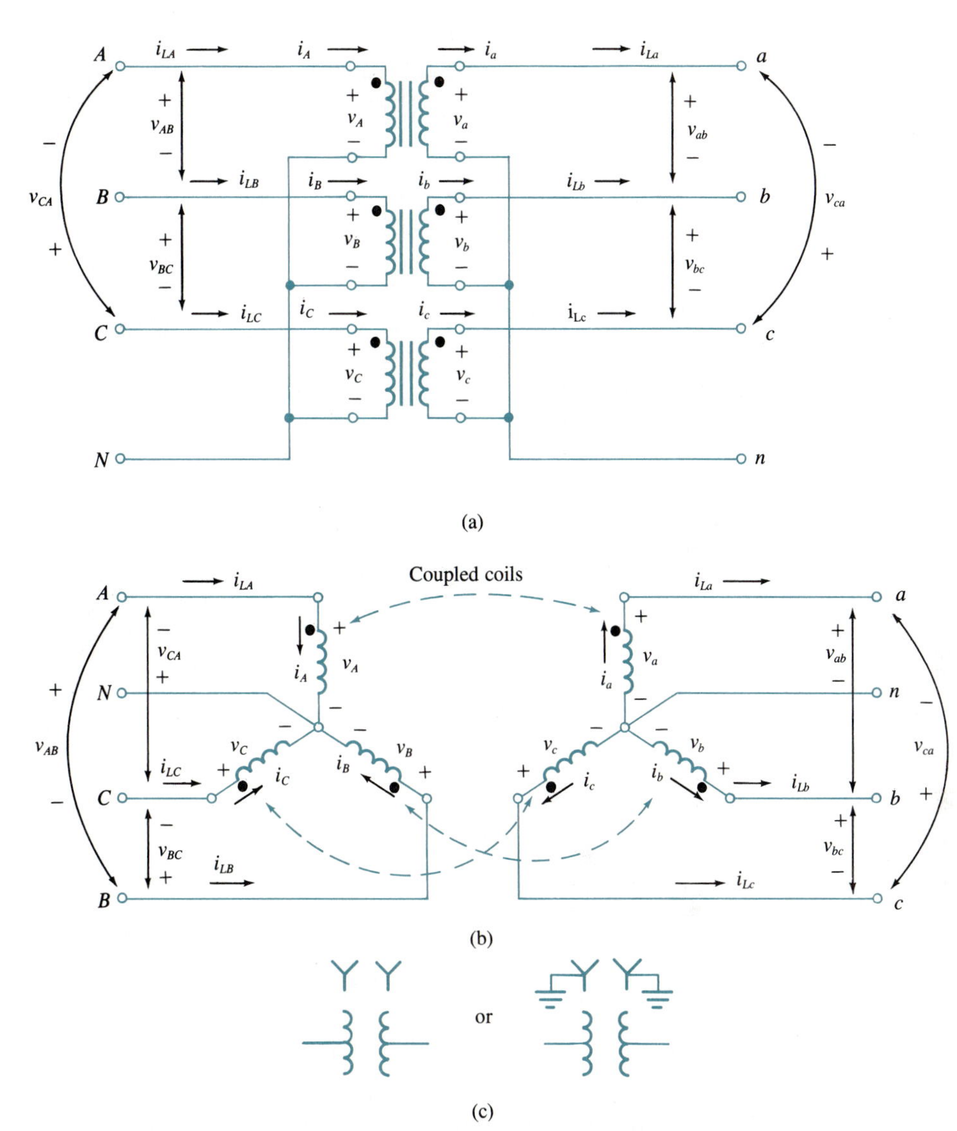

FIG. 17.3-1. (a) A three-phase transformer comprising three single-phase transformers connected in a Y-Y configuration, (b) a second representation, and (c) symbols representing the transformer.

high-voltage applications and in some four-wire systems to deliver power to industrial users. The main disadvantage stems from transformer behavior when third harmonics of the fundamental frequency are generated because of practical nonlinearities in the magnetization characteristics of the core. Even though some remedies exist,† other transformer connections overcome these problems more easily, so that few Y-Y transformers are used in practice.

† Third, or *tertiary*, windings are sometimes added to each of the three transformers and are delta-connected. This technique can remove the third-harmonic problem but adds to the transformer's cost and reduces its reliability.

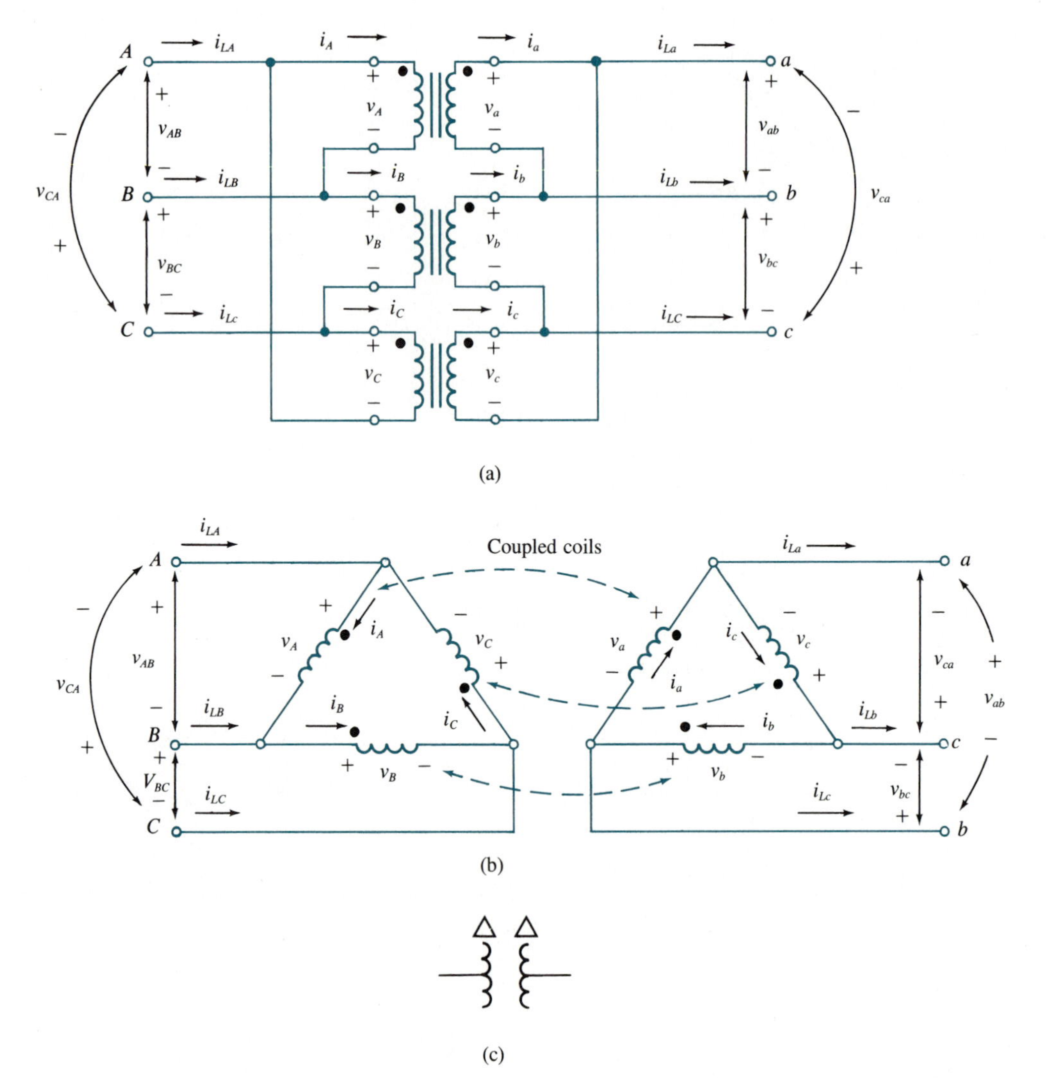

FIG. 17.3-2.
(a) A three-phase transformer comprising three single-phase transformers connected in a Δ-Δ configuration, (b) a second representation, and (c) symbols representing the transformer.

Delta-Delta Transformer

The delta-delta transformer of Fig. 17.3-2 does not suffer from the third-harmonic problems found in the Y-Y connection. The closed form of the delta provides a path for third-harmonic currents to circulate so that they do not flow into the power system.

Analysis of the Δ-Δ transformer is straightforward. Line voltages follow definitions of Fig. 17.3-2 and the use of (17.3-1):

$$v_{ab} = v_a = \frac{v_A}{n} = \frac{v_{AB}}{n} \tag{17.3-4a}$$

$$v_{bc} = v_b = \frac{v_B}{n} = \frac{v_{BC}}{n} \tag{17.3-4b}$$

$$v_{ca} = v_c = \frac{v_C}{n} = \frac{v_{CA}}{n} \tag{17.3-4c}$$

Next, we apply (17.2-6) to (17.3-1), once for the secondary Δ and once for the primary, to obtain

$$i_{La} = ni_{LA} \tag{17.3-5a}$$

$$i_{Lb} = ni_{LB} \tag{17.3-5b}$$

$$i_{Lc} = ni_{LC} \tag{17.3-5c}$$

These equations show that all secondary line voltages and currents are in phase with their respective primary quantities.

The Δ-Δ transformer is often found in medium- and lower-voltage applications because its insulation must withstand the full line-to-line voltages. While this fact may be considered a disadvantage, it also has a major advantage. If one of the three transformers is lost (by open circuits or removal), the system will still operate properly if the load being served is slightly less than 58% of full capacity.† The mode of operation is called an *open-delta connection*. It allows easy servicing or replacement of single transformers when the 3-ϕ unit is implemented with three 1-ϕ transformers. It also allows power companies to provide full 3-ϕ service with later growth capability in service by adding a third transformer.

EXAMPLE 17.3-1

Assume a 3-ϕ, Δ-Δ transformer is to step down 13.8 kV to 480 V in a balanced system for which all loads are resistive. If the system is operating at full capacity of 100 kVA, we find the primary and secondary line currents and the resistance of a typical leg of a Δ-connected load.

Because of the resistive balanced load, the product of the rms voltage (480 V) and rms current (I_{sec}) of a secondary line must equal $\frac{1}{3}$ of the total power. Thus, $I_{sec} = 100 \times 10^3/(3)(480) \approx 69.44$ A. The transformer ratio is $n = 13{,}800/480 = 28.75$, so the primary's rms line current is $I_{pri} = I_{sec}/n \approx 69.44/28.75 \approx 2.42$ A. In

† D. R. Brown, and E. P. Hamilton III, *Electromechanical Energy Conversion*, Macmillan, New York, 1984, p. 469.

the Δ load $\frac{1}{3}$ the total power is expended in each leg. If I_Δ is the rms current in a leg, then $I_\Delta = 100 \times 10^3/(3)(480) \approx 69.44$ A. The leg's resistance becomes $R_\Delta = 480/I_\Delta \approx 480/69.44 \approx 6.91\ \Omega$.

Wye-Delta Transformer

Figure 17.3-3 defines the wye-delta transformer. It is particularly suited for stepping down high voltages. The high voltage is applied to the Y to take advantage of its lower insulation needs compared with an equivalent Δ connection. Other advantages of the Y-Δ transformer are as follows: It provides a neutral for grounding on the high-voltage side (as required in most high-voltage transmission systems); it provides a path through the Δ for third-harmonic currents so that they do not propagate through the power system; equipment connected on the low-voltage Δ are given some protection from lightning strikes on the high-voltage side;† it provides some stability to unbalanced loads (Δ partially redistributes imbalances).

The largest disadvantage of the Y-Δ transformer concerns the phase shifts it introduces into its secondary voltages and currents. From Fig. 17.3-3 and (17.3-1), which applies to the individual transformers, we have, for a balanced load,

$$v_{ab} = v_a = \frac{v_A}{n} \qquad i_{LA} = i_A = \frac{i_a}{n} \tag{17.3-6a}$$

$$v_{bc} = v_b = \frac{v_B}{n} \qquad i_{LB} = i_B = \frac{i_b}{n} \tag{17.3-6b}$$

$$v_{ca} = v_c = \frac{v_C}{n} \qquad i_{LC} = i_C = \frac{i_c}{n} \tag{17.3-6c}$$

To relate line voltages, we apply (17.2-4), with capital letter subscripts, to the Y of the primary to write (17.3-6) as

$$v_{ab} = \frac{v_A}{n} = \frac{1}{n\sqrt{3}}\, e^{-j\pi/6} v_{AB} \tag{17.3-7a}$$

$$v_{bc} = \frac{v_B}{n} = \frac{1}{n\sqrt{3}}\, e^{-j\pi/6} v_{BC} \tag{17.3-7b}$$

$$v_{ca} = \frac{v_C}{n} = \frac{1}{n\sqrt{3}}\, e^{-j\pi/6} v_{CA} \tag{17.3-7c}$$

† Brown and Hamilton, op. cit., p. 467.

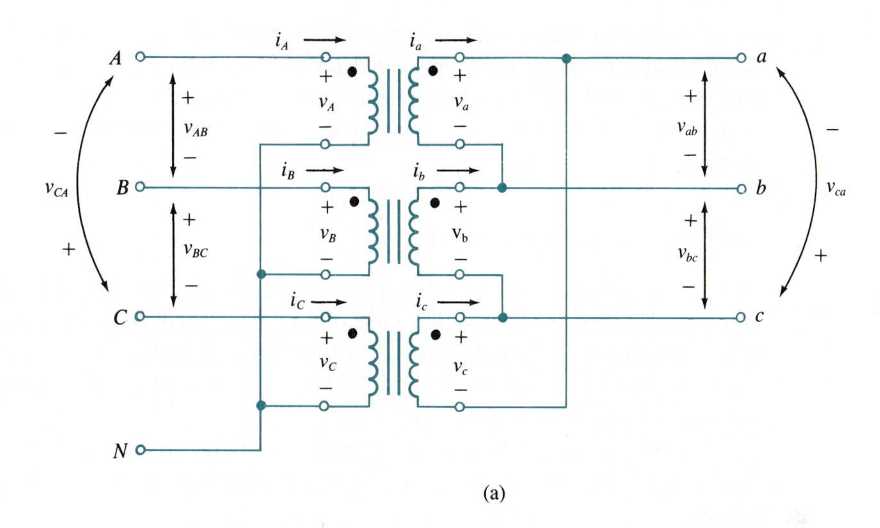

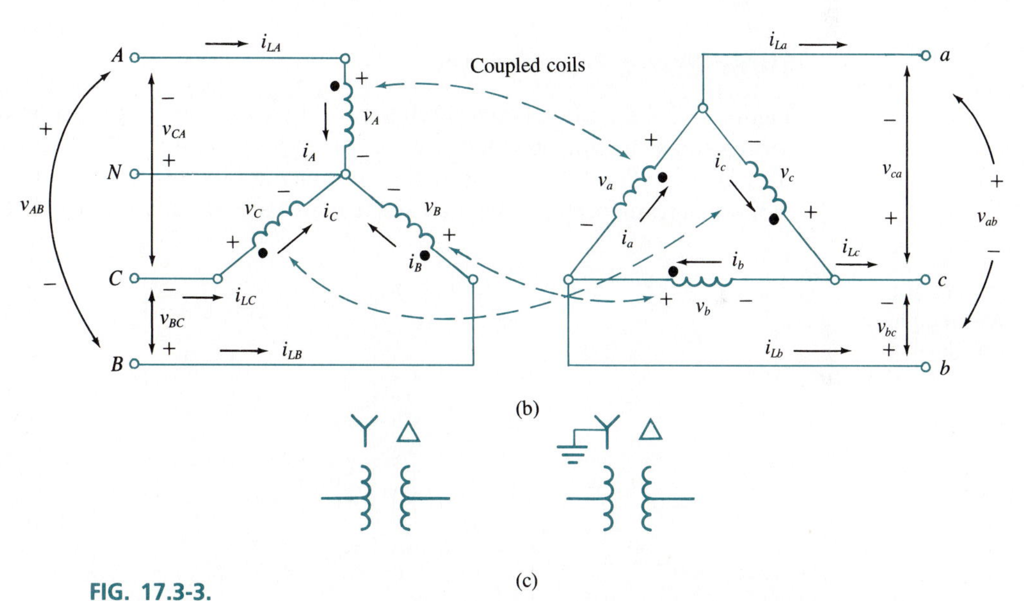

(c)

FIG. 17.3-3.
(a) A three-phase transformer comprising three single-phase transformers connected in a Y-Δ configuration, (b) a second representation, and (c) symbols representing the transformer.

To find line currents, we substitute (17.3-6) into (17.2-6), which applies to the Δ circuit:

$$i_{La} = n\sqrt{3}\, e^{-j\pi/6} i_{LA} \tag{17.3-8a}$$

$$i_{Lb} = n\sqrt{3}\, e^{-j\pi/6} i_{LB} \tag{17.3-8b}$$

$$i_{Lc} = n\sqrt{3}\, e^{-j\pi/6} i_{LC} \tag{17.3-8c}$$

for a balanced load. These results indicate that secondary line voltages and currents are, respectively, phase-shifted by $-\pi/6$ ($-30°$) relative to their primary line-voltage and -current counterparts. By a different connection the phase can be $+30°$. Thus, care must be exercised when paralleling transformers in power systems.

Delta-Wye Transformer

The delta-wye transformer has a primary defined by the left side of Fig. 17.3-2(b) and a secondary as in the right side of Fig. 17.3-1(b). Line voltages are given by (17.3-7) if the $\sqrt{3}$ is moved to the numerator and the phase is changed to $+\pi/6$. Line currents are given by (17.3-8) if the $\sqrt{3}$ is moved to the denominator and the phase is changed to $+\pi/6$.

Transformers used to step voltage up to transmission level are often Δ-Y. Advantages and disadvantages are essentially the same as for the Y-Δ transformer.

Three-Phase Transformer

Figure 17.3-4 illustrates how three separate 1-ϕ transformers can be combined to form a single 3-ϕ transformer. Each core shares a common flux-return path in one of four legs. The total flux ϕ is the sum of the fluxes ϕ_1, ϕ_2, and ϕ_3 due to the individual transformers. Because these fluxes are directly related to the primary's

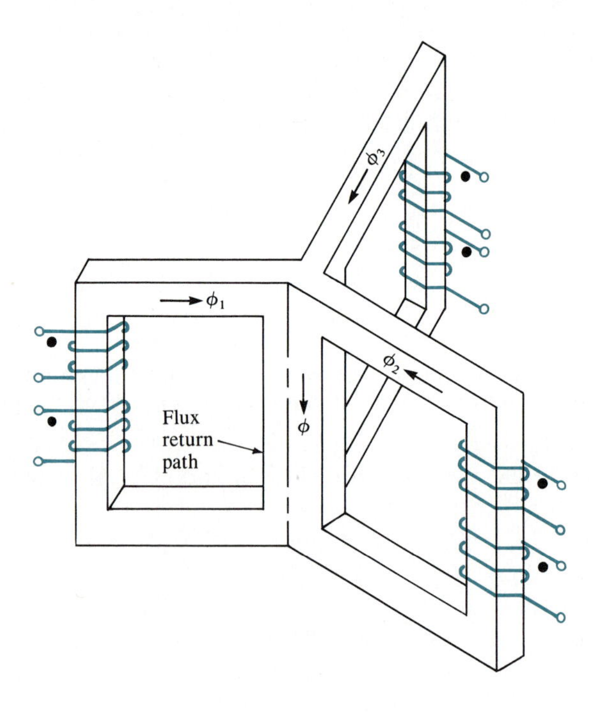

FIG. 17.3-4. A three-phase transformer of the core type.

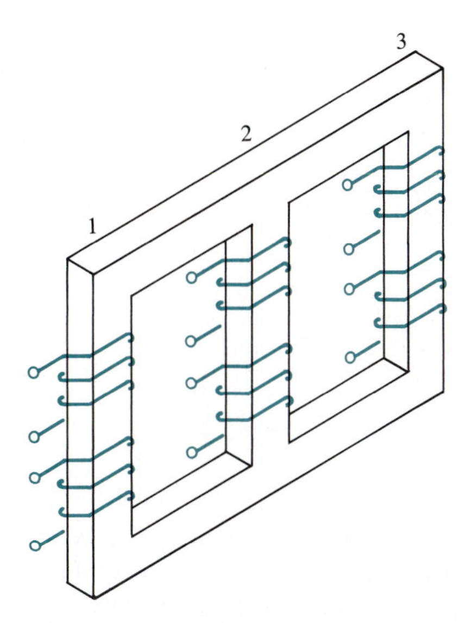

FIG. 17.3-5. A practical three-phase transformer.

exciting voltages, which have the phase shifts 0, $-2\pi/3$, and $-4\pi/3$, their sum will be zero for a balanced load. This fact means the fourth leg is unnecessary. It can be omitted, which leads to the practical construction of Fig. 17.3-5. In both cases the internal connections necessary to give a particular type of transformer (Δ-Δ, Δ-Y, etc.) are not shown, for simplicity.

For the same power rating a single 3-ϕ transformer is usually smaller, weighs less, is more efficient, and costs less than three individual transformers. It also needs fewer external connections. These advantages must be compared against its disadvantages. It is more expensive to replace due to the failure of any one winding. It is typically larger and heavier than one of the three individual transformers, so it is more difficult to replace.

17.4 POWER TRANSMISSION LINES

Three-phase power in the power grid is typically conveyed by a three-wire arrangement called a transmission line. In the most common application the wires are hung overhead from a tower usually made of metal, wood, or reinforced concrete. In some cases, transmission lines are placed underground, but such lines are not in widespread use because of high cost and several severe technical limitations; their length rarely exceeds a few kilometers.

An example of the type of tower often used for 765-kV transmission (rms, line-to-line voltage) is sketched in Fig. 17.4-1. One or more small wires (two in the illustration) atop the tower are for lightning protection. Most lightning strikes will hit these wires, which carry the current in two directions, ultimately to ground

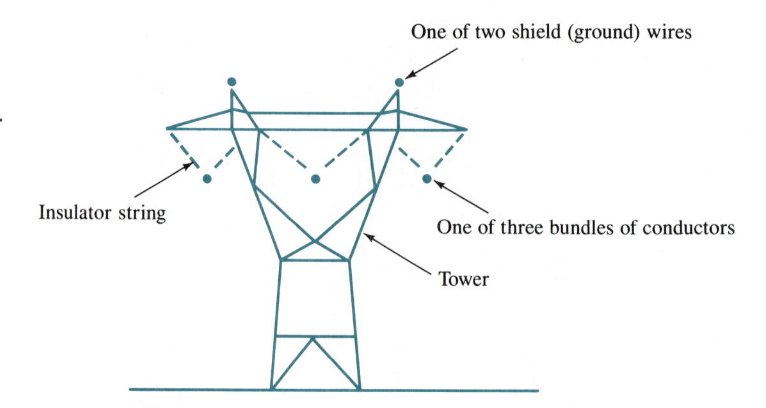

FIG. 17.4-1. A typical tower for a 765-kV transmission line.

through the towers. The three "wires" of the transmission line are hung from the tower by long strings of insulators. Towers come in a variety of shapes and sizes. They usually suspend the three lines in a horizontal plane, but some stack the three phases in the vertical plane. Others arrange the wires in a triangular pattern. Some towers carry *two* transmission lines (six power wires plus lightning-protection wires).

The "wires," one for each of the three phases, can consist of only one conductor at lower voltages or a *bundle* of conductors at the higher voltages (above about 230 kV). Bundles having from one to four conductors are typical, but higher numbers have been used. Although not a necessity, bundles usually have round conductors of nearly the same radius separated by equal amounts, as sketched in Fig. 17.4-2. The conductors of a bundle share the total current of a phase. Bundling increases the effective radius of the line's conductor and reduces the electric field strength near the conductors, which reduces corona† and energy losses due to corona. The disadvantages of bundling are increased cost, added clearance needed in the structure, a more complicated suspension, and increased ice and wind loading in many climates.

In spite of having a lower conductivity than copper, aluminum is almost exclusively used in conductors because of its lighter weight and lower cost. For strength in the long lines draped between towers, a bare stranded aluminum conductor is usually placed over a stranded steel core. It is called an ACSR (aluminum-conductor steel-reinforced) line. ACSR lines typically have overall diameters of up to about 40 mm and a current capacity of up to about 1380 A.

Transmission Line Parameters

Even though 60 Hz is a very low frequency of operation, long power lines (over about 240 km) must be analyzed by use of wave concepts (as in Sec. 14.1). The

† Corona is a luminous discharge that occurs near electric wires due to ionization of the surrounding air caused by excessive electric fields.

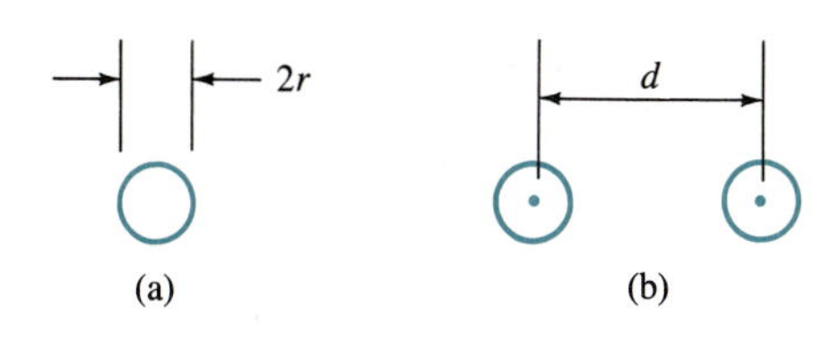

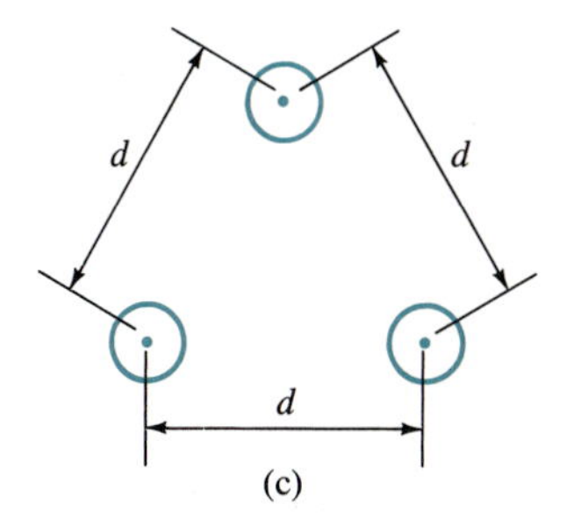

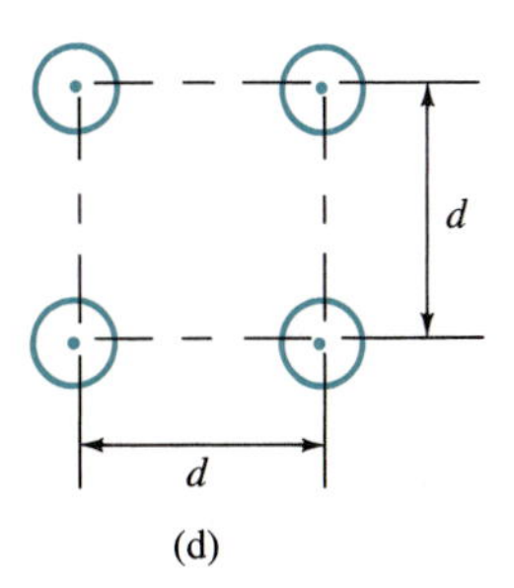

FIG. 17.4-2. Conductor bundles with (a) one, (b) two, (c) three, and (d) four conductors, all of radius r and separation d.

line is treated as a distributed-parameter device; the parameters that define its behavior are L, the series inductance, C, the shunt capacitance, R, the series resistance, and G, the shunt conductance. These parameters are specified on a per-unit-length basis. As will be described in Sec. 17.6, the power transmission line can usually be analyzed by study of only one of the three phases, because all behave in a similar manner (for a balanced system). Therefore, the parameters L, C, R, and G are also given on a per-phase basis.

Under normal operating conditions G, which represents the resistive leakage between a phase and ground, has negligible effect on performance and is customarily neglected ($G = 0$ assumed). Resistance R is best determined from manufacturer's data. For a 3-ϕ line with conductor bundles labeled a, b, and c, and distances between bundle centers of D_{ab}, D_{bc}, and D_{ca}, as shown in Fig. 17.4-3, the values of L and C are given by†

$$L \text{ (per phase)} = (2 \times 10^{-7}) \ln \left[\frac{(D_{ab}D_{bc}D_{ca})^{1/3}}{r_{\text{eq}}}\right] \quad \text{H/m} \tag{17.4-1}$$

where

$$r_{\text{eq}} = \begin{cases} 0.7788r & \text{1 conductor per bundle} \quad (17.4\text{-}2a) \\ (0.7788rd)^{1/2} & \text{2 conductors per bundle} \quad (17.4\text{-}2b) \\ (0.7788rd^2)^{1/3} & \text{3 conductors per bundle} \quad (17.4\text{-}2c) \\ 1.09(0.7788rd^3)^{1/4} & \text{4 conductors per bundle} \quad (17.4\text{-}2d) \end{cases}$$

† R. D. Shultz and R. A. Smith, *Introduction to Electric Power Engineering*, Harper & Row, New York, 1985, chap. 5.

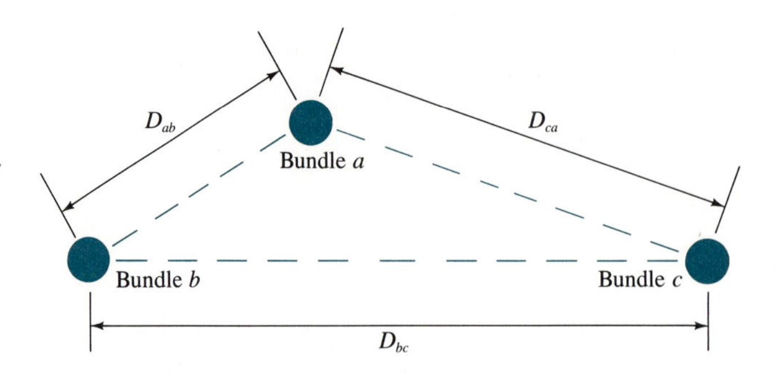

FIG. 17.4-3. Bundle separations in a three-phase transmission line.

and

$$C \text{ (per phase)} = \frac{55.56 \times 10^{-12}}{\ln [(D_{ab}D_{bc}D_{ca})^{1/3}/r'_{\text{eq}}]} \quad \text{F/m} \qquad (17.4\text{-}3)$$

with

$$r'_{\text{eq}} = \begin{cases} r & \text{1 conductor per bundle} & (17.4\text{-}4a) \\ (rd)^{1/2} & \text{2 conductors per bundle} & (17.4\text{-}4b) \\ (rd^2)^{1/3} & \text{3 conductors per bundle} & (17.4\text{-}4c) \\ 1.09(rd^3)^{1/4} & \text{4 conductors per bundle} & (17.4\text{-}4d) \end{cases}$$

These last four equations assume all conductors in a bundle are round with radius r, separation d, and uniform current in the wires. Also assumed is a transposed-line configuration.†

EXAMPLE 17.4-1

Assume $r = 1.24$ cm in a 138-kV power line having one conductor per phase with $D_{ab} = D_{bc} = D_{ca} = 2.8$ m. Find L and C. From (17.4-1) and (17.4-3):

$$L \text{ (per phase)} = (2 \times 10^{-7}) \ln \left[\frac{2.8}{0.7788(0.0124)}\right] \approx 1.13 \ \mu\text{H/m}$$

$$C \text{ (per phase)} = \frac{55.56 \times 10^{-12}}{\ln (2.8/0.0124)} \approx 10.25 \times 10^{-12} \text{ F/m}$$

† A transposed line periodically changes positions of the wires of each phase so that all wires occupy each position over about one-third of a given distance. This arrangement makes a nonsymmetrical line appear symmetrical.

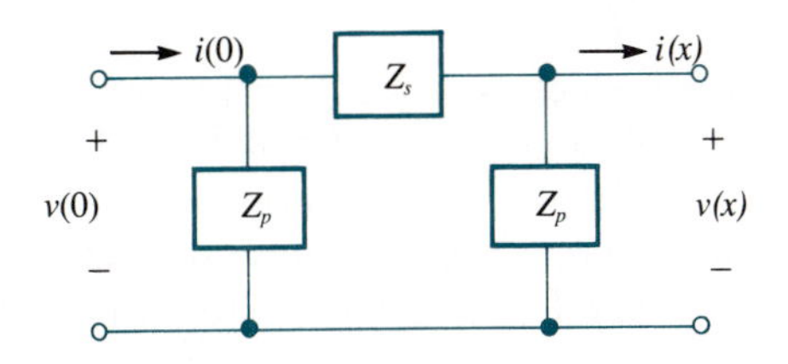

FIG. 17-4-4. An equivalent circuit for one phase of a three-phase transmission line segment of length x.

Transmission Line Models

It can be shown† that the circuit of Fig. 17.4-4 is an accurate model of any one phase of a three-phase transmission line. It relates the voltage and current at one point on the line (left end of the circuit) to the voltage and current at another point a distance x further down the line. The impedances Z_s and Z_p are given by

$$Z_s = Z_w \sinh(\gamma x) \qquad (17.4\text{-}5)$$

$$Z_p = \frac{Z_w}{\tanh(\gamma x/2)} \qquad (17.4\text{-}6)$$

where

$$Z_w = \left(\frac{R + j\omega L}{G + j\omega C}\right)^{1/2} \quad \Omega \qquad (17.4\text{-}7)$$

$$\gamma = [(R + j\omega L)(G + j\omega C)]^{1/2} \quad \text{m}^{-1} \qquad (17.4\text{-}8)$$

Of course, L and C are given by (17.4-1) and (17.4-3), respectively, and G can be ignored in most cases.

The equivalent circuit assumes that voltage and currents are expressed in complex form and applies even for long lines ($x > 240$ km). However, for medium-length lines, where 80 km $< x <$ 240 km, it can be shown that $|\gamma x| \ll 1$; so the impedances of Fig. 17.4-4 reduce to

$$Z_s \approx (R + j\omega L)x \quad \text{medium-length line} \qquad (17.4\text{-}9)$$

$$Z_p \approx \frac{2}{(G + j\omega C)x} \quad \text{medium-length line} \qquad (17.4\text{-}10)$$

Finally, for short lines, where $x < 80$ km, C and G may both be ignored, and

$$Z_s \approx (R + j\omega L)x \quad \text{short line} \qquad (17.4\text{-}11)$$

$$Z_p \approx \infty \quad \text{short line} \qquad (17.4\text{-}12)$$

† O. I. Elgerd, *Electric Energy Systems Theory, An Introduction*, 2d ed., McGraw-Hill, New York, 1982, pp. 204–205.

EXAMPLE 17.4-2

A 765-kV transmission line for which $D_{ab} = D_{bc} = 13.7$ m and $D_{ca} = 27.4$ m uses bundled wires with four round conductors of radius 18.1 mm. Conductors are separated by $d = 45.7$ cm, and the resistance of a single conductor at 60 Hz can be taken as 0.05 Ω/km. We find the impedances of the equivalent circuit of Fig. 17.4-4 for a line of length 120 km. Since four conductors form a bundle,

$$R = \frac{0.05}{4} = 0.0125 \ \Omega/\text{km} \qquad \text{(per phase)}$$

From (17.4-1) and (17.4-2d):

$$L \text{ (per phase)} = (2 \times 10^{-7}) \ln \left\{ \frac{2^{1/3}(13.7)(1/1.09)}{[0.7788(0.0181)(0.457)^3]^{1/4}} \right\}$$
$$\approx 8.83 \times 10^{-7} \text{ H/m}$$

Equations (17.4-3) and (17.4-4d) give

$$C \text{ (per phase)} = \frac{55.56 \times 10^{-12}}{\ln \{2^{1/3}(13.7)/1.09[0.0181(0.457)^3]^{1/4}\}}$$
$$\approx 12.76 \times 10^{-12} \text{ F/m}$$

From (17.4-8) and (17.4-7):

$$\gamma = \{[1.25 \times 10^{-5} + j2\pi(60)(8.83 \times 10^{-7})][0 + j2\pi(60)(12.76 \times 10^{-12})]\}^{1/2}$$
$$= [(1.25 + j33.29)(10^{-5})(0 + j4.81 \times 10^{-9})]^{1/2}$$
$$\approx (0.238 + j12.66) \times 10^{-7} \text{ m}^{-1}$$

$$Z_w = \left[\frac{(1.25 + j33.29)(10^{-5})}{j4.81 \times 10^{-9}}\right]^{1/2} = 263.12 - j4.938 \ \Omega$$

Next,

$$\tanh \frac{\gamma x}{2} = \tanh (0.00143 + j0.076) \approx 0.0762 \exp (j1.55)$$

$$\sinh \gamma x = \sinh (0.00286 + j0.152) \approx 0.151 \exp (j1.55)$$

Finally, from (17.4-9) and (17.4-10);

$$Z_s \approx [(1.25 \times 10^{-5}) + j2\pi(60)(8.83 \times 10^{-7})](1.2 \times 10^5)$$
$$= (1.5 + j33.29) \ \Omega$$

$$Z_p \approx \frac{2/(1.2 \times 10^5)}{0 + j2\pi(60)(12.76 \times 10^{-12})} \approx -j3465 \ \Omega$$

The impedance Z_p can clearly be neglected in this line.

17.5 OTHER POWER SYSTEM COMPONENTS

Generators, transformers, and transmission lines are not the only components needed to form an effective power system. Others, such as fuses, surge arresters, switches, and circuit breakers, are needed to provide protection from faults and to aid in maintenance. A *fault* is often associated with a short circuit at some point in the system, but it can also refer to other abnormal conditions, such as current overloads, over- and undervoltages, and frequency changes. Voltage-regulating and power-factor-correcting devices are also needed to provide quality of service under normal conditions.

Fuses

A fuse is a relatively low-cost component designed to open (disconnect) a current line under fault conditions. In high-power applications a fusible material, such as silver, copper, aluminum, or some alloy, melts at an appropriate overcurrent condition. After the metal separation an arc typically maintains the connection. Several types of fuses are available that differ in the manner of arc suppression. One of the fastest-acting uses a silver fusible element embedded in sand; it can interrupt current in less than one-half of a cycle of the 60-Hz current. A disadvantage of a fuse is that it must be replaced once it performs its function.

Arc suppression is less of a problem at lower-power applications. It was not too many years ago that homes used only fuses for circuit protection. Current values of 15, 20, and 30 A were common. In recent times the fuse has been displaced by the circuit breaker because of its ability to be reset without replacement.

Circuit Breakers and Switches

In high-power applications a *circuit breaker* is a large three-pole device usually designed to interrupt a 3-ϕ circuit on command from a control circuit. The circuit senses when and where a fault or abnormal condition exists and opens the appropriate circuit breaker to isolate the problem. Circuit breakers can also be operated manually or by a remotely located operator for system maintenance.

Arcing occurs when a circuit breaker opens. Arc suppression is accomplished by both mechanical design and choice of media (air, oil, vacuum, and sulfur hexafloride, or SF_6) in which the arc occurs. Typical times required from break command to zero current is about two cycles of 60 Hz. Steady-state current ratings of 8000 A are available, and interrupt-current ratings are up to 80 kA at 230 kV and 63 kA at 500 kV.†

† H. M. Rustabakke (ed.), *Electric Utility Systems and Practices*, 4th ed., Wiley, New York, 1983, p. 172.

Low-power circuit breakers, as used in home wiring or inside some small motors (the garbage disposal, for example), use thermal or magnetic devices to trip a switch open when current exceeds a specified value. These breakers may be manually reset after the fault has been corrected.

It should be noted that a circuit breaker is a device meant to interrupt a circuit *under load*, as contrasted with a *switch*, the name often reserved for an interrupt device that operates under reduced or no load. Switches are helpful in isolating parts of a system or in removing equipment for repair.

Surge Arresters

A surge arrester is a component connected between a power line and ground. Its purpose is to divert large surge currents (from a lightning strike, for example) to ground without allowing excessive voltages on the line. It is essentially a nonlinear resistance. For line voltages below a value called the *protected voltage*, the device permits only a small leakage current. As the voltage of a surge approaches the protected level, current increases rapidly to shunt the undesired excess to ground. After the surge passes, the arrester returns to its low-leakage state. Surge arresters are important in keeping surge voltages from damaging transformers and other system components at substations.

Voltage and Power Factor Control

Utilities must maintain the voltage of its customers within an acceptable range (typically 110–126 V for a nominally 120-V home system). This task is usually performed in a substation by a *step-voltage regulator*. It consists of a step-down transformer with a multitap secondary winding. Voltage is adjusted by tap selection. Voltage-sensing and -control devices are used to automatically regulate voltage.

Other sources of voltage variation in a power system are the load and the system itself. Voltage will vary with the load's reactive power and position in the system. Load compensation amounts to reducing the net reactive power through power factor correction. Because most industrial loads are inductive, correction occurs when an equivalent capacitive reactive power is added. The effect of location in the system can be compensated for by correcting for transmission line shunt capacitance and series inductance. Several corrective approaches are possible; most use either series or shunt capacitors, shunt inductors, or synchronous motors with no shaft load.

The bulk of power factor correction is by addition of shunt capacitors at small distribution substations and at distribution feeders on poles. Some degree of automatic adjustment with varying load conditions can be implemented at substations. A smaller amount of shunt capacitor correction occurs at larger substations and at power transmission levels.

Synchronous motors with no shaft loads are able to present either capacitive or inductive reactive power to its power line. Use of these motors is effective in power factor correction for large loads (to about 250 Mvar).

Series capacitors are sometimes used at extra high voltages (over 345 kV) to compensate for series impedances of transmission lines. Similarly, shunt inductances are used to compensate for the line's shunt capacitance.

EXAMPLE 17.5-1

A medium-sized motor has a per-phase impedance of $4.19 + j2.92\ \Omega$ when connected to a 3-ϕ, 480-V line. The per-phase voltage of the line is therefore $480/\sqrt{3} \approx 277$ V rms. We determine the shunt capacitor needed to fully correct this motor's power factor as indicated in Fig. 17.5-1. For complete compensation Z_{in} must be real. Impedance Z_{in} is first found:

$$Z_{\text{in}} = \frac{jX_C(4.19 + j2.92)}{jX_C + 4.19 + j2.92} = \frac{4.19X_C^2 + jX_C(26.08 + 2.92X_C)}{4.19^2 + (2.92 + X_C)^2}$$

For the imaginary part to be zero we require $X_C = -8.93\ \Omega$. At 60 Hz the capacitance is $C = [2\pi(60)(8.93)]^{-1} = 297\ \mu\text{F}$. With compensation $Z_{\text{in}} = 6.22\ \Omega$; the rms current per phase is $277/6.22 = 44.5$ A. Before compensation the motor's impedance magnitude per phase was $[(4.19)^2 + (2.92)^2]^{1/2} = 5.11\ \Omega$, which corresponded to an rms-current magnitude of $277/5.11 = 54.2$ A. The current drain on the power system has been reduced by a factor of $44.5/54.2 = 0.82$, or 18%.

17.6 PER-PHASE ANALYSIS

Power systems are usually analyzed on a per-phase basis. That is, voltages, currents, and powers are computed for only one phase of a 3-ϕ system under the assumption that the system is balanced. For analysis performed on phase a (of a sequence abc) the voltages and currents in phase b would equal those of phase a

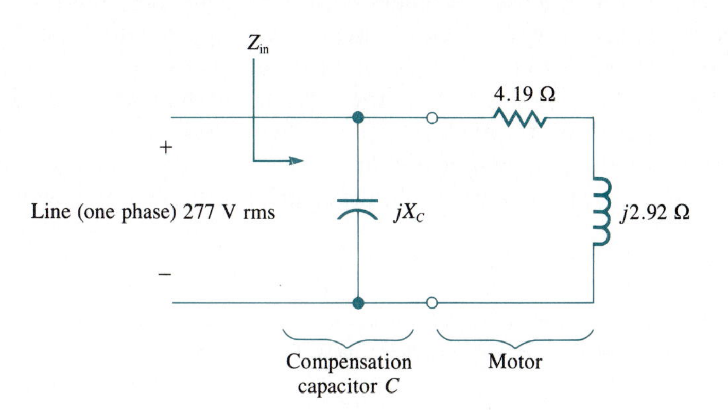

FIG. 17.5-1. Shunt capacitance power factor compensation of the motor of Example 17.5-1.

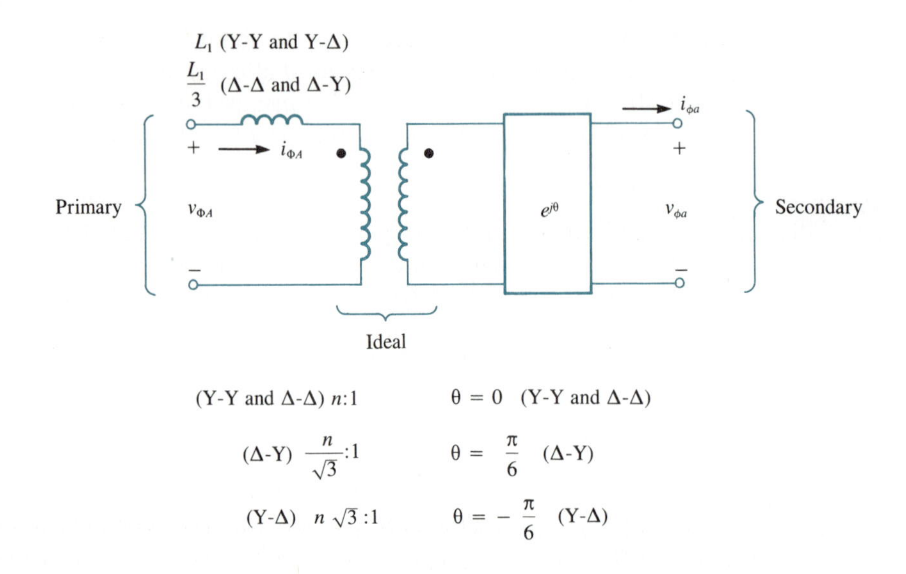

FIG. 17.6-1. Per-phase equivalent circuits for power transformers.

except for an additional phase shift of $-2\pi/3$ ($-120°$). Phase c would be the same as a except for a phase shift of $-4\pi/3$ ($-240°$). Powers would be equal in all phases, so total power is 3 times that of the per-phase analysis. In a complicated system of many sources and loads, superposition can be invoked to obtain currents and voltages.

To apply per-phase analysis, we first convert all delta-connected sources, transformers, and loads to equivalent wye connections. The equivalent circuit of one phase is then drawn and analyzed. The conversion of sources follows the use of (17.2-4) and is straightforward. Similarly, load conversions result from use of (17.2-11). Proper representations for transformers are less obvious.

Practical modeling of power transformers is well beyond our limited scope. In fact, the resulting models are relatively complex except when losses are small. Fortunately, with modern transformers the ohmic losses can often be completely neglected. When losses in magnetic flux linkages are small (the usual case), the equivalent circuit (per phase) is approximately an inductance† in series with an ideal transformer followed by an appropriate phase shift, as shown in Fig. 17.6-1. The phase shift is added to both the current and the voltage at the secondary of the ideal transformer. Inductance L_1 is a "leakage inductance" associated with the primary winding of each of the three winding pairs. When the primary of the 3-ϕ transformer is Δ-connected, the inductance of the Y equivalent representation is $L_1/3$.

† The inductance represents losses in magnetic flux linkages.

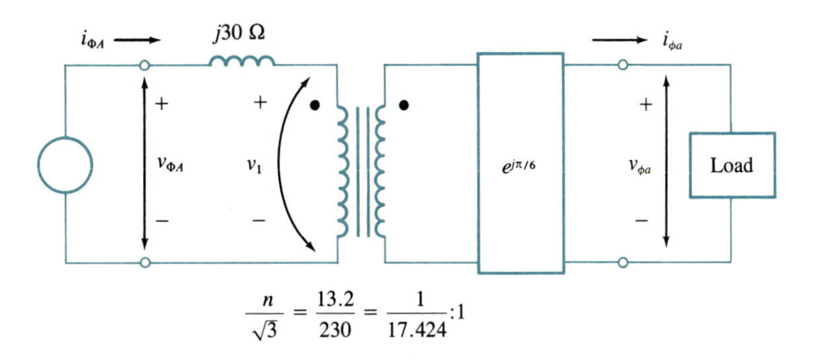

FIG. 17.6-2. The equivalent circuit applicable to Example 17.6-1.

EXAMPLE 17.6-1

A 3-ϕ generator of terminal voltage 13.2 kV delivers 180 kW of average power at power factor 0.9 (lagging current) to the Δ-connected primary of a Δ-Y transformer. The transformer has a 90-Ω leakage reactance and provides nominal voltage step-up from 13.2 to 230 kV, where a balanced secondary load absorbs power. We find the system's equivalent circuit per phase, various voltages and currents, and the load's impedance and power factor.

Figure 17.6-1 for a Δ-Y transformer is combined with a generator and load to give the per-phase equivalent circuit of Fig. 17.6-2. Since 13.2 kV is the generator's rms line voltage, its equivalent Y's phase voltage is $13.2/\sqrt{3}$ kV rms. The peak voltage is $\sqrt{2}(13.2)/\sqrt{3}$ kV, so

$$v_{\Phi A} = \frac{\sqrt{2}(13.2 \times 10^3)}{\sqrt{3}} e^{j\omega t} \qquad \text{V}$$

The angle of the power factor at the generator is $\theta = \cos^{-1} 0.9 = 0.144\pi\ (25.84°)$. Total reactive power is $(180 \times 10^3)\sin(\theta)/0.9 = 87.178$ kvar. Total complex power of the generator is 180 kW + j87.178 kvar = 200 exp ($j0.144\pi$) kVA. The complex power per phase, denoted by P_Φ, is one-third the total, or

$$P_\Phi = 60 \text{ kW} + j29.059 \text{ kvar} = \frac{200}{3} e^{j0.144\pi} \qquad \text{kVA}$$

We now use (4.5-7) to solve for $i_{\Phi A}$:

$$i_{\Phi A} = \frac{2P_\Phi^*}{v_{\Phi A}^*} = \frac{2(200)e^{-j0.144\pi + j\omega t}}{3\sqrt{2}(13.2)/\sqrt{3}}$$

$$= \frac{\sqrt{2}(200)}{\sqrt{3}(13.2)} e^{j\omega t - j0.144\pi} \qquad \text{A}$$

The transformer's turns ratio is 13.2 kV, the primary winding's voltage divided by the secondary's phase voltage of $230/\sqrt{3}$ kV, or

$$n = \frac{13.2\sqrt{3}}{230}$$

Load current now follows from Fig. 17.6-2:

$$i_{\phi a} = i_{\Phi A} \frac{n}{\sqrt{3}} e^{j\pi/6} = \frac{\sqrt{2}(200)}{\sqrt{3}(230)} e^{j\omega t + j0.0227\pi}$$

Next,

$$\begin{aligned} v_1 &= v_{\Phi A} - j30 i_{\Phi A} \\ &= \left[\frac{\sqrt{2}(13.2 \times 10^3)}{\sqrt{3}} + \frac{30\sqrt{2}(200)}{\sqrt{3}(13.2)} e^{-j\pi/2 - j0.144\pi}\right] e^{j\omega t} \\ &= (10.62 \times 10^3)\, e^{j\omega t - j0.01\pi} \end{aligned}$$

$$v_{\phi a} = v_1 \frac{\sqrt{3}}{n} e^{j\pi/6} = (185.0 \times 10^3) e^{j\omega t + j0.157\pi}$$

Load impedance, denoted by Z_L, is

$$Z_L = \frac{v_{\phi a}}{i_{\phi a}} = \frac{(185.0 \times 10^3) e^{j0.157\pi}}{0.71 e^{j0.0227\pi}} = 260{,}563 e^{j0.134\pi}\ \Omega$$

The load's power factor is cos $0.134\pi = 0.913$, the cosine of the angle of Z_L (see Sec. 4.5). A further calculation of Re $(v_{\phi a} i^*_{\phi a}/2)$ shows that all the generator's real power flows in the load.

17.7 POWER DISTRIBUTION

The manner in which power is delivered to a load depends on the type of customer being served. We discuss several typical cases.

Industrial Loads

Large mills, factories, and other industrial operations may present large demands on a power system. These large loads sometimes are satisfied by having one or more "customer" bulk-power substations at the industrial site (Fig. 17.1-1). Transformers in these substations are then selected according to the special needs of the customer.

Commercial Loads

For relatively isolated loads, such as a large commercial building, shopping center, or hospital, power is taken from a line arriving from a distribution substation (Fig. 17.1-1). The line is often referred to as the *primary network* or *primary feed*. The voltage of the primary service is reduced to a lower level by *distribution transformers*, whose outputs are called the *secondary network*. The secondary network is often 3-ϕ, four-wire service at 480Y/277 V.† This type of service is known as a *spot network*.

For commercial loads spread over large areas, as in large cities downtown, the secondary voltage is often 3-ϕ, four-wire at 208Y/120 V provided over a *grid network*.

Residential, Suburban, and Rural Loads

For residential, suburban, and rural loads, the primary service is from a distribution substation that feeds a general area, with lateral branches to cover the area. The primary feed is 3-ϕ four-wire service with 12,470Y/7200 V, 24,900Y/14,400 V, and 34,500Y/19,920 V being typical voltages. Lateral-branch feeders are sometimes 3-ϕ four-wire service, but more often they are single-phase taken by branching from one line and the neutral. Primary-feed lines are often overhead, but many modern residential and suburban services are now underground. Single-phase branches to rural areas can be long (30 km or more) and use voltages of 7.2 or 14.4 kV.

Distribution transformers reduce the primary network's voltage to 1-ϕ, three-wire secondary service to homes at 120/240 V. Figure 17.7-1 sketches the most important functions. A 3-ϕ, four-wire primary feeder provides the secondary service to the house through a distribution transformer, which may serve more than one house. Three wires labeled *A*, *B*, and *N* enter the house through a power meter (not illustrated), which registers the cumulative energy consumption. Wires *A* and *B* are each at 120 V rms relative to the neutral *N*. Line *B* is the negative of *A*, so that 240 V exists between lines *A* and *B*. The neutral is grounded at the service entrance by driving a copper rod into the ground.

Inside the house a power distribution panel is the central point where various branch circuits (there can easily be up to 30 or more of these) feed power to various locations. Each branch circuit has its own circuit breaker to protect (cut out) on short circuits or overloads. Breakers with 20- to 30-A capacities are typical. All branches are also controlled by main circuit breakers that are typically of 100- to 200-A capacity. Most branch circuits are 120 V and feed to various outlets, lights, and other devices (ceiling fans, etc.). Some branch circuits are 240 V to serve heavier loads, such as hot-water heaters, clothes dryers, and electric stoves.

In a modern residence each 120-V branch circuit in Fig. 17.7-1 includes a third grounded wire that originates from the neutral bus in the distribution panel. This wire (not shown in the figure) is used to reduce shock hazards by providing a direct path to ground for appliances and other electrical apparatus. In a similar manner, the 240-V circuits are also three-wire.

---

† This is standard notation, where the first number (480) is the line-to-line voltage from a Y-connected transformer secondary, and the second (277) is the line-to-neutral voltage.

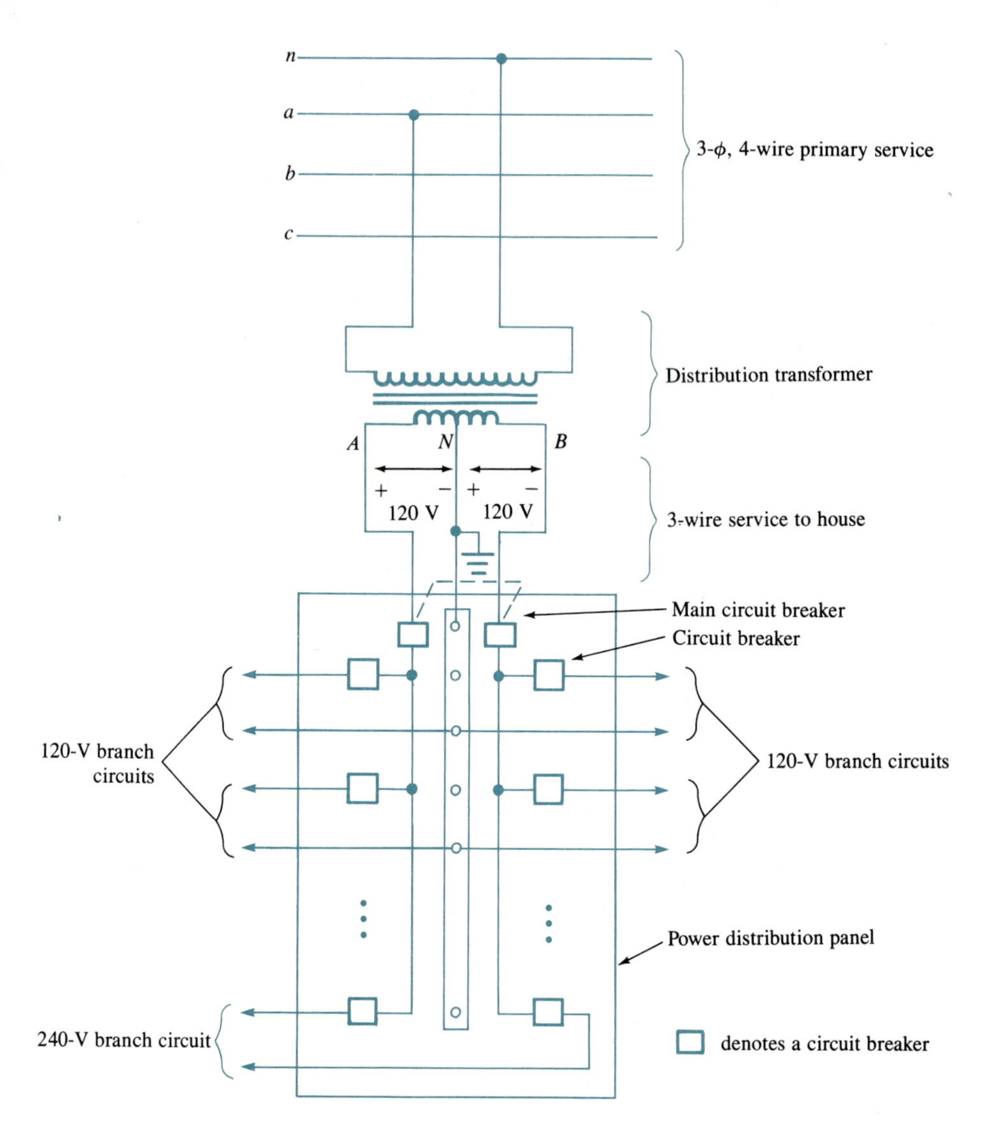

FIG. 17.7-1. Three-wire, single-phase service for household wiring as derived from a three-phase, four-wire primary service.

PROBLEMS

17-1. If electric power consumption increased exactly 2.5% each year, precisely how many years pass before consumption doubles?

17-2. A transmission line at 69 kV is converted to 765 kV. Assume that the same wires are used and that the loads are resistive and draw the same average power. By what fraction will the line losses in the 765-kV line be lower than in the 69-kV line?

17-3. Suppose a generator has 32 poles. What shaft speed will produce 60-Hz power?

17-4. Sketch (versus time) the real voltages described by (17.2-2).

★ **17-5.** Prove that only two phase sequences (*abc* and *acb*) are possible from a three-phase generator.

17-6. What maximum magnitude of current is expected to flow in each phase winding of a three-phase generator if its phase voltage is 7620 V rms and its rated capacity is 100 MVA?

★ **17-7.** Show that $i_n = 0$ in Fig. 17.2-1(a) if (17.2-2) applies.

17-8. In Fig. 17.2-1(a) voltages are given by (17.2-2), but the load impedances are Z_L, $Z_L/1.1$, and $Z_L/0.9$, respectively, that support currents i_a, i_b, and i_c. Find $i_n = i_a + i_b + i_c$ and compare the result with individual currents.

17-9. Show that (17.2-9) is true.

★ **17-10.** Show that (17.2-10) satisfies (17.2-8a).

17-11. A wye-connected load has impedances $Z_Y = 100 + j12\ \Omega$. What are the equivalent delta-connected load impedances?

17-12. In Fig. 17.2-3(a), $Z_a = 150\ \Omega$, $Z_b = 75 - j15\ \Omega$, and $Z_c = 120 + j45\ \Omega$. Find Z_1, Z_2, and Z_3 of the delta load in Fig. 17.2-3(b).

17-13. A delta-connected load has impedances $Z_3 = 12\ \Omega$, $Z_2 = 9\ \Omega$, and $Z_1 = 11\ \Omega$. What are the corresponding impedances of an equivalent wye-connected load?

17-14. An extra-high-voltage transmission line in a power system can be treated as a 3-ϕ source for which the line voltage is 625 kV. If the line provides a real power of 1000 MW to a balanced load with a lagging power factor of 0.85, what is the load's impedance? What magnitude of line current (per phase) flows?

17-15. Suppose a Y-connected generator has an open-circuited rms voltage of 13.8 kV (line to line). Its internal resistance can be considered to be negligible and its synchronous reactance is 5 Ω per phase. What minimum resistance (per leg) is required in a balanced wye-connected resistive load to prevent the generator from providing more than its rated capacity of 15 MVA? What real power is dissipated in this load?

17-16. A Y-connected generator of 15-kV rms line voltage feeds a 69-kV rms transmission line through a transformer connected Y-Y. What is the transformer ratio?

17-17. A Y-Y transformer, as in Fig. 17.3-1(b), has a balanced Y-connected load on its secondary with leg impedances Z_Y. For a transformer ratio *n*, find the leg impedance of the equivalent balanced Y-connected load presented to any generator that feeds the primary.

17-18. The generator of Problem 17-16 is connected to a Δ-Δ transformer for which $n = 1/15.34$. What is the line voltage of the transformer's secondary?

17-19. Work Problem 17-18 except assume (a) Y-Δ and (b) Δ-Y transformers.

17-20. The primaries of two Δ-Y transformers are connected in parallel to a 34.5-kV distribution line. The secondary line voltages of transformers 1 and 2 are 480 and 208 V, respectively. They both have Y-connected balanced loads with leg impedances of $6.3 + j2.8\ \Omega$ and $4.0 + j1.2\ \Omega$, respectively. Find the complex power delivered to each of the two loads and to the input to the transformers. What is the per-leg impedance of the Y equivalent load on the transmission line? Assume ideal transformers.

17-21. If an identical extra conductor per phase is added to the line of Example 17.4-1 to form a two-conductor bundle of separation 15 cm, determine the new values of L (per phase) and C (per phase).

17-22. Bundles of a 345-kV transmission line are at the corners of an equilateral triangle with sides of 6 m. Three conductors of radius 15 mm are separated by 22 cm. Find L (per phase) and C (per phase).

17-23. A transmission line is defined by per-phase quantities $R = 0.07\ \Omega/\text{km}$, $L = 1.1$ mH/km, $G \approx 0$, and $C = 9.2$ pF/m. Determine Z_p and Z_s in Fig. 17.4-4 for a line of length 450 km.

★ **17-24.** A generator provides power to a load over a 3-ϕ system having the one-line diagram of Fig. P17-24. If the transformers are assumed ideal, find (a) the complex power, (b) the real power, and (c) the power factor of the load seen by the generator. Use per-phase analysis and assume that the line is as defined in Example 17.4-2.

FIG. P17-24.

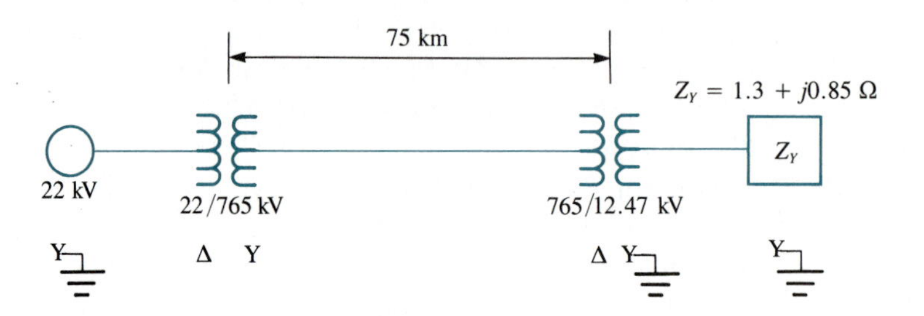

17-25. Assume a house has seven impedances $140 + j10\ \Omega$ in parallel on line A in Fig. 17.7-1. Three more of these impedances are in parallel but on terminal B. An impedance of 11.5 Ω is tied between A and B (a hot-water heater). What real power flows in each of the impedances? What complex power and real power flow into the house?

CHAPTER 18

Direct and Alternating Current Machinery

18.0 INTRODUCTION

Electric machines are electromechanical devices that convert mechanical energy to electric energy (a generator) or vice versa (a motor). They are the workhorses of modern society. Although alternating current (ac) motors and generators account for the bulk of the machinery in use, there are many applications where direct current (dc) machines are desirable, even necessary. Ac generators are available to generate power levels up to millions of watts. Motors can be found with power ratings from a very small fraction of a horsepower up to thousands of horsepower.†

In the following sections we review some of the more important aspects of motors and generators, both dc and ac.

18.1 DIRECT CURRENT (DC) GENERATORS

All generators (ac and dc) depend on the basic principle that a conductor moving through a magnetic field will have a voltage induced into it. In the usual case the conductor is a length of wire. Figure 18.1-1(a) illustrates the principle, where a straight wire of length l between ends a, b is moving at speed v in a direction

† One horsepower (hp) equals 745.7 W or 550 ft·lb/s.

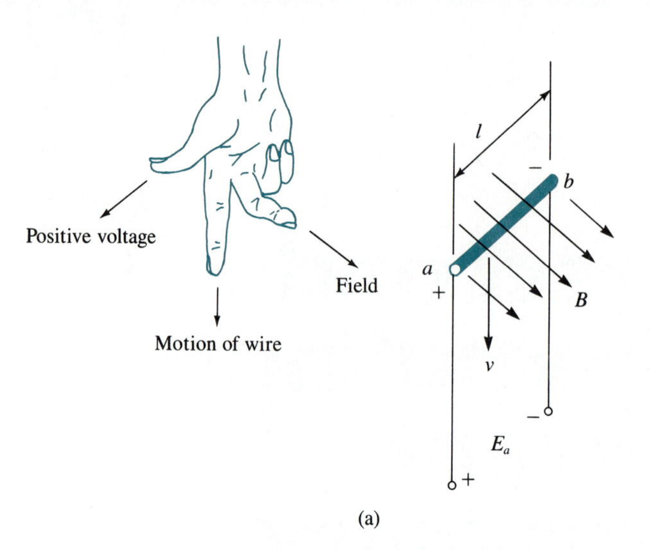

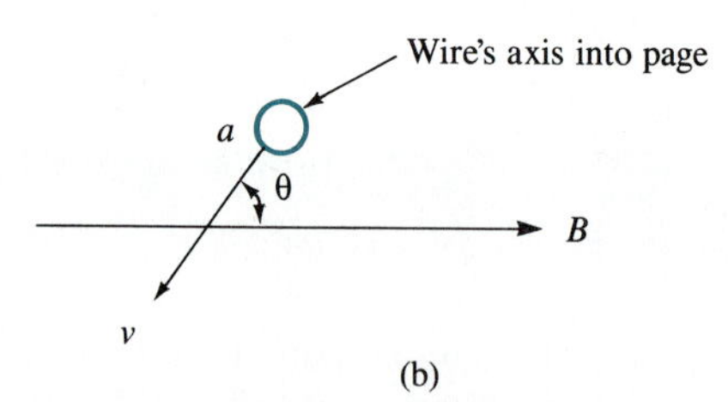

FIG. 18.1-1. (a) The voltage induced in a wire as given by the right-hand rule. (b) A wire with motion not perpendicular to the field direction.

perpendicular to a magnetic field of flux density B. The wire's axis is also perpendicular to the field and to its direction of motion. The induced voltage will be positive at terminal a relative to terminal b. Its magnitude is vBl. Its direction can be determined by the *right-hand rule*. It says that if the thumb, index, and middle fingers of the right hand are set to point in mutually perpendicular directions, with the index finger in the direction of the wire's motion and the middle finger pointing toward the field's direction, the thumb will point in the direction of positive voltage.

More generally, if the wire's motion is not perpendicular to the field's direction but makes an angle θ relative to it, as shown in Fig. 18.1-1(b), the induced voltage is

$$E_a = vBl \sin \theta \tag{18.1-1}$$

Note that motion parallel to the field ($\theta = 0$) results in *no* induced voltage.

Elementary Generator Operation

Use of the foregoing basic principle allows an elementary dc generator to be constructed, as sketched in Fig. 18.1-2(a). The ends of a one-turn coil of wire are attached to a conducting ring split so that it has two small gaps $G1$ and $G2$. This

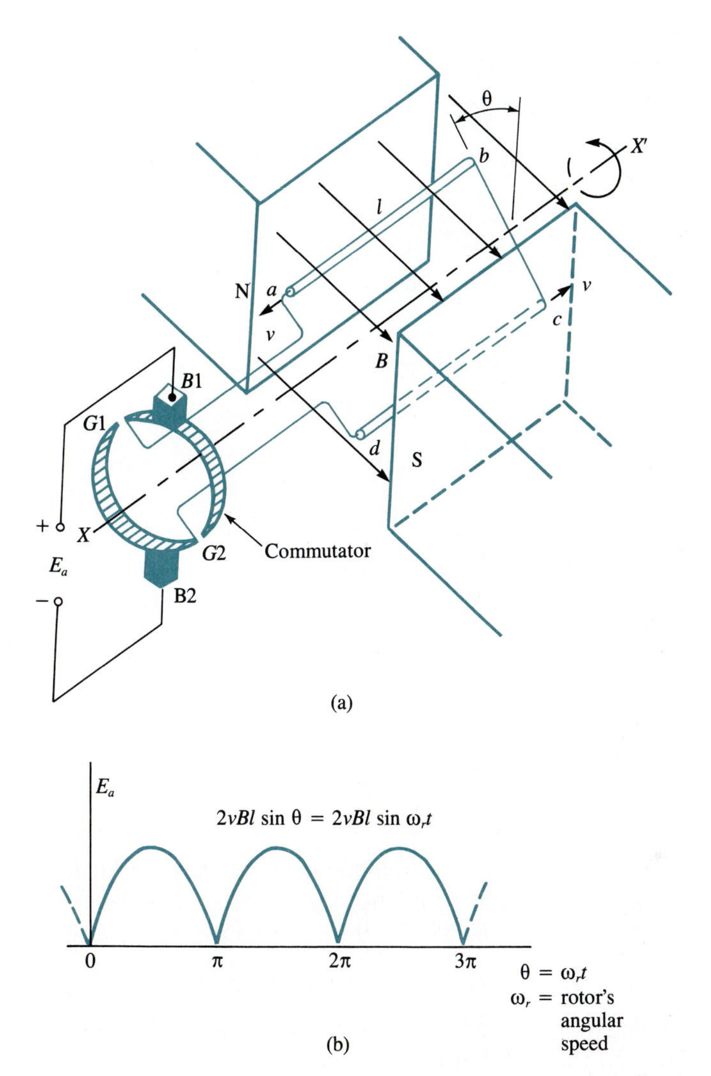

FIG. 18.1-2. (a) A simple dc generator and (b) its output voltage.

ring is called a *commutator* with two *segments*. Two sliding conducting contacts $B1$ and $B2$, called *brushes*, form the output terminals of the generator. The commutator and wire coil are attached to a mechanical structure, called a *rotor* (not shown), that rotates counterclockwise as viewed from the commutator end. Rotation is about axis XX'. The rotor turns inside a magnetic field established by permanent magnets. The structure that provides the fixed field is called the *stator* since it does not rotate. Our simple generator has a stator with one north (N) and one south (S) pole and is called a *two-pole generator*.

When $\theta = 0$ in Fig. 18.1-2(a), conductor *ab* is moving (tangentially) parallel to the field, so $\sin\theta = 0$ and its induced voltage is zero. The same is true of conductor *cd* on the other side of the coil. As θ increases from zero, both

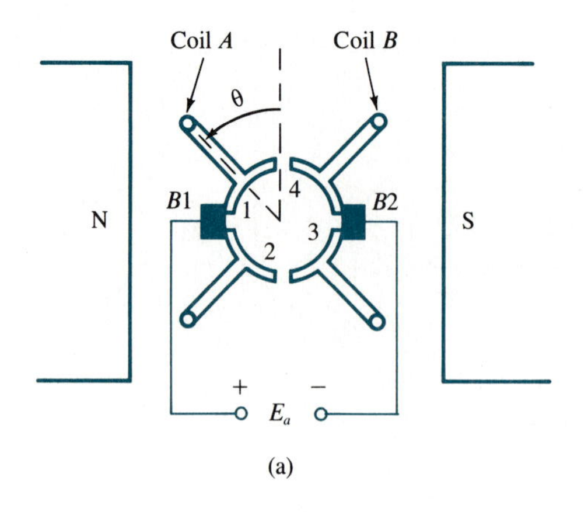

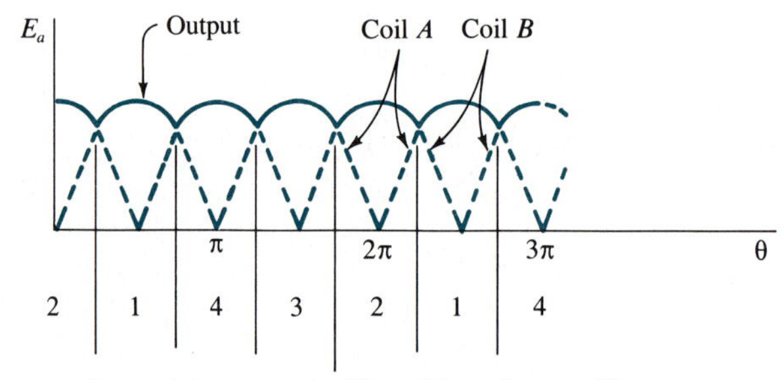

FIG. 18.1-3. (a) A simple two-pole dc generator with two one-turn windings and (b) its output voltage.

conductors cut the magnetic field and begin to have induced voltage according to (18.1-1). By the right-hand rule the voltages are positive at *a* and *c* relative to *b* and *d*, respectively. In other words, they add so that the output voltage is

$$E_a = 2vBl \sin \theta \tag{18.1-2}$$

or just twice that induced in one of the wires. As $0 \le \theta \le \pi$, $B1$ remains connected to point *a* on the coil and $B2$ is connected to point *d*. Output follows (18.1-1), as shown in Fig. 18.1-2(b).

When $\theta = \pi$, $B1$ is at gap $G2$ and $B2$ is at $G1$. As θ becomes larger than π, the brushes switch segments on the commutator. The result is to reverse the sides of the windings attached to $B1$ and $B2$. Therefore, for $\pi \le \theta \le 2\pi$, behavior is identical to that for $0 \le \theta \le \pi$. The overall output of Fig. 18.1-2(b) has a dc component but also has many unwanted harmonics, in accordance with Fourier analysis.

Practical generators have several methods by which the ripple in the generated voltage can be reduced. One effective method relies on additional windings and further subdivision of the commutator into additional segments. Figure 18.1-3(a) depicts the addition of one winding. The commutator now has four segments. Brushes make transitions between segments while the voltages of the

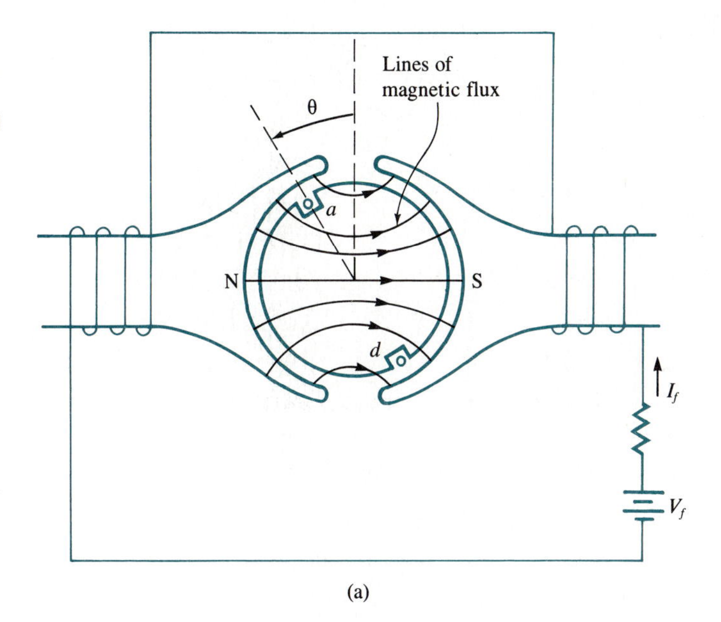

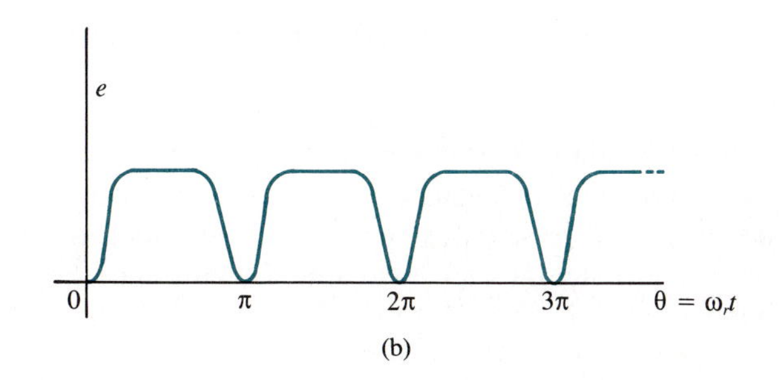

FIG. 18.1-4. (a) Shaped poles in a two-pole dc generator and (b) the output voltage for a single-armature coil.

windings are at $2vBl/\sqrt{2}$ for the assumed constant magnetic field, as sketched in Fig. 18.1-3(b). Ripple has been greatly reduced. Additional windings and commutator segments can reduce the ripple even further.

In real generators the rotor windings are mounted in slots cut into a cylindrical magnetic rotor structure called the *armature*. The presence of the highly magnetic material near the poles, in conjunction with shaping of the poles, as sketched in Fig. 18.1-4(a), results in a nearly constant magnetic flux density in the gap between the poles and armature. The resulting output voltage is given in Fig. 18.1-4(b).

In all but the smaller-power dc generators the stator's magnetic field is provided by an electromagnet, as depicted in Fig. 18.1-4(a). The amount of flux density (B) is determined by the field current, which is sometimes provided by an external source.

Modern dc generators often use multiple poles (an even positive integer) and

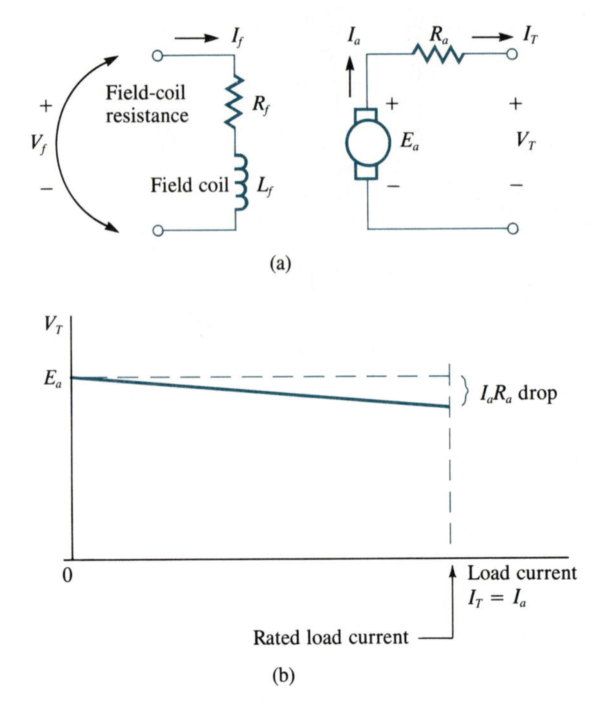

FIG. 18.1-5. (a) The equivalent circuit of a dc generator and (b) its terminal-voltage behavior with load current.

have multiple windings that can be wound in several ways. These details, as well as other refinements in design, are beyond our scope. For our purposes it is sufficient to gain a feeling for the manner in which these devices work and to understand that machines are available that produce nearly constant dc voltage (low ripple).

Equivalent Circuit

For most purposes the equivalent circuit of a dc generator is as sketched in Fig. 18.1-5(a). The terms R_f and L_f represent, respectively, the resistance and inductance of the field coils.† The armature circuit comprises a Thevènin equivalent source of voltage E_a and resistance R_a. The way a dc generator behaves is related to the way the field winding is excited. Broadly, we can classify dc generators as *separately excited* or *internally (self-) excited.*

† Even though the inductance plays no role in the dc circuit in steady-state conditions, it is included in the equivalent circuit as a reminder that F_f is caused by a winding in the real generator.

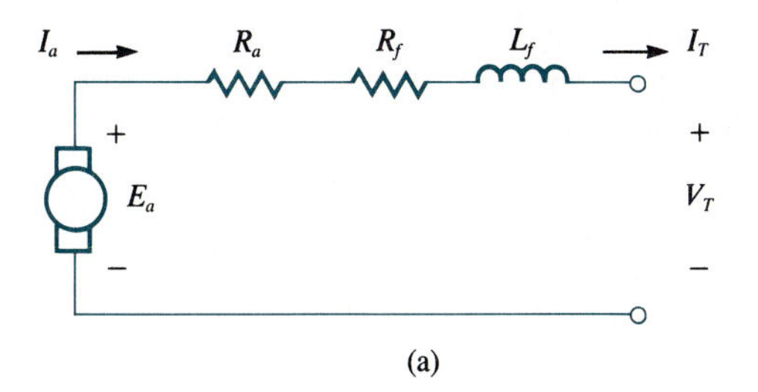

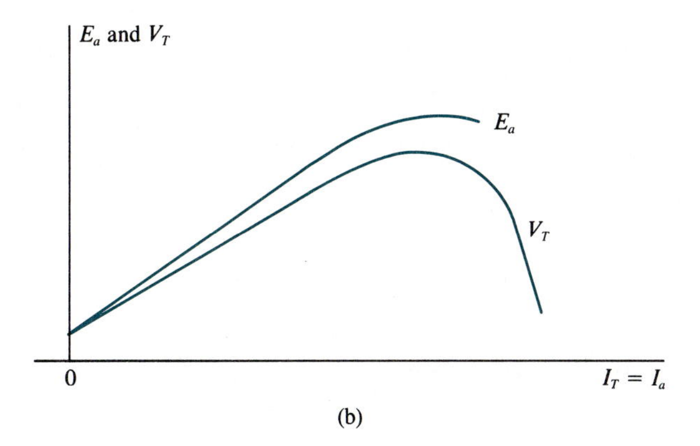

FIG. 18.1-6. (a) The equivalent circuit of a series dc generator and (b) its terminal-voltage behavior with load current.

Separate Excitation

Under separate excitation of the field winding, current I_f is constant and independent of load. Generated voltage E_a depends on I_f but is fixed for all loads. As a function of load current, the terminal voltage behaves as in Fig. 18.1-5(b).†

Series Excitation

In the series-excited generator, armature current is passed through the field winding to provide field excitation. The equivalent circuit of Fig. 18.1-6(a) applies. If no load current flows, there is no field flux except for the residual flux of the magnetic structure of the poles. Thus E_a is small. As load current increases, magnetic flux density and E_a increase. This trend continues until the pole and armature materials go into magnetic saturation (which usually occurs above the machine's current rating). At this point E_a becomes constant. Further increases in load current I_T causes mainly additional voltage drop in the armature resistance

† The curves assume that the generator has *compensation windings* to counteract some practical effects. Without these windings the droop in Fig. 18.1-5(b) is more severe and falls more rapidly as load current approaches the rated value.

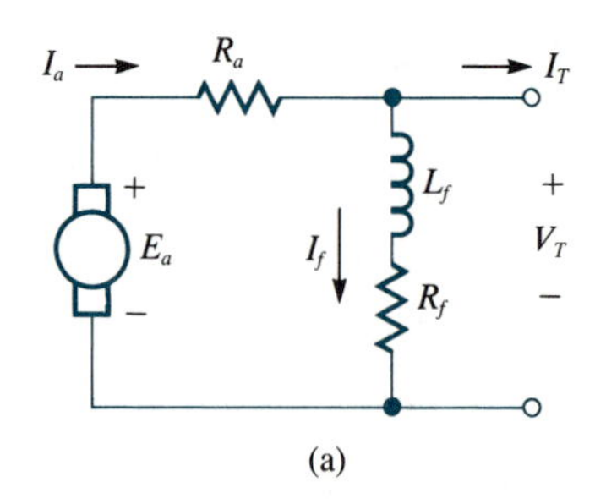

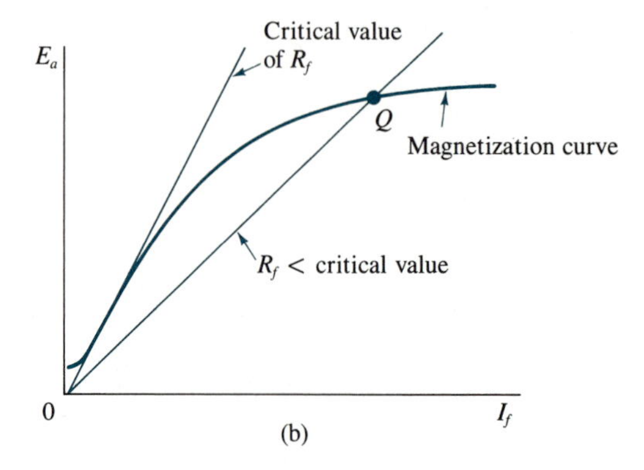

FIG. 18.1-7. (a) The equivalent circuit of a shunt dc generator and (b) its magnetization curve and field coil's load line.

R_a, and terminal (load) voltage begins to decrease. These effects are sketched in Fig. 18.1-6(b). Dc generators with series excitation are not widely used except in specialized applications such as arc welding and street cars.†

Shunt Excitation

In the shunt type of internal (self-) excitation the field winding is in parallel (shunt) with the armature, as sketched in Fig. 18.1-7(a). When the generator is initially started, the voltage E_a is small and entirely due to residual magnetism in the field poles. This voltage causes a small current I_f to flow in the field winding, which increases magnetic flux, thereby increasing E_a, and so on. The induced voltage increases with I_f according to the *magnetization curve* of Fig. 18.1-7(b). This process, called *buildup*, stabilizes when E_a equals the voltage required to exactly cause the current I_f that corresponds to E_a [point Q in Fig. 18.1-7(b)].

Since the steady-state value of E_a depends on R_f through I_f ($I_f \approx E_a/R_f$), some control of generator voltage can be obtained by adjustment of R_f. A practical arrangement is to place an additional resistance in series with the field coil; in this case R_f represents the total shunt resistance. If R_f is above a value called the *critical resistance* [Fig. 18.1-7(b)], buildup will not occur.

† A. N. Chaston, *Electric Machinery*, Prentice-Hall, Englewood Cliffs, N.J., 1986, p. 255.

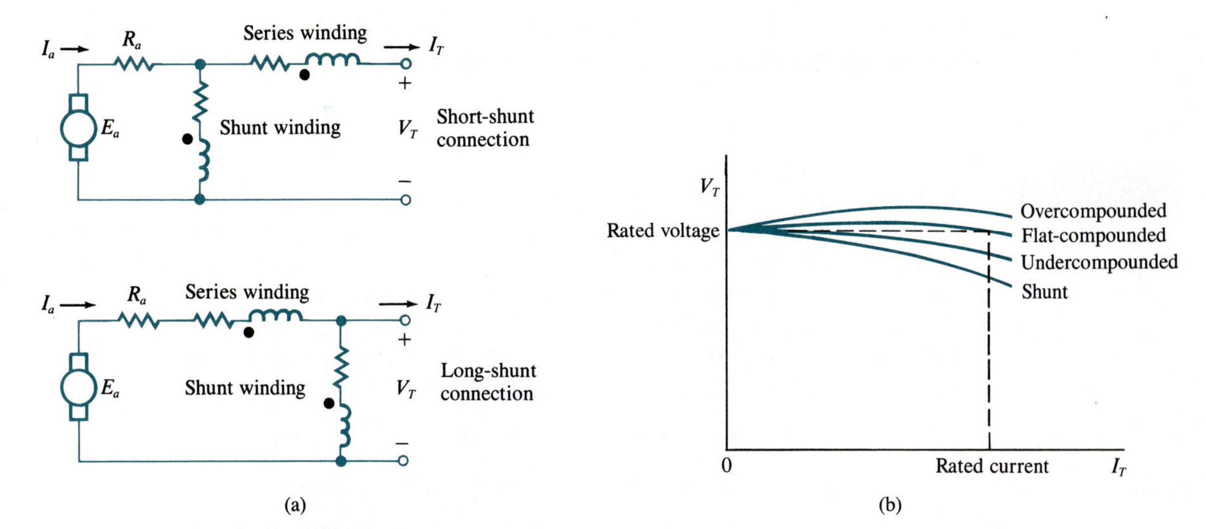

FIG. 18.1-8.
(a) Short-shunt and long-shunt versions of a cumulatively compounded dc generator and (b) its load curves.

After buildup, the terminal (load) voltage of the shunt generator decreases more rapidly with load current than in the separately excited generator [Fig. 18.1-5(b)]. Even so, the drop in the shunt machine is usually small, so that it can be used in applications requiring nearly constant voltage. An advantage of the shunt generator is that E_a drops rapidly to zero as load current exceeds the rated value. This feature provides a built-in protection in the event of a short-circuited load.

Compound Connections

Placing two sets of field windings in a dc generator gives a combination of series and shunt self-excitation that produces good regulation of voltage with load variations. Two implementations are possible, as sketched in Fig. 18.1-8(a); there is little practical difference between them. When the two windings produce fluxes that aid each other, the field is called *cumulatively wound*.†

Performance curves are given in Fig. 18.1-8(b). The undercompounded case occurs when the series coil has only small effect (small number of turns). As the series coil is made more dominant, a condition is reached called *flat compounding*, where the no-load and rated-load voltages are equal with a slight rise for in-between loads. With still more series-coil effect the generator becomes *overcompounded*, having a rising voltage with load current.

Although flat compounding is most often desired to maintain nearly constant voltage, overcompounding has one advantage. Where a load is some distance

---

† *Differentially* wound fields have nonaiding magnetic flux. We shall omit discussion of this generator because of its poor terminal-voltage characteristics.

from the generator, overcompounding can compensate for the increasing voltage drop on the feeder line caused by increasing load.

EXAMPLE 18.1-1

A cumulatively compounded dc generator has a long-shunt connection, series- and shunt-winding resistances of 0.02 and 50 Ω, respectively, and an armature resistance of 0.1 Ω; it delivers 86 A at 120 V to a load. We find the load's power and the power losses in the armature and field windings.

Load power is 120(86) = 10.32 kW. The shunt winding's current is $\frac{120}{50}$ = 2.4 A, so its power loss is $2.4^2(50)$ = 288 W. Armature current becomes 86 + 2.4 = 88.4 A, so armature- and series-field-winding losses become

$$\text{Armature:} \qquad 88.4^2(0.1) \approx 781.5 \text{ W}$$

$$\text{Series winding:} \qquad 88.4^2(0.02) \approx 156.3 \text{ W}$$

If a loss of around 600 W is allowed for losses in the magnetic and rotational structures, the total losses are 600 + 156.3 + 781.5 + 288 = 1825.8 W. Thus, this generator has a *power efficiency* η of

$$\eta = \frac{10{,}320}{10{,}320 + 1825.8} = 0.8497 \qquad \text{or } 84.97\%$$

18.2 DC MOTORS

Most motors (ac and dc) depend on the basic principle that a current-carrying conductor embedded in a magnetic field will have a force exerted on it. Figure 18.2-1 illustrates this principle for a single straight wire of length l between ends a and b. The wire carries a current I_a toward b from a and has an axis that is perpendicular to the lines of a magnetic field having flux density B. The force F that acts on the wire will have a magnitude $I_a Bl$; its direction is given by a *left-hand rule*,† where thumb, index, and middle fingers are all extended in mutually perpendicular directions. With the index finger pointing in the current's direction and the thumb in the direction of magnetic field, the force is in the direction of the middle finger.

Elementary Motor Operation

Dc motors are constructed similarly to dc generators. In fact, a dc machine can act as either a motor or a generator. Because of this fact, we consider and discuss the simple machines of Figs. 18.1-2, 18.1-3, and 18.1-4 as motors.

† A right-hand rule is possible (Problem 18-11) but it is more difficult to illustrate in Fig. 18.2-1.

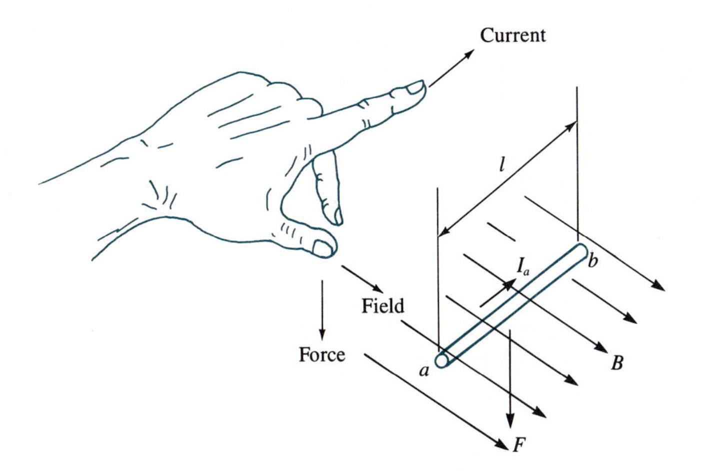

FIG. 18.2-1. The force on a current-carrying wire in a magnetic field as given by the left-hand rule.

The system of Fig. 18.1-2(a) becomes a motor by simply applying a dc voltage between brushes $B1$ and $B2$. If $B1$ is the positive terminal of an applied voltage V_T, a current I_a flows in the wires from a to b and from c to d. According to the left-hand rule, a force $F = I_aBl$ acts on each wire, as shown in Fig. 18.2-2(a). If the wires are at a radius r from the axis of rotation, the torque due to both wires is such as to cause counterclockwise shaft rotation. The torque† T_a is

$$T_a = 2(I_aBl)r \sin \theta \tag{18.2-1}$$

This armature torque is sometimes referred to as the *induced torque*, analogous to induced voltage in a generator.

Under action of the torque the armature of Fig. 18.1-2(a) will rotate. Torque is defined by (18.2-1) until θ exceeds π. For $\pi \leq \theta \leq 2\pi$ the action of the commutator is to reverse the direction of current in the coil to maintain the counterclockwise torque. For $\theta > 2\pi$ the process repeats in multiples of 2π. The overall behavior of T_a versus θ will have the form of half-cycles of a sinusoid, as in Fig. 18.2-2(b).

Real motors remove the pulsating nature of the torque present in our elementary motor. Multiple windings on a magnetic armature and multiple, shaped poles all contribute to making motors provide smooth torque with time. These improvements are directly analogous to improvements in voltage ripple in generators (Fig. 18.1-3 and 18.1-4).

Note that when a positive voltage is applied to $B1$ in the machine of Fig. 18.1-2(a), the direction of rotation is the same as when the machine is used as a generator to produce a positive voltage at $B1$. In fact, once a dc motor begins to rotate, it also acts as a generator having a generated voltage E_a (called the counter-electromotive force, or counter-emf) that is in opposition to the applied voltage. It is the counter-emf that limits the armature current to precisely the

† Recall that torque is the work involved in rotational motion. It equals the product of the turning force and the arm length (from turning axis) taken perpendicular to the direction of the line of force.

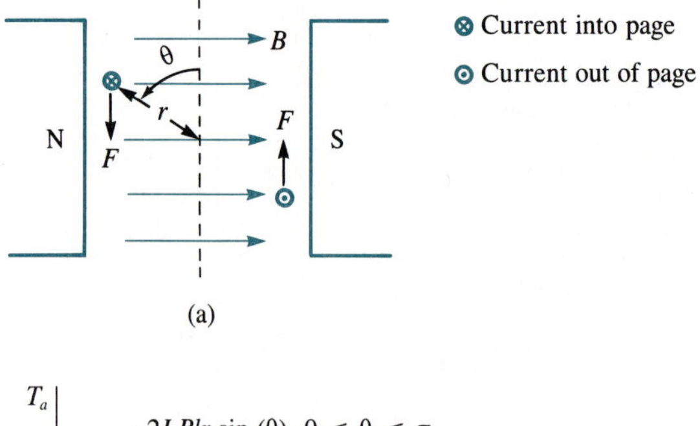

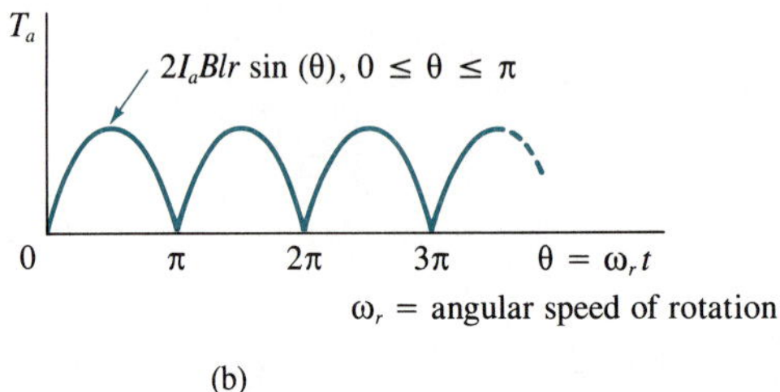

FIG. 18.2-2. (a) Forces on wires in a one-turn armature coil of an elementary motor and (b) torque output versus shaft angle.

amount needed to make the input electric power equal the sum of the load's power and power losses in the motor.

Equivalent Circuit

The equivalent circuit of a dc motor is exactly the same as in Fig. 18.1-5(a) for the dc generator, except that the direction of the armature current is reversed. Some of the most important motor relationships can be justified by plausible arguments centered around our elementary motor. On starting with (18.2-1) and recognizing that the sin θ term is replaced by unity in a well-designed motor, we see that torque is proportional to armature current and a magnetic flux (product of flux density and area). Thus, for real motors

$$T_a = K\Phi I_a \tag{18.2-2}$$

where K is a proportionality constant related to mechanical structure and Φ is the flux per pole. For our elementary generator $K\Phi = 2Blr$. Similarly, we note that $v = r\omega_r$ and write (18.1-2) as $E_a = 2Blr\omega_r$; so by direct extension to a real motor,

$$E_a = K\Phi\omega_r \tag{18.2-3}$$

Here ω_r is the rotor's shaft speed (radians per second). From the equivalent circuit we also have

$$V_T = E_a + I_aR_a \tag{18.2-4}$$

These last three equations plus the magnetization curve (plot of E_a versus I_f) allow most dc motors to be analyzed.

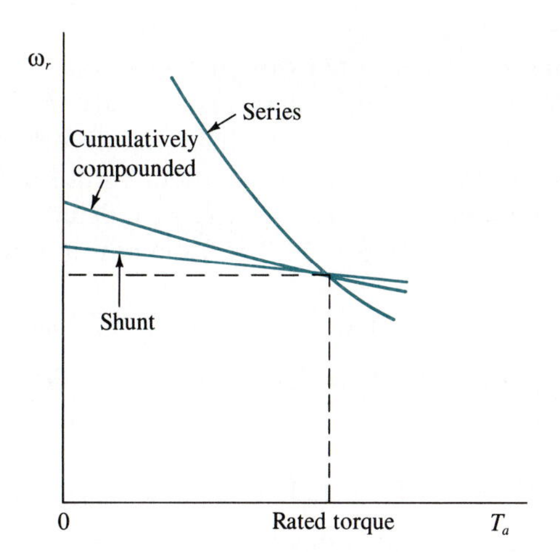

FIG. 18.2-3. Motor speed versus torque for several dc motors.

Series, Shunt, and Compound Motors

Dc motors may be connected in all the ways described previously for generators. Motor performance is usually shown as a plot of shaft speed versus load torque. Figure 18.2-3 sketches relative performances of series, shunt, and cumulative-compound motors that all have the same full-load ratings.

The shunt motor is useful in many industrial applications where high starting torque is not needed, as in fans and blowers. It can be operated over a range of speeds (two-to-one, or more) by varying field current (with a rheostat). Care must be taken never to operate the machine with the field winding open, however, because dangerous speeds can occur (called *runaway*). Runaway can also occur in some ordinary shunt motors subjected to load changes. In such motors the speed-torque curve in Fig. 18.2-3 begins to rise at higher loads. An increase in torque leads to higher speed and still higher torque, etc. More expensive shunt motors have special compensation windings to solve this problem. Ordinary shunt machines can add a small amount of cumulative compounding to remove the effect; they are called *stabilized-shunt motors*.

Dc motors can be operated in a separately excited mode. However, since it and the shunt motor both have constant field current, both connections give essentially the same performance under similar conditions (same machine, same field current). For this reason the separately excited motor is not shown in Fig. 18.2-3. By using (18.2-2) and (18.2-3) to eliminate I_a and E_a from (18.2-4), we find the speed characteristic of these two types of motors:

$$\omega_r = \frac{V_T}{K\Phi} - \frac{R_a T_a}{(K\Phi)^2} \qquad (18.2\text{-}5)$$

It is a linearly decreasing function of torque.

EXAMPLE 18.2-1

A long shunt-connected dc motor is fed from a 120-V source (V_T). Its armature and field-circuit resistances are 1.4 and 100 Ω, respectively. At full shaft load I_a = 10 A and $\omega_r/2\pi$ = 30 Hz (1800 rpm). At zero shaft load (free running) I_a = 0.8 A. We find the free-running speed and line power required at full load.

For V_T = 120 V, field current is $I_f = V_T/R_f = 120/100 = 1.2$ A. Total line current is $I_T = I_f + I_a = 1.2 + 10.0 = 11.2$ A. Line power at full load is $V_T I_T$ = 120(11.2) = 1344 W.

To find no-load speed, we start with (18.2-4). At full load

$$E_{a2} = V_T - I_{a2}R_a = 120 - 10(1.4) = 106 \text{ V}$$

At no load

$$E_{a1} = V_T - I_{a1}R_a = 120 - 0.8(1.4) = 118.88 \text{ V}$$

From (18.2-3) with constant machine flux

$$\frac{E_{a2}}{E_{a1}} = \frac{\omega_{r2}}{\omega_{r1}}$$

so

$$\frac{\omega_{r1}}{2\pi} = \frac{\omega_{r2}E_{a1}}{2\pi E_{a2}} = 30\,\frac{118.88}{106} = 33.65 \text{ Hz} \qquad \text{or 2019 rpm}$$

Places where applied power is dissipated in a dc motor is inferred from (18.2-4). On multiplication by I_a we have

$$V_T I_a = E_a I_a + I_a^2 R_a \tag{18.2-6}$$

The quantity $V_T I_a$ is the power provided to the armature circuit; it plus the field winding's power, which is $V_T I_f$ in the shunt motor, becomes the total power provided *to* the motor. The term $I_a^2 R_a$ is the power lost in the armature circuit as heat. The term $E_a I_a$ represents the power converted to mechanical power. Since mechanical torque (energy) per unit time is mechanical (shaft) power, denoted by P_s, we write

$$P_s = E_a I_a = T_a \omega_r \tag{18.2-7}$$

In terms of shaft horsepower (hp) we rewrite (18.2-7) as

$$\text{hp} = \frac{T_a \omega_r}{745.7} \tag{18.2-8}$$

EXAMPLE 8.2-2

The power components in the motor of Example 18.2-1 at full load are as follows:

Shaft power: $P_s = E_a I_a = 106(10) = 1060 \text{ W} = \frac{1060}{745.7} \text{ hp} \approx 1.42 \text{ hp}$

Armature loss: $I_a^2 R_a = 10^2(1.4) = 140$ W

Field loss: $V_T I_f = 120(1.2) = 144$ W

Total input power $= V_T(I_a + I_f) = 120(11.2) = 1344$ W

Note that the power efficiency of this motor is $1060/1344 \approx 0.789$ (or 78.9%).

Series-connected motors are often found in applications where high starting torque is needed and speed variations are not objectionable. Starter motors in automobiles, tractors and other machinery, and electric railway and car motors are common applications. Behavior of a somewhat idealized series motor can be developed by substitution of E_a from (18.2-4) into (18.2-3) and then elimination of Φ and I_a by use of (18.2-2) and the assumption that Φ is proportional to $I_f = I_a$† according to

$$\Phi = cI_a \tag{18.2-9}$$

where c is a constant. The result is

$$\omega_r = \frac{V_T}{\sqrt{KcT_a}} - \frac{R_a + R_f}{Kc} \tag{18.2-10}$$

Thus, shaft speed varies inversely as the square root of torque. This behavior means that a series motor should *never* be operated without load; otherwise, excessive speed and machine destruction could occur. In fact, the connection between motor and load should never involve a belt or other device that could break. At the other extreme the motor provides high torque at start and low speeds.

The cumulatively compounded motor combines some of the qualities of both series and shunt motors. It can give good starting torque without overspeeding at no load. Generally, the compound motor is very versatile and finds considerable use. Its main disadvantage is its expense.

Generally, a dc motor can serve also as a dc generator. However, a cumulatively compounded dc motor will become differentially compounded when serving as a generator, and a differentially compounded motor becomes a cumulatively compounded generator.‡ Since differentially compounded motors and generators have significant problems, this condition may restrict the dual-purpose use of compounded machines.

Starting DC Motors

When a dc motor starts from rest, its counter-emf E_a is zero and armature current is very large from (18.2-4). So that armature burnout is prevented, modern dc

---

† For an unsaturated field circuit (18.2-9) is approximately true. Most machines operate near saturation at full load, so our results apply more to performance interpretation than analysis.

‡ S. J. Chapman, *Electric Machinery Fundamentals*, McGraw-Hill, New York, 1985, p. 312.

motors use a starting resistor in series with the armature to limit starting currents. Automatic circuits progressively remove portions of the resistance so that speed increases until resistance is completely removed at some prescribed time. As part of these automatic circuits, relays are usually present to guard against loss of field current. If a loss occurs, the circuits disconnect the motor from the source.

18.3 THREE-PHASE SYNCHRONOUS GENERATORS

For a number of practical reasons, ac generators are constructed differently than dc generators. The ac machine typically has the field windings on the *rotor*, and the output (armature) windings are placed on the *stator*. This arrangement provides for better cooling and mechanical stability than in the dc machine, where the armature and field are on the rotor and stator, respectively. In addition, ac machines require no commutators, so their power levels can be much larger than those of dc generators.

Some discussion of three-phase (3-ϕ) ac generators was given in Sec. 17.2. Here we present only enough additional detail for the reader to have a plausible idea of operation.

Elementary Three-Phase Generator

An elementary three-phase generator is sketched in cross section in Fig. 18.3-1. The rotor consists of a permanent magnet of maximum flux density B_{max} along its axis. Flux density is assumed to decrease cosinusoidally with angle from the north

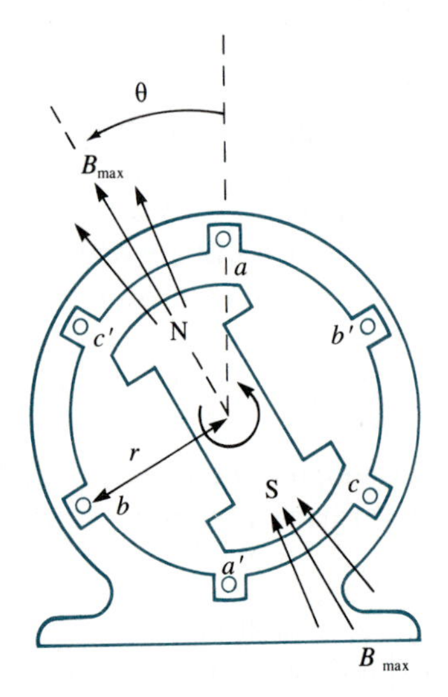

FIG. 18.3-1. An elementary three-phase generator with a two-pole rotor.

pole's maximum. The flux density normal to one side of a one-turn coil at a is then $B_{max} \cos \theta$. A similar flux density occurs at the other side of the coil at a'. Two other one-turn coils b, b' and c, c' are spaced in increments of $2\pi/3$ (120°) around the stator. The rotor is driven counterclockwise by the prime mover, and $\theta = 0$ at time $t = 0$.

The right-hand rule of voltage induction applies if we note carefully that the index finger points in the wire's direction of motion *relative to the field*. The wire at point a in Fig. 18.3-1 moves to the *right* relative to the field, so the wire's induced voltage is positive at a and negative at a'. Its amplitude is $2v(B_{max} \cos \theta)l$, where l is the length of one side of the coil. For wires a radial distance r from the rotational axis, the speed of the field passing the wire is

$$v = r\omega_r \tag{18.3-1}$$

where ω_r is the rotor's angular rate. The total voltage induced in a one-turn coil becomes

$$e_a = 2rlB_{max}\omega_r \cos \theta \tag{18.3-2}$$

Finally, if we allow the coil to have N_c turns per phase windings and define

$$A = 2rl \tag{18.3-3}$$

$$\Phi = B_{max}A = B_{max}2rl \tag{18.3-4}$$

the total voltage induced into the phase winding is

$$E_a = \sqrt{2}\left(\frac{N_c\Phi\omega_r}{\sqrt{2}}\right) \cos \theta = \sqrt{2}(\sqrt{2}\, \pi N_c\Phi f) \cos \omega t \tag{18.3-5}$$

The quantity $\sqrt{2}\, \pi N_c \Phi f$ is the rms voltage generated in the phase winding, and $\theta = \omega t = 2\pi f t$.

Other Three-Phase Generators

Generators often use more than two poles. Figure 18.3-2 illustrates four-pole (a) and six-pole (b) generators. In the four-pole unit the winding of phase a starts at wire 1 (point a) at the top, emerges in the rear, and connects there to wire 4, which emerges at the top and connects to wire 7, which connects to wire 10 in the rear; wire 10 emerges at the top at a' to complete the winding's path. A similar behavior occurs in the six-pole machine involving wires 1, 4, 7, 10, 13, and 16.

Phase winding b progresses through wires 3, 6, 9, and 12 in the four-pole generator and through 3, 6, 9, 12, 15, and 18 in the six-pole machine. Winding c begins at point c in each machine and uses the remaining wires in similar fashion.

Clearly, there are p wires in each phase winding in a p-pole machine when only one wire is present in a stator slot. If N_c conductors are placed in each slot by making N_c turns through the entire phase winding, there are $N_c p$ conductors per phase that generate voltages in series. From the right-hand rule the rms voltage generated is $N_c p r \omega_r B_{max} l/\sqrt{2} = \sqrt{2}\, \pi N_c \Phi f$, which is the same as in (18.3-5) for

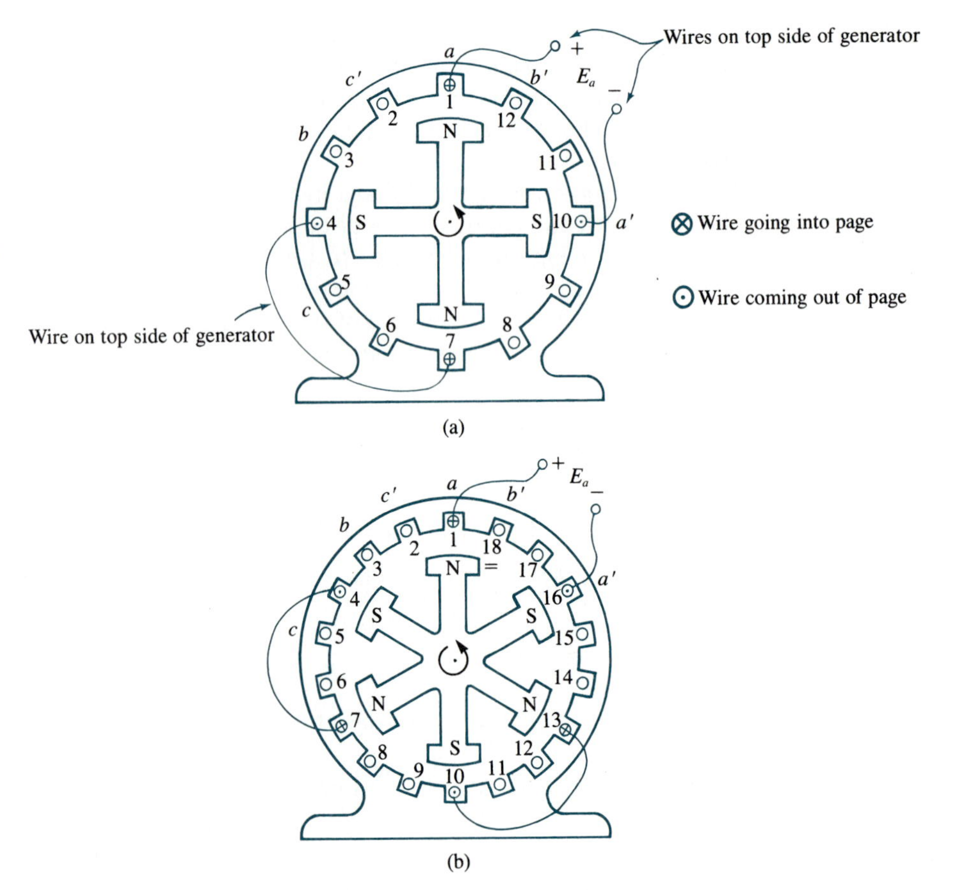

FIG. 18.3-2. Three-phase generators with (a) four poles and (b) six poles. Only one phase winding is illustrated in each case (having one conductor per slot), but wires are shown for all three phases.

the two-pole generator, since $\omega_r/2\pi = f_r = 2f/p = \omega/(\pi p)$.

In a more realistic machine, windings are more complicated than in the simple schemes of Fig. 18.3-2 and 18.3-1. Windings are typically distributed over the periphery of the stator. This spatial distribution reduces the generated voltage by a small amount that is accounted for by a *winding factor*, k_w, with a value typically in the range 0.85 to 0.95. Generated voltage of phase a becomes

$$E_a = \sqrt{2}(\sqrt{2}\ \pi k_w N_c \Phi f) \cos \omega t = E_s \cos \omega t \tag{18.3-6}$$

For the other phases

$$E_b = E_s \cos\left(\omega t - \frac{2\pi}{3}\right) \tag{18.3-7}$$

$$E_c = E_s \cos\left(\omega t - \frac{4\pi}{3}\right) \tag{18.3-8}$$

Except in very low-power machines, the rotor's poles are not permanent magnets but electromagnets. Poles as in Fig. 18.3-1 and 18.3-2 are called *salient poles* and are mainly used in slower machines, such as in a hydroelectric plant. Higher-speed two- and four-pole machines usually have round (cylindrical) rotors with field windings in slots over the entire surface. The field coil is fed dc via slip rings, as described in Sec. 17.2. For details on rotor excitation the reader is referred to that section.

The three-phase generator is formed when the three-phase windings are connected in wye or delta form (Sec. 17.2). Its equivalent circuit per phase is a resistance in series with a reactance, where the resistance is often negligible (Fig. 17.2-4).

18.4 THREE-PHASE SYNCHRONOUS MOTOR

A three-phase synchronous motor has essentially the same construction as a three-phase synchronous generator. In fact, the generator can operate as a motor and vice versa. As noted in Chapter 17, the word *synchronous* means that angular frequency ω of the electric voltages is some exact multiple of the shaft's angular speed ω_s,† according to (17.2-1):

$$\omega = \frac{p\omega_s}{2} \tag{18.4-1}$$

Thus, a $p = 2$-pole synchronous motor operated on a 60-Hz power line will have a shaft speed of 3600 rpm. Even for reasonable load variations, this speed stays constant, unlike the speed of the dc motor studied earlier.

Constant shaft speed is important for many industrial applications and is one of the main advantages of the synchronous motor. It is especially attractive for higher-power (over 50 hp, or 37.3 kW), lower-speed (below 500 rpm), directly connected loads such as compressors, crushers, grinders, and conveyor lines, where it can be more economical than other forms of motors.

The larger-power motors typically require dc excitation for the rotor's field, which allows for the possibility of adjusting field current. For a given shaft load there is a value of field current (called *normal* excitation) where the motor's power factor is unity. Field currents above or below this value (for the same load) cause the machine's power factor to become leading (capacitive) or lagging (inductive), respectively. These two conditions are known as *overexcitation* and *underexcitation*, respectively. The ability to adjust the power factor is an important advantage of the synchronous motor. It allows the machine not only to perform a function (work) but also to adjust for the poor power factor of other types of machinery

---

† We use ω_s to denote a shaft speed that is synchronous, as opposed to the earlier notation ω_r for a possibly nonsynchronous shaft speed.

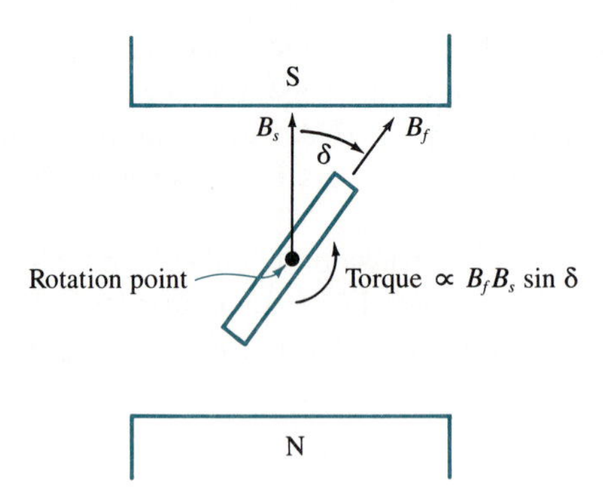

FIG. 18.4-1. The alignment of magnetic fields.

that cannot be adjusted. Some synchronous motors are even operated without load specifically for power factor correction in heavy industries.

Operation–Rotating Magnetic Fields

Operation of a synchronous motor is based on the principle that magnetic fields tend to align themselves. The small permanent magnet with magnetic field B_f, as shown in Fig. 18.4-1, will move to align its field with that of the larger magnet of field B_s. When the fields are aligned, angle δ is zero and the small magnet is fixed. However, if the fields are displaced by angle δ, a torque proportional to $B_f B_s \sin \delta$ will tend to pull the small magnet back into alignment. Clearly, if the fields are once aligned and the larger magnet begins to revolve around its rotation point (to cause the field B_s to rotate in space), the small magnet will follow. If a small "load" is attached to the axis of the small magnet, a small value of *fixed* angle δ will develop so that the torque required to keep B_f in synchronism with B_s exactly equals the torque of the load. A simple synchronous motor has been developed.

Operation of our simple motor depends intimately on the creation of the rotating fields B_s and B_f. Field B_f is established by the dc-excited field circuit on the rotor of a real motor. The second rotating field B_s is obtained electrically from a fixed stator, as we next describe.

Consider three-phase windings on the stator of a machine as sketched in Fig. 18.4-2(a). Let the windings be excited by a 3-ϕ source with sequence *abc*. The magnetic fields of the three windings will be directed in space as shown in Fig. 18.4-2(b). The horizontal component B_x of the total field is the sum of the three horizontal components:

$$\begin{aligned} B_x &= -B_{\max} \cos(\theta) + B_{\max} \cos\left(\theta - \frac{4\pi}{3}\right) \cos\left(\frac{\pi}{3}\right) + B_{\max} \cos\left(\theta - \frac{2\pi}{3}\right) \cos\left(\frac{\pi}{3}\right) \\ &= -\tfrac{3}{2} B_{\max} \cos\theta \end{aligned} \qquad (18.4\text{-}2)$$

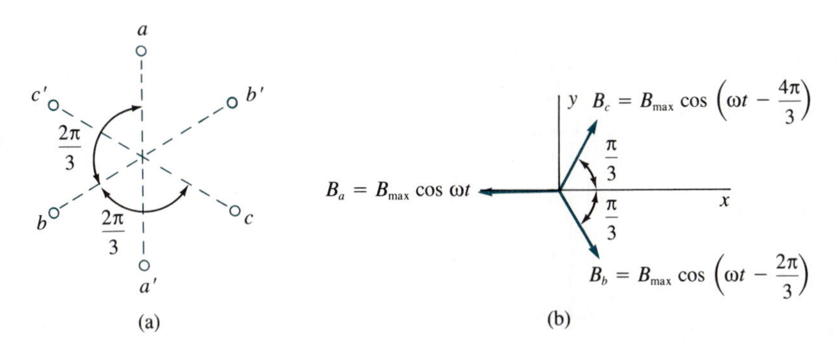

FIG. 18.4-2. (a) Stator windings of a three-phase motor and (b) the physical directions of the magnetic fields due to the windings.

These two equations describe a magnetic field that rotates in space. Its direction is defined by the rotation angle $\theta = \omega t$, and its amplitude is *constant* at $\sqrt{B_x^2 + B_y^2} = 3B_{max}/2$.

We may visualize the synchronous motor as operating such that its rotor revolves in a way that makes its field *chase* the rotating field of the stator. In proper (synchronized) operation the two fields rotate at exactly the same rate, but the rotor's field lags by exactly the correct angle δ to cause the torque required by the load. Under normal loads torque is provided with no speed loss. There is a maximum, called the *pullout torque*, that causes the machine to cease function and can create excessive vibration. Pullout is typically 2 or 3 times larger than the machine's rated value, and it increases with field current.

Starting Synchronous Motors

Without special design 3-ϕ synchronous motors will not start by simply applying line power. The reason is that the stator's magnetic field rotates so fast relative to the rotor (which starts from zero speed) that both clockwise and counterclockwise torques are produced that average to zero. In some high-power motors external prime movers first bring the motor to synchronous speed and then are replaced by the load. In most motors *damper windings*, also known as *amortisseur windings*, are included that aid starting.

A damper winding consists of conducting bars laid in notches cut into the rotor. The bars are shorted at each end of the rotor, as sketched in Fig. 18.4-3 for a two-pole motor. During starting the field winding is shorted (by automatic starting devices) through a resistance to limit field current. The rotating stator field induces a current into a damper winding that establishes a magnetic field; it reacts

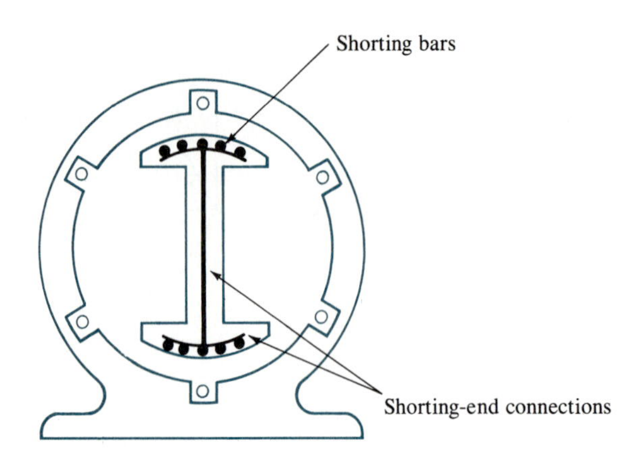

FIG. 18.4-3. A cross section of a two-pole motor's rotor with damper windings.

with the stator's field in such a way that a net starting torque will exist (see Problem 18-19). Because some relative motion must exist between the stator's field and the damper winding for the induced field to exist, the rotor can never reach synchronous speed. It can get close enough, however, for the regular field current to complete the process.

18.5 THREE-PHASE INDUCTION MOTOR

In the previous section we found that a three-phase motor could run only on damper windings. The motor could not reach synchronous speed, but it would operate. The possibility obviously exists to eliminate the rotor's main field windings altogether if a motor with nonsynchronous speed is acceptable. Such motors are called *induction motors*.

The typical induction motor has a rotor with shorting bars around its entire periphery. It resembles the exercise cages in which squirrels or hamsters run and is called a *squirrel-cage rotor* (Fig. 18.5-1).† Squirrel-cage induction motors are rugged, require little maintenance, and are relatively inexpensive. These advantages explain why they satisfy most industrial requirements (in the 5- to 10,000-hp range) and are widely used. In machines where the stator is designed for a rotor with p poles, the squirrel-cage rotor acts to develop the necessary p poles.

Let ω_r represent the angular shaft speed of an induction motor. *Slip* is defined

† Induction motors exist with three-phase windings on the rotor. We do not discuss these here. See D. V. Richardson and A. J. Caisse, Jr., *Rotating Electric Machinery and Transformer Technology*, 3d ed., Prentice-Hall, Englewood Cliffs, N.J., 1987, p. 455.

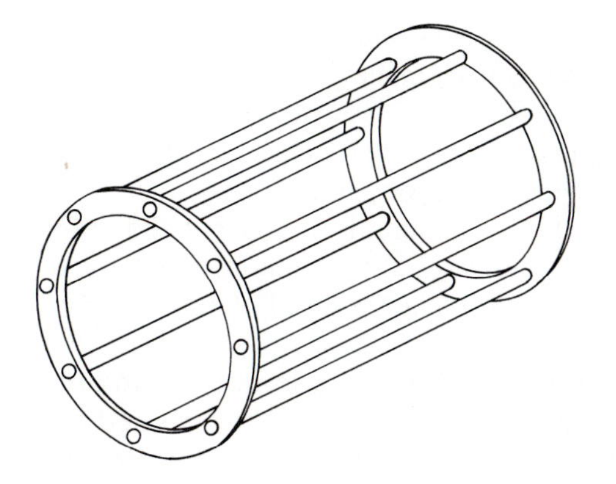

FIG. 18.5-1. A squirrel cage for an induction motor's rotor.

as the fraction of synchronous speed ω_s:

$$s = \text{slip} = \frac{\omega_s - \omega_r}{\omega_s} \tag{18.5-1}$$

or

$$\omega_r = \omega_s(1 - s) \tag{18.5-2}$$

Several classes of induction motors have been established by the National Electrical Manufacturers Association (NEMA). The approximate relative performances of classes A, B, C, and D are sketched in Fig. 18.5-2. The class A motor has the smallest slip at full load (0.02 to 0.05) of the various classes. It also has slightly higher starting torque (about 10% more) and higher pullout torque (the maximum torque the machine can provide) than a class B motor but has a very high starting current (about $\frac{1}{3}$ more than class B).

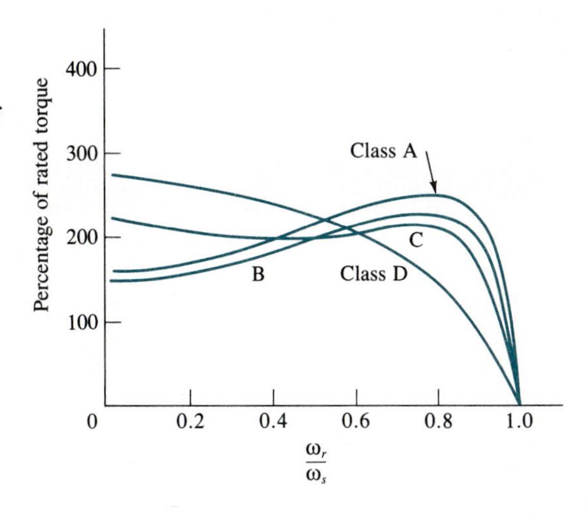

FIG. 18.5-2. Torque-speed curves for several classes of induction motor.

In recent times class B motors have been replacing class A machines, mainly because they offer significantly smaller starting current (about 5 times full-load current) but having only slightly poorer slip (0.03 to 0.05), less starting torque (about 10% less), and less pullout torque. Class B motors enjoy wide use and can legitimately be called general-purpose machines.

Class C motors are able to maintain reasonable slip at full load (<0.05) while giving high starting torque (200 to 250% of rated torque) and relatively low starting current (around 4 to 5 times full-load rating). Pullout torque is slightly less than for classes A or B.

Class D motors provide high starting torque (up to about 300% of full-load torque) and low starting current (about 3 to 8 times rated current). These advantages are gained at the expense of large slip (often over 0.1).

Design of the squirrel-cage rotors accounts for most of the variation in motor performance (Fig. 18.5-2). Class A rotors have large low-resistance bars near the rotor's surface, while large bars are placed deeper in the rotor of class B motors. Small high-resistance bars near the rotor's surface are characteristic of class D motors. The class C rotor combines small bars near the surface, to gain low-speed torque, with deeper larger bars (forming a *double*-squirrel-cage rotor) to maintain higher-speed characteristics somewhat like those of class B.

EXAMPLE 18.5-1

A 10-pole three-phase induction motor delivers full-rated power of 100 hp to a load, and the motor's shaft speed is 698 rpm. We find the motor's synchronous speed, its slip, and the load's torque when the power-source frequency is 60 Hz.

From (18.4-1), $\omega_s/2\pi = 2f/p = 2(60)/10 = 12$ Hz (or 720 rpm). From (18.5-1)

$$s = 1 - \frac{\omega_r}{\omega_s} = 1 - \frac{698}{720} \approx 0.0306$$

Finally, (18.2-8) relates mechanical torque to the load's power: $T_a = 745.7\,(\text{hp})/\omega_r = 745.7(100)/(24\pi) \approx 989$ N·m.

18.6 SINGLE-PHASE INDUCTION MOTORS

Whereas three-phase synchronous and induction motors are the workhorses of heavy industry, the workhorses in small industry, businesses, and homes are single-phase induction motors. Since they must operate from a single-phase source, they cannot use the unique properties of three-phase windings to create rotating magnetic fields, as needed for motor operation. Although beyond our scope, it can be shown that an induction motor comprising a single-phase stator winding and a squirrel-cage rotor will function. Such a motor only functions if

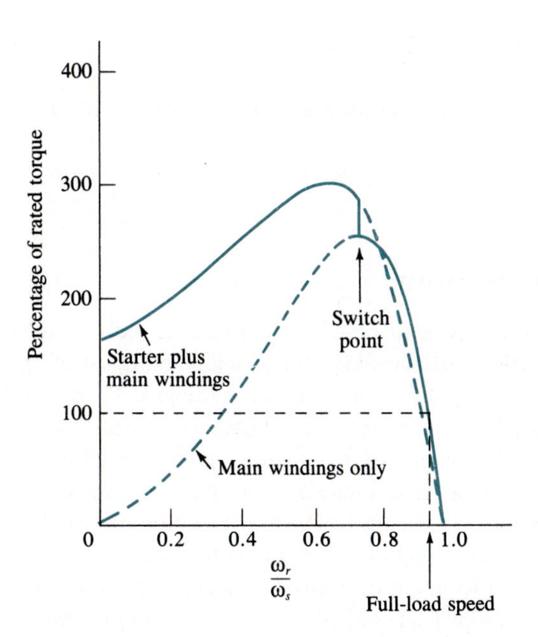

FIG. 18.6-1. Torque-speed curves for a split-phase induction motor.

initially started by some auxiliary method, since it has no starting torque at zero rotor speed.

Single-phase induction motors are usually classified according to the schemes used to start the machine. We discuss three types.

Split-Phase Motors

By addition of a second (or auxiliary) *starter winding*, in parallel with the main winding, a split-phase motor is formed. The extra winding has fewer turns of smaller wire such that it appears mainly resistive. The main winding appears mostly inductive. Starter-winding current leads main-winding current as a result. Current phasing in time, combined with a physical displacement of coils, leads to a rotating-stator magnetic field and starting torque.

The torque provided by the starter winding at low speeds is adequate to get the motor up to a speed where the main windings function. At roughly 75% of synchronous speed a centrifugal switch cuts out the starter winding, and operation is by main windings only. Without the cutout the starter windings may burn out.† Figure 18.6-1 shows typical performance curves. Start torques of from about 150 to 200% of rated torque are possible. At a time near the time when the torque due to the main winding is maximum (about $\omega_r = 0.75\omega_s$), the starter winding is

† Switch failure leading to a burned-out starter winding is one of the principal causes of failure in single-phase induction motors, which are otherwise rugged and reliable machines under normal conditions.

removed. The main winding continues to accelerate the shaft until load torque is achieved.

Split-phase motors have moderate starting torque and reasonably low starting current (5 to 7 times rated current). Rotation direction can be reversed by reversing the auxiliary winding's connections but leaving the main winding unchanged.

Capacitor-Type Motors

The starting torque of the split-phase motor can be increased by placing a capacitor in series with the start winding. By proper selection of the capacitor's size, starting torque can be over 3 times the rated torque. Typical performance of this form of motor, called a *capacitor-start motor*, is illustrated in Fig. 18.6-2. Applications are in machines where high starting torque and low starting current are desirable, as in washing machines, dryers, air conditioners, pumps, and compressors. The capacitor-start motor is more expensive than the split-phase machine and typically has ratings from about $\frac{1}{6}$ to 2 hp.

It is also possible to use a capacitor in series with the auxiliary winding on a permanent basis. This design, called a *permanent-capacitor motor*, is more efficient and has a better power factor than the regular single-phase induction motor at normal loads. It also does not require a switch, which makes it reliable. Starting torque is not large (below 100% of rating), but starting current is low. It has relatively high efficiency and a good power factor at rated load. Typical ratings are from about $\frac{1}{6}$ to $\frac{3}{4}$ hp. A typical torque-speed characteristic is shown in Fig. 18.6-3.

The good running characteristics of the permanent-capacitor motor can be maintained while achieving high starting torque by adding a second (switched)

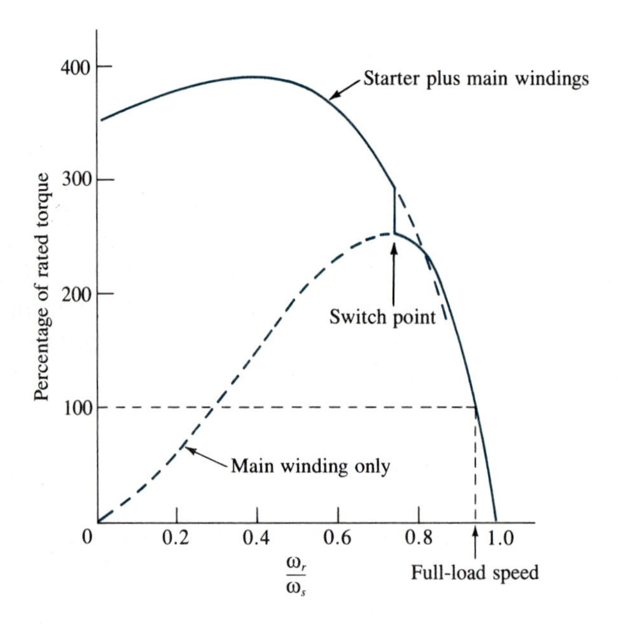

FIG. 18.6-2. Torque-speed curves for a capacitor-start induction motor.

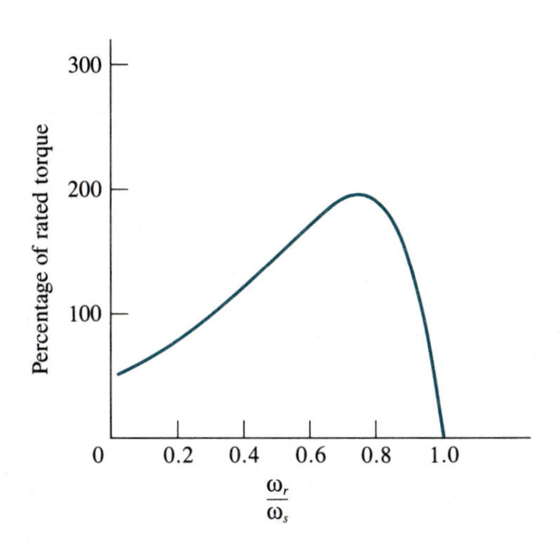

FIG. 18.6-3. Torque-speed curves for a permanent-capacitor induction motor.

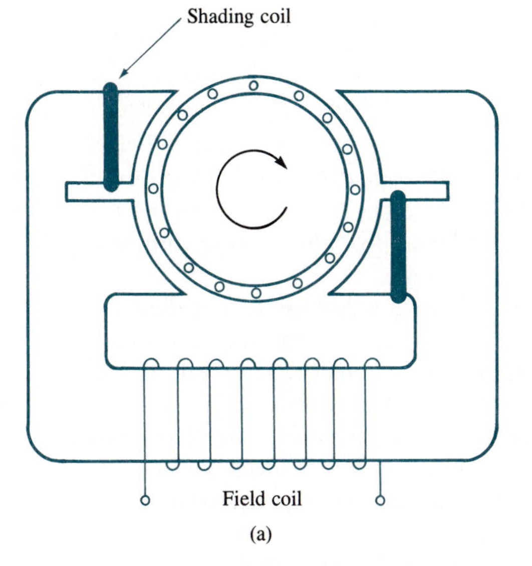

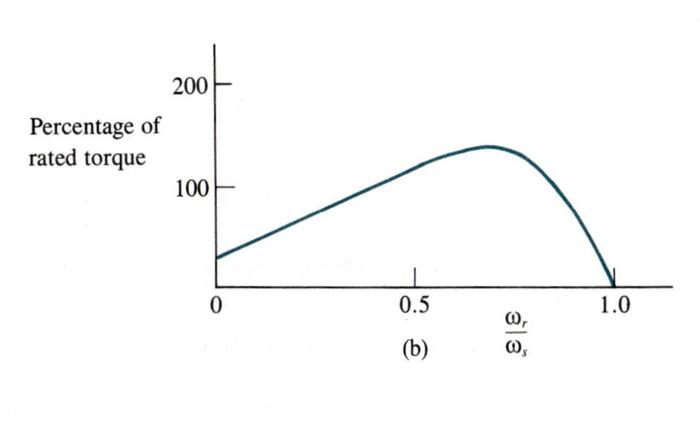

FIG. 18.6-4. (a) A shaded-pole motor and (b) its torque-speed characteristic.

capacitor. Called a *capacitor-start capacitor-run* motor, its torque-speed curves are similar to those of Fig. 18.6-2. Starting and running characteristics can be as good as any of the motors discussed in this section, but this motor may also be the most expensive.

Shaded-Pole Motors

It is also possible to obtain motor action with only a single stator winding. The *shaded-pole motor* uses a squirrel-cage rotor and splits the poles of the stator, as shown in Fig. 18.6-4(a). One or more shorting rings of copper or another good conductor are placed around part of each pole, creating what is called a *shaded pole*. The flux caused by the stator's coil and that caused by current induced into the shorting coils react in a complicated way (not described here) but sufficient to develop torque and cause rotation. Rotation is directed toward the shaded part of a pole from the unshaded part.

Shaded-pole motors are characterized by low cost but also have low starting torque and high slip [Fig. 18.6-4(b)]. Nevertheless, it is the standard general-purpose motor of choice in the subfractional horsepower range (less than $\frac{1}{20}$ hp) because of its low cost. It is not typically found with ratings above about $\frac{1}{6}$ hp. In the range from $\frac{1}{20}$ to $\frac{1}{6}$ hp it must compete with the split-phase motor.

18.7 OTHER TYPES OF MOTORS

We mention two other types of motors before closing this chapter.

Universal Motor

If the polarity of the voltage applied to a series-connected dc motor is changed, the directions of both armature current and field flux change, but the direction of the torque is not changed, as indicated by (18.2-2). This fact allows the series dc motor to operate on an ac voltage. To the motor, the voltage appears to be equivalent to a source of pulsating "dc" voltage.

Motors designed to operate on either an ac or a dc source are called *universal motors*. Eddy-current (power) losses in the magnetic materials are reduced by making both the rotor and the stator of laminated construction (built up from thin sheets of materials). The torque-speed characteristic with ac excitation is similar to the curve for dc operation. Advantages are low cost and the ability to operate at speeds above 3600 rpm (the maximum speed of other ac machines). In some applications its ability to control speed with applied voltage (speed increases with an increase in voltage) is an advantage. The motor's disadvantages are increased maintenance and reduced reliability because of the commutator and brushes.

Universal motors range from $\frac{1}{1000}$ to about 1 hp. They are found in many home appliances and portable drills and saws. In these applications the combination of motor losses and losses in the gear train that the motor is usually driving are enough that speeds at no-load are not excessive.

Hysteresis Motor

The *hysteresis motor* is rather interesting because its rotor has no windings at all, not even damper windings. It consists of a shaft-mounted, solid, cylindrical, nonmagnetic material (such as brass) inside a cylindrical shell of special magnetic material having a smooth surface. The stator generally can be of any form that

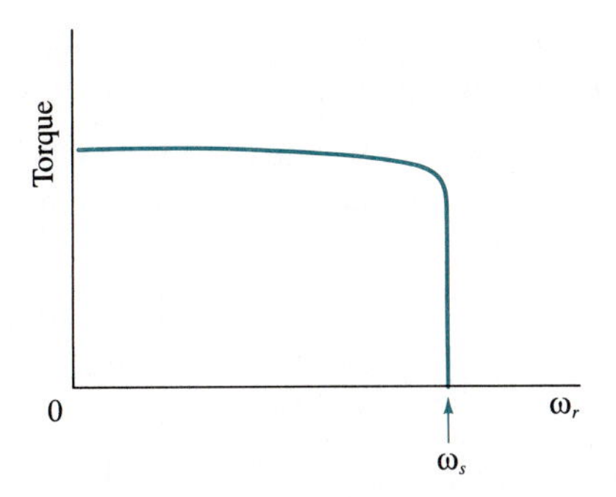

FIG. 18.7-1. Hysteresis motor's torque-speed characteristic.

produces a rotating magnetic field. Split-phase, permanent-capacitor, and shaded-pole choices are typical, with the latter two being popular.

The stator's field induces a magnetic field in the rotor. The rotating stator's field "drags" the rotor's field around with it. Because of hysteresis (a nonlinear magnetization present in magnetic materials), the rotor's field lags the stator's field by a fixed angle related to the rotor's material, and torque is developed. Thus, torque is nearly constant with speed. The motor quickly accelerates until synchronous operation is achieved. The torque-speed curve for the hysteresis motor is shown in Fig. 18.7-1.

Because of their synchronous operation, hysteresis motors are extensively used in electric clocks. They exhibit very smooth operation and are even useful in gyroscopes for inertial guidance systems. They can be found with ratings up to about $\frac{1}{8}$ hp (about 100 W) but usually do not exceed a few watts.

PROBLEMS

18-1. For the wire in Fig. 18.1-1, $l = 10$ cm, $B = 0.4$ T, and $v = 4.7$ m/s. What is the maximum induced voltage E_a that can occur? For an angle $\theta = \pi/3$, what is E_a?

18-2. In the elementary dc generator of Fig. 18.1-2(a), $l = 12$ cm, $r = 4$ cm, $B = 0.4$ T, and $v = 5$ m/s. What is the shaft's speed in rpm? Sketch the output voltage and label its peak amplitude.

18-3. Explain how Fig. 18.1-2(b) would change if the simple generator had N turns of wire instead of one.

18-4. If the output of the generator of Fig. 18.1-2 is filtered to remove all periodic components in its voltage, what is the dc output voltage?

18-5. Work Problem 18-4 except for the generator of Fig. 18.1-3.

18-6. A separately excited dc generator with compensation windings has a load rating of 12.5 kW at 250 V and an armature resistance of 0.35 Ω. What is the terminal voltage with no load attached?

18-7. Explain why the shunt dc generator's load voltage decreases more rapidly with load current than that of the separately excited generator.

18-8. A shunt-connected dc generator provides 120 V to a load of 20 Ω. If R_a = 0.8 Ω and R_f = 240 Ω, find (a) the load's power, (b) the field current, and (c) E_a.

18-9. A shunt-connected dc generator with R_a = 0.04 Ω and R_f = 80 Ω delivers full-rated power of 200 kW to a load when the field winding draws 6 A. Find (a) V_T, (b) E_a, and (c) the total power dissipated in the generator.

18-10. A long-shunt dc generator for which R_a = 0.04 Ω and E_a = 235 V delivers 50 kW at 220 V to a load when the shunt field current is 1.8 A. What are the resistances of the shunt and series fields?

18-11. Define a right-hand rule for Fig. 18.2-1.

18-12. What force acts on a wire 15 cm long in a field B = 20 T when I_a = 1.2 A, as defined in Fig. 18.2-1?

18-13. Sketch the torque versus θ for the elementary motor of Fig. 18.2-2(a) if r = 2.5 cm and the other parameters are as given in Problem 18-12.

18-14. The field resistance of a shunt-connected dc motor is 150 Ω. The applied voltage and line current are 240 V and 66.6 A, respectively. If the armature resistance is 0.12 Ω and shaft speed is 1750 rpm, find (a) the back emf E_a, (b) the power delivered to the load, in watts and in horsepower, and (c) the load's torque.

★ **18-15.** A shunt-connected dc motor for which R_a = 0.3 Ω is connected to a 50-V line. Its armature current is measured to be 12 A when the shaft's torque is 3.55 N·m at a speed of 1500 rpm. When the motor's load is removed, the torque required to maintain shaft losses (friction) is 0.0355 N·m. What is its no-load shaft speed?

18-16. A three-phase, Y-connected generator is to provide direct 480Y/277-V service to an industrial load. If it has N_c = 5 turns per phase winding, k_w = 0.9, rotor length l = 1.0 m, and radius r = 0.2 m, what value of maximum flux density B_{max} is required for each rotor pole?

18-17. A large 36-pole synchronous motor uses input power at 60 Hz. What is its shaft speed?

18-18. A three-phase, two-pole synchronous motor uses a 60-Hz power line. It develops a maximum magnetic flux density B_{max} = 40 T per phase winding. The rotor develops an axial flux density B_f = 2 T per pole. The load causes a shaft angle δ = π/8 rad. What is the load's torque? What power (watts) is delivered to the load?

★ **18-19.** For four directions of the stator field (up, left, down, and right) in Fig. 18.4-3, show that the interaction of the field of the amortisseur winding always gives starting torque in the same direction.

18-20. A three-phase induction motor for a 60-Hz line has four poles and a slip of 0.03. What is its shaft speed?

18-21. A three-phase induction motor is to be selected to drive a load requiring 2.5 times as much starting torque as running torque (flywheel with a viscous load). What NEMA class of motor is best for this load?

APPENDIX

Software Package

A.0 INTRODUCTION

The use of computers to facilitate the learning process is integrated into this text. A diskette containing two different programs, (1) FRESP (frequency response) and (2) NUMCON (numbers conversions) is available to the instructor. The two programs were written in BASIC for IBM personal computers and IBM-compatible computers; however, with minor changes these programs should run on any system. Both programs are interactive and menu-driven programs. This appendix serves as the manual for those wishing to use the problem solving software package. The software is provided free of charge to the instructor as part of the solutions manual with permission to make as many copies as needed for class use. We provide here a brief but, for our purposes, adequate description of each program and procedures for using them.

A.1 PROGRAM DESCRIPTIONS

FRESP (Frequency Response)

FRESP is an interactive program that allows the user to do the following:

1. Find the roots of a polynomial.
2. Find the poles and zeros of a transfer function.
3. Determine the frequency response by using a Bode plot.

The transfer function is defined in terms of its numerator and denominator polynomial coefficients. Once a transfer function has been defined, the program computes its poles and zeros and displays them on the screen. The user may choose to send the result to the printer.

Once the poles and zeros of a transfer function have been computed, plots for the magnitude and phase versus frequency can be obtained. In order to obtain the plot in the frequency range of interest, the user specifies the upper and lower limits for the frequency (fhi and flo), phase angle (phi and plo), and magnitude (mhi and mlo) parameters. Once these parameters have been defined, only those limits that change on subsequent plots need be changed. The x-axis of the plot is logarithmic, and the number of decades per plot is automatically determined by using the upper and lower limits of the frequency range. Once the plots have been displayed on the screen, the user may select to have the results printed in a tabular form or in a graphic form as viewed on the screen.

Even though effort was made to make this program efficient and user-friendly, the program has the following limitations:

1. The maximum degree for any polynomial is 19.
2. The minimum degree for any polynomial is 0.
3. The maximum number of decades per plot is 5.
4. The printout of the plot can only be obtained by using the "Print Screen" option.

NUMCON (Numbers Conversion)

NUMCON is an interactive menu-driven program that allows the user to perform the following number conversions:

1. Decimal to binary, octal, or hexadecimal.
2. Binary to decimal.
3. Octal to decimal.
4. Hexadecimal to decimal.

Integers as well as floating-point numbers may be converted. The largest number that can be converted is 10E10 and the smallest is 10E − 10. Fractional parts are truncated to 16 places when converted.

A.2 SYSTEM CONFIGURATION

The two programs will run on an IBM PC, XT, AT, PS/2, or any other compatible PCs with 512K bytes of memory or more. The floating-point math coprocessor (8087, 80287, or 80387) is optional. However, if it is present, the programs will run at full speed; otherwise, they will run 5–15 times slower. Any color adaptor and color-display monitor may be used. Any printer may be used.

This software package runs under DOS version 3.0 or higher. It is recommended that the system be booted with a CONFIG.SYS file, which contains the following two statements:

Files = 10

Buffers = 10

A.3 SOFTWARE INSTALLATION

This section describes the procedure for installing the software package. This procedure allows the user to copy the **"COMMAND.COM"**, **"GRAPHICS.COM"**, and **"BASICA.COM"** from the user's DOS diskette, along with the two programs from the software diskette, to form a system disk. The installation is described for both a two-disk system and a hard-disk system.

Two-Disk System

To begin the two-drive installation, the following items are needed:

1. The software-package original diskette.
2. A DOS diskette.
3. Two blank diskettes.

The procedure is as follows:

1. Insert the DOS diskette in drive A and boot up the system.
2. When the DOS prompt **"A>"** appears, insert one of the blank diskettes into drive B.
3. To format the system disk and copy the **"COMMAND.COM"** type

   Format B:/S Press† **ENTER**

4. When formatting is complete, the following message will appear:

   Format another (Y/N)?

5. Type

   N Press **ENTER**

† "Press ENTER" is our way of saying "hit the ENTER key" on the keyboard.

6. To copy the **"GRAPHICS.COM"** and the **"BASICA.COM"**, type:

   Copy **GRAPHICS.COM** **b:*.*** Press **ENTER**
   Copy **BASICA.COM** **b:*.*** Press **ENTER**

7. Now, remove the DOS diskette from drive A, and insert the software-package original diskette in its place. At the **"A>"** prompt, type

   B: Press **ENTER**

8. To copy the two programs from the software-package original diskette, at the **"B>"** prompt, type

   Copy **A:*.*** **b:** Press **ENTER**

9. Remove the software-package original diskette from drive A and store it for safekeeping.
10. Insert the second blank disk in drive A, and at the **"B>"** prompt, type

    Format A: Press **ENTER**

11. When formatting is complete, type

    N Press **ENTER**

12. Remove the diskette from drive A and label it as the ‘‘DATA’’ diskette.
13. Remove the diskette from drive B and label it as the ‘‘SOFPAC’’ diskette.

Hard-Disk System

1. Boot up the system from the hard disk.
2. If your DOS prompt does not display the current drive and path, then type

   PROMPT PG Press **ENTER**

3. When the DOS prompt **"C:\>"** appears, create a new directory for the software package. We will call this directory **"SOFPAC"**. Type

   MD **SOFPAC** Press **ENTER**

4. To make **"SOFPAC"** the default directory, type

   CD **SOFPAC** Press **ENTER**

5. Now, insert the software-package original diskette in drive A, and at the **"C:\SOFPAC>"** prompt, type

   Copy A:*.* Press **ENTER**

6. Remove the software-package original diskette from drive A and store it for safekeeping.
7. Provide a path for **"BASICA.COM"** and **"GRAPHICS.COM"** utilities. For example, if these programs are in the **"C:\DOS"** subdirectory, then you would type

   Path **C:\DOS** Press **ENTER**

A.4 RUNNING SOFPAC

SOFPAC consists of two programs: FRESP and NUMCON. For a two-drive system, insert the "SOFPAC" diskette in drive A and "DATA" diskette in drive B. For a hard-disk system, make sure that the "SOFPAC" directory is the default directory. The remainder of this section will describe the step-by-step procedure for the execution of each of the two different programs.

FRESP (Frequency Response)

1. At the DOS prompt (i.e., >), type

   Graphics Press **ENTER**

2. Now, you are ready to run FRESP. Type

   Basica **FRESP** Press **ENTER**

3. The screen will be cleared and the following message will be displayed:

TRANSFER FUNCTION PROGRAM

MAIN MENU

This program will find the poles and zeros of a transfer function and draw its Bode plot.

Do you wish to continue (y/n)?

4. Type **Y** to begin a session.
5. Now, you should be prompted to enter the degree of the numerator, as follows:

TRANSFER FUNCTION PROGRAM

ENTRY SCREEN

Input the degree of the numerator?

6. Type an integer for the degree of the numerator and press **ENTER**.
7. Type the coefficients for the polynomial in descending order. For example, for $s^2 + 2s + 3$, you would type

   1 **2** **3** Press **ENTER**

8. If you entered the data in step 7 correctly, enter **Y** for the **"Is this correct (Y/N)?"** prompt; otherwise, enter **N** and edit the values as instructed.
9. Repeat steps 6, 7, and 8 for the degree and coefficients of the denominator.
10. Once the poles and zeros are computed and displayed, press any key to view the Bode plots.

11. The screen should be cleared, and a logarithmic chart is drawn. If you wish to obtain the Bode plots for the magnitude and the phase, press **ENTER**; otherwise, type **Y** to return to the main screen.
12. Assuming that you selected to view the Bode plots, you will be prompted to enter the values for **flo**, **fhi**, **mlo**, **mhi**, **plo**, and **phi**, where

    flo = lower limit of frequency range, rad/s
    fhi = upper limit of frequency range, rad/s
    mlo = lower limit of magnitude value, dB
    mhi = upper limit of magnitude value, dB
    plo = lower limit of phase angle, degrees
    phi = upper limit of phase angle, degrees

13. Once these values are entered, two graphs will be plotted. The cyan-colored curve represents the magnitude, and the white-colored curve represents the phase.
14. If you wish to obtain a printout of the magnitude and phase responses versus frequency in a tabular form, enter **Y** for the **"Print Table of Result (y/n)?"** prompt; otherwise, enter **N** and hit the **ENTER** key.
15. To obtain a printout of the Bode plots for the magnitude and phase, hit the **Print Screen** function key.
16. To return to the main menu, type **Y**.
17. To terminate this session, type **N**; and at the basic prompt **"OK"**, type

    SYSTEM Press **ENTER**

    Now, you should be back under the control of your DOS commands.

NUMCON (Numbers Conversion)

1. At the DOS prompt, type

    BASICA NUMCON Press **ENTER**

2. The screen will be cleared and the following menu will be displayed:

    1) Decimal to binary **5) Binary to decimal**
    2) Decimal to octal **6) Octal to decimal**
    3) Decimal to hexadecimal **7) Hexadecimal to decimal**
    4) Quit

    Enter number of choice and press ENTER

3. Enter your choice of conversion and follow the instructions.
4. To terminate this session from the main menu, type

    4 Press **ENTER**

5. The screen should be cleared, and when the basic prompt **"OK"** is displayed, type

   SYSTEM Press **ENTER**

   Now you should be back under the control of your DOS commands.

APPENDIX B

PSpice

B.0 INTRODUCTION

PSpice is an electric circuit simulator. It will calculate the behavior of electric circuits; that is, PSpice will calculate, print, and/or plot the response over time to different inputs, the response to different frequencies, and other information about a circuit. PSpice can be thought of as a "software breadboard." Using PSpice, we can perform the same measurements as we can with an actual circuit and many others that would not be feasible with a breadboard.

PSpice is the PC student version of the widely used SPICE (Simulation Program with Integrated Circuit Emphasis) program, which was developed at the University of California at Berkeley in the early 1970s. An educational demonstration version of PSpice can be obtained from

MicroSim Corp.
20 Fairbanks
Irving, Ca. 92719-9905

We provide in this appendix a brief but, for our purposes, adequate overview of PSpice.

B.1 CIRCUIT DESCRIPTION

The information required to describe and analyze a circuit is entered one statement per line in an ASCII text file. This file lists each active- and passive-circuit element, describes their interconnections using node numbers, specifies the type of analysis desired, and specifies the type of desired output data.

B.2 TYPES OF STATEMENTS

The types of PSpice statements that are required to specify the problem and the form of the solution may be subdivided into the following five categories:

1. Title and comments.
2. Data statements.
3. Command or control statements.
4. Output specification and format.
5. END statement.

A sample circuit and its complete PSpice program are shown in Fig. B.2-1. This program computes and prints the voltage at node 1, the voltage drop between nodes 1 and 2, and the loop current. This example is provided to illustrate the different statements, which are discussed in the following subsections.

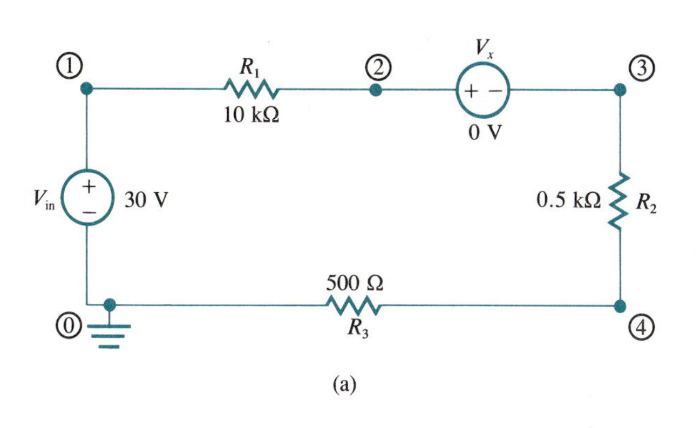

(a)

```
DC analysis sample PSpice program.

* This line is a comments line; the line above is
* a title statement.

R1    1   2   10k
R2    3   4   .5k
R3    4   0   .5k
Vin   1   0    30
Vx    2   3     0

* The above 5 statements are data statements
* that describe the circuit.

.PRINT DC V(1), V(1,2), I(Vx)

* The above statement prints the voltage at node
* 1, the voltage between nodes 1 and 2, and the
* current flowing through Vx.

.END
```

(b)

FIG. B.2-1.
(a) A sample circuit and (b) a corresponding sample PSpice program.

Title and Comment Statements

Every PSpice program must begin with a title statement that starts in column 1 of the first line of the file (a blank line may be used as a title). If the title requires more than one line, a second line can be used provided that the second line starts with a "+" in column 1. The title can be any user-defined text and will appear in the output.

PSpice ignores all lines that start with an asterisk (*). Therefore, asterisks are used to include comments. Comments are very important in documenting our programs. Comments will also appear as part of the output listing. In addition, a semicolon (;) may be used to insert a comment on any line. PSpice ignores everything to the right of a semicolon on any given line.

Data Statements

PSpice writes and solves the nodal equations of a circuit. Each active and passive element of the circuit is specified by using two nodes. The nodes are arbitrarily numbered by the user, except that node 0 is always the reference node to which the voltage of other nodes are referenced.

Circuit passive elements such as resistors (*R*), capacitors (*C*), and inductors (*L*) are specified between nodes by using the following general forms:

```
RXXXXXXX    n+    n-    value
CXXXXXXX    n+    n-    value <IC = init_val>
LXXXXXXX    n+    n-    value <IC = init_val>
```

where

XXXXXXX = any seven alphanumeric characters

n+ = positive assigned direction node

n− = negative assigned direction node

value = nonzero value of the element

init_val = nonzero initial-voltage or -current values for the capacitor or the inductor, respectively

The "<" and ">" are used to indicate that the enclosed parameter is optional and need not be specified.

The units for *R*, *C*, and *L* are ohm, farad, and henry, respectively. The value of the element may be an integer, a floating-point number, a floating-point number followed by exponent, or one of the following scale factors:

Scale Factor	Value
T	1E12
G	1E09
M	1E06
k	1E03
m	1E-3
u	1E-6
n	1E-9
p	1E-12
f	1E-15

Circuit active elements (i.e., sources) such as an independent voltage source (*V*), independent current source (*I*), voltage-dependent voltage source (*E*), current-dependent voltage source (*H*), voltage-dependent current source (*G*), and current-dependent current source (*F*) are also specified between nodes, using the following general forms:

VXXXXXXX n+ n− <DC value> <AC acmag acphase>
IXXXXXXX n+ n− <DC value> <AC acmag acphase>
EXXXXXXX n+ n− cn+ cn− value
GXXXXXXX n+ n− cn+ cn− value
FXXXXXXX n+ n− cvname value
HXXXXXXX n+ n− cvname value

where

DC = dc-type source

AC = ac-type source

acmag = magnitude of the ac-type source

acphase = phase of the ac-type source, degrees

cn+ = positive node of the controlling voltage source

cn− = negative node of the controlling voltage source

cvname = name of the voltage source through which the controlling current flows

The type of the independent voltage and current sources could be either dc or ac or both. For dc sources, the value could optionally be preceded by "DC". For ac sources, the values for the magnitude and the phase will go after "AC".

Fig. B.2-2 shows circuit elements in different node topologies with their equivalent PSpice data statements.

FIG. B.2-2. Different PSpice elements' specifications.

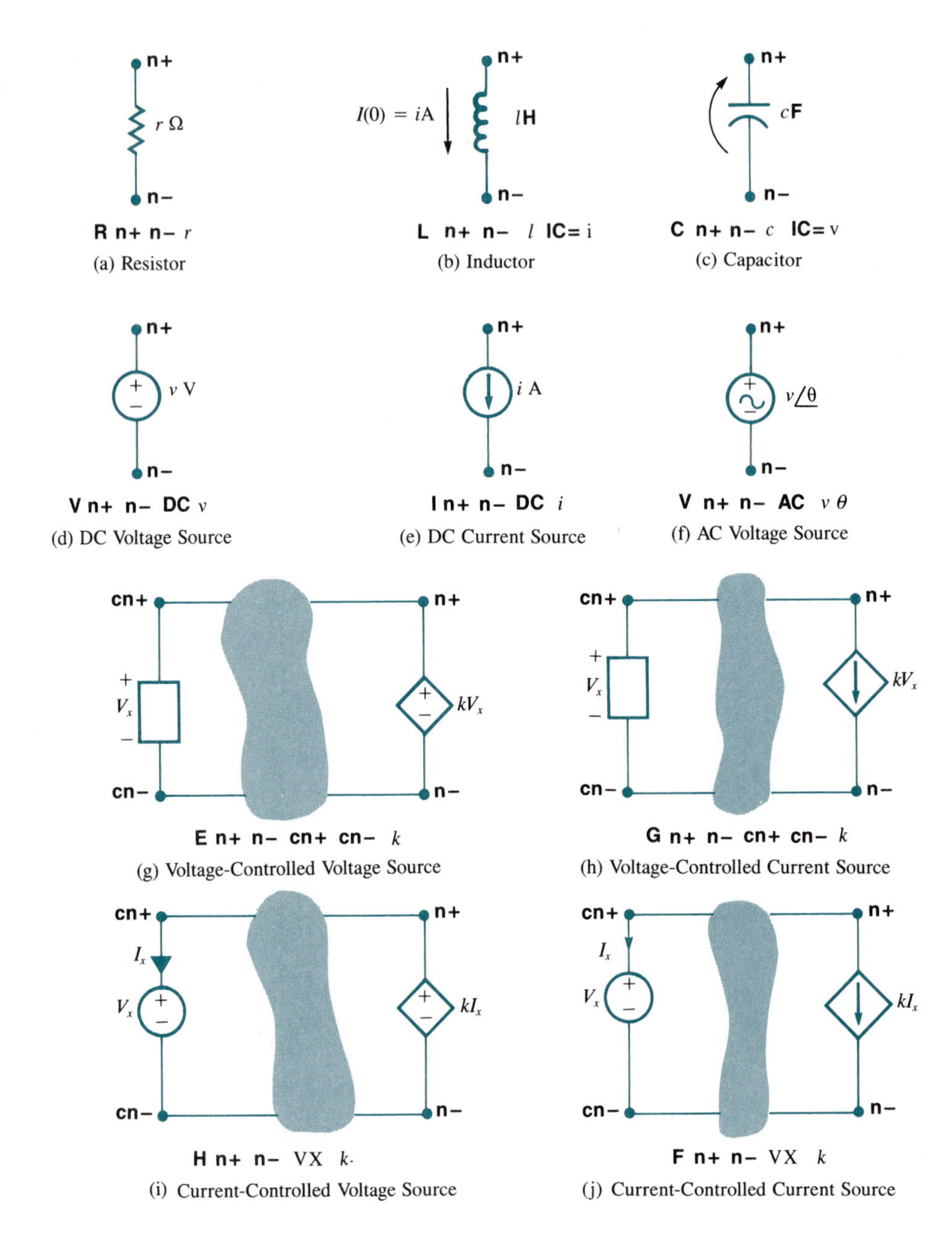

Command (Control) Statements

Once a circuit has been described by using the data statements, the command statements are used to specify the type of analysis to be performed on the given circuit. All PSpice command statements are preceded by a period (.). Even though

PSpice can perform numerous analysis types, we will limit our discussion only to dc, ac, and transient analyses.

DC Analysis

Dc analysis is the default analysis, which computes the dc node voltages and branch currents. In addition, PSpice can perform this analysis over a range of values for different dc voltage and current sources. This command statement has the following format:

.DC s_name init_value fin_value step

where

s_name = voltage- or current-input source name

init_value = initial value of the input source, V or A

fin_value = final value of the input source, V or A

step = incremental step value of the input source, V or A

For example, the command statement

.DC VX 10 15 1

will compute the dc node voltages and branch currents for all values of VX between 10 and 15 V, in increments of 1 V.

AC Analysis

Ac analysis is the frequency-response analysis that computes all the ac node voltages and branch currents over a range of frequencies. This command statement has the following format:

.AC variation_type num_pts init_freq fin_freq

In the above statement, ''variation_type'' specifies how the frequency of the input source is to be varied. The frequency variation can be either linear (LIN), octave† (OCT), or decade‡ (DEC). For linear variation, the number of points, plotted between the initial frequency (init_freq) in hertz and the final frequency (fin_freq) in hertz, will be equal to the value given by num_pts. For octave variation, num_pts specifies the number of points to be plotted per octave. For decade, the num_pts specifies the number of points to be plotted per decade. For example, the statement:

.AC LIN 4 100 1000

---

† A twofold increase in frequency.

‡ A tenfold increase in frequency.

will compute the ac node voltages and branch currents at frequencies of 100, 400, 700, and 1000 Hz. The statement

.AC DEC 4 100 1000

will compute the ac node voltages and branch currents at frequencies of 100, 178, 316, 562, and 1000 Hz. Finally, the statement

.AC OCT 4 100 1000

will compute the ac node voltages and branch currents at frequencies of 100, 119, 141, 168, 200, 238, 283, 336, . . . , 800, and 951 Hz.

Transient Analysis

Transient analysis is the time response of the circuit. In this analysis, one or more time-varying input sources may be specified, and the voltages and currents of the circuit's time response are calculated. The transient-analysis command statement specifies the time interval over which the transient analysis is performed. This command statement has the following format:

.TRANS step t_stop t_start **<UIC>**

where

step = time interval between successive time values, s

t_stop = stop time, s

t_start = start time, s

<UIC> = use initial condition flag

The transient analysis will always be performed from $t = 0$ to t_stop; however, only the values between t_start and t_stop will be produced as outputs. The UIC is an optional flag that, when specified, instructs PSpice to use the initial conditions of the inductors and the capacitors in the transient analysis. For example, the command statement

.TRANS .005 .1 .05 UIC

will evaluate the transient analysis by using the initial conditions of the capacitors and the inductors. The transient analyses evaluated between $t = 0.05$ s and $t = 0.1$ s will be printed or plotted at 0.005-s intervals.

Output Specification and Format

Once the circuit has been described and the desired analyses have been designated, different outputs can be specified. The output can be a single value, a tabular listing, or a plot. The command statement, for either a single value or a tabular-listing printout, has the following format:

.PRINT analy_type OUT1 OUT2 OUT3 . . . OUT8

where analy_type is the type of analysis (DC, AC, or TRANS) and OUT1, OUT2, OUT3, . . . , OUT8 are the output variables that can be specified. The output variables can represent node voltages (*V*) or branch currents (*I*). The dc node voltage has the following format:

V(n+, n−)

which specifies the voltage drop between n+ and n−. If n− and the preceding comma are omitted, n− is assumed to be the reference (i.e., ground) node. The dc branch current has the following format:

I(VXXXXXXX)

where VXXXXXXX is the name of the independent voltage source through which the current *I* flows. For example, the print statement

.PRINT DC V(5) V(2, 4) I(VIN)

will print the dc voltage at node 5 with respect to ground, the dc voltage drop between nodes 2 and 4, and the dc current flowing through the voltage source VIN.

For ac analysis, five different outputs can be specified. These outputs can be accessed by appending one of the following prefixes to the letter V or I in the output statement:

Prefix	Type
R	Real part
I	Imaginary part
M	Magnitude
P	Phase (in degrees)
DB	20 log(magnitude) or dB

For example, the statement

.PRINT AC VR(5) VI(5) IM(VX) IP(VX)

will print the real and imaginary values of the voltage at node 5 and the magnitude and phase angle of the current flowing through the voltage source VX.

The format for the plot statement is exactly the same as that for the print statement. We simply use PLOT instead of PRINT. For example, the PLOT statement

.PLOT AC VM(5) IM(VX)

will plot the magnitude of the voltage at node 5 and the magnitude of the current flowing through VX.

Finally, the width of the outputs can be controlled by using the WIDTH command, which has the following statement:

.WIDTH **OUT**=XX

where XX equals the number of columns desired. You can use either 80 or 132 columns. The 80-column format is best when you are viewing plots on a terminal screen. The 132-column format is best when a hard copy of the output is desired. However, the printer must be set up for 17 CPI (characters per inch) when you use the 132-column format.

The END Statement

Every PSpice program must end with an END statement, which signals the end of input data that describes the circuit and the type of analyses and outputs desired. The END statement has the following format:

.END

An input file may contain more than one circuit description and its desired analyses. However, each circuit and its commands are marked by the END statement.

B.3 TIME-VARYING INDEPENDENT SOURCES

The general form for the independent-voltage-source or -current-source data statement has the following form:

Name n+ n− <DC value> <AC value> <transient value>

where Name = VXXXXXXX or IXXXXXXX

The dc value, when specified, will be used for the dc-operating-point analysis and the dc sweep. Similarly, the ac value will be used for ac analysis. The default values for both dc and ac values are 0s. The transient value, when specified, will be used only for transient analysis. The transient value has different forms for different time-varying input sources. We describe here the formats of data statements for *pulse-*, *piecewise-linear*, *sinusoidal-*, and *exponential*-waveform input sources.

Pulse-Waveform Source

The pulse waveform for a voltage source V_{in} is shown in Fig. B.3-1(a), and its PSpice data statement has the following format:

Name n+ n− PULSE (V_i V_p T_d T_r T_f T_w PER)

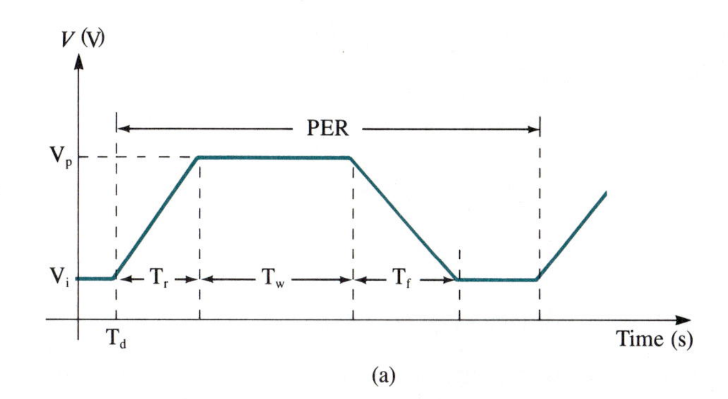

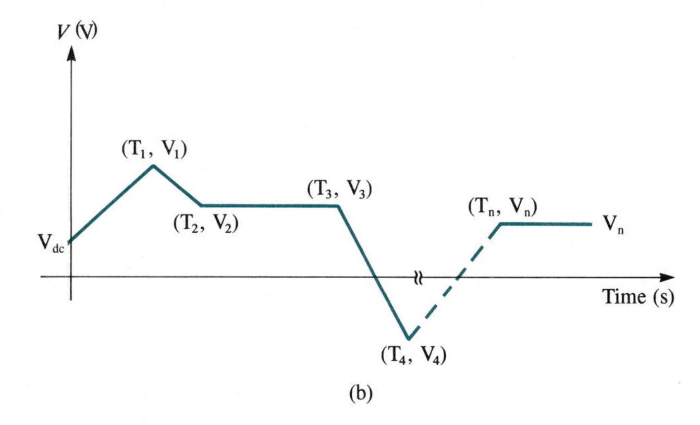

FIG. B.3-1. Independent-source waveforms: (a) pulse and (b) piecewise linear.

where

$$V_i = \text{initial voltage, V}$$
$$V_p = \text{value of source during pulse, V}$$
$$T_d = \text{delay between } t = 0 \text{ and start of pulse, s}$$
$$T_r = \text{rise time, s}$$
$$T_f = \text{fall time, s}$$
$$T_w = \text{pulse width, s}$$
$$\text{PER} = \text{period, s}$$

The terms T_r and T_f must not be zero. For example, the statement

```
VIN     2     3     PULSE (0 5 0 1n 1n 5m 7m)
```

describes a square-wave voltage source V_{in} that is connected between nodes 2 and 3 and has the following parameters:

$$\text{Initial voltage} = 0 \text{ V}$$
$$\text{Pulsed voltage} = 5 \text{ V}$$

$$\text{Delay time} = 0 \text{ s}$$
$$\text{Rise time} = 1 \text{ ns}$$
$$\text{Fall time} = 1 \text{ ns}$$
$$\text{Pulse width} = 5 \text{ ms}$$
$$\text{Signal period} = 7 \text{ ms}$$

Piecewise-Linear-Waveform Source

The piecewise-linear waveform for a voltage source V_{in} is shown in Fig. B.3-1(b), and its PSpice data statement has the following format:

VIN n+ n− PWL (T_1 V_1 T_2 V_2 · · · T_n V_n)

where $T_1 < T_2 < \cdots < T_n$ and each pair of values (T_i, V_i), $1 \leq i \leq n$, specifies the value of the source to be V_i at time T_i. Values between T_i and T_{i+1} are computed by using linear interpolation.

Sine-Waveform Source

The general equation for an exponentially damped sine waveform for a voltage source V_{in} is as follows:

$$V_{in}(t) = V_{off} + V_p \sin\left(2\pi f(t - T_d) - \frac{\theta}{360}\right)e^{-(t-T_d)\mathrm{DF}}$$

where

$$V_{off} = \text{offset voltage, V}$$
$$V_p = \text{peak amplitude, V}$$
$$f = \text{frequency, Hz}$$
$$T_d = \text{delay between } t = 0 \text{ and start of signal, s}$$
$$\mathrm{DF} = \text{damping factor, s}^{-1}$$
$$\theta = \text{phase shift, degrees}$$

The PSpice data statement for the above equation is

VIN n+ n− SIN (V_{off} V_p f T_d DF θ)

For example, the statement

VIN 2 3 SIN (0 .5 100k 1m .5 45)

describes a sine-waveform voltage source that is connected between nodes 2 and 3. This voltage starts at 0 V and remains at that level for 1 ms. Then the voltage becomes an exponentially damped sine waveform with $V_p = 0.5$ V, $f = 100$ kHz, DF $= 0.5$ s^{-1}, and $\theta = 45°$.

Exponential-Waveform Source

The general format of a PSpice data statement for an exponential-waveform voltage source VIN is as follows:

VIN n+ n− EXP (V_i V_p T_r T_{rc} T_f T_{fc})

where

$$V_i = \text{initial voltage, V}$$
$$V_p = \text{peak voltage, V}$$
$$T_r = \text{rise delay time, s}$$
$$T_{rc} = \text{rise time constant, s}$$
$$T_f = \text{fall delay time, s}$$
$$T_{fc} = \text{fall time constant, s}$$

For example, the statement

VIN 2 3 EXP (0 .5 .2u 30u 60u 40u)

describes a voltage source V_{in} that is connected between nodes 2 and 3. This voltage starts at 0 V and remains at 0 for 0.2 μs. At $t = T_r = 0.2$ μs, the voltage starts to increase exponentially with a time constant T_{rc} of 30 μs to a peak value V_p of 0.5 V. At $t = T_f = 60$ μs, the voltage starts to decrease exponentially with a time constant T_{fc} of 40 μs.

B.4 EXAMPLES

In this section, we present complete examples for each type of the PSpice analyses discussed in the previous sections. These examples have been completely tested and the output listings from the computer are included.

EXAMPLE B.4-1

DC Analysis

Write a PSpice program to determine the output voltage V_o and the loop current I for the circuit shown in Fig. P3-31.

Solution

The circuit is redrawn in Fig. B.4-1 and is labeled arbitrarily for PSpice analysis. Observe that since the current I through the 1500-Ω resistor is one of the desired output variables, a 0-V voltage source is placed in series with the 1500 Ω. This 0-V

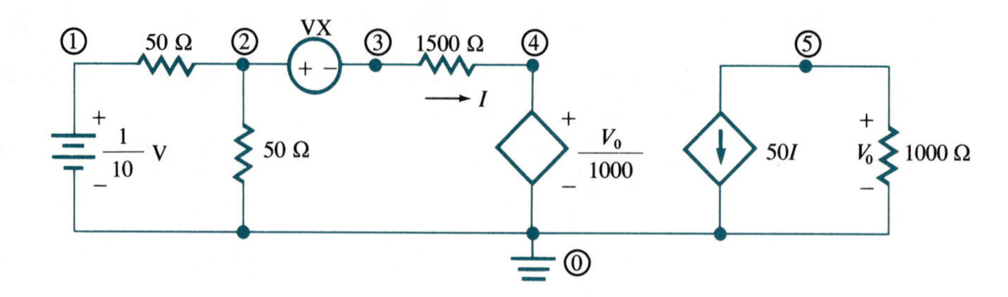

FIG. B.4-1. The circuit in Fig. P3-31 redrawn for PSpice analysis.

source is used as an ammeter. The PSpice program is as follows:

```
DC analysis example
*Calculate the output voltage Vo and the loop current I for the
*circuit given in Fig. B.4-1.
*
*Circuit Description
*
V1    1    0    DC      .1
R1    1    2    50
R2    2    0    50
VX    2    3    DC      0
R3    3    4    1500
E     4    0    5       0     0.001
F     5    0    VX      50
RL    5    0    1000
*
*Command Statements
*
.DC        V1     .1     .1     1
.PRINT DC     V(5)     I(VX)
.WIDTH=132
.END
```

The output listing produced by the above program includes the following:

```
**************************************************************************
V1              V(5)            I(VX)
1.00E-01        -1.695E+00      3.390E-05
```

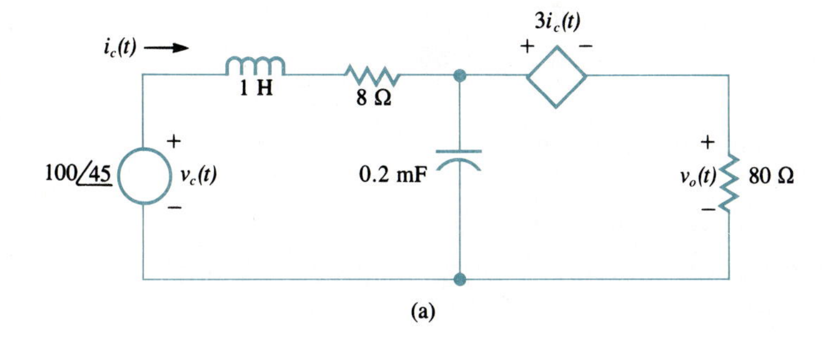

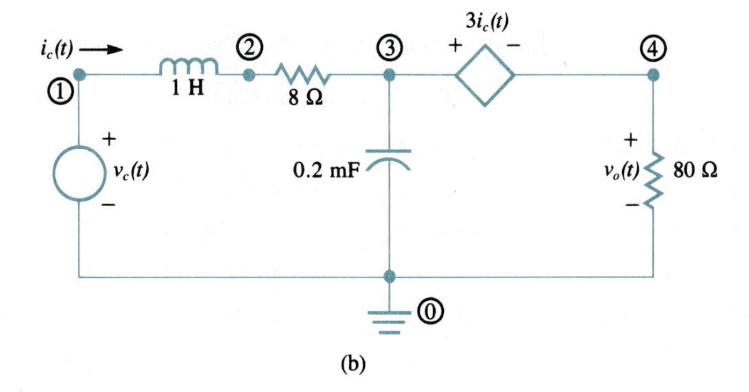

FIG. B.4-2. An ac analysis example: (a) the original circuit and (b) the circuit for PSpice.

EXAMPLE B.4-2

AC Analysis

Write a PSpice program to determine the magnitudes and phase angles of both the output voltage $v_o(t)$ and the source current $i_c(t)$ for the circuit of Fig. B.4-2(a). The voltage source of this circuit is an ac source with a frequency f that varies linearly from 60 to 70 Hz. Produce a listing for different frequencies.

Solution

The circuit is redrawn for PSpice analysis in Fig. B.4-2(b). The PSpice program is as follows:

```
AC ANALYSIS EXAMPLE
*
*Circuit Description
*
VC    1    0    AC      100    45
L     1    2    1
R     2    3    8
C     3    0    0.2m
```

```
H       3    4    VC    3
RL      4    0    80
*
*Command Statements
*
.AC     LIN    10    60    70
.PRINT AC    VM(4)    VP(4)    IM(VC)    IP(VC)
.WIDTH=132
.END
```

The output listing produced by the above program includes the following:

FREQ	VM(4)	VP(4)	IM(VC)	IP(VC)
6.000E+01	7.956E−01	−4.403E+01	2.652E−01	1.362E+02
6.111E+01	7.812E−01	−4.405E+01	2.604E−01	1.362E+02
6.222E+01	7.672E−01	−4.406E+01	2.557E−01	1.362E+02
6.333E+01	7.538E−01	−4.408E+01	2.513E−01	1.362E+02
6.444E+01	7.408E−01	−4.410E+01	2.469E−01	1.361E+02
6.556E+01	7.282E−01	−4.411E+01	2.427E−01	1.361E+02
6.667E+01	7.161E−01	−4.413E+01	2.387E−01	1.361E+02
6.778E+01	7.044E−01	−4.414E+01	2.348E−01	1.361E+02
6.889E+01	6.930E−01	−4.415E+01	2.310E−01	1.361E+02
7.000E+01	6.820E−01	−4.417E+01	2.273E−01	1.360E+02

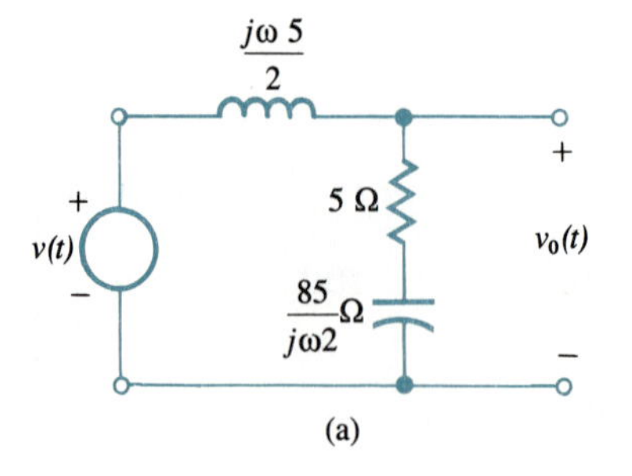

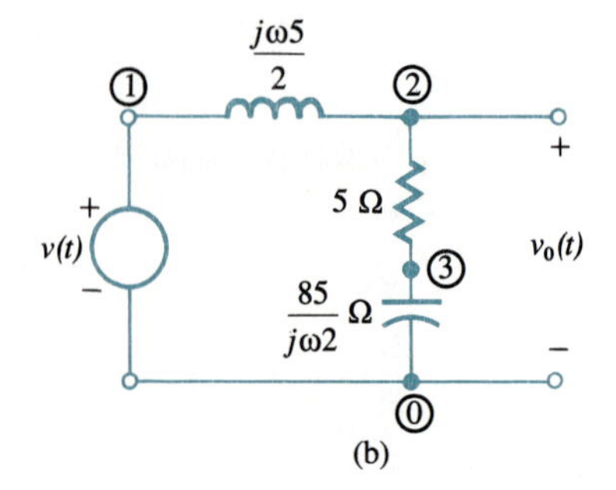

FIG. B.4-3. A transient-analysis example: (a) the original circuit, (b) the circuit for PSpice, and (c) the PSpice plot of V(2).

```
TRANSIENT ANALYSIS EXAMPLE
• TRANSIENT ANALYSIS                    TEMPERATURE = 27.000 DEG C

TIME          V(2)  .000D+00    5.000D-01    1.000D+00    1.500D+00    2.000D+00
.000D+00     .000D+00
1.000D-01    1.193D-01
2.000D-01    4.138D-01
3.000D-01    7.343D-01
4.000D-01    1.032D+00
5.000D-01    1.278D+00
6.000D-01    1.445D+00
7.000D-01    1.516D+00
8.000D-01    1.489D+00
9.000D-01    1.399D+00
1.000D+00    1.269D+00
1.100D+00    1.121D+00
1.200D+00    9.783D-01
1.300D+00    8.620D-01
1.400D+00    7.913D-01
1.500D+00    7.627D-01
1.600D+00    7.722D-01
1.700D+00    8.127D-01
1.800D+00    8.745D-01
1.900D+00    9.458D-01
2.000D+00    1.012D+00
2.100D+00    1.064D+00
2.200D+00    1.098D+00
2.300D+00    1.112D+00
2.400D+00    1.108D+00
2.500D+00    1.088D+00
2.600D+00    1.058D+00
2.700D+00    1.025D+00
2.800D+00    9.942D-01
2.900D+00    9.693D-01
3.000D+00    9.527D-01
3.100D+00    9.462D-01
3.200D+00    9.497D-01
3.300D+00    9.595D-01
3.400D+00    9.733D-01
3.500D+00    9.887D-01
3.600D+00    1.003D+00
3.700D+00    1.015D+00
3.800D+00    1.022D+00
3.900D+00    1.025D+00
4.000D+00    1.023D+00
4.100D+00    1.019D+00
4.200D+00    1.012D+00
```

(c)

EXAMPLE B.4-3

Transient Analysis

For the circuit shown in Fig. B.4-3(a), write a PSpice program that will plot the output voltage $v_o(t)$ in 0.1-s intervals over a 6-s range starting at $t = 0$ s. The input voltage for this circuit is a unit step function.

Solution

A unit-step-function source can be represented in PSpice as a pulse waveform with a large pulse width. The circuit is redrawn for PSpice analysis in Fig. B.4-3(b). The PSpice program for plotting the output voltage is as follows:

```
TRANSIENT ANALYSIS EXAMPLE
*
*Circuit Description
*
VIN     1     0     PULSE (0 1 0 0 0 950 1000)
R1      2     3     5
L       1     2     2.5
C       3     0     0.0235
*
*Command Statements
*
.TRAN         .1    6
.WIDTH=132
.PLOT         TRAN V(2)
.END
```

The output produced by the above program includes the plot shown in Fig. B.4-3(c). These results may also be compared with Fig. 5.2-3.

APPENDIX C

Useful Mathematical Quantities

C.1 TRIGONOMETRIC IDENTITIES

$$\cos(x \pm y) = \cos(x)\cos(y) \mp \sin(x)\sin(y) \quad \text{(C-1)}$$

$$\sin(x \pm y) = \sin(x)\cos(y) \pm \cos(x)\sin(y) \quad \text{(C-2)}$$

$$\cos\left(x \pm \frac{\pi}{2}\right) = \mp \sin(x) \quad \text{(C-3)}$$

$$\sin\left(x \pm \frac{\pi}{2}\right) = \pm \cos(x) \quad \text{(C-4)}$$

$$\cos(2x) = \cos^2(x) - \sin^2(x) \quad \text{(C-5)}$$

$$\sin(2x) = 2\sin(x)\cos(x) \quad \text{(C-6)}$$

$$2\cos(x) = e^{jx} + e^{-jx} \quad \text{(C-7)}$$

$$2j\sin(x) = e^{jx} - e^{-jx} \quad \text{(C-8)}$$

$$2\cos(x)\cos(y) = \cos(x-y) + \cos(x+y) \quad \text{(C-9)}$$

$$2\sin(x)\sin(y) = \cos(x-y) - \cos(x+y) \quad \text{(C-10)}$$

$$2\sin(x)\cos(y) = \sin(x-y) + \sin(x+y) \quad \text{(C-11)}$$

$$2\cos^2(x) = 1 + \cos(2x) \quad \text{(C-12)}$$

$$2\sin^2(x) = 1 - \cos(2x) \quad \text{(C-13)}$$

$$4\cos^3(x) = 3\cos(x) + \cos(3x) \quad \text{(C-14)}$$

$$4\sin^3(x) = 3\sin(x) - \sin(3x) \tag{C-15}$$

$$8\cos^4(x) = 3 + 4\cos(2x) + \cos(4x) \tag{C-16}$$

$$8\sin^4(x) = 3 - 4\cos(2x) + \cos(4x) \tag{C-17}$$

$$A\cos(x) - B\sin(x) = R\cos(x + \theta) \tag{C-18}$$

where

$$R = \sqrt{A^2 + B^2} \tag{C-19a}$$

$$\theta = \tan^{-1}\frac{B}{A} \tag{C-19b}$$

$$A = R\cos\theta \tag{C-19c}$$

$$B = R\sin\theta \tag{C-19d}$$

C.2 INDEFINITE INTEGRALS

Rational Algebraic Functions

$$\int (a + bx)^n\,dx = \frac{(a + bx)^{n+1}}{b(n + 1)} \qquad 0 < n \tag{C-20}$$

$$\int \frac{dx}{a + bx} = \frac{1}{b}\ln|a + bx| \tag{C-21}$$

$$\int \frac{dx}{(a + bx)^n} = \frac{-1}{(n - 1)(b)(a + bx)^{n-1}} \qquad 1 < n \tag{C-22}$$

$$\begin{aligned} \int \frac{dx}{c + bx + ax^2} &= \frac{2}{\sqrt{4ac - b^2}}\tan^{-1}\left(\frac{2ax + b}{\sqrt{4ac - b^2}}\right) && b^2 < 4ac \\ &= \frac{1}{\sqrt{b^2 - 4ac}}\ln\left|\frac{2ax + b - \sqrt{b^2 - 4ac}}{2ax + b + \sqrt{b^2 - 4ac}}\right| && b^2 > 4ac \\ &= \frac{-2}{2ax + b} && b^2 = 4ac \end{aligned} \tag{C-23}$$

$$\int \frac{x\,dx}{c + bx + ax^2} = \frac{1}{2a}\ln|ax^2 + bx + c| - \frac{b}{2a}\int \frac{dx}{c + bx + ax^2} \tag{C-24}$$

$$\int \frac{dx}{a^2 + b^2x^2} = \frac{1}{ab}\tan^{-1}\frac{bx}{a} \tag{C-25}$$

$$\int \frac{x\,dx}{a^2 + x^2} = \frac{1}{2}\ln(a^2 + x^2) \tag{C-26}$$

$$\int \frac{x^2\,dx}{a^2 + x^2} = x - a \tan^{-1}\frac{x}{a} \tag{C-27}$$

$$\int \frac{dx}{(a^2 + x^2)^2} = \frac{x}{2a^2(a^2 + x^2)} + \frac{1}{2a^3}\tan^{-1}\frac{x}{a} \tag{C-28}$$

$$\int \frac{x\,dx}{(a^2 + x^2)^2} = \frac{-1}{2(a^2 + x^2)} \tag{C-29}$$

$$\int \frac{x^2\,dx}{(a^2 + x^2)^2} = \frac{-x}{2(a^2 + x^2)} + \frac{1}{2a}\tan^{-1}\frac{x}{a} \tag{C-30}$$

$$\int \frac{dx}{(a^2 + x^2)^3} = \frac{x}{4a^2(a^2 + x^2)^2} + \frac{3x}{8a^4(a^2 + x^2)} + \frac{3}{8a^5}\tan^{-1}\frac{x}{a} \tag{C-31}$$

$$\int \frac{x^2\,dx}{(a^2 + x^2)^3} = \frac{-x}{4(a^2 + x^2)^2} + \frac{x}{8a^2(a^2 + x^2)} + \frac{1}{8a^3}\tan^{-1}\frac{x}{a} \tag{C-32}$$

$$\int \frac{x^4\,dx}{(a^2 + x^2)^3} = \frac{a^2x}{4(a^2 + x^2)^2} - \frac{5x}{8(a^2 + x^2)} + \frac{3}{8a}\tan^{-1}\frac{x}{a} \tag{C-33}$$

$$\int \frac{dx}{(a^2 + x^2)^4} = \frac{x}{6a^2(a^2 + x^2)^3} + \frac{5x}{24a^4(a^2 + x^2)^2} + \frac{5x}{16a^6(a^2 + x^2)} + \frac{5}{16a^7}\tan^{-1}\frac{x}{a} \tag{C-34}$$

$$\int \frac{x^2\,dx}{(a^2 + x^2)^4} = \frac{-x}{6(a^2 + x^2)^3} + \frac{x}{24a^2(a^2 + x^2)^2} + \frac{x}{16a^4(a^2 + x^2)} + \frac{1}{16a^5}\tan^{-1}\frac{x}{a} \tag{C-35}$$

$$\int \frac{x^4\,dx}{(a^2 + x^2)^4} = \frac{a^2x}{6(a^2 + x^2)^3} - \frac{7x}{24(a^2 + x^2)^2} + \frac{x}{16a^2(a^2 + x^2)} + \frac{1}{16a^3}\tan^{-1}\frac{x}{a} \tag{C-36}$$

$$\int \frac{dx}{a^4 + x^4} = \frac{1}{4a^3\sqrt{2}}\ln\left(\frac{x^2 + ax\sqrt{2} + a^2}{x^2 - ax\sqrt{2} + a^2}\right) + \frac{1}{2a^3\sqrt{2}}\tan^{-1}\left(\frac{ax\sqrt{2}}{a^2 - x^2}\right) \tag{C-37}$$

$$\int \frac{x^2\,dx}{a^4 + x^4} = -\frac{1}{4a\sqrt{2}}\ln\left(\frac{x^2 + ax\sqrt{2} + a^2}{x^2 - ax\sqrt{2} + a^2}\right) + \frac{1}{2a\sqrt{2}}\tan^{-1}\left(\frac{ax\sqrt{2}}{a^2 - x^2}\right) \tag{C-38}$$

Trigonometric Functions

$$\int \cos(x)\,dx = \sin(x) \tag{C-39}$$

$$\int x\cos(x)\,dx = \cos(x) + x\sin(x) \tag{C-40}$$

$$\int x^2\cos(x)\,dx = 2x\cos(x) + (x^2 - 2)\sin(x) \tag{C-41}$$

$$\int \sin(x)\,dx = -\cos(x) \tag{C-42}$$

$$\int x \sin(x)\, dx = \sin(x) - x \cos(x) \tag{C-43}$$

$$\int x^2 \sin(x)\, dx = 2x \sin(x) - (x^2 - 2) \cos(x) \tag{C-44}$$

Exponential Functions

$$\int e^{ax}\, dx = \frac{e^{ax}}{a} \qquad a \text{ real or complex} \tag{C-45}$$

$$\int x e^{ax}\, dx = e^{ax}\left(\frac{x}{a} - \frac{1}{a^2}\right) \qquad a \text{ real or complex} \tag{C-46}$$

$$\int x^2 e^{ax}\, dx = e^{ax}\left(\frac{x^2}{a} - \frac{2x}{a^2} + \frac{2}{a^3}\right) \qquad a \text{ real or complex} \tag{C-47}$$

$$\int x^3 e^{ax}\, dx = e^{ax}\left(\frac{x^3}{a} - \frac{3x^2}{a^2} + \frac{6x}{a^3} - \frac{6}{a^4}\right) \qquad a \text{ real or complex} \tag{C-48}$$

$$\int e^{ax} \sin(x)\, dx = \frac{e^{ax}}{a^2 + 1}[a \sin(x) - \cos(x)] \tag{C-49}$$

$$\int e^{ax} \cos(x)\, dx = \frac{e^{ax}}{a^2 + 1}[a \cos(x) + \sin(x)] \tag{C-50}$$

C.3 DEFINITE INTEGRALS

$$\int_{-\infty}^{\infty} e^{-a^2x^2 + bx}\, dx = \frac{\sqrt{\pi}}{a} e^{b^2/(4a^2)} \qquad a > 0 \tag{C-51}$$

$$\int_0^{\infty} x^2 e^{-x^2}\, dx = \frac{\sqrt{\pi}}{4} \tag{C-52}$$

$$\int_0^{\infty} Sa(x)\, dx = \int_0^{\infty} \frac{\sin(x)}{x}\, dx = \frac{\pi}{2} \tag{C-53}$$

$$\int_0^{\infty} Sa^2(x)\, dx = \frac{\pi}{2} \tag{C-54}$$

C.4 FINITE SERIES

$$\sum_{n=1}^{N} n = \frac{N(N + 1)}{2} \tag{C-55}$$

$$\sum_{n=1}^{N} n^2 = \frac{N(N+1)(2N+1)}{6} \tag{C-56}$$

$$\sum_{n=1}^{N} n^3 = \frac{N^2(N+1)^2}{4} \tag{C-57}$$

$$\sum_{n=0}^{N} x^n = \frac{x^{N+1}-1}{x-1} \tag{C-58}$$

$$\sum_{n=0}^{N} \frac{N!}{n!(N-n)!} x^n y^{N-n} = (x+y)^N \tag{C-59}$$

$$\sum_{n=0}^{N} e^{j(\theta+n\phi)} = \frac{\sin[(N+1)\phi/2]}{\sin(\phi/2)} e^{j[\theta+(N\phi/2)]} \tag{C-60}$$

$$\sum_{n=0}^{N} \binom{N}{n} = \sum_{n=0}^{N} \frac{N!}{n!(N-n)!} = 2^N \tag{C-61}$$

Answers to Selected Problems

CHAPTER 1

1-1. $i = 0.87$ A

1-4. $1.092 \times 10^{20} |\cos(50t)|$ electrons per second.

1-8. $F = -1.08 \times 10^7$ N

1-13. $P_{av} = 1404$ W

1-16. $I = 9.333 \times 10^{-3}$ A

1-22. $Q = 0.01$ C

1-26. $Q = 102.04$ μC

1-29. $L = 4.2$ cm

1-33. $L = 2.8$ mH

1-38. $k = 0.559$

1-44. 7 branches, 5 nodes, 7 loops, and 3 meshes

CHAPTER 2

2-2. Rate = 2 V/s

2-6. $P_{av} = \frac{A^2}{2R} + \frac{B^2}{2R}$; yes

2-13. **(a)** $\cos 18$

(b) $\frac{1}{22}$

(c) $e^{-4} \cos 9$

(d) $0.5e^{-7}$

2-18. $W = \alpha(\sqrt{2} - 1)^{1/6}$

2-21. Nyquist rate = 40 MHz

CHAPTER 3

3-2. $V_{ab} = 103$ V, with terminal a most positive

3-5. $I_1 = 3$ A, $I_2 = 5$ A, $I_3 = -2$ A, $V_1 = 54$ V, $V_2 = 125$ V, and $V_3 = -16$ V; both batteries provide power

3-9. $V_1 = \frac{16}{13}$ V, $V_2 = \frac{-510}{39}$ V, $I_{4\Omega} = \frac{4}{13}$ A, $I_{4\Omega} = \frac{4}{13}$ A, $I_{10\Omega} = \frac{-51}{39}$ A, $I_{10\Omega} = \frac{-51}{39}$ A, and $I_{6\Omega} = \frac{31}{13}$ A

3-13. $I_1 = \frac{33}{54}$ A and $I_2 = \frac{14}{45}$ A

3-15. $V_1 = \frac{1440}{109}$ V

3-16. $V_0 = (15{,}310/803)V_a + (16{,}695/5621)V_b$

3-20. Dc current source of 4 A in parallel with $\frac{8}{3}$ Ω

3-28. $I_1 = \frac{-21}{52}$ A, $I_2 = \frac{-165}{442}$ A, and $I_3 = \frac{87}{442}$ A

3-31. $V_0 = \frac{-100}{59}$ V

3-35. $V_1 = \frac{99}{26}$ V, $V_2 = \frac{1299}{221}$ V, and $V_3 = \frac{1494}{221}$ V

CHAPTER 4

4-4. $\omega = \dfrac{10^5}{9}$ rad/s

4-6. $L = \dfrac{600}{7\pi}$ μH

4-11. $Z_{ab} = 3.1 - j17.05$ Ω

4-14. $Z = -j30$ Ω

4-19. A source of voltage $47i_c(t)$ in series with $47 + j15$ Ω

4-24. $v_{0c} \approx 2.71e^{j\omega t + j1.42}$

4-28. $P_{av} = \left(\dfrac{5}{118^2}\right)[25 - 24\cos(\theta_1 - \theta_2)]$

4-32. P_{av} (in 4 S) $= \frac{18{,}944}{1573}$ W, P_{av} (in 2 S) $= \frac{9216}{1573}$ W, $P_c = \dfrac{28{,}160 - j12{,}672}{1573}$ VA

4-35. $X_{added} = \frac{-1043}{221}$ Ω

CHAPTER 5

5-1. $v_o(t) = u(t)(1 - e^{-Rt/L})$

5-3. $v_o(t) = \delta(t) - u(t)\left(\dfrac{R}{L}\right)e^{-Rt/L}$

5-13. $V(\omega) = \dfrac{-j(\pi/\alpha)^{1/2}}{4\alpha^2}\,\omega\left[3 - \left(\dfrac{\omega^2}{2\alpha}\right)\right]e^{-\omega^2/(2\alpha)}$

5-16. **(a)** $H(s) = \dfrac{\frac{-2}{3}}{s+2} + \dfrac{\frac{8}{3}}{s+8}$

(b) $h(t) = \frac{-2}{3}u(t)e^{-2t} + \frac{8}{3}u(t)e^{-8t}$

5-20. **(a)** $H(\omega) = \dfrac{2}{3 + j\omega}$

(b) $V(\omega) = \dfrac{6}{(3 + j\omega)^3}$

(c) $V_o(\omega) = \dfrac{12}{(3 + j\omega)^4}$

(d) $v_o(t) = 2u(t)t^3e^{-3t}$

5-25. **(a)** 0

(b) $\frac{40}{1587}$

(c) $\frac{4}{3}$

5-27. $v_o(t) = \frac{6}{11}[\mathrm{Sgn}(t) - 2u(t)e^{-t/2}]$

CHAPTER 6

6-2. $\sigma = 331.35\ (\Omega\cdot\mathrm{m})^{-1}$, $\rho = 30.179 \times 10^{-4}\ \Omega\cdot\mathrm{m}$; results differ from those of Example 6.1-2 by a factor of $\dfrac{1}{3.529}$

6-7. $\dfrac{i_D}{I_0} = 10^2, 10^4, 10^6$; $V_\gamma = 0.239, 0.476, 0.715$; for $V_\gamma = 0.6$, $\dfrac{i_D}{I_0} = 1.091 \times 10^5$

6-11. $R \geq 735\ \Omega$, $R_{L,\min} \geq 514.9\ \Omega$

6-16. Load variation = $\pm 1.928\%$ of nominal

6-22. $RC = 0.4167$

6-25. $\beta = 65.67$, $i_E = 1.6$ mA, and $i_C = 1.58$ mA

6-29. **(a)** $\alpha = 0.9$

(b) $i_B = 0.08$ mA

(c) $i_{SE} = 9.545 \times 10^{-16}$ A

(d) $\beta = 9$

6-35. **(a)** $v_{DS} = 1.0$ V, $i_D = \frac{2}{3}$ mA

(b) $v_{DS} = 2.0$ V, $i_D = \frac{8}{3}$ mA

6-39. $V_{DS} = 17.5$ V

6-46. $K = 0.3 \times 10^{-3}\ \mathrm{A/V^2}$

CHAPTER 7

7-1. $R_B \approx 56{,}500\ \Omega$, $R_C = \frac{6}{13}\ \text{k}\Omega$

7-6. $I_1 = 0.673$ mA, $I_2 = 0.617$ mA, and $I_{EQ} = \frac{654}{108}$ mA

7-10. $R_S = 78.21\ \Omega$, $R_1 = 33{,}891\ \Omega$, $R_2 = 7821\ \Omega$, $R_D = 103.26\ \Omega$, and $V_G = 6$ V

7-15. $A_{v1} = -139.57$ and $A_i = -18.76$

7-18. $A_{v1} = -3.748$ and $A_i = -5.62$

7-22. **(a)** $R_i = 6941\ \Omega$
(b) $R_{\text{in}} = 4627\ \Omega$
(c) $A_{v1} = 0.961$
(d) $A_i = 16.46$

7-28. **(a)** $R_i = 32.95\ \Omega$
(b) $R_E = 57.9\ \Omega$
(c) $A_{v1} = 25.24$
(d) $A_i = 0.095$

7-31. **(a)** $g_m = 0.0298$ S, $I_{DSS} = 0.0977$ A
(b) $R_{S1} = 34.54\ \Omega$

7-36. **(a)** $R_{\text{in}} = 169.9\ \Omega$
(b) $A_{v1} = 28.62$
(c) $A_i = 0.324$

7-38. **(a)** $\omega_H = 6\pi \times 10^4$ rad/s
(b) $\omega_L = 130\pi$ rad/s
(c) $A_{v0} \approx 20$

7-41. $Q = 53.5$, $L = 0.2433\ \mu\text{H}$, and $C = 909.5$ pF

CHAPTER 8

8-1. $R_1 = \frac{154}{153}\ \text{k}\Omega$, $\dfrac{v_0}{v_{\text{in}}} = -15.83$

8-4. Voltage indication $= V_0 = \dfrac{R_2}{100}$, so $R_2 = 100V_0$

8-10. $R_f = 4100\ \Omega$

8-14. $R_{\text{in}} = 2R_1$

8-16. $v_0 = 6.016v_a - 5.016v_b$

8-20. Arbitrarily choose $R_B = R_2$; then $R_A = R_1$, $C_A = \dfrac{R_3C}{R_1}$, and

$$C_B = \frac{(R_2 + R_3)C}{R_2}$$

8-24. (a) $R_3 = 4396\ \Omega$

(b) $v_0 = -0.68$ mV

8-29. Gain margin (theoretically) $= \infty$, phase margin $= 55.66°$

8-31. $\omega_3 = 25\pi$ rad/s

8-35. $\sigma_M = 2.45 \times 10^{-4}$ N·m/(rad/s), $K_M = 3.401\ (\text{V}\cdot\text{s})^{-1}$, and $\omega_L = 12.25$ rad/s

8-38. $K_D = 2.387\ \mu$V/(rad/s)

8-46. $\omega_r = 1001\omega_L \approx 2\pi \times 10^6$ rad/s

CHAPTER 9

9-6. The dc part of v_0 is independent of the diode resistances given and equals $\dfrac{2V_{\text{in, peak}}}{\pi}$, where $V_{\text{in, peak}}$ = peak amplitude of $v_{\text{in}}(t)$

9-8. Fraction of T_0 in conduction $= 0.133$, $V_{\text{ripple}} = \dfrac{V_{\text{peak}}}{9.6}$

9-12. $\dfrac{\Delta v_0}{\Delta v_1} \approx 0.001537$

9-15. $v_0 = -12.69$ V, -12.063 V, -11.437 V, and 6.05 V, for $v_{\text{in}} = -3$ V, -1 V, $+1$ V, and 5 V, respectively

9-19. $C = \dfrac{10^{-7}}{10\pi\sqrt{6}}$ F, $R_2 > 145$ kΩ required

9-23. (a) Yes

(b) $L = 3.166\ \mu$H

9-27. $R_{B1} = 288.5\ \Omega$, $R_{B2} = 25.97$ kΩ, $R_{C1} = 100\ \Omega$, $R_{C2} = 100\ \Omega$; take output as v_{C1}

9-28. $f_0 < 5$ kHz, $R_B = \dfrac{10^4}{\ln 2}\ \Omega$

9-32. $R = \dfrac{10{,}000}{8.4}\ \Omega$

CHAPTER 10

10-1. (a) $(91)_{10}$

(b) $(.15625)_{10}$

(c) $(29.625)_{10}$

10-4. **(a)** $(177777)_8$
(b) $(.1)_8$
(c) $(573.2)_8$

10-7. **(a)** $(109506)_{10}$
(b) $(0.05541992)_{10}$
(c) $(598.4453)_{10}$

10-10. **(b)** $(.0000110111000101)_2$

10-13. **(a)** $(1110010011)_2$
(c) $(001000011)_2$
(e) $(11110101111010)_2$
(g) Quotient = $(10001)_2$ and remainder = $(0010)_2$

10-15. **(a)** $(11001010)_2$
(c) $(1.111100)_2$

10-16. **(c)** $(00111000)_2$
(d) $(10110101)_2$

10-20. **(a)**

A	B	F
0	0	0
0	1	1
1	0	1
1	1	0

10-21. **(c)** $F = A \cdot B + B \cdot C$

10-23. **(c)** $F = \overline{A} \cdot B + B \cdot \overline{D} + \overline{A} \cdot \overline{C} \cdot \overline{D}$

10-28. **(a)** $F_1 = A \cdot \overline{B} + \overline{B} \cdot C$
(c) $F_3 = (X + Y) \cdot (Y + Z)$

10-31. **(a)** $A \oplus B \oplus C \oplus D$
(c) No

CHAPTER 11

11-2.

S	R	Q	$\overline{Q}$
0	0	not allowed	
0	1	1	0
1	0	0	1
1	1	Q_p	$\overline{Q}_p$

11-4.

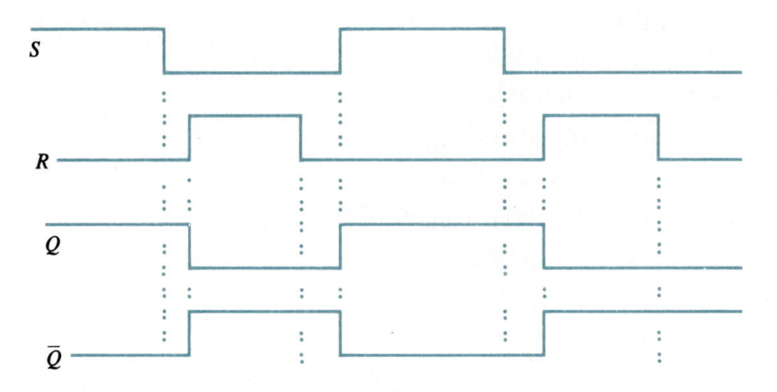

11-10. $(1101)_2$

11-13. 15 flip-flops

11-16. 16 flip-flops

11-17. 36.80 books per tape

11-19. **(a)** 18.75 V

(b) 0

(c) 0.625 V

11-23. 10 kΩ

CHAPTER 12

12-2. Hardware: ALU, CU, memory, and I/Os; software: system and user

12-11. 20 lines

12-17. **(a)** $(252)_{10}$

(b) $(253)_{10}$

12-20. The size of the op code field

12-24. **(a)** CY = 1, Z = 0, AC = 0, P = 1, and S = 1

(b) CY = 0, Z = 0, AC = 1, P = 0, and S = 1

12-26. (A) = (B) = (L) = $(100)_{10}$; (D) = 0; (E) = $(250)_{10}$; (H) = $(66)_{10}$; (memory location 250) = 100

12-27. **(b)** **LXI H,** 5599H

SHLD 100

12-31. $A_7 = A_0 = 1$; $A_4 = A_3 = 0$; A_1 and A_2 are complemented

12-35.

```
MOV A, B
ORA A
RAR
```

RAR

RAR

12-39. **LDA** 2000

OUT 200

See Fig. 12.5-3 for the block diagram

12-41. Advantages: saves memory, saves user space, provides clarity; disadvantage: slower than inline coding

CHAPTER 13

13-3. **(a)** Public data network

(b) Private communication network

(c) Remote-access communication network

13-7. **(a)** Data-link layer

(c) Physical layer

(e) Application layer

(g) Transport layer

13-11. **(a)** Twisted pairs of wire

(b) Coaxial cables

(c) Fiber-optics cables

13-16. 1024

13-20. $(01001000111)_2$

13-27. By providing tables of the token-passing order for each station

13-31.

	Token Ring	Token Bus
Topology	Ring	Bus
Media	Twisted pair	Broadband cable
Encoding	Differential manchester	Manchester

13-33. Simultaneous transmissions on the same cable; frequency-division multiplexing (FDM)

13-35.

	Strobe	Parity	b_6	b_5	b_4	b_3	b_2	b_1	b_0
B	1	0	1	0	0	0	0	1	0
C	1	1	1	0	0	0	0	1	1

13-38. 660 characters

13-42. **(a)** When each signaling represents 1 bit

(b) When each signaling represents 2 bits

CHAPTER 14

14-2. $25.975\ \Omega \leq Z_L \leq 96.248\ \Omega$

14-7. $f_c = 6.656$ GHz

14-10. From (14.2-9), $Z_0 = 25.25\ \Omega$; from (14.2-10), $Z_0 = 24.51\ \Omega$

14-13. Theoretical: 256.76 MHz $< f <$ 513.52 Mhz; practical: 320.95 MHz $\leq f \leq$ 487.85 MHz

14-18. Attenuation$|_c$ = 0.1292 dB/100 m at $1.25f_c$; attenuation$|_c$ = 0.0875 dB/100 m at $1.90f_c$

14-22. $A_e \approx 18.72\ \text{cm}^2$

14-25. **(a)** $G_D \approx 187.8$

(b) $\theta_B \approx 11.1°$, $\phi_B \approx 13.0°$

(c) Yes

14-29. From Fig. 14.7-3 attenuation (oxygen + water vapor) $\approx$ 2.8 dB

14-33. $dN_a = 4.0 \times 10^{-18}$ W

14-36. $T_s = 160.2$ K

CHAPTER 15

15-1. 107

15-5. **(a)** $\eta_{\text{AM}} \approx 0.104$

(b) $P_c = 484$ W, $P_f = 56.25$ W, $P_{\text{AM}} = 540.25$ W

15-8. $\left(\frac{S_o}{N_o}\right)_{\text{AM}} \approx 1995$

15-11. $W_3 \leq 50.97$ kHz

15-16. $\left(\frac{S_o}{N_o}\right)_{\text{DSB}} = 5 \times 10^4$, unchanged

15-20. **(a)** $\left(\frac{S_i}{N_i}\right)_{\text{FM}} \geq 853.3$

(b) Yes

15-23. $K_{\text{cr}} = 5$

15-25. From (15.6-8): $R_{\text{FM}} = 2.26$; from (15.6-9): $R_{\text{FM}} = 1.98$

★15-28. Bandwidth $\approx$ 5.44 MHz

CHAPTER 16

16-2. $f_s = 9.0$ MHz

16-6. $|f(t)|_{\max} = 3.2$ V

16-9. **(a)** $|f(t)|_{\text{max}} = 2.56$ V

(b) $K_{\text{cr}} \leq 2.73$

16-11. $N_b = 5$ in Problem 16-4, and $N_b = 7$ in Problem 16-9

16-18. $\frac{\mathcal{N}_0}{2} = 0.985 \times 10^{-6}$ V²/Hz

16-20. $\left(\frac{S_o}{N_o}\right)_{\text{PCM}} = 13{,}081$; operation is below threshold

16-24. $P_e = 3.862 \times 10^{-5}$ and $E_b = 3.146$ μJ

16-26. $\frac{E_b}{\mathcal{N}_0} = 23.61$

16-28. New amplitude = 2.071 V

CHAPTER 17

17-1. 28.07 years

17-2. Fraction = 8.135×10^{-3}

17-6. Maximum (peak) magnitude = 6186 A

17-11. $Z_\Delta = 300 + j36\ \Omega$

17-15. $R_\Delta \geq 38.09\ \Omega$ and load power = 1.638 MW

17-19. **(a)** 132.8 kV rms

(b) 398.3 kV rms

17-21. L (per phase) = 8.63×10^{-7} H/m and C (per phase) = 13.26×10^{-12} F/m

17-25. Power in one impedance of $140 + j10\ \Omega$ = 102.3 W; power in 11.5-Ω impedance = 5009 W; complex house power = $6032 + j73.1$ var; real house power = 6032 W

CHAPTER 18

18-1. $E_a(\text{max}) = 0.188$ V, $E_a\left(\text{for } \frac{\pi}{3}\right) = 0.163$ V

18-4. $E_{\text{dc}} = 0.637$ times the peak voltage

18-8. **(a)** $P_L = 720$ W

(b) $I_f = 0.5$ A

(c) $E_a = 125.2$ V

18-12. $F = 3.6$ N

★18-15. Speed = 1615 rpm

18-18. Torque = 45.92 N·M and power = 17.31 kW

18-21. Class D

Index